Springer Tracts in Modern Physics

Volume 262

Springer Tracts in Modern Physics

Springer Tracts in Modern Physics provides comprehensive and critical reviews of topics of current interest in physics. The following fields are emphasized: Elementary Particle Physics, Condensed Matter Physics, Light Matter Interaction, Atomic and Molecular Physics, Complex Systems, Fundamental Astrophysics.

Suitable reviews of other fields can also be accepted. The editors encourage prospective authors to correspond with them in advance of submitting a manuscript. For reviews of topics belonging to the above mentioned fields, they should address the responsible editor as listed below.

Special offer: For all clients with a print standing order we offer free access to the electronic volumes of the Series published in the current year.

More information about this series at http://www.springer.com/series/426

Kamakhya Prasad Ghatak

Einstein's Photoemission

Emission from Heavily-Doped Quantized Structures

Springer

Kamakhya Prasad Ghatak
Electronics and Communication
Engineering
National Institute of Technology
Agartala
India

ISSN 0081-3869 ISSN 1615-0430 (electronic)
ISBN 978-3-319-11187-2 ISBN 978-3-319-11188-9 (eBook)
DOI 10.1007/978-3-319-11188-9

Library of Congress Control Number: 2014953237

Springer Cham Heidelberg New York Dordrecht London

Printed on acid-free paper

Springer is part of Springer Science+Business Media (www.springer.com)

Professors P.T. Landsberg, P.N. Butcher,
K. Seeger and J.S. Blakemore are now living
in a wonderful world from where the concept
of return visa to this globe called
Earth permanently stays in the band gap
regime. The long interaction with them
for more than three decades through personal
contacts and letters has transformed
an unsophisticated engineer to the present
author. I earnestly believe that they still
enjoy the academic output of my research
group and I write them mental letters
and dedicate this monograph to them but
cannot express my gratitude since I do not
know the present addresses of the said
four first class semiconductors physicists.

Kamakhya Prasad Ghatak

Preface

The unification of the concept of the asymmetry of the wave vector space of the charge carriers in semiconductors with the modern techniques of fabricating nano-structured materials such as molecular beam epitaxy, metal organic chemical vapor deposition, fine line lithography and other modern fabrication techniques in one, two and three dimensions (such as quantum wells (QWs), Doping super-lattices, inversion and accumulation layers, quantum well super-lattices, carbon nano-tubes, quantum wires, quantum wire super-lattices, magnetic quantization, magneto size quantization, quantum dots, magneto inversion and accumulation layers, magneto quantum well super-lattices, magneto NIPIs, quantum dot super-lattices and other field aided low dimensional electronic systems) spawns not only useful quantum effect devices but also unearth new concepts in the realm of low dimensional solid state electronics and related disciplines. These semiconductor nanostructures occupy a central position in the entire arena of condensed matter science, materials science, computational and theoretical nano-science and technology, semiconductor optoelectronics, quantized structures and semiconductor physics in general by their own right and find extensive applications in quantum registers, quantum switches, quantum sensors, hetero-junction field-effect, quantum logic gates, quantum well and quantum wire transistors, quantum cascade lasers, high-frequency microwave circuits, high-speed digital networks, high-resolution terahertz spectroscopy, advanced integrated circuits, super-lattice photo-oscillator, super-lattice photo-cathodes, resonant tunneling diodes and transistors, super-lattice coolers, thermo-electric devices, thin film transistors, micro-optical systems, intermediate-band solar cells, high performance infrared imaging systems, band-pass filters, optical modulators, thermal sensors, optical switching systems, single electron/molecule electronics, nano-tube based diodes, and other nano-electronic devices. Knowledge regarding these quantized structures may be gained from original research contributions in scientific journals, various patents, proceedings of the conferences, review articles, and different research monographs [1] respectively. Mathematician Simmons rightfully tells us [2] that the mathematical knowledge is said to be doubling in every 10 years, and in this context, we can also envision extrapolation of the Moore's law by projecting it in the perspective of the advancement of new research and analyses, in turn, generating novel concepts particularly in the entire arena of materials science in general [3].

Although many new effects in quantized structures have already been reported, the interest for further research of other aspects of such quantum-confined materials is becoming increasingly important. One such significant property is Einstein's Photoemission (EP) which is a physical phenomenon and occupies a singular position in the whole arena of Modern Physics and related disciplines in general and whose importance has already been established since the inception of Einstein's photoelectric effect (for which Einstein won Nobel Prize in 1921), which in recent years finds extensive applications in modern optoelectronics, characterization and investigation of condensed matter systems, photoemission spectroscopy and related aspects in connection with the investigations of the optical properties of nano-structures [4–8]. Interest in low dimensional silicon nanostructures also grew up and gained momentum, after the discovery of room temperature photoluminescence and electroluminescence of silicon nano-wires in porous silicon [4]. Work on ultrathin layers of $SiSiO_2$ super-lattices resulting into visible light emission at room temperature clearly exhibited low dimensional quantum confinement effect [5] and one of the most popular techniques for analyzing the low dimensional structures is to employ photoemission techniques. Recent observation of room temperature photoluminescence and electro luminescence in porous silicon has stimulated vigorous research activities in silicon nanostructures [6].

In this context, it may be noted that the available reports on the said areas [4–7] cannot afford to cover even an entire chapter regarding the EP from heavily doped (HD) quantized structures and incidentally the second book of the present research group devoted solely to the elementary study of EP [8] from optoelectronic materials and their nanostructures does not even contain a paragraph regarding the EP from HD Quantized Structures. The EP depends on the density-of-states (DOS) function which, in turn, is significantly affected by the different carrier energy spectra of different semiconductors having various band structures. In recent years, various energy wave vector dispersion relations of the carriers of different materials have been proposed [9] which have created the interest in studying the EP from HD materials and their quantized counterparts. The present monograph solely investigates the EP from HD quantized structures of non-linear optical, III-V, II-VI, Gallium Phosphide, Germanium, Platinum Antimonide, stressed, IV-VI, Lead Germanium Telluride, Tellurium, II-V, Zinc and Cadmium diphosphides, Bismuth Telluride, III-V, II-VI, IV-VI and HgTe/CdTe quantum well HD super-lattices with graded interfaces under magnetic quantization, III-V, II-VI, IV-VI and HgTe/CdTe HD effective mass super-lattices under magnetic quantization, quantum confined effective mass super-lattices and super-lattices of HD optoelectronic materials with graded interfaces on the basis of the newly derived appropriate respective HD dispersion relation in each case. Incidentally, even after 20 years of continuous effort, we see that the complete investigation of the EP comprising of the whole set of the HD materials and allied sciences is really a sea and permanently enjoys the domain of impossibility theorems.

It is well-known that the classical equation of the photo-emitted current density is [10] $J = \left[4\pi\alpha_0 e m_c g_v (k_B T)^2 / h^3\right] \exp[(hv - \phi)/(k_B T)]$ (where α_0 $e, m_c, g_v, k_B, T, h, h\upsilon$ and ϕ and are the probability of photoemission, electron charge, effective electron mass at the edge of the conduction band, valley degeneracy, the Boltzmann constant, temperature, the Planck constant, incident photon energy along z-axis and work function respectively). The afore-mentioned equation is valid for both the charge carriers and in this conventional form it appears that, the photoemission changes with the effective mass, temperature, work function and the incident photon energy respectively. This relation holds only under the condition of carrier non-degeneracy.

The EP has different values for different materials and varies with doping and with external fields which creates quantization of the wave-vector space of the carriers leading to various types of quantized structures. The nature of these variations has been studied in [4–35] and some of the significant features are as follow:

1. The EP from bulk materials increases with the increase in doping.
2. The EP exhibits oscillatory dependence with inverse quantizing magnetic field because of the Shubnikov de Haas (SdH) effect.
3. The EP changes significantly with the magnitude of the externally applied quantizing electric field in electronic materials.
4. The EP from quantum confined Bismuth, nonlinear optical, III-V, II-VI and IV-VI materials oscillate with nano-thickness in various manners which are totally band structure dependent.
5. The nature of variations is significantly influenced by the energy band constants of various materials having different band structures.
6. The EP has significantly different values in quantum confined semiconductor super-lattices and various other quantized structures.

It is well known that heavy doping and carrier degeneracy are the keys to unlock the important properties of semiconductors and they are especially instrumental in dictating the characteristics of Ohomic contacts and Schottky contacts respectively [36]. It is an amazing fact that although the heavily doped semiconductors (HDS) have been investigated in the literature but the study of the carrier transport in such materials through proper formulation of the Boltzmann transport equation which needs in turn, the corresponding HD carrier energy spectra *is still one of the open research problems*.

It is well known that the band tails are being formed in the forbidden zone of HDS and can be explained by the overlapping of the impurity band with the conduction and valence bands [37]. Kane [38] and Bonch Bruevich [39] have independently derived the theory of band tailing for semiconductors having unperturbed parabolic energy bands. Kane's model [38] was used to explain the experimental results on tunneling [40] and the optical absorption edges [41, 42] in this context. Halperin and Lax [43] developed a model for band tailing applicable only to the deep tailing states. Although Kane's concept is often used in the literature for the investigation of band tailing [44, 45], it may be noted that this model

[38, 46] suffers from serious assumptions in the sense that the local impurity potential is assumed to be small and slowly varying in space coordinates [45]. In this respect, the local impurity potential may be assumed to be a constant. In order to avoid these approximations, we have developed in this book, the electron energy spectra for HDS for studying the EP based on the concept of the variation of the kinetic energy [37, 45] of the electron with the local point in space coordinates. This kinetic energy is then averaged over the entire region of variation using a Gaussian type potential energy. On the basis of the *E–k* dispersion relation, we have obtained the electron statistics for different HDS for the purpose of numerical computation of the respective EPs. It may be noted that, a more general treatment of many-body theory for the DOS of HDS merges with one-electron theory under macroscopic conditions [37]. Also, the experimental results for the Fermi energy and others are the average effect of this macroscopic case. So, the present treatment of the one-electron system is more applicable to the experimental point of view and it is also easy to understand the overall effect in such a case [47]. In a HDS, each impurity atom is surrounded by the electrons, assuming a regular distribution of atoms, and it is screened independently [44, 46, 48]. The interaction energy between electrons and impurities is known as the impurity screening potential. This energy is determined by the inter-impurity distance and the screening radius, which is known as the screening length. The screening radius changes with the electron concentration and the effective mass. Furthermore, these entities are important for HDS in characterizing the semiconductor properties [49, 50] and the modern electronic devices [44, 51]. The works on Fermi energy and the screening length in an n-type GaAs have already been initiated in the literature [52, 53], based on Kane's model. Incidentally, the limitations of Kane's model [38, 45], as mentioned above, are also present in their studies.

At this point, it may be noted that many band tail models are proposed using the Gaussian distribution of the impurity potential variation [38, 45]. From the very start, we have used the Gaussian band tails to obtain the *exact E–k dispersion relations* in HD non-linear optical, III-V, II-VI, Gallium Phosphide, Germanium, Platinum Antimonide, stressed, IV-VI, Lead Germanium Telluride, Tellurium, II-V, Zinc and Cadmium diphosphides, Bismuth Telluride, III-V, II-VI, IV-VI and HgTe/CdTe quantum well HD super-lattices with graded interfaces under magnetic quantization, III-V, II-VI, IV-VI and HgTe/CdTe HD effective mass super-lattices under magnetic quantization, quantum confined effective mass super-lattices and super-lattices of HD optoelectronic materials with graded interfaces respectively. Our method is not at all related with the DOS technique as used in the aforementioned works. From the electron energy spectrum, one can obtain the DOS but the DOS technique, as used in the literature cannot provide the *E–k* dispersion relation. *Therefore, our study is more fundamental than those in the existing literature, because the Boltzmann transport equation, which controls the study of the charge transport properties of the semiconductor devices, can be solved if and only if the E–k dispersion relation is known.* We wish to note that the Gaussian function for the impurity potential distribution has been used by many authors. It has been widely used since 1963 when Kane first proposed it and we will also use the Gaussian distribution for the present study.

This book, is divided into two parts (the first and second parts contain four and ten chapters respectively) and four Appendices, is partially based on our on-going researches on the EP from HDS from 1990 and an attempt has been made to present a cross section of the EP from wide range of HDS and their quantized-structures with varying carrier energy spectra under various physical conditions. The first chapter deals with the influence of quantum confinement on the EP from non-parabolic HDS and at first we study the EP from QWs of HD nonlinear optical materials on the basis of a generalized electron dispersion law introducing the anisotropies of the effective masses and the spin orbit splitting constants respectively together with the inclusion of the crystal field splitting within the framework of the *k.p* formalism. *We will observe that the complex electron dispersion law in HDS instead of real one occurs from the existence of the essential poles in the corresponding electron energy spectrum in the absence of band tails.* It may be noted that the complex band structures have already been studied for bulk semiconductors and super lattices without heavy doping [54] and bears no relationship with the complex electron dispersion law as formulated in this book. The physical picture behind the existence of the complex energy spectrum in heavily doped nonlinear optical semiconductors is the interaction of the impurity atoms in the tails with the splitting constants of the valance bands. The more is the interaction, the more the prominence of the complex part than the other case. In the absence of band tails, there is no interaction of the impurity atoms in the tails with the spin orbit constants and consequently, the complex part vanishes. Besides, the complex spectra are not related to same evanescent modes in the band tails and the conduction bands. One important consequence of the HDS forming band tails is that *the effective mass exists in the forbidden zone, which is impossible without the effect of band tailing. In the absence of band tails, the effective mass in the band gap of semiconductors is infinity. Besides, depending on the type of the unperturbed carrier energy spectrum, the new forbidden zone will appear within the normal energy band gap for HDS.*

The results of HD III-V (e.g. InAs, InSb, GaAs etc.), ternary (e.g. $Hg_{1-x}Cd_xTe$), quaternary (e.g. $In_{1-x}Ga_xAs_{1-y}P_y$ lattice matched to InP) compounds form a special case of our generalized analysis under certain limiting conditions as stated already. The EP from HD QWs of II-VI, IV-VI, stressed Kane type semiconductors, Te, GaP, $PtSb_2$,, Bi_2Te_3, Ge, and GaSb has also been investigated by formulating the respective appropriate HD energy band structure. The importance of the afore-mentioned semiconductors has also been described in the same chapter. In the absence of band tails and under the condition of extreme carrier degeneracy together with certain limiting conditions, all the results for all the EPs from all the HD QWs of Chap. 1 get simplified into the form [10] $J_{2D} = (\alpha_0 e g_v/2\hbar d_z^2) \sum_{n_{zmin}}^{n_{zmax}} n_z[E_{F2D} - \frac{\hbar^2}{2m_c}(n_z\pi/d_z)^2]$ (where d_z is the film thickness along z direction, n_z is the size quantum number along z direction and E_{F2D} is the Fermi energy in the presence of size quantization as measured from the edge of the conduction band in the vertically upward direction in the absence of any quantization) exhibiting the necessary mathematical compatibility test. In Chaps. 2 and 3 the EP from nano wires (NWs) and quantum boxes (QBs) of all the materials of Chap. 1 have respectively been investigated.

With the advent of modern experimental techniques of fabricating nano-materials, it is possible to grow semiconductor super-lattices (SLs) composed of alternative layers of two different degenerate layers with controlled thickness [55]. These structures have found wide applications in many new devices such as photodiodes [56], photo-resistors [57], transistors [58], light emitters [59], tunneling devices [60], etc. [61–72]. The investigations of the physical properties of narrow gap SLs have increased extensively, since they are important for optoelectronic devices and also since the quality of hetero-structures involving narrow gap materials has been greatly improved. It is well known that Keldysh [73] first suggested the fundamental concept of a super-lattice (SL), although it was successfully experimental realized by Esaki and Tsu [74]. The importance of SLs in the field of nano-electronics has already been described in [75–77]. The most extensively studied III-V SL is the one consisting of alternate layers of GaAs and $Ga_{1-x}Al_xAs$ owing to the relative ease of fabrication. The GaAs layers forms quantum wells and $Ga_{1-x}Al_xAs$ form potential barriers. The III-V SL's are attractive for the realization of high speed electronic and optoelectronic devices [78]. In addition to SLs with usual structure, SLs with more complex structures such as II-VI [79], IV-VI [80] and HgTe/CdTe [81] SL's have also been proposed. The IV-VI SLs exhibit quite different properties as compared to the III-V SL due to the peculiar band structure of the constituent materials [82]. The epitaxial growth of II-VI SL is a relatively recent development and the primary motivation for studying the mentioned SLs made of materials with the large band gap is in their potential for optoelectronic operation in the blue [82]. HgTe/CdTe SL's have raised a great deal of attention since 1979, when as a promising new materials for long wavelength infrared detectors and other electro-optical applications [83]. Interest in Hg-based SL's has been further increased as new properties with potential device applications were revealed [84]. These features arise from the unique zero band gap material HgTe [85] and the direct band gap semiconductor CdTe which can be described by the three band mode of Kane [86]. The combination of the aforementioned materials with specified dispersion relation makes HgTe/CdTe SL very attractive, especially because of the possibility to tailor the material properties for various applications by varying the energy band constants of the SLs. In addition to it, for effective mass SLs, the electronic sub-bands appear continually in real space [87].

We note that all the aforementioned SLs have been proposed with the assumption that the interfaces between the layers are sharply defined, of zero thickness, i.e., devoid of any interface effects. The SL potential distribution may be then considered as a one dimensional array of rectangular potential wells. The aforementioned advanced experimental techniques may produce SLs with physical interfaces between the two materials crystallo-graphically abrupt; adjoining their interface will change at least on an atomic scale. As the potential form changes from a well (barrier) to a barrier (well), an intermediate potential region exists for the electrons. The influence of finite thickness of the interfaces on the electron dispersion law is very important, since the electron energy spectrum governs the electron transport in SLs.

In this context, it may be noted that the effects of quantizing magnetic field (B) on the band structures of compound semiconductors are most striking than that of the parabolic one and are easily observed in experiments. A number of interesting physical features originate from the significant changes in the basic energy wave vector relation of the carriers caused by the magnetic field. The valuable information could also be obtained from experiments under magnetic quantization regarding the important physical properties such as Fermi energy and effective masses of the carriers, which affect almost all the transport properties of the electron devices [88] of various materials having different carrier dispersion relations [89]. In Chap. 4, the magneto EP from III-V, II-VI, IV-VI, HgTe/CdTe and strained layer quantum well heavily doped super-lattices (QWHDSLs) with graded interfaces will be studied. Besides the magneto EP from III-V, II-VI, IV-VI, HgTe/CdTe and strained layer quantum well HD effective mass super-lattices respectively has been explored and the same from the quantum dots of the aforementioned HD SLs has further been investigated in the same chapter.

It is worth remarking that, in the methods as given in the literature, the physics of photoemission has been incorporated in the lower limit of the photoemission integral and assuming that the band structure of the bulk materials becomes an invariant quantity in the presence of photo-excitation necessary for Einstein's photoelectric effect. The basic band structure of semiconductors changes in the presence of intense external light waves in a fundamental way, which has been incorporated mathematically through the expressions of the DOS function on the basis of a newly formulated electron dispersion law and the velocity along the direction of photoemission respectively in addition to the appropriate fixation of the lower limit of the photo-emission integral for the purpose of investigating the EP. The second part of the book investigates the EP from HD III-V semiconductors and their quantized counter parts. In Chap. 5, we study the EP from HD Kane type semiconductors on the basis of the newly formulated electron energy spectrum in the presence of intense light waves. An important concept highly relevant to the measurement of band-gap in HD electronic materials in the presence of external photo-excitation has also been discussed in this perspective. Under the conditions of extreme degeneracy, the invariant band structure concept in the presence of light waves and certain other limiting constraints all the results of this chapter for the EP assumes the well-known form [10] $J = (2\pi\alpha_0 e m_c g_v / h)(v - v_0)^2$, ($v_0$ is the threshold frequency) which indicates the fact current density is independent of temperature and when the energy of light quantum is much greater than the work function the material, the condition of extreme degeneracy is reached.

In Chap. 6, the EP has been investigated under magnetic quantization from HD Kane type materials on the basis of the concept as presented in Chap. 5. Chapter 7 covers the study of the EP from QWs, NWs and QBs of HD optoelectronic materials as an extension of the new dispersion relations of the bulk HD materials as investigated in Chap. 5. In Chap. 8, the magneto EP from HD effective mass super lattices, quantum well, quantum well wire, and quantum dot HD effective mass super-lattices have been investigated by formulating the appropriate electron dispersion laws. The experimental aspects of EP are extremely wide and it is virtually impossible even to highlight the major developments in a chapter. For the purpose of condensed presentation, the experimental aspects of EP from different nano-structured materials have been discussed in Chap. 9 which also contains few important related applications of the content of this book. The Chap. 10 contains the conclusion and the scope for future research.

The Appendix A studies the EP from HDnonlinear optical, III-V, IV-VI, stressed compounds, n-Te,n-GaP, $PtSb_2$, *Bismuth Telluriden-Ge,Gallium AntimonideII-V semiconductors and Lead Germanium Telluride under magnetic quantization respectively.* In this *Appendix we shall observe that the EEM depends on Landau quantum number in addition toFermi energyand the other system constants due to the specific band structures of theHD materialstogether with the fact EEM exist in the band gap due to the presence of finite scattering potential as noted already. Thus we present a very simplified analysis of the EP from HD non-parabolic semiconductors under magnetic quantization, which is a big topic of research by its own right.*

In Appendix B, the magneto EP from HD III-V, II-VI, IV-VI, HgTe/CdTe and strained layer super-lattices with graded interfaces and the HD effective mass super-lattices of the aforementioned materials have been investigated.

It is worth remarking that the influence of crossed electric and quantizing magnetic fields on the transport properties of semiconductors having various band structures are relatively less investigated as compared with the corresponding magnetic quantization, although, the cross-fields are fundamental with respect to the addition of new physics and the related experimental findings. It is well known that in the presence of electric field (E_o) along x-axis and the quantizing magnetic field along z-axis, the dispersion relations of the conduction electrons in semiconductors become modified and for which the electron moves in both the z and y directions. The motion along y-direction is purely due to the presence of E_0 along x-axis and in the absence of electric field, the effective electron mass along y-axis tends to infinity which indicates the fact that the electron motion along y-axis is forbidden. The effective electron mass of the isotropic, bulk semiconductors having parabolic energy bands exhibits mass anisotropy in the presence of cross fields and this anisotropy depends on the electron energy, the magnetic quantum number, the electric and the magnetic fields respectively, although, the effective electron mass along z-axis is a constant quantity. In 1966, Zawadzki and Lax [90] formulated the electron dispersion law for III-V semiconductors in accordance with the two band model of Kane under cross fields configuration which generates the interest to study this particular topic of solid state science in general [91]. *The Appendix C investigates the EP under cross-field configuration from HD nonlinear optical, III-V, II-VI, IV-VI and stressed Kane type semiconductors respectively. This appendix also tells us that the EEM in all the cases is a function of the finite scattering potential, the magnetic quantum number and the Fermi energy even for HD semiconductors whose bulk electrons in the absence of band tails are defined by the parabolic energy bands.*

With the advent of nano-devices, the build-in electric field becomes so large that the electron energy spectrum changes fundamentally instead of being invariant and the Appendix D investigates the EP under intense electric field from bulk specimens of HD III-V, ternary and quaternary semiconductors. This appendix also explores the influence of electric field on the EP on the basis of HD new dispersion law in for QWs, NWs, QBs, under magnetic quantization, QWs under magnetic quantization and effective mass HD super-lattices under magnetic quantization.

In these four Appendices no graphs together with results and discussions are being presented since we feel that the readers should not lose a chance to enjoy the complex computer algorithm to investigate the EP in the respective case generating new physics and thereby transforming each Appendix into a short monograph by considering various other important materials having different dispersion relations.

It is needless to say that this monograph is based on the *'iceberg principle'* [92] and the rest of which will be explored by the researchers of different appropriate fields. Since, there is no existing report devoted solely to the study of EP from HD quantized structures to the best of our knowledge, we hope that the present book will a useful reference source for the present and the next generation of the readers and the researchers of materials and allied sciences in general. Since the production of error free first edition of any book from every point of view is a permanent member in the domain of impossibility theorems, therefore in spite of our joint concentrated efforts for couple of years together with the seasoned team of Springer, the same stands very true for this monograph also. Various expressions and a few chapters of this book have been appearing for the first time in printed form. The suggestions from the readers for the development of the book will be highly appreciated for the purpose of inclusion in the future edition, if any. In this book, from chapter one to till the end, we have presented *300 open research problems* for the graduate students, Ph.D. aspirants, researchers, engineers in this pinpointed research topic. We strongly hope that alert readers of this monograph will not only solve the said problems by removing all the mathematical approximations and establishing the appropriate uniqueness conditions, but also will generate new research problems both theoretical and experimental and, thereby, transforming this monograph into a solid book. Incidentally, our readers after reading this book will easily understand that how little is presented and how much more is yet to be investigated in this exciting topic which is the signature of coexistence of new physics, advanced mathematics combined with the inner fire for performing creative researches in this context from the young scientists since like Kikoin [93] we feel that *A young scientist is no good if his teacher learns nothing from him and gives his teacher nothing to be proud of.* We emphatically write that the problems presented here form the integral part of this book and will be useful for the readers to initiate their own contributions on the EP from HDS and their quantized counter parts since like Sakurai [94] we firmly believe *The reader who has read the book but cannot do the exercise has learned nothing.* It is nice to note that if we assign the alphabets A to Z, the positive integers from 1 to 26, chronologically, then the word *ATTITUDE* receives the perfect score 100 and is the vital quality needed from the readers since *attitude* is the ladder on which all the other virtues mount.

In this monograph, we have investigated various dispersion relations of different HD quantized structures and the corresponding carrier statistics to study the concentration dependence of the EP from HD quantum confined materials. *Besides, the expressions of effective electron mass and the sub-band energy have been formulated throughout this monograph as a collateral study, for the purpose of in-depth investigations of the said important pinpointed research topics.* Thus, in this book, the readers will get much information regarding the influence of quantization in HD low dimensional materials having different band structures. For the enhancement of the materials aspect, we have considered various materials having the same dispersion relation to study the influence of energy band constants of the different HDS on EP. Although the name of the book is extreme specific, from the content, one can easily infer that it should be useful in graduate courses on materials science, condensed matter physics, solid states electronics, nano-science and technology and solid-state sciences and devices in many Universities and the Institutions in addition to both Ph.D. students and researchers in the aforementioned fields. *Last but not the least, we do hope that our humble effort will kindle the desire to delve deeper into this fascinating and deep topic by any one engaged in materials research and device development either in academics or in industries.*

References

1. K.P. Ghatak, S. Bhattacharya, *Debye Screening Length: Effects of Nanostructured Materials*. Springer Tracts in Modern Physics, vol. 255 (Springer, Heidelberg, 2014); S. Bhattacharya, K. P. Ghatak, *Fowler-Nordheim Field Emission: Effects in Semiconductor Nanostructures*. Springer Series in Solid state Sciences, vol. 170 (Springer, Heidelberg, 2012); S. Bhattacharya, K.P. Ghatak, *Effective Electron Mass in Low Dimensional Semiconductors*. Springer Series in Materials Sciences, vol. 167, (Springer, Heidelberg, 2013); K.P. Ghatak, S. Bhattacharya, *Thermo Electric Power in Nano structured Materials Strong Magnetic Fields*. Springer Series in Materials Science, vol. 137 (Springer, Heidelberg, 2010); K.P. Ghatak, S. Bhattacharya, D. De, *Einstein Relation in Compound Semiconductors and Their Nanostructures*. Springer Series in Materials Science vol. 116 (Springer, Heidelberg, 2009)
2. G.E. Simmons, *Differential Equations with Applications and Historical Notes*. International Series in Pure and Applied Mathematics (McGraw-Hill, New York, 1991)
3. H. Huff (ed.), *Into the Nano Era-Moore's Law beyond Planner Silicon CMOS*. Springer Series in Materials Science, vol. 106 (Springer, Berlin, 2009)
4. L.T. Canham, Appl. Phys. Lett. **57**, 1046 (1990)
5. Z.H. Lu, D.J. Lockwood, J.M. Baribeam, Nature **378**, 258 (1995)
6. A.G. Cullis, L.T. Canham, P.D.O. Calocott, J. Appl. Phys. **82**, 909 (1997)
7. M. Cardona, L. Ley, *Photoemission in Solids 1 and 2*, Topics in Applied Physics, vols. 26, 27 (Springer, Berlin, 1978); S. Hüfner, *Photoelectron Spectroscopy* (Springer, Berlin, 2003); S. Hüfner, (ed.), *Very High Resolution Photoelectron Spectroscopy*, Lecture Notes in Physics, vol. 715 (Springer, Berlin, 2007); D.W. Lynch, C.G. Olson, *Photoemission Studies of High-*

Temperature Superconductors (Cambridge University Press, Cambridge, 1999); B. Feuerbacher, B. Fitton, R.F. Willis (eds.), *Photoemission and the Electronic Properties of Surfaces* (Wiley, New York, 1978); W. Schattke, M.A.V. Hove, *Solid-State Photoemission and Related Methods: Theory and Experiment* (Wiley, Weinheim, 2003); V.V. Afanas'ev, *Internal Photoemission Spectroscopy: Principles and Applications* (Elsevier, North Holland, 2010); D. J. Lockwood, *Light Emission in Silicon Based Materials and Devices*, ed. by H.S. Nalwa vol. 2 (Academic Press, San Diego, 2001)

8. K.P. Ghatak, D. De, S. Bhattacharya, *Photoemission from Optoelectronic Materials and Their Nanostructures*. Springer Series in Nanostructure Science and Technology (Springer, Heidelberg, 2009)
9. B. Mitra, K.P. Ghatak, Phys. Letts. A **141**, 81 (1989); B. Mitra, K.P. Ghatak, Solid-state Electronics **32**, 810 (1989); K.P. Ghatak, B. Mitra, Phys. Letts. A **156**, 233 (1991); S. Choudhury, L.J. Singh, K.P. Ghatak, Nanotechnology **15**, 180 (2004); K.P. Ghatak, S.N. Biswas, Appl. Phys. **70**, 299 (1991); K.P. Ghatak et al., Appl. Phys. **23**, 241 (1980); K.P. Ghatak, D.K. Basu, B. Nag, J. Phys. Chem. Sol. **58**, 133 (1997); P.K. Chakraborty, G.C. Datta, K.P. Ghatak, Physica B **339**, 198 (2003); K.P. Ghatak, A. Ghoshal, B. Mitra, Il Nuovo Cimento D **14**, 903 (1992); B. Mitra, A. Ghoshal, K.P. Ghatak, Phys. Stat. Sol. B **154**, K147 (1989); K.P. Ghatak, M. Mondal, J. Magn. Magn. Mater. **74**, 203 (1988); K.P. Ghatak et al., Ann. Phys. **17**, 195 (2008); K.P. Ghatak, B. Mitra Physica Scripta **42**, 103 (1990); B. Mitra, K.P. Ghatak, Solid-state Electronics **32**, 515 (1989); K.P. Ghatak et al., J. Appl. Phys. **103**, 094314 (2008); B. Nag, K.P. Ghatak, Phys. Scr. **54**, 657 (1996); M. Mondal, S. Bhattacharya, K.P. Ghatak, Appl. Phys. A **42**, 331 (1987); K.P. Ghatak, M. Monda, l Zeitschrift für Physik B Condensed Matter **64**, 223 (1986); A.N. Chakravarti et al., Appl. Phys. A **26**, 165 (1981)
10. R.K. Pathria, *Statistical Mechanics*, 2nd edn. (Butterworth-Heinemann, Oxford, 1996)
11. D. De et al., Beilstein J. Nanotech. **2**, 339 (2012); D. De et al., Super-lattices Microstruct. **47**, 377 (2010); K.P. Ghatak, M. Mondal, S.N. Biswas, J. Appl. Phys. **68**, 3032 (1990); M. Mondal, S. Banik, K.P. Ghatak, J. Low Temp. Phys. **74**, 423 (1989); B. Mitra, A. Ghoshal, K. P. Ghatak, Phys. Status Solidi B **150**, K67 (1988); K.P. Ghatak et al., J. Nonlinear Opt. Quantum Opt. **13**, 267 (1995); L. Torres, L. Lopez-Diaz, J. Iniguez, Appl. Phys. Lett. **73**, 3766, (1996); A.Y. Toporov, R.M. Langford, A.K. Petford-Long, Appl. Phys. Lett. **77**, 3063 (2000) ; L. Torres et al., J. Appl. Phys. **85**, 6208 (1999); S. Tiwari, S. Tiwari, Cryst. Res. Technol. **41**, 78 (2006); R. Houdré et al., Phys. Rev. Lett. **55**, 734 (1985)
12. K.P. Ghatak et al., Phys. B **403**, 2116 (2008); S.M. Adhikari, K.P. Ghatak, J. Adv. Phys. **2**, 130 (2013); D. De et al., Adv. Sci. Eng. Med. **4**, 211 (2012); A. Kumar et al., J. Comput. Theor. Nanosci. **7**, 115 (2010); R. Houdré et al., Surf. Sci. **168**, 538 (1986); T.C. Chiang, R. Ludeke, D.E. Eastman, Phys. Rev. B **25**, 6518 (1982); S.P. Svensson et al., J. Vac. Sci. Technol. B **2**, 235 (1984); S.F. Alvarado, F. Ciccacci, M. Campagna, Appl. Phys. Letts. **39**, 615 (1981); L. Fleming et al., Vac. Sci. Technol. B **22**, 2000 (2004); H. Shimoda et al., Phys. Rev. Lett. **88**, 015502 (2002); M. Maillard, P. Monchicourt, M.P. Pileni, Chem. Phys. Lett. **380**, 704 (2003)
13. B. Mitra, K.P. Ghatak, Phys. Scr. **40**, 776 (1989); D. De, S. Bhattacharya, K.P. Ghatak, International work shop in physics of semiconductor devices, IEEE **897** (2007); S. Bhattacharya et al., J. Comput. Theor. Nanosci. **3**, 280 (2006)
14. I. Matsuda et al., Phys. Rev. B **73**, 113311 (2006); V.L. Colvin, A.P. Alivisatos, J.G. Tobin, Phys. Rev. Lett. **66**, 2786 (1991); B. Schroeter, K. Komlev, W. Richter, Mater. Sci. Eng. B **88**, 259 (2002); G. F. Bertsch, N. Van Giai, N. Vinh Mau, Phys. Rev. A **61**, 033202 (2000)
15. K.P. Ghatak, S.N. Biswas, Nonlinear Opt. **4**, 39 (1993)
16. K.P. Ghatak, S.N. Biswas, *Growth and Characterization of Materials for Infrared Detectors and Nonlinear Optical Switches*, SPIE, vol. 1484 (USA, 1991), p. 136
17. K.P. Ghatak, B. De, *Polymeric Materials for Integrated Optics and Information Storage, Materials Research Society (MRS) Symposium Proceedings*, MRS Spring Meeting, 1991, vol. 228, p. 237; L.J. Heyderman, F. Nolting, Quitmann, Appl. Phys. Lett. **83**, 1797 (2003); R.P. Cowburn, A.O. Adeyeye, J.A.C. Bland, Appl. Phys. Lett. **70**, 2309 (1997)

18. K.P. Ghatak, B. Nag, G. Majumdar, *Strained Layer Expitaxy: Materials, Processing, and Device Applications, MRS Symposium Proceedings*, MRS Spring Meeting, 1995, vol. 379, p. 85
19. K.P. Ghatak, *High Speed Phenomena in Photonic Materials and Optical Bistability*, SPIE, USA, 1990, vol. 1280, p. 53
20. K.P. Ghatak, *Long Wave Length Semiconductor Devices, Materials and Processes Symposium Proceedings, MRS Symposium Proceedings, MRS Spring Meeting*, 1990, vol. 216, p. 469
21. K.P. Ghatak, A. Ghoshal, S. Bhattacharyya, *Nonlinear Optical Materials and Devices for Photonic Switching* (SPIE, Belington, 1990) vol. 1216, p. 282
22. K.P. Ghatak, *Nonlinear Optics III* (SPIE, Bellingham, 1992) vol. 1626, p. 115
23. K.P. Ghatak, A. Ghoshal, B. De, *Optoelectronic Devices and Applications* (SPIE, Bellingham, 1990) vol. 1338, p. 111
24. K.P. Ghatak, S.N. Biswas, in *Proceedings of the Society of Photo-optical and Instrumentation Engineers*, Nonlinear Optics II, (SPIE, Belington, 1991) vol. 1409, p. 28; K.P. Ghatak, *Process Module Metrology* (SPIE, Belington, 1992) vol. 1594, p. 110; K.P. Ghatak, *International Conference on the Application and Theory of Periodic Structures* (SPIE, Belington, 1991) vol. 1545, p. 282
25. K.P. Ghatak, M. Mondal, Solid State Electron. **31**, 1561 (1988)
26. K.P. Ghatak, M. Mondal, J. Appl. Phys. **69**, 1666 (1991)
27. C. Majumdar, A.B. Maity, A.N. Chakravarti, Phys. Status Solidi B **140**, K7 (1987)
28. C. Majumdar, A.B. Maity, A.N. Chakravarti, Phys. Status Solidi B **141**, K35 (1987)
29. N.R. Das, K.K. Ghosh, D. Ghoshal, Phys. Status Solidi B **197**, 97 (1996)
30. C. Majumdar, A.B. Maity, A.N. Chakravarti, Phys. Status Solidi B **144**, K13 (1987)
31. N.R. Das, A.N. Chakravarti, Phys. Status Solidi B **176**, 335 (1993)
32. S. Sen, N.R. Das, A.N. Chakravarti, J. Phys.: Condens. Matter **19**, 186205 (2007); N.R. Das, S. Ghosh, A.N. Chakravarti, Phys. Status Solidi B **174**, 45 (1992)
33. A.B. Maity, C. Majumdar, A.N. Chakravarti, Phys. Status Solidi B **144**, K93, (1987)
34. A.B. Maity, C. Majumdar, A.N. Chakravarti, Phys. Status Solidi B **149**, 565 (1988)
35. A.V.D. Ziel, *Solid State Physical Electronics* (Prentice Hall, Inc. Eaglewood Cliffs, 1957); A. Modinos, *Field, Thermionic and Secondary Electron Emission Spectroscopy* (Plenum Press, New York, 1984)
36. S.N. Mohammad, J. Appl. Phys. **95**, 4856 (2004); V.K. Arora, Appl. Phys. Lett. **80**, 3763 (2002); S.N. Mohammad, J. Appl. Phys. **95**, 7940 (2004); S.N. Mohammad, Phil. Mag. **84**, 2559 (2004); S.N. Mohammad, J. Appl. Phys. **97**, 063703 (2005); K. Suzue et al., J. Appl. Phys. **80**, 4467 (1996); S.N. Mohammad et al., Electron. Lett. **32**, 598 (1996); Z. Fan et al., J. Electron. Matter. **25**, 1703 (1996); C. Lu et al., J. Appl. Phys. **91**, 9216 (2002); S.N. Mohammad, J. Appl. Phys. **63**, 1614 (1988); S.N. Mohammad, J. Appl. Phys. **68**, 1710 (1990)
37. R.K. Willardson, *Semiconductors and Semimetals*, ed. by A.C. Beer (Academic Press, New York, 1966) p. 102
38. E.O. Kane, Phys. Rev. **131**, 79 (1963); Phys. Rev. B **139**, 343 (1965)
39. V.L. Bonch Bruevich, Sov. Phys. Solid State **4**, 53 (1963)
40. R.A. Logan, A.G. Chynoweth, Phys. Rev. **131**, 89 (1963)
41. C.J. Hwang, J. Appl. Phys. **40**, 3731 (1969)
42. J.I. Pankove, Phys. Rev. A **130**, 2059 (1965)
43. B.I. Halperin, M. Lax, Phys. Rev. **148**, 722 (1966)
44. R.A. Abram, G.J. Rees , B.L.H. Wilson, Adv. Phys. **27**, 799 (1978)
45. B.I. Shklovskii, A.L. Efros, *Electronic Properties of Doped Semiconductors*. Springer series in Solid State Sciences, vol. 45 (Springer, Berlin, 1984)
46. E.O. Kane, Solid State Electron. **28**, 3 (1985)
47. P.K. Chakraborty, J.C. Biswas, J. Appl. Phys. **82**, 3328 (1997)
48. B.R. Nag, *Electron Transport in Compound Semiconductors. Springer Series in Solid state Sciences*, vol. 11 (Springer, Heidelberg, 1980)
49. P.E. Schmid, Phys. Rev. B **23**, 5531 (1981)

50. G.E. Jellison Jr. et al., Phys. Rev. Lett. **46**, 1414 (1981)
51. V.I. Fistul, *Heavily Doped Semiconductors* Chap. 7 (Plenum, New York, 1969)
52. C.J. Hwang, J. Appl. Phys. **41**, 2668 (1970); W. Sritrakool, H.R. Glyde, V. Sa Yakanit, Can. J. Phys. **60**, 373 (1982); H. Ikoma, J. Phys. Soc. Japan, **27**, 514 (1969)
53. A.N. Chakravarti et al., Appl. Phys. **25**, 105 (1981); A.N. Chakravarti et al. Z. für Physik B Conden. Matter. **47**, 149 (1982); M. Mondal, N. Chattopadhyay, K.P. Ghatak, J. Low. Temp. Phys. **66**, 131 (1987); M. Mondal, K.P. Ghatak, Phys. Lett. A **131**, 529 (1988); K.P. Ghatak, A. Ghoshal, B. Mitra, Il Nuovo Cimento. **D 13**, 867 (1991); K.P. Ghatak, A. Ghoshal, Phys. Status Solidi A **126**, K53 (1991); E.A. Aruhanov et al., Sov. Phys. Semiconduct. **15**, 828 (1981); P.N. Hai et al., Appl. Phys. Letts. **77**, 1843 (2000); D.P. DiVincenzo, E.J. Mele, Phys. Rev. B **29**, 1685 (1984); P. Perlin et al., Appl. Phys. Letts. **68**, 1114 (1996); F. Masia et al., Phys. Rev. B **73**, 073201 (2006); I.M. Tsidilkovski, *Band Structures of Semiconductor*. (Pergamon Press, London, 1982); K.P. Ghatak, M. Mondal, Z. F. Physik B **69**, 471 (1988); B. Mitra et al., Nonlinear Opt. Quantum Opts. **12**, 83 (1995); K.P. Ghatak, S.N. Biswas, *Proceedings of SPIE* **1484**, 149 (1991); M. Mondal, K.P. Ghatak, *Graphite Intercalation Compounds: Science and Applications*, MRS Proceedings, ed. by M. Endo, M.S. Dresselhaus, G. Dresselhaus, MRS Fall Meeting, EA **16**, 173 (1988); K.P. Ghatak, M. Mondal, Z. fur, Nature A **41**, 881 (1986); C.C. Wu, C.J. Lin, J. Low Temp. Phys. **57**, 469 (1984); M.H. Chen, C.C. Wu, C.J. Lin, J. Low Temp. Phys. **55**, 127 (1984)
54. J.N. Schulman, Y.C. Chang, Phys. Rev. B **24**, 4445 (1981)
55. K.P. Ghatak, J. Mukhopadhyay J.P. Banerjee, *SPIE Proceedings Series*, **4746**, 1292 (2002); K.P. Ghatak et al., Il Nuovo Cimento. D **20**, 227 (1998); K.P. Ghatak, B. De, *Materials Research Society Symposium Proceedings* **300**, 513 (1993); K.P. Ghatak, B. Mitra, Il Nuovo Cimento. D **15**, 97 (1993); K.P. Ghatak et al., *Proceedings of the Society of Photo-Optical Instrumentation Engineers* **1308**, 356 (1990); B. Mitra, K.P Ghatak, Phys. Lett. A **146**, 357 (1990); B. Mitra, K.P. Ghatak, Phys. Lett. A **142**, 401 (1989); B. Mitra, K.P. Ghatak, Phys. Status Solidi B **149**, K117 (1988); S. Bhattacharyya, K.P. Ghatak, S. Biswas, OE/Fibers' 87, Inter. Soc. Opt. Photon. 73 (1987); K.P. Ghatak, A.N. Chakravarti, Phys. Status Solidi B **117**, 707 (1983)
56. F. Capasso, *Semiconductors and Semimetals* vol. 22 (Academic Press, New York, 1985), p. 2
57. F. Capassoet et al., Appl. Phys. Letts. **47**, 420 (1985)
58. F. Capasso, R.A. Kiehl, J. Appl. Phys. **58**, 1366 (1985)
59. K. Ploog, G.H. Doheler, Adv. Phys. **32**, 285 (1983)
60. F. Capasso, K. Mohammed, A.Y. Cho, Appl. Phys. Lett. 478 (1986)
61. R. Grill, C. Metzner, G.H. Döhler, Phys. Rev. B **63**, 235316 (2001); Phys. Rev. B **61**, 15614 (2000)
62. A.R. Kost et al.. J. Appl. Phys. **99**, 023501 (2006)
63. A.G. Smirnov, D.V. Ushakov, V.K. Kononenko, *Proceedings of SPIE*, **4706**, 70 (2002)
64. D.V. Ushakov, V.K. Kononenko, I.S. Manak, *Proceedings of SPIE*, **4358**, 171 (2001)
65. J.Z. Wang et al., Phys. Rev. B **62**, 6956 (2000)
66. A.R. Kost et al., Appl. Phys. Lett. **63**, 3494 (1993)
67. S. Bastola, S.J. Chua, S.J. Xu, J. Appl. Phys. **83**, 1476 (1998)
68. Z.J. Yang, E.M. Garmire, D. Doctor, J. Appl. Phys. **82**, 3874 (1997)
69. G.H. Avetisyan et al., *Proceedings of SPIE*, **2694**, 216 (1996)
70. U. Pfeiffer et al., Appl. Phys. Lett. **68**, 1838 (1996)
71. H.L. Vaghjiani et al., J. Appl. Phys. **76**, 4407 (1994)
72. P. Kiesel et al., In *Proceedings of SPIE*, vol. 85, (1993), p. 278
73. L.V. Keldysh, Sov. Phys. Solid State **4**, 1658 (1962)
74. L. Esaki, R. Tsu, IBM J. Res. Dev. **14**, 61 (1970)
75. R. Tsu, *Super-lattices to Nanoelectronics* (Elsevier, 2005); E.L. Ivchenko, G. Pikus, *Superlattices and Other Heterostructures* (Springer, Berlin, 1995)
76. G. Bastard, *Wave Mechanics Applied to Heterostructures*, (Editions de Physique, Les Ulis, France, 1990)

77. K.P. Ghatak, B. Mitra, Int. J. Electron. **72**, 541 (1992); S.N. Biswas, K.P. Ghatak, Int. J. Electron. **70**, 125 (1991); B. Mitra, K.P. Ghatak, Phys. Status Solidi B **149**, K117 (1988); K. P. Ghatak, B. Mitra, A. Ghoshal, Phys. Status Solidi B **154**, K121 (1989)
78. K.V. Vaidyanathan et al., Solid State Electron. **26**, 717 (1983)
79. B.A. Wilson, IEEE, J. Quantum Electron. **24**, 1763 (1988)
80. M. Krichbaum et al., IEEE, J. Quantum Electron. **24**, 717 (1988)
81. J.N. Schulman, T.C. McGill, Appl. Phys. Letts. **34**, 663 (1979)
82. H. Kinoshita, T. Sakashita, H. Fajiyasu, J. Appl. Phys. **52**, 2869 (1981)
83. L. Ghenin et al., Phys. Rev. B **39**, 1419 (1989)
84. C.A. Hoffman et al., Phys. Rev. B **39**, 5208 (1989)
85. V.A. Yakovlev, Sov. Phys. Semiconductors **13**, 692 (1979)
86. E.O. Kane, J. Phys. Chem. Solids **1**, 249 (1957)
87. H. Sasaki, Phys. Rev. B **30**, 7016 (1984)
88. S. Debbarma, K.P. Ghatak, Rev. Theor. Sci. (2014 in press)
89. A.N. Chakravarti et al., Phys. Status Solidi B **109**, 705 (1982); K.P. Ghatak et al., Phys. Status Solidi B **110**, 323 (1982); K.P. Ghatak et al., Phys. Scr. **75**, 820 (2007); A.N. Chakravarti et al., Phys. Status Solidi B **112**, 75 (1982); M. Mondal, K. P. Ghatak, Phys. Status Solidi B **120**, K63 (1983); M. Mondal, K.P. Ghatak, Acta Phys. Pol. A **66**, 539 (1984); M. Mondal, K. P. Ghatak, Acta Phys. Pol. A **66**, 47 (1984); M. Mondal, K.P. Ghatak, Phys. Status Solidi B **126**, K47 (1984); K.P. Ghatak, Il Nuovo Cimento D **14**, 1187 (1992); B. Nag, K.P. Ghatak, J. Phys. Chem. Solids **59**, 713 (1998); P.K. Chakraborty, K.P. Ghatak, J. Phys. D: Appl. Phys. **32**, 2438 (1999); P.K. Chakraborty, S.K. Biswas, K.P. Ghatak, Physica B **352**, 111 (2004)
90. W. Zawadzki, B. Lax, Phys. Rev. Lett. **16**, 1001 (1966)
91. K.P. Ghatak et al., Nonlinear Opt. Quantum Opt. **16**, 241 (1996); K.P. Ghatak et al., Nonlinear Opt. Quantum Opt. **16**, 167 (1996); K.P. Ghatak et al., Nonlinear Opt. Quantum Opt. **16**, 9 (1996); S. Bhattaacharya et al., J. Comput. Theor. Nanosci. **3**, 423 (2006); M. Mondal, K.P. Ghatak, Ann. Phys. **46**, 502 (1989); K.P. Ghatak, OE Fiber-DL, 435 (1991); K.P. Ghatak, B. Mitra, Int. J. Electron. Theor. Exp. **70**, 343 (1991); B. Mitra, K.P. Ghatak, Phys. Lett. A **137**, 413 (1989); M. Mondal, K.P. Ghatak, Ann. Phys. **501**, 502 (1989); M. Mondal, K.P. Ghatak, Phys. Status Solidi B **147**, K179 (1988); S. Biswas, N. Chattopadhyay, K.P. Ghatak, Int. Soc. Opt. Photonics, *Proceedings of the Society Photo-Optical Instrumentation Engineering*, vol. 836 (USA, 1987), p. 175
92. A. Pais, J. *Robert Oppenheimer* (Oxford University Press, Oxford, 2006), p. 18
93. I.K. Kikoin, *Science for Everyone: Encounters with Physicists and Physics* (Mir Publishers, Moscow, 1989), p. 154
94. J.J. Sakurai, *Advanced Quantum Mechanics* (Addison-Wesley Publishing Company, Reading, 1967)

Acknowledgments

I am grateful to A.N. Chakravarti, my thesis advisor and an *ardent advocator of the concept of theoretical minimum of Landau*, who convinced a 21 years old Circuit theorist that theoretical materials science, in general, is the hidden dual dance of quantum mechanics, statistical mechanics together with advanced mathematical techniques, and even to appreciate the incredible beauty, he not only placed a stiff note for me to derive all the equations in the Monumental Course of Theoretical Physics, the Classics of Landau–Lifshitz together with the two-volume classics of Morse-Feshbach 40 years ago but also forced me to stay in the creative zone of research. I express my gratitude to Late B.R. Nag of the Institute of Radio-physics and Electronics of the University of Calcutta, to whom I am ever grateful as a student and research worker, the Semiconductor Grandmaster of India for his three books on semiconductor science in general and more than 200 research papers (many of them are absolutely in *honor's class*) which still fire my imagination. I am grateful to Late S.C. Dasgupta and M. Mitra (both of them were formidable Applied Mathematicians with deep physical insight) of the Department of Mathematics of the then Bengal Engineering College, Shibpur, Howrah (presently Indian Institute of Engineering Science and Technology) for teaching me deeply the various methods of different branches of Applied Mathematics with special emphasis to *analytic number theory* when I was pursuing the bachelor degree in the branch of Electronics and Telecommunication Engineering 45 years ago. I am indebted to Late C.K. Majumdar of the Department of Physics of the University of Calcutta to light the fire for Theoretical Physics. I am grateful to P.N. Singh, the Ex-Chairman of NIT Agartala for his encouragement and enthusiastic support and the Director G. Mugeraya for his keen academic insight and creation of congenial atmosphere to promote research and higher learning among the faculty members bypassing all the apparent difficulties.

I am grateful to all my friends and colleagues for the last 40 years from my college days till date for forming my inner fire to solve independently the problems from five volumes Berkeley Physics Course, three volumes Sakurai, ten volumes of Addision-Wesley Modular Series on Solid State Devices, two volumes Cohen-Tannoudji et.al., three volumes Edwards, three volumes Carslaw, six volumes Guillemin, three volumes Tuttle, two volumes Scott, two volumes Budak, five volumes Chen, five volumes Smirnov.... It is curious to note that they are insisting me in the real sense of the term to verify all the entries of Gradshteyn-Ryzhik and three volumes Bateman manuscript project for the last 40 years. It may be noted that academic output of a nice scholar is the response of a good quality RC coupled amplifier in the mid-band zone whereas the same for a fine research worker is the Dirac's delta function. Incidentally, I can safely say that I belong to neither as a consequence of the academic surroundings which this life presents to me. I express my warm gratitude to H.L. Hartnagel, D. Bimberg, W.L. Freeman and W. Schommers for various academic interactions spanning over the last three decades. I am extremely grateful to Dr. Hari Singh Nalwa, the President and CEO of American Scientific Publishers (www.aspbs.com) an ardent supporter of the work of my research group. I remember the sweet memories of P.N. Robson, I.M. Stephenson, V.S. Letokhov, J. Bodnar and K.F. Brennan with true reverence for inspiring me with the ebullient idea that the publications of the research monographs from internationally reputed Publishers containing innovative concepts is of prime importance to excel in creative research activity.

Late P.T. Landsberg often expressed his desire to write a foreword for this book in 2007 but due to previous other heavy academic and related commitments I was unable to write this present monograph and his wife Mrs. Sophia Landsberg in her letter of desperate sadness dated January 12, 2009 informed me that from 2008 onward due to Alzheimer's disease her husband had to give up scientific works. In 21st December, 1974, A.N. Chakravarti, my mentor in my first interaction with him emphatically energized me that I must master "Semiconductor Statistics" (Pergamon Press, London, 1962) by J.S. Blakemore for my initiation in semiconductor physics. Later on I came in deep touch with K. Seeger, the well-known author of the book "Semiconductor Physics" (Springer Series in Solid State Sciences, vol. 40, Springer-Verlag, Germany, 1982). The solid scientist Late P. N. Butcher has been a steady hidden force since 1983 before his sad demise with respect to our scripting the series in band structure dependent properties of nano-structured materials. He insisted me repeatedly regarding it and to tune with his high rigorous academic standard, myself with my prominent members of my research group wrote [1] as the first one, [2] as the second one, [3] as the third one, [4] as the fourth one, [5] as the fifth one, [6] as the sixth one, [7] as the seventh one and the present monograph as the eighth one. Both P.T. Landsberg and P.N. Butcher visited the Institute of Radio Physics and Electronics and immediately being a young research scholar I formed permanent covalent bonds with them. Consequently at the fag end of my academic life, I dedicate this monograph to the said four Semiconductor Grand Masters as a token of my regards and gratitude to them individually for the long academic interactions lasting for more than three decades.

I offer special thanks to Late N. Guhachoudhury of Jadavpur University for instilling in me the thought that the *academic output = ((desire X determination X dedication)–(false enhanced self ego pretending like a true friend although a real unrecognizable foe))* although a thank you falls in the band gap regime for my beloved better half See who really formed the backbone of my long unperturbed research career, since in accordance with "Sanatan" Hindu Dharma, the fusion of marriage has transformed us to form a single entity, where the individuality is being lost. I am grateful to all the members of my research group (from 1983 till date) for not only quantum confining me in the infinitely deep quantum wells of *Ramanujan and Rabindranath* but also inspiring me to teach quantum mechanics and related topics from the eight volumes classics of Greiner et al.

From the flash back of my memory I wish to offer my indebtedness to M. Mondal, the first member of my research team who in 1983 joined with me to complete his Ph.D. work under my guidance when R.K. Poddar, the then Vice-chancellor of the University of Calcutta selected me as a Lecturer (presently Assistant Professor) of the famous Department of Radio Physics and Electronics. In 1987, S.K. Sen, the then Vice-chancellor of Jadavpur University accepted me as the Reader (presently Associate Professor) in the Department of Electronics and Telecommunication Engineering and since then a flood of young researchers (more than 12 in number consisting of B. Mitra, A. Ghoshal, D. Bhattacharya, A.R. Ghatak,…) around me created my research team and insisted me to work with them at the @ of 16 hours per day including holidays in different directions of research for the purpose of my creative conversion from an ordinary engineer to a 360° research scientist and consequently I enjoyed the high rate of research publications in different reputed International journals in various aspects of band structure dependent properties of quantized structures. It is nice to note that the said young talented researchers obtained their respective Ph.D. degree under my direct supervision. Incidentally in 1994, R.K. Basu, the then Vice-chancellor of the University of Calcutta selected me as a Professor in the Department of Electronic Science and another flood of research over helmed me in a new direction of materials science. The persons responsible for this change include S. Datta, S. Sengupta, A. Ali…. The 11th and 12th names of this golden series are S. Bhattacharya and D. De respectively who, in turn formed permanent covalent bond with me not only with respect to research (S. Bhattacharya and D. De are respectively the co-authors of seven and two Monographs in different series of Springer) but also in all aspects of life in general. Recently, this teacher-student relationship has been interchanged with respect to these two creative young persons and both of them are not only the first two hidden authors of this monograph but being the experts of numerical simulations of electron transport in nano-devices they also carried out the complex computer programming for the plots as used here.

It is curious to note that after serving 18 years as a Professor in the Department of Electronic Science, in 2012 P.K. Bose, the then Director of the National Institute of Technology, Agartala requested me to join as a Professor in the Department of Electronics and Communication Engineering. Being my life-long friend, I accepted his offer and more than ten young scholars around me are again directing my research in the last phase of my life in an altogether new direction. In my 30+ years of teaching life (I have the wide experience of teaching Physics, Mathematics, Applied Mechanics(from engineering statics up to nonlinear mechanics including indeterminate structures) and 70 % of the courses of Electronics and Telecommunication and Electrical Engineering respectively) and 40+ years of research life (mostly in Materials Science, Nano-science and Number Theory), I have finally realized that the research without teaching is body without blood and teaching without research is body without brain although my all time hero, creatively prolific number theorist Godfrey Harold Hardy in his classic book entitled "A Mathematician's Apology" (Cambridge University Press, 1990) tells us "I hate teaching".

I must express my gratitude to S. Debbarma, of Computer Science and Engineering Department of my present working place, one of the strong members of my research group, for offering important suggestions for the condensed presentation of this monograph. I offer special thanks to K. Sarkar, P. Bhardwaj, N. Debbarma, R. Das and other members of my research team for placing their combined effort towards the development of this book in the DO-LOOP of a computer and critically reading the manuscript in its last phase before sending it to C. Ascheron, Executive Editor Physics, Springer-Verlag. Last but not the least; I am grateful for ever to our life long time tested friend S. Sanyal, Principal, Lake School for Girls, Kolkata for not only motivating me at rather turbulent moments of my academic carrier but also instilling in me the concept that *the ratio of total accumulated knowledge in my chosen field of research to my own contribution in my topic of research tends to infinity at any time*.

As always, myself with the members of my research team are grateful to C. Ascheron, Executive Editor Physics, Springer Verlag, in the real sense of the term for his inspiration and priceless technical assistance from the very start of our first monograph from Springer. I am grateful to Peter Wölfle, Editor Condensed matter theory of Springer Tract in Modern Physics for accepting this book. We owe a lot to A. Duhm, Associate Editor Physics, Springer, and E. Suer, assistant to Ascheron. Naturally, the author is responsible for non-imaginative shortcomings. I firmly believe that our Mother Nature has propelled this Project in her own unseen way in spite of several insurmountable obstacles.

Agartala, India, July 2014 Kamakhya Prasad Ghatak

References

1. K.P. Ghatak, S. Bhattacharya, D. De, *Einstein Relation in Compound Semiconductors and their Heterostructures*. Springer Series in Materials Science, vol. 116 (Springer, Heidelberg, 2009)
2. K.P. Ghatak, D. De, S. Bhattacharya, *Photoemission from Optoelectronic Materials and their Nanostructures*. Springer Series in Nanostructure Science and Technology (Springer, Heidelberg, 2009)
3. K.P. Ghatak, S. Bhattacharya, *Thermo Electric Power in Nanostructured Materials Strong Magnetic Fields*. Springer Series in Materials Science, vol. 137 (Springer, Heidelberg, 2010)
4. S. Bhattacharya, K.P. Ghatak, *Fowler-Nordheim Field Emission: Effects in Semiconductor Nanostructures*. Springer Series in Solid state Sciences, vol. 170 (Springer, Heidelberg, 2012)
5. S. Bhattacharya, K.P. Ghatak, *Effective Electron Mass in Low Dimensional Semiconductors*. Springer Series in Materials Sciences, vol. 167 (Springer, Heidelberg, 2013)
6. K.P. Ghatak, S. Bhattacharya, *Debye Screening Length: Effects of Nanostructured Materials*. Springer Tracts in Modern Physics, vol. 255 (Springer, Heidelberg, 2014)
7. K.P. Ghatak, S. Bhattacharya, *Heavily Doped 2D Quantized Structures and the Einstein Relation*. Springer Tracts in Modern Physics, vol. 260 (Springer, Heidelberg, 2014 in press)

Contents

Part II The EP from HD III-V Semiconductors and Their Quantized Counter Parts

About the Author

Professor Kamakhya Prasad Ghatak born in India in 1953, obtained the B.E degree in Electronics and Telecommunication Engineering from the then Bengal Engineering College Shibpur (Presently Indian Institute of Engineering Science and Technology) of the University of Calcutta in 1974, the M.Tech degree from the Institute of Radio Physics and Electronics of the University of Calcutta in 1976. He obtained the Ph.D. (Tech) degree from the University of Calcutta in 1988 on the basis of 27 published research papers in International peer-reviewed S.C.I. Journals which is still a record in the said Institute. He joined as Lecturer in the Institute of Radio Physics and Electronics of the University of Calcutta in 1983, Reader in the Department of Electronics and Telecommunication Engineering of Jadavpur University in 1987 and Professor in the Department of Electronic Science of the University of Calcutta in 1994. After serving 18 years as a Professor in the said Department, from March 2012, he has joined as a Professor in the Department of Electronics and Communication Engineering of National Institute of Technology, Agartala, Tripura and presently acting as Dean (Faculty Welfare) of the said Institute. He was at the top of the merit list in all the cases respectively. Professor K.P. Ghatak is the First Recipient of the Degree of Doctor of Engineering of Jadavpur University in 1991 since the University inception in 1955 and in the same year he received the Indian National Science Academy visiting fellowship to IIT-Kharagpur. He is the principal co-author of more than 200 research papers on Semiconductor Nano-science in eminent S.C.I. Journals and more than 50 research papers in the Proceedings of SPIE and MRS of USA and many of his papers are being cited many times. Professor Ghatak is the invited Speaker of SPIE, MRS, etc., the referee, Editor and Associate Editor of different eminent Journals. He is the supervisor of more than two dozens of Ph.D. candidates in various aspects of materials and nano-sciences and many of them are working as Professor, Associate Professor and Assistant Professor in different Universities and reputed Academic Institutions. He is the principal co-author of the SEVEN research monographs among them three from Springer series in Materials Science (Vols. 116, 137, 167), one form Springer Series in Nanostructure Science and Technology, one form Springer Series in Solid-State Sciences (Vol. 170), and two form Springer Tracts in

Modern Physics (vols. 255, 260). The All Indian Council for Technical Education has selected the first Research and Development project in his life for the best project award in Electronics and second best research project award considering all the branches of Engineering for the year 2006. In 2011, the University Grant Commission recommended a research project to him and placed him at the top in the list of awardees. His present teaching interests are Non-Linear Network Analysis and Synthesis, Non-Linear Devices, Non-Linear Mechanics, Non-Linear Integral and Partial Differential Equations and present research interests are nano science and number theory respectively. His brief CV has been enlisted in many biographical references of USA and UK. Professor Ghatak is the firm believer of the principle of theoretical minimum of Landau and his vision, mission and passion is to stay deeply in solitude. For more information, please log in to http://www.amazon.com/Kamakhya-Prasad-Ghatak/e/B003B09OEY.

Symbols

α	Band nonparabolicity parameter
a	The lattice constant
a_0, b_0	The widths of the barrier and the well for super-lattice structures
A_0	The amplitude of the light wave
$\vec{A}$	The vector potential
$A(E, n_z)$	The area of the constant energy 2D wave vector space for ultrathin films
B	Quantizing magnetic field
B_2	The momentum matrix element
b	Bandwidth
c	Velocity of light
C_1	Conduction band deformation potential
C_2	A constant which describes the strain interaction between the conduction and valance bands
ΔC_{44}	Second order elastic constant
ΔC_{456}	Third order elastic constant
δ	Crystal field splitting constant
Δ_o	Interface width
$\Delta\left(\frac{1}{B}\right)$	Period of SdH oscillation
d_0	Super-lattice period
$D_0(E)$	Density-of-states (DOS) function
$D_B(E)$	DOS function in magnetic quantization
$D_B(E, \lambda)$	DOS function under the presence of light waves
d_x, d_y, d_z	Nano thickness along the x, y and z-directions
$\Delta_\parallel$	Spin-orbit splitting constants parallel
$\Delta_\perp$	Spin-orbit splitting constants perpendicular to the C-axis
Δ	Isotropic spin-orbit splitting constant
d^3k	Differential volume of the k space
$\in$	Energy as measured from the center of the band gap
ε	Trace of the strain tensor

ε_0	Permittivity of free space
ε_∞	Semiconductor permittivity in the high frequency limit
ε_{sc}	Semiconductor permittivity
ΔE_g	Increased band gap
$\lvert e \rvert$	Magnitude of electron charge
E	Total energy of the carrier
E_o, ζ_0	Electric field
E_g	Band gap
E_i	Energy of the carrier in the ith band
E_{ki}	Kinetic energy of the carrier in the ith band
E_F	Fermi energy
E_{FB}	Fermi energy in the presence of magnetic quantization
E_n	Landau sub band energy
E_{Fs}	Fermi energy in the presence of size quantization
$\bar{E}_{Fn}$	Fermi energy for nipis
$'E_{FSL}$	Fermi energy in super-lattices
$\vec{\varepsilon}_s$	Polarization vector
E_{FQWSL}	Fermi energy in quantum wire super-lattices with graded interfaces
E_{FL}	Fermi energy in the presence of light waves
$E_{F_{EL}}$	Fermi energyunder quantizing magnetic field in the presence of light waves
E_{F2DL}	2D Fermi energy in the presence of light waves
E_{F1DL}	1D Fermi energy in the presence of light waves
E_{g0}	Un-perturbed band-gap
$Erfc$	Complementary error function
Erf	Error function
E_{F_h}	Fermi energy of HD materials
$\bar{E}_{hd}$	Electron energy within the band gap
F_s	Surface electric field
$F(V)$	Gaussian distribution of the impurity potential
$F_j(\eta)$	One parameter Fermi-Dirac integral of order j
f_0	Equilibrium Fermi-Dirac distribution function of the total carriers
f_{0i}	Equilibrium Fermi-Dirac distribution function of the carriers in the ith band
g_v	Valley degeneracy
G	Thermoelectric power under classically large magnetic field
G_0	Deformation potential constant
g^*	Magnitude of the band edge g-factor
h	Planck's constant
$\hat{H}$	Hamiltonian
$\hat{H}'$	Perturbed Hamiltonian
$H(E - E_n)$	Heaviside step function

$\hat{i}$, $\hat{j}$ and $\hat{k}$	Orthogonal triads
i	Imaginary unit
I	Light intensity
j_{cl}	Conduction current contributed by the carriers of the ith band
k	Magnitude of the wave vector of the carrier
k_B	Boltzmann's constant
λ	Wavelength of the light
$\bar{\lambda}_0$	Splitting of the two spin-states by the spin-orbit coupling and the crystalline field
$\bar{l}, \bar{m}, \bar{n}$	Matrix elements of the strain perturbation operator
L_x, L_z	Sample length along x and z directions
L_0	Super-lattices period length
L_D	Debye screening length
m_1	Effective carrier masses at the band-edge along x direction
m_2	Effective carrier masses at the band-edge along y direction
m_3	The effective carrier masses at the band-edge along z direction
m_2'	Effective- mass tensor component at the top of the valence band (for electrons) or at the bottom of the conduction band (for holes)
m_i^*	Effective mass of the ith charge carrier in the ith band
$m_{\parallel}^*$	Longitudinal effective electron masses at the edge of the conduction band
$m_{\perp}^*$	Transverse effective electron masses at the edge of the conduction band
m_c	Isotropic effective electron masses at the edge of the conduction band
$m_{\perp,1}^*, m_{\parallel,1}^*$	Transverse and longitudinal effective electron masses at the edge of the conduction band for the first material in super-lattice
m_r	Reduced mass
m_v	Effective mass of the heavy hole at the top of the valance band in the absence of any field
n	Landau quantum number
n_x, n_y, n_z	Size quantum numbers along the x, y and z-directions
n_{1D}, n_{2D}	1D and 2D carrier concentration
n_{2Ds}, n_{2Dw}	2D surface electron concentration under strong and weak electric field
$\bar{n}_{2Ds}$, $\bar{n}_{2Dw}$	Surface electron concentration under the strong and weak electric field quantum limit
n_i	Mini-band index for nipi structures
$N_{nipi}(E)$	DOS function for nipi structures
$N_{2DT}(E)$	2D DOS function
$N_{2D}(E, \lambda)$	2D DOS function in the presence of light waves
$N_{1D}(E, \lambda)$	1D DOS function in the presence of light waves

n_0	Total electron concentration
$\bar{n}_0$	Electron concentration in the electric quantum limit
n_i	Carrier concentration in the ith band
P	Isotropic momentum matrix element
P_n	Available noise power
$P_{\parallel}$	Momentum matrix elements parallel to the direction of crystal axis
$P_{\perp}$	Momentum matrix elements perpendicular to the direction of crystal axis
$\bar{r}$	Position vector
S_i	Zeros of the Airy function
$\vec{s}_0$	Momentum vector of the incident photon
t	Time scale
t_c	Tight binding parameter Surface electric field
T	Absolute temperature
$\tau_i(E)$	Relaxation time of the carriers in the ith band
$u_1\left(\vec{k},\vec{r}\right)$, $u_2\left(\vec{k},\vec{r}\right)$	Doubly degenerate wave functions
$V(E)$	Volume of k space
V_0	Potential barrier encountered by the electron
$V(\vec{r})$	Crystal potential
x, y	Alloy compositions
z_t	Classical turning point
μ_i	Mobility of the carriers in the ith band
μ	Average mobility of the carriers
$\zeta(2r)$	Zeta function of order 2 r
$\Gamma(j+1)$	Complete Gamma function
η	Normalized Fermi energy
η_g	Impurity scattering potential
ω_0	Cyclotron resonance frequency
θ	Angle
μ_0	Bohr magnetron
ω	Angular frequency of light wave
$\uparrow', \downarrow'$	Spin up and down function

Part I
Influence of Quantum Confinement on the EP from Non-Parabolic Semiconductors

Chapter 1
The EP from Quantum Wells (QWs) of Heavily Doped (HD) Non-parabolic Semiconductors

1.1 Introduction

In recent years, with the advent of fine lithographical methods [1–3] molecular beam epitaxy [4], organometallic vapor-phase epitaxy [5], and other experimental techniques, the restriction of the motion of the carriers of bulk materials in one (QWs, doping super-lattices, accumulation, and inversion layers), two (nanowires) and three (quantum dots, magneto-size quantized systems, magneto inversion layers, magneto accumulation layers, quantum dot super-lattices, magneto QW super-lattices, and magneto doping superlattices) dimensions have in the last few years, attracted much attention not only for their potential in uncovering new phenomena in nano-science but also for their interesting quantum device applications [6–9]. In QWs, the restriction of the motion of the carriers in the direction normal to the film (say, the z direction) may be viewed as carrier confinement in an infinitely deep 1D rectangular potential well, leading to quantization [known as quantum size effect (QSE)] of the wave vector of the carriers along the direction of the potential well, allowing 2D carrier transport parallel to the surface of the film representing new physical features not exhibited in bulk semiconductors [10–14]. The low-dimensional hetero-structures based on various materials are widely investigated because of the enhancement of carrier mobility [15]. These properties make such structures suitable for applications in QWs lasers [16], hetero-junction FETs [17, 18], high-speed digital networks [19–22], high-frequency microwave circuits [23], optical modulators [24], optical switching systems [25], and other devices. The constant energy 3D wave-vector space of bulk semiconductors becomes 2D wave-vector surface in QWs due to dimensional quantization. Thus, the concept of reduction of symmetry of the wave-vector space and its consequence can unlock the physics of low-dimensional structures. In this chapter, we study the EP from QWs of HD non-parabolic semiconductors having different band structures in the presence of Gaussian band tails. At first we shall investigate the EP from QWs of HD nonlinear optical compounds which are being used in nonlinear optics

K.P. Ghatak, *Einstein's Photoemission*, Springer Tracts in Modern Physics 262,
DOI 10.1007/978-3-319-11188-9_1

and light emitting diodes [26]. The quasi-cubic model can be used to investigate the symmetric properties of both the bands at the zone center of wave vector space of the same compound. Including the anisotropic crystal potential in the Hamiltonian, and special features of the nonlinear optical compounds, Kildal [27] formulated the electron dispersion law under the assumptions of isotropic momentum matrix element and the isotropic spin-orbit splitting constant, respectively, although the anisotropies in the two aforementioned band constants are the significant physical features of the said materials [28–30]. In Sect. 1.2.1, the EP from QWs of HD nonlinear optical semiconductors has been investigated on the basis of newly formulated HD dispersion relation of the said compound by considering the combined influence of the anisotropies of the said energy band constants together with the inclusion of the crystal field splitting respectively within the framework of $\vec{k} \cdot \vec{p}$ formalism. The III-V compounds find applications in infrared detectors [31], quantum dot light emitting diodes [32], quantum cascade lasers [33], QWs wires [34], optoelectronic sensors [35], high electron mobility transistors [36], etc. The electron energy spectrum of III-V semiconductors can be described by the three- and two-band models of Kane [37–39], together with the models of Stillman et al. [40], Newson and Kurobe [41] and, Palik et al. [42] respectively. In this context it may be noted that the ternary and quaternary compounds enjoy the singular position in the entire spectrum of optoelectronic materials. The ternary alloy $Hg_{1-x}Cd_xTe$ is a classic narrow gap compound. The band gap of this ternary alloy can be varied to cover the spectral range from 0.8 to over 30 μm [43] by adjusting the alloy composition. $Hg_{1-x}Cd_xTe$ finds extensive applications in infrared detector materials and photovoltaic detector arrays in the 8–12 μm wave bands [44]. The above uses have generated the $Hg_{1-x}Cd_xTe$ technology for the experimental realization of high mobility single crystal with specially prepared surfaces. The same compound has emerged to be the optimum choice for illuminating the narrow sub-band physics because the relevant material constants can easily be experimentally measured [45]. Besides, the quaternary alloy $In_{1-x}Ga_xAs_yP_{1-y}$ lattice matched to InP, also finds wide use in the fabrication of avalanche photo-detectors [46], hetero-junction lasers [47], light emitting diodes [48] and avalanche photodiodes [49], field effect transistors, detectors, switches, modulators, solar cells, filters, and new types of integrated optical devices are made from the quaternary systems [50]. It may be noted that all types of band models as discussed for III-V semiconductors are also applicable for ternary and quaternary compounds. In Sect. 1.2.2, the EP from QWs of HD III-V, ternary and quaternary semiconductors has been studied in accordance with the corresponding HD formulation of the band structure and the simplified results for wide gap materials having parabolic energy bands under certain limiting conditions have further been demonstrated as a special case in the absence of band-tails and thus confirming the compatibility test. The II-VI semiconductors are being used in nano-ribbons, blue green diode lasers, photosensitive thin films, infrared detectors, ultra-high-speed bipolar transistors, fiber optic communications, microwave devices, solar cells, semiconductor gamma-ray detector arrays, semiconductor detector gamma camera and allow for a greater density of data storage on optically addressed compact discs [51–58]. The carrier energy spectra in II-VI compounds are defined by the Hopfield model [59] where the splitting of the two-spin states by the spin-orbit coupling and the crystalline field has been taken into account. The Sect. 1.2.3 contains the investigation of the EP from QWs of HD II-VI compounds.

Lead Chalcogenides (PbTe, PbSe, and PbS) are IV-VI non-parabolic semiconductors whose studies over several decades have been motivated by their importance in infrared IR detectors, lasers, light-emitting devices, photo-voltaic, and high temperature thermo-electrics [60–64]. PbTe, in particular, is the end compound of several ternary and quaternary high performance high temperature thermoelectric materials [65–69]. It has been used not only as bulk but also as films [70–73], QWs [74] super-lattices [75, 76] nanowires [77] and colloidal and embedded nanocrystals [78–81], and PbTe films doped with various impurities have also been investigated [82–89]. These studies revealed some of the interesting features that had been seen in bulk PbTe, such as Fermi level pinning and, in the case of superconductivity [90]. In Sect. 1.2.4, the 2D EP from QWs of HD IV-VI semiconductors has been studied taking PbTe, PbSe, and PbS as examples. The stressed semiconductors are being investigated for strained silicon transistors, quantum cascade lasers, semiconductor strain gages, thermal detectors, and strained-layer structures [91–94]. The EP from QWs of HD stressed compounds (taking stressed n-InSb as an example) has been investigated in Sect. 1.2.5 The vacuum deposited Tellurium (Te) has been used as the semiconductor layer in thin-film transistors (TFT) [95] which is being used in CO_2 laser detectors [96], electronic imaging, strain sensitive devices [97, 98], and multichannel Bragg cell [99]. Section 1.2.6 contains the investigation of EP from QWs of HD Tellurium. The n-Gallium Phosphide (n-GaP) is being used in quantum dot light emitting diode [100], high efficiency yellow solid state lamps, light sources, high peak current pulse for high gain tubes. The green and yellow light emitting diodes made of nitrogen-doped n-GaP possess a longer device life at high drive currents [101–103]. In Sect. 1.2.7, the EP from QWs of HD n-GaP has been studied. The Platinum Antimonide ($PtSb_2$), finds application in device miniaturization, colloidal nanoparticle synthesis, sensors and detector materials and thermo-photovoltaic devices [104–106]. Section 1.2.8 explores the EP from QWs of HD $PtSb_2$. Bismuth telluride (Bi_2Te_3) was first identified as a material for thermoelectric refrigeration in 1954 [107] and its physical properties were later improved by the addition of bismuth selenide and antimony telluride to form solid solutions. The alloys of Bi_2Te_3 are useful compounds for the thermoelectric industry and have been investigated in the literature [108–112]. In Sect. 1.2.9, the EP from QWs of HD Bi_2Te_3 has been considered. The usefulness of elemental semiconductor Germanium is already well known since the inception of transistor technology and, it is also being used in memory circuits, single photon detectors, single photon avalanche diode, ultrafast optical switch, THz lasers and THz spectrometers [113–116]. In Sect. 1.2.10, the EP has been studied from QWs of HD Ge. Gallium Antimonide (GaSb) finds applications in the fiber optic transmission window, hetero-junctions, and QWs. A complementary hetero-junction field effect transistor in which the channels for the p-FET device and the n-FET device forming the complementary FET are formed from GaSb. The band gap energy of GaSb makes it suitable for low power operation [117–122]. In Sect. 1.2.11, the EP from QWs of HD GaSb has been studied. Section 1.3 contains the result and discussions pertaining to this chapter. The last Sect. 1.4 contains open research problems.

1.2 Theoretical Background

1.2.1 The EP from QWs of HD Non-linear Optical Semiconductors

The form of k. p matrix for nonlinear optical compounds can be expressed extending Bodnar [28] as

$$H = \begin{bmatrix} H_1 & H_2 \\ H_2^+ & H_1 \end{bmatrix} \tag{1.1}$$

where,

$$H_1 \equiv \begin{bmatrix} E_{g_0} & 0 & P_{\|}k_z & 0 \\ 0 & (-2\Delta_{\|}/3) & (\sqrt{2}\Delta_{\perp}/3) & 0 \\ P_{\|}k_z & (\sqrt{2}\Delta_{\perp}/3) & -(\delta+\frac{1}{3}\Delta_{\|}) & 0 \\ 0 & 0 & 0 & 0 \end{bmatrix}, \quad H_2 \equiv \begin{bmatrix} 0 & -f_{,+} & 0 & f_{,-} \\ f_{,+} & 0 & 0 & 0 \\ 0 & 0 & 0 & 0 \\ f_{,+} & 0 & 0 & 0 \end{bmatrix}$$

in which E_{g_0} is the band gap in the absence of any field, $P_{\|}$ and $P_{\perp}$ are the momentum matrix elements parallel and perpendicular to the direction of crystal axis respectively, δ is the crystal field splitting constant, $\Delta_{\|}$ and $\Delta_{\perp}$ are the spin-orbit splitting constants parallel and perpendicular to the C-axis respectively, $f_{,\pm} \equiv \left(P_{\perp}/\sqrt{2}\right)\left(k_x \pm ik_y\right)$ and $i = \sqrt{-1}$. Thus, neglecting the contribution of the higher bands and the free electron term, the diagonalization of the above matrix leads to the dispersion relation of the conduction electrons in bulk specimens of nonlinear optical semiconductors as

$$\gamma(E) = f_1(E)k_s^2 + f_2(E)k_z^2 \tag{1.2}$$

where

$$\gamma(E) \equiv E(E+E_{g_0})\left[(E+E_{g_0})(E+E_{g_0}+\Delta_{\|}) + \delta\left(E+E_{g_0}+\frac{2}{3}\Delta_{\|}\right) + \frac{2}{9}\left(\Delta_{\|}^2 - \Delta_{\perp}^2\right)\right],$$

E is the total energy of the electron as measured from the edge of the conduction band in the vertically upward direction in the absence of any quantization, $k_s^2 = k_x^2 + k_y^2$,

$$f_1(E) \equiv \frac{\hbar^2 E_{g0}\left(E_{g0} + \Delta_\perp\right)}{\left[2m_\perp^*\left(E_{g0} + \frac{2}{3}\Delta_\perp\right)\right]}\left[\delta\left(E + E_{g0} + \frac{1}{3}\Delta_\parallel\right) + \left(E + E_{g0}\right)\right.$$
$$\left. \times \left(E + E_{g0} + \frac{2}{3}\Delta_\parallel\right) + \frac{1}{9}\left(\Delta_\parallel^2 - \Delta_\perp^2\right)\right],$$

$$f_2(E) \equiv \frac{\hbar^2 E_{g0}\left(E_{g0} + \Delta_\parallel\right)}{\left[2m_\parallel^*\left(E_{g0} + \frac{2}{3}\Delta_\parallel\right)\right]}\left[\left(E + E_{g0}\right)\left(E + E_{g0} + \frac{2}{3}\Delta_\parallel\right)\right],$$

ħ = h/2π, h is Planck's constant and $m_\parallel^*$ and $m_\perp^*$ are the longitudinal and transverse effective electron masses at the edge of the conduction band respectively.

Thus the generalized unperturbed electron energy spectrum for the bulk specimens of the nonlinear optical materials in the absence of band tails can be expressed following (1.2) as

$$\frac{\hbar^2 k_z^2}{2m_\parallel^*} + \left(\frac{b_\parallel}{b_\perp}\frac{c_\perp}{c_\parallel}\right)\frac{\hbar^2 k_s^2}{2m_\perp^*} = \left\{\frac{E(\alpha E+1)(b_\parallel E+1)}{(c_\parallel E+1)} + \frac{\alpha b_\parallel}{c_\parallel}\left[\delta E + \frac{2}{9}\left(\Delta_\parallel^2 - \Delta_\perp^2\right)\right] - \left(\frac{2}{9}\right)\frac{\alpha b_\parallel}{c_\parallel}\frac{\left(\Delta_\parallel^2 - \Delta_\perp^2\right)}{(c_\parallel E+1)}\right\}$$
$$- \left(\frac{\hbar^2 k_s^2}{2m_\perp^*}\right)\left\{\left(\frac{b_\parallel}{b_\perp}\frac{c_\perp}{c_\parallel}\right)\left[\left(\frac{\delta}{2} + \frac{\Delta_\parallel^2 - \Delta_\perp^2}{6\Delta_\parallel}\right)\frac{\alpha_\parallel}{\alpha_\parallel E+1} + \left(\frac{\delta}{2} - \left\{\frac{\Delta_\parallel^2 - \Delta_\perp^2}{6\Delta_\parallel}\right\}\right)\frac{c_\parallel}{c_\parallel E+1}\right]\right\} \tag{1.3}$$

where,

$$b_\parallel \equiv 1/(E_g + \Delta_\parallel), c_\perp \equiv 1/\left(E_g + \frac{2}{3}\Delta_\perp\right), b_\perp \equiv 1/(E_g + \Delta_\perp),$$
$$c_\parallel \equiv 1/\left(E_g + \frac{2}{3}\Delta_\parallel\right) \text{ and } \alpha \equiv 1/E_g$$

The Gaussian distribution *F(V)* of the impurity potential is given by [123, 124]

$$F(V) = \left(\pi\eta_g^2\right)^{-1/2}\exp\left(-V/\eta_g^2\right) \tag{1.4}$$

where, η_g is the impurity scattering potential. It appears from (1.4) that the variance parameter η_g is not equal to zero, but the mean value is zero. Further, the impurities are assumed to be uncorrelated and the band mixing effect has been neglected in this simplified theoretical formalism.

We have to average the kinetic energy in the order to obtain the E-k dispersion relation in nonlinear optical materials in the presence of band tails. Using the (1.3) and (1.4), we get

$$
\begin{aligned}
&\left[\frac{\hbar^2 k_z^2}{2m_\parallel^*}\int_{-\infty}^{E} F(V)dV\right] + \left[\left(\frac{b_\parallel}{b_\perp}\frac{c_\perp}{c_\parallel}\right)\frac{\hbar^2 k_s^2}{2m_\perp^*}\int_{-\infty}^{E} F(V)dV\right] \\
&= \left\{\int_{-\infty}^{E}\frac{(E-V)[\alpha(E-V)+1][b_\parallel(E-V)+1]}{[c_\parallel(E-V)+1]}F(V)dV\right. \\
&+\frac{\alpha b_\parallel}{c_\parallel}\left[\delta\int_{-\infty}^{E}(E-V)F(V)dV+\frac{2}{9}\left(\Delta_\parallel^2-\Delta_\perp^2\right)\int_{-\infty}^{E}F(V)dV\right] \\
&\left.-\left(\frac{2}{9}\right)\frac{\alpha b_\parallel}{c_\parallel}\left(\Delta_\parallel^2-\Delta_\perp^2\right)\int_{-\infty}^{E}\frac{F(V)dV}{[c_\parallel(E-V)+1]}\right\} \\
&-\left(\frac{\hbar^2 k_s^2}{2m_\perp^*}\right)\left\{\left[\left(\frac{b_\parallel}{b_\perp}\frac{c_\perp}{c_\parallel}\right)\left(\frac{\delta}{2}+\frac{\Delta_\parallel^2-\Delta_\perp^2}{6\Delta_\parallel}\right)\alpha\int_{-\infty}^{E}\frac{F(V)dV}{[\alpha(E-V)+1]}\right.\right. \\
&\left.\left.+\left(\frac{\delta}{2}-\frac{\Delta_\parallel^2-\Delta_\perp^2}{6\Delta_\parallel}\right)c_\parallel\int_{-\infty}^{E}\frac{F(V)dV}{(c_\parallel(E-V)+1)}\right]\right\}
\end{aligned}
\tag{1.5}
$$

The (1.5) can be rewritten as [125–128]

$$
\begin{aligned}
&\frac{\hbar^2 k_z^2}{2m_\parallel^*}I(1)+\left(\frac{b_\parallel}{b_\perp}\frac{c_\perp}{c_\parallel}\right)\frac{\hbar^2 k_s^2}{2m_\perp^*}I(1) \\
&\quad=\left\{I_3(c_\parallel)+\frac{\alpha b_\parallel}{c_\parallel}\left[\delta I(4)+\frac{2}{9}\left(\Delta_\parallel^2-\Delta_\perp^2\right)I(1)\right]-\left(\frac{2}{9}\right)\frac{\alpha b_\parallel}{c_\parallel}\left(\Delta_\parallel^2-\Delta_\perp^2\right)I_6(c_\parallel)\right\} \\
&\quad-\left(\frac{\hbar^2 k_s^2}{2m_\perp^*}\right)\left\{\left(\frac{b_\parallel}{b_\perp}\frac{c_\perp}{c_\parallel}\right)\left[\left(\frac{\delta}{2}+\frac{\Delta_\parallel^2-\Delta_\perp^2}{6\Delta_\parallel}\right)\alpha I(\alpha)+\left(\frac{\delta}{2}-\left\{\frac{\Delta_\parallel^2-\Delta_\perp^2}{6\Delta_\parallel}\right\}\right)c_\parallel I(c_\parallel)\right]\right\}
\end{aligned}
\tag{1.6}
$$

where,

$$
I(1)\equiv\int_{-\infty}^{E}F(V)dV \tag{1.7}
$$

$$
I_3(c_\parallel)\equiv\int_{-\infty}^{E}\frac{(E-V)[\alpha(E-V)+1][b_\parallel(E-V)+1]}{[c_\parallel(E-V)+1]}F(V)dV \tag{1.8}
$$

$$I(4) \equiv \int_{-\infty}^{E} (E - V)F(V)dV \tag{1.9}$$

$$I(\alpha) \equiv \int_{-\infty}^{E} \frac{F(V)dV}{[\alpha(E - V) + 1]} \tag{1.10}$$

Substituting $E - V \equiv x$ and $x/\eta_g \equiv t_0$, we get from (1.7)

$$I(1) = \left(\exp\left(-E^2/\eta_g^2\right)/\sqrt{\pi}\right) \int_0^{\infty} \exp\left[-t_0^2 + \left(2Et_0/\eta_g\right)\right] dt_0$$

Thus,

$$I(1) = \left[\frac{1 + Erf(E/\eta_g)}{2}\right] \tag{1.11}$$

where, $Erf(E/\eta_g)$ is the error function of (E/η_g).

From (1.9), one can write

$$\begin{aligned} I(4) &= \left(1/\eta_g\sqrt{\pi}\right) \int_{-\infty}^{E} (E - V) \exp\left(-V^2/\eta_g^2\right) dV \\ &= \frac{E}{2}[1 + Erf(E/\eta_g)] - \left\{\frac{1}{\sqrt{\pi\eta_g^2}} \int_{-\infty}^{E} V \exp\left(-V^2/\eta_g^2\right) dV\right\} \end{aligned} \tag{1.12}$$

After computing this simple integration, one obtains

Thus,

$$I(4) = \eta_g \exp\left(-E/\eta_g^2\right)\left(2\sqrt{\pi}\right)^{-1} + \frac{E}{2}\left(1 + Erf(E/\eta_g)\right) = \gamma_0\left(E, \eta_g\right) \tag{1.13}$$

From (1.10), we can write

$$I(\alpha) = \frac{1}{\sqrt{\pi\eta_g^2}} \int_{-\infty}^{E} \frac{\exp\left(-V^2/\eta_g^2\right) dV}{[\alpha(E - V) + 1]} \tag{1.14}$$

When,

$$V \to \pm\infty, \frac{1}{[\alpha(E - V) + 1]} \to 0 \text{ and } \exp\left(-V^2/\eta_g^2\right) \to 0;$$

Thus (1.14) can be expressed as

$$I(\alpha) = \left(1/\alpha\eta_g\sqrt{\pi}\right) \int_{-\infty}^{\infty} \exp(-t^2)(u - t)^{-1} dt \tag{1.15}$$

where,

$$\frac{V}{\eta_g} \equiv t \text{ and } u \equiv \left(\frac{1 + \alpha E}{\alpha\eta}\right).$$

It is well known that [129, 130]

$$W(Z) = (i/\pi) \int_{-\infty}^{\infty} (Z - t)^{-1} \exp(-t^2) dt \tag{1.16}$$

In which $i = \sqrt{-1}$ and Z, in general, is a complex number.
We also know [129, 130],

$$W(Z) = \exp(-Z^2) Erfc(-iZ) \tag{1.17}$$

where,

$$Erfc(Z) \equiv 1 - Erf(Z).$$

Thus,

$$Erfc(-iu) = 1 - Erf(-iu)$$

Since,

$$Erf(-iu) = -Erf(iu)$$

Therefore,

$$Erfc(-iu) = 1 + Erf(iu).$$

Thus,

$$I(\alpha) = \left[-i\sqrt{\pi}/\alpha\eta_g\right] \exp(-u^2)[1 + Erf(iu)] \tag{1.18}$$

We also know that [129, 130]

$$Erf(x+iy) = Erf(x) + \left(\frac{e^{-x^2}}{2\pi x}\right)\left[(1-\cos(2xy)) + i\sin(2xy) + \frac{2}{\pi}e^{-x^2}\sum_{p=1}^{\infty}\frac{\exp(-p^2/4)}{(p^2+4x^2)}\right] \times \left[f_p(x,y) + ig_p(x,y) + \varepsilon(x,y)\right] \tag{1.19}$$

where,

$$f_p(x,y) \equiv [2x - 2x\cosh(py)\cos(2xy) + p\sinh(py)\sin(2xy)],$$
$$g_p(x,y) \equiv [2x\cosh(py)\sin(2xy) + p\sinh(py)\cos(2xy)],$$
$$|\varepsilon(x,y)| \approx 10^{-16}|Erf(x+iy)|$$

Substituting $x = 0$ and $y = u$ in (1.19), one obtains,

$$Erf(iu) = \left(\frac{2i}{\pi}\right)\sum_{p=1}^{\infty}\left\{\frac{\exp(-p^2/4)}{p}\sinh(pu)\right\} \tag{1.20}$$

Therefore, one can write

$$I(\alpha) = C_{21}(\alpha, E, \eta_g) - iD_{21}(\alpha, E, \eta_g) \tag{1.21}$$

where,

$$C_{21}(\alpha, E, \eta_g) \equiv \left[\frac{2}{\alpha\eta_g\sqrt{\pi}}\right]\exp(-u^2)\left[\sum_{P=1}^{\infty}\left\{\frac{\exp(-p^2/4)}{p}\sinh(pu)\right\}\right] \text{ and}$$
$$D_{21}(\alpha, E, \eta_g) \equiv \left[\frac{\sqrt{\pi}}{\alpha\eta_g}\exp(-u^2)\right].$$

The (1.21) consists of both real and imaginary parts and therefore, $I(\alpha)$ is complex, which can also be proved by using the method of analytic continuation of the subject Complex Analysis.

The integral $I_3(c_{\|})$ in (1.8) can be written as

$$I_3(c_{\|}) = \left(\frac{\alpha b_{\|}}{c_{\|}}\right) I(5) + \left(\frac{\alpha c_{\|} + b_{\|}c_{\|} - \alpha b_{\|}}{c_{\|}^2}\right) I(4) + \frac{1}{c_{\|}}\left(1 - \frac{\alpha}{c_{\|}}\right)\left(1 - \frac{b_{\|}}{c_{\|}}\right) I(1) - \left\{\frac{1}{c_{\|}}\left(1 - \frac{\alpha}{c_{\|}}\right)\left(1 - \frac{b_{\|}}{c_{\|}}\right) I(c_{\|})\right\} \quad (1.22)$$

where

$$I(5) \equiv \int_{-\infty}^{E} (E - V)^2 F(V) dV \quad (1.23)$$

From (1.23) one can write

$$I(5) = \frac{1}{\sqrt{\pi \eta_g^2}}\left[E^2 \int_{-\infty}^{E} \exp\left(\frac{-V^2}{\eta_g^2}\right) dV - 2E \int_{-\infty}^{E} V \exp\left(\frac{-V^2}{\eta_g^2}\right) dV + \int_{-\infty}^{E} V^2 \exp\left(\frac{-V^2}{\eta_g^2}\right) dV\right]$$

The evaluations of the component integrals lead us to write

$$I(5) = \frac{\eta_g E}{2\sqrt{\pi}} \exp\left(\frac{-E^2}{\eta_g^2}\right) + \frac{1}{4}\left(\eta_g^2 + 2E^2\right)\left[1 + Erf\left(\frac{E}{\eta_g}\right)\right] = \theta_0(E, \eta_g) \quad (1.24)$$

Thus combining the aforementioned equations, $I_3(c_{\|})$ can be expressed as

$$I_3(c_{\|}) = A_{21}(E, \eta_g) + iB_{21}(E, \eta_g) \quad (1.25)$$

where,

$$\begin{aligned} A_{21}(E, \eta) \equiv & \left[\frac{\alpha b_{\|}}{c_{\|}}\left[\frac{\eta_g E}{2\sqrt{\pi}} \exp\left(\frac{-E^2}{\eta_g^2}\right) + \frac{1}{4}\left(\eta_g^2 + 2E^2\right)\left\{1 + Erf\left(\frac{E}{\eta_g}\right)\right\}\right]\right] \\ & + \left[\frac{\alpha c_{\|} + b_{\|}c_{\|} - \alpha b_{\|}}{c_{\|}^2}\right]\left\{\frac{E}{2}[1 + Erf(E/\eta_g)] + \frac{\eta_g \exp\left(-E^2/\eta_g^2\right)}{2\sqrt{\pi}}\right\} \\ & + \frac{1}{c_{\|}}\left(1 - \frac{\alpha}{c_{\|}}\right)\left(1 - \frac{b_{\|}}{c_{\|}}\right)\frac{1}{2}[1 + Erf(E/\eta_g)] - \left\{\frac{2}{c_{\|}^2 \eta_g \sqrt{\pi}}\left(1 - \frac{\alpha}{c_{\|}}\right)\left(1 - \frac{b_{\|}}{c_{\|}}\right)\exp(-u_1^2)\right\} \\ & \times \left[\sum_{p=1}^{\infty}\left\{\frac{\exp(-p^2/4)}{p} \sinh(pu_1)\right\}\right], \end{aligned}$$

$$u_1 \equiv \left[\frac{1 + c_{\|}E}{c_{\|}\eta_g}\right] \text{ and } B_{21}(E, \eta_g) \equiv \frac{\sqrt{\pi}}{c_{\|}^2 \eta_g}\left(1 - \frac{\alpha}{c_{\|}}\right)\left(1 - \frac{b_{\|}}{c_{\|}}\right)\exp(-u_1^2).$$

Therefore, the combination of all the appropriate integrals together with algebraic manipulations leads to the expression of the dispersion relation of the conduction electrons of HD nonlinear optical materials forming Gaussian band tails as

$$\frac{\hbar^2 k_z^2}{2m_{\parallel}^* T_{21}(E,\eta_g)} + \frac{\hbar^2 k_s^2}{2m_{\perp}^* T_{22}(E,\eta_g)} = 1 \tag{1.26}$$

where, $T_{21}(E,\eta_g)$ and $T_{22}(E,\eta_g)$ have both real and complex parts and are given by

$$T_{21}(E,\eta_g) \equiv [T_{27}(E,\eta_g) + iT_{28}(E,\eta_g)], T_{27}(E,\eta_g) \equiv \left[\frac{T_{23}(E,\eta_g)}{T_5(E,\eta_g)}\right],$$

$$T_{23}(E,\eta_g) \equiv \left[A_{21}(E,\eta_g) + \frac{\alpha b_{\parallel}}{c_{\parallel}}\left[\delta\gamma_0(E,\eta_g) + \frac{1}{9}\left(\Delta_{\parallel}^2 - \Delta_{\perp}^2\right)[1 + Erf(E/\eta_g)]\right]\right.$$
$$\left. - \left\{\frac{2}{9}\left(\frac{\alpha b_{\parallel}}{c_{\parallel}}\right)\left(\Delta_{\parallel}^2 - \Delta_{\perp}^2\right) G_{21}(c_{\parallel}, E, \eta_g)\right\}\right],$$

$$G_{21}(E,\eta_g) \equiv \frac{2}{c_{\parallel}\eta_g\sqrt{\pi}}\exp(-u_1^2)\sum_{p=1}^{\infty}\left\{\frac{\exp(-p^2/4)}{p}\sinh(pu_1)\right\},$$

$$T_5(E,\eta_g) \equiv \frac{1}{2}[1 + Erf(E/\eta_g)],$$

$$T_{28}(E,\eta_g) \equiv \left[\frac{T_{24}(E,\eta_g)}{T_5(E,\eta_g)}\right], T_{24}(E,\eta_g) \equiv \left[B_{21}(E,\eta_g) + \frac{2}{9}\frac{\alpha b_{\parallel}}{c_{\parallel}}\left(\Delta_{\parallel}^2 - \Delta_{\perp}^2\right) H_{21}(c_{\parallel}, E, \eta_g)\right],$$

$$H_{21}(c_{\parallel}, E, \eta_g) \equiv \left[\frac{\sqrt{\pi}}{\eta_g c_{\parallel}}\exp(-u_1^2)\right], T_{22}(E,\eta_g) \equiv [T_{29}(E,\eta_g) + iT_{30}(E,\eta_g)],$$

$$T_{29}(E,\eta_g) \equiv \frac{T_{23}(E,\eta_g)T_{25}(E,\eta_g) - T_{24}(E,\eta_g)T_{26}(E,\eta_g)}{\left[(T_{25}(E,\eta_g))^2 + (T_{26}(E,\eta_g))^2\right]},$$

$$T_{25}(E,\eta_g) \equiv \left[\left(\frac{b_{\parallel}}{b_{\perp}}\frac{c_{\perp}}{c_{\parallel}}\right)\frac{1}{2}\left[1 + Erf\left(\frac{E}{\eta_g}\right)\right] + \left(\frac{b_{\parallel}}{b_{\perp}}\frac{c_{\perp}}{c_{\parallel}}\right)\left(\frac{\delta}{2} + \left[\frac{\Delta_{\parallel}^2 - \Delta_{\perp}^2}{6\Delta_{\parallel}}\right]\right)\alpha_{\parallel} C_{21}(\alpha_{\parallel}, E, \eta_g)\right.$$
$$\left. + \left(\frac{b_{\parallel}c_{\perp}}{b_{\perp}}\right)\left(\frac{\delta}{2} - \left[\frac{\Delta_{\parallel}^2 - \Delta_{\perp}^2}{6\Delta_{\parallel}}\right]\right) G_{21}(\alpha_{\parallel}, E, \eta_g),\right]$$

$$C_{21}(\alpha, E, \eta_g) \equiv \left[\frac{2}{\alpha\sqrt{\pi}\eta_g}\exp(-u^2)\left[\sum_{p=1}^{\infty}\frac{\exp(-p^2/4)}{p}\sinh(pu)\right]\right],$$

$$T_{26}(E,\eta_g) \equiv \left(\frac{b_{\parallel}}{b_{\perp}}\frac{c_{\perp}}{c_{\parallel}}\right)\left(\frac{\delta}{2} - \frac{\Delta_{\parallel}^2 - \Delta_{\perp}^2}{6\Delta_{\parallel}}\right)\alpha D_{21}(\alpha, E, \eta_g) + \frac{b_{\parallel}c_{\perp}}{b_{\perp}}\left(\frac{\delta}{2} - \frac{\Delta_{\parallel}^2 - \Delta_{\perp}^2}{6\Delta_{\parallel}}\right) H_{21}(c_{\parallel}, E, \eta_g)$$

and $$T_{30}(E,\eta_g) \equiv \frac{T_{24}(E,\eta_g)T_{25}(E,\eta_g) + T_{23}(E,\eta_g)T_{26}(E,\eta_g)}{\left[(T_{25}(E,\eta_g))^2 + (T_{26}(E,\eta_g))^2\right]}$$

From (1.26), it appears that the energy spectrum in HD nonlinear optical semiconductors is complex. **The complex nature of the electron dispersion law in HD semiconductors occurs from the existence of the essential poles in the corresponding electron energy spectrum in the absence of band tails.** It may be noted that the complex band structures have already been studied for bulk semiconductors and super lattices without heavy doping [131, 132] and bears no relationship with the complex electron dispersion law as indicated by (1.26). The physical picture behind the formulation of the complex energy spectrum in HDS is the interaction of the impurity atoms in the tails with the splitting constants of the valance bands. More is the interaction; more is the prominence of the complex part than the other case. In the absence of band tails, $\eta_g \to 0$, and there is no interaction of the impurity atoms in the tails with the spin orbit constants. As a result, there exist no complex energy spectrum and (1.26) gets converted into (1.2) when $\eta_g \to 0$. Besides, the complex spectra are not related to same evanescent modes in the band tails and the conduction bands.

It is interesting to note that the single important concept in the whole spectra of materials and allied sciences is the effective electron mass which is in disguise in the apparently simple (1.26), and can, briefly be described as follows:

Effective electron mass: The effective mass of the carriers in semiconductors, being connected with the mobility, is known to be one of the most important physical quantities, used for the analysis of electron devices under different operating conditions [133]. The carrier degeneracy in semiconductors influences the effective mass when it is energy dependent. Under degenerate conditions, only the electrons at the Fermi surface of n-type semiconductors participate in the conduction process and hence, the effective mass of the electrons corresponding to the Fermi level (EEM) would be of interest in electron transport under such conditions. The Fermi energy is again determined by the electron energy spectrum and the carrier statistics and therefore, these two features would determine the dependence of the effective electron mass in degenerate n-type semiconductors under the degree of carrier degeneracy. In recent years, various energy wave vector dispersion relations have been proposed [134–146] which have created the interest in studying the effective mass in such materials under external conditions. It has, therefore, different values in different materials and varies with electron concentration, with the magnitude of the reciprocal quantizing magnetic field under magnetic quantization, with the quantizing electric field as in inversion layers, with the nano-thickness as in UFs and nano wires and with superlattice period as in the quantum confined superlattices of small gap semiconductors with graded interfaces having various carrier energy spectra [147–178].

The transverse and the longitudinal EEMs at the Fermi energy (E_{F_h}) of HD nonlinear optical materials can, respectively, be expressed as

$$m_\perp^*(E_{F_h}, \eta_g) = m_\perp^* \{T_{29}(E, \eta_g)\}'\Big|_{E=E_{F_h}} \tag{1.27}$$

and

$$m_\parallel^*(E_{F_h}, \eta_g) = m_\parallel^* \{T_{27}(E, \eta_g)\}'\Big|_{E=E_{F_h}} \tag{1.28}$$

where E_{F_h} is the Fermi energy of HDS in the presence of band tails as measured from the edge of the conduction band in the vertically upward direction in the absence of band tails and the primes denote the differentiations of the differentiable functions with respect to Fermi energy in the appropriate case.

In the absence of band tails $\eta_g \to 0$ and we get

$$m_\perp^*(E_F, 0) = \frac{\hbar^2}{2}\left[\frac{\psi_2(E)\{\psi_1(E)\}' - \psi_1(E)\{\psi_2(E)\}'}{\{\psi_2(E)\}^2}\right]\Bigg|_{E=E_F} \tag{1.29}$$

and

$$m_\parallel^*(E_F, 0) = \frac{\hbar^2}{2}\left[\frac{\psi_3(E)\{\psi_1(E)\}' - \{\psi_1(E)\}\{\psi_3(E)\}'}{\{\psi_3(E)\}^2}\right]\Bigg|_{E=E_F} \tag{1.30}$$

where E_F is the Fermi energy as measured from the edge of the conduction band in the vertically upward direction in the absence of any perturbation, $\psi_1(E) = \gamma(E), \psi_2(E) = f_1(E)$ and $\psi_3(E) = f_2(E)$.

Comparing the aforementioned equations, one can infer that **the effective masses exist in the forbidden zone, which is impossible without the effect of band tailing. For semiconductors, in the absence of band tails the effective mass in the band gap is infinity**.

The DOS function is given by

$$N_{HD}(E, \eta_g) = \frac{2g_v m_\perp^* \sqrt{2m_\parallel^*}}{3\pi^2\hbar^3} R_{11}(E, \eta_g)\cos[\psi_{11}(E, \eta_g)] \tag{1.31a}$$

where,

$$R_{11}(E,\eta_g) \equiv \left[\left[\{T_{29}(E,\eta_g)\}'\sqrt{x(E,\eta_g)} + \frac{T_{29}(E,\eta_g)\{x(E,\eta_g)\}'}{2\sqrt{x(E,\eta_g)}} - \{T_{30}(E,\eta_g)\}'\sqrt{y(E,\eta_g)} - \frac{T_{30}(E,\eta_g)\{y(E,\eta_g)\}'}{2\sqrt{y(E,\eta_g)}}\right]^2 + \left[\{T_{29}(E,\eta_g)\}'\sqrt{y(E,\eta_g)} + \frac{T_{29}(E,\eta_g)\{y(E,\eta_g)\}'}{2\sqrt{y(E,\eta_g)}} + \{T_{30}(E,\eta_g)\}'\sqrt{x(E,\eta_g)} - \frac{T_{30}(E,\eta_g)\{x(E,\eta_g)\}'}{2\sqrt{x(E,\eta_g)}}\right]^2\right]^{1/2},$$

$$x(E,\eta_g) \equiv \frac{1}{2}\left[T_{27}(E,\eta_g) + \sqrt{\{T_{27}(E,\eta_g)\}^2 + \{T_{28}(E,\eta_g)\}^2}\right],$$

$$y(E,\eta_g) \equiv \frac{1}{2}\left[\sqrt{\{T_{27}(E,\eta_g)\}^2 + \{T_{28}(E,\eta_g)\}^2} - T_{27}(E,\eta_g)\right] \text{ and}$$

$$\psi_{11}(E,\eta_g) \equiv \tan^{-1}\left[\left[\{T_{29}(E,\eta_g)\}'\sqrt{y(E,\eta_g)} + \frac{T_{29}(E,\eta_g)}{2\sqrt{y(E,\eta_g)}} + \{T_{30}(E,\eta_g)\}'\sqrt{x(E,\eta_g)} + \frac{T_{30}\{x(E,\eta_g)\}'}{2\sqrt{x(E,\eta_g)}}\right] \times \left[\{T_{29}(E,\eta_g)\}'\sqrt{x(E,\eta_g)} + \frac{T_{29}(E,\eta_g)\{x(E,\eta_g)\}'}{2\sqrt{x(E,\eta_g)}} - \{T_{30}(E,\eta_g)\}'\sqrt{y(E,\eta_g)} + \frac{T_{30}\{y(E,\eta_g)\}'}{2\sqrt{y(E,\eta_g)}}\right]^{-1}\right].$$

The oscillatory nature of the DOS for HD nonlinear optical materials is apparent from (1.31a). For, $\psi_{11}(E,\eta_g) \geq \pi$, the cosine function becomes negative leading to the negative values of the DOS. The electrons cannot exist for the negative values of the DOS and therefore, this region is forbidden for electrons, which indicates that in the band tail, **there appears a new forbidden zone in addition to the normal band gap of the semiconductor.**

The use of (1.31a) leads to the expression of the electron concentration as

$$n_0 = \frac{2g_v m_\perp^* \sqrt{2m_\parallel^*}}{3\pi^2\hbar^3}\left[I_{11}\left(E_{F_h}, \eta_g\right) + \sum_{r=1}^{s} L(r)[I_{11}\left(E_{F_h}, \eta_g\right)]\right] \tag{1.31b}$$

where, $I_{11}\left(E_{F_h}, \eta_g\right) \equiv \left[T_{29}\left(E, \eta_g\right)\sqrt{x\left(E_{F_h}, \eta_g\right)} - T_{30}\left(E_{F_h}, \eta_g\right)\sqrt{y\left(E_{F_h}, \eta_g\right)}\right]$, $L(r) = 2(k_B T)^{2r}(1 - 2^{1-2r})\xi(2r)\frac{\partial^{2r}}{\partial(\text{Fermi energy})^{2r}}$, r is the set of real positive integers whose upper s and $\xi(2r)$ is the Zeta function of order 2r [129, 130].

The consequence of the photoelectric effect is the creation of the concept of photoelectric current density (J) which, can, in turn, be written through the photoemission integral (P_I) as [7]

$$J = \frac{\alpha_0 e}{4}(P_I) \tag{1.31c}$$

where,

$$P_I = \int_{E_0}^{\infty} N\left(E', \eta_g\right) v_z\left(E', \eta_g\right) f(E) dE' \tag{1.31d}$$

$E_0 \equiv \xi_i + W - h\nu$, ξ_i *is obtained by substituting* $k_x = 0, k_y = 0$ *and* $k_z = 0$ *in the dispersion relation*, W is the electron affinity, $E' \equiv E - E_0$, E is the total energy of the electron as measured from the edge of the conduction band in the vertically upward direction in the absence of any quantization, $N\left(E', \eta_g\right)$ is the DOS function at $E = E'$, $v_z\left(E', \eta_g\right)$ is the velocity of the emitted electron along z-axis when, $E = E'$,$f(E)$ is the Fermi Dirac occupation probability factor and can be written in this case as

$$f(E) = \left[1 + \exp\frac{\left(E - E_{F_h}\right)}{k_B T}\right]^{-1}. \tag{1.31e}$$

Thus combining the appropriate equations, the EP from HD non-linear materials can be expressed as

$$J = M_1 \text{ Real part of } \int_{E_0}^{\infty} N_1\left(\mathrm{E}, \eta_g\right) \mathrm{dE}' \tag{1.31f}$$

where

$$M_1 = \frac{\alpha_0 e g_v m_\perp^*}{3\pi^2\hbar^3} \text{ and } N_1(E,\eta_g) = \frac{\sqrt{T_{21}(\mathrm{E}',\eta_g)}}{T'_{21}(\mathrm{E}',\eta_g)} R_{11}(\mathrm{E}',\eta_g)\cos[\psi_{11}(\mathrm{E}',\eta_g)]\mathrm{f}(\mathrm{E})$$

For dimensional quantization along z-direction, the dispersion relation of the 2D electrons in this case can be written following (1.26) as

$$\frac{\hbar^2(\mathrm{n}_z\pi/\mathrm{d}_z)^2}{2m_\parallel^* T_{21}(E,\eta_g)} + \frac{\hbar^2 k_s^2}{2m_\perp^* T_{22}(E,\eta_g)} = 1 \tag{1.32}$$

where, $n_z (= 1, 2, 3, \ldots)$ and d_z are the size quantum number and the nano-thickness along the z-direction respectively.

The general expression of the total 2D DOS $(N_{2DT}(E))$ can, in general, be expressed as

$$N_{2DT}(E) = \frac{2g_v}{(2\pi)^2} \sum_{n_z=1}^{n_{z\max}} \frac{\partial A(E,n_z)}{\partial E} H(E - E_{n_z}) \tag{1.33}$$

where, g_v is the valley degeneracy, $A(E,n_z)$ is the area of the constant energy 2D wave vector space and in this case it is for QWs, $H(E - E_{n_z})$ is the Heaviside step function and E_{n_z} is the corresponding sub-band energy. Using (1.32) and (1.33), the expression of the $N_{2DT}(E)$ for QWs of HD nonlinear optical semiconductors can be written as

$$N_{2DT}(E) = \frac{m_\perp^* g_v}{\pi\hbar^2} \sum_{n_z=1}^{n_{z\max}} T'_{1D}(E,\eta_g,\mathrm{n}_z) H(\mathrm{E} - \mathrm{E}_{n_z D1}) \tag{1.34}$$

where, $T_{1D}(E,\eta_g,\mathrm{n}_z) = [1 - \frac{\hbar^2(\mathrm{n}_z\pi/\mathrm{d}_z)^2}{2m_\parallel^* T_{21}(E,\eta_g)}] T_{22}(E,\eta_g)$ and the sub band energies $E_{n_z D1}$ in this case is given by the following equation

$$\frac{\hbar^2(\mathrm{n}_z\pi/\mathrm{d}_z)^2}{2m_\parallel^* T_{21}(E_{n_z D1},\eta_g)} = 1 \tag{1.35}$$

Thus we observe that both the total DOS and sub-band energies of QWs of HD nonlinear optical semiconductors are complex due to the presence of the pole in energy axis of the corresponding materials in the absence of band tails.

The EEM in this case is given by

$$m^*(E_{F1HD}, \eta_g, \mathrm{n}_z) = m_\perp^*[\text{Real part of } T_{1D}'(E_{F1HD}, \eta_g, \mathrm{n}_z)] \tag{1.36}$$

Thus, we observe that the EEM is the function of size quantum number and the Fermi energy due to the combined influence of the crystal filed splitting constant and the anisotropic spin-orbit splitting constants respectively. Besides it is a function of η_g due to which the EEM exists in the band gap, which is otherwise impossible.

Combining (1.34) with the Fermi-Dirac occupation probability factor, integrating between E_{n_zD1} to infinity and applying the generalized Sommerfeld's lemma [179], the 2D carrier statistics in this case assumes the form

$$n_{2D} = \frac{m_\perp^* g_v}{\pi\hbar^2} \sum_{n_z=1}^{n_z\max} [\text{Real part of}[T_{1D}(E_{F1HD}, \eta_g, \mathrm{n}_z) + T_{2D}(E_{F1HD}, \eta_g, \mathrm{n}_z)]] \tag{1.37}$$

where, $T_{2D}(E_{F1HD}, \eta_g, \mathrm{n}_z) = \sum_{r=1}^{s} L(r)[T_{1D}(E_{F1HD}, \eta_g, \mathrm{n}_z)]$, E_{F1HD} is the Fermi energy in the presence of size quantization of the QWs of HD non-linear optical materials as measured from the edge of the conduction band in the vertically upward direction in the absence of any perturbation.

The photoelectric current density in QWs can, in general be written as

$$J_{2D} = \frac{\alpha_o e}{2d_z} \sum_{n_{z\min}}^{n_{z\max}} \int_{E_{n_z}}^{\infty} N_{2D}(E) f(E) v_z(E_{n_z}) dE \tag{1.38a}$$

where $N_{2D}(E)$ is the density-of-states function per sub band, $v_z(E_{n_z})$ is the velocity of the electron in the n_z^{th} sub band, the factor $\frac{1}{2}$ originates owing to the fact that only half of the electron will migrate towards the surface and escape [36].

Thus combining the appropriate equations the J_{2D} in this case assumes the form

$$J_{2D} = M_2 \,\text{Real part of} \sum_{n_{z\min}}^{n_{z\max}} \int_{\mathrm{E}_{n_zD1}}^{\infty} N_2(\mathrm{E}, \eta_g, \mathrm{n}_z) \mathrm{dE} \tag{1.38b}$$

where

$$n_{z_{\min}} \geq \frac{\sqrt{2m_\parallel^*}}{\hbar\pi} d_z \sqrt{T_{27}(\mathrm{W} - \mathrm{h}\nu, \eta_g)}, \quad M_2 = \frac{\alpha_0 e g_v}{d_z \pi\hbar^2} \left(\frac{m_\parallel^*}{2}\right)^{\frac{1}{2}}$$

and

$$N_2(\mathrm{E}, \eta_g, \mathrm{n}_z) = \frac{\sqrt{T_{21}(\mathrm{E}_{n_zD1}, \eta_g)}}{T'_{21}(E_{n_zD1}, \eta_g)} \mathrm{f(E)} T'_{1D}(\mathrm{E}, \eta_g, \mathrm{n}_z)$$

Therefore using (1.37) and (1.38b) we can study the EP in this case.

In the absence of band-tails, the 2D dispersion relation the EEM in the x-y plane at the Fermi level, the total 2D DOS, the sub-band energy $E_{n_{z1}}$, the surface electron concentration per unit area and the EP for QWs of non-linear optical materials in the absence of band tails can, respectively, be written as

$$\psi_1(E) = \psi_2(E)k_s^2 + \psi_3(E)(n_z\pi/d_z)^2 \tag{1.39}$$

$$m^*(E_{Fs}, n_z) = \left(\frac{\hbar^2}{2}\right)[\psi_2(E_{Fs})]^{-2}\left[\psi_2(E_{Fs})\left\{\{\psi_1(E_{Fs})\}' - \{\psi_3(E_{Fs})\}'\left(\frac{n_z\pi}{d_z}\right)^2\right\} - \left\{\psi_1(E_{Fs}) - \psi_3(E_{Fs})\left(\frac{n_z\pi}{d_z}\right)^2\right\}\{\psi_2(E_{Fs})\}'\right] \tag{1.40}$$

$$N_{2DT}(E) = \left(\frac{g_v}{2\pi}\right)\sum_{n_z=1}^{n_{z\max}}[\psi_2(E)]^{-2}\left[\psi_2(E)\left\{\{\psi_1(E)\}' - \{\psi_3(E)\}'\left(\frac{n_z\pi}{d_z}\right)^2\right\} - \left\{\psi_1(E) - \psi_3(E)\left(\frac{n_z\pi}{d_z}\right)^2\right\}\{\psi_2(E)\}'\right]H\left(E - E_{n_{z1}}\right) \tag{1.41}$$

$$\psi_1(E_{n_{z1}}) = \psi_2(E_{n_{z1}})(n_z\pi/d_z)^2 \tag{1.42}$$

$$n_{2D} = \frac{g_v}{2\pi}\sum_{n_z=1}^{n_{z\max}}[T_{51}(E_{Fs}, n_z) + T_{52}(E_{Fs}, n_z)] \tag{1.43}$$

$$J_{2D} = M_3 \sum_{n_{z\min}}^{n_{z\max}} \int_{E_{n_{z1}}}^{\infty} N_3(\mathrm{E}, \mathrm{n}_z)\mathrm{dE} \tag{1.44}$$

where

$$M_3 = \frac{\alpha_0 e g_v}{2\pi\hbar d_z}, N_3(\mathrm{E}, \mathrm{n}_z) = [[\frac{\psi_1(E_{n_{z1}})}{\psi_3(E_{n_{z1}})}]^{1/2}[\frac{\psi_3^2(E_{n_{z1}})}{\psi_3(E_{n_{z1}})\psi'_1(E_{n_{z1}}) - \psi'_3(E_{n_{z1}})\psi_1(E_{n_{z1}})}]$$
$$\times [\psi_2(E)]^{-2}[\psi_2(E)\{\{\psi_1(E)'\} - \{\psi_3(E)\}'(\frac{n_z\pi}{d_z})^2\} - \{\psi_1(E) - \psi_3(E)(\frac{n_z\pi}{d_z})^2\}\{\psi_2(E)\}']]\mathrm{f(E)},$$

$\psi_1(\mathrm{E}) = \gamma(\mathrm{E}), \psi_2(\mathrm{E}) = \mathrm{f}_1(\mathrm{E}), \psi_3(\mathrm{E}) = \mathrm{f}_2(\mathrm{E})$, sub band energies E_{Fs} is the Fermi energy in the 2-D sized quantized material in the presence of size quantization and in the absence of band-tails as measured from the edge of the conduction band in the vertically upward direction in the absence of any quantization,

$$T_{51}(E_{Fs}, n_z) \equiv \left[\frac{\psi_1(E_{Fs}) - \psi_3(E_{Fs})(n_z\pi/d_z)^2}{\psi_2(E_{Fs})}\right] \text{ and}$$

$$T_{52}(E_{Fs}, n_z) \equiv \sum_{r=1}^{s} L(r)[T_{51}(E_{Fs}, n_z)].$$

In the absence of band-tails, the DOS for bulk specimens of non-linear optical semiconductors is given by

$$D_0(E) = g_v(3\pi^2)^{-1}\psi_4(E) \tag{1.45}$$

$$\psi_4(E) \equiv \left[\frac{3}{2}\frac{\sqrt{\psi_1(E)}[\psi_1(E)]'}{\psi_2(E)\sqrt{\psi_3(E)}} - \frac{[\psi_2(E)]'[\psi_1(E)]^{3/2}}{[\psi_2(E)]^2\sqrt{\psi_3(E)}} - \frac{1}{2}\frac{[\psi_3(E)]'[\psi_1(E)]^{3/2}}{\psi_2(E)[\psi_3(E)]^{3/2}}\right],$$

$$[\psi_1(E)]' \equiv \left[(2E + E_g)\psi_1(E)\left[E(E + E_g)\right]^{-1} + E(E + E_g)\left(2E + 2E_g + \delta + \Delta_{\|}\right)\right],$$

$$[\psi_2(E)]' \equiv \left[2m_\perp^*\left(E_g + \frac{2}{3}\Delta_\perp\right)\right]^{-1}\left[\hbar^2 E_g(E_g + \Delta_\perp)\right]\left[\delta + 2E + 2E_g + \frac{2}{3}\Delta_{\|}\right]$$

$$\text{and } [\psi_3(E)]' \equiv \left[2m_{\|}^*\left(E_g + \frac{2}{3}\Delta_{\|}\right)\right]^{-1}\left[\hbar^2 E_g(E_g + \Delta_{\|})\right]\left[2E + 2E_g + \frac{2}{3}\Delta_{\|}\right]$$

Combining (1.45) with the Fermi-Dirac occupation probability factor and using the generalized Sommerfeld's lemma, the electron concentration can be written as

$$n_0 = g_v(3\pi^2)^{-1}[M(E_F) + N(E_F)] \tag{1.46a}$$

where, $M(E_F) \equiv \left[\frac{[\psi_1(E_F)]^{\frac{3}{2}}}{\psi_2(E_F)\sqrt{\psi_3(E_F)}}\right]$, E_F is the Fermi energy of the bulk specimen in the absence of band tailsfermi energy as measured from the edge of the conduction band in the vertically upward direction and $N(E_F) \equiv \sum_{r=1}^{s} L(r)M(E_F)$

In this case the EP is given by

$$J = M_4 \int_{E_0}^{\infty} N_4(E)dE' \tag{1.47}$$

where

$$M_4 = \frac{\alpha_0 e g_v}{6\pi^2\hbar} \text{ and } N_4(E) = \psi_4(E')f(E) \times \left(\frac{\psi_1(E')}{\psi_3(E')}\right)^{1/2}\left[\frac{\psi_3^2(E')}{\psi_3(E')\psi_1'(E') - \psi_3'(E')\psi_1(E')}\right].$$

1.2.2 *The EP from QWs of HD III-V Semiconductors*

The dispersion relation of the conduction electrons of III-V compounds are described by the models of Kane (both three and two bands) [38, 156, 157], Stillman et al. [40] and Palik et al. [42] respectively. For the purpose of complete and coherent presentation and relative comparison, the EP from QWs of HD III-V semiconductors have also been investigated in accordance with the aforementioned different dispersion relations as follows:

(a) The Three Band Model of Kane

Under the conditions, $\delta = 0, \Delta_{\parallel} = \Delta_{\perp} = \Delta$ (isotropic spin orbit splitting constant) and $m_{\parallel}^* = m_{\perp}^* = m_c$ (isotropic effective electron mass at the edge of the conduction band), (1.2) gets simplified as

$$\frac{\hbar^2 k^2}{2m_c} = I_{11}(E), \quad I_{11}(E) \equiv \frac{E(E + E_{g0})(E + E_{g_0} + \Delta)(E_{g_0} + \frac{2}{3}\Delta)}{E_{g_0}(E_{g_0} + \Delta)(E + E_{g_0} + \frac{2}{3}\Delta)} \tag{1.48}$$

which is known as the three band model of Kane [38, 39] and is often used to investigate the physical properties of III-V materials. Under the said conditions, the HD electron dispersion law in this case can be written from (1.26) as

$$\frac{\hbar^2 k^2}{2m_c} = T_{31}(E, \eta_g) + iT_{32}(E, \eta_g) \tag{1.49}$$

where,

$$T_{31}(E,\eta_g) \equiv \left(\frac{2}{1+Erf(E/\eta_g)}\right)\left[\frac{\alpha b}{c}\theta_0(E,\eta_g) + \left[\frac{\alpha c + bc - \alpha b}{c^2}\right]\gamma_0(E,\eta_g)\right.$$
$$+\frac{1}{c}\left(1-\frac{\alpha}{c}\right)\left(1-\frac{b}{c}\right)\frac{1}{2}\left[1+Erf\left(\frac{E}{\eta_g}\right)\right]$$
$$\left.-\frac{1}{c}\left(1-\frac{\alpha}{c}\right)\left(1-\frac{b}{c}\right)\frac{2}{c\eta_g\sqrt{\pi}}\exp(-u_2^2)\left[\sum_{p=1}^{\infty}\frac{\exp(-p^2/4)}{p}\sinh(pu_2)\right]\right],$$
$$b \equiv (E_g+\Delta)^{-1}, c \equiv \left(E+\frac{2}{3}\Delta\right)^{-1}, u_2 \equiv \frac{1+cE}{c\eta_g} \text{ and } T_{32}(E,\eta_g)$$
$$\equiv \left(\frac{2}{1+Erf(E/\eta_g)}\right)\frac{1}{c}\left(1-\frac{\alpha}{c}\right)\left(1-\frac{b}{c}\right)\frac{\sqrt{\pi}}{c\eta_g}\exp(-u_2^2).$$

Thus, the complex energy spectrum occurs due to the term $T_{32}(E,\eta_g)$ and this imaginary band is quite different from the forbidden energy band.

The EEM at the Fermi level is given by

$$m^*(E_{F_h},\eta_g) = m_c\{T_{31}(E,\eta_g)\}'\Big|_{E=E_{F_h}} \tag{1.50}$$

Thus, the EEM in HD III-V, ternary and quaternary materials exists in the band gap, which is the new attribute of the theory of band tailing.

In the absence of band tails, $\eta_g \to 0$ and the EEM assumes the form

$$m^*(E_F) = m_c\{I_{11}(E)\}'\big|_{E=E_F} \tag{1.51}$$

The DOS function in this case can be written as

$$N_{HD}(E,\eta_g) = \frac{g_v}{3\pi^2}\left(\frac{2m_c}{\hbar^2}\right)^{3/2} R_{21}(E,\eta_g)\cos[\vartheta_{21}(E,\eta_g)] \tag{1.52}$$

where,

$$R_{21}(E,\eta_g) \equiv \left[\frac{\left[\{\alpha_{11}(E,\eta_g)\}'\right]^2}{4\alpha_{11}(E,\eta_g)} + \frac{\left[\{\beta_{11}(E,\eta_g)\}'\right]^2}{4\beta_{11}(E,\eta_g)}\right]^{1/2},$$

$$\alpha_{11}(E,\eta_g) \equiv \frac{1}{2}\left[T_{33}(E,\eta_g) + \sqrt{\{T_{33}(E,\eta_g)\}^2 + \{T_{34}(E,\eta_g)\}^2}\right],$$

$$T_{33}(E,\eta_g) \equiv \left[\{T_{31}(E,\eta_g)\}^3 - 3T_{31}(E,\eta_g)\{T_{32}(E,\eta_g)\}^2\right],$$

$$T_{34}(E,\eta_g) \equiv \left[3T_{32}(E,\eta_g)\{T_{31}(E,\eta_g)\}^2 - \{T_{32}(E,\eta_g)\}^3\right],$$

$$\beta_{11}(E,\eta_g) \equiv \frac{1}{2}\left[\sqrt{\{T_{33}(E,\eta_g)\}^2 + \{T_{34}(E,\eta_g)\}^2} - T_{33}(E,\eta_g)\right] \text{ and}$$

$$\vartheta_{21}(E,\eta_g) \equiv \tan^{-1}\left[\frac{\{\beta_{11}(E,\eta_g)\}'}{\{\alpha_{11}(E,\eta_g)\}'}\sqrt{\frac{\alpha_{11}(E,\eta_g)}{\beta_{11}(E,\eta_g)}}\right].$$

Thus, the oscillatory DOS function becomes negative for $\vartheta_{21}(E,\eta_g) \geq \pi$ and a new forbidden zone will appear in addition to the normal band gap.

The electron concentration can be expressed as

$$n_0 = \frac{g_v}{3\pi^2}\left(\frac{2m_c}{\hbar^2}\right)^{3/2}\left[\bar{I}_{111e}(E_{F_h},\eta_g) + \sum_{r=1}^{s} L(r)[\bar{I}_{111e}(E,\eta_g)]\right] \tag{1.53}$$

where,

$$\bar{I}_{111e}(E_{F_h},\eta_g) = \{\gamma_2(E_{F_h},\eta_g)\}^{3/2}$$

The EP in this case is given by

$$J = M_5 \,\text{Real part of} \int_{\mathrm{E}_0}^{\infty} N_5(\mathrm{E},\eta_g)\mathrm{dE}' \tag{1.54}$$

where

$$M_5 = \frac{\alpha_0 e g_v m_c}{12\pi^2\hbar^3} \text{ and } N_5(\mathrm{E},\eta_g)$$

$$= \frac{\sqrt{T_{31}(\mathrm{E}',\eta_g) + iT_{32}(\mathrm{E}',\eta_g)}}{T'_{31}(\mathrm{E}',\eta_g) + iT'_{32}(\mathrm{E}',\eta_g)} R_{21}(\mathrm{E}',\eta_g)\cos[\vartheta_{21}(\mathrm{E}',\eta_g)]\mathrm{f(E)}$$

For dimensional quantization along z-direction, the dispersion relation of the 2D electrons in this case can be written following (1.49) as

$$\frac{\hbar^2(\mathrm{n}_z\pi/\mathrm{d}_z)^2}{2m_c}+\frac{\hbar^2(k_s)^2}{2m_c}=T_{31}(E,\eta_g)+iT_{32}(E,\eta_g) \tag{1.55}$$

The expression of the $N_{2DT}(E)$ in this case assumes the form

$$N_{2DT}(E)=\frac{m_c g_v}{\pi\hbar^2}\sum_{n_z=1}^{n_{z\max}}T'_{5D}(E,\eta_g,\mathrm{n}_z)H(\mathrm{E}-\mathrm{E}_{n_zD5}) \tag{1.56}$$

where,

$$T_{5D}(E,\eta_g,\mathrm{n}_z)=[T_{31}(E,\eta_g)+iT_{32}(E,\eta_g)-\hbar^2(\mathrm{n}_z\pi/\mathrm{d}_z)^2(2m_c)^{-1}]$$

and the sub band energies E_{n_zD5} in this case given by

$$\{\hbar^2(n_z/d_z)^2\}(2m_c)^{-1}=T_{31}(E_{n_zD5},\eta_g) \tag{1.57}$$

Thus we observe that both the total DOS in QWs of HD III-V compounds and the sub band energies are complex due to the presence of the pole in energy axis of the corresponding materials in the absence of band tails.

The EEM in this case is given by

$$m^*(E_{F1HD},\eta_g,\mathrm{n}_z)=m_c[T'_{31}(E_{F1HD},\eta_g,\mathrm{n}_z)] \tag{1.58}$$

Therefore under the same conditions as used in obtaining (1.48) from (1.2), the 2D carrier statistics in this case can be written by using the same conditions from (1.37) as

$$n_{2D}=\frac{m_c g_v}{\pi\hbar^2}\sum_{n_z=1}^{n_{z\max}}[\text{Real part of }[\mathrm{T}_{5D}(E_{F1HD},\eta_g,\mathrm{n}_z)+T_{6D}(E_{F1HD},\eta_g,\mathrm{n}_z)]] \tag{1.59}$$

where,

$$T_{6D}(E_{F1HD},\eta_g,\mathrm{n}_z)=\sum_{r=1}^{s}L(r)[T_{5D}(E_{F1HD},\eta_g,\mathrm{n}_z)],$$

The EP in this is given by

$$J_{2D}=M_6\,\text{Real part of}\sum_{n_{z\min}}^{n_{z\max}}\int_{\mathrm{E}_{n_zD5}}^{\infty}N_6(\mathrm{E},\eta_g,\mathrm{n}_z)\mathrm{dE} \tag{1.60a}$$

where

$$M_6 = \frac{\alpha_0 e g_v m_c}{2\pi\hbar^2 d_z}\left(\frac{2}{m_c}\right)^{1/2} \text{ and } N_6(\mathrm{E}, \eta_g, \mathrm{n}_z) = \left[\frac{\sqrt{T_{31}(\mathrm{E}_{n_z D5}, \eta_g) + iT_{32}(\mathrm{E}_{n_z D5}, \eta_g)}}{T'_{31}(\mathrm{E}_{n_z D5}, \eta_g) + iT'_{32}(\mathrm{E}_{n_z D5}, \eta_g)} \mathrm{T}'_{5D}(\mathrm{E}, \eta_g, \mathrm{n}_z)\mathrm{f}(\mathrm{E})\right]$$

In the absence of band tails, the 2D dispersion relation, EEM in the x-y plane at the Fermi level, the total 2D DOS, the sub-band energy, the electron concentration and the EP for QWs of III-V materials assume the following forms

$$\frac{\hbar^2 k_s^2}{2m_c} + \frac{\hbar^2}{2m_c}(n_z\pi/d_z)^2 = I_{11}(E) \tag{1.60b}$$

$$m^*(E_{Fs}) = m_c\{I_{11}(E_{Fs})\}' \tag{1.61}$$

It is worth noting that the EEM in this case is a function of Fermi energy alone and is independent of size quantum number.

$$N_{2DT}(E) = \left(\frac{m_c g_v}{\pi\hbar^2}\right)\sum_{n_z=1}^{n_{z\max}}\left\{[I_{11}(E)]' H\left(E - E_{n_{z_2}}\right)\right\} \tag{1.62}$$

where, the sub-band energies $E_{n_{z_2}}$ can be expressed as

$$I_{11}(E_{n_{z_2}}) = \frac{\hbar^2}{2m_c}(n_z\pi/d_z)^2 \tag{1.63}$$

$$n_{2D} = \frac{m_c g_v}{\pi\hbar^2}\sum_{n_z=1}^{n_{z\max}}[T_{53}(E_{Fs}, n_z) + T_{54}(E_{Fs}, n_z)] \tag{1.64}$$

$$J_{2D} = M_7 \sum_{n_{z\min}}^{n_{z\max}} \int_{E_{n_{z2}}}^{\infty} N_7(\mathrm{E}, \mathrm{n}_z)\mathrm{dE} \tag{1.65}$$

where

$$T_{53}(E_{Fs}, n_z) \equiv \left[I_{11}(E_{Fs}) - \frac{\hbar^2}{2m_c}\left(\frac{n_z\pi}{d_z}\right)^2\right], T_{54}(E_{Fs}, n_z) \equiv \sum_{r=1}^{s} L(r)T_{53}(E_{Fs}, n_z),$$

$$M_7 = \frac{\alpha_0 e g_v}{2\pi\hbar^2 d_z}\left(\frac{m_c}{2}\right)^{1/2} \text{ and } N_7(\mathrm{E}, \mathrm{n}_z) = I'_{11}(\mathrm{E})\mathrm{f}(\mathrm{E})\frac{\sqrt{I_{11}(\mathrm{E}_{n_{z2}})}}{I'_{11}(\mathrm{E}_{n_{z2}})}$$

In the absence of band tails, the DOS function, the electron concentration, and the EP in bulk III-V, ternary and quaternary materials in accordance with the unperturbed three band model of Kane assume the following forms

$$D_0(E) = 4\pi g_v\left(\frac{2m_c}{h^2}\right)^{3/2}\sqrt{I_{11}(E)}\left[I'_{11}(E)\right] \tag{1.66}$$

$$n_0 = \frac{g_v}{3\pi^2}\left(\frac{2m_c}{\hbar^2}\right)^{3/2}\left[M_1(E_F) + N_1(E_F)\right] \tag{1.67}$$

and

$$J = \left[\frac{4\pi e m_c (K_BT)^2\alpha_0 g_v}{h^3}\right]\left[\frac{\left(1+\frac{2}{3}\alpha\Delta\right)}{(1+\alpha\Delta)}\left[2\alpha k_BTF_2(\eta_0) + \left(1+2\alpha E_0+\frac{1}{3}\alpha\Delta\right)F_1(\eta_0)\right.\right.$$
$$\left.\left.+\frac{E_0+\alpha E_0^2+\frac{1}{3}\alpha\Delta E_0}{k_BT}F_{-1}(\eta_0) + a_0\left[\ln\left|\frac{\overline{a_0}+\eta_0}{\overline{a_0}}\right| + \phi(\eta_0)\right]\right]\right] \tag{1.68}$$

where,

$$I'_{11}(E) \equiv I_{11}(E)\left[\frac{1}{E} + \frac{1}{E+E_g} + \frac{1}{E+E_g+\Delta} - \frac{1}{E+E_g+\frac{2}{3}\Delta}\right], M_1(E_F) \equiv \left[I_{11}(E_F)\right]^{3/2},$$

$$N_1(E_F) \equiv \sum_{r=1}^{s} L(r)M_1(E_F), a_0 = \frac{E_0+E_{g_0}+\frac{2}{3}\Delta}{k_BT}, \overline{a_0} = \frac{2}{9}\frac{\Delta^2}{(k_BT)^2}\left(1+\frac{2}{3}\alpha\Delta\right),$$

$$\phi(\eta_0) = \sum_{r=1}^{S_0} 2\left(1-2^{1-2r}\right)\zeta(2r)\frac{(-1)^{2r-1}(2r-1)!}{(\overline{a_0}+\eta_0)^{2r}}, \eta_0 = \frac{hv-\phi}{k_BT}$$

and $F_j(\eta)$ is the one parameter Fermi-Dirac integral of order j which can be written [180−182] as

$$F_j(\eta) = \left(\frac{1}{\Gamma(j+1)}\right)\int_0^\infty \frac{x^j dx}{1+\exp(x-\eta)}, \quad j > -1 \tag{1.69}$$

or for all j, analytically continued as a complex contour integral around the negative x-axis

$$F_j(\eta) = \left(\frac{\Gamma(-j)}{2\pi\sqrt{-1}}\right)\int_{-\infty}^{+0} \frac{x^j dx}{1+\exp(-x-\eta)} \tag{1.70}$$

where η is the dimensionless quantity and x is independent variable,

Under the inequalities $\Delta \gg E_{g0}$ or $\Delta \ll E_{g_0}$, (1.48) can be expressed as

$$E(1+\alpha E) = \frac{\hbar^2 k^2}{2m_c} \tag{1.71}$$

where $\alpha \equiv \left(E_{g_0}\right)^{-1}$ and is known as band non-parabolicity.

It may be noted that (1.71) is the well-known two band model of Kane and is used in the literature to study the physical properties of those III-V and opto-electronic materials whose energy band structures obey the aforementioned inequalities.

The dispersion relation in HD III-V, ternary and quaternary materials whose energy spectrum in the absence of band tails obeys the two band model of Kane as defined by (1.71), can be written as

$$\frac{\hbar^2 k^2}{2m_c} = \gamma_2\left(E, \eta_g\right) \tag{1.72}$$

where,

$$\gamma_2\left(E, \eta_g\right) \equiv \left[\frac{2}{1+Erf\left(E/\eta_g\right)}\right]\left[\gamma_0\left(E, \eta_g\right) + \alpha\theta_0\left(E, \eta_g\right)\right].$$

The EEM in this case can be written as

$$m^*\left(E_{F_h}, \eta_g\right) = m_c\left\{\gamma_2\left(E, \eta_g\right)\right\}'\Big|_{E=E_{F_h}} \tag{1.73a}$$

Thus, one again observes that the EEM in this case exists in the band gap.

In the absence of band tails, $\eta_g \to 0$ and the EEM assumes the well-known form

$$m^*(E_F) = m_c\{1+2\alpha E\}|_{E=E_F} \tag{1.73b}$$

The DOS function in this case can be written as

$$N_{HD}\left(E, \eta_g\right) = \frac{gv}{2\pi^2}\left(\frac{2m_c}{\hbar^2}\right)^{3/2}\sqrt{\gamma_2\left(E, \eta_g\right)}\left\{\gamma_2\left(E, \eta_g\right)\right\}' \tag{1.73c}$$

Since, the poles of the original two band Kane model are at infinity and and no finite poles with respect to energy, therefore the HD counterpart will be totally real and the complex band vanishes.

The electron concentration is given by

$$n_0 = \frac{g_v}{3\pi^2}\left(\frac{2m_c}{\hbar^2}\right)^{3/2}\left[\bar{I}_{111}(E_{F_h},\eta_g) + \sum_{r=1}^{s} L(r)[\bar{I}_{111}(E_{F_h},\eta_g)]\right] \tag{1.74}$$

where,

$$\bar{I}_{111}(E_{F_h},\eta_g) = \{\gamma_2(E_{F_h},\eta_g)\}^{3/2}$$

The EP in this case is given by

$$J = M_7 \int_{\mathrm{E}_0}^{\infty} N_7(\mathrm{E},\eta_g)\mathrm{dE}' \tag{1.75}$$

where

$$M_7 = \frac{\alpha_0 e g_v m_c}{2\pi^2\hbar^3} \text{ and } N_7(\mathrm{E},\eta_g) = \gamma_2(\mathrm{E}',\eta_g)\mathrm{f}(\mathrm{E})$$

For dimensional quantization along z-direction, the dispersion relation of the 2D electrons in this case can be written following (1.70) as

$$\frac{\hbar^2(\mathrm{n}_z\pi/\mathrm{d}_z)^2}{2m_c} + \frac{\hbar^2(k_s)^2}{2m_c} = \gamma_2(E,\eta_g) \tag{1.76}$$

The expression of the $N_{2DT}(E)$ in this case can be written as

$$N_{2DT}(E) = \frac{m_c g_v}{\pi\hbar^2}\sum_{n_z=1}^{n_{z\max}} T'_{7D}(E,\eta_g,\mathrm{n}_z)H(\mathrm{E}-\mathrm{E}_{n_zD7}) \tag{1.77}$$

where,

$$T_{7D}(E,\eta_g,\mathrm{n}_z) = [\gamma_2(E,\eta_g) - \hbar^2(\mathrm{n}_z\pi/\mathrm{d}_z)^2(2m_c)^{-1}],$$

The sub-band energies E_{n_zD7} in this case given by

$$\left\{\hbar^2(\mathrm{n}_z\pi/\mathrm{d}_z)^2\right\}(2m_c)^{-1} = \gamma_2(E_{\mathrm{n}_z\mathrm{D}7},\eta_g) \tag{1.78}$$

Thus, we observe that both the total DOS and sub-band energies of QWs of HD III-V compounds in accordance with two band model of Kane are not at all complex since the dispersion relation in accordance with the said model has no pole in the finite complex plane.

The EEM in this case is given by

$$m^*(E_{F1HD}, \eta_g, \mathrm{n}_z) = m_c[\gamma_2'(E_{F1HD}, \eta_g, \mathrm{n}_z)] \tag{1.79}$$

Therefore under the same conditions as used in obtaining (1.48) from (1.2), the 2D carrier statistics in this case can be written by using the same conditions from (1.77) as

$$n_{2D} = \frac{m_c g_v}{\pi\hbar^2} \sum_{n_z=1}^{n_{z\max}} [T_{7D}(E_{F1HD}, \eta_g, \mathrm{n}_z) + T_{8D}(E_{F1HD}, \eta_g, n_z)] \tag{1.80}$$

where,

$$T_{8D}(E_{F1HD}, \eta_g, \mathrm{n}_z) = \sum_{r=1}^{s} L(r)[T_{7D}(E_{F1HD}, \eta_g, \mathrm{n}_z)],$$

The EP in this case is given by

$$J_{2D} = M_8 \sum_{n_{z\min}}^{n_{z\max}} \int_{\mathrm{E}_{n_z D7}}^{\infty} N_8(\mathrm{E}, \eta_{g_0}, \mathrm{n}_z)\mathrm{dE} \tag{1.81}$$

where

$$M_8 = \frac{\alpha_0 e g_v m_c}{2\pi d_z \hbar^2}(\frac{2}{m_c})^{1/2} \text{ and } N_8(\mathrm{E}, \eta_g, \mathrm{n}_z) = T_{7D}'(\mathrm{E}, \eta_g, \mathrm{n}_z)\mathrm{f(E)} \times \frac{\gamma_2(\mathrm{E}_{n_z D7}, \eta_g)}{\gamma_2'(\mathrm{E}_{n_z D7}, \eta_g)}$$

Under the inequalities $\Delta \gg E_{g_0}$ or, $\Delta \ll E_{g_0}$ (1.60a), (1.160b) assumes the form

$$E(1 + \alpha E) = \frac{\hbar^2 k_s^2}{2m_c} + \frac{\hbar^2}{2m_c}\left(\frac{n_z \pi}{d_z}\right)^2 \tag{1.81a}$$

The EEM can be written from (1.81a) as

$$m^*(E_{F_s}) = m_c(1 + 2\alpha E_{F_s}) \tag{1.81b}$$

The total 2D DOS function assumes the form

$$N_{2DT}(E) = \frac{m_c g_v}{\pi\hbar^2} \sum_{n_z=1}^{n_{z\max}} (1 + 2\alpha E) H\left(E - E_{n_{z3}}\right) \tag{1.82}$$

where, the sub-band energy $(E_{n_{z3}})$ can be expressed as

$$\frac{\hbar^2}{2m_c}(n_z\pi/d_z)^2 = E_{n_{z3}}\left(1+\alpha E_{n_{z3}}\right) \tag{1.83}$$

The 2D electron statistics can be written as

$$\begin{aligned} n_{2D} &= \frac{m_c g_v}{\pi\hbar^2}\sum_{n_z=1}^{n_{z\max}}\int_{E_{n_3}}^{\infty}\frac{(1+2\alpha E)dE}{1+\exp\left(\frac{E-E_{F_s}}{k_BT}\right)} \\ &= \frac{m_c k_B T g_v}{\pi\hbar^2}\sum_{n_z=1}^{n_{z\max}}\left[(1+2\alpha E_{n_{z3}})F_0(\eta_{n_1})+2\alpha k_BTF_1(\eta_{n_1})\right] \end{aligned} \tag{1.84}$$

where,

$$\eta_{n_1} \equiv (E_{Fs}-E_{n_{z3}})/k_BT$$

The EP in this case the can be written as

$$\begin{aligned} J_{2D} = {} & [\frac{\alpha_0 e g_v m_c}{2\pi\hbar^2 d_z}(\frac{2}{m_c})^{1/2}][\sum_{n_{z\min}}^{n_{z\max}}\frac{\sqrt{E_{z_{\max}}(1+\alpha E_{n_z3})}}{(1+2\alpha E_{n_z3})}[(1+2\alpha E_{n_z3})F_0(\eta_{n_1}) \\ & +2\alpha k_BTF_0(\eta_{n_1})]] \end{aligned} \tag{1.85}$$

The forms of the DOS, the electron statistics and the EP for bulk specimens of III-V materials in the absence of band tails whose energy band structures are defined by the two-band model of Kane can, respectively, be written as

$$D_0(E) = 4\pi g_v\left(\frac{2m_c}{h^2}\right)^{3/2}\sqrt{I_{11e}(E)}\left[I'_{11e}(E)\right] \tag{1.86}$$

$$n_0 = \frac{g_v}{3\pi^2}\left(\frac{2m_c}{\hbar^2}\right)^{3/2}[M_2(E_F)+N_2(E_F)] \tag{1.89}$$

and

$$J = \frac{4\pi\alpha_0 e m_c (k_BT)^2}{h^3}[F_1(\eta_0)+2\alpha k_B T F_2(\eta_0)] \tag{1.90}$$

where,

$$I_{11e}(E) \equiv E(1+\alpha E), I'_{11e}(E) \equiv (1+2\alpha E), M_2(E_F) \equiv [I_{11e}(E_F)]^{3/2}$$

and

$$N_2(E_F) \equiv \sum_{r=1}^{s} L(r) M_2(E_F)$$

Under the constraints $\Delta \gg E_g$ or $\Delta \ll E_g$ together with the inequality $\alpha E_F \ll 1$, the (1.89) assumes the forms as

$$n_0 = g_v N_c \left[F_{1/2}(\eta) + \left(\frac{15\alpha k_B T}{4} \right) F_{3/2}(\eta) \right] \tag{1.91}$$

where,

$$N_C \equiv 2 \left(\frac{2\pi m_c k_B T}{h^2} \right)^{3/2}$$

and

$$\eta \equiv \frac{E_F}{k_B T}$$

The dispersion relation in HDS whose energy spectrum in the absence of band tails obeys the parabolic energy bands (1.69) is given by

$$\frac{\hbar^2 k^2}{2m_c} = \gamma_3(E, \eta_g) \tag{1.92}$$

where,

$$\gamma_3(E, \eta_g) = \left[\frac{2}{(1 + Erf(E/\eta_g))} \right] \gamma_0(E, \eta_g). \tag{1.93}$$

Since the dispersion relation in accordance with the said model is an all zero function with no pole in the finite complex plane, therefore the HD counterpart will be totally real, which is also apparent form the expression (1.92).

The EEM in this case can be written as

$$m^*\left(E_{F_h}, \eta_g\right) = m_c\left\{\gamma_3\left(E_{F_h}, \eta_g\right)\right\}' \tag{1.94}$$

In the absence of band tails, $\eta_g \to 0$ and the EEM assumes the form

$$m^*(E_F) = m_c \tag{1.95}$$

It is well-known that the EEM in unperturbed parabolic energy bands is a constant quantity in general excluding cross-fields configuration. However, the same mass in the corresponding HD bulk counterpart becomes a complicated function of Fermi energy and the impurity potential together with the fact that the EEM also exists in the band gap solely due to the presence of finite η_g.

The DOS function in this case can be written as

$$N_{HD}\left(E, \eta_g\right) = \frac{g_v}{2\pi^2}\left(\frac{2m_c}{\hbar^2}\right)^{3/2}\sqrt{\gamma_3\left(E, \eta_g\right)}\left\{\gamma_3\left(E, \eta_g\right)\right\}' \tag{1.96}$$

The electron concentration is given by

$$n_0 = \frac{g_v}{3\pi^2}\left(\frac{2m_c}{\hbar^2}\right)^{3/2}\left[\bar{I}_{113}\left(E_{F_h}, \eta_g\right) + \sum_{r=1}^{s} L(r)[\bar{I}_{113}\left(E_{F_h}, \eta_g\right)]\right] \tag{1.97}$$

where,

$$\bar{I}_{113}\left(E, \eta_g\right) = \left\{\gamma_3\left(E_{F_h}, \eta_g\right)\right\}^{3/2}$$

The EP in this case is given by

$$J = M_7\int_{\mathrm{E}_0}^{\infty} N_8(\mathrm{E}, \eta_g)\mathrm{dE}' \tag{1.98}$$

where

$$N_8(\mathrm{E}, \eta_g) = \gamma_3(\mathrm{E}', \eta_g)\mathrm{f}(\mathrm{E})$$

For dimensional quantization along z-direction, the dispersion relation of the 2D electrons in this case can be written following (1.93) as

$$\frac{\hbar^2(\mathrm{n}_z\pi/\mathrm{d}_z)^2}{2m_c} + \frac{\hbar^2(k_s)^2}{2m_c} = \gamma_3\left(E, \eta_g\right) \tag{1.99}$$

the expression of the $N_{2DT}(E)$ in this case can be written as

$$N_{2DT}(E) = \frac{m_c g_v}{\pi\hbar^2} \sum_{n_z=1}^{n_{z\max}} T'_{9D}(E, \eta_g, \mathrm{n}_z) H(\mathrm{E} - \mathrm{E}_{n_z D9}) \tag{1.100}$$

where,

$$T_{9D}(E, \eta_g, \mathrm{n}_z) = [\gamma_3(E, \eta_g) - \hbar^2 (n_z \pi/\mathrm{d}_z)^2 (2m_c)^{-1}].$$

The sub band energies $E_{n_z D9}$ in this case given by

$$\{\hbar (n_z \pi/d_z)^2\}(2m_c)^{-1} = \gamma_3(E_{n_z D9}, \eta_g) \tag{1.101}$$

The EEM in this case can be written as

$$m^*(E_{F1HD}, \eta_g, n_z) = m_c [\gamma'_3(E_{F1HD}, \eta_g)] \tag{1.102}$$

Therefore under the same conditions as used in obtaining (1.48) from (1.2), the 2D carrier statistics in this case can be written by using the same conditions from (1.77) as

$$n_{2D} = \frac{m_c g_v}{\pi\hbar^2} \sum_{n_z=1}^{n_{z\max}} [T_{9D}(E_{F1HD}, \eta_g, \mathrm{n}_z) + T_{10D}(E_{F1HD}, \eta_g, \mathrm{n}_z)] \tag{1.103}$$

where,

$$T_{10D}(E_{F1HD}, \eta_g, \mathrm{n}_z) = \sum_{r=1}^{s} L(r) [T_{9D}(E_{F1HD}, \eta_g, \mathrm{n}_z)],$$

The EP in this case is given by

$$J_{2D} = M_8 \sum_{n_{z\min}}^{n_{z\max}} \int_{\mathrm{E}_{n_z D9}}^{\infty} N_9(\mathrm{E}, \eta_{g0}, \mathrm{n}_z) \mathrm{dE} \tag{1.104a}$$

where

$$N_9(\mathrm{E}, \eta_g, \mathrm{n}_z) = T'_{9D}(\mathrm{E}, \eta_g, \mathrm{n}_z) \mathrm{f(E)} \times \frac{\gamma_3(\mathrm{E}_{n_z D9}, \eta_g)}{\gamma'_3(\mathrm{E}_{n_z D9}, \eta_g)}$$

Under the condition $\alpha \to 0$, the expressions of total 2D DOS, for semiconductors without forming band tails whose bulk electrons are defined by the isotropic parabolic energy bands can, be written as

$$N_{2DT}(E) = \frac{m_c g_v}{\pi \hbar^2} \sum_{n_z=1}^{n_{z\max}} H\left(E - E_{n_{zp}}\right) \tag{1.104b}$$

The sub-band energy $\left(E_{n_{zp}}\right)$, the n_{2D} and the EP can, respectively, be expressed as

$$E_{n_{zp}} = \frac{\hbar^2}{2m_c}\left(\frac{n_z \pi}{d_z}\right)^2 \tag{1.105}$$

$$n_{2D} = \frac{m_c k_B T g_v}{\pi \hbar^2} \sum_{n_z=1}^{n_{z\max}} F_0(\eta_{n_2}) \tag{1.106a}$$

$$J_{2D} = \frac{\alpha_o e k_B T g_v}{2\hbar d_z^2} \sum_{n_{z\min}}^{n_{z\max}} n_z \left[F_0\left(\eta_{n_2}\right)\right], \tag{1.106b}$$

where,

$$n_{z_{\min}} \geq \left(\frac{d_z}{\pi}\right)\left(\frac{\sqrt{2m_c}}{\hbar}\right)(W - h\nu)^{1/2} \text{ and } \eta_{n_2} \equiv \frac{1}{k_B T}\left[E_{Fs} - \frac{\hbar^2}{2m_c}\left(\frac{n_z \pi}{d_z}\right)^2\right]$$

Converting the summation over n_z to the integration over n_z, (1.106b) gets transformed to the well-known relation as [38, 39]

$$J = \frac{4\pi \alpha_0 e m_c g_v (k_B T)^2}{h^3} F_1(\eta_0) \tag{1.106c}$$

This indirect test not only exhibits the mathematical compatibility of our formulation but also shows the fact that our simple analysis is a more generalized one, since one can obtain the corresponding results for relatively wide gap 2D materials having parabolic energy bands under certain limiting conditions from our present derivation.

(b) The Model of Stillman et al.

In accordance with the model of Stillman et al. [40], the electron dispersion law of III-V materials assumes the form

$$E = \overline{t_{11}} k^2 - \overline{t_{12}} k^4 \tag{1.107}$$

where,

$$\overline{t_{11}} \equiv \frac{\hbar^2}{2m_c};$$

$$\bar{t}_{12} \equiv \left(1 - \frac{m_c}{m_0}\right)^2 \left(\frac{\hbar^2}{2m_c}\right)^2 \left[\left(3E_{g_0} + 4\Delta + \frac{2\Delta^2}{E_{g_0}}\right) \cdot \left\{\left(E_{g_0} + \Delta\right)\left(2\Delta + 3E_{g_0}\right)\right\}^{-1}\right]$$

and m_0 is the free electron mass

In the presence of band tails, (1.107) gets transformed as

$$\frac{\hbar^2 k^2}{2m_c} = I_{12}(E, \eta_g) \quad (1.108)$$

where,

$$I_{12}(E, \eta_g) = a_{11}[1 - (1 - a_{12}\gamma_3(E\eta_g))^{\frac{1}{2}}], a_{11} \equiv \left(\frac{\hbar^2 \bar{t}_{11}}{4m_c \bar{t}_{12}}\right) \text{ and } a_{12} \equiv \frac{4\bar{t}_{12}}{\bar{t}_{11}^2}$$

The EEM can be written as

$$m^*\left(E_{F_h}, \eta_g\right) = m_c\left\{I_{12}\left(E_{F_h}, \eta_g\right)\right\}' \quad (1.109)$$

The DOS function in this case can be written as

$$N_{HD}\left(E, \eta_g\right) = \frac{g_v}{2\pi^2}\left(\frac{2m_c}{\hbar^2}\right)^{3/2} \sqrt{I_{12}\left(E, \eta_g\right)}\left\{I_{12}\left(E, \eta_g\right)\right\}' \quad (1.110)$$

The electron concentration is given by

$$n_0 = \frac{g_v}{3\pi^2}\left(\frac{2m_c}{\hbar^2}\right)^{3/2}\left[\bar{I}_{121}\left(E_{F_h}, \eta_g\right) + \sum_{r=1}^{s} L(r)[\bar{I}_{121}\left(E, \eta_g\right)]\right] \quad (1.111)$$

where,

$$\bar{I}_{121}\left(E_{F_h}, \eta_g\right) = \left\{I_{12}\left(E_{F_h}, \eta_g\right)\right\}^{3/2}$$

The EP in this is given by

$$J = M_7 \int_{\mathrm{E}_0}^{\infty} N_9(\mathrm{E}, \eta_g)\mathrm{dE}' \quad (1.112)$$

where

$$N_9(\mathrm{E}, \eta_g) = I_{12}(\mathrm{E}', \eta_g)\mathrm{f}(\mathrm{E})$$

For dimensional quantization along z-direction, the dispersion relation of the 2D electrons in this case can be written following (1.108) as

$$\frac{\hbar^2(\mathrm{n}_z\pi/\mathrm{d}_z)^2}{2m_c} + \frac{\hbar^2(k_s)^2}{2m_c} = I_{12}(E, \eta_g) \tag{1.113}$$

the expression of the $N_{2DT}(E)$ in this case can be written as

$$N_{2DT}(E) = \frac{m_c g_v}{\pi\hbar^2} \sum_{n_z=1}^{n_{zmzx}} T'_{11D}(E, \eta_g, \mathrm{n}_z) H(\mathrm{E} - \mathrm{E}_{n_zD11}) \tag{1.114}$$

where,

$$T_{11D}(E, \eta_g, \mathrm{n}_z) = [I_{12}(E, \eta_g) - \hbar^2(\mathrm{n}_z\pi/\mathrm{d}_z)^2(2m_c)^{-1}],$$

The sub band energies E_{n_zD11} in this case given by

$$\left\{\hbar^2(\mathrm{n}_z\pi/d_z)^2\right\}(2m_c)^{-1} = I_{12}(\mathrm{E}_{n_zD11}, \eta_g) \tag{1.115}$$

The EEM in this case assumes the form

$$m^*(E_{F1HD}, \eta_g, \mathrm{n}_z) = m_c[I'_{12}(E_{F1HD}, \eta_g, \mathrm{n}_z)] \tag{1.116}$$

The 2-D electron statistics in this case can be written as

$$n_{2D} = \frac{m_c g_v}{\pi\hbar^2} \sum_{n_z=1}^{n_{z\max}} [T_{11D}(E_{F1HD}, \eta_g, \mathrm{n}_z) + T_{12D}(E_{F1HD}, \eta_g, \mathrm{n}_z)] \tag{1.117a}$$

where,

$$T_{12D}(E_{F1HD}, \eta_g, \mathrm{n}_z) = \sum_{r=1}^{s} L(r)[T_{11D}(E_{F1HD}, \eta_g, \mathrm{n}_z)],$$

The EP in this is given by

$$J_{2D} = M_8 \sum_{n_{z\min}}^{n_{z\max}} \int_{\mathrm{E}_{n_zD11}}^{\infty} N_{10}(\mathrm{E}, \eta_g, \mathrm{n}_z)\mathrm{dE} \tag{1.117b}$$

where

$$N_9(\mathrm{E}, \eta_g, \mathrm{n}_z) = T'_{11D}(E, \eta_g, \mathrm{n}_z)\mathrm{f}(\mathrm{E}) \times \frac{I_{12}(\mathrm{E}_{n_zD11}, \eta_g)}{I'_{12}(E_{n_zD11}, \eta_g)}$$

For unperturbed material, the 2-D EEM can be expressed as

$$m^*(E_{Fs}) = m_c\{I_{12}(E_{Fs})\}' \quad (1.118)$$

where

$$I_{12}(E) \equiv a_{11}[1 - a_{12}(E))^{\frac{1}{2}}]$$

It appears that the EEM in this case is a function of Fermi energy alone and is independent of size quantum number.

The total 2D DOS function in the absence of band tails in this case can be written as

$$N_{2DT}(E) = \left(\frac{m_c g_v}{\pi\hbar^2}\right) \sum_{n_z=1}^{n_{z\max}} \left\{ [I_{12}(E)]' H\left(E - E_{n_{z3}}\right) \right\} \quad (1.119)$$

where, the sub band energies $E_{n_{z3}}$ can be expressed as

$$I_{12}(E_{n_{z3}}) = \frac{\hbar^2}{2m_c}(n_z\pi/d_z)^2 \quad (1.120)$$

The 2D electron concentration assumes the form

$$n_{2D} = \frac{m_c g_v}{\pi\hbar^2} \sum_{n_z=1}^{n_{z\max}} [T_{55}(E_{Fs}, n_z) + T_{56}(E_{Fs}, n_z)] \quad (1.121)$$

where

$$T_{55}(E_{Fs}, n_z) \equiv \left[I_{12}(E_{Fs}) - \frac{\hbar^2}{2m_c}\left(\frac{n_z\pi}{d_z}\right)^2\right] \text{ and } T_{56}(E_{Fs}, n_z) \equiv \sum_{r=1}^{s} L(r) T_{55}(E_{Fs}, n_z)$$

The EP in this case is given by

$$J_{2D} = \frac{\alpha_0 e g_v m_c}{2\pi\hbar^2 d_z}\left(\frac{2}{m_c}\right)^{1/2}\left[\sum_{n_{z\min}}^{n_{z\max}} \left[\frac{\sqrt{I_{12}(E_{n_z3})}}{I'_{12}(E_{n_z3})}\right] \int_{\mathrm{E}_{n_z3}}^{\infty} I'_{12}(\mathrm{E})\mathrm{f}(\mathrm{E})\mathrm{dE}\right] \quad (1.122a)$$

The expression of electron concentration for bulk specimens of III-V semiconductors (in the absence of band tails) can be written in accordance with the model of Stillman et al. as

$$n_0 = \frac{g_v}{3\pi^2}\left(\frac{2m_c}{\hbar^2}\right)^{3/2}[M_{A_{10}}(E_F) + N_{A_{10}}(E_F)] \tag{1.122b}$$

where,

$$M_{A_{10}}(E_F) = [I_{12}(E_F)]^{3/2} \text{ and } N_{A_{10}}(E_F) = \sum_{r=1}^{s} L(r)[M_{A_{10}}(E_F)]$$

The EP in this case can be expressed as

$$J = \frac{4\pi\alpha_0 e g_v m_c}{h^3}\int_{E_0}^{\infty} I_{12}(E')\mathrm{f(E)dE'} \tag{1.122c}$$

(c) Model of Palik et al.

The energy spectrum of the conduction electrons in III-V semiconductors up to the fourth order in effective mass theory, taking into account the interactions of heavy hole, light hole and the split-off holes can be expressed in accordance with the model of Palik et al. [42] as

$$E = \frac{\hbar^2 k^2}{2m_c} - \bar{B}_{11}k^4 \tag{1.123}$$

where

$$\bar{B}_{11} = \left[\frac{\hbar^4}{4E_{g_0}(m_c)^2}\right]\left[\frac{1 + \frac{x_{11}^2}{2}}{1 + \frac{x_{11}}{2}}\right](1 - y_{11})^2,\ x_{11} = \left[1 + \left(\frac{\Delta}{E_{g_0}}\right)\right]^{-1} \text{ and } y_{11} = \frac{m_c}{m_o}$$

The (1.123) gets simplified as

$$\frac{\hbar^2 k^2}{2m_c} = I_{13}(E) \tag{1.124}$$

where

$$I_{13}(E) = \bar{b}_{12}\left[\bar{a}_{12} - \left((\bar{a}_{12})^2 - 4E\bar{B}_{11}\right)^{1/2}\right],\quad \bar{a}_{12} = \left(\frac{\hbar^2}{2m_c}\right)$$

and $\bar{b}_{12} = \left[\frac{\bar{a}_{12}}{2\bar{B}_{11}}\right]$

Under the condition of heavy doping forming Gaussian band tails, (1.124) assumes the form

$$\frac{\hbar^2 k^2}{2m_c} = I_{13}(E, \eta_g) \tag{1.125}$$

where,

$$I_{13}(E, \eta_g) = \bar{b}_{12}[\bar{a}_{12} - ((\bar{a}_{12})^2 - 4\bar{B}_{11}\gamma_3(E, \eta_g))^{1/2}]$$

The EEM can be written as

$$m^*(E_{F_h}, \eta_g) = m_c\{I_{13}(E_{F_h}, \eta_g)\}' \tag{1.126}$$

The DOS function in this case can be expressed as

$$N_{HD}(E, \eta_g) = \frac{g_v}{2\pi^2}\left(\frac{2m_c}{\hbar^2}\right)^{3/2}\sqrt{I_{13}(E, \eta_g)}\{I_{13}(E, \eta_g)\}' \tag{1.127}$$

Since, the original band model in this case is a no pole function, in the finite complex plane therefore, the HD counterpart will be totally real and the complex band vanishes.

The electron concentration is given by

$$n_0 = \frac{g_v}{3\pi^2}\left(\frac{2m_c}{\hbar^2}\right)^{3/2}\left[\bar{I}_{123}(E_{F_h}, \eta_g) + \sum_{r=1}^{s} L(r)[\bar{I}_{123}(E_{F_h}, \eta_g)]\right] \tag{1.128}$$

where,

$$\bar{I}_{123}(E_{F_h}, \eta_g) = \{I_{13}(E_{F_h}, \eta_g)\}^{3/2}$$

The EP in this is given by

$$J = M_7 \int_{\mathrm{E}_0}^{\infty} N_{11}(E, \eta_g)\mathrm{dE}' \tag{1.129}$$

where

$$N_{11}(E, \eta_g) = I_{13}(E', \eta_g)\mathrm{f(E)}$$

For dimensional quantization along z-direction, the dispersion relation of the 2D electrons in this case can be written following (1.108) as

$$\frac{\hbar^2 (\mathrm{n}_z \pi / d_z)^2}{2m_c} + \frac{\hbar^2 (k_s)^2}{2m_c} = I_{13}(E, \eta_g) \tag{1.130}$$

the expression of the $N_{2DT}(E)$ in this case can be written as

$$N_{2DT}(E) = \frac{m_c g_v}{\pi \hbar^2} \sum_{n_z}^{n_{z\max}} T'_{13D}(E, \eta_g, n_z) H(E - E_{n_z D13}) \tag{1.131}$$

where,

$$T_{13D}(E, \eta_g, \mathrm{n}_z) = [I_{13}(E, \eta_g) - \hbar^2 (\mathrm{n}_z \pi / \mathrm{d}_z)^2 (2m_c)^{-1}],$$

The sub band energies $E_{n_z D13}$ in this case given by

$$\{\hbar (n_z \pi / d_z)^2\} (2m_c)^{-1} = I_{13}(E_{n_z D13}, \eta_g) \tag{1.132}$$

The EEM in this case can be expressed as

$$m^*(E_{F1HD}, \eta_g, n_z) = m_c [I'_{13}(E_{F1HD}, \eta_g, n_z)] \tag{1.133}$$

The 2-D electron statistics in this case can be written as

$$n_{2D} = \frac{m_c g_v}{\pi \hbar^2} \sum_{n_z=1}^{n_z \max} [T_{13D}(E_{F1HD}, \eta_g, \mathrm{n}_z) + T_{14D}(E_{F1HD}, \eta_g, \mathrm{n}_z)] \tag{1.134}$$

where,

$$T_{14D}(E_{F1HD}, \eta_g, \mathrm{n}_z) = \sum_{r=1}^{s} L(r) [T_{13D}(E_{F1HD}, \eta_g, \mathrm{n}_z)],$$

The EP in this is given by

$$J_{2D} = M_8 \sum_{n_{z\min}}^{n_{z\max}} \int_{\mathrm{E}_{n_z D13}}^{\infty} N_{13}(\mathrm{E}, \eta_{g0}, \mathrm{n}_z) \mathrm{dE} \tag{1.135a}$$

where

$$N_{13}(\mathrm{E},\eta_g,\mathrm{n}_z)=T'_{13D}(\mathrm{E},\eta_g,\mathrm{n}_z)\mathrm{f}(\mathrm{E})\times\frac{I_{13}(E_{n_{zD13}},\eta_g)}{I'_{13}(\mathrm{E}_{n_{zD13}},\eta_g)}$$

The 2D electron dispersion relation in the absence of band tails this case assumes the form

$$\frac{\hbar^2k_s^2}{2m_c}+\frac{\hbar^2}{2m_c}(n_z\pi/d_z)^2=I_{13}(E) \tag{1.135b}$$

The EEM in this case can be written from (1.135b) as

$$m^*(E_{F_S})=m_c[I_{13}(E_{F_S})]' \tag{1.135c}$$

The total 2D DOS function can be written as

$$N_{2DT}(E)=\left(\frac{m_cg_v}{\pi\hbar^2}\right)\sum_{n_z=1}^{n_{z\max}}\left\{[I_{13}(E)]'H\left(E-E_{n_{z4}}\right)\right\} \tag{1.136}$$

where, the sub band energies $E_{n_{z4}}$ can be expressed as

$$I_{13}(E_{n_{z4}})=\frac{\hbar^2}{2m_c}(n_z\pi/d_z)^2 \tag{1.137}$$

The 2D electron concentration assumes the form

$$n_{2D}=\frac{m_cg_v}{\pi\hbar^2}\sum_{n_z=1}^{n_{z\max}}[T_{57}(E_{Fs},n_z)+T_{58}(E_{Fs},n_z)] \tag{1.138}$$

where

$$T_{57}(E_{Fs},n_z)\equiv\left[I_{13}(E_{Fs})-\frac{\hbar^2}{2m_c}\left(\frac{n_z\pi}{d_z}\right)^2\right]\text{ and}$$

$$T_{58}(E_{Fs},n_z)\equiv\sum_{r=1}^{s}L(r)T_{57}(E_{Fs},n_z)$$

The EP in this case is given by

$$J_{2D}=\frac{\alpha_0eg_vm_c}{2\pi\hbar^2d_z}\left(\frac{2}{m_c}\right)^{1/2}\left[\sum_{n_{z\min}}^{n_{z\max}}\left[\frac{\sqrt{I_{13}(\mathrm{E}_{n_{z4}})}}{I'_{13}(\mathrm{E}_{n_{z4}})}\right]\int_{\mathrm{E}_{n_{z4}}}^{\infty}I'_{13}(\mathrm{E})\mathrm{f}(\mathrm{E})\mathrm{dE}\right] \tag{1.139a}$$

The expression of electron concentration for bulk specimens of III-V semiconductors (in the absence of band tails) can be written in accordance with the model of Stillman et al. as

$$n_0 = \frac{g_v}{3\pi^2}\left(\frac{2m_c}{\hbar^2}\right)^{3/2}[\bar{M}_{A_{10}}(E_F) + \bar{N}_{A_{10}}(E_F)] \tag{1.139b}$$

where,

$$\bar{M}_{A_{10}}(E_F) = [I_{13}(E_F)]^{3/2} \text{ and } \bar{N}_{A_{10}}(E_F) = \sum_{r=1}^{s} L(r)[\bar{M}_{A_{10}}(E_F)]$$

The EP in this case can be expressed as

$$J = \frac{4\pi\alpha_0 e g_v m_c}{h^3}\int_{E_0}^{\infty} I_{13}(E')\mathrm{f(E)dE'}. \tag{1.139c}$$

1.2.3 The EP in QWs of HD II-VI Semiconductors

The carrier energy spectra in bulk specimens of II-VI compounds in accordance with Hopfield model [59] can be written as

$$E = a_o' k_s^2 + b_o' k_z^2 \pm \bar{\lambda}_o k_s \tag{1.140}$$

where $a_0' \equiv \hbar^2/2m_\perp^*, b_0' \equiv \hbar^2/2m_\parallel^*$, and $\bar{\lambda}_0$ represents the splitting of the two-spin states by the spin orbit coupling and the crystalline field.

Therefore the dispersion relation of the carriers in HD II-VI materials in the presence of Gaussian band tails can be expressed as

$$\gamma_3(E,\eta_g) = a_0' k_s^2 + b_0' k_z^2 \pm \bar{\lambda}_0 k_s \tag{1.141}$$

Thus, the energy spectrum in this case is real since the corresponding E-k relation in the absence of band tails as given by (1.141) is a no pole function in the finite complex plane.

The transverse and the longitudinal EEMs masses are, respectively, given by

$$m_\perp^*(E_{F_h},\eta_g) = m_\perp^*\{\gamma_3(E,\eta_g)\}'\left[1 \pm \left(\frac{\bar{\lambda}_0}{\sqrt{(\bar{\lambda}_0)^2 + 4a_0'\gamma_3(E,\eta_g)}}\right)\right]\Bigg|_{E=E_{F_h}} \tag{1.142}$$

and

$$m_{\parallel}^{*}(E_{F_h},\eta_g) = m_{\parallel}^{*}\{\gamma_3(E,\eta_g)\}'\Big|_{E=E_{F_h}} \tag{1.143}$$

Thus the transverse EEM in HD II-VI semiconductors is a function of electron energy and is double valued due to the presence of $\bar{\lambda}_0$ and due to heavy doping the same mass exists in the band gap.

In the absence of band tails, $\eta_g \to 0$, we get

$$m_{\perp}^{*}(E_F) = m_{\perp}^{*}\left[1 \pm \left(\frac{\bar{\lambda}_0}{\sqrt{(\bar{\lambda}_0)^2 + 4a_0'E}}\right)\right]\Bigg|_{E=E_F} \tag{1.144}$$

and

$$m_{\parallel}^{*}(E_F) = m_{\parallel}^{*} \tag{1.145}$$

The volume in k-space as enclosed (1.141) can be expressed as

$$V(E,\eta_g) = \frac{4\pi}{3a_0'\sqrt{b_0'}}\left[\{\gamma_3(E,\eta_g)\}^{3/2} + \frac{3}{8}\frac{(\bar{\lambda}_0)^2\sqrt{\gamma_3(E,\eta_g)}}{a_0'} \pm \left(\frac{3}{4}\frac{\bar{\lambda}_0}{\sqrt{a_0'}}\right)\right.$$
$$\left.\times\left(\gamma_3(E,\eta_g) + \frac{(\bar{\lambda}_0)^2}{4a_0'}\right)\sin^{-1}\left[\frac{\sqrt{\gamma_3(E,\eta_g)}}{\sqrt{\gamma_3(E,\eta_g) + \frac{(\bar{\lambda}_0)^2}{4a_0'}}}\right]\right] \tag{1.146}$$

Therefore, the electron concentration can be written as

$$n_0 = \frac{g_v}{3\pi^2 a_0'\sqrt{b_0'}}\left[\bar{I}_{124}(E_{F_h},\eta_g) + \sum_{r=1}^{s} L(r)[\bar{I}_{124}(E_{F_h},\eta_g)]\right] \tag{1.147}$$

where,

$$\bar{I}_{124}(E_{F_h},\eta_g) = \left[\{\gamma_3(E_{F_h},\eta_g)\}^{3/2} + \frac{3}{8}\frac{(\bar{\lambda}_0)^2\sqrt{\gamma_3(E_{F_h},\eta_g)}}{a_0'}\right]$$

The EP in this case is given by

$$J = \frac{\alpha_0 e g_v}{12\pi^2 a_0' \sqrt{b_0'}} \left(\frac{2}{m_{11}}\right)^{1/2} \int_{E_0}^{\infty} \frac{\sqrt{\gamma_3(E', \eta_g)}}{\gamma_3'(E', \eta_g)} [\bar{I}_{124}(E', \eta_g)]' \mathrm{f(E)dE'} \tag{1.148}$$

The dispersion relation of the conduction electrons of QWs of HD II-VI materials for dimensional quantization along z-direction can be written following (1.141) as

$$\gamma_3(E, \eta_g) = a_0' k_s^2 + b_0' \left(\frac{\pi n_z}{d_z}\right)^2 \pm \bar{\lambda}_0 k_s \tag{1.149}$$

The EEM can be expressed following (1.149) as

$$m^*(E_{F1HD}, n_z, \eta_g) = m_\perp^* [1 \mp \frac{(\bar{\lambda}_0)\gamma_3'(E_{F1HD}, \eta_g)}{[(\bar{\lambda}_0)^2 - 4a_0' b_0' \left(\frac{n_z \pi}{d_z}\right)^2 + 4a_0' \gamma_3(E_{F1HD}, \eta_g)]^{1/2}} \tag{1.150}$$

Thus we observe that the doubled valued effective mass in 2-D QWs of HD II-VI materials is a function of Fermi energy, size quantum number and the screening potential respectively together with the fact that the same mass exists in the band gap due to the sole presence of the splitting of the two-spin states by the spin orbit coupling and the crystalline field.

The sub-band energy in this case is given by

$$\gamma_3(E_{n_Z D14}, \eta_g) = b_0' \left(\frac{\pi n_z}{d_z}\right)^2 \tag{1.151}$$

The surface electron concentration at low temperatures assumes the form

$$n_{2D} = \frac{g_v m_\perp^*}{\pi \hbar^2} \sum_{n_Z=1}^{n_{Z\max}} \left(\gamma_3(E_{F1HD}, \eta_g) - E_{n_Z D14} + (\bar{\lambda}_0)^2 m_\perp^* \hbar^{-2}\right) \tag{1.152}$$

The EP in this cases given by

$$J_{2D} = \frac{\alpha_0 e g_v}{\pi d_z \hbar^2} m_\perp^* \left(\frac{2}{m_{11}^*}\right)^{1/2} [\sum_{n_{z\min}}^{n_{z\max}} [\frac{\sqrt{\gamma_3(E_{n_z D14}, \eta_g)}}{\gamma_3'(E_{n_{n_z D14}}, \eta_s)}] \int_{E_{n_z D14}}^{\infty} \gamma_3(E, \eta_g) f(E) dE] \tag{1.153a}$$

The dispersion relation of the conduction electrons of QWs of II-VI materials for dimensional quantization along z-direction in the absence of band tails can be written following (1.140) as

$$E = a_0' k_s^2 + b_0' \left(\frac{n_z \pi}{d_z}\right)^2 \pm \bar{\lambda}_0 k_s \tag{1.153b}$$

Using (1.153b), the EEM in this case can be written as

$$m^*(E_{Fs}, n_z) = m_\perp^* [1 \mp \frac{(\bar{\lambda}_0)}{[(\bar{\lambda}_0)^2 - 4a_0' b_0' \left(\frac{n_z \pi}{d_z}\right)^2 + 4a_0' E_{Fs}]^{1/2}}] \tag{1.154}$$

The sub-band energy $E_{n_{z5}}$ assumes the form

$$E_{n_z 5} = b_0' (n_z \pi / d_z)^2 \tag{1.155}$$

The area of constant energy 2D quantized surface in this case is given by where

$$A_\pm(E, n_z) = \left[\frac{\pi}{2(a_0')^2}\left[(\overline{\lambda_0})^2 + 2a_0'\left(E - E_{n_{z5}}\right) \pm \overline{\lambda_0}\left[(\overline{\lambda_0})^2 + 4a_0'(E - E_{n_{z5}})\right]^{1/2}\right]\right]$$

The surface electron concentration can be expressed in this case as

$$n_{2D} = \frac{-2g_v}{2(2\pi)^2} \sum_{n_z=1}^{n_{z\max}} \int_{E_{n_{z5}}}^{\infty} [A_+(E_{Fs}, n_z) + A_-(E_{Fs}, n_z)] \frac{\partial}{\partial E} \{f_0(E)\} dE \tag{1.156}$$

where $f_0(E)$ is the Fermi-Dirac occupation probability factor.

From (1.156) we get

$$n_{2D} = \frac{g_v m_\perp^* k_B T}{\pi \hbar^2} \sum_{n_z=1}^{n_{z\max}} F_0(\eta_{n_{z8}}) \tag{1.157}$$

where

$$\eta_{n_{z8}} = \left(E_{Fs} - E_{n_{z5}} + (\bar{\lambda})^2 m_\perp^* \hbar^{-2}\right)(k_B T)^{-1}$$

Therefore the EP is given by

$$J_{2D} = k_B T \frac{\alpha_0 e g_v}{\pi d_z \hbar^2} m_\perp^* \left(\frac{2}{m_{11}^*}\right)^{1/2} [\sum_{n_{z\min}}^{n_{z\max}} [\sqrt{E_{n_z D5}}] \mathrm{F}_1(\eta_{n_z 8})]. \tag{1.158}$$

1.2.4 The EP from QWs of HD IV-VI Semiconductors

(a) The dispersion relation of the conduction electrons in IV-VI semiconductors can be expressed in accordance with Dimmock [183] as

$$\left[\bar{\varepsilon}-\frac{E_{g0}}{2}-\frac{\hbar^2k_s^2}{2m_t^-}-\frac{\hbar^2k_z^2}{2m_l^-}\right]\left[\bar{\varepsilon}+\frac{E_{g0}}{2}+\frac{\hbar^2k_s^2}{2m_t^-}+\frac{\hbar^2k_z^2}{2m_l^-}\right]=P_\perp^2k_s^2+P_\parallel^2k_z^2 \quad (1.159)$$

where, $\bar{\varepsilon}$ is the energy as measured from the center of the band gap E_{g0}, $m_t^\pm$ and $m_l^\pm$ represent the contributions to the transverse and longitudinal effective masses of the external L_6^+ and L_6^- bands arising from the $\vec{k}\cdot\vec{p}$ perturbations with the other bands taken to the second order.

Substituting, $P_\perp^2 \equiv \left(\hbar^2E_g/2m_t^*\right), P_\parallel^2 \equiv \left(\frac{\hbar^2E_g}{2m_l^*}\right)$ and $\bar{\varepsilon} \equiv [E+(\frac{E_g}{2})]$ (where, m_t^* and m_l^* are the transverse and the longitudinal effective masses at k = 0), (1.159) gets transformed as

$$\left[E-\frac{\hbar^2k_s^2}{2m_t^-}-\frac{\hbar^2k_z^2}{2m_l^-}\right]\left[1+\alpha E+\alpha\frac{\hbar^2k_s^2}{2m_t^+}+\alpha\frac{\hbar^2k_z^2}{2m_l^+}\right]=\frac{\hbar^2k_s^2}{2m_t^*}+\frac{\hbar^2k_z^2}{2m_l^*} \quad (1.160)$$

From (1.160), we can write

$$\frac{\alpha\hbar^4k_s^4}{4m_t^+m_t^-}+\hbar^2k_s^2\left[\left(\frac{1}{2m_t^*}-\frac{1}{2m_t^-}\right)+\alpha E\left(\frac{1}{2m_t^-}-\frac{1}{2m_t^+}\right)+\frac{\alpha\hbar^2k_z^2}{4m_l^-m_t^+}\right]$$
$$+\left[\left(\frac{\hbar^2k_z^2}{2m_l^*}+\frac{\hbar^2k_z^2}{2m_l^-}\right)+\frac{\alpha E}{2}\hbar^2k_z^2\left(\frac{1}{m_l^-}-\frac{1}{m_l^+}\right)+\frac{\alpha\hbar^4k_z^4}{4m_l^+m_t^-}-E(1+\alpha E)\right]=0 \quad (1.161)$$

Using (1.161), the dispersion relation of the conduction electrons in HD IV-VI materials can be expressed as

$$\frac{\alpha\hbar^4k_s^4}{4m_t^+m_l^-}Z_0(E,\eta_g)+\hbar^2k_s^2\left[\lambda_{71}(E,\eta_g)k_z^2+\lambda_{72}(E,\eta_g)\right]$$
$$+\left[\lambda_{73}(E,\eta_g)k_z^2+\lambda_{74}(E,\eta_g)k_z^4-\lambda_{75}(E,\eta_g)\right]=0 \quad (1.162)$$

where,

$$Z_0(E,\eta_g) \equiv \frac{1}{2}\left[1 + Erf\left(\frac{E}{\eta_g}\right)\right], \lambda_{70}(E,\eta_g) \equiv \frac{\alpha}{4m_t^+ m_t^-} Z_0(E,\eta_g)$$

$$\lambda_{71}(E,\eta_g) \equiv \left[\frac{\alpha\hbar^2}{4m_t^- m_l^+} Z_0(E,\eta_g) + \frac{\alpha\hbar^2}{4m_l^- m_t^+} Z_0(E,\eta_g)\right],$$

$$\lambda_{72}(E,\eta_g) \equiv \left[\left(\frac{1}{2m_t^*} - \frac{1}{2m_t^-}\right) Z_0(E,\eta_g) + \alpha\left(\frac{1}{2m_t^-} - \frac{1}{2m_t^+}\right)\gamma_0(E,\eta_g)\right],$$

$$\lambda_{73}(E,\eta_g) \equiv \left[\left(\frac{\hbar^2}{2m_l^*} + \frac{\hbar^2}{2m_l^-}\right) Z_0(E,\eta_g) + \frac{\alpha\hbar^2}{2}\left(\frac{1}{m_l^-} - \frac{1}{2m_l^+}\right)\gamma_0(E,\eta_g)\right],$$

$$\lambda_{74}(E,\eta_g) \equiv \frac{\alpha\hbar^4 Z_0(E,\eta_g)}{4m_l^+ m_l^-} \text{ and } \lambda_{75}(E,\eta_g) \equiv \left[\gamma_0(E,\eta_g) + \alpha\theta_0(E,\eta_g)\right]$$

Thus, the energy spectrum in this case is real since the corresponding dispersion relation in the absence of band tails as given by (1.162) is a pole-less function with respect to energy axis in the finite complex plane.

The respective transverse and the longitudinal EEMs' in this case can be written as

$$m_\perp^*(E_{F_h},\eta_g) = \{2Z_0(E,\eta_g)\}^{-2}\left[Z_0(E,\eta_g)\left[-\{\lambda_{72}(E,\eta_g)\}' + \frac{\{\lambda_{78}(E,\eta_g)\}'}{2\sqrt{\lambda_{78}(E,\eta_g)}}\right] - \{Z_0(E,\eta_g)\}'\left[-\lambda_{72}(E,\eta_g) + \sqrt{\lambda_{78}(E,\eta_g)}\right]\right]\Bigg|_{E=E_{F_h}} \tag{1.163}$$

where,

$$\lambda_{78}(E,\eta_g) \equiv \left[4\lambda_{70}(E,\eta_g)\lambda_{75}(E,\eta_g)\right]$$

and

$$m_\parallel^*(E_{F_h},\eta_g) = \frac{\hbar^2}{4}\left[-\{\lambda_{84}(E,\eta_g)\}' + \frac{\{\lambda_{84}(E,\eta_g)\}'\lambda_{84}(E,\eta_g) + 2\{\lambda_{85}(E,\eta_g)\}'}{\sqrt{(\lambda_{84}(E,\eta_g))^2 + 4\lambda_{85}(E,\eta_g)}}\right]\Bigg|_{E=E_{F_h}} \tag{1.164}$$

in which,

$$\lambda_{84}(E,\eta_g) \equiv \frac{\lambda_{73}(E,\eta_g)}{\lambda_{74}(E,\eta_g)} \text{ and } \lambda_{85}(E,\eta_g) \equiv \frac{\lambda_{75}(E,\eta_g)}{\lambda_{74}(E,\eta_g)}$$

Thus, we can see that the both the EEMs' in this case exist in the band gap. In the absence of band tails, $\eta_g \to 0$, we get

$$m_{\perp}^{*}(E_F) = \frac{\hbar^2}{2}\left[-\{\alpha_{11}(E)\}' + \frac{\alpha_{511}\{T_{311}(E)\}'}{2\sqrt{T_{311}(E)}}\right]\Bigg|_{E=E_F} \quad (1.165)$$

where

$$\alpha_{11}(E) \equiv \frac{2m_t^+ m_t^-}{\alpha\hbar^2}\alpha_{211}(E), \alpha_{211}(E) \equiv \left[\frac{1}{2m_t^*} - \frac{\alpha E}{2m_t^+} + \frac{1+\alpha E}{2m_t^-}\right], \alpha_{511} \equiv \frac{2m_t^+ m_t^-}{\alpha\hbar^2}\omega_{11}$$

$$(\omega_{11}) \equiv \left[\frac{\alpha^2}{16}\left[\frac{1}{m_t^- m_l^+} + \frac{1}{m_l^- m_t^+}\right]^2 - \frac{\alpha^2}{4m_t^- m_t^+ m_l^- m_l^+}\right]^{1/2}, \quad T_{311}(E) \equiv \frac{\omega_{311}(E)}{(\omega_{11})^2},$$

$$\omega_{311}(E) \equiv \left[\frac{\alpha E(1+\alpha E)}{m_t^+ m_t^-} + \left[\frac{1}{2m_t^*} - \left(\frac{\alpha E}{2m_t^+}\right) + \frac{(1+\alpha E)}{2m_t^-}\right]^2\right]$$

and

$$m_{\|}^{*}(E_F) = \left(\frac{m_l^+ m_l^-}{\alpha}\right)\left[\left(\frac{\alpha}{2m_l^+} - \frac{\alpha}{2m_l^-}\right) + \frac{1}{2}\left\{\frac{2\left[\frac{1}{2m_l^*} + \frac{1+\alpha E}{2m_l^-} - \frac{\alpha E}{2m_l^+}\right]\left(\frac{\alpha}{2m_l^-} - \frac{\alpha}{2m_l^+}\right) + \frac{\alpha(1+2\alpha E)}{m_l^- m_l^+}}{\left[\left[\frac{1}{2m_l^*} + \frac{1+\alpha E}{2m_l^-} - \frac{\alpha E}{2m_l^+}\right]^2 + \frac{\alpha E(1+\alpha E)}{m_l^- m_l^+}\right]^{1/2}}\right\}\right]\Bigg|_{E=E_F} \quad (1.166)$$

The volume in k-space as enclosed by (1.162) can be written through the integral as

$$V(E,\eta_g) = 2\pi \int_0^{\lambda_{86}(E,\eta_g)} \left[-[\lambda_{79}(E,\eta_g)k_z^2 + \lambda_{80}(E,\eta_g)] + \sqrt{\lambda_{81}(E,\eta_g)k_z^4 + \lambda_{82}(E,\eta_g)k_z^2 + \lambda_{83}(E,\eta_g)}\right] dk_z \quad (1.167)$$

where,

$$\lambda_{86}(E,\eta_g) \equiv \left[\frac{\sqrt{[\lambda_{84}(E,\eta_g)]^2+4\lambda_{85}(E,\eta_g)}-\lambda_{84}(E,\eta_g)}{2}\right]^{1/2}, \quad \lambda_{79}(E,\eta_g) \equiv \frac{\lambda_{71}(E,\eta_g)}{2\hbar^2 Z_0(E,\eta_g)}$$

$$\lambda_{81}(E,\eta_g) \equiv \frac{\lambda_{76}(E,\eta_g)}{4\hbar^4[Z_0(E,\eta_g)]^2}, \quad \lambda_{76}(E,\eta_g) \equiv [\lambda_{71}(E,\eta_g)]^2, \lambda_{76}(E,\eta_g) \equiv [\lambda_{71}(E,\eta_g)]^2$$

$$\lambda_{77}(E,\eta_g) \equiv [2\lambda_{71}(E,\eta_g)\lambda_{72}(E,\eta_g) - 4\lambda_{70}(E,\eta_g)\lambda_{73}(E,\eta_g) - 4\lambda_0(E,\eta_g)\lambda_{74}(E,\eta_g)],$$

$$\lambda_{83}(E,\eta_g) \equiv \frac{\lambda_{78}(E,\eta_g)}{9\hbar^4[Z_0(E,\eta_g)]^2} \text{ and } \lambda_{78}(E,\eta_g) \equiv [4\lambda_{70}(E,\eta_g)\lambda_{75}(E,\eta_g)]$$

Thus,

$$V(E,\eta_g) = [\lambda_{87}(E,\eta_g)] \int_0^{\lambda_{86}(E,\eta_g)} \left[\sqrt{k_z^4 + \lambda_{88}(E,\eta_g)k_z^2 + \lambda_{89}(E,\eta_g)} - \lambda_{90}(E,\eta_g)\right] dk_z \tag{1.168}$$

where,

$$\lambda_{87}(E,\eta_g) \equiv 2\pi\sqrt{\lambda_{81}(E,\eta_g)}, \lambda_{88}(E,\eta_g) \equiv \frac{\lambda_{82}(E,\eta_g)}{\lambda_{81}(E,\eta_g)}, \lambda_{89}(E,\eta_g) \equiv \frac{\lambda_{83}(E,\eta_g)}{\lambda_{81}(E,\eta_g)}$$

and

$$\lambda_{90}(E,\eta_g) \equiv 2\pi\left[\frac{\lambda_{79}(E,\eta_g)\{\lambda_{86}(E,\eta_g)\}^3}{3} + \lambda_{80}(E,\eta_g)\lambda_{89}(E,\eta_g)\right].$$

The (1.168) can be written as

$$V(E,\eta_g) = [\lambda_{87}(E,\eta_g)\lambda_{95}(E,\eta_g) - \lambda_{90}(E,\eta_g)] \tag{1.169}$$

in which,

$$\lambda_{95}(E,\eta_g) \equiv \left[\frac{\lambda_{91}(E,\eta_g)}{3}[-E_i[\lambda_{93}(E,\eta_g),\lambda_{94}(E,\eta_g)]\right.$$
$$\times\left[\{\lambda_{91}(E,\eta_g)\}^2+\{\lambda_{92}(E,\eta_g)\}^2+2\{\lambda_{92}(E,\eta_g)\}^2F_i[\lambda_{93}(E,\eta_g),\lambda_{94}(E,\eta_g)]\right]$$
$$+\left(\frac{\{\lambda_{86}(E,\eta_g)\}}{3}\right)\left[\{\lambda_{86}(E,\eta_g)\}^2+\{\lambda_{91}(E,\eta_g)\}^2+2\{\lambda_{92}(E,\eta_g)\}^2\right]$$
$$\left.\times\left[\left[\{\lambda_{91}(E,\eta_g)\}^2+\{\lambda_{86}(E,\eta_g)\}^2\right]^{1/2}\left[\{\lambda_{92}(E,\eta_g)\}^2+\{\lambda_{86}(E,\eta_g)\}^2\right]^{-1/2}\right]\right],$$

$$\{\lambda_{91}(E,\eta_g)\}^2 \equiv \frac{1}{2}\left[\sqrt{\{\lambda_{88}(E,\eta_g)\}^2-4\lambda_{89}(E,\eta_g)}+\lambda_{88}(E,\eta_g)\right], E_i[\lambda_{93}(E,\eta_g),\lambda_{94}(E,\eta_g)]$$

is the incomplete elliptic integral of the 2nd kind and is given by [129, 130],

$$E_i[\lambda_{93}(E,\eta_g),\lambda_{94}(E,\eta_g)] \equiv \int_0^{\lambda_{93}(E,\eta_g)}\left[\left\{1-\{\lambda_{94}(E,\eta_g)\}^2\sin^2\xi\right\}^{1/2}\right]d\xi,$$

ξ is the variable of integration in this case,

$$\lambda_{93}(E,\eta_g) \equiv \tan^{-1}\left[\frac{\lambda_{86}(E,\eta_g)}{\lambda_{92}(E,\eta_g)}\right], \{\lambda_{92}(E,\eta_g)\}^2 \equiv \frac{1}{2}\left[\lambda_{88}(E,\eta_g)-\sqrt{\{\lambda_{88}(E,\eta_g)\}^2-4\lambda_{89}(E,\eta_g)}\right],$$

$$\lambda_{94}(E,\eta_g) \equiv \frac{\sqrt{\{\lambda_{91}(E,\eta_g)\}^2-\{\lambda_{92}(E,\eta_g)\}^2}}{\lambda_{91}(E,\eta_g)}, F_i[\lambda_{93}(E,\eta_g),\lambda_{94}(E,\eta_g)]$$

is the incomplete elliptic integral of the 1st kind and is given by [129, 130],

$$F_i[\lambda_{93}(E,\eta_g),\lambda_{94}(E,\eta_g)] \equiv \int_0^{\lambda_{93}(E,\eta_g)}\left[\left\{1-\{\lambda_{94}(E,\eta_g)\}^2\sin^2\xi\right\}^{1/2}\right]d\xi.$$

The DOS function in this case is given by

$$N_{HD}(E,\eta_g) = \frac{g_v}{4\pi^3}\left[\{\lambda_{87}(E,\eta_g)\}'\lambda_{95}(E,\eta_g)+\{\lambda_{95}(E,\eta_g)\}'\lambda_{87}(E,\eta_g)-\{\lambda_{90}(E,\eta_g)\}'\right] \tag{1.170}$$

Therefore the electron concentration can be expressed as

$$n_0 = \frac{g_v}{4\pi^3}\left[\bar{I}_{125}(E_{F_h}, \eta_g) + \sum_{r=1}^{s} L(r)[\bar{I}_{125}(E_{F_h}, \eta_g)]\right] \tag{1.171}$$

where,

$$\bar{I}_{125}(E_{F_h}, \eta_g) = [\{\lambda_{87}(E_{F_h}, \eta_g)\}\lambda_{95}(E_{F_h}, \eta_g) - \{\lambda_{90}(E_{F_h}, \eta_g)\}]$$

The EP in the case is given by

$$J = \frac{\alpha_0 e g_v}{8\pi^3\hbar}\int_{E_0}^{\infty}\frac{f(E)\sqrt{\lambda_{100}(E', \eta_s)}}{\lambda'_{100}(E', \eta_g)}\lambda'_{101}(E', \eta_s)dE' \tag{1.172}$$

where,

$$\lambda_{100}(E, \eta_g) = [-\lambda_{73}(E, \eta_g) + \sqrt{\lambda_{73}^2(E, \eta_g) + 4\lambda_{74}(E, \eta_g)\lambda_{75}(E, \eta_g)}][2\lambda_{74}(E, \eta_g)]^{-1}$$
and $\lambda_{101}(E, \eta_g) = [\lambda_{87}(E, \eta_g)\lambda_{95}(E, \eta_g) - \lambda_{90}(E, \eta_g)]$

The 2D dispersion relation of the conduction electrons in QWs of IV-VI materials in the absence of band tails for the dimensional quantization along z direction can be expressed as

$$\begin{aligned} &E(1+\alpha E) + \alpha E\left(\frac{\hbar^2 k_x^2}{2x_4} + \frac{\hbar^2 k_y^2}{2x_5}\right) + \alpha E\frac{\hbar^2}{2x_6}\left(\frac{n_z\pi}{d_z}\right)^2 - (1+\alpha E)\left(\frac{\hbar^2 k_x^2}{2x_1} + \frac{\hbar^2 k_y^2}{2x_2}\right) \\ &- \alpha\left(\frac{\hbar^2 k_x^2}{2x_1} + \frac{\hbar^2 k_y^2}{2x_2}\right)\left(\frac{\hbar^2 k_x^2}{2x_4} + \frac{\hbar^2 k_y^2}{2x_5}\right) - \alpha\left(\frac{\hbar^2 k_x^2}{2x_1} + \frac{\hbar^2 k_y^2}{2x_2}\right)\frac{\hbar^2}{2x_6}\left(\frac{n_z\pi}{d_z}\right)^2 - (1+\alpha E)\frac{\hbar^2}{2x_3}\left(\frac{n_z\pi}{d_z}\right)^2 \\ &- \alpha\frac{\hbar^2}{2x_3}\left(\frac{n_z\pi}{d_z}\right)^2\left(\frac{\hbar^2 k_x^2}{2x_4} + \frac{\hbar^2 k_y^2}{2x_5}\right) - \alpha\frac{\hbar^2}{2x_3}\left(\frac{n_z\pi}{d_z}\right)^2\frac{\hbar^2}{2x_6}\left(\frac{n_z\pi}{d_z}\right)^2 = \frac{\hbar^2 k_x^2}{2m_1} + \frac{\hbar^2 k_y^2}{2m_2} + \frac{\hbar^2}{2m_3}\left(\frac{n_z\pi}{d_z}\right)^2 \end{aligned} \tag{1.173}$$

where

$$x_4 = m_t^+, x_5 = \frac{m_t^+ + 2m_l^+}{3}, x_6 = \frac{3m_t^+ m_l^+}{2m_l^+ + m_t^+}, x_1 = m_t^-, x_2 = \frac{m_t^- + 2m_l^-}{3}, x_3 = \frac{3m_t^- m_l^-}{2m_l^- + m_t^-},$$
$$m_1 = m_t^*, m_2 = \frac{m_t^* + 2m_l^*}{3} \text{ and } m_3 = \frac{3m_l^* m_l^*}{m_t^* + 2m_l^*}.$$

Therefore, the HD 2-D dispersion relation In this case assumes the form

$$\gamma_2(E,\eta_g)+\alpha\gamma_3(E,\eta_g)\left(\frac{\hbar^2k_x^2}{2x_4}+\frac{\hbar^2k_y^2}{2x_5}\right)+\alpha\gamma_3(E,\eta_g)\frac{\hbar^2}{2x_6}\left(\frac{n_z\pi}{d_z}\right)^2-(1+\alpha\gamma_3(E,\eta_g))\left(\frac{\hbar^2k_x^2}{2x_1}+\frac{\hbar^2k_y^2}{2x_2}\right)$$
$$-\alpha\left(\frac{\hbar^2k_x^2}{2x_1}+\frac{\hbar^2k_y^2}{2x_2}\right)\left(\frac{\hbar^2k_x^2}{2x_4}+\frac{\hbar^2k_y^2}{2x_5}\right)-\alpha\left(\frac{\hbar^2k_x^2}{2x_1}+\frac{\hbar^2k_y^2}{2x_2}\right)\frac{\hbar^2}{2x_6}\left(\frac{n_z\pi}{d_z}\right)^2-(1+\alpha\gamma_3(E,\eta_g))\frac{\hbar^2}{2x_3}\left(\frac{n_z\pi}{d_z}\right)^2$$
$$-\alpha\frac{\hbar^2}{2x_3}\left(\frac{n_z\pi}{d_z}\right)^2\left(\frac{\hbar^2k_x^2}{2x_4}+\frac{\hbar^2k_y^2}{2x_5}\right)-\alpha\frac{\hbar^2}{2x_3}\left(\frac{n_z\pi}{d_z}\right)^2\frac{\hbar^2}{2x_6}\left(\frac{n_z\pi}{d_z}\right)^2=\frac{\hbar^2k_x^2}{2m_1}+\frac{\hbar^2k_y^2}{2m_2}+\frac{\hbar^2}{2m_3}\left(\frac{n_z\pi}{d_z}\right)^2 \tag{1.174}$$

Substituting, $k_x = r\cos\theta$ and $k_y = r\sin\theta$ (where r and θ are 2D polar coordinates in 2D wave vector space) in (1.174), we can write

$$r^4\left[\alpha\frac{1}{4}\left(\frac{\hbar^2\cos^2\theta}{x_1}+\frac{\hbar^2\sin^2\theta}{x_2}\right)\left(\frac{\hbar^2\cos^2\theta}{x_4}+\frac{\hbar^2\sin^2\theta}{x_5}\right)\right]+r^2\frac{1}{2}\left[\left(\frac{\hbar^2\cos^2\theta}{m_1}+\frac{\hbar^2\sin^2\theta}{m_2}\right)\right.$$
$$+\alpha\frac{\hbar^2}{2x_3}\left(\frac{n_z\pi}{d_z}\right)^2\left(\frac{\hbar^2\cos^2\theta}{x_4}+\frac{\hbar^2\sin^2\theta}{x_5}\right)+\alpha\left(\frac{\hbar^2\cos^2\theta}{x_1}+\frac{\hbar^2\sin^2\theta}{x_2}\right)\frac{\hbar^2}{2x_6}\left(\frac{n_z\pi}{d_z}\right)^2$$
$$\left.+\hbar^2(1+\alpha\gamma_3(E,\eta_g))\left(\frac{\cos^2\theta}{x_1}+\frac{\sin^2\theta}{x_2}\right)-\hbar^2\alpha\gamma_3(E,\eta_g)\left(\frac{\cos^2\theta}{x_4}+\frac{\sin^2\theta}{x_5}\right)\right]-\left[\gamma_2(E,\eta_g)\right.$$
$$\left.+\alpha\gamma_3(E,\eta_g)\frac{\hbar^2}{2x_6}\left(\frac{n_z\pi}{d_z}\right)^2-(1+\alpha\gamma_3(E,\eta_g))\frac{\hbar^2}{2x_3}\left(\frac{n_z\pi}{d_z}\right)^2-\alpha\left(\frac{\hbar^4}{4x_3x_6}\left(\frac{n_z\pi}{d_z}\right)^4\right)\right]=0 \tag{1.175}$$

The area $A(E,n_z)$ of the 2D wave vector space can be expressed as

$$A(E,n_z)=\bar{J}_1-\bar{J}_2 \tag{1.176}$$

where

$$\bar{J}_1\equiv 2\int_0^{\pi/2}\frac{c_1}{b_1}d\theta \tag{1.177}$$

and

$$\bar{J}_2\equiv 2\int_0^{\pi/2}\frac{ac_1^2}{b_1^3}d\theta \tag{1.178}$$

in which,

$$a \equiv \left[\alpha\left(\frac{\hbar^4}{4}\right)\left(\frac{\cos^2\theta}{x_1}+\frac{\sin^2\theta}{x_2}\right)\left(\frac{\cos^2\theta}{x_4}+\frac{\sin^2\theta}{x_5}\right)\right],$$

$$\begin{aligned} b_1 \equiv \left(\frac{\hbar^2}{2}\right)\Bigg[&\left(\frac{\cos^2\theta}{m_1}+\frac{\sin^2\theta}{m_2}\right)+\alpha\left(\frac{\hbar^2}{2x_3}\right)\left(\frac{n_z\pi}{d_z}\right)^2\left(\frac{\cos^2\theta}{x_4}+\frac{\sin^2\theta}{x_5}\right) \\ &+\alpha\left(\frac{\hbar^2}{2x_6}\right)\left(\frac{n_z\pi}{d_z}\right)^2\left(\frac{\cos^2\theta}{m_1}+\frac{\sin^2\theta}{m_2}\right) \\ &+(1+\alpha\gamma_3(E,\eta_g))\left(\frac{\cos^2\theta}{x_1}+\frac{\sin^2\theta}{x_2}\right)-\alpha\gamma_3(E,\eta_g)\left(\frac{\cos^2\theta}{x_4}+\frac{\sin^2\theta}{x_5}\right)\Bigg] \end{aligned}$$

and

$$c_1 \equiv \left[\gamma_2(E,\eta_g)+\alpha\gamma_3(E,\eta_g)\left(\frac{\hbar^2}{2x_6}\right)\left(\frac{n_z\pi}{d_z}\right)^2-(1+\alpha\gamma_3(E,\eta_g))\left(\frac{\hbar^2}{2x_3}\right)\left(\frac{n_z\pi}{d_z}\right)^2-\alpha\left(\frac{\hbar^4}{4x_3x_6}\right)\left(\frac{n_z\pi}{d_z}\right)^4\right]$$

The (1.177) can be expressed as

$$\bar{J}_1 = 2\int_0^{\pi/2}\frac{t_{31}(E,n_z)d\theta}{A_{11}(E,n_z)\cos^2\theta+B_{11}(E,n_z)\sin^2\theta}$$

where,

$$\begin{aligned} t_{31}(E,n_z) &\equiv c_1, A_{11}(E,n_z) \equiv \frac{\hbar^2}{2m_1}t_{11}(E,n_z), \\ t_{11}(E,n_z) &\equiv \left[1+m_1\left[\frac{1}{x_4}\frac{\alpha\hbar^2}{2x_3}\left(\frac{n_z\pi}{d_z}\right)^2+\frac{\alpha\hbar^2}{2x_1x_6}\left(\frac{n_z\pi}{d_z}\right)^2+\frac{1+\alpha\gamma_2(E,\eta_g)}{x_1}-\frac{\alpha\gamma_3(E,\eta_g)}{x_4}\right]\right] \\ B_{11}(E,n_z) &\equiv \frac{\hbar^2}{2m_2}t_{21}(E,n_z) \text{ and} \\ t_{21}(E,n_z) &\equiv \left[1+m_2\left[\frac{\alpha\hbar^2}{2x_3x_5}\left(\frac{n_z\pi}{d_z}\right)^2+\frac{\alpha\hbar^2}{2x_2x_6}\left(\frac{n_z\pi}{d_z}\right)^2+\frac{1+\alpha\gamma_3(E,\eta_g)}{x_2}-\frac{\alpha\gamma_3(E,\eta_g)}{x_5}\right]\right]. \end{aligned}$$

Performing the integration, we get

$$\bar{J}_1 = \pi t_{31}(E,n_z)[A_{11}(E,n_z)B_{11}(E,n_z)]^{-1/2} \tag{1.179}$$

From (1.178) we can write

$$\bar{J}_2 = \frac{\alpha t_{31}^2(E,n_z)\hbar^4}{2B_{11}^3(E,n_z)} I \tag{1.180}$$

where,

$$I \equiv \int_0^\infty \frac{(a_1 + a_2 z^2)(a_3 + a_4 z^2)dz}{\left[(\bar{a})^2 + z^2\right]^3}, (\bar{a})^2 = \left(\frac{A_{11}(E,n_z)}{B_{11}(E,n_z)}\right), \tag{1.181}$$

in which $a_1 \equiv \frac{1}{x_1}, a_2 \equiv \frac{1}{x_2}, z = \tan\theta, \theta$ is a new variable, $a_3 \equiv \frac{1}{x_4}, a_4 \equiv \frac{1}{x_5}$ and $(\bar{a})^2 \equiv \left(\frac{A_1(E,n_z)}{B_1(E,n_z)}\right)$.

The use of the Residue theorem leads to the evaluation of the integral in (1.181) as

$$I = \frac{\pi}{4\bar{a}}[a_1 a_4 + 3a_2 a_4], \tag{1.182}$$

Therefore, the 2D area of the 2D wave vector space can be written as

$$A_{HD}(E,n_z) = \frac{\pi t_{31}(E,n_z)}{\sqrt{A_{11}(E,n_z)B_{11}(E,n_z)}}\left[1 - \frac{1}{x_5}\left(\frac{1}{x_1} + \frac{3}{x_2}\right)\frac{\alpha t_{31}(E,n_z)\hbar^4}{8B_{11}^2(E,n_z)}\right] \tag{1.183}$$

The EEM for the HD QWs of IV-VI materials can thus be written as

$$m^*(E,n_z) = \frac{\hbar^2}{2}[\theta_{5HD}(E,n_z)]\Big|_{E=E_{F1HD}} \tag{1.184}$$

where,

$$\begin{aligned}
\theta_{5HD}(E,n_z) \equiv & \left[1 - \frac{1}{x_5}\left(\frac{1}{x_1} + \frac{3}{x_2}\right)\frac{\alpha t_{31}(E,n_z)\hbar^4}{8[B_{11}(E,n_z)]^2}\right][A_{11}(E,n_z)B_{11}(E,n_z)]^{-1} \\
& \times \left[\sqrt{A_{11}(E,n_z)B_{11}(E,n_z)}\{t_{31}(E,n_z)\}' - t_{31}(E,n_z)\right. \\
& \left.\times\left\{\frac{1}{2}\{A_{11}(E,n_z)\}'\left[\frac{B_{11}(E,n_z)}{A_{11}(E,n_z)}\right]^{1/2} + \frac{1}{2}\{B_{11}(E,n_z)\}'\left[\frac{A_{11}(E,n_z)}{B_{11}(E,n_z)}\right]^{1/2}\right\}\right] \\
& - \frac{1}{8}\frac{t_{31}(E,n_z)\alpha\hbar^4}{\sqrt{A_{11}(E,n_z)B_{11}(E,n_z)}}\frac{1}{x_5}\left(\frac{1}{x_1} + \frac{3}{x_2}\right)[B_{11}(E,n_z)]^{-4} \\
& \times \left[\{B_{11}(E,n_z)\}^2\{t_{31}(E,n_z)\}' - 2B_{11}(E,n_z)\{B_{11}(E,n_z)\}' t_{31}(E,n_z)\right].
\end{aligned}$$

Thus, the EEM is a function of Fermi energy and the quantum number due to the band non-parabolicity.

The total DOS function can be written as

$$N_{2DT}(E) = \left(\frac{g_v}{2\pi}\right) \sum_{n_z=1}^{n_{z\max}} \theta_{5HD}(E, n_z) H\left(E - E_{n_{z7HD}}\right) \tag{1.185}$$

where the sub-band energy $\left(E_{n_{z7}}\right)$ in this case can be written as

$$\gamma_2(E_{n_{z7HD}}, \eta_g) + \alpha\gamma_3(E_{n_{z7HD}}, \eta_g)\frac{\hbar^2}{2x_6}\left(\frac{n_z\pi}{d_z}\right)^2 - \left(1 + \alpha\gamma_3(E_{n_{z7HD}}, \eta_g)\right)\frac{\hbar^2}{2x_3}\left(\frac{n_z\pi}{d_z}\right)^2$$
$$- \alpha\frac{\hbar^2}{2x_3}\left(\frac{n_z\pi}{d_z}\right)^2\frac{\hbar^2}{2x_6}\left(\frac{n_z\pi}{d_z}\right)^2 - \left[\frac{\hbar^2}{2m_3}\left(\frac{n_z\pi}{d_z}\right)^2\right] = 0 \tag{1.186}$$

The use (1.185) leads to the expression of 2D electron statistics as

$$n_{2D} = \frac{g_v}{2\pi}\sum_{n_z=1}^{n_{z\max}}[T_{55HD}(E_{F1HD}, n_z) + T_{56HD}(E_{F1HD}, n_z)] \tag{1.187}$$

where

$$T_{55HD}(E_{F1HD}, n_z) \equiv \frac{A_{HD}(E_{F1HD}, n_z)}{\pi} \text{ and}$$
$$T_{56HD}(E_{F1HD}, n_z) \equiv \sum_{r=1}^{s} L(r) T_{55HD}(E_{F1HD}, n_z).$$

The EP in this case is given by

$$J_{2D} = \frac{\alpha_0 e g_v}{h d_z}\left[\sum_{n_{z\min}}^{n_{z\max}}\left[\frac{\sqrt{\gamma_{12}(E_{n_{z7HD}}, \eta_g)}}{\gamma'_{12}(E_{n_z 7HD}, \eta_g)}\right]\left[\int_{E_{n_z 7HD}}^{\infty}\theta_{5HD}(E, n_z) f(E) dE\right]\right]$$

where,

$$\gamma_{12}(E_{n_z 7HD}, \eta_g) = [-\gamma_{10}(E_{n_z 7HD}, \eta_g) + \sqrt{\gamma_{10}^2(E_{n_z 7HD}, \eta_g) + 4\gamma_{11}(E_{n_z 7HD}, \eta_g)}]/2,$$
$$\gamma_{10}(E_{n_z 7HD}, \eta_g) = \left[\frac{\alpha\hbar^4}{4x_3x_6}\right]^{-1}\left[\frac{\hbar^2}{2m_3} + (1 + \alpha\gamma_3(E_{n_z 7HD}, \eta_g))\frac{\hbar^2}{2x_3} - \alpha\gamma_3(E_{n_z 7HD}, \eta_g)\frac{\hbar^2}{2x_6}\right] \text{ and}$$
$$\gamma_{11}(E_{n_z 7HD}, \eta_g) = \left[\frac{\alpha\hbar^4}{4x_3x_6}\right]^{-1}[\gamma_2(E_{n_z 7HD}, \eta_g)]$$

In the absence of band-tails the EEM in QWs of IV-VI materials can be written as

$$m^*(E,n_z)=\frac{\hbar^2}{2}\left[\theta_5(E,n_z)\right]\Big|_{E=E_{Fs}} \tag{1.188}$$

where,

$$\begin{aligned}\theta_5(E,n_z)\equiv&\left[1-\frac{1}{x_5}\left(\frac{1}{x_1}+\frac{3}{x_2}\right)\frac{\alpha t_{30}(E,n_z)\hbar^4}{8[B_{10}(E,n_z)]^2}\right][A_{10}(E,n_z)B_{10}(E,n_z)]^{-1}\\&\times\left[\sqrt{A_{10}(E,n_z)B_{10}(E,n_z)}\{t_{30}(E,n_z)\}'-t_{30}(E,n_z)\right.\\&\left.\times\left\{\frac{1}{2}\{A_{10}(E,n_z)\}'\left[\frac{B_{10}(E,n_z)}{A_{10}(E,n_z)}\right]^{1/2}+\frac{1}{2}\{B_{10}(E,n_z)\}'\left[\frac{A_{10}(E,n_z)}{B_{10}(E,n_z)}\right]^{1/2}\right\}\right]\\&-\frac{1}{8}\frac{t_{30}(E,n_z)\alpha\hbar^4}{\sqrt{A_{10}(E,n_z)B_{10}(E,n_z)}}\frac{1}{x_5}\left(\frac{1}{x_1}+\frac{3}{x_2}\right)[B_{10}(E,n_z)]^{-4}\\&\times\left[\{B_{10}(E,n_z)\}^2\{t_{30}(E,n_z)\}'-2B_{10}(E,n_z)\{B_{10}(E,n_z)\}'t_{30}(E,n_z)\right]\end{aligned}$$

where

$$\begin{aligned}t_{30}(E,n_z)&\equiv c_0,\\c_0&\equiv\left[E(1+\alpha E)+\alpha E\left(\frac{\hbar^2}{2x_6}\right)\left(\frac{n_z\pi}{d_z}\right)^2\right.\\&\left.\quad-(1+\alpha E)\left(\frac{\hbar^2}{2x_3}\right)\left(\frac{n_z\pi}{d_z}\right)^2-\alpha\left(\frac{\hbar^4}{4x_3x_6}\right)\left(\frac{n_z\pi}{d_z}\right)^4\right],\\A_{10}(E,n_z)&\equiv\frac{\hbar^2}{2m_1}t_{10}(E,n_z),\\t_{10}(E,n_z)&\equiv\left[1+m_1\left[\frac{1}{x_4}\frac{\alpha\hbar^2}{2x_3}\left(\frac{n_z\pi}{d_z}\right)^2+\frac{\alpha\hbar^2}{2x_1x_6}\left(\frac{n_z\pi}{d_z}\right)^2+\frac{1+\alpha E}{x_1}-\frac{\alpha E}{x_4}\right]\right],\\B_{10}(E,n_z)&\equiv\frac{\hbar^2}{2m_2}t_{20}(E,n_z)\text{and}\\t_{20}(E,n_z)&\equiv\left[1+m_2\left[\frac{\alpha\hbar^2}{2x_3x_5}\left(\frac{n_z\pi}{d_z}\right)^2+\frac{\alpha\hbar^2}{2x_2x_6}\left(\frac{n_z\pi}{d_z}\right)^2+\frac{1+\alpha E}{x_2}-\frac{\alpha E}{x_5}\right]\right]\end{aligned}$$

Thus, the EEM is a function of Fermi energy and the quantum number due to the band non-parabolicity.

The total DOS function can be written as

$$N_{2DT}(E) = \left(\frac{g_v}{2\pi}\right) \sum_{n_z=1}^{n_{z\max}} \theta_5(E, n_z) H\left(E - E_{n_{z7}}\right) \quad (1.189)$$

where the sub-band energy $\left(E_{n_{z7}}\right)$ in this case can be written as

$$E_{n_{z7}}\left(1 + \alpha E_{n_{z7}}\right) + \alpha E_{n_{z7}} \frac{\hbar^2}{2x_6}\left(\frac{n_z\pi}{d_z}\right)^2 - \left(1 + \alpha E_{n_{z7}}\right)\frac{\hbar^2}{2x_3}\left(\frac{n_z\pi}{d_z}\right)^2 \\ - \alpha\frac{\hbar^2}{2x_3}\left(\frac{n_z\pi}{d_z}\right)^2 \frac{\hbar^2}{2x_6}\left(\frac{n_z\pi}{d_z}\right)^2 - \left[\frac{\hbar^2}{2m_3}\left(\frac{n_z\pi}{d_z}\right)^2\right] = 0 \quad (1.190)$$

In the absence of band-tails, the expression of 2D electron statistics can be written as

$$n_{2D} = \frac{g_v}{2\pi} \sum_{n_z=1}^{n_{z\max}} \left[T_{550}(E_{Fs}, n_z) + T_{560}(E_{Fs}, n_z)\right] \quad (1.191)$$

where,

$$T_{550}(E_{Fs}, n_z) \equiv \frac{A_0(E_{Fs}, n_z)}{\pi}, A_0(E, n_z) \\ = \frac{\pi t_{30}(E, n_z)}{\sqrt{A_{10}(E, n_z) B_{10}(E, n_z)}}\left[1 - \frac{1}{x_5}\left(\frac{1}{x_1} + \frac{3}{x_2}\right)\frac{\alpha t_{30}(E, n_z)\hbar^4}{8B_{10}^2(E, n_z)}\right], \text{ and}$$

$$T_{50}(E_{Fs}, n_z) \equiv \sum_{r=1}^{s} L(r) T_{50}(E_{Fs}, n_z)$$

$$J_{2D} = \frac{\alpha_0 e g_v}{2\pi\hbar d_z} \sum_{n_{z\min}}^{n_{z\max}} \frac{\sqrt{\gamma_{51}\left(E_{n_{z7}}\right)}}{\gamma'_{51}\left(E_{n_{z7}}\right)} \int_{E_{n_{z7}}}^{\infty} \theta(\mathrm{E}, \mathrm{n}_z) \mathrm{f}(\mathrm{E}) \mathrm{dE} \quad (1.192)$$

where

$$\gamma_{51}(\mathrm{E}) = \frac{-\gamma_{50} + \sqrt{\gamma_{50}^2 + 4\gamma_{51}}}{2}$$

$$\gamma_{50}(\mathrm{E}) = \left[\frac{\alpha\hbar^4}{4x_3x_6}\right]^{-1}\left[\frac{1}{2x_3} - \frac{\alpha E}{2x_6} + \frac{1 + \alpha E}{2x_3}\right],$$

$$\gamma_{51}(\mathrm{E}) = \left[\frac{\alpha\hbar^4}{4x_3x_6}\right]^{-1} E(1 + \alpha E)$$

For bulk specimens of IV-VI materials, the expressions of electron concentration and the EP assume the forms

$$n_0 = \left(\frac{g_v}{2\pi^2}\right)[M_{A_4}(E_{F_b}) + N_{A_4}(E_{F_b})] \tag{1.193}$$

$$J = \frac{\alpha_0 e g_v}{4\pi^2\hbar}\int_{E_0}^{\infty}\frac{\sqrt{I_{17}(E')}}{I'_{17}(E')}M'_{A4}(E')f(E)dE' \tag{1.194a}$$

where

$$M_{A_4}(E_{F_b}) = \left[\alpha_5 J_{A_1}(E_{F_b}) - \alpha_3(E_{F_b})\bar{\tau}_{A_1}(E_{F_b}) - \frac{\alpha_4}{3}[\bar{\tau}_{A_1}(E_{F_b})]^3\right], \alpha_5 = \left[\frac{2m_t^+ m_t^-}{\alpha\hbar^2}\omega_{A_1}\right],$$

$$\omega_{A_1} = \left[\frac{\alpha^2}{16}\left[\frac{1}{m_t^- m_l^+} + \frac{1}{m_l^- m_t^+}\right]^2 - \frac{\alpha^2}{4m_t^+ m_t^- m_l^- m_l^+}\right],$$

$$J_{A_1}(E_{F_b}) = \frac{A_A(E_{F_b})}{3}\left[-(A_A^2(E_{F_b}) + B_A^2(E_{F_b}))E(\lambda, q) + 2B_A^2(E_{F_b})F(\lambda, q)\right] + \frac{\bar{\tau}_{A1}(E_{F_b})}{3}$$
$$\times\left[(\bar{\tau}_{A_1}(E_{F_b}))^2 + A_A^2(E_{F_b}) + 2B_A^2(E_{F_b})\right]$$
$$\times\left[A_A^2(E_{F_b}) + \bar{\tau}_A^2(E_{F_b})\right]^{1/2}\left[B_A^2(E_{F_b}) + \bar{\tau}_{A_1}^2(E_{F_b})\right]^{-1/2}$$

$$\lambda = \tan^{-1}\frac{\bar{\tau}_A(E_{F_b})}{B_A(E_{F_b})}, q = \left[\frac{\sqrt{A_A^2(E_{F_b}) - B_A^2(E_{F_b})}}{A_A(E_{F_b})}\right]$$

$$A_A(E_{F_b}) = \left[\tau_{A_2}(E_{F_b}) + \sqrt{\tau_{A2}^2(E_{F_b}) - 4\tau_{A_3}(E_{F_b})}\right]^{1/2}/\sqrt{2},$$

$$B_A(E_{F_b}) = \left[\tau_{A_2}(E_{F_b}) - \sqrt{\tau_{A_2}^2(E_{F_b}) - 4\tau_{A_3}(E_{F_b})}\right]^{1/2}/\sqrt{2},$$

$$\tau_{A_2}(E_{F_b}) = \frac{\omega_{A_2}(E_{F_b})}{\omega_{A_1}^2}, \tau_{A3}(E_{F_b}) = \frac{\omega_{A_3}(E_{F_b})}{\omega_{A1}^2},$$

$$\omega_{A_2}(E_{F_b}) = \left[\frac{\alpha}{2}\left[\frac{1}{2m_t^*} - \frac{\alpha\cdot E_{F_b}}{2m_t^+} + \frac{1+\alpha\cdot E_{F_b}}{2m_t^-}\right]\cdot\left[\frac{1}{m_t^- m_l^+} + \frac{1}{m_l^- m_t^+}\right]\right.$$
$$\left. - \frac{\alpha}{m_t^+ m_t^-}\left[\frac{1}{2m_l^*} + \frac{\alpha\cdot E_{F_b}}{2m_l^+} + \frac{1+\alpha\cdot E_{F_b}}{2m_l^-}\right]\right]$$

$$\omega_{A_3}(E_{F_b}) = \left[\frac{\alpha\cdot E_{F_b}(1+\alpha\cdot E_{F_b})}{m_t^+ m_t^-} + \left[\frac{1}{2m_t^*} - \frac{\alpha\cdot E_{F_b}}{2m_t^+} + \frac{1+\alpha\cdot E_{F_b}}{2m_t^-}\right]^2\right],$$

$$\alpha_2(E_{F_b}) = \left[\frac{1}{2m_t^*} - \frac{\alpha\cdot E_{F_b}}{2m_t^+} + \frac{1+\alpha\cdot E_{F_b}}{2m_t^-}\right], \alpha_3 = \frac{\alpha\hbar^2}{4}\left[\frac{1}{m_t^- m_l^+} + \frac{1}{m_l^- m_t^+}\right],$$

$$\tau_{A_1}(E_{F_b}) = \left[\frac{2m_l^+ m_l^-}{\alpha\hbar^2}\right]^{1/2}\left[-\left[\frac{1}{2m_l^*} + \frac{1+\alpha\cdot E_{F_b}}{m_l^-} - \frac{\alpha\cdot E_{F_b}}{2m_l^+}\right]\right.$$
$$\left. + \left[\left[\frac{1}{2m_l^*} + \frac{1+\alpha\cdot E_{F_b}}{m_l^-} - \frac{\alpha\cdot E_{F_b}}{2m_l^+}\right]^2 + \frac{\alpha\cdot E_{F_b}(1+\alpha\cdot E_{F_b})}{m_l^- m_l^+}\right]^{1/2}\right]^{1/2}$$

$E(\lambda, q)$ is the incomplete Elliptic integral of second kind, $F(\lambda, q)$ is the incomplete Elliptic integral of first kind,

$$N_{A_4}(E_{F_b}) = \sum_{r=1}^{s} L(r)[M_{A_4}(E_{F_b})],$$

$$I_{17}(E') = \frac{-I_{15}(E') + \sqrt{I_{15}^2(E') + 4I_{16}(E')}}{2},$$

$$I_{15}(E') = \frac{2m_l^+ m_t^-}{\alpha \hbar^2}[\frac{1}{m_l^*} + \frac{1}{m_l^-} + \alpha E'(\frac{1}{m_l^-} - \frac{1}{m_l^+})] \text{ and}$$

$$I_{16}(E') = \frac{4m_l^+ m_t^-}{\alpha \hbar^4} E'(1 + \alpha E')$$

(b) The dispersion relation of the conduction electrons in bulk specimens of IV-VI semiconductors in accordance with the model of Bangert and Kastner is given by

$$\omega_1(\mathrm{E})k_s^2 + \omega_2(\mathrm{E})k_z^2 = 1 \tag{1.194b}$$

where

$$\omega_1(\mathrm{E}) = (2E)^{-1}[\frac{(\bar{\mathrm{R}})^2}{\mathrm{E}_{g0}(1 + \alpha_1 \mathrm{E})} + \frac{(\bar{\mathrm{S}})^2}{\Delta_c'(1 + \alpha_2 \mathrm{E})} + \frac{(\bar{\mathrm{Q}})^2}{\Delta_c'(1 + \alpha_3 \mathrm{E})}]$$

and

$$\omega_2(\mathrm{E}) = (2E)^{-1}[\frac{(\bar{\mathrm{A}})^2}{\mathrm{E}_{g_0}(1 + \alpha_1 \mathrm{E})} + \frac{(\bar{\mathrm{S}} + \bar{\mathrm{Q}})^2}{\Delta_c''(1 + \alpha_3 \mathrm{E})}],$$

$$(\bar{R})^2 = 2.3 \times 10^{-10}(ev_m)^2, (\bar{S})^2 = 4.6(\bar{R})^2,$$

$$\alpha_1 = \frac{1}{\mathrm{E}_{g_0}}, \alpha_2 = \frac{1}{\Delta_c'}, \alpha_3 = \frac{1}{\Delta_c''},$$

$$\Delta_c'' = 3.28\,\mathrm{eV}, \quad \Delta_c' = 3.07\,\mathrm{eV}, \quad (\bar{Q})^2 = 1.3(\bar{R})^2, \quad (\bar{A})^2 = 0.8 \times 10^{-4}(\mathrm{eVm})^2$$

The electron energy spectrum in heavily doped IV-VI materials in accordance with this model can be expressed by using the methods as given in this chapter as

$$\begin{aligned} 2I(4) = {} & k_s^2[\{c_1(\alpha_1, E, E_g) - iD_1(\alpha_1 E, E_g)\}\frac{(\bar{R})^2}{E_{g_0}} + \{c_2(\alpha_2, E, E_g) - iD_2(\alpha_2, E, E_g)\}\frac{(\bar{S})^2}{\Delta_c'} \\ & + \{c_3(\alpha_3, E, E_g) - iD_3(\alpha_3, E, E_g)\}\frac{(\bar{Q})^2}{\Delta_c''}] + k_z^2[\frac{2(\bar{A})^2}{E_{g_0}}\{c_1(\alpha_1, E, E_g) - iD_1(\alpha_1, E, E_g)\} \\ & + \frac{(\bar{S} + \bar{Q})^2}{\Delta_c''}\{c_3(\alpha_3, E, E_g) - iD_3(\alpha_3, E, E_g)\}] \end{aligned} \tag{1.194c}$$

where

$$\alpha_1 = \frac{1}{E_g}, \alpha_2 = \frac{1}{\Delta_c'}, \alpha_3 = \frac{1}{\Delta_c''}, G_i = \frac{1 + \alpha_i E}{\eta_g \alpha_i},$$

$$c_i(\alpha_i, E, \eta_g) = \left[\frac{2}{\alpha_i \eta_g \sqrt{\pi}}\right] \exp(-u_i^2) \times \left[\sum_{p=1}^{\infty} \{\exp(-p^2/4)(\sinh(pu_i))\} p^{-1}\right]$$

$$i = 1, 2, 3 \quad D_i(\alpha_i, E, \eta_g) = \left[\frac{\sqrt{\pi}}{\alpha_i \eta_g}\right] \exp(-u_i^2),$$

Therefore (1.194c) can be written as,

$$F_1(E, \eta_g) k_s^2 + F_2(E, \eta_g) k_z^2 = 1 \qquad (1.194\text{d})$$

where,

$$F_1(E, \eta_g) = [2\gamma_0(E, \eta_g)]^{-1} [\frac{(\bar{R})^2}{E_g} \{C_1(\alpha_1 E, E_g) - iD_1(\alpha_1, E, E_g)\}$$
$$+ \frac{(\bar{S})^2}{\Delta_c'} \{C_2(\alpha_2, E, E_g) - iD_2(\alpha_2, E, E_g\})$$
$$+ \frac{(\bar{Q})^2}{\Delta_c''} \{C_3(\alpha_3, E, E_g) - iD_3(\alpha_3, E, E_g)\}] \text{ and}$$

$$F_2(E, \eta_g) = [2\gamma_0(E, \eta_g)]^{-1} [\frac{2(\bar{A})^2}{E_g} \{C_1(\alpha_1, E, \eta_g) - iD_1(\alpha_1, E, \eta_g)\}$$
$$+ \frac{(\bar{S} + \bar{Q})^2}{\Delta_c''} \{C_3(\alpha_3, E, E_g) - iD_3(\alpha_3, E, E_g)\}]$$

Since $F_1(E, \eta_g)$ and $F_2(E, \eta_g)$ are complex, the energy spectrum is also complex in the presence of Gaussian band tails.

The EEMs can be written as

$$m_\perp^*(E_{F_h}, \eta_g) = (\frac{\hbar^2}{2}) \text{ Real part of } (\frac{F_1'(E_{F_h}, \eta_g)}{F_1^2(E_{F_h}, \eta_g)}) \qquad (1.194\text{e})$$

$$m_{11}^*(E_{F_h}, \eta_g) = (\frac{\hbar^2}{2}) \text{ Real part of } (\frac{F_2'(E_{F_h}, \eta_g)}{F_2^2(E_{F_h}, \eta_g)}) \qquad (1.194\text{f})$$

It appears then that, the evolution of the masses needs an expression of the carrier concentration, which in turn is determined by the DOS function.

The DOS function in this case can be expressed as

$$N_{HD}(\mathrm{E},\eta_g) = \frac{g_v}{3\pi^2} F_3'(\mathrm{E},\eta_g),\; F_3(\mathrm{E},\eta_g) = [F_1(\mathrm{E},\,\eta_g)\sqrt{F_2(\mathrm{E},\eta_g)}]^{-1} \tag{1.194g}$$

The electron concentration is given by

$$n_0 = \frac{g_v}{3\pi^2}\text{Real part of } [F_3(\mathrm{E}_{F_h},\eta_g) + \sum_{r=1}^{s} L(\mathrm{r})[F_3(\mathrm{E}_{F_h},\eta_g)]] \tag{1.194h}$$

The EP can be written as

$$J = \frac{\alpha_0 e g_v}{6\pi^2\hbar}\text{Real part of } \int_{\mathrm{E}_0}^{\infty} F_5(\mathrm{E}',\eta_g)F_3'(\mathrm{E}',\eta_g)\mathrm{f(E)dE}' \tag{1.194i}$$

where

$$F_5(\mathrm{E}',\eta_g) = \frac{F_2^2(\mathrm{E}',\eta_g)}{F_2'(\mathrm{E}',\eta_g)\sqrt{F_2(\mathrm{E}',\eta_g)}}$$

The 2D dispersion relation in this case assumes the form

$$k_s^2 = F_6(E,\eta_g,n_z) \tag{1.194j}$$

where,

$$F_6(E,\eta_g,n_z) = [\frac{[1 - F_2(E,\eta_g)(n_z\pi/d_z)^2]}{F_1(E,\eta_g)}]$$

The EEM in this case is given by

$$m^*(E_{F1HD},\eta_g,n_z) = \frac{\hbar^2}{2}\text{Real part of } [\mathrm{F}_6'(E_{F1HD},\eta_g,n_z)] \tag{1.194k}$$

The total DOS function can be written as

$$N_{2DT}(\mathrm{E}) = \frac{g_v}{2\pi}\sum_{n_z=1}^{n_{z\max}} F_6'(E,\eta_g,n_z)H(E - E_{n_{z71HD}}) \tag{1.194l}$$

where $E_{n_{z71HD}}$ the quantized energy in this case and is given by

$$1 = F_2(E_{n_{z71HD}}, \eta_g)(\pi n_z/d_z)^2 \quad (1.194\text{m})$$

The surface electron concentration can be expressed as

$$n_0 = \frac{g_v}{2\pi} \text{Real part of } [\sum_{n_t=1}^{n_{z\max}} [F_6(\mathrm{E}_{F1HD}, \eta_g, n_z) + \sum_{r=1}^{s} L(r)[F_6(\mathrm{E}_{F1HD}, \eta_g, n_z)]]] \quad (1.194\text{n})$$

The EP in this case is given by

$$J_{2D} = \frac{\alpha_0 e g_v}{h d_z} [\sum_{n_{z\min}}^{n_{z\max}} [\frac{F_2^2(E_{n_{z71HD}}, \eta_g)}{F_2'(En_{z71HD}, \eta_g)\sqrt{F_2(E_{n_{z71HD}}, \eta_g)}}][\int_{E_{n_{z71HD}}}^{\infty} F_6'(E, \eta_g, \mathrm{n}_z) f(E) \mathrm{dE}]] \quad (1.194\text{o})$$

In the absence of band-tails the EEMs can be written as

$$m_{\perp}^*(E_F) = (\frac{\hbar^2}{2})(\frac{F_{11}'(E_F)}{F_{11}^2(E_F)}) \quad (1.194\text{p})$$

$$m_{11}^*(E_F) = (\frac{\hbar^2}{2})(\frac{F_{12}'(E_F)}{F_{12}^2(E_F)}) \quad (1.194\text{q})$$

where

$$F_{11}(E) = [\frac{(\bar{\mathrm{R}})^2}{E_{go}(1+\alpha_1 E)} + \frac{(\bar{S})^2}{\Delta_c'(1+\alpha_2 E)} + \frac{(\bar{Q})^2}{\Delta_{go}''(1+\alpha_3 E)}][2E]^{-1}$$

and

$$F_{12}(E) = [\frac{(\bar{A})^2}{E_{go}(1+\alpha_1 E)} + \frac{(\bar{S}+\bar{Q})^2}{\Delta_{go}''(1+\alpha_3 E)}][2E]^{-1}$$

It appears then that, the evolution of the masses needs an expression of the carrier concentration, which in turn is determined by the DOS function.

The DOS function in this case can be expressed as

$$N(\mathrm{E}) = \frac{g_v}{3\pi^2} F_{13}'(\mathrm{E}),\ F_{13}(\mathrm{E}) = [F_{11}(\mathrm{E})\sqrt{F_{12}(\mathrm{E})}]^{-1} \quad (1.194\text{r})$$

The electron concentration is given by

$$n_0 = \frac{g_v}{3\pi^2}[F_{13}(E_F) + \sum_{r=1}^{s} L(r)[F_{13}(E_F)]] \tag{1.194s}$$

The EP can be written as

$$J = \frac{\alpha_0 e g_v}{6\pi^2 \hbar} \int_{\mathrm{E}_0}^{\infty} F_{15}(\mathrm{E}')F'_{13}(\mathrm{E}')\mathrm{f}(\mathrm{E})\mathrm{dE}' \tag{1.194t}$$

where

$$F_{15}(\mathrm{E}') = \frac{F_{12}^2(\mathrm{E}')}{F'_{12}(\mathrm{E}')\sqrt{F_{12}(\mathrm{E}')}}$$

In the absence of band-tails, the 2D dispersion relation in this case assumes the form

$$k_s^2 = F_{16}(E, n_z) \tag{1.194u}$$

where,

$$F_{16}(E, n_z) = [\frac{[1 - F_{12}(E)(n_z\pi/d_z)^2]}{F_{11}(E)}]$$

The EEM in this case is given by

$$m^*(E_{F_s}, n_z) = \frac{\hbar^2}{2}[F'_{16}(E_{F_s}, n_z)] \tag{1.194v}$$

The total DOS function can be written as

$$N_{2DT}(\mathrm{E}) = \frac{g_v}{2\pi}\sum_{n_z=1}^{n_{z\max}} F'_{16}(E, n_z)H(E - E_{n_{z711}}) \tag{1.194w}$$

where $E_{n_{z711}}$ is the quantized energy in this case and is given by

$$1 = F_{12}(E_{n_{z711}}, \eta_g)(\pi n_z/d_z)^2 \tag{1.194x}$$

The surface electron concentration can be expressed as

$$n_0 = \frac{g_v}{2\pi}[\sum_{n_z}^{n_{z\max}} [F_{16}(E_{F_s}, \mathrm{n}_z) + \sum_{r=1}^{s} L(r)[F_{16}(E_{F_s}, \mathrm{n}_z)]]] \tag{1.194y}$$

The EP in this case is given by

$$J_{2D} = \frac{\alpha_0 e g_v}{h d_z}[\sum_{n_{z\min}}^{n_{z\max}} [\frac{F_{12}^2(E_{n_{z711}})}{F'_{12}(E_{n_{z711}})\sqrt{F_{12}(E_{n_{z711}})}}][\int_{E_{n_{z711}}}^{\infty} F'_{16}(E, \mathrm{n}_z) f(E) \mathrm{d}E]]. \tag{1.194z}$$

1.2.5 The EP from QWs of HD Stressed Kane Type Semiconductors

The electron energy spectrum in stressed Kane type semiconductors can be written [184–186] as

$$\left(\frac{k_x}{\bar{a}_0(E)}\right)^2 + \left(\frac{k_y}{\bar{b}_0(E)}\right)^2 + \left(\frac{k_z}{\bar{c}_0(E)}\right)^2 = 1 \tag{1.195}$$

where, $[\bar{a}_0(E)]^2 \equiv \frac{\bar{K}_0(E)}{\bar{A}_0(E) + \frac{1}{2}\bar{D}_0(E)}$, $\bar{K}_0(E) \equiv \left[E - C_1\varepsilon - \frac{2C_2^2\varepsilon_{xy}^2}{3E'_g}\right]\left(\frac{3E'_g}{2B_2^2}\right)$, C_1 is the conduction band deformation potential, ε is the trace of the strain tensor $\hat{\varepsilon}$ which can be written as $\hat{\varepsilon} = \begin{bmatrix} \varepsilon_{xx} & \varepsilon_{xy} & 0 \\ \varepsilon_{xy} & \varepsilon_{yy} & 0 \\ 0 & 0 & \varepsilon_{zz} \end{bmatrix}$, C_2 is a constant which describes the strain interaction between the conduction and valance bands, $E'_g \equiv E_g + E - C_1\varepsilon$, B_2 is the momentum matrix element,

$$\bar{A}_0(E) \equiv \left[1 - \frac{(\bar{a}_0 + C_1)}{E'_g} + \frac{3\bar{b}_0\varepsilon_{xx}}{2E'_g} - \frac{\bar{b}_0\varepsilon}{2E'_g}\right],$$

$$\bar{a}_0 \equiv -\frac{1}{3}(\bar{b}_0 + 2\bar{m}),$$

$$\bar{b}_0 \equiv \frac{1}{3}(\bar{l} - \bar{m}),\ \bar{d}_0 \equiv \frac{2\bar{n}}{\sqrt{3}},$$

$\bar{l}, \bar{m}, \bar{n}$ are the matrix elements of the strain perturbation operator, $\bar{D}_0(E) \equiv (\bar{d}_0\sqrt{3})\frac{\varepsilon_{xy}}{E'_g}$,

$$[\bar{b}_0(E)]^2 \equiv \frac{\bar{K}_0(E)}{\bar{A}_0(E) - \frac{1}{2}\bar{D}_0(E)}, \quad [\bar{c}_0(E)]^2 \equiv \frac{\bar{K}_0(E)}{\bar{L}_0(E)} \text{ and}$$

$$\bar{L}_0(E) \equiv \left[1 - \frac{(\bar{a}_0 + C_1)}{E'_g} + \frac{3\bar{b}_0\varepsilon_{zz}}{E'_g} - \frac{\bar{b}_0\varepsilon}{2E'_g}\right]$$

The use of (1.195) can be written as

$$(E - \alpha_1)k_x^2 + (E - \alpha_2)k_y^2 + (E - \alpha_3)k_z^2 = t_1E^3 - t_2E^2 + t_3E + t_4 \tag{1.196a}$$

where

$$\alpha_1 \equiv \left[E_g - C_1\varepsilon - (\bar{a}_0 + C_1)\varepsilon + \frac{3}{2}\bar{b}_0\varepsilon_{xx} - \frac{\bar{b}_0}{2}\varepsilon + \left(\sqrt{3}/2\right)\varepsilon_{xy}\bar{d}_0\right],$$

$$\alpha_2 \equiv \left[E_g - C_1\varepsilon - (\bar{a}_0 + C_1)\varepsilon + \frac{3}{2}\bar{b}_0\varepsilon_{xx} - \frac{\bar{b}_0}{2}\varepsilon - \left(\sqrt{3}/2\right)\varepsilon_{xy}\bar{d}_0\right],$$

$$\alpha_3 \equiv \left[E_g - C_1\varepsilon - (\bar{a}_0 + C_1)\varepsilon + \frac{3}{2}\bar{b}_0\varepsilon_{zz} - \frac{\bar{b}_0}{2}\varepsilon\right],$$

$$t_1 \equiv \left(3/2B_2^2\right),$$

$$t_2 \equiv \left(1/2B_2^2\right)\left[6(E_g - C_1\varepsilon) + 3C_1\varepsilon\right],$$

$$t_3 \equiv \left(1/2B_2^2\right)\left[3(E_g - C_1\varepsilon)^2 + 6C_1\varepsilon(E_g - C_1\varepsilon) - 2C_2^2\varepsilon_{xy}^2\right] \text{ and}$$

$$t_4 \equiv \left(1/2B_2^2\right)\left[-3C_1\varepsilon(E_g - C_1\varepsilon)^2 + 2C_2^2\varepsilon_{xy}^2\right].$$

The (1.196a) can be written as

$$Ek^2 - T_{17}k_x^2 - T_{27}k_y^2 - T_{37}k_z^2 = \left[q_{67}E^3 - R_{67}E^2 + V_{67}E + \rho_{67}\right] \tag{1.196b}$$

where,

$$\mathrm{T}_{17} = \alpha_1, \ \mathrm{T}_{27} = \alpha_2, \ \mathrm{T}_{37} = \alpha_3, \ \mathrm{t}_1 = \mathrm{q}_{67}, \ \mathrm{t}_2 = \mathrm{R}_{67}, \ \mathrm{t}_3 = \mathrm{V}_{67} \text{ and } \mathrm{t}_4 = \rho_{67}$$

Under the condition of heavy doping, (1.196b) can be written as

$$\begin{aligned} &I(4)k^2 - T_{17}I(1)k_x^2 - T_{27}I(1)k_y^2 - T_{37}k_z^2I(1) \\ &\quad = \left[q_{67}I(6) - R_{67}I(5) + V_{67}I(4) + \rho_{67}I(1)\right] \end{aligned} \tag{1.196c}$$

where,

$$I(6) = \int_{-\infty}^{E} (E - V)^3 F(V)dV \quad (1.197)$$

The (1.197) can be written as

$$I(6) = E^3 I(1) - 3E^2 I(7) + 3EI(8) - I(9) \quad (1.198)$$

In which,

$$I(7) = \int_{-\infty}^{E} VF(V)dV \quad (1.199)$$

$$I(8) = \int_{-\infty}^{E} V^2 F(V)dV \quad (1.200)$$

$$I(9) = \int_{-\infty}^{E} V^3 F(V)dV \quad (1.201)$$

Using (1.4), together with simple algebraic manipulations, one obtains

$$I(7) = \frac{-\eta_g}{2\sqrt{\pi}} \exp\left(\frac{-E^2}{\eta_g^2}\right) \quad (1.202)$$

$$I(8) = \frac{\eta_g^2}{4}\left[1 + Erf\left(\frac{E}{\eta_g}\right)\right] \quad (1.203)$$

and

$$I(9) = \frac{-\eta_g^3}{2\sqrt{\pi}} \exp\left(\frac{-E^2}{\eta_g^2}\right)\left[1 + \frac{E^2}{\eta_g^2}\right] \quad (1.204)$$

Thus (1.197) can be written as

$$I(6) = \left[\frac{E}{2}\left[1 + Erf\left(\frac{E}{\eta_g}\right)\right]\left[E^2 + \frac{3}{2}\eta_g^2\right] + \frac{\eta_g}{2\sqrt{\pi}} \exp\left(\frac{-E^2}{\eta_g^2}\right)\left[4E^2 + \eta_g^2\right]\right] \quad (1.205)$$

Thus, combining the appropriate equations, the dispersion relations of the conduction electrons in HD stressed materials can be expressed as

$$P_{11}(E,\eta_g)k_x^2 + Q_{11}(E,\eta_g)k_y^2 + S_{11}(E,\eta_g)k_z^2 = 1 \tag{1.206}$$

where,

$$P_{11}(E,\eta_g) \equiv \left[\frac{\gamma_0(E,\eta_g) - (T_{17}/2)[1 + Erf(E/\eta_g)]}{\Delta_{14}(E,\eta_g)}\right],$$

$$\Delta_{14}(E,\eta_g) \equiv \left[q_{67}\left\{\frac{E}{2}\left[1 + Erf\left(\frac{E}{\eta_g}\right)\right]\left[E^2 + \frac{3}{2}\eta_g^2\right] + \frac{\eta_g}{2\sqrt{\pi}}\exp\left(\frac{-E^2}{\eta_g^2}\right)\left[4E^2 + \eta_g^2\right]\right\}\right.$$
$$\left. - R_{67}\theta_0(E,\eta_g) + V_{67}\gamma_0(E,\eta_g) + \frac{\rho_{67}}{2}[1 + Erf(E/\eta_g)]\right],$$

$$Q_{11}(E,\eta_g) \equiv \left[\frac{\gamma_0(E,\eta_g) - (T_{27}/2)[1 + Erf(E,\eta_g)]}{\Delta_{14}(E,\eta_g)}\right] \text{ and}$$

$$S_{11}(E,\eta_g) \equiv \left[\frac{\gamma_0(E,\eta_g) - (T_{37}/2)[1 + Erf(E,\eta_g)]}{\Delta_{14}(E,\eta_g)}\right].$$

Thus, the energy spectrum in this case is real since the dispersion relation of the corresponding materials in the absence of band tails as given by (1.195) has no poles in the finite complex plane.

The EEMs along x, y and z directions in this case can be written as

$$m_{xx}^*(E_{F_h},\eta_g) = \frac{\hbar^2}{2}\left[[\gamma_0(E_{F_h},\eta_g) - (T_{17}/2)[1 + Erf(E_{F_h}/\eta_g)]]^{-2}\right.$$
$$\times\left[\{\Delta_{14}(E_{F_h},\eta_g)\}'[\gamma_0(E_{F_h},\eta_g) - (T_{17}/2)[1 + Erf(E_{F_h},\eta_g)]]\right]$$
$$\left. - \Delta_{14}(E_{F_h},\eta_g)\left[\frac{1}{2}\left[1 + Erf\left(\frac{E_{F_h}}{\eta_g}\right)\right] - \left\{\frac{T_{17}}{\eta_g\sqrt{\pi}}\exp\left(\frac{-E_{F_h}^2}{\eta_g^2}\right)\right\}\right]\right] \tag{1.207}$$

$$m_{yy}^*(E_{F_h},\eta_g) = \frac{\hbar^2}{2}\left[[\gamma_0(E_{F_h},\eta_g) - (T_{27}/2)[1 + Erf(E_{F_h}/\eta_g)]]^{-2}\right.$$
$$\times\left[\{\Delta_{14}(E_{F_h},\eta_g)\}'[\gamma_0(E_{F_h},\eta_g) - (T_{27}/2)[1 + Erf(E_{F_h}/\eta_g)]]\right]$$
$$\left. - \Delta_{14}(E_{F_h},\eta_g)\left[\frac{1}{2}\left[1 + Erf\left(\frac{E_{F_h}}{\eta_g}\right)\right] - \left\{\frac{T_{27}}{\eta_g\sqrt{\pi}}\exp\left(\frac{-E_{F_h}^2}{\eta_g^2}\right)\right\}\right]\right] \tag{1.208}$$

and

$$m_{zz}^{*}(E_{F_h},\eta_g) = \frac{\hbar^2}{2}\left[[\gamma_0(E_{F_h},\eta_g) - (T_{37}/2)[1+Erf(E_{F_h}/\eta_g)]]^{-2}\right.$$
$$\times\left[\{\Delta_{14}(E_{F_h},\eta_g)\}'[\gamma_0(E_{F_h},\eta_g) - (T_{37}/2)[1+Erf(E_{F_h}/\eta_g)]]\right]$$
$$\left.-\Delta_{14}(E_{F_h},\eta_g)\left[\frac{1}{2}\left[1+Erf\left(\frac{E_{F_h}}{\eta_g}\right)\right] - \left\{\frac{T_{37}}{\eta_g\sqrt{\pi}}\exp\left(\frac{-E_{F_h}^2}{\eta_g^2}\right)\right\}\right]\right] \tag{1.209}$$

Thus, we can see that the EEMs in this case exist within the band gap.
In the absence of band tails, $\eta_g \to 0$ we get

$$m_{xx}^{*}(E_F) = \hbar^2\bar{a}_0(E_F)\{\bar{a}_0(E_F)\}' \tag{1.210}$$

$$m_{xx}^{*}(E_F) = \hbar^2\bar{b}_0(E_F)\{\bar{b}_0(E_F)\}' \tag{1.211}$$

and

$$m_{xx}^{*}(E_F) = \hbar^2\bar{c}_0(E_F)\{\bar{c}_0(E_F)\}' \tag{1.212}$$

The DOS function in this case can be written as

$$N_{HD}(E,\eta_g) = \frac{g_v}{3\pi^2}\Delta_{100}(E,\eta_g)$$

where

$$\Delta_{100}(E,\eta_g) = \{\Delta_{15}(E,\eta_g)\}^{-2}\left[\frac{3}{2}\{\Delta_{15}(E,\eta_g)\}\sqrt{\Delta_{14}(E,\eta_g)}\{\Delta_{14}(E,\eta_g)\}'\right.$$
$$\left.-\{\Delta_{14}(E,\eta_g)\}^{3/2}\{\Delta_{15}(E,\eta_g)\}'\right] \text{ and}$$
$$\Delta_{15}(E,\eta_g) \equiv [[\gamma_0(E,\eta_g) - (T_{17}/2)[1+Erf(E/\eta_g)]][\gamma_0(E,\eta_g) - (T_{27}/2)[1+Erf(E,\eta_g)]]$$
$$\times[\gamma_0(E,\eta_g) - (T_{37}/2)[1+Erf(E/\eta_g)]]]^{1/2} \tag{1.213}$$

Using (1.213), the electron concentration at can be written as

$$n_0 = \frac{g_v}{3\pi^2}\left[\bar{I}_{126}(E_{F_h},\eta_g) + \sum_{r=1}^{s} L(r)[\bar{I}_{126}(E_{F_h},\eta_g)]\right] \tag{1.214}$$

where

$$\bar{I}_{126}(E_{F_h}, \eta_g) = [\frac{\{\Delta_{14}(E_{F_h}, \eta_g)\}^{3/2}}{\Delta_{15}(E_{F_h}, \eta_g)}]$$

The EP in this case is given by

$$J = \frac{\alpha_0 e g_v}{6\pi^2 \hbar}[\int_{E_0}^{\infty} \frac{\Delta_{100}(\mathrm{E}', \eta_g)\mathrm{S}_{11}^2(\mathrm{E}', \eta_g)}{\mathrm{S}_{11}'(\mathrm{E}', \eta_g)\sqrt{\mathrm{S}_{11}(\mathrm{E}', \eta_g)}} \mathrm{f(E)dE'}] \tag{1.215}$$

The dispersion relation of the conduction electrons in HD QWs of Kane type semiconductors can be written as

$$P_{11}(E, \eta_g)k_x^2 + Q_{11}(E, \eta_g)k_y^2 + S_{11}(E, \eta_g)(\pi n_z/dz)^2 = 1 \tag{1.216}$$

The EEM can be expressed as

$$m^*(E_{F1HD}, \eta_g, n_z) = \frac{\hbar^2}{2} A_{56}'(E_{F1HD}, \eta_g, n_z)$$

where,

$$A_{56}(E, \eta_g, n_z) = \frac{\pi\left[1 - S_{11}(E, \eta_g)(\eta_z \pi/d_z)^2\right]}{\sqrt{P_{11}(E, \eta_g)Q_{11}(E, \eta_g)}} \tag{1.217}$$

From (1.217), it appears that the EEM is a function of Fermi energy, and size quantum number and the same mass exists in the band gap.

Thus, the total 2D DOS function can be expressed as

$$N_{2DT}(E) = \left(\frac{g_v}{2\pi}\right) \sum_{n_z=1}^{n_{z\max}} A_{56}'(E_{F1HD}, \eta_g, n_z) \tag{1.218}$$

The sub band energies $\left(E_{n_{zg}HD}\right)$ are given by

$$S_{11}\left(E_{n_{zg}HD}, \eta_g\right)(\pi n_z/d_z)^2 = 1 \tag{1.219}$$

The 2D surface electron concentration per unit area for QWs of stressed HD Kane type compounds can be written as

$$n_{2D} = \frac{g_v}{2\pi}\sum_{n_z=1}^{n_{z\max}}\left[T_{57HD}(E_{Fs1HD},\eta_g,n_z) + T_{58HD}(E_{Fs1HD},\eta_g,n_z)\right] \tag{1.220a}$$

where,

$$T_{57}(E_{Fs1HD},\eta_g,n_z) \equiv A_{56}\left(E_{F1HD},\eta_g,n_z\right)$$

and

$$T_{58HD}(E_{F1HD},\eta_g,n_z) \equiv \sum_{r=1}^{s} L(r)T_{57HD}(E_{F1HD},\eta_g,n_z).$$

The EP in this case is given by

$$J_{2D} = \frac{\alpha_0 e g_v}{hd_z}[\sum_{n_{z\min}}^{n_{z\max}}[\frac{S_{11}^2(E_{n_{z8HD}},\eta_g)}{S_{11}'(E_{n_{z8HD}},\eta_g)\sqrt{S_{11}(E_{n_{z8HD}},\eta_g)}}]][\int_{E_{n_{z8HD}}}^{\infty} A_{56}'(E,\eta_g,n_z)f(E)\mathrm{dE}] \tag{1.220b}$$

In the absence of band tails, the 2D electron energy spectrum in QWs of stressed materials assumes the form

$$\frac{k_x^2}{[\bar{a}_0(E)]^2} + \frac{k_y^2}{[\bar{b}_0(E)]^2} + \frac{1}{[\bar{c}_0(E)]^2}(n_z\pi/d_z)^2 = 1 \tag{1.221}$$

The area of 2D wave vector space enclosed by (1.221) can be written as

$$A(E,n_z) = \pi P^2(E,n_z)\bar{a}_0(E)\bar{b}_0(E)$$

where

$$P^2(E,n_z) = [1 - [n_z\pi/d_z\bar{c}_0(E)]^2].$$

From (1.221), the EEM can be written as

$$m^*(E_{F_s},n_z) = \frac{\hbar^2}{2}[P^2(E_{F_s},n_z)\bar{a}_0(E_{F_s})\bar{b}_0(E_{F_s})]' \tag{1.222}$$

Thus, the total 2D DOS function can be expressed as

$$N_{2DT}(E) = \left(\frac{g_v}{2\pi}\right) \sum_{n_z=1}^{n_{z\max}} \theta_6(E, n_z) H(E - E_{n_{z11}}) \tag{1.223}$$

in which,

$$\theta_6(E, n_z) = \Big[2P(E, n_z)\{P(E, n_z)\}' \bar{a}_0(E) \bar{b}_0(E) + \{P(E, n_z)\}^2 \{\bar{a}_0(E)\}' \bar{b}_0(E) + \{P(E, n_z)\}^2 \{\bar{b}_0(E)\}' \bar{a}_0(E)\Big]$$

The sub band energies $(E_{n_{z11}})$ are given by

$$\bar{c}_0(E_{n_{z11}}) = n_z \pi / d_z \tag{1.224}$$

The 2D surface electron concentration per unit area for QWs of stressed Kane type compounds can be written as

$$n_{2D} = \frac{g_v}{2\pi} \sum_{n_z=1}^{n_{z\max}} [T_{61}(E_{Fs}, n_z) + T_{62}(E_{Fs}, n_z)] \tag{1.225}$$

where

$$T_{61}(E_{Fs}, n_z) \equiv \left[P^2(E_{Fs}, n_z) \bar{a}_0(E_{Fs}) \bar{b}_0(E_{Fs})\right] \text{and}$$

$$T_{62}(E_{Fs}, n_z) \equiv \sum_{r=1}^{s} L(r) T_{61}(E_{Fs}, n_z).$$

The EP in this case is given by

$$J_{2D} = \frac{\alpha_0 e g_v}{2h d_z} [\sum_{n_{z\min}}^{n_{z\max}} [\bar{c}_0(\mathrm{E}_{n_{z11}})]' [\int_{\mathrm{E}_{n_{z11}}}^{\infty} \theta_6(\mathrm{E}, \mathrm{n}_z) f(\mathrm{E}) d\mathrm{E}]] \tag{1.226}$$

The DOS function for bulk specimens of stressed Kane type semiconductors in the absence of band tail can be written as

$$D_0(E) = g_v (3\pi^2)^{-1} \bar{T}_0(E) \tag{1.227}$$

where

$$\overline{T}_0(E) = \left[\bar{a}_0(E)\bar{b}_0(E)[\bar{c}_0(E)]' + \bar{a}_0(E)[\bar{b}_0(E)]'\bar{c}_0(E) + [\bar{a}_0(E)]'\bar{b}_0(E)\bar{c}_0(E)\right]$$

Combining (1.227) with the Fermi-Dirac occupation probability factor and using the generalized Sommerfeld lemma the electron concentration in this case can be expressed as

$$n_0 = g_v(3\pi^2)^{-1}[M_4(E_F) + N_4(E_F)] \tag{1.228}$$

where,

$$M_4(E_F) \equiv [\bar{a}_0(E_F)\bar{b}_0(E_F)\bar{c}_0(E_F)] \text{ and}$$
$$N_4(E_F) \equiv \sum_{r=1}^{s} L(r)M_4(E_F).$$

The EP in this case is given by

$$J = \frac{\alpha_0 e g_v}{12\pi^2\hbar}\int_{E_0}^{\infty} \bar{T}_0(E')[\bar{c}_0(E')]'\text{f(E)dE}'. \tag{1.229}$$

1.2.6 The EP from QWs of HD Te

The dispersion relation of the conduction electrons in Te can be expressed as [187]

$$E = \psi_1 k_z^2 + \psi_2 k_s^2 \pm [\psi_3^2 k_z^2 + \psi_4^2 k_s^2]^{1/2} \tag{1.230}$$

where, the values of the system constants are given in Table 1.1.

The carrier energy spectrum in HD Te can be written as

$$\gamma_3(E,\eta_g) = \psi_1 k_z^2 + \psi_2 k_s^2 \pm [\psi_3^2 k_z^2 + \psi_4^2 k_s^2]^{1/2} \tag{1.231}$$

The EEMs along k_z and k_s directions assume the forms

$$m_z^*(E_{F_h},\eta_g) = \frac{\hbar^2}{2\psi_1}\left[1 - \frac{\psi_3}{\sqrt{\psi_3^2 + 4\psi_1\gamma_3(E_{F_h},\eta_g)}}\right]\gamma_3'(E_{F_h},\eta_g) \tag{1.232}$$

Table 1.1 The numerical values of the energy band constants of few materials

	Materials	Numerical values of the energy band constants
1	The conduction electrons of n-Cadmium Germanium Arsenide can be described by three types of band models	1. The values of the energy band constants in accordance with the generalized electron dispersion relation of nonlinear optical materials are as follows $E_{g0} = 0.57\,\text{eV}, \Delta_{\parallel} = 0.30\,\text{eV}, \Delta_{\perp} = 0.30\,\text{eV}, m_{\parallel}^{*} = 0.034m_0, m_{\perp}^{*} = 0.039m_0$, $T = 4\,K$, $\delta = -0.21\,\text{eV}$, $g_v = 1$, $\varepsilon_{sc} = 18.4\varepsilon_0$ (ε_{sc} and ε_0 $T = 4\,K$, are the permittivity of the semiconductor material and free space respectively) and $W = 4\,\text{eV}$
		2. In accordance with the three band model of Kane, the spectrum constants are given by $\Delta = (\Delta_{\parallel} + \Delta_{\perp})/2 = 0.33\,\text{eV}$, $E_{g0} = 0.57\,\text{eV}$, and $\delta = 0\,\text{eV}$
		3. In accordance with two band model of Kane, $E_{g0} = 0.57\,\text{eV}$ and $m_c = 0.0365\,m_0$
2	n-Indium Arsenide	The values $E_{g_0} = 0.3\,\text{eV}$, $\Delta = 0.43\,\text{eV}$, $m_c = 0.026m_0$, $g_v = 1$, $\varepsilon_{sc} = 12.25\varepsilon_0$ and $W = 5.06\,\text{eV}$ are valid for three band model of Kane
3	n-Gallium Arsenide	The values $E_{g_0} = 1.55\,\text{eV}$, $\Delta = 0.35\,\text{eV}$, $m_c = 0.07\,m_0$, $g_v = 1$, $\varepsilon_{sc} = 12.9\varepsilon_0$ and $W = 4.07\,\text{eV}$ are valid for three band model of Kane. The values $\alpha_{13} = -1.97 \times 10^{-37}\,\text{eVm}^4$ and $a_{15} = -2.3 \times 10^{-34}\,\text{eVm}^4$. The values $\alpha_{11} = -2132 \times 10^{-40}\,\text{eVm}^4$, $\alpha_{12} = -9030 \times 10^{-50}\,\text{eVm}^5$, $\beta_{11} = -2493 \times 10^{-40}\,\text{eVm}^4$, $\beta_{12} = 12594 \times 10^{-50}\,\text{eVm}^5$, $\gamma_{11} = 30 \times 10^{-30}\text{eVm}^3$, $\gamma_{12} = -154 \times 10^{-42}\text{eVm}^4$
4	n-Gallium Aluminum Arsenide	$E_{g0} = (1.424 + 1.266x + 0.26x^2)\,\text{eV}, \Delta = (0.34 - 0.5x)\,\text{eV}$ $m_c = [0.066 + 0.088x]m_0$, $g_v = 1$, $\varepsilon_{sc} = [13.18 - 3.12x]\varepsilon_0$ and $W = (3.64 - 0.14x)\text{eV}$
5	n-Mercury Cadmium Telluride	$E_{g_0} = (-0.302 + 1.93x + 5.35 \times 10^{-4}(1 - 2x)T - 0.810x^2 + 0.832x^3)\text{eV}$, $\Delta = (0.63 + 0.24x - 0.27x^2)\text{eV}$, $m_c = 0.1m_0E_{g_0}(\text{eV})^{-1}$, $g_v = 1, m_c = 0.048m_0$ $\varepsilon_{sc} = [20.262 - 14.812x + 5.22795x^2]\varepsilon_0$ and $W = (4.23 - 0.813(E_{g_0} - 0.083))\text{eV}$
6	n-Indium Gallium Arsenide Phosphide lattice matched to Indium Phosphide	$E_{g_0} = (1.337 - 0.73y + 0.13y^2)\text{eV}$, $\Delta = (0.114 + 0.26y - 0.22y^2)\text{eV}$, $m_c = (0.08 - 0.039y)m_0$, $y = (0.1896 - 0.4052x)/(0.1896 - 0.0123x)$, $g_v = 1$, $\varepsilon_{sc} = [10.65 + 0.1320y]\varepsilon_0$ and $W(x, y) = [5.06(1 - x)y + 4.38(1 - x)(1 - y) + 3.64xy + 3.75\{x(1 - y)\}]\text{eV}$
7	n-Indium Antimonide	$E_{g_0} = 0.2352\,\text{eV}$, $\Delta = 0.81\,\text{eV}$, $m_c = 0.01359m_0$, $g_v = 1$, $\varepsilon_{sc} = 15.56\varepsilon_0$ and $W = 4.72\,\text{eV}$
8	n-Gallium Antimonide	The values of $E_{g_0} = 0.81\,\text{eV}$, $\Delta = 0.80\,eV, P = 9.48 \times 10^{-10}\text{eVm}$, $\bar{\varsigma}_0 = -2.1$, $\bar{v}_0 = -1.49$, $\bar{\omega}_0 = 0.42$, $g_v = 1$ and $\varepsilon_{sc} = 15.85\varepsilon_0$ are valid for the model of Seiler et al. The values $\bar{E}_1 = 1.024\,\text{eV}$, $\bar{E}_2 = 0\,\text{eV}$, $\bar{E}_3 = -1.132\,\text{eV}$, $\bar{E}_4 = -0.05\,\text{eV}$, $\bar{E}_5 = 1.107\,\text{eV}$, $\bar{E}_6 = -0.113\,\text{eV}$ and $\bar{E}_7 = -0.0072\,\text{eV}$ are valid for the model of Zhang

(continued)

Table 1.1 (continued)

	Materials	Numerical values of the energy band constants
9	n-Cadmium Sulphide	$m_{\parallel}^{*} = 0.7m_0$, $m_{\perp}^{*} = 1.5m_0$, $\bar{\lambda}_0 = 1.4 \times 10^{-8}$ eVm, $g_v = 1$, $\varepsilon_{sc} = 15.5\varepsilon_0$ and $W = 4.5$ eV
10	n-Lead Telluride	The values $m_t^{-} = 0.070m_0$, $m_l^{-} = 0.54m_0$, $m_t^{+} = 0.010m_0$, $m_l^{+} = 1.4m_0$, $P_{\parallel} = 141$meV nm, $P_{\perp} = 486$ meV nm, $E_{g_0} = 190$ meV, $g_v = 4$, $\varepsilon_{sc} = 33\varepsilon_0$ and $W = 4.6$ eV are valid for the Dimmock model. The values $(\bar{R})^2 = 2.3 \times 10^{-19}(\text{eVm})^2$, $E_{g_0} = 0.16$ eV, $(\bar{s})^2 = 4.6(\bar{R})^2$, $\Delta_c' = 3.07$ eV, $(\bar{Q})^2 = 1.3(\bar{R})^2$, $\Delta_c'' = 3.28$ eV, $(\bar{A})^2 = 0.83 \times 10^{-19}(\text{eVm})^2$ and $W = 4.21$ eV are valid for the model of Bangert and Kastner. The values $m_{tv} = 0.0965m_0$, $m_{lv} = 1.33m_0$, $m_{tc} = 0.088m_0$, $m_{lc} = 0.83m_0$ are valid for the model of Foley et al. The values $m_1 = 0.0239m_0$, $m_2 = 0.024m_0$, $m_2' = 0.31m_0$, $m_3 = 0.24m_0$ are valid for the Cohen model
11	Stressed n-Indium Antimonide	The values $m_c = 0.048m_0$, $E_{g_0} = 0.081$ eV, $B_2 = 9 \times 10^{-10}$ eVm, $C_1^c = 3$ eV, $C_2^c = 2$ eV, $a_0 = -10$ eV, $b_0 = -1.7$ eV, $d = -4.4$ eV, $S_{xx} = 0.6 \times 10^{-3}(\text{kbar})^{-1}$, $S_{yy} = 0.42 \times 10^{-3}(\text{kbar})^{-1}$, $S_{zz} = 0.39 \times 10^{-3}(\text{kbar})^{-1}$, $S_{xy} = 0.5 \times 10^{-3}(\text{kbar})^{-1}$, $\varepsilon_{xx} = \sigma S_{xx}$, $\varepsilon_{yy} = \sigma S_{yy}$, $\varepsilon_{zz} = \sigma S_{zz}$, $\varepsilon_{xy} = \sigma S_{xy}$, σ is the stress in kilobar, $g_v = 1$ are valid for the model of Seiler et al.
12	Bismuth	$E_{g_0} = 0.0153$ eV, $m_1 = 0.00194m_0$, $m_2 = 0.313m_0$, $m_3 = 0.00246m_0$, $m_2' = 0.36m_0$, $g_v = 3$, $g_s = 2$, $M_2 = 0.128m_0$, $M_2' = 0.80m_0$ and $W = 4.34$ eV
13	Mercury Telluride	$m_v^{*} = 0.028m_0$, $g_v = 1$, $\varepsilon_{\infty} = 15.2\varepsilon_0$ and $W = 5.5$ eV
14	Platinum Antimonide	For valence bands, along $\langle 111 \rangle$ direction, $W = 5.5$ eV, $l_1 = 1.09$ eV, $v_1 = 0.17$ eV, $\bar{\bar{n}} = 0.22$ eV, $\bar{\bar{a}} = 0.643$ nm, $I_0 = 0.30(\text{eV})^2$, $\delta_0' = 0.33$ eV, $g_v = 8$, $\varepsilon_{sc} = 30\varepsilon_0$ and $\phi \approx 3.0$ eV
15	n-Gallium Phosphide	$m_{\parallel}^{*} = 0.92m_0$, $m_{\perp}^{*} = 0.25m_0$, $k_0 = 1.7 \times 10^{19} m^{-1}$, $\lvert V_G \rvert = 0.21$ eV, $g_v = 6$, $g_s = 2$ and $W = 3.75$ eV
16	Germanium	$E_{g_0} = 0.785$ eV, $m_{\parallel}^{*} = 1.57m_0$, $m_{\perp}^{*} = 0.0807m_0$ and $W = 4.14$ eV
17	Tellurium	The values $A_6 = 6.7 \times 10^{-16}$meVm^2, $A_7 = 4.2 \times 10^{-16}$ meV m^2, $A_8 = (6 \times 10^{-8}$ meV m$^2)$ and $A_9 = (3.6 \times 10^{-8}$ meV m$)^2$ are valid for the model of Bouat et al. The values $t_1 = 0.06315$ eV, $t_2 = -10.0\hbar^2/2m_0$, $t_3 = -5.55\hbar^2/2m_0$, $t_4 = 0.3 \times 10^{-36}$eVm^4, $t_5 = 0.3 \times 10^{-36}$eVm^4, $t_6 = -5.55\hbar^2/2m_0$, $t_7 = 6.18 \times 10^{-20}(\text{eVm})^2$ and $W = 1.9708$ eV are valid for the model of Ortenberg and Button

(continued)

Table 1.1 (continued)

	Materials	Numerical values of the energy band constants
18	Lead Germanium Telluride	The values $\bar{E}_{g_0} = 0.21\,\text{eV}$, $g_v = 4$ and $\phi \approx 6\,\text{eV}$ are valid for the model of Vassilev
19	Cadmium Antimonide	The values $A_{10} = -4.65 \times 10^{-19}\text{eVm}^2$, $A_{11} = -2.035 \times 10^{-19}\text{eVm}^2$, $A_{12} = -5.12 \times 10^{-19}\text{eVm}^2$, $A_{13} = -0.25 \times 10^{-10}\text{eVm}^2$, $A_{14} = 1.42 \times 10^{-19}\text{eVm}^2$, $A_{15} = 0.405 \times 10^{-19}\text{eVm}^2$, $A_{16} = -4.07 \times 10^{-19}\text{eVm}^2$, $A_{17} = 3.22 \times 10^{-10}\text{eVm}$, $A_{18} = 1.69 \times 10^{-20}(\text{eVm})^2$, $A_{19} = 0.070\,\text{eV}$ and $\phi \approx 2\,\text{eV}$ are valid for the model of Yamada
20	Cadmium Diphosphide	The values $\beta_1 = 8.6 \times 10^{-21}\text{eVm}^2$, $\beta_2 = 1.8 \times 10^{-21}(\text{eVm})^2$, $\beta_4 = 0.0825\,\text{eV}$, $\beta_5 = -1.9 \times 10^{-19}\text{eVm}^2$ and $\phi \approx 5\,\text{eV}$ are valid for the model of Chuiko
21	Zinc Diphosphide	The values $\beta_1 = 8.7 \times 10^{-21}\text{eVm}^2$, $\beta_2 = 1.9 \times 10^{-21}(\text{eVm})^2$, $\beta_4 = 0.0875\,\text{eV}$, $\beta_5 = -1.9 \times 10^{-19}\text{eVm}^2$ and $W \approx 3.9\,\text{eV}$ are valid for the model of Chuiko
22	Bismuth Telluride	The values $E_{g_0} = 0.145\,\text{eV}$, $\alpha_{11} = 3.25$, $\alpha_{22} = 4.81$, $\alpha_{33} = 9.02$, $\alpha_{23} = 4.15$, $g_s = 1$, $g_v = 6$ and $\phi = 5.3\,\text{eV}$ are valid for the model of Stordeur et al.
23	Antimony	The values $\bar{\alpha}_{11} = 16.7$, $\bar{\alpha}_{22} = 5.98$, $\bar{\alpha}_{33} = 11.61$, $\bar{\alpha}_{23} = 7.54$ and $W = 4.63\,\text{eV}$ are valid for the model of Ketterson
24	Zinc Selenide	$m_2^* = 0.16m_0$, $\Delta_2 = 0.42\,\text{eV}$, $E_{g_{02}} = 2.82\,\text{eV}$ and $W = 3.2\,\text{eV}$
25	Lead Selenide	$m_t^- = 0.23m_0$, $m_l^- = 0.32m_0$, $m_t^+ = 0.115m_0$, $m_l^+ = 0.303m_0$, $P_\parallel \approx 138\,\text{meV nm}$, $P_\perp \approx 471\text{meV nm}$, $E_{g_0} = 0.28\,\text{eV}$, $\varepsilon_{sc} = 21.0\varepsilon_0$ and $W = 4.2\,\text{eV}$

and

$$m_s^*(E_{F_h},\eta_g)=\frac{\hbar^2}{2\psi_2}\left[1-\frac{\psi_4}{\sqrt{\psi_4^2+4\psi_2\gamma_3(E_{F_h},\eta_g)}}\right]\gamma_3'(E_{F_h},\eta_g) \quad (1.233)$$

The investigations of EEMs require the expression of electron concentration, which can be written from (1.231) as

$$n_0=\frac{g_v}{3\pi^2}\left[t_{1HD}(E_{F_h},\eta_g)+t_{2HD}(E_{F_h},\eta_g)\right] \quad (1.234a)$$

where,

$$t_{1HD}(E_{F_h},\eta_g)=\left[3\psi_{5HD}(E_{F_h},\eta_g)\Gamma_{3HD}(E_{F_h},\eta_g)-\psi_6\Gamma_{3HD}^3(E_{F_h},\eta_g)\right],$$

$$\psi_{5HD}(E_{F_h},\eta_g)=\left[\frac{\gamma_3(E_{F_h},\eta_g)}{\psi_2}+\frac{\psi_4^2}{2\psi_2^2}\right],$$

$$\Gamma_{3HD}(E_{F_h},\eta_g)=\frac{\sqrt{\psi_3^2+4\psi_1\gamma_3(E_{F_h},\eta_g)}}{2\psi_1},$$

$$\psi_6=\frac{\psi_1}{\psi_2} \text{ and } t_{2HD}(E_{F_h},\eta_g)=\sum_{r=1}^{s}L(r)t_{1HD}(E_{F_h},\eta_g)$$

$$J=\frac{\alpha_0 e g_v}{12\pi^2\hbar}\left[\int_{\mathrm{E}_0}^{\infty}\frac{\sqrt{\psi_2^2+4\psi_1\gamma_3(\mathrm{E}',\eta_g)}}{\gamma_3'(\mathrm{E}',\eta_g)}t_{1HD}'(\mathrm{E}',\eta_g)f(\mathrm{E})d\mathrm{E}'\right] \quad (1.1234b)$$

The 2D electron energy spectrum in HD QW of Te can be written using (1.230) as

$$k_s^2=\psi_{5HD}(E,\eta_g)-\psi_6\left(\frac{\pi n_z}{d_z}\right)^2\pm\psi_7\left[\psi_{8HD}^2(E,\eta_g)-\left(\frac{\pi n_z}{d_z}\right)^2\right]^{1/2} \quad (1.235)$$

where,

$$\psi_7=\frac{\psi_4\sqrt{\psi_1}}{\psi_2^{3/2}} \text{ and } \psi_{8HD}^2(E,\eta_g)=\left[\frac{\psi_4^4+4\gamma_3(E,\eta_g)\psi_2\psi_4^2+4\psi_2^2\psi_3^2}{4\psi_1\psi_2\psi_4^2}\right]$$

The EEM in this case is given by

$$m^{*}(E_{F1HD},\eta_g,n_z)=\frac{\hbar^2}{2}\left[\psi'_{5HD}(E_{F1HD},\eta_g)\pm\frac{\psi_{8HD}(E_{F1HD},\eta_g)\psi'_{8HD}(E_{F1HD},\eta_g)}{\sqrt{\psi^2_{8HD}(E_{F1HD},\eta_g)-(\pi n_z/d_z)^2}}\right] \tag{1.236}$$

The total DOS function in this case can be expressed as

$$N_{2DT}(E)=\frac{g_v}{\pi}\sum_{n_z=1}^{n_{z\max}}\psi'_{5HD}(E,\eta_g)H(E-E_{n_{z59HD}}) \tag{1.237}$$

where $E_{n_{z59HD}}$ is the lowest positive root of the equation

$$\psi_{5HD}(E_{n_{z59HD}},\eta_g)-\psi_6\left(\frac{\pi n_z}{d_z}\right)^2\pm\psi_7\left[\psi^2_{8HD}(E_{n_{z59HD}},\eta_g)-\left(\frac{\pi n_z}{d_z}\right)^2\right]^{1/2}=0 \tag{1.238}$$

The surface electron concentration is given by

$$n_{2D}=\frac{g_v}{\pi}\sum_{n_z=1}^{n_{z\max}}\left[t_{1HDTe}(E_{F1HD},\eta_g,n_z)+t_{2HDTe}(E_{F1HD},\eta_g,n_z)\right] \tag{1.239a}$$

where,

$$t_{1HDTe}(E_{F1HD},\eta_g,n_z)=\left[\psi_{5HD}(E_{F1HD},\eta_g,n_z)-\psi_6\left(\frac{\pi n_z}{d_z}\right)^2\right]\text{ and}$$

$$t_{2HDTe}(E_{F1HD},\eta_g,n_z)=\sum_{r=1}^{s}L(r)\left[t_{1HDTe}(E_{F1HD},\eta_g,n_z)\right]$$

The EP in this case is given by

$$J_{2D}=\frac{\alpha_0 e g_v}{hd_z}\Big[\sum_{n_{z\min}}^{n_{z\max}}\Big[\frac{\sqrt{\psi_2^2+4\psi_1\gamma_3(\mathrm{E}_{n_{z59HD}},\eta_g)}}{2\psi_1\gamma'_3(\mathrm{E}_{n_{z59HD}},\eta_g)}\Big]\int_{\mathrm{E}_{n_{z59HD}}}^{\infty}[\psi'_{5HD}(\mathrm{E},\eta_g)f(\mathrm{E})d\mathrm{E}']\Big] \tag{1.239b}$$

The 2D electron energy spectrum in QWs of Te in the absence of band tails assumes the form

$$k_s^2 = \psi_5(E) - \psi_6\left(\frac{\pi n_z}{d_z}\right)^2 \pm \psi_7[\psi_8^2(E) - \left(\frac{\pi n_z}{d_z}\right)^2]^{1/2} \quad (1.240)$$

where,

$$\psi_5(E) = \left[\frac{E}{\psi_2} + \frac{\psi_4^2}{2\psi_2^2}\right] \text{ and } \psi_8^2(E) = \frac{\psi_4^4 + 4E\psi_2\psi_4^2 + 4\psi_2^2\psi_3^2}{4\psi_1\psi_2\psi_4^2}$$

Thus, the total 2D DOS function can be expressed as

$$N_{2DT}(E) = \left(\frac{g_v}{\pi}\right)\sum_{n_z=1}^{n_{z\max}} t'_{40}(E, n_z)H(E - E_{n_{z12}}) \quad (1.241)$$

where,

$$t_{40}(E, n_z) = [\psi_5(E) - \psi_6\left(\frac{\pi n_z}{d_z}\right)^2 \pm \psi_7[\psi_8^2(E) - \left(\frac{\pi n_z}{d_z}\right)^2]^{1/2}]^{1/2}$$

The sub-band energies ($E_{n_{z12}}$) are given by

$$E_{n_{z12}} = \psi_1(n_z\pi/d_z)^2 \pm \psi_3(n_z\pi/d_z) \quad (1.242a)$$

Using (1.240) the EEM can be expressed as

$$m^*(E_{F_s}, n_z) = \frac{\hbar^2}{2} t'_{40}(E_{F_s}, n_z) \quad (1.242b)$$

The 2D surface electron concentration per unit area for QWs of Te can be written as

$$n_{2D} = \frac{g_v}{\pi}\sum_{n_z=1}^{n_{z\max}} [t_{40}(E_{Fs}, n_z) + t_{41}(E_{Fs}, n_z)] \quad (1.243)$$

where

$$t_{41}(E_{Fs}, \eta_z) \equiv \sum_{r=1}^{s} L(r) t_{40}(E_{Fs}, n_z).$$

The EP in this case is given by

$$J_{2D} = \frac{\alpha_0 e g_v}{2\psi_1 h d_z}[\sum_{n_{z\min}}^{n_{z\max}} [\sqrt{\psi_2^2 + 4\psi_1 \mathrm{E}_{n_z 12}}] \int_{\mathrm{E}_{n_z 12}}^{\infty} [t'_{40}(\mathrm{E}, \mathrm{n}_z) f(\mathrm{E}) \mathrm{dE}']]] \tag{1.244}$$

The electron concentration and the EP for bulk specimens of Te in the absence of band tails can, respectively, be expressed as

$$n_0 = \frac{g_v}{3\pi^2}\left[\mathrm{M}_g(\mathrm{E}_F) + \mathrm{N}_g(\mathrm{E}_F)\right] \tag{1.245}$$

and

$$J = \frac{\alpha_0 e g_v}{24\psi_1 \pi^2 \hbar} \int_{E_0}^{\infty} \sqrt{\psi_2^2 + 4\psi_1 E'} M_9(\mathrm{E}') \mathrm{f}(\mathrm{E}) \mathrm{dE}' \tag{1.246}$$

where,

$$M_9(E_F) = \left[3\psi_5(E_F)\Gamma_3(E_F) - \psi_6 \Gamma_3^3(E_F)\right], \quad \psi_5(E_F) = \left[\frac{E_F}{\psi_2} + \frac{\psi_4^2}{2\psi_2^2}\right] \text{ and}$$

$$\Gamma_3(E_F) = [2\psi_1]^{-1}\left[\sqrt{\psi_3^2 + 4\psi_1 E_F} - \psi_3\right] \text{and} N_9(E_F) \equiv \sum_{r=1}^{s} L(r) M_9(E_F).$$

1.2.7 The EP from QWs of HD Gallium Phosphide

The energy spectrum of the conduction electrons in n-GaP can be written as [188]

$$E = \frac{\hbar^2 k_s^2}{2m_\perp^*} + \frac{\hbar^2}{2m_\parallel^*}\left[\bar{A}' k_s^2 + k_z^2\right] - \left[\frac{\hbar^4 k_0^2}{m_\parallel^{*2}}(k_s^2 + k_z^2) + |V_G|^2\right]^{1/2} + |V_G| \tag{1.247}$$

where, k_0 and $|V_G|$ are constants of the energy spectrum and $\bar{A}' = 1$.

The dispersion relation of the conduction electrons in HD n-GaP can be expressed as

$$\gamma_3(E, \eta_g) = \frac{\hbar^2 k_s^2}{2m_\perp^*} + \frac{\hbar^2}{2m_\parallel^*}\left[\bar{A}' k_s^2 + k_z^2\right] - \left[\frac{\hbar^4 k_0^2}{m_\parallel^{*2}}(k_s^2 + k_z^2) + |V_G|^2\right]^{1/2} - |V_G| \tag{1.248}$$

The EEMs assume the forms as

$$m_z^*(E_{F_h},\eta_g) = \frac{\hbar^2\gamma_3'(E_{F_h},\eta_g)}{b}\Big[1 \pm (C+bD)\big[C^2+4bD^2+4bC\gamma_3(E_{F_h},\eta_g) - 4bCD + 4b^2\gamma_3(E_{F_h},\eta_g)D\big]^{-1/2}\Big] \quad (1.249)$$

and

$$m_s^*(E_{F_h},\eta_g) = \frac{\hbar^2}{2}\left[t_{11}\gamma_3'(E_{F_h},\eta_g) - t_{41}t_5'(E_{F_h},\eta_g)\right] \quad (1.250)$$

where,

$b = \frac{\hbar^2}{2m_\parallel^*}$, $C = (\hbar^2 k_0/m_\parallel^*)^2$, $D = |V_G|$, $t_{11} = \frac{1}{a}$, $a = \frac{\hbar^2}{2m_\perp^*} + \bar{A}'b$, $t_{41} = \frac{\sqrt{g_3}}{2a^2}$,
$g_3 = (4abc + 4a^2c)$, $t_5^2(E_{F_h},\eta_g) = [g_2 - 4aC\gamma_3(E_{F_h},\eta_g)](g_3)^{-1}$, $g_2 = (4a^2b^2 + C^2 + 4aCD)$

The electron concentration can be expressed as

$$n_0 = \frac{g_v}{4\pi^2}\left[\bar{I}_{127}(E_{F_h},\eta_g) + \sum_{r=1}^{s} L(r)[\bar{I}_{127}(E_{F_h},\eta_g)]\right] \quad (1.251)$$

where

$$\begin{aligned}
\bar{I}_{127}(E_{F_h},\eta_g) &= [M_{1HD}(E_{F_h},\eta_g)] \\
M_{1HD}(E_{F_h},\eta_g) &= [2(t_{11}\gamma_3(E_{F_h},\eta_g) + t_{21})\sqrt{t_{81} + t_{91}\gamma_3(E_{F_h},\eta_g)} \\
&\quad + (t_{31}/3)\theta_{,-}^3(E_{F_h},\eta_g) + (t_{41}/2) \\
&\quad \times [\theta_{,-}(E_{F_h},\eta_g)\sqrt{\theta_-^2(E_{F_h},\eta_g) + t_5(E_{F_h},\eta_g)} \\
&\quad - \sqrt{t_5(E_{F_h},\eta_g)}] + (t_{41}t_5(E_{F_h},\eta_g)/2) \\
&\quad \ln\left|\frac{\theta_{,-}(E_{F_h},\eta_g) + \sqrt{\theta_-^2(E_{F_h},\eta_g) + t_5(E_{F_h},\eta_g)}}{\sqrt{t_5(E_{F_h},\eta_g)}}\right|]], \\
t_{21} &= \frac{g_1}{2a^2}, \; g_1 = -(C + 2aD), \\
t_{81} &= [t_{41}^4 + 4t_{41}^2 t_{21}t_{31} + (4t_{31}^2 t_{41}^2 g_2)(g_3)^{-1}], \; t_{31} = \frac{b}{a}, \\
t_{91} &= [4t_{11}t_{31}t_{41}^2 + 8t_{11}t_{21}t_{31}^2 - (16t_{31}^2 t_{41}^2 aC)(g_3)^{-1}], \\
\theta_-(E_{F_h},\eta_g) &= (t_{31}\sqrt{2})^{-1}[t_{61} + t_{71}\gamma_3(E_{F_h},\eta_g) - \sqrt{t_{81} + t_{91}\gamma_3(E_{F_h},\eta_g)}], \\
t_{61} &= (t_{41}^2 + 2t_{21}t_{31}) \text{ and } t_{71} = (2t_{11}t_{31})
\end{aligned}$$

The EP in this case is given by

$$J = \frac{\alpha_0 e g_v}{4\pi^2\sqrt{m_{11}^*}} \int_{E_0}^{\infty} \frac{[\bar{I}_{127}(E', \eta_g)]' \mathrm{f(E)\,dE'}}{\gamma_{100}'(E', \eta_g)} \tag{1.252}$$

where

$$\gamma_{100}(E', \eta_g) = [2\gamma_3(E', \eta_g) + 2\mathrm{V}_G - \frac{4\hbar^4 k_0^4}{m_{\|}^*} - [4V_G^2 + \frac{4\hbar^2 k_0^2}{m_{\|}^{*2}} - \frac{8V_G\hbar^2 k_0^2}{m_{\|}^*} - \frac{4\gamma_3(E', \eta_g)\hbar^2 k_0^2}{m_{\|}^*}]^{1/2}]^{1/2}$$

The 2D dispersion relation in QW of HD GaP can be expressed following (1.248) as

$$k_s^2 = t_{11}\gamma_3(E, \eta_g) + t_{21} - t_{31}(\frac{\pi n_z}{d_z})^2 - t_{41}[(\frac{\pi n_z}{d_z})^2 + t_5^2(E, \eta_g)]^{1/2} \tag{1.253}$$

The EEM in this case can be written following (1.253) as

$$m^*(E_{F1HD}, \eta_g, n_z) = \frac{\hbar^2}{2}\left[t_{11}\gamma_3'(E_{F1HD}, \eta_g) - t_{41}t_5(E_{F1HD}, \eta_g)t_5'(E_{F1HD}, \eta_g) \times [(\frac{\pi n_z}{d_z})^2 + t_5^2(E_{F1HD}, \eta_g)]^{-1/2}\right] \tag{1.254}$$

The total DOS function up to Fermi level assumes the form

$$N_{2DT}((E_{F1HD}, \eta_g) = \frac{g_v}{2\pi}\sum_{n_{z=1}}^{n_{z\,\max}} [t_{11}\gamma_3'(E_{F1HD}, \eta_g) - t_{41}t_5(E_{F1HD}, \eta_g)t_5'(E_{F1HD}, \eta_g) [(\frac{\pi n_z}{d_z})^2 + t_5^2(E_{F1HD}, \eta_g)]^{-1/2} H(E_{F1HD} - E_{n_{z8THD}}) \tag{1.255}$$

where, $E_{n_{z8THD}}$ is given by the equation

$$t_{11}\gamma_3(E_{n_{z8THD}}, \eta_g) + t_{21} - t_{31}(\frac{\pi n_z}{d_z})^2 - t_{41}[(\frac{\pi n_z}{d_z})^2 + t_5^2(E_{n_{z8THD}}, \eta_g)]^{1/2} = 0 \tag{1.256}$$

The surface electron concentration in QW of HD n-GaP can be written as

$$n_5 = \frac{g_v}{\pi} \sum_{n_z=1}^{n_{z\max}} \left[t_{3HDGaP}(E_{F1HD}, \eta_g, n_z) + t_{4HDGaP}(E_{F1HD}, \eta_g, n_z) \right] \tag{1.257a}$$

where,

$$t_{1HDGaP}(E_{F1HD}, \eta_g, n_z) = t_{11}\gamma_3(E_{F1HD}, \eta_g, n_z) + t_{21} - t_{31}\left(\frac{\pi n_z}{d_z}\right)^2 - t_{41}\left[\left(\frac{\pi n_z}{d_z}\right)^2 + t_5^2(E_{F1HD}, \eta_g, n_z)\right]^{1/2}$$

$$t_{4HDGaP}(E_{F1HD}, \eta_g, n_z) = \sum_{r=1}^{s} L(r)\left[t_{3HDGaP}(E_{F1HD}, \eta_g, n_z)\right]$$

The EP in this case is given by

$$J_{2D} = \frac{\alpha_0 e g_v}{4 d_z \pi \sqrt{m_\parallel^*}} \left[\sum_{n_{z\min}}^{n_{z\max}} [\gamma'_{101}(E_{n_{z8THD}}, \eta_g, n_z)]^{-1} \left[\int_{E_{n_{z8THD}}}^{\infty} \gamma_{101}(E, \eta_g, n_z) f(E) dE \right] \right] \tag{1.257b}$$

where

$$\gamma_{101}(E, \eta_g, n_z) = \left[t_{11}\gamma'_3(E, \eta_g) - t_{41} t_5(E, \eta_g) t'_5(E, \eta_g) \left[\left(\frac{\pi n_z}{d_z}\right)^2 + t_5^2(E, \eta_g) \right]^{-1/2} \right]$$

The 2D electron dispersion relation in size-quantized n-GaP in the absence of band tails assumes the form

$$E = a k_s^2 + C(n_z \pi / d_z)^2 + |V_G| - \left[D k_s^2 + |V_G|^2 + D(n_z \pi / d_z)^2 \right]^{1/2} \tag{1.258}$$

The sub-band energy $(E_{n_{z13}})$ are given by

$$E_{n_{z13}} = C(\pi n_z / d_z)^2 + |V_G| - \left[|V_G|^2 + D(\pi n_z / d_z)^2 \right]^{1/2} \tag{1.259}$$

The (1.258) can be expressed as

$$k_s^2 = t_{42}(E, n_z) \tag{1.260}$$

in which,

$$t_{42}(E,n_z) \equiv \left[\{2a(E-t_1)+D\} - \left\{[2a(E-t_1)+D]^2 - 4a^2\left[(E-t_1)^2 - t_2\right]\right\}^{1/2}\right],$$
$$t_1 \equiv |V_G| + C(\pi n_z/d_z)^2 \text{ and } t_2 \equiv |V_G|^2 + D(\pi n_z/d_z)^2.$$

The total DOS function is given by

$$N_{2DT}(E) = \frac{g_v}{4\pi a^2}\sum_{n_z=1}^{n_{z\max}} [t'_{42}(E,n_z)]H(E-E_{n_{z13}}) \tag{1.261a}$$

Using (1.260) the EEM can be expressed as

$$m^*(E_{F_s},n_z) = \frac{\hbar^2}{2} t'_{42}(E_{F_s},n_z) \tag{1.261b}$$

The electron statistics in QWs in n-GaP assumes the form

$$\begin{aligned} n_{2D} &= [\left(\frac{g_v}{4\pi a^2}\right)\sum_{n_z=1}^{n_{z\max}} [t_{42}(E_{Fs},n_z) + t_{43}(E_{Fs},n_z)]] \\ t_{43}(E_{Fs},n_z) &= \sum_{r=1}^{s} L(r)[t_{42}(E_{Fs},n_z) \end{aligned} \tag{1.262}$$

where,

$$t_{43}(E_{Fs},n_z) = \sum_{r=1}^{s} L(r)[t_{42}(E_{Fs},n_z)$$

The EP in this case is given by

$$J_{2D} = \frac{\alpha_0 e g_v}{8d_z\pi a^2\sqrt{m_\|^*}}[\sum_{n_{z_{\min}}}^{n_{z\max}} [\gamma'_{102}(E_{n_{z13}})]^{-1}[\int_{E_{n_{z13}}}^{\infty} t'_{42}(E,n_z)f(E)dE]] \tag{1.263}$$

where

$$\gamma_{102}(E_{n_{z13}}) = [2E_{n_{z13}} + 2V_G - \frac{2\hbar^2 k_0}{m_\|^*} - [4V_G^2 + \frac{4\hbar^4 k_0^4}{m_\|^{*2}} - \frac{8V_G\hbar^2k_0^2}{m_\|^*} - \frac{4E_{n_{z13}}\hbar^2k_0^2}{m_\|^*}]^{1/2}]^{1/2}$$

The EEMs in bulk specimens of n-GaP in the absence of band tails can be written as

$$m_s^*(E_F) = \frac{\hbar^2}{2}[t_{11} - t_{41}t_5'(E_F)] \tag{1.264}$$

and

$$m_z^*(E_F) = \frac{\hbar^2}{b}[1 - C[4bCE_F + 4b^2D^2 + C^2 - 4bCD]^{-1/2}] \tag{1.265}$$

where

$$t_5(E_F) = [\frac{g_2 - 4aCE_F}{g_3}]^{1/2}$$

The electron concentration and the EPin this case assume the forms

$$n_0 = \frac{g_v}{4\pi^2}[M_1(E_F) + N_1(E_F)] \tag{1.266}$$

$$J = \frac{\alpha_0 e g_v}{16\pi^2\sqrt{m_\parallel^*}} \int_{E_0}^{\infty} \frac{M_1'(E')}{\gamma_{102}'(E')} f(E)dE' \tag{1.267}$$

where

$$\begin{aligned}
M_1(E_F) = {} & [2(t_{11}E_F + t_{21})\sqrt{t_{91}E_F + t_{81}} \\
& + \frac{t_{31}}{3}\phi^3(E_F) + \frac{t_{41}}{2}[\phi(E_F)\sqrt{\phi^2(E_F) + t_5(E_F)} \\
& + \frac{t_{41}t_5(E_F)}{2}[\ln\left|\frac{\phi(E_F) + \sqrt{\phi^2(E_F) + t_5(E_F)}}{\sqrt{t_5(E_F)}}\right|]],
\end{aligned}$$

$$\phi(E_F) = (t_{31}\sqrt{2})^{-1}[t_{61} + E_F t_{71} - [t_{81} + t_{91}E_F]^{1/2}]$$

$$N_1(E_F) = \sum_{r=1}^{s_0}[L(r)M_1(E_F)]$$

and

$$\gamma_{102}(E') = [2E' + 2V_G - \frac{2\hbar^2 k_0^2}{m_\parallel^*} - [4V_G^2 + \frac{4\hbar^2 k_0^4}{m_\parallel^{*2}} - \frac{8V_G\hbar^2 k_0^2}{m_\parallel^*} - \frac{4E'\hbar^2 k_0^2}{m_\parallel^*}]^{1/2}]^{1/2}.$$

1.2.8 The EP in QWs of HD Platinum Antimonide

The dispersion relation for the n-type $PtSb_2$ can be written as [189]

$$[E + \lambda_0 \frac{a^2}{4} k^2 - l k_s^2 \frac{a^2}{4}][E + \delta_0 - v \frac{a^2}{4} k^2 - n' k_s^2 \frac{a^2}{4}] = I(\frac{a^4}{16}) k^4 \quad (1.268)$$

The (1.268) assumes the form

$$[E + \omega_1 k_s^2 + \omega_2 k_z^2][E + \delta_0 + \omega_3 k_s^2 - \omega_4 k_z^2] = I_1 (k_s^2 + k_z^2)^2 \quad (1.269)$$

where

$$\omega_1 = [\lambda_0 \frac{a^2}{4} + l \frac{a^2}{4}],\ \omega_2 = \lambda_0 \frac{a^2}{4},\ \omega_3 = [n' \frac{a^2}{4} - \upsilon \frac{a^2}{4}],\ \omega_4 = \upsilon \frac{a^2}{4},\ I_1 = I(\frac{a^2}{4})^2,$$

λ_0, l, δ_0, v, n' and a are the band constants.

The carrier dispersion law in HD $PtSb_2$ can be written as

$$T_{11} k_s^4 - k_s^2 [T_{21}(E, \eta_g) - T_{31} k_z^2] + [T_{41} k_z^4 - T_{51}(E, \eta_g) k_z^2 - T_{61}(E, \eta_g)] = 0 \quad (1.270)$$

where,

$$\begin{aligned} T_{11} &= (I_1 - \omega_2 \omega_3),\ T_{21}(E, \eta_g) = [\omega_1 \delta_0 + \omega_1 \gamma_3(E, \eta_g) + \omega_3 \gamma_3(E, \eta_g)], \\ T_{31} &= [2I_1 + \omega_2 \omega_4 - \omega_2 \omega_3],\ T_{41} = [2I_1 + \omega_2 \omega_4], \\ T_{51}(E, \eta_g) &= [\omega_2 \gamma_0(E, \eta_g) - \omega_4 \gamma_3(E, \eta_g) + \omega_2 \gamma_3(E, \eta_g)], \\ T_{61}(E, \eta_g) &= [\gamma_8(E, \eta_g) + \gamma_0 \gamma_3(E, \eta_g)] \text{and} \\ \gamma_8(E, \eta_g) &= 2\theta_0(E, \eta_g)[1 + Erf(E/\eta_g)]^{-1} \end{aligned}$$

The EEMs are given by

$$\begin{aligned} m_s^*(E_{F_h}, \eta_g) = \frac{\hbar^2}{2T_{11}} [T'_{21}(E_{F_h}, \eta_g) \\ + \frac{(T_{21}(E_{F_h}, \eta_g) T'_{21}(E_{F_h}, \eta_g) + 2T_{11} T'_{61}(E_{F_h}, \eta_g))}{\sqrt{T_{21}^2(E_{F_h}, \eta_g) + 4T_{11} T_{61}(E_{F_h}, \eta_g)}}] \text{ and} \end{aligned} \quad (1.271)$$

$$\begin{aligned} m_s^*(E_{F_h}, \eta_g) = (\frac{\hbar^2}{2T_{41}})[T'_{51}(E_{F_h}, \eta_g) + [T_{51}(E_{F_h}, \eta_g) T'_{51}(E_{F_h}, \eta_g) + 2T_{41} T'_{61}(E_{F_h}, \eta_g)] \\ \times [T_{51}^2(E_{F_h}, \eta_g) + 4T_{41} T_{61}(E_{F_h}, \eta_g)]^{-1/2}] \end{aligned} \quad (1.272)$$

The electron concentration assumes the form

$$n_0 = \frac{g_v}{3\pi^2}\left[\bar{I}_{128}(E_{F_h},\eta_g) + \sum_{r=1}^{s} L(r)[\bar{I}_{128}(E_{F_h},\eta_g)]\right] \quad (1.273a)$$

where,

$$\bar{I}_{128}(E_{F_h},\eta_g) = [M_{6HD}(E_{F_h},\eta_g)],$$

$$M_{6HD}(E_{F_h},\eta_g) = [T_{91HD}(E_{F_h},\eta_g)\rho_{2HD}(E_{F_h},\eta_g) - T_{101}\frac{\rho^3_{2HD}(E_{F_h},\eta_g)}{3} - T_{11}J_3(E_{F_h},\eta_g)]$$

$$T_{91HD}(E_{F_h},\eta_g) = \frac{T_{21}(E_{F_h},\eta_g)}{2T_{11}},$$

$$\rho_{2HD}(E_{F_h},\eta_g) = [(2T_{41})^{-1}[T_{51}(E_{F_h},\eta_g) + \sqrt{T^2_{51}(E_{F_h},\eta_g) + 4T_{41}T_{61}(E_{F_h},\eta_g)}]]^{1/2}$$

$$T_{101} = [T_{31}/2T_{11}],$$

$$J_3(E_{F_h},\eta_g) = \frac{\rho_{2HD}(E_{F_h},\eta_g)}{3}[[A^2_{3HD}(E_{F_h},\eta_g) + B^2_{3HD}(E_{F_h},\eta_g)]E_0(\eta(E_{F_h},\eta_g),t(E_{F_h},\eta_g)) - [A^2_{3HD}(E_{F_h},\eta_g) - B^2_{3HD}(E_{F_h},\eta_g)]F_0(\eta(E_{F_h},\eta_g),t(E_{F_h},\eta_g))] + \frac{\rho_{2HD}(E_{F_h},\eta_g)}{3}[(A^2_{3HD}(E_{F_h},\eta_g) - \rho^2_{2HD}(E_{F_h},\eta_g))(B^2_{3HD}(E_{F_h},\eta_g) - \rho^2_{2HD}(E_{F_h},\eta_g))]^{1/2},$$

$E_0(\eta(E_{F_h},\eta_g),t(E_{F_h},\eta_g))$ and $F_0(\eta(E_{F_h},\eta_g),t(E_{F_h},\eta_g))$ are the incomplete elliptic integrals of second and first respectively,

$$A^2_{3HD}(E_{F_h},\eta_g) = \frac{1}{2}[T_{12}(E_{F_h},\eta_g) + \sqrt{T^2_{12}(E_{F_h},\eta_g) - 4T_{13}(E_{F_h},\eta_g)}],$$

$$T_{12}(E_{F_h},\eta_g) = [T_7(E_{F_h},\eta_g)/\overline{T}_{61}]\overline{T}_{61} = [T^2_{31} - 4T_{11}T_{41}],$$

$$T_7(E_{F_h},\eta_g) = [2T_{31}T_{21}(E_{F_h},\eta_g) - 4T_{11}T_{51}(E_{F_h},\eta_g)],$$

$$T_{13}(E_{F_h},\eta_g) = (T_8(E_{F_h},\eta_g)/\overline{T}_8),$$

$$T_8(E_{F_h},\eta_g) = [T^2_{21}(E_{F_h},\eta_g) + 4T_{11}T_{61}(E_{F_h},\eta_g)],$$

$$B^2_{3HD}(E_{F_h},\eta_g) = \frac{1}{2}[T_{12}(E_{F_h},\eta_g) - \sqrt{T^2_{12}(E_{F_h},\eta_g) - 4T_{13}(E_{F_h},\eta_g)}],$$

$$\overline{T}_{11} = [\sqrt{\overline{T}_{61}}/2T_{11}]t(E_{F_h},\eta_g) = [B_3(E_{F_h},\eta_g)/A_3(E_{F_h},\eta_g)],$$

$$\eta(E_{F_h},\eta_g) = \sin^{-1}[\frac{\rho_2(E_{F_h},\eta_g)}{B_3(E_{F_h},\eta_g)}]$$

The EP in this case is given by

$$J = \frac{\alpha_0 e g_v}{6\pi^2 \hbar} \int_{E_0}^{\infty} \frac{\sqrt{U_{100}(E', \eta_g)}[\bar{\text{I}}_{128}(E', \eta_g)]' f(E) dE'}{U'_{100}(E', \eta_g)} \tag{1.273b}$$

where

$$U_{100}(E', \eta_g) = (2\text{T}_{41})^{-1}[\text{T}_{51}(E', \eta_g) + \sqrt{T_{51}^2(E', \eta_g) + 4\text{T}_{41}\text{T}_{61}(E', \eta_g)}]$$

From (1.270) the dispersion relation in QWs of HD $PtSb_2$ can be expressed as

$$T_{11}k_s^4 - P_{1HD}(E, \eta_g, n_z)k_s^2 + P_{2HD}(E, \eta_g, n_z) = 0 \tag{1.274}$$

where,

$$P_{1HD}(E, \eta_g, n_z) = [T_{21}(E_{F_h}, \eta_g) - T_{31}(\pi n_z/d_z)]$$
$$P_{2HD}(E, \eta_g, n_z) = [T_{41}(\pi n_z/d_z)^4 - T_{51}(E_{F_h}, \eta_g)(\pi n_z/d_z)^2 - T_{61}(E_{F_h}, \eta_g)]$$

(1.274) can be written as

$$k_s^2 = A_{60}(E, \eta_g, n_z) \tag{1.275}$$

where,

$$A_{60}(E, \eta_g, n_z) = [P_{1HD}(E, \eta_g, n_z) - \sqrt{P_{1HD}^2(E, \eta_g, n_z) - 4T_{11}P_{2HD}(E, \eta_g, n_z)}]$$

The EEM assumes the form

$$m^*(E_{F1HD}, \eta_g, n_z) = \frac{\hbar^2}{2} A'_{60}(E_{F1HD}, \eta_g, n_z) \tag{1.276}$$

The surface electron concentration is given by

$$n_0 = \frac{g_v}{2\pi} \sum_{n_z=1}^{n_{z\max}} [A_{60}(E_{F1HD}, \eta_g, n_z) + B_{60}(E_{F1HD}, \eta_g, n_z)] \tag{1.277a}$$

where,

$$B_{60}(E_{F1HD}, \eta_g, n_z) = \sum_{r=1}^{s_0} L(r)[A_{60}(E_{F1HD}, \eta_g, n_z)]$$

The EP in this case is given by

$$J_{2D} = \frac{\alpha_0 e g_v}{2hd_z} \sum_{n_{z\min}}^{n_{z\max}} \sqrt{\frac{U_{100}(E_{n_{z100}}, \eta_g)}{U'_{100}(E_{n_{z100}}, \eta_g)}} \int_{E_{n_{z100}}}^{\infty} A'_{60}(E, \eta_g, n_z) f(E) dE \tag{1.277b}$$

where $E_{n_{z100}}$ is the lowest positive root of the equation

$$P_{2HD}(E_{n_{z100}}, \eta_g, n_z) = 0 \tag{1.277c}$$

From (1.269), we can write the expression of the 2D dispersion law in QWs of n-$PtSb_2$, in the absence of band tails as

$$k_s^2 = t_{44}(E, n_z) \tag{1.278}$$

where,

$$t_{44}(E, n_z) = [2A_9]^{-1}\left[-A_{10}(E, n_z) + \sqrt{A_{10}^2(E, n_z) + 4A_9 A_{11}(E, n_z)}\right],$$

$$A_9 \equiv [I_1 + \omega_1\omega_3], A_{10}(E, n_z) \equiv \left[\omega_3 E + \omega_1\left\{E + \delta_0 - \omega_4\left(\frac{\pi n_z}{d_z}\right)^2\right\} + \omega_2\omega_3\left(\frac{\pi n_z}{d_z}\right)^2 + 2I_1\left(\frac{\pi n_z}{d_z}\right)^4\right]$$

and

$$A_{11}(E, n_z) \equiv \left[E\left[E + \delta_0 - \omega_4\left(\frac{\pi n_z}{d_z}\right)^2\right] + \omega_2\left(\frac{\pi n_z}{d_z}\right)^2 \times \left[E + \delta_0 - \omega_4\left(\frac{\pi n_z}{d_z}\right)^2\right] - I_1\left(\frac{\pi n_z}{d_z}\right)^4\right]$$

The area of k_s space can be expressed as

$$A(E, n_z) = \pi t_{44}(E, n_z) \tag{1.279}$$

The total DOS function assumes the form

$$N_{2DT}(E) = \frac{g_v}{2\pi} \sum_{n_z=1}^{n_{z\max}} [t'_{44}(E, n_z)] H(E - E_{n_{z14}}) \tag{1.280}$$

where the quantized levels $E_{n_{z14}}$ can be expressed through the equation

$$E_{n_{z14}} = (2)^{-1}\left[-\left[\omega_2\left(\frac{\pi n_z}{d_z}\right)^2+\delta_0-\omega_4\left(\frac{\pi n_z}{d_z}\right)^2\right] + \left\{\left[\omega_2\left(\frac{\pi n_z}{d_z}\right)^2+\delta_0-\omega_4\left(\frac{\pi n_z}{d_z}\right)^2\right]^2+4\left[I_1\left(\frac{\pi n_z}{d_z}\right)^4+\omega_2\omega_4\left(\frac{\pi n_z}{d_z}\right)^4-\omega_2\delta_0\left(\frac{\pi n_z}{d_z}\right)^2\right]\right\}^{1/2}\right] \quad (1.281a)$$

Using (1.278), the EEM in this case can be written as

$$m^*(E_{F_s}, n_z) = \frac{\hbar^2}{2}t'_{44}(E_{F_s}, n_z) \quad (1.281b)$$

The electron statistics can be written as

$$n_{2D} = \frac{g_v}{2\pi}\sum_{n_z=1}^{n_{z\max}}[t_{44}(E_{F_s}, n_z) + t_{45}(E_{F_s}, n_z)] \quad (1.282)$$

where

$$t_{45}(E_{F2D}, n_z) \equiv \sum_{r=1}^{s} L(r)[t_{44}(E_{F2D}, n_z)]$$

The EP in this case is given by

$$J_{2D} = \frac{\alpha_0 e g_v}{d_z h}\sum_{n_{z\min}}^{n_{z\max}}\frac{\sqrt{\omega_{100}(E_{n_{z14}})}}{\omega'_{100}(E_{n_{z14}})}\int_{E_{n_{z14}}}^{\infty} t'_{44}(E, n_z)f(\mathrm{E})\mathrm{dE} \quad (1.283)$$

where

$$\omega_{100}(E) = [[[(\omega_2 E + \delta_0\omega_2 - \omega_4 E)^2 + 4(\omega_2\omega_4 + I_1)(E^2 + \delta_0 E)^{1/2} + (\omega_2 E + \delta_0\omega_2 - \omega_4 E)][2(\omega_2\omega_4 + I_1)]^{-1}].$$

1.2.9 The EP from QWs of HD Bismuth Telluride

The dispersion relation of the conduction electrons in Bi_2Te_3 can be written as [190–192]

$$E(1+\alpha E) = \bar{\omega}_1 k_x^2 + \bar{\omega}_2 k_y^2 + \bar{\omega}_3 k_z^2 + 2\bar{\omega}_4 k_z k_y \tag{1.284}$$

where

$$\overline{\omega}_1 = \frac{\hbar^2}{2m_0}\overline{\alpha}_{11},\ \overline{\omega}_2 = \frac{\hbar^2}{2m_0}\overline{\alpha}_{22},\ \overline{\omega}_3 = \frac{\hbar^2}{2m_0}\overline{\alpha}_{33},\ \overline{\omega}_4 = \frac{\hbar^2}{2m_0}\overline{\alpha}_{23}$$

in which $\overline{\alpha}_{11}$, $\overline{\alpha}_{22}$, $\overline{\alpha}_{33}$ and $\overline{\alpha}_{23}$ are system constants.

The dispersion relation in HD Bi_2Te_3 assumes the form

$$\gamma_2(E,\eta_g) = \bar{\omega}_1 k_x^2 + \bar{\omega}_2 k_y^2 + \bar{\omega}_3 k_z^2 + 2\bar{\omega}_4 k_z k_y \tag{1.285}$$

The EEMs can, respectively, be expressed as

$$m_x^*(E_{F_h},\eta_g) = \frac{\hbar^2}{2\bar{w}_1}\gamma_2'(E_{F_h},\eta_g) \tag{1.286}$$

$$m_y^*(E_{F_h},\eta_g) = \frac{\hbar^2}{2\bar{w}_2}\gamma_2'(E_{F_h},\eta_g) \tag{1.287}$$

$$m_z^*(E_{F_h},\eta_g) = \frac{\hbar^2}{2\bar{w}_3}\gamma_2'(E_{F_h},\eta_g) \tag{1.288}$$

The DOS function in this case is given by

$$N(E) = 4\pi g_v\left(\frac{2m_0}{h^2}\right)^{3/2}\frac{\sqrt{\gamma_2(E,\eta_g)}\gamma_2'(E,\eta_g)}{\sqrt{\alpha_{11}\alpha_{22}\alpha_{33} - 4\alpha_{11}\alpha_{23}^2}} \tag{1.289}$$

Thus combining (1.289) with the Fermi Dirac occupation probability factor, the electron concentration can be written as

$$n_0 = \frac{g_v}{3\pi^2}\left(\frac{2m_0}{h^2}\right)^{3/2}(\alpha_{11}\alpha_{22}\alpha_{33} - 4\alpha_{11}\alpha_{23}^2)^{-1/2}[U_{1HD}(E_{F_h},\eta_g) + U_{2HD}(E_{F_h},\eta_g)] \tag{1.290a}$$

where,

$$U_{1HD}(E_{F_h},\eta_g) = [\gamma_2(E_{F_h},\eta_g)]^{3/2},\ U_{2HD}(E_{F_h},\eta_g) = \sum_{r=1}^{s} L(r)[U_{1HD}(E_{F_h},\eta_g)]$$

The EP in this case is given by

$$J = \frac{2\pi\alpha_0 e g_v \sqrt{\bar{\omega}_3}}{\hbar}[\sqrt{\alpha_{11}\alpha_{22}\alpha_{33} - 4\alpha_{11}\alpha_{23}^2}]^{-1}(\frac{2m_0}{\hbar^2})^{3/2}\int_{E_0}^{\infty}\gamma_2(\mathrm{E}',\eta_g)f(\mathrm{E})\mathrm{dE}' \tag{1.290b}$$

The dispersion relation in QWs of HD Bi_2Te_3 can be expressed as

$$\gamma_2(E,\eta_g) = \bar{\omega}_1(\frac{\pi n_x}{d_x})^2 + \bar{\omega}_2 k_y^2 + \bar{\omega}_3 k_z^2 + 2\bar{\omega}_4 k_z k_y \tag{1.291}$$

The EEM can be expressed as

$$m^*(E_{F1HD},\eta_g) = \frac{m_0}{\sqrt{\alpha_{11}\alpha_{33} - 4\alpha_{23}^2}}\gamma_2'(E_{F1HD},\eta_g) \tag{1.292}$$

The surface electron concentration can be written as

$$n_{2D} = \frac{g_v}{2\pi}[\sum_{n_z=1}^{n_{z\max}} R_{60}(\mathrm{E}_{F1HD},\eta_g,\mathrm{n}_z) + R_{61}(\mathrm{E}_{F1HD},\eta_g,\mathrm{n}_z)] \tag{1.293a}$$

$$R_{60}(\mathrm{E}_{F1HD},\eta_g,\mathrm{n}_z) = \frac{1}{\sqrt{\alpha_{11}\alpha_{33} - 4\alpha_{23}^2}}[\frac{2m_0\gamma_2'(E_{F1HD},\eta_g)}{\hbar^2} - \frac{2m_0}{\hbar^2}(\frac{\pi n_z}{d_x})\bar{\alpha}_{11}] \text{ and}$$

$$R_{61}(E_{F1HD},\eta_g,\mathrm{n}_z) = \sum_{r=1}^{s} L(r)[R_{60}(E_{F1HD},\eta_g,\mathrm{n}_z)]$$

The EP in this case is given by

$$J = \frac{\alpha_0 e g_v \sqrt{\overline{\omega}_1}}{h d_z}\sum_{n_{z\min}}^{n_{z\max}}\frac{\sqrt{\gamma_2(\mathrm{E}_{n_{z90HD}},\eta_g)}}{\gamma_2'(\mathrm{E}_{n_{z90HD}},\eta_g)}\int_{\mathrm{E}_{n_{z90HD}}}^{\infty} R_{60}'(\mathrm{E},\eta_g,n_z)f(\mathrm{E})\mathrm{dE} \tag{1.293b}$$

The 2D electron dispersion law in QWs of Bi_2Te_3 in the absence of band tails assumes the form

$$E(1+\alpha E) = \overline{\omega}_1(\frac{n_x\pi}{d_x})^2 + \overline{\omega}_2 k_y^2 + \overline{\omega}_3 k_z^2 + 2\overline{\omega}_4 k_z k_y \tag{1.294}$$

The area of the ellipse is given by

$$A_n(E, n_x) = \frac{\pi}{\sqrt{\bar{\alpha}_{22}\bar{\alpha}_{33} - 4\bar{\alpha}_{23}^2}}\left[\frac{2m_0E(1+\alpha E)}{\hbar^2} - \bar{\omega}_1\left(\frac{n_x\pi}{d_x}\right)^2\right] \tag{1.295}$$

The total DOS function assumes the form

$$N_{2DT}(E) = \frac{g_v m_0}{\pi\hbar^2\sqrt{\bar{\alpha}_{22}\bar{\alpha}_{33} - 4\bar{\alpha}_{23}^2}}\sum_{n_x=1}^{n_{x\max}}(1+2\alpha E)H(E - E_{n_{z15}}) \tag{1.296}$$

where, $(E_{n_{z15}})$ can be expressed through the equation

$$E_{n_{z15}}(1+\alpha E_{n_{z15}}) = \bar{\omega}_1\left(\frac{n_x\pi}{d_x}\right)^2 \tag{1.297a}$$

The EEM in this case assumes the form as

$$m^*(E_{F_s}) = \frac{m_0(1+2\alpha(E_{F_s}))}{\sqrt{\bar{\alpha}_{22}\bar{\alpha}_{33} - 4\bar{\alpha}_{23}^2}} \tag{1.297b}$$

The electron concentration can be written as

$$n_{2D} = \frac{k_BTg_v}{\pi\hbar^2}\left(\frac{m_0}{\sqrt{\bar{\alpha}_{22}\bar{\alpha}_{33} - 4\bar{\alpha}_{23}^2}}\right)\sum_{n_x=1}^{n_{x\max}}\left[(1+2\alpha E)F_0(\eta_{n_{15}}) + 2\alpha k_BTF_1(\eta_{n_{15}})\right] \tag{1.298}$$

where,

$$\eta_{n_{15}} = \frac{E_{Fs} - E_{n_{z15}}}{k_BT}$$

The EP in this case is given by

$$J_{2D} = \frac{\alpha_0 e g_v(k_B\mathrm{T})\sqrt{\omega_1}}{(\pi\hbar^3 d_z)\sqrt{\alpha_{22}\alpha_{23} - 4\alpha_{23}}^2}\left[\sum_{n_{z\min}}^{n_{z\max}}\frac{\sqrt{E_{n_{z15}}(1+\alpha E_{n_{z15}})}}{(1+2\alpha E_{n_{z15}})}\left[(1+2\alpha E_{n_{z15}})\mathrm{F}_0(\eta_{n_{15}}) + 2\alpha k_B\mathrm{T}\mathrm{F}_1(\eta_{n_{15}})\right]\right]. \tag{1.299}$$

1.2.10 The EP from QWs of HD Germanium

It is well known that the conduction electrons of n-Ge obey two different types of dispersion laws since band non-parabolicity has been included in two different ways as given in the literature [193–195].

(a) The energy spectrum of the conduction electrons in bulk specimens of n-Ge can be expressed in accordance with Cardona et al. [193, 194] as

$$E = -\frac{E_{g_0}}{2} + \frac{\hbar^2 k_z^2}{2m_{\|}^*} + \left[\frac{E_{g_0}^2}{4} + E_{g_0} k_s^2 \left(\frac{\hbar^2}{2m_{\perp}^*}\right)\right]^{1/2} \tag{1.300}$$

where in this case $m_{\|}^*$ and $m_{\perp}^*$ are the longitudinal and transverse effective masses along $\langle 111 \rangle$ direction at the edge of the conduction band respectively

The (1.300) can be written as

$$\frac{\hbar^2 k_s^2}{2m_{\perp}^*} = E(1+\alpha E) + \alpha\left(\frac{\hbar^2 k_z^2}{2m_{\|}^*}\right)^2 - (1+2\alpha E)\left(\frac{\hbar^2 k_z^2}{2m_{\|}^*}\right) \tag{1.301}$$

The dispersion relation under the condition of heavy doping can be expressed from (1.301) as

$$\frac{\hbar^2 k_s^2}{2m_{\perp}^*} = \gamma_2(E,\eta_g) + \alpha\left(\frac{\hbar^2 k_z^2}{2m_{\|}^*}\right)^2 - (1+2\alpha\gamma_3(E,\eta_g))\frac{\hbar^2 k_z^2}{2m_{\|}^*} \tag{1.302}$$

The EEMs can be written as

$$m_s^*(E_{Fh},\eta_g) = m_{\perp}^* \gamma_2'(E_{Fh},\eta_g) \tag{1.303}$$

and

$$m_z^*(E_{Fh},\eta_g) = m_{\|}^*[\gamma_3'(E_{Fh},\eta_g) - \frac{[\gamma_3'(E_{Fh},\eta_g)[1+2\alpha\gamma_3(E_{Fh},\eta_g)] - \gamma_2'(E_{Fh},\eta_g)]}{\sqrt{[1+2\alpha\gamma_3(E_{Fh},\eta_g)]^2 - 4\alpha\gamma_2(E_{Fh},\eta_g)}}] \tag{1.304}$$

The electron concentration can be written as

$$n_0 = \frac{8\pi g_v m_{\perp}^* \sqrt{2m_{\|}^*}}{h^3}\left[\bar{I}_{129}(E_{F_h},\eta_g) + \sum_{r=1}^{s} L(r)[\bar{I}_{129}(E_{F_h},\eta_g)]\right] \tag{1.305a}$$

where,

$$\bar{I}_{129}(E_{F_h},\eta_g) = [M_{8HD}(E_{F_h},\eta_g)],$$

$$M_{8HD}(E_{F_h},\eta_g) = [\gamma_3(E_{F_h},\eta_g)]^{1/2}[\gamma_2(E_{F_h},\eta_g) + \frac{\alpha}{5}\gamma_3^2(E_{F_h},\eta_g)] - \frac{\gamma_3(E_{F_h},\eta_g)}{3}[1 + 2\alpha\gamma_3(E_{F_h},\eta_g)]$$

$$J = \frac{4\pi\alpha_0 e g_v}{h^3} m_\perp^* \sqrt{2m_\parallel^*} \int_{E_0}^{\infty} [\overline{I_{129}}(\mathrm{E}',\eta_g)]' \frac{\sqrt{\gamma_{200}(\mathrm{E}',\eta_g)}}{\gamma'_{200}(\mathrm{E}',\eta_g)} f(\mathrm{E})\mathrm{dE}' \tag{1.305b}$$

where

$$\gamma_{200}(\mathrm{E},\eta_g) = [2\alpha(\frac{\hbar^2}{2m_\parallel^*})^2]^{-1} \cdot [[\frac{\hbar^2}{2m_\parallel^*}(1+2\alpha\gamma_3(\mathrm{E},\eta_g))] - [[\frac{\hbar^2}{2m_\parallel^*}(1+2\alpha\gamma_3(\mathrm{E},\eta_g))]^2 - 4\gamma_2(\mathrm{E},\eta_g)(\frac{\hbar^2}{2m_\parallel^*})^2]^{1/2}]$$

In the presence of size quantization, the dispersion law in QW of HD Ge can be written following (1.302) as

$$\frac{\hbar^2 k_s^2}{2m_\parallel^*} = \gamma_2(E,\eta_g) + \alpha\left(\frac{\hbar^2 (n_z\pi/d_z)^2}{2m_\parallel^*}\right)^2 - (1+2\alpha\gamma_3(E,\eta_g))\frac{\hbar^2 (n_z\pi/d_z)^2}{2m_\parallel^*} \tag{1.306a}$$

The EEM assumes the form

$$m_s^*(E_{F1HD},\eta_g,n_z) = m_\perp^*[\gamma_2'(E_{F1HD},\eta_g) - \frac{\alpha\hbar^2}{m_\parallel^*}(\frac{n_z\pi}{d_z})^2\gamma_3'(E_{F_h},\eta_g)] \tag{1.306b}$$

The surface electron concentration per unit area is given by

$$n_{2D} = \frac{g_v m_\perp^*}{\pi\hbar^2}\sum_{n_z=1}^{n_{z\max}}[R_1(E_{F1HD},\eta_g,n_z) + S_1(E_{F1HD},\eta_g,n_z)] \tag{1.307a}$$

where,

$$R_1(E_{F1HD},\eta_g,n_z) = [\gamma_2(E_{F1HD},\eta_g) + \alpha\left(\frac{\hbar^2 (n_z\pi/d_z)^2}{2m_\parallel^*}\right)^2 - (1+2\alpha\gamma_3(E_{F1HD},\eta_g))\frac{\hbar^2 (n_z\pi/d_z)^2}{2m_\parallel^*}]$$

and

$$S_1(E_{F1HD},\eta_g,n_z)=\sum_{r=1}^{S}L(r)[R_1(E_{F1HD},\eta_g,n_z)]$$

The EP in this case is given by

$$J_{2D}=\frac{\alpha_0 e g_v m_\perp^*}{d_z\pi\hbar^3}\sum_{n_{z\min}}^{n_{z\max}}\frac{\sqrt{\gamma_{200}(\mathrm{E}_{n_{z100HD}},\eta_g)}}{\gamma_{200}'(\mathrm{E}_{n_{z100HD}},\eta_g)}\int_{\mathrm{E}_{n_{z100HD}}}^{\infty}R_1'(E,\eta_g,n_z)f(\mathrm{E})\mathrm{dE}\qquad(1.307\mathrm{b})$$

where $\mathrm{E}_{n_{z100HD}}$ is the lowest positive root of the equation

$$\gamma_2(\mathrm{E}_{n_{z100HD}},\eta_g)+\alpha[\frac{\hbar^2}{2m_\parallel^*}(\frac{n_z\pi}{d_z})^2]^2-[1+2\alpha\gamma_3(\mathrm{E}_{n_{z100HD}},\eta_g)]\frac{\hbar^2}{2m_\parallel^*}(\frac{n_z\pi}{d_z})^2=0$$

In the presence of size quantization along k_z direction, the 2D dispersion relation of the conduction relations in QWs of n-Ge in the absence of band tails can be written by extending the method as given in [187] as

$$\frac{\hbar^2k_x^2}{2m_1^*}+\frac{\hbar^2k_y^2}{2m_2^*}=\gamma(E,n_z)\qquad(1.308)$$

where,

$$m_1^*\equiv m_\perp^*,m_2^*=\frac{m_\perp^*+2m_\parallel^*}{3},$$

$$\gamma(E,n_z)\equiv\left[E(1+\alpha E)-(1+2\alpha E)\frac{\hbar^2}{2m_3^*}\left(\frac{n_z\pi}{d_z}\right)^2+\alpha\left[\frac{\hbar^2}{2m_3^*}\left(\frac{n_z\pi}{d_z}\right)^2\right]^2\right]$$

and

$$m_3^*=\frac{3m_\parallel^*m_\perp^*}{2m_\parallel^*+m_\perp^*}.$$

The area of ellipse of the 2D surface as given by (1.308) can be written as

$$A(E,n_z)=\frac{2\pi\sqrt{m_1^*m_2^*}}{\hbar^2}\gamma(E,n_z)\qquad(1.309\mathrm{a})$$

The EEM in this case can be written as

$$m^*(E_{F_s}, n_z) = (\sqrt{m_1^* m_2^*})[\gamma(E_{F_s}, n_z)]' \tag{1.309b}$$

The DOS function per sub-band can be expressed as

$$N_{2D}(E) = \frac{4\sqrt{m_1^* m_2^*}}{\pi\hbar^2}\left[1 + 2\alpha E - 2\alpha\left(\frac{\hbar^2}{2m_3^*}\left(\frac{\pi n_z}{d_z}\right)^2\right)\right] \tag{1.310}$$

The total DOS function is given by

$$N_{2DT}(E) = \frac{4}{\pi\hbar^2}\sqrt{m_1^* m_2^*}\sum_{n_r=1}^{n_{z\max}}\left[1 + 2\alpha E - 2\alpha\left(\frac{\hbar^2}{2m_3^*}\left(\frac{\pi n_z}{d_z}\right)^2\right)\right]H(E - E_{n_{z16}}) \tag{1.311}$$

where, $E_{n_{z16}}$ is the positive root of the following equation

$$E_{n_{z16}}(1 + \alpha E_{n_{z16}}) - (1 + 2\alpha E_{n_{z16}})\left(\frac{\hbar^2}{2m_3^*}\left(\frac{\pi n_z}{d_z}\right)^2\right) + \alpha\left(\frac{\hbar^2}{2m_3^*}\left(\frac{\pi n_z}{d_z}\right)^2\right)^2 = 0 \tag{1.312}$$

Thus combining (1.311) with the Fermi Dirac occupation probability factor, the 2D electron statistics in this case can be written as

$$n_{2D} = \frac{4\sqrt{m_1^* m_2^*}k_B T}{\pi\hbar^2}\sum_{n_r=1}^{n_{z\max}}[(\mathrm{A}_1(\mathrm{n}_z) + 2\alpha\eta_{n_{z16}})\mathrm{F}_0(\eta_{n_{z16}}) + 2\alpha k_B T\,\mathrm{F}_1(\eta_{n_{z16}})] \tag{1.313}$$

where

$$A_1(n_z) \equiv \left[1 + 2\alpha(\hbar^2/2m_3^*)(\pi n_z/d_z)^2\right] \text{ and } \eta_{n_{z16}} \equiv \frac{1}{k_B T}\left[E_{F2D} - E_{n_{z16}}\right].$$

The EP in this case is given by

$$J_{2D} = \frac{\alpha_0 e}{d_z\hbar}\frac{4k_B T\sqrt{m_1^* m_2^*}}{\pi\hbar}[\sum_{n_{z\min}}^{n_{z\max}}\frac{\sqrt{\gamma_{201}(\mathrm{E}_{n_{z16}})}}{\gamma'_{201}(\mathrm{E}_{n_{z16}})}[(\mathrm{A}_1(\mathrm{n}_z) + 2\alpha\eta_{n_{z16}})\mathrm{F}_0(\eta_{n_{z16}}) + 2\alpha k_B T\,\mathrm{F}_1(\eta_{n_{z16}})]] \tag{1.314}$$

The expressions of EEMs' in bulk specimens of Ge in the absence of band tails can be written following (1.301) as

$$m_z^*(E_F) = m_{\|}^* \tag{1.315}$$

$$m_s^*(E_F) = m_{\perp}^*(1 + 2\alpha E_F) \tag{1.316}$$

The DOS function for bulk specimens of Ge in the absence of band tails can be written following (1.301) as

$$N(E) = 4\pi g_v \left(\frac{2m_D^*}{h^2}\right)^{\frac{3}{2}} \left[E^{\frac{1}{2}} - \frac{5}{6}\alpha E^{\frac{3}{2}} + \frac{18\alpha}{5}\left(\frac{m_{11}^*}{\hbar^2}\right)^2 E^{7/2}\right]; m_D = \left(m_{\perp}^{*^2} \cdot m_{\|}^*\right)^{\frac{1}{3}} \tag{1.317}$$

Using (1.317), the electron concentration in bulk specimens of Ge can be written as

$$n_0 = N_{c1}\left[F_{\frac{1}{2}}(\eta) - \frac{5}{4}\alpha k_B T F_{\frac{3}{2}}(\eta) + \frac{189}{4}\alpha k_B T\left(\frac{m_{11}^* k_B T}{\hbar^2}\right)^2 F_{\frac{7}{2}}(\eta)\right];$$
$$N_{c1} = 2g_v\left(\frac{2\pi m_D^* k_B T}{h^2}\right)^{\frac{3}{2}} \tag{1.318}$$

The EP in this case is given by

$$J = \frac{4\alpha_0 e g_v \pi}{h^3}(m_{\|}^*)^{-1/2}(m_D)^{3/2}[\mathrm{F}_0(\eta_0) - \frac{5}{3}\alpha \mathrm{k}_B^2\mathrm{T}^2\mathrm{F}_2(\eta_0) + \frac{108\alpha \mathrm{k}_B^3\mathrm{T}^3}{5}(\frac{m_{\|}^*}{\hbar^2})^2\mathrm{F}_3(\eta_0)] \tag{1.319}$$

(b) The dispersion relation of the conduction electron in bulk specimens of n-Ge can be expressed in accordance with the model of Wang and Ressler [195] can be written as

$$E = \frac{\hbar^2 k_z^2}{2m_{\|}^*} + \frac{\hbar^2 k_s^2}{2m_{\perp}^*} - \bar{\alpha}_4\left(\frac{\hbar^2 k_z^2}{2m_{\perp}^*}\right)^2 - \bar{\alpha}_5\left(\frac{\hbar^2 k_s^2}{2m_{\perp}^*}\right)\left(\frac{\hbar^2 k_z^2}{2m_{\|}^*}\right) - \bar{\alpha}_6\left(\frac{\hbar^2 k_z^2}{2m_{\|}^*}\right)^2 \tag{1.320}$$

where,

$$\bar{\alpha}_4 = \beta_4(\frac{2m_{\perp}^*}{\hbar^2}), \beta_4 = 1.4\beta_5,$$

$$\beta_5 = \frac{\alpha\hbar^4}{4}[(m_{\perp}^*)^{-1} - (m_0)^{-1}]^2, \bar{\alpha}_5 = \bar{\alpha}_7(\frac{4m_{\perp}^* m_{\|}^*}{\hbar^4}), \bar{\alpha}_7 = 0.8\beta_5 \text{ and } \bar{\alpha}_6$$
$$= (0.005\beta_5)(\frac{2m_{\|}^*}{\hbar^2})^2$$

The energy spectrum under the condition of heavy doping can be written as

$$\gamma_3(E,\eta_g) = \frac{\hbar^2 k_z^2}{2m_{\|}^*} + \frac{\hbar^2 k_s^2}{2m_{\perp}^*} - \bar{\alpha}_4\left(\frac{\hbar^2 k_s^2}{2m_{\perp}^*}\right)^2 - \bar{\alpha}_5\left(\frac{\hbar^2 k_s^2}{2m_{\perp}^*}\right)\left(\frac{\hbar^2 k_z^2}{2m_{\|}^*}\right) - \bar{\alpha}_6\left(\frac{\hbar^2 k_z^2}{2m_{\|}^*}\right)^2 \tag{1.321a}$$

The sub band energy and the 2D Dispersion relation can respectively be expressed as

$$[\bar{\alpha}_8 - \bar{\alpha}_3(\frac{n_z\pi}{d_z})^2 - \bar{\alpha}_{10}[(\frac{n_z\pi}{d_z})^4 + \bar{\alpha}_{11}(\frac{n_z\pi}{d_z})^2 + \bar{\alpha}_{11}(E_{n_{zHD24}},\eta_g)]^{1/2}] = 0 \tag{1.321b}$$

$$\frac{\hbar^2 k_s^2}{2m_{\perp}^*} = \overline{\alpha_8} - \overline{\alpha_9} k_z^2 - \overline{\alpha_{10}}\left[k_z^4 + \overline{\alpha_{11}} k_z^2 + \overline{\alpha_{12}}(E,\eta_g)\right]^{1/2} \tag{1.321c}$$

where

$$\overline{\alpha_8} = \frac{1}{2\overline{\alpha_4}}, \quad \overline{\alpha_9} = \frac{\overline{\alpha_5}}{2\overline{\alpha_4}}(\frac{\hbar^2}{2m_{\|}^*}), \quad \overline{\alpha_{10}} = \frac{1}{2\overline{\alpha_4}}(\frac{\hbar^2}{2m_{\|}^*})\sqrt{\overline{\alpha_5}^2 - 4\overline{\alpha_4}\overline{\alpha_6}},$$

$$\overline{\alpha_{11}} = \frac{2m_{\|}^*}{\hbar^2}\left[\frac{4\overline{\alpha_4} - 2\overline{\alpha_5}}{\overline{\alpha_5}^2 - 4\overline{\alpha_4}\overline{\alpha_6}}\right] \text{ and } \overline{\alpha_{12}}(E,\eta_g) = (\frac{2m_{\|}^*}{\hbar^2})\left[\frac{(1 - 4\overline{\alpha_4}\gamma_3(E,\eta_g))}{\overline{\alpha_5}^2 - 4\overline{\alpha_4}\overline{\alpha_6}}\right]$$

The EEMs' can be written as

$$m_z^*(E_{Fh},\eta_g) = [\frac{m_{\|}^*\gamma_3'(E_{Fh},\eta_g)}{\sqrt{1 - 4\bar{\alpha}_6\gamma_3(E_{Fh},\eta_g)}}] \tag{1.322}$$

$$m_{\perp}^*(E_{Fh},\eta_g) = [\frac{m_{\perp}^*\gamma_3'(E_{Fh},\eta_g)}{\sqrt{1 - 4\bar{\alpha}_4\gamma_3(E_{Fh},\eta_g)}}] \tag{1.323}$$

The electron concentration in HD Ge in accordance with the model of Wang and Ressler can be expressed as

$$n_0 = \frac{m_{\perp}^* g_v}{\pi^2\hbar^2}[I_3(E_{F_h},\eta_s) + I_4(E_{F_h},\eta_s)] \tag{1.324a}$$

where

$$I_3(E_{F_h},\eta_s)=\left[\overline{\alpha_8}\rho_{10}(E_{F_h},\eta_g)-\frac{\overline{\alpha_9}}{3}\rho_{10}^3(E_{F_h},\eta_s)-\overline{\alpha_{10}}J_{10}(E_{F_h},\eta_s)\right]$$

$$\rho_{10}(E_{F_h},\eta_s)=\frac{1}{\hbar}\left[\frac{m_{\|}^*}{\overline{\alpha_6}}\right]^{\frac{1}{2}}\left[1-\sqrt{1-4\overline{\alpha_6}\gamma_3(E_{F_h},\eta_s)}\right]^{\frac{1}{2}}$$

$$J_{10}(E_{F_h},\eta_g)=\frac{\bar{A}_1(E_{F_h},\eta_g)}{3}[-E_0(\lambda(E_{F_h},\eta_g),q(E_{F_h},\eta_g)).$$
$$[\bar{A}_1^2(E_{F_h},\eta_g)+\bar{B}_1^2(E_{F_h},\eta_g)]+2\bar{B}_1^2(E_{F_h},\eta_g)F_0(\lambda(E_{F_h},\eta_g),q(E_{F_h},\eta_g))]$$
$$+\frac{\bar{A}_1(E_{F_h},\eta_g)}{3}[\rho_{10}^2(E_{F_h},\eta_g)+\bar{A}_1^2(E_{F_h},\eta_g)+2\bar{B}_1^2(E_{F_h},\eta_g)]$$
$$\left[\frac{\bar{A}_1^2(E_{F_h},\eta_g)+\rho_{10}^2(E_{F_h},\eta_g)}{\bar{B}_1^2(E_{F_h},\eta_g)+\rho_{10}^2(E_{F_h},\eta_g)}\right]^{\frac{1}{2}},\bar{A}_1^2(E_{F_h},\eta_g)=\frac{1}{2}[\bar{\alpha}_{11}+\sqrt{\bar{\alpha}_{11}^2-4\bar{\alpha}_{12}^2(E_{F_h},\eta_g)}],$$

$$\bar{B}_1^2(E_{F_h},\eta_g)=\frac{1}{2}[\bar{\alpha}_{11}-\sqrt{\bar{\alpha}_{11}^2-4\bar{\alpha}_{12}^2(E_{F_h},\eta_g)}],$$

$$\lambda(E_{F_h},\eta_g)=\tan^{-1}[\frac{\rho_{10}(E_{F_h},\eta_g)}{\bar{B}_1(E_{F_h},\eta_g)}],$$

$$q(E_{F_h},\eta_g)=\left[\left[\frac{\bar{A}_1^2(E_{F_h},\eta_g)-\bar{B}_1^2(E_{F_h},\eta_g)}{\bar{A}_1^2(E_{F_h},\eta_g)}\right]\right]$$

and

$$I_4(E_{F_h},\eta_g)=\sum_{r=1}^{s}L(r)[I_3(E_{F_h},\eta_s)]$$

The EP in this case is given by

$$J=\frac{\alpha_0 e g_v m_{\perp}^*}{2\pi\hbar^3}\int_{E_0}^{\infty}I_3'(\mathrm{E}',\eta_g)\frac{\sqrt{\gamma_{203}(\mathrm{E}',\eta_g)}}{\gamma_{203}'(\mathrm{E}',\eta_g)}f(\mathrm{E})\mathrm{dE}' \tag{1.324b}$$

where

$$\gamma_{203}(\mathrm{E},\eta_g)=[2\overline{\alpha_6}(\frac{\hbar^2}{2m*})^2]^{-1}[\frac{\hbar^2}{2m_{\|}^*}-\sqrt{(\frac{\hbar^2}{2m*})^2-4\overline{\alpha_6}(\frac{\hbar^2}{2m*})\gamma_3(\mathrm{E},\eta_g)}]$$

The dispersion relation in QW of HD Ge can be written as

$$\frac{\hbar^2k_s^2}{2m_{\perp}^*}=\overline{\alpha_8}-\overline{\alpha_9}(\frac{n_z\pi}{d_z})^2-\overline{\alpha_{10}}\left[(\frac{n_z\pi}{d_z})^4+\overline{\alpha_{11}}(\frac{n_z\pi}{d_z})^2+\overline{\alpha_{12}}(E,\eta_g)\right]^{1/2}, \tag{1.325}$$

The (1.325) can be expressed as

$$\frac{\hbar^2 k_s^2}{2m_\perp^*} = A_{75}(E, \eta_g, n_z) \tag{1.326}$$

where,

$$A_{75}(E, \eta_g, n_z) = [\bar{\alpha}_8 - \bar{\alpha}_9(\frac{n_z \pi}{d_z})^2 - \bar{\alpha}_{10}[(\frac{n_z \pi}{d_z})^4 + \bar{\alpha}_{11}(\frac{n_z \pi}{d_z})^2 + \bar{\alpha}_{12}(E, \eta_g)]^{1/2}]$$

The EEM is given by

$$m_s^*(E_{F1HD}, \eta_g, n_z) = m_\perp^* A'_{75}(E_{F1HD}, \eta_g, n_z) \tag{1.327}$$

The electron concentration per unit area assumes the form

$$n_{2D} = \frac{m_\perp^* g_v}{\pi \hbar^2} \sum_{n_z=1}^{n_{z\max}} [A_{75}(E_{F1HD}, \eta_g, n_z) + A_{76}(E_{F1HD}, \eta_g, n_z)] \tag{1.328a}$$

where,

$$A_{76}(E_{F1HD}, \eta_g, n_z) = \sum_{r=1}^{s} L(r)[A_{75}(E_{F1HD}, \eta_g, n_z)]$$

The EP in this case is given by

$$J_{2D} = \frac{\alpha_0 e g_v m_\perp^*}{d_z \pi \hbar^3} \sum_{n_{z\min}}^{n_{z\max}} \frac{\sqrt{\delta_{203}(\mathrm{E}_{n_{z205HD}}, \eta_g)}}{\delta'_{203}(\mathrm{E}_{n_{z205HD}}, \eta_g)} \int_{\mathrm{E}_{n_{z205HD}}}^{\infty} A'_{75}(\mathrm{E}, \eta_g, n_z) f(\mathrm{E}) d\mathrm{E} \tag{1.328b}$$

where $\mathrm{E}_{n_{z205HD}}$ is the lowest positive root of the equation

$$A_{75}(\mathrm{E}_{n_{z205HD}}, \eta_g, \mathrm{n}_z) = 0 \tag{1.328c}$$

and

$$\delta_{203}(\mathrm{E}, \eta_g) = [2]^{-1}[\delta_{201} = \sqrt{\delta_{201}^2 - 4\delta_{202}(\mathrm{E}, \eta_g)}], \delta_{201}(\mathrm{E}, \eta_g) = \frac{[2\overline{\alpha_8 \alpha_9} + (\overline{\alpha_{10}})^3]}{[(\overline{\alpha_9})^2 - (\overline{\alpha_{10}}^2)]}$$

and

$$\delta_{202}(\mathrm{E},\eta_g) = \frac{[(\overline{\alpha_8})^2 - \overline{\alpha_{12}}(\mathrm{E},\eta_g)(\overline{\alpha_{10}})^2]}{[(\overline{\alpha_9})^2 - (\overline{\alpha_{10}}^2)]}$$

The 2D dispersion law in the absence of band tails can be expressed as

$$E = A_5(n_z) + A_6(n_z)\beta - \bar{\alpha}_4\beta^2 \tag{1.329}$$

where,

$$A_5(n_z) = \frac{\hbar^2}{2m_3^*}(\frac{\pi n_z}{d_z})^2[1 - \bar{\alpha}_6\frac{\hbar^2}{2m_3^*}(\frac{\pi n_z}{d_z})^2],\ A_6(n_z) = [1 - \bar{\alpha}_5(\frac{\hbar^2}{2m_3^*})(\frac{\pi n_z}{d_z})^2]$$

and

$$\beta \equiv \frac{\hbar^2 k_x^2}{2m_1^*} + \frac{\hbar^2 k_y^2}{2m_2^*}.$$

The (1.329) can be written as

$$\frac{\hbar^2 k_x^2}{2m_1^*} + \frac{\hbar^2 k_y^2}{2m_2^*} = I_1(E,n_z) \tag{1.330}$$

where,

$$I_1(E,n_z) = (2\bar{\alpha}_4)^{-1}[A_6(n_z) - [A_6^2(n_z) - 4\bar{\alpha}_4 E + 4\bar{\alpha}_4 A_5(n_z)]^{1/2}]$$

From (1.330), the area of the 2D k_s-space is given by

$$A(E,n_z) = \frac{2\pi\sqrt{m_1^* m_2^*}}{\hbar^2} I_1(E,n_z) \tag{1.331a}$$

Using (1.331a) in this case can be expressed as

$$m^*(E_{F_s},n_z) = (\sqrt{m_1^* m_2^*})[I_1(E_{F_z},n_z)]' \tag{1.331b}$$

The DOS function per sub-band can be written as

$$N_{2D}(E) = \frac{4}{\pi}\frac{\sqrt{m_1^* m_2^*}}{\hbar^2}\{I_1(E,n_z)\}' \tag{1.332}$$

where

$$\{I_1(E,n_z)\}' \equiv \frac{\partial}{\partial E}[I_1(E,n_z)]$$

The total DOS function assumes the form

$$N_{2DT}(E) = \frac{4\sqrt{m_1^* m_2^*}}{\pi\hbar^2}\sum_{n_z=1}^{n_{z\max}}\{I_1(E,n_z)\}'H(E-E_{n_{z17}}) \tag{1.333}$$

where, the sub-band energy $\left(E_{n_{z17}}\right)$ are given by

$$E_{n_{z17}} = (\frac{\hbar^2}{2m_3^*})(\frac{n_z\pi}{d_z})^2[1-\bar{\alpha}_6(\frac{\hbar^2}{2m_3^*})(\frac{n_z\pi}{d_z})^2] \tag{1.334}$$

The electron statistics can be written as

$$n_{2D} = \frac{4\sqrt{m_1^* m_2^*}}{\pi\hbar^2}\sum_{n_z=1}^{n_{z\max}} t_{46}(E_{Fs},n_z)+t_{47}(E_{Fs},n_z) \tag{1.335}$$

where

$$t_{46}(E_{Fs},n_z) \equiv I_1(E_{Fs},n_z),\ t_{47}(E_{Fs},n_z) \equiv \sum_{r=1}^{s} L(r)(t_{46}(E_{Fs},n_z))$$

The EP in this case is given by

$$J_{2D} = \frac{4\alpha_0 e\sqrt{m_1^* m_2^*}}{d_z\pi\hbar^3}[\sum_{n_{z\min}}^{n_{z\max}}\sqrt{\frac{\delta_{204}(\mathrm{E}_{n_{z17}})}{\delta'_{204}(\mathrm{E}_{n_{z17}})}}[\int_{\mathrm{E}_{n_{z17}}}^{\infty} I_1'(\mathrm{E},\mathrm{n}_z)\mathrm{f}(\mathrm{E})\mathrm{dE}]] \tag{1.336}$$

where

$$\delta_{204}(\mathrm{E}) = [2\overline{\alpha_6}(\frac{\hbar^2}{2m_3^*})^2]^{-1}[\frac{\hbar^2}{2m_3^*}-\sqrt{(\frac{\hbar^2}{2m_3^*})^2-4\bar{\alpha}_6(\frac{\hbar^2}{2m_3^*})E}].$$

1.2.11 The EP from QWs of HD Gallium Antimonide

The dispersion relation of the conduction electrons in n-GaSb can be written as [196]

$$E = \frac{\hbar^2 k^2}{2m_0} - \frac{\bar{E}'_{go}}{2} + \frac{\bar{E}'_{go}}{2}[1 + \frac{\hbar^2 k^2}{\bar{E}'_{go}}(\frac{1}{m_c} - \frac{1}{m_0})]^{\frac{1}{2}} \tag{1.337}$$

where

$$\bar{E}'_{go} = [E_{go} + \frac{5.10^{-5}T^2}{2(112+T)}]\text{eV}$$

The (1.337) can be expressed as

$$\frac{\hbar^2 k^2}{2m_c} = I_{36}(E) \tag{1.338}$$

where

$$\begin{aligned} I_{36}(E) = {} & [E + \bar{E}'_{g0} - (m_c/m_0)(\bar{E}'_{g0}/2) \\ & - [(\bar{E}'_{g0}/2)^2 + [((\bar{E}'_{g0})^2/2)(1 - (m_c/m_0))] \\ & + [(\bar{E}'_{g0}/2)(1 - (m_c/m_0))]^2 + E\bar{E}'_{g0}(1 - (m_c/m_0))]^{1/2}] \end{aligned}$$

Under the condition of heavy doping (1.338) assumes the form

$$\frac{\hbar^2 k^2}{2m_c} = I_{36}(E, \eta_g) \tag{1.339}$$

where,

$$\begin{aligned} I_{36}(E,\eta_g) = {} & [\gamma_3(E,\eta_g) + E'_g - \frac{m_c}{m_0}.\frac{E'_g}{2} - [(\frac{E'_g}{2})^2 + [\frac{E'_g}{2}(1 - \frac{m_c}{m_0})]^2 + (\frac{E'_g}{2})^2(1 - \frac{m_c}{m_0}) \\ & + \gamma_3(E,\eta_g)E'_g(1 - \frac{m_c}{m_0})]^{1/2}] \end{aligned}$$

The EEM can be written as

$$m^*(E_{F_h}, \eta_g) = m_c\{I_{36}(E_{F_h}, \eta_g)\}' \tag{1.340}$$

The DOS function in this case can be written as

$$N_{HD}(E,\eta_g) = \frac{g_v}{2\pi^2}\left(\frac{2m_c}{\hbar^2}\right)^{3/2}\sqrt{I_{36}(E,\eta_g)}\{I_{36}(E,\eta_g)\}' \tag{1.341}$$

Since, the original band model in this case is a no pole function, therefore, the HD counterpart will be totally real, and the complex band vanishes.

The electron concentration is given by

$$n_0 = \frac{g_v}{3\pi^2}\left(\frac{2m_c}{\hbar^2}\right)^{3/2}\left[\{I_{36}(E_{F_h},\eta_g)\}^{3/2} + \sum_{r=1}^{s} L(r)[\{I_{36}(E_{F_h},\eta_g)\}^{3/2}]\right] \quad (1.342)$$

The EP in this case is given by

$$J = \frac{4\alpha_0 \pi e g_v m_c}{h^3}\int_{E_0}^{\infty} I_{36}(E',\eta_g)f(E)dE' \quad (1.343)$$

For dimensional quantization along z-direction, the dispersion relation of the 2D electrons in QWs of HD GaSb can be written following (1.339) as

$$\frac{\hbar^2(n_z\pi/d_z)^2}{2m_c} + \frac{\hbar^2(k_s)^2}{2m_c} = I_{36}(E,\eta_g) \quad (1.344)$$

The expression of the $N_{2DT}(E)$ in this case can be written as

$$N_{2DT}(E) = \frac{m_c g_v}{\pi\hbar^2}\sum_{n_z=1}^{n_{z\max}} T'_{119D}(E,\eta_g,n_z)H(E - E_{n_zD119}) \quad (1.345)$$

where,

$$T_{119D}(E,\eta_g,n_z) = [I_{36}(E,\eta_g) - \hbar^2(n_z\pi/d_z)^2(2m_c)^{-1}],$$

The sub band energies E_{n_zD119} in this case given by

$$\left\{\hbar^2(n_z\pi/d_z)^2\right\}(2m_c)^{-1} = I_{36}(E_{n_zD119},\eta_g) \quad (1.346)$$

The EEM in this case assumes the form

$$m^*(E_{F1HD},\eta_g,n_z) = m_c[I'_{36}(E_{F1HD},\eta_g,n_z)] \quad (1.347)$$

The 2-D electron statistics in this case can be written as

$$n_{2D} = \frac{m_c g_v}{\pi\hbar^2}\sum_{n_z=1}^{n_{z\max}}[T_{119D}(E_{F1HD},\eta_g,n_z) + T_{119D}(E_{F1HD},\eta_g,n_z)] \quad (1.348a)$$

where,

$$T_{129D}(E_{F1HD},\eta_g,\mathrm{n}_z)\sum_{r=1}^{s}L(r)[T_{119D}(E_{F1HD},\eta_g,\mathrm{n}_z)],$$

The EP in this case is given by

$$J_{2D}=\frac{\alpha_0 e g_v\sqrt{m_c}}{\sqrt{2}d_z\pi\hbar^3}[\sum_{n_{z\min}}^{n_{z\max}}\frac{\sqrt{I_{36}(\mathrm{E}_{n_{zD119}},\eta_g)}}{I'_{36}(\mathrm{E}_{n_{zD119}},\eta_g)}[\int_{\mathrm{E}_{n_{zD119}}}^{\infty}T'_{119D}(\mathrm{E},\eta_g,n_z)\mathrm{f(E)dE}]] \quad (1.348\mathrm{b})$$

The total 2D DOS function in the absence of band tails in this case can be written as

$$N_{2DT}(E)=(\frac{m_c g_v}{\pi\hbar^2})\sum_{n_z=1}^{n_{\max}}\{[I_{36}(E)]'H(E-E_{n_{z44}})\} \quad (1.349)$$

where, the sub-band energies $E_{n_{z44}}$ can be expressed as

$$I_{36}(E_{n_{z44}})=\frac{\hbar^2}{2m_c}(\pi n_z/d_z)^2 \quad (1.350\mathrm{a})$$

The EEM in this case can be written as

$$m^*(E_{F_s})=(m_c)[I_{36}(E_{F_s})]' \quad (1.350\mathrm{b})$$

The 2D carrier concentration assumes the form

$$n_{2D}=(\frac{m_c g_v}{\pi\hbar^2})\sum_{n_z=1}^{n_{\max}}[\overline{\mathrm{T}}_{55}(E_{Fs},n_z)+\overline{\mathrm{T}}_{56}(E_{Fs},n_z)] \quad (1.351)$$

where

$$\overline{\mathrm{T}}_{55}(E_{Fs},n_z)=[I_{36}(E_{Fs})-\frac{\hbar^2}{2m_c}(\pi n_z/d_z)^2]\text{ and }\overline{\mathrm{T}}_{56}(E_{Fs},n_z)=\sum_{r=1}^{s}L(r)[\overline{\mathrm{T}}_{55}(E_{Fs},n_z)]$$

The EP in this case is given by

$$J_{2D}=\frac{\alpha_0 e g_v\sqrt{m_c}}{\sqrt{2}d_z\pi\hbar^2}[\sum_{n_{z\min}}^{n_{z\max}}\frac{\sqrt{I_{36}(\mathrm{E}_{n_{z44}})}}{I'_{36}(\mathrm{E}_{n_{z44}})}[\int_{\mathrm{E}_{n_{z44}}}^{\infty}I'_{36}(\mathrm{E})\mathrm{f(E)dE}]] \quad (1.352)$$

The expression of electron concentration for bulk specimens of GaSb (in the absence of band tails) can be expressed as

$$n_0 = \frac{g_v}{3\pi^2}\left(\frac{2m_c}{\hbar^2}\right)^{3/2}[\bar{M}_{A_{10}}(E_F) + \bar{N}_{A_{10}}(E_F)] \tag{1.353}$$

where,

$$\bar{M}_{A_{10}}(E_F) = [I_{36}(E_F)]^{3/2} \text{ and } \bar{N}_{A_{10}}(E_F) = \sum_{r=1}^{s} L(r)[\bar{M}_{A_{10}}(E_F)]$$

The EP in this case is given by

$$J = \frac{\alpha_0 e}{4}\frac{g_v}{3\pi^2}\sqrt{\frac{2}{m_c}}(\frac{2m_c}{\hbar^2})^{3/2}[\int_{\mathrm{E}'}^{\infty}\frac{\sqrt{I_{36}(\mathrm{E}')}}{I'_{36}(\mathrm{E}')}[\overline{\mathrm{M}}_{A100}(\mathrm{E}')]'\mathrm{f}(\mathrm{E})\mathrm{dE}']] \tag{1.354}$$

Thus, we can summarize the whole mathematical background in the following way.

In this chapter, we have investigated the 3D and 2D EPs from HD bulk and QWs of non-linear optical materials on the basis of a newly formulated electron dispersion law considering the anisotropies of the effective electron masses, the spin orbit splitting constants and the influence of crystal field splitting within the framework of **k.p** formalism. The results for 3D and 2D EPs from HD bulk and QWs of III-V, ternary and quaternary compounds in accordance with the three and two band models of Kane form a special case of our generalized analysis. We have also studied the EP in accordance with the models of Stillman et al. and Palik et al. respectively since these models find use to describe the electron energy spectrum of the aforesaid materials. The 3D and 2D EPs has also been derived for HD bulk and QWs of II-VI, IV-VI, stressed materials, Te, n-GaP, p-$PtSb_2$, Bi_2Te_3, n-Ge and n-GaSb compounds[sub]Band structure by using the models of Hopfield, Dimmock, Bangert and Kastner, Seiler, Bouat and Thuillier, Rees, Emtage, Kohler, Cardona, Wang et al. and Mathur et al. respectively and transforming each and on the basis of the appropriate carrier energy spectra. The well-known expressions of the EPs in the absence of band tails for wide gap materials have been obtained as special cases of our generalized analysis under certain limiting conditions. This indirect test not only exhibits the mathematical compatibility of our formulation but also shows the fact that our simple analysis is a more generalized one, since one can obtain the corresponding results for relatively wide gap materials having parabolic energy bands under certain limiting conditions from our present derivation.

1.3 Results and Discussion

Using the appropriate equations and taking the energy band constants as given in Table 1.1, we have plotted the normalized EP from QWs of HD $CdGeAs_2$ (an example of nonlinear optical materials) as a function of d_z as shown in plot (a) of Fig. 1.1, in which the plot (b) corresponds to $\delta = 0$. The plot (c) has been drawn in accordance with the three band model of Kane and the plot (d) refers to the two band model of Kane together with the plot (e) exhibiting the variation in accordance with the parabolic energy bands for the overall assessments of the energy band constants on the EP in this case. The Fig. 1.2 exhibits the plots of the normalized EP from QWs of HD $CdGeAs_2$ as a function of the normalized incident photon energy for all cases Figs. 1.2 and 1.3 shows the dependence of the said variable on the normalized electron degeneracy for all cases of Fig. 1.2.

The normalized EP from QWs of HD n-InAs (an example of III-V materials) in accordance with the three and two band models of Kane as functions of film thickness, normalized incident photon energy and the normalized electron degeneracy have, respectively, been presented in Figs. 1.4, 1.5 and 1.6. The Figs. 1.7, 1.8 and 1.9 exhibit the variations of normalized EP from QWs of HD n-InSb as functions of film thickness, normalized incident photon energy and the normalized electron degeneracy respectively. The variations of the normalized EP from QWs of HD CdS (an example of II-VI materials) as functions of thickness, normalized

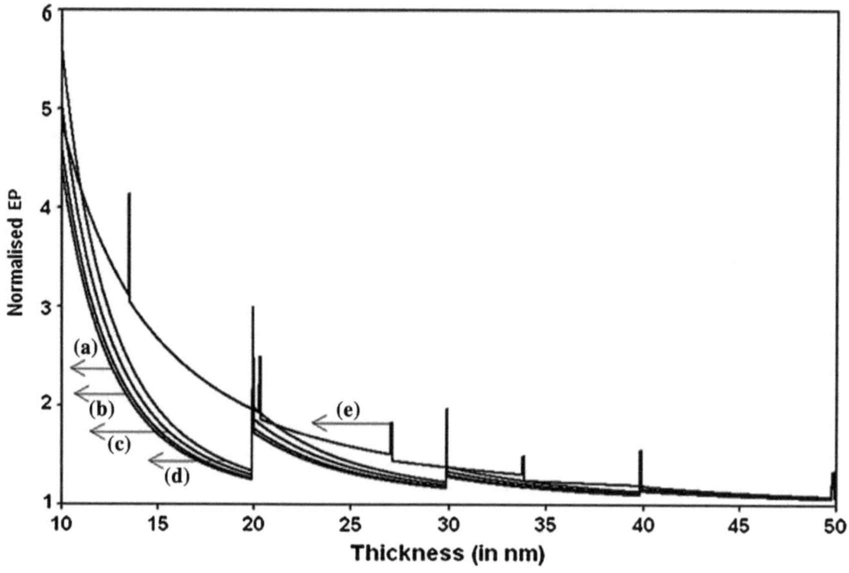

Fig. 1.1 Plot of the normalized EP from QWs of HD $CdGeAs_2$ as a function of d_z in accordance with **a** generalized band model, **b** $\delta = 0$, **c** the three-band model of Kane, **d** the two band model of Kane, and **e** the parabolic energy bands

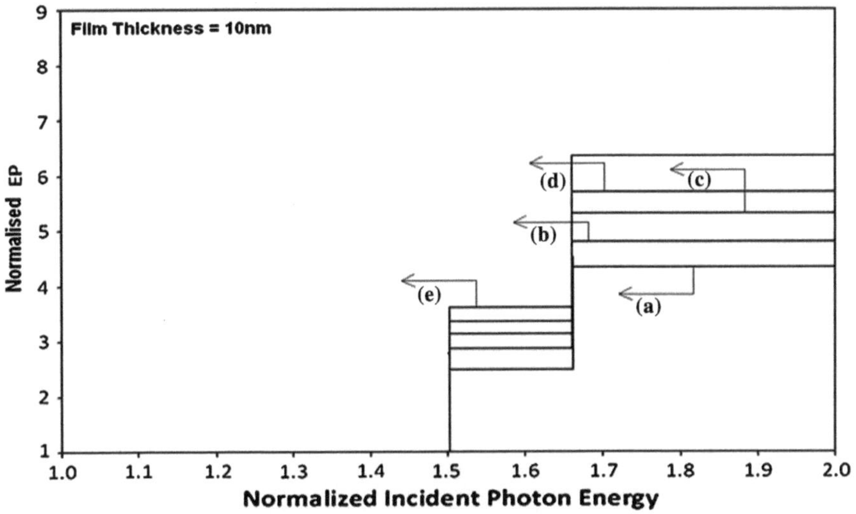

Fig. 1.2 Plot of the normalized EP from QWs of HD $CdGeAs_2$ as a function of normalized incident photon energy for all cases of Fig. 1.1

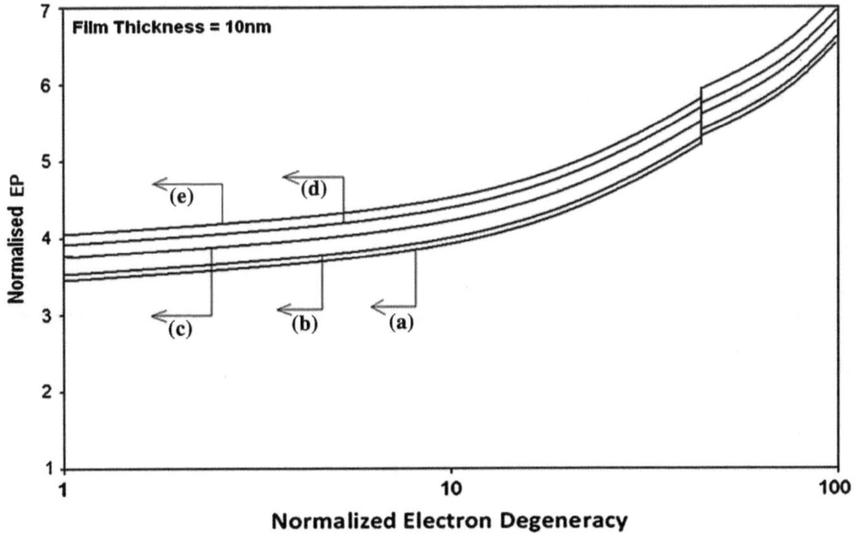

Fig. 1.3 Plot of the normalized EP from QWs of HD $CdGeAs_2$ as a function of normalized electron degeneracy for all cases of Fig. 1.1

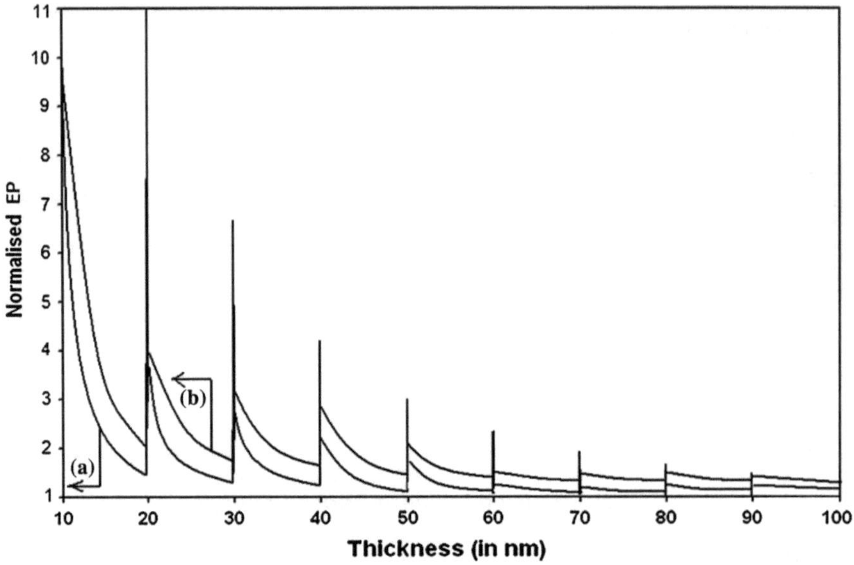

Fig. 1.4 Plot of the normalized EP from QWs of HD n-InAs as a function of d_z in accordance with **a** the three band model of Kane and **b** the two band model of Kane

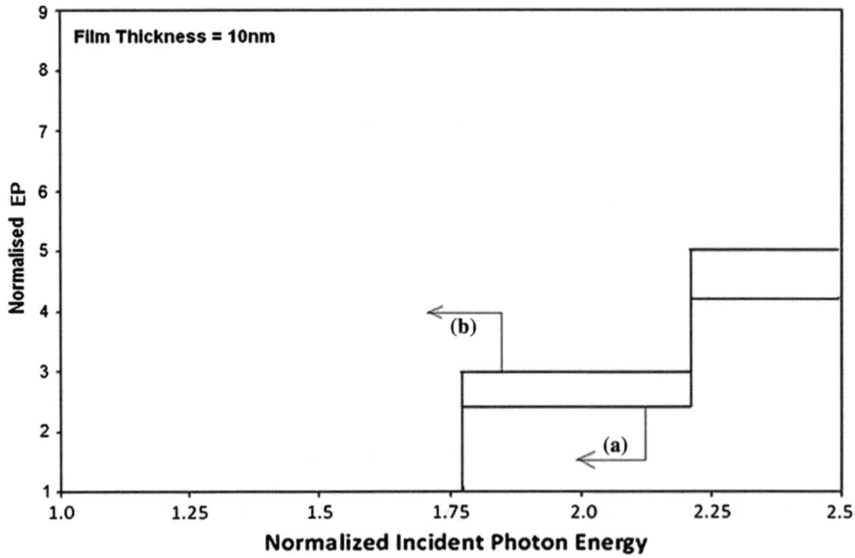

Fig. 1.5 Plot of the normalized EP from QWs of HD n-InAs as a function of normalized incident photon energy in accordance with **a** the three band model of Kane and **b** the two band model of Kane

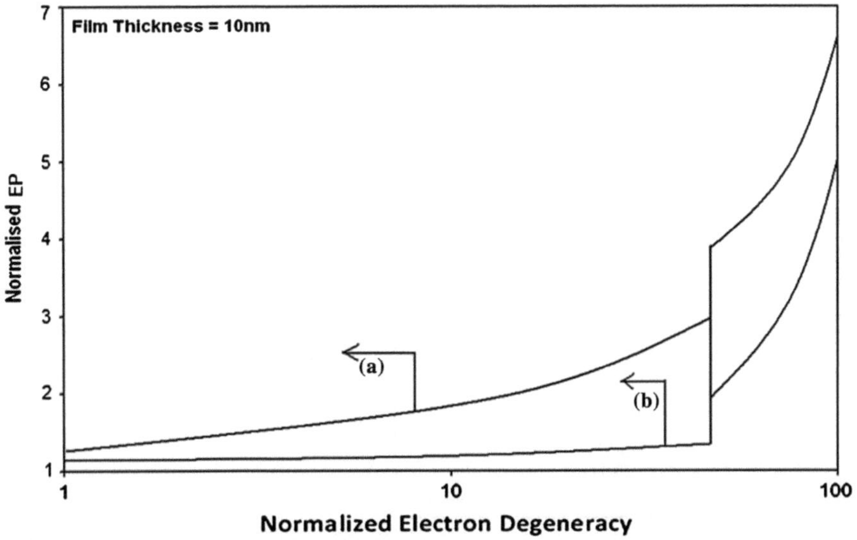

Fig. 1.6 Plot of the normalized EP from QWs of HD n-InAs as a function of normalized electron degeneracy in accordance with **a** the three band model of Kane and **b** the two band model of Kane

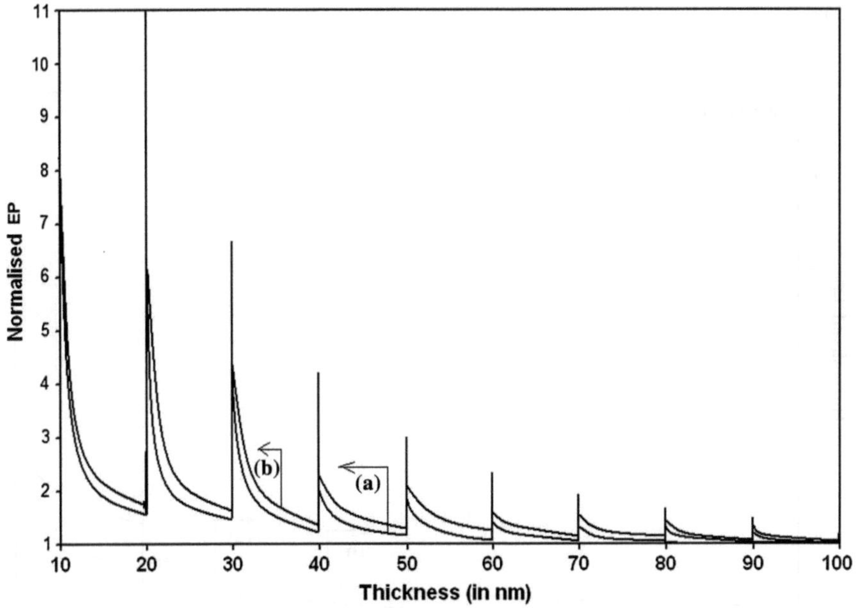

Fig. 1.7 Plot of the normalized EP from QWs of HD n-InSb as a function of d_z in accordance with **a** the three band model of Kane and **b** the two band model of Kane

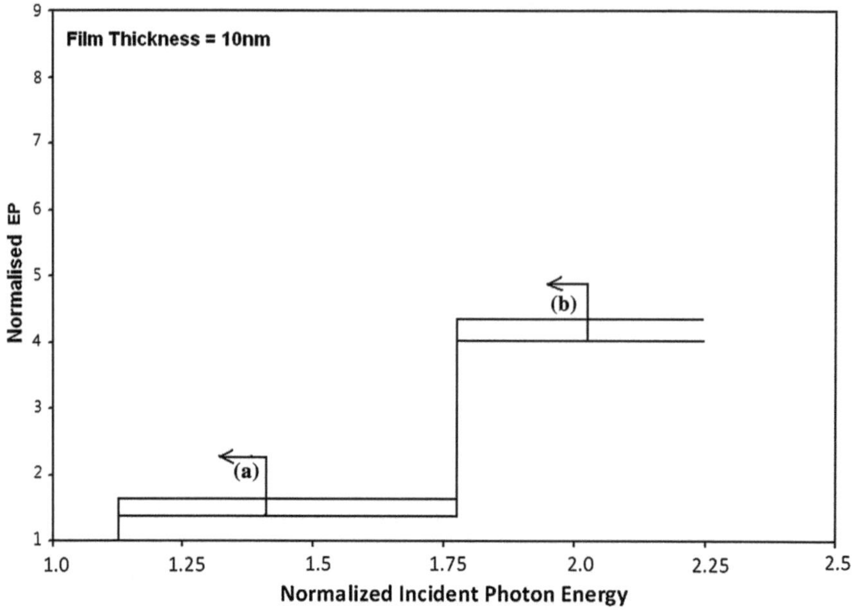

Fig. 1.8 Plot of the normalized EP from QWs of HD n-InSb as a function of normalized incident photon energy in accordance with **a** the three band model of Kane and **b** the two band model of Kane

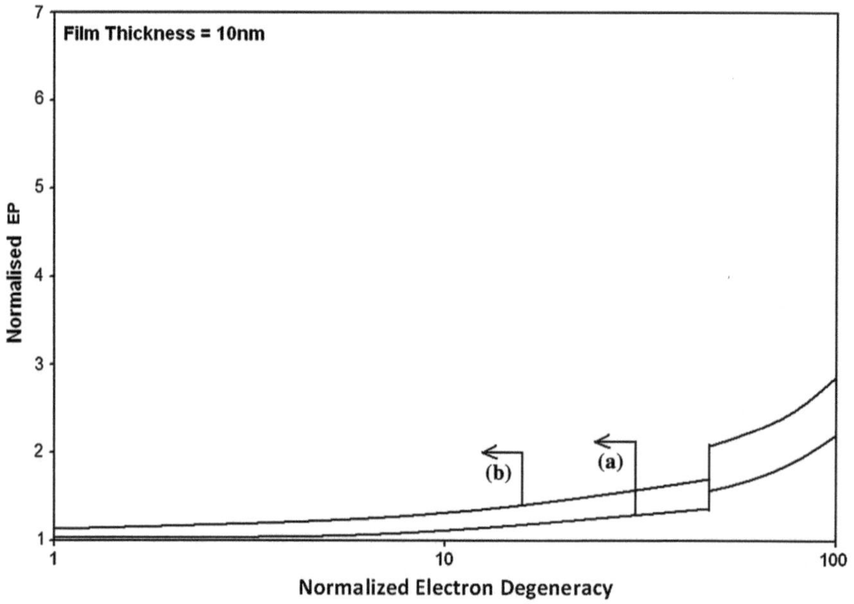

Fig. 1.9 Plot of the normalized EP from QWs of HD n-InSb as a function of normalized electron degeneracy in accordance with **a** the three band model of Kane and **b** the two band model of Kane

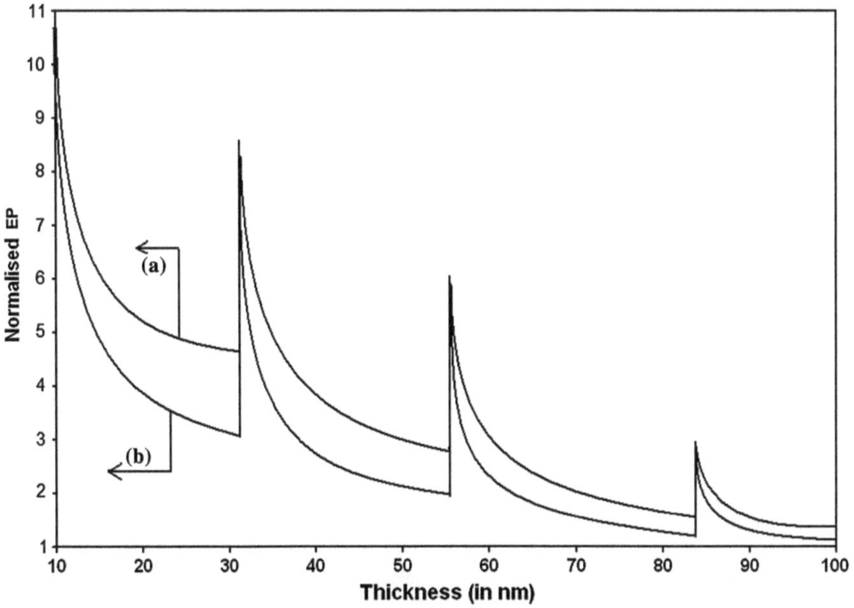

Fig. 1.10 Plot of the normalized EP from QWs of HD CdS as a function of d_z with **a** $\bar{\lambda}_0 = 0$ and **b** $\bar{\lambda}_0 = 0$

incident photon energy and normalized electron degeneracy have respectively been drawn in Figs. 1.10, 1.11 and 1.12, where the plots for $\bar{\lambda}_0 = 0$ have further been drawn for the purpose of assessing the influence of the splitting of the two-spin states by the spin orbit coupling and the crystalline field. The thickness, normalized photon energy and the normalized electron degeneracy dependences of normalized EP from QWs of HD GaP have been shown in Figs. 1.13, 1.14 and 1.15 respectively. The dependence of normalized EP with reference to the aforementioned variables from QWs of HD n-Ge and $PtSb_2$, has been shown in Figs. 1.16, 1.17, 1.18, 1.19, 1.20 and 1.21 in accordance with the models of Cardona et al.,Wang and Ressler and Emtage respectively. Figures 1.22, 1.23 and 1.24 manifest the variations of the normalized EP from QWs of HD stressed n-InSb as functions of the film thickness, normalized incident photon energy and the normalized electron degeneracy respectively. The Figs. 1.25, 1.26, 1.27 exhibit the normalized EP from QWs of HD IV-VI materials as functions of film thickness, normalized incident photon energy and normalized electron degeneracy.

The influence of quantum confinement is immediately apparent from Figs. 1.1, 1.4, 1.7, 1.10, 1.13, 1.16, 1.19, 1.22 and 1.25 since the EP depends strongly on the thickness of the quantum-confined materials in contrast with the corresponding bulk specimens. The EP decreases with increasing film thickness in an oscillatory way with different numerical magnitudes for QWs of HD materials. It appears from the aforementioned figures that the EP exhibits spikes for particular values of film

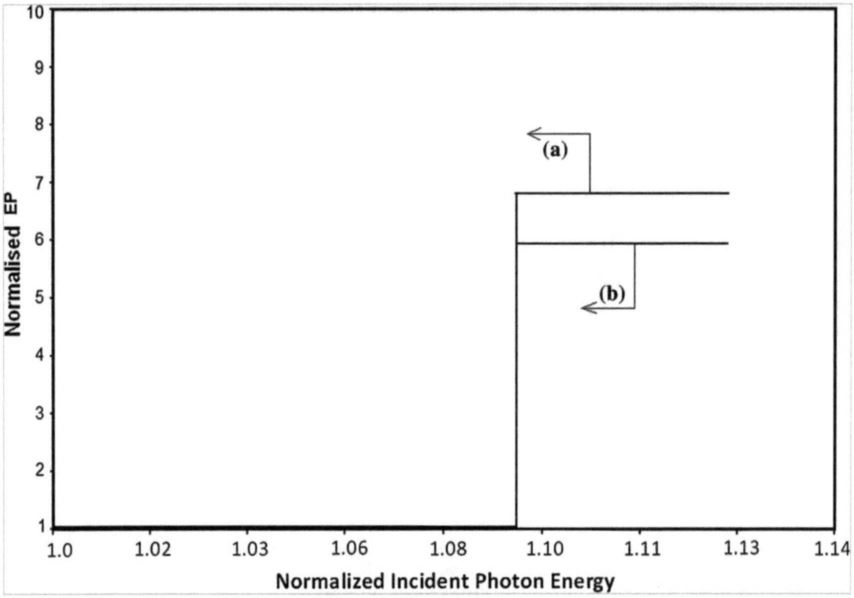

Fig. 1.11 Plot of the normalized EP from QWs of HD CdS as a function of normalized incident photon energy with **a** $\bar{\lambda}_0 = 0$ and **b** $\bar{\lambda}_0 = 0$

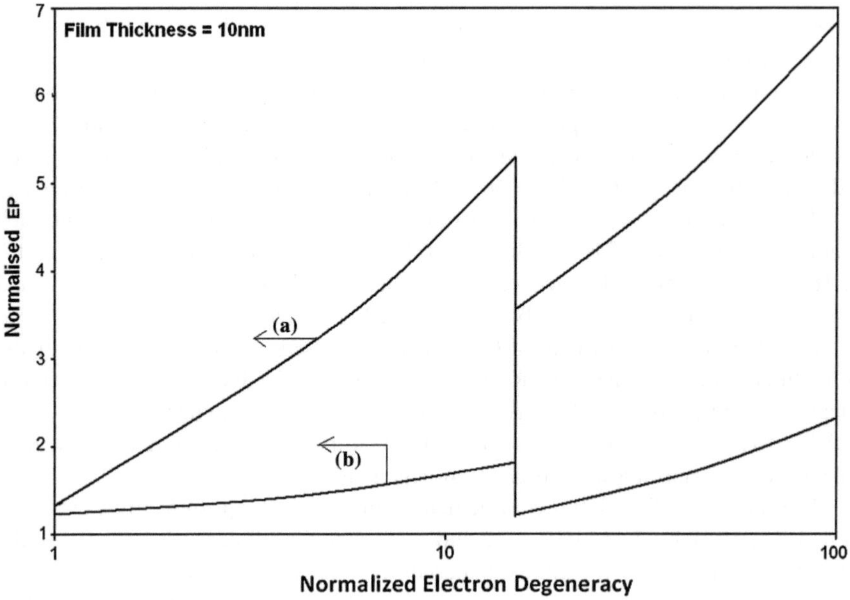

Fig. 1.12 Plot of the normalized EP from QWs of HD CdS as a function of normalized electron degeneracy with **a** $\bar{\lambda}_0 = 0$ and **b** $\bar{\lambda}_0 = 0$

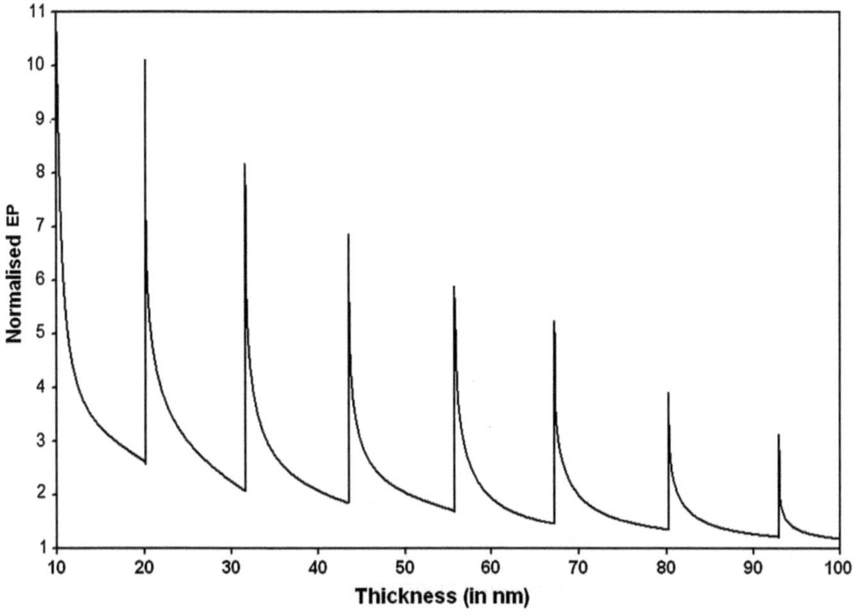

Fig. 1.13 Plot of the normalized EP from QWs of HD n-GaP as a function of d_z

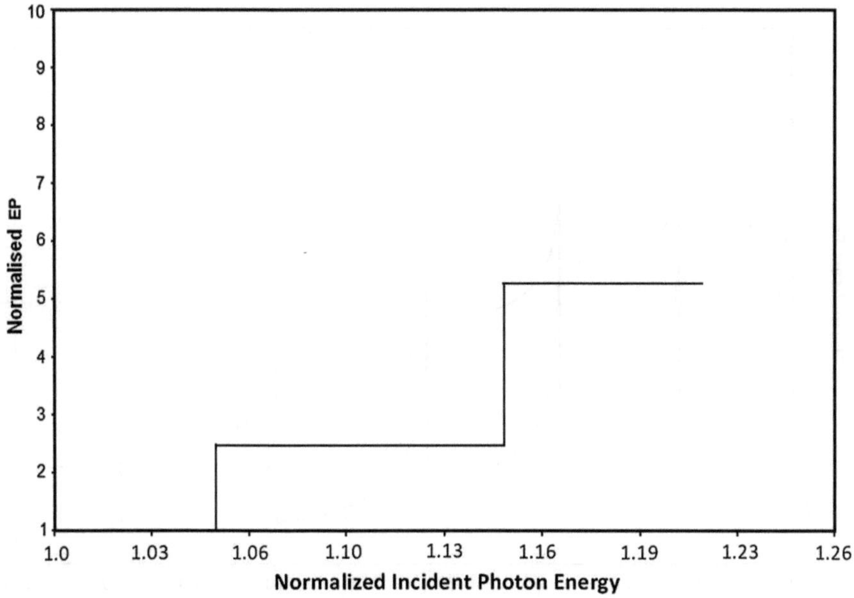

Fig. 1.14 Plot of the normalized EP from QWs of HD n-GaP as a function of normalized incident photon energy

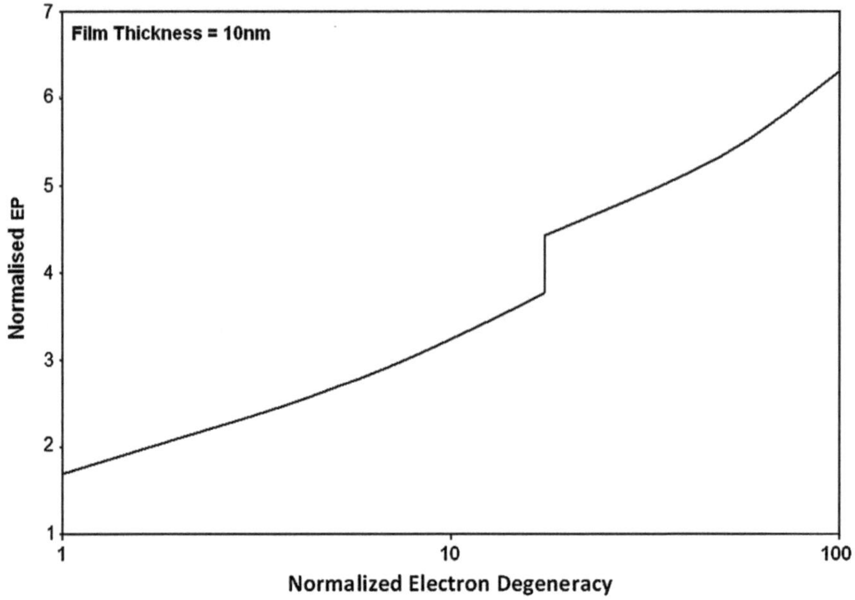

Fig. 1.15 Plot of the normalized EP from QWs of HD n-GaP as a function of normalized electron degeneracy

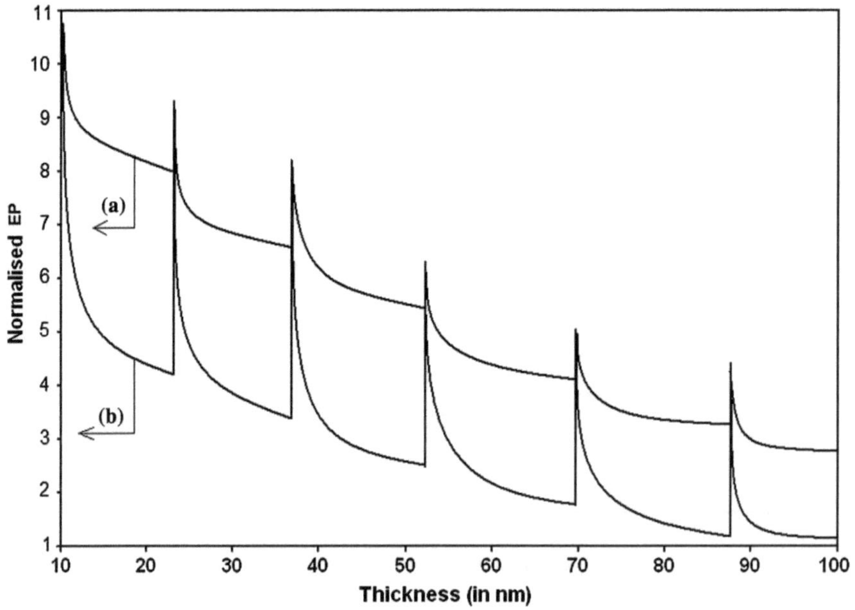

Fig. 1.16 Plot of the normalized EP from QWs of HD n-Ge as a function of thickness in accordance with **a** Cardona et al. and **b** Wang et al.

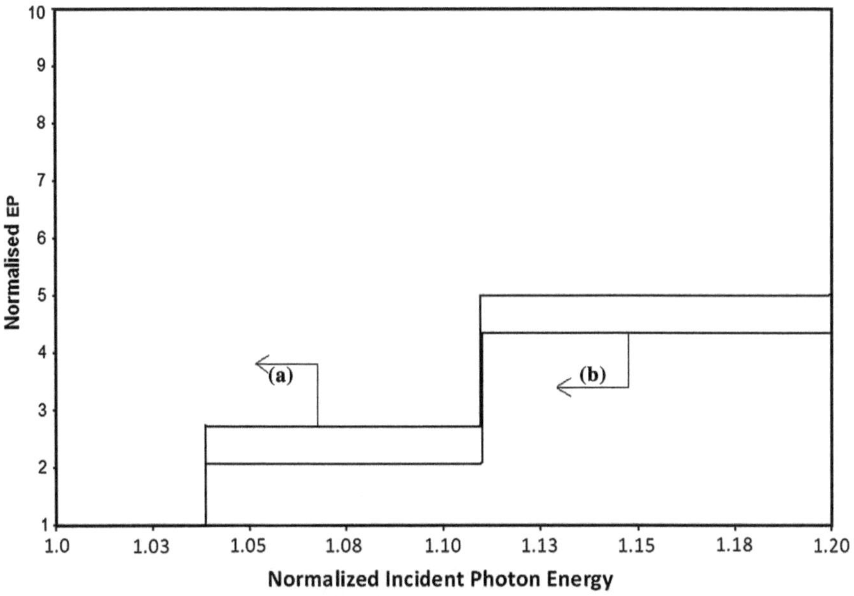

Fig. 1.17 Plot of the normalized EP from QWs of HD n-Ge as a function of normalized incident photon energy for both the cases of Fig. 1.16

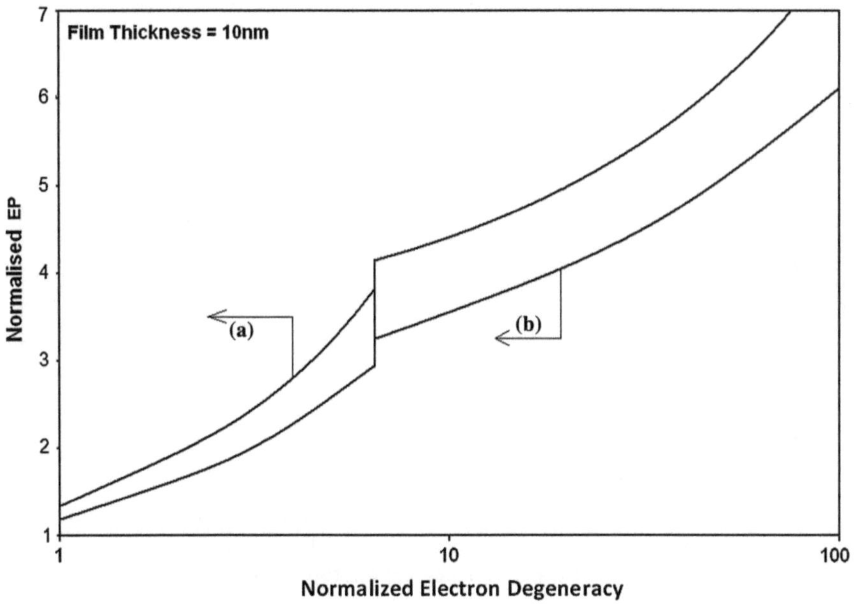

Fig. 1.18 Plot of the normalized EP from QWs of HD n-Ge as a function of normalized electron degeneracy for both the cases of Fig. 1.16

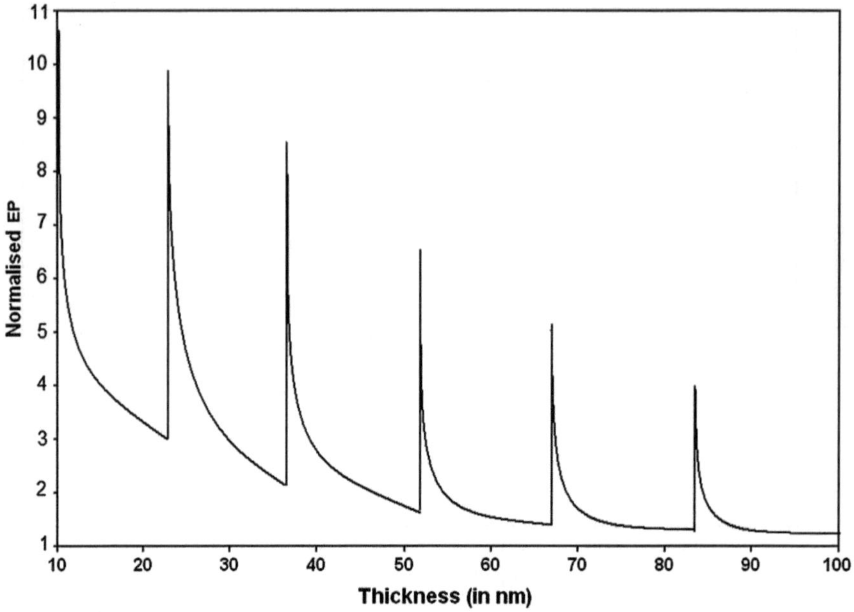

Fig. 1.19 Plot of the normalized EP from QWs of HD n-$PtSb_2$ as a function of thickness

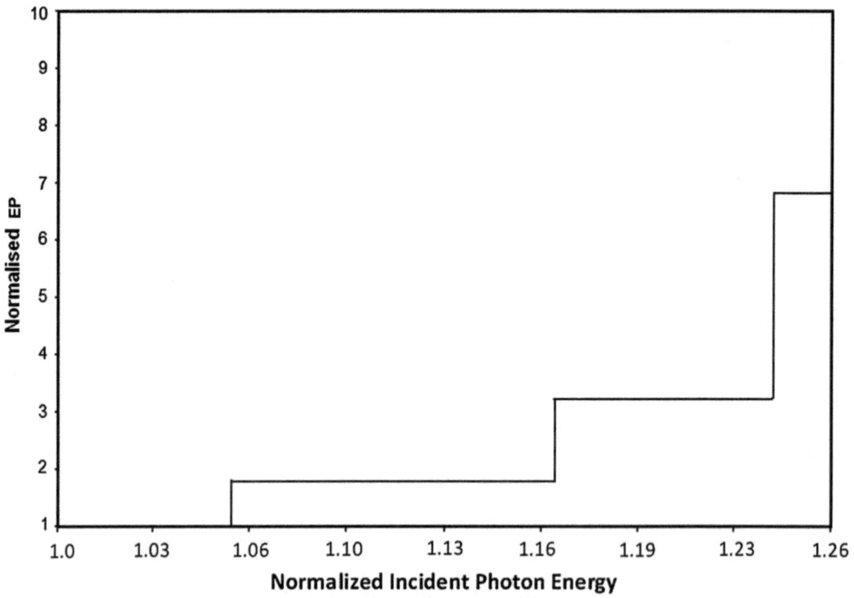

Fig. 1.20 Plot of the normalized EP from QWs of HD n-$PtSb_2$ as a function of normalized incident photon energy

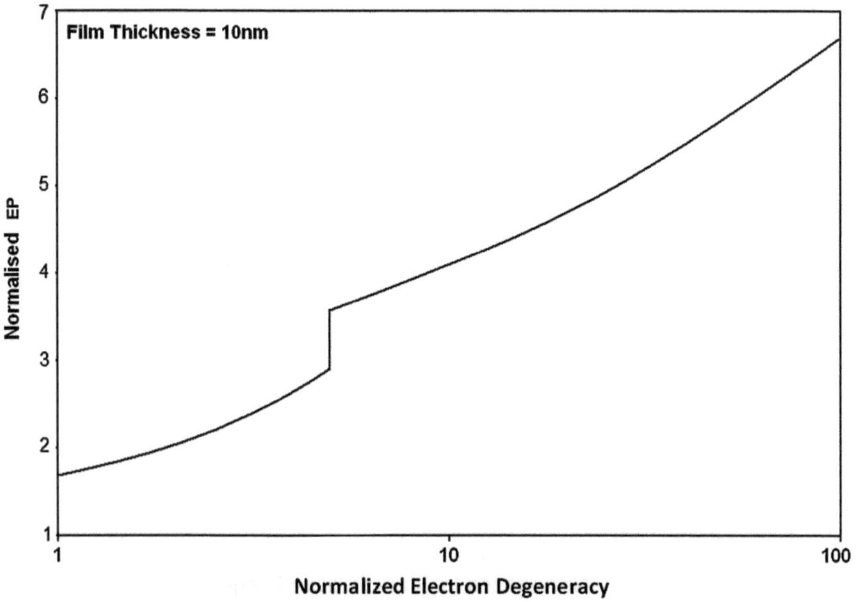

Fig. 1.21 Plot of the normalized EP from QWs of HD n-$PtSb_2$ as a function of normalized electron degeneracy

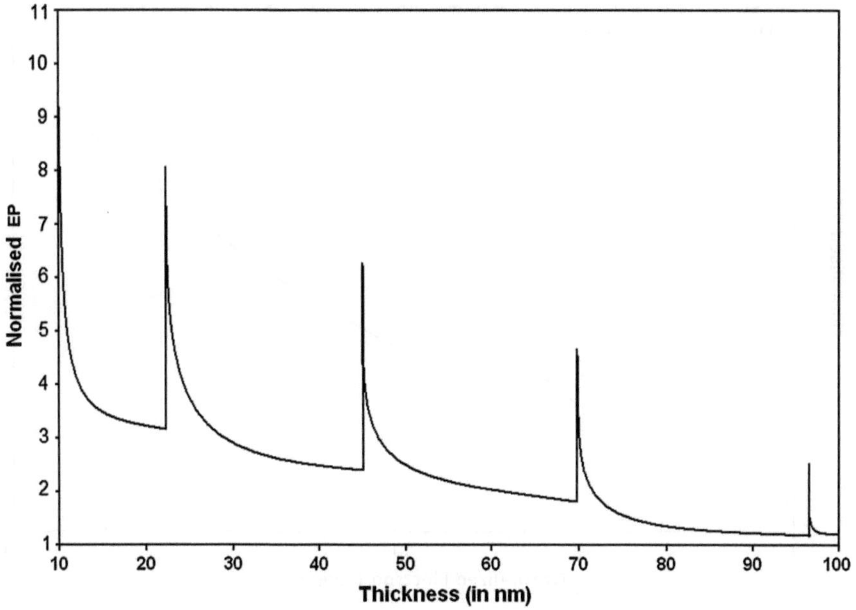

Fig. 1.22 Plot of the normalized EP from QWs of HD stressed n-InSb as a function of film thickness

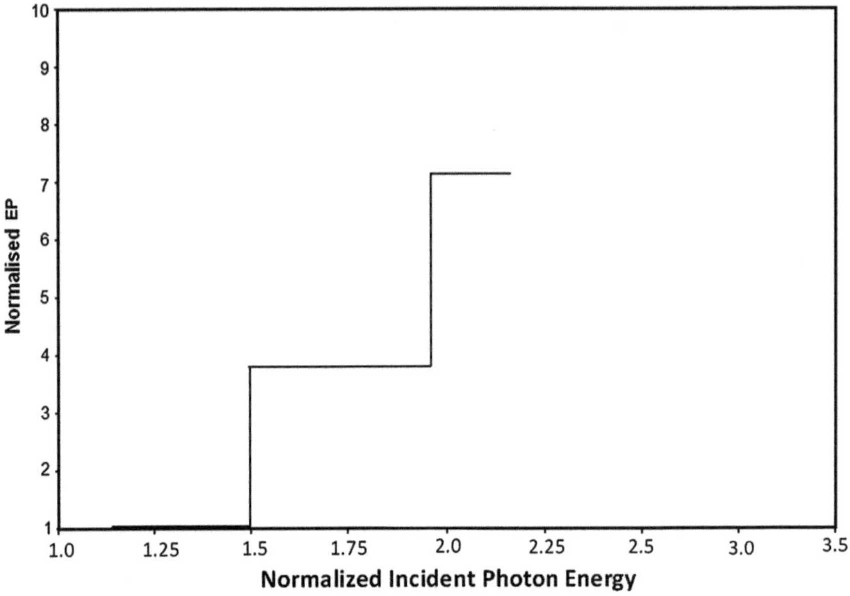

Fig. 1.23 Plot of the normalized EP from QWs of HD stressed n-InSb as a function of normalized incident photon energy

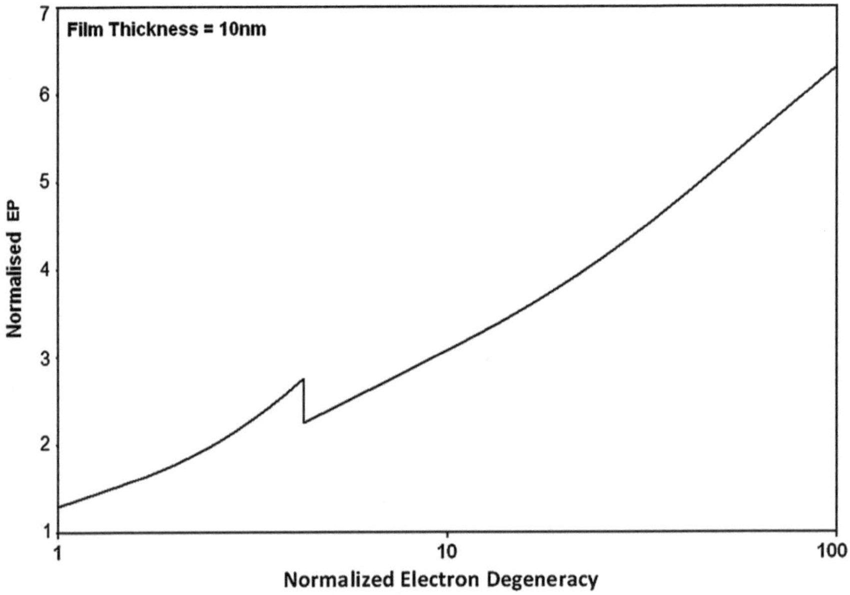

Fig. 1.24 Plot of the normalized EP from QWs of HD stressed n-InSb as a function normalized electron degeneracy

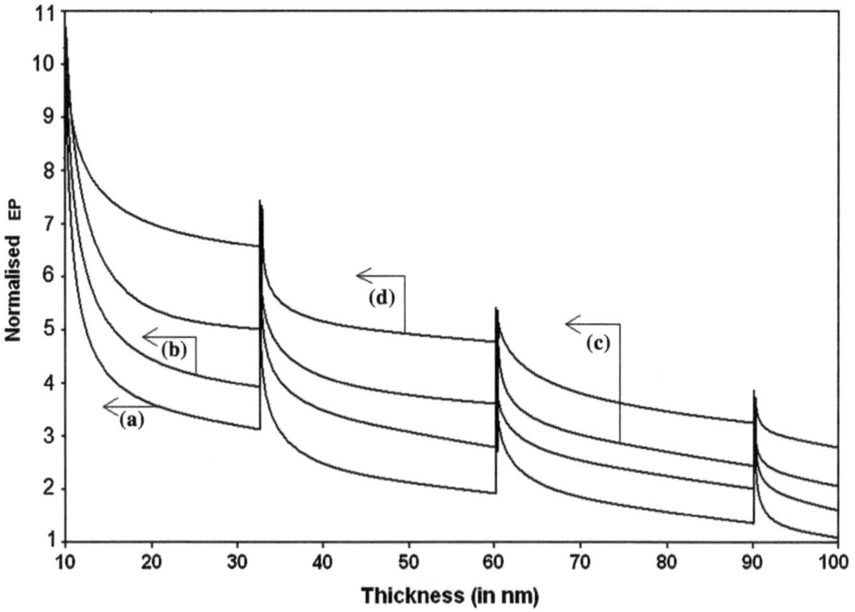

Fig. 1.25 Plot of the normalized EP from QWs of HD PbTe as a function of film thickness in accordance with the models of **a** the Dimmok and **b** the Bangert and Kastner respectively. The plots **c** and **d** exhibit the same for PbSe

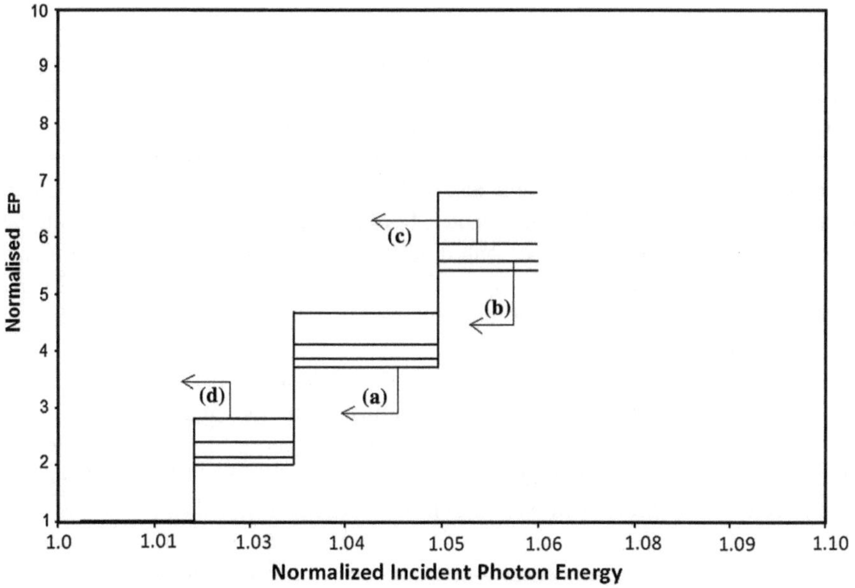

Fig. 1.26 Plot of the normalized EP from QWs of HD PbTe as a function of incident photon energy in accordance with the models of **a** the Dimmok and **b** the Bangert and Kastner respectively. The plots **c** and **d** exhibit the same for PbSe

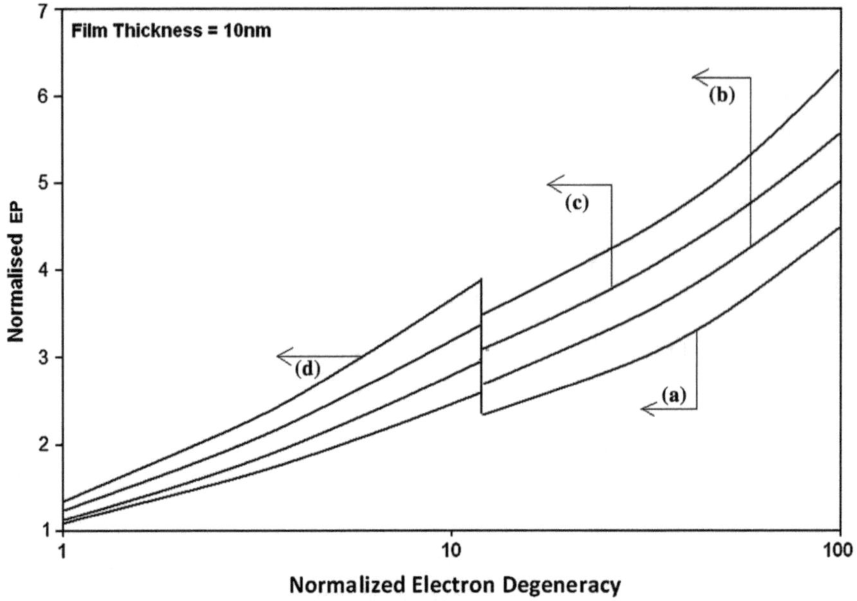

Fig. 1.27 Plot of the normalized EP from QWs of HD PbTe as a function of electron degeneracy in accordance with the models of **a** the Dimmok and **b** the Bangert and Kastner respectively. The plots **c** and **d** exhibit the same for PbSe

thickness which, in turn, depends on the particular band structure of the specific material. Moreover, the photoemission from QWs of HD compounds can become several orders of magnitude larger than of bulk specimens of the same materials, which is also a direct signature of quantum confinement. This oscillatory dependence will be less and less prominent with increasing film thickness. It appears from Figs. 1.3, 1.6, 1.9, 1.12, 1.15, 1.18, 1.21, 1.24 and 1.27 that the EP increases with increasing carrier degeneracy and also exhibits spikes for all types of quantum confinement as considered in this chapter. For bulk specimens of the same material, the EP will be found to increase continuously with increasing electron degeneracy in a non-oscillatory manner. The Figs. 1.2, 1.5, 1.8, 1.11, 1.14, 1.17, 1.20, 1.23 and 1.26 illustrate the dependence of the EP from quantum-confined HD materials on the normalized incident photon energy.

The EP increases with increasing photon energy in a step like manner for all the figures. The appearance of the discrete jumps in all the figures is due to the redistribution of the electrons among the quantized energy levels when the size quantum number corresponding to the highest occupied level changes from one fixed value to the others. With varying electron degeneracy, a change is reflected in the EP through the redistribution of the electrons among the size-quantized levels. It may be noted that at the transition zone from one sub band to another, the height of the peaks between any two sub-bands decreases with the increasing in the degree of quantum confinement and is clearly shown in all the curves. It should be noted that although, the EP varies in various manners with all the variables as evident from all the figures, the rates of variations are totally band-structure dependent. The influence of the energy band models on the EP from various types of HD quantum-confined materials can also be assessed from the plots. With different sets of energy band parameters, different numerical values of the EP will be obtained though the nature of variations of the same as shown here would be similar for the other types of materials and the simplified analysis of this chapter exhibits the basic qualitative features of the EP phenomena from such compounds. Another important point in this context is the existence of the effective mass within the forbidden zone, which is impossible without the formation of band tails.

It is an amazing fact that the study of the carrier transport in HD quantized materials through proper formulation of the Boltzmann transport equation which needs in turn, the corresponding HD carrier energy spectra is still one of the open research problems.

It may be noted that with the advent of MBE and other experimental techniques, it is possible to fabricate quantum-confined structures with an almost defect-free surface. The numerical computations have been performed using the fact that the probability of photon absorption in direct band-gap compounds is close to unity. If the direction normal to the film was taken differently from that as assumed in this work, the expressions for the EP from QWs of HD materials would be different analytically, since the basic dispersion relations for many materials are anisotropic. In formulating the generalized electron energy spectrum for non-linear optical materials, we have considered the crystal-field splitting parameter, the anisotropies in the momentum-matrix elements, and the spin-orbit splitting parameters, respectively. In the absence of the crystal field splitting parameter together with the assumptions of isotropic effective electron mass and isotropic spin orbit splitting, our basic relation as given by (1.2) converts into the well-known three-band Kane model and is valid for III-V compounds, in general. It should be used as such for studying the electronic properties of n-InAs where the spin-orbit splitting parameter (Δ) is of the order of band gap (E_g). For many important materials $\Delta \gg E_g$ and under this inequality, the three band model of Kane assumes the form $E(1 + EE_g^{-1}) = \hbar^2 k^2 / 2m_c$ which is the well-known two-band Kane model. Also under the condition, $E_g \to \infty$, the above equation gets simplified to the well-known form of parabolic energy bands as $E = \hbar^2 k^2 / 2m_c$. It is important to note that under certain limiting conditions, all the results for all the models as derived here have transformed into the well-known expression of the 2D EP for size quantized materials having parabolic bands. We have not considered other types of compounds or external physical variables for numerical computations in order to keep the presentation brief. With different sets of energy band constants, we shall get different numerical values of the 2D EP though the nature of variations of the 2D EP as shown here would be similar for the other types of materials and the simplified analysis of this chapter exhibits the basic qualitative features of the 2D EP for such compounds.

Our method is not at all related to the DOS technique as used in the literature. From the E-k dispersion relation, we can obtain the DOS, but the DOS technique as used in the literature cannot provide the E-k dispersion relation. Therefore, our study is more fundamental than those of the existing literature because the Boltzmann transport equation, which controls the study of the charge transport properties of semiconductor devices, can be solved if and only if the E-k dispersion relation is known. We wish to note that we have not considered the many body effects in this simplified theoretical formalism due to the lack of availability in the literature of proper analytical techniques for including them for the generalized systems as considered in this chapter. Our simplified approach will be useful for the purpose of comparison when methods of tackling the formidable problem after inclusion of the many body effects for the present generalized systems appear. It is worth remarking in this context that from our simple theory under certain limiting conditions we get the well-known result of the EP from wide gap materials having parabolic energy bands. The inclusion of the said effects would certainly increase the accuracy of the results, although the qualitative features of the 2D EP in QWs of HD materials discussed in this chapter would not change in the presence of the aforementioned effects. The influence of energy band models and the various band constants on the EP for different materials can also be studied from all the Figures of this chapter. One important concept of this chapter is the presence of poles in the finite complex plane in the dispersion relation of the materials in the absence of band tails creates the complex energy spectrum in the corresponding HD samples. Besides, from the DOS function in this case, it appears that a new forbidden zone has been created in addition to the normal band gap of the semiconductor. If the basic dispersion relation in the absence of band tails contains no poles in the finite complex plane, the corresponding HD energy band spectrum will be real, although it may be the complicated functions of exponential and error functions and deviate considerably from that in the absence of band tailing.

The numerical results presented in this chapter would be different for other materials but the nature of variation would be unaltered. The theoretical results as given here would be useful in analyzing various other experimental data related to this phenomenon. We must note that the study of transport phenomena and the formulation of the electronic properties of HD nano-compounds are based on the dispersion relations in such materials. It is worth remarking that this simplified formulation exhibits the basic qualitative features of 2D EP from 2D materials. The basic objective of this chapter is not solely to demonstrate the influence of quantum confinement on the 2D EP from QWs of HD non-parabolic materials but also to formulate the appropriate electron statistics in the most generalized form, since the transport and other phenomena in HD 2D materials having different band structures and the derivation of the expressions of many important electronic properties are based on the temperature-dependent electron statistics in such compounds. Finally, we can write that the analysis as presented in this chapter can be used to investigate the Burstein Moss shift, the carrier contribution to the elastic constants, the specific heat, activity coefficient, reflection coefficient, Hall coefficient, plasma frequency, various scattering mechanisms and other different transport coefficients of modern

HD non-parabolic quantum confined HD devices operated under different external conditions having varying band structures.

1.4 Open Research Problems

The problems under these sections of this monograph are by far the most important part for the readers and few open research problems are presented from this chapter till end. The numerical values of the energy band constants for various semiconductors are given in Table 1.1 *for the related computer simulations.*

(R.1.1) Investigate the EP for the HD bulk semiconductors whose respective dispersion relations of the carriers in the absence of band tails and any externally applied field are given below:

(a) The electron dispersion law in n-GaP can be written as [197]

$$E = \frac{\hbar^2 k_s^2}{2m_{\|}^*} + \frac{\hbar^2 k_s^2}{2m_{\perp}^*} \mp \frac{\bar{\Delta}}{2} \pm \left[\left(\frac{\bar{\Delta}}{2} \right)^2 + P_1 k_z^2 + D_1 k_x^2 k_y^2 \right]^{1/2} \tag{R.1.1}$$

where, $\bar{\Delta} = 335\,\text{meV}$, $P_1 = 2 \times 10^{-10}\text{eVm}$, $D_1 = P_1 a_1$ and $a_1 = 5.4 \times 10^{-10}\text{m}$.

(b) The dispersion relation for the conduction electrons for IV-VI semiconductors can also be described by the models of Cohen [198], McClure and Choi [199], Bangert et al. [200] and Foley et al. [201] respectively.

(i) In accordance with Cohen [198], the dispersion law of the carriers is given by

$$E(1 + \alpha E) = \frac{p_x^2}{2m_1} + \frac{p_z^2}{2m_3} = \frac{\alpha E p_y^2}{2m_2'} + \left(\frac{\alpha p_y^4}{4m_2 m_2'} \right) + \frac{p_y^2}{2m_2}(1 + \alpha E) \tag{R.1.2}$$

where m_1, m_2 and m_3 are the effective carrier masses at the band-edge along x, y and z directions respectively and m_2' is the effective- mass tensor component at the top of the valence band (for electrons) or at the bottom of the conduction band (for holes).

(ii) The carrier energy spectra can be written, following McClure and Choi [199] as

$$E(1+\alpha E)=\frac{p_x^2}{2m_1}+\frac{p_y^2}{2m_2}+\frac{p_z^2}{2m_3}+\frac{p_y^2}{2m_2}\alpha E\left\{1-\left(\frac{m_2}{m_2'}\right)\right\}+\frac{p_y^4\alpha}{4m_2m_2'}-\frac{\alpha p_x^2p_y^2}{4m_1m_2}-\frac{\alpha p_y^2p_z^2}{4m_2m_3} \quad \text{(R.1.3)}$$

(iii) The carrier energy spectrum of IV-VI semiconductors in accordance with Foley et al. [201] can be written as

$$E+\frac{E_g}{2}=E_-(k)+\left[\left[E_+(k)+\frac{E_g}{2}\right]^2+P_\perp^2k_s^2+P_\parallel^2k_z^2\right]^{1/2} \quad \text{(R.1.4)}$$

where, $E_+(k)=\frac{\hbar^2k_s^2}{2m_\perp^+}+\frac{\hbar^2k_z^2}{2m_\parallel^+}$, $E_-(k)=\frac{\hbar^2k_s^2}{2m_\perp^-}+\frac{\hbar^2k_z^2}{2m_\parallel^-}$ represents the contribution from the interaction of the conduction and the valance band edge states with the more distant bands and the free electron term, $\frac{1}{m_\perp^\pm}=\frac{1}{2}\left[\frac{1}{m_{tc}}\pm\frac{1}{m_{tv}}\right]$, $\frac{1}{m_\parallel^\pm}=\frac{1}{2}\left[\frac{1}{m_{1c}}\pm\frac{1}{m_{1v}}\right]$,

For n-PbTe

$$\mathrm{P}_\perp=4.61\times10^{-10}\,\mathrm{eVm},\ \mathrm{P}_\parallel=4.61\times10^{-10}\,\mathrm{eVm},\ \frac{m_0}{m_{tv}}=10.36,\ \frac{m_0}{m_{tv}}=0.75,\ \frac{m_0}{m_{tc}}=11.36,\ \frac{m_0}{m_{1c}}=1.20\ \text{and}\ g_v=4$$

(c) The hole energy spectrum of p-type zero-gap semiconductors (e.g. HgTe) is given by Ivanov-Omskii et al. [202]

$$E=\frac{\hbar^2k^2}{2m_v^*}+\frac{3e^2}{128\varepsilon_\infty}k-\left(\frac{2E_B}{\pi}\right)\ln\left|\frac{k}{k_0}\right| \quad \text{(R.1.5)}$$

where m_v^* is the effective mass of the hole at the top of the valence band, $E_B\equiv\frac{m_0e^2}{2\hbar^2\varepsilon_\infty^2}$ and $k_0\equiv\frac{m_0e^2}{\hbar^2\varepsilon_\infty^2}$.

(d) The conduction electrons of n-GaSb obey the following two dispersion relations:

(i) In accordance with the model of Seiler et al. [203]

$$E=\left[-\frac{E_g}{2}+\frac{E_g}{2}[1+\alpha_4k^2]^{1/2}+\frac{\bar{\varsigma}_0\hbar^2k^2}{2m_0}+\frac{\bar{v}_0f_1(k)\hbar^2}{2m_0}\pm\frac{\bar{\omega}_0f_2(k)\hbar^2}{2m_0}\right] \quad \text{(R.1.6)}$$

where $\alpha_4\equiv4P^2\left(E_g+\frac{2}{3}\Delta\right)\left[E_g^2(E_g+\Delta)\right]^{-1}$, P is the isotropic momentum matrix element, $f_1(k)\equiv k^{-2}\left[k_x^2k_y^2+k_y^2k_z^2+k_z^2k_x^2\right]$

represents the warping of the Fermi surface, $f_2(k) \equiv \left[\left\{k^2\left(k_x^2k_y^2 + k_y^2k_z^2 + k_z^2k_x^2\right) - 9k_x^2k_y^2k_z^2\right\}^{1/2} k^{-1}\right]$ represents the inversion asymmetry splitting of the conduction bandand $\bar{\varsigma}_0, \bar{v}_0$ and $\bar{\omega}_0$ represent the constants of the electron spectrum in this case.

(ii) In accordance with the model of Zhang et al. [204]

$$E = \left[E_2^{(1)} + E_2^{(2)}K_{4,1}\right]k^2 + \left[E_4^{(1)} + E_4^{(2)}K_{4,1}\right]k^4 + k^6\left[E_6^{(1)} + E_6^{(2)}K_{4,1} + E_6^{(3)}K_{6,1}\right] \quad \text{(R.1.7)}$$

where

$$K_{4,1} \equiv \frac{5}{4}\sqrt{21}\left[\frac{k_x^4 + k_y^4 + k_z^4}{k^4} - \frac{3}{5}\right],$$

$$K_{6,1} \equiv \sqrt{\frac{639,639}{32}}\left[\frac{k_x^2k_y^2k_z^2}{k^6} + \frac{1}{22}\left(\frac{k_x^4 + k_y^4 + k_z^4}{k^4} - \frac{3}{5}\right) - \frac{1}{105}\right],$$

the coefficients are in eV, the values of k are $10\left(\frac{a}{2\pi}\right)$ times those of k in atomic units (a is the lattice constant), $E_2^{(1)} = 1.0239620$, $E_2^{(2)} = 0$, $E_4^{(1)} = -1.1320772$, $E_4^{(2)} = 0.05658$, $E_6^{(1)} = 1.1072073$, $E_6^{(2)} = -0.1134024$ and $E_6^{(3)} = -0.0072275$.

(e) In addition to the well-known band models of III-V semiconductors as discussed in this monograph, the conduction electrons of such compounds obey the following three dispersion relations:

(i) In accordance with the model of Rossler [205]

$$E = \frac{\hbar^2k^2}{2m^*} + \bar{\alpha}_{10}k^4 + \bar{\beta}_{10}\left[k_x^2k_y^2 + k_y^2k_z^2 + k_z^2k_x^2\right] \pm\bar{\gamma}_{10}\left[k^2\left(k_x^2k_y^2 + k_y^2k_z^2 + k_z^2k_x^2\right) - 9k_x^2k_y^2k_z^2\right]^{1/2} \quad \text{(R.1.8)}$$

where, $\bar{\alpha}_{10} = \bar{\alpha}_{11} + \bar{\alpha}_{12}k$, $\bar{\beta}_{10} = \bar{\beta}_{11} + \bar{\beta}_{12}k$ and $\bar{\gamma}_{10} = \bar{\gamma}_{11} + \bar{\gamma}_{12}k$, in which, $\bar{\alpha}_{11} = -2,132 \times 10^{-40}\,\text{eVm}^4$, $\bar{\alpha}_{12} = 9,030 \times 10^{-50}\,\text{eVm}^5$, $\bar{\beta}_{11} = -2,493 \times 10^{-40}\,\text{eVm}^4$, $\bar{\beta}_{12} = 12,594 \times 10^{-50}\,\text{eVm}^5$, $\bar{\gamma}_{11} = 30 \times 10^{-30}\,\text{eVm}^3$ and $\bar{\gamma}_{12} = -154 \times 10^{-42}\,\text{eVm}^4$.

(ii) In accordance with Johnson and Dickey [206], the electron energy spectrum assumes the form

$$E = -\frac{E_g}{2} + \frac{\hbar^2 k^2}{2}\left[\frac{1}{m_0} + \frac{1}{m_{\gamma b}}\right] + \frac{E_g}{2}\left[1 + 4\frac{\hbar^2 k^2 \bar{f}_1(E)}{2m_c' \; E_g}\right]^{1/2}$$

where, $\frac{m_0}{m_c'} \equiv P^2\left[\frac{\left(E_g+\frac{2\Delta}{3}\right)}{E_g\left(E_g+\Delta\right)}\right]$, $\bar{f}_1(E) \equiv \frac{\left(E_g+\Delta\right)\left(E+E_g+\frac{2\Delta}{3}\right)}{\left(E_g+\frac{2\Delta}{3}\right)\left(E+E_g+\Delta\right)}$, $m_c' = 0.139 m_0$ and $m_{\gamma b} = \left[\frac{1}{m_c'} - \frac{2}{m_0}\right]^{-1}$.

(iii) In accordance with Agafonov et al. [207], the electron energy spectrum can be written as

$$E = \frac{\bar{\eta} - E_g}{2}\left[1 - \frac{\hbar^2 k^2}{2\bar{\eta} m_c}\left\{\frac{D\sqrt{3} - 3\bar{B}}{2\left(\frac{\hbar^2}{2m_c}\right)}\right\}\left[\frac{k_x^4 + k_y^4 + k_z}{k^4}\right]\right] \quad \text{(R.1.9)}$$

where, $\bar{\eta} \equiv \left(E_g^2 + \frac{8}{3}P^2 k^2\right)^{1/2}$, $\bar{B} \equiv -21\frac{\hbar}{2m_0}$ and $D \equiv -40\left(\frac{\hbar^2}{2m_0}\right)$.

(f) The dispersion relation of the carriers in n-type Pb_{1-x}-Ga_xTe with x = 0.01 can be written following Vassilev [208] as

$$\begin{aligned}&\left[E - 0.606k_s^2 - 0.0722k_z^2\right]\left[E + \bar{E}_g + 0.411k_s^2 + 0.0377k_z^2\right]\\ &= 0.23k_s^2 + 0.02k_z^2 \pm \left[0.06\bar{E}_g + 0.061k_s^2 + 0.0066k_z^2\right]k_s\end{aligned} \quad \text{(R.1.10)}$$

where, $\bar{E}_g (= 0.21\,\text{eV})$ is the energy gap for the transition point, the zero of the energy E is at the edge of the conduction band of the Γ point of the Brillouin zone and is measured positively upwards, k_x, k_y and k_z are in the units of 10^9m^{-1}.

(g) The energy spectrum of the carriers in the two higher valance bands and the single lower valance band of Te can, respectively, be expressed as [209]

$$\begin{aligned}\bar{E} =& A_{10}k_z^2 + B_{10}k_s^2\\ &\pm \left[\Delta_{10}^2 + (\beta_{10}k_z)^2\right]^{1/2} \text{ and } \bar{E} = \Delta_{\parallel} + A_{10}k_z^2 + B_{10}k_s^2 \pm \beta_{10}k_z\end{aligned} \quad \text{(R.1.11)}$$

where, $\bar{E}$ is the energy of the hole as measured from the top of the valance and within it, $A_{10} = 3.77 \times 10^{-19} \text{eVm}^2$, $B_{10} = 3.57 \times 10^{-19} \text{eVm}^2$, $\Delta_{10} = 0.628\,\text{eV}$, $(\beta_{10})^2 = 6 \times 10^{-20}\,\text{eVm}^2$ and $\Delta_{\parallel} = 1{,}004 \times 10^{-5}\,\text{eV}$ are the spectrum constants.

(h) The dispersion relation of the holes in p-InSb can be written in accordance with Cunningham [210] as

$$\bar{E} = c_4(1+\gamma_4 f_4)k^2 \pm \frac{1}{3}\left[2\sqrt{2}\sqrt{c_4}\sqrt{16+5\gamma_4}\sqrt{E_4}g_4 k\right] \quad \text{(R.1.12)}$$

where, $c_4 \equiv \frac{\hbar^2}{2m_0} + \theta_4$, $\theta_4 \equiv 4.7\frac{\hbar^2}{2m_0}$, $\gamma_4 \equiv \frac{b_4}{c_4}$, $b_4 \equiv \frac{3}{2}b_5 + 2\theta_4$, $b_5 \equiv 2.4\frac{\hbar^2}{2m_0}$, $f_4 \equiv \frac{1}{4}\left[\sin^2 2\theta + \sin^4\theta \sin^2 2\phi\right]$, θ is measured from the positive z-axis, ϕ is measured from positive x-axis, $g_4 \equiv \sin\theta\left[\cos^2\theta + \frac{1}{4}\sin^4\theta \sin^2 2\phi\right]$ and $E_4 = 5\times 10^{-4}$eV.

(i) The energy spectrum of the valance bands of CuCl in accordance with Yekimov et al. [211] can be written as

$$E_h = (\gamma_6 - 2\gamma_7)\frac{\hbar^2 k^2}{2m_0} \quad \text{(R.1.13)}$$

and

$$E_{l,s} = (\gamma_6+\gamma_7)\frac{\hbar^2k^2}{2m_0} - \frac{\Delta_1}{2} \pm \left[\frac{\Delta_1^2}{4} + \gamma_7\Delta_1\frac{\hbar^2k^2}{2m_0} + 9\left(\frac{\gamma_7\hbar^2k^2}{2m_0}\right)^2\right]^{1/2} \quad \text{(R.1.14)}$$

where,

$$\gamma_6 = 0.53,\ \gamma_7 = 0.07,\ \Delta_1 = 70\,\text{meV}.$$

(j) In the presence of stress, χ_6 along $\langle 000\rangle$ and $\langle 111\rangle$ directions, the energy spectra of the holes in semiconductors having diamond structure valance bands can be respectively expressed following Roman et al. [212] as

$$E = A_6k^2 \pm \left[\bar{B}_7^2k^4 + \delta_6^2 + B_7\delta_6\left(2k_z^2 - k_s^2\right)\right]^{1/2} \quad \text{(R.1.15)}$$

and

$$E = A_6k^2 \pm \left[\bar{B}_7^2k^4 + \delta_7^2 + \frac{D_6}{\sqrt{3}}\delta_7\left(2k_z^2 - k_s^2\right)\right]^{1/2} \quad \text{(R.1.16)}$$

where, A_6, B_7, D_6 and C_6 are inverse mass band parameters in which $\delta_6 \equiv l_7(\bar{S}_{11} - \bar{S}_{12})\chi_6$, $\bar{S}_{ij}$ are the usual elastic compliance constants, $\bar{B}_7^2 \equiv \left(B_7^2 + \frac{c_6^2}{5}\right)$ and $\delta_7 \equiv \left(\frac{d_8 S_{44}}{2\sqrt{3}}\right)\chi_6$. For gray tin, $d_8 = -4.1$ eV, $l_7 = -2.3$ eV, $A_6 = 19.2\frac{\hbar^2}{2m_0}$, $B_7 = 26.3\frac{\hbar^2}{2m_0}$, $D_6 = 31\frac{\hbar^2}{2m_0}$ and $c_6^2 = -1112\frac{\hbar^2}{2m_0}$.

(k) The dispersion relation of the carriers of cadmium and zinc diphosphides are given by [213]

$$E = \left[\beta_1 + \frac{\beta_2\beta_3(k)}{8\beta_4}\right]k^2 \pm \left\{\left[\beta_4\beta_3(k)x\left(\beta_5 - \frac{\beta_2\beta_3(k)}{8\beta_4}\right)k^2\right] + 8\beta_4^2\left(1 - \frac{\beta_3^2(k)}{4}\right) - \beta_2\left(1 - \frac{\beta_3^2(k)}{4}\right)k^2\right\}^{1/2} \quad \text{(R.1.17)}$$

where $\beta_1, \beta_2, \beta_4\, and\, \beta_5$ are system constants and $\beta_3(k) = \frac{k_x^2 + k_y^2 - 2k_z^2}{k^2}$.

(R.1.2) Investigate the EP for bulk specimens of the HD semiconductors in the presences of Gaussian, exponential, Kane, Halperian, Lax and Bonch-Burevich types of band tails [38, 39] for all systems whose unperturbed carrier energy spectra are defined in R.1.1.

(R.1.3) Investigate the EP for QWs of all the HD semiconductors as considered in R.1.2.

(R.1.4) Investigate the EP for HD bulk specimens of the negative refractive index, organic, magnetic and other advanced optical materials in the presence of an arbitrarily oriented alternating electric field.

(R.1.5) Investigate the EP for the QWs of HD negative refractive index, organic, magnetic and other advanced optical materials in the presence of an arbitrarily oriented alternating electric field.

(R.1.6) Investigate the EP for the multiple QWs of HD materials whose unperturbed carrier energy spectra are defined in R.1.1.

(R.1.7) Investigate the EP for all the appropriate HD low dimensional systems of this chapter in the presence of finite potential wells.

(R.1.8) Investigate the EP for all the appropriate HD low dimensional systems of this chapter in the presence of parabolic potential wells.

(R.1.9) Investigate the EP for all the appropriate HD systems of this chapter forming quantum rings.

(R.1.10) Investigate the EP for all the above appropriate problems in the presence of elliptical Hill and quantum square rings.

(R.1.11) Investigate the EP for parabolic cylindrical HD low dimensional systems in the presence of an arbitrarily oriented alternating electric field for all the HD materials whose unperturbed carrier energy spectra are defined in R.1.1.

(R.1.12) Investigate the EP for HD low dimensional systems of the negative refractive index and other advanced optical materials in the presence of an arbitrarily oriented alternating electric field and non-uniform light waves.

(R.1.13) Investigate the EP for triangular HD low dimensional systems of the negative refractive index, organic, magnetic and other advanced optical materials in the presence of an arbitrarily oriented alternating electric field in the presence of strain.

(R.1.14) Investigate the EP in HD quantum wires of non-parabolic semiconductors as discussed in this chapter.

(R.1.15) Investigate the EP for all the problems of (R.1.13) in the presence of arbitrarily oriented magnetic field.

(R.1.16) Investigate the EP for all the problems of (R.1.13) in the presence of alternating electric field.

(R.1.17) Investigate the EP for all the problems of (R.1.13) in the presence of alternating magnetic fieldAlternating magnetic field.

(R.1.18) Investigate the EP for all the problems of (R.1.13) in the presence of crossed electric field and quantizing magnetic fields.

(R.1.19) Investigate the EP for all the problems of (R.1.13) in the presence of crossed alternating electric field and alternating quantizing magnetic fields.

(R.1.20)
(a) Investigate the EP for HD low dimensional systems of the negative refractive index, organic, magnetic and other advanced optical materials in the presence of an arbitrarily oriented alternating electric field considering many body effects
(b) Investigate all the appropriate problems of this chapter for a Dirac electron.

(R.1.21) Investigate all the appropriate problems of this chapter by including the many body, image force, broadening and hot carrier effects respectively.

(R.1.22) Investigate all the appropriate problems of this chapter by removing all the mathematical approximations and establishing the respective appropriate uniqueness conditions.

References

1. R.K. Pathria, *Statistical Mechanics*, 2nd edn. (Butterworth-Heinemann, Oxford, 1996)
2. P.M. Petroff, A.C. Gossard, W. Wiegmann, Appl. Phys. Lett. **45**, 620 (1984)
3. J.M. Gaines, P.M. Petroff, H. Kroemar, R.J. Simes, R.S. Geels, J.H. English, J. Vac. Sci. Technol. B **6**, 1378 (1988)
4. J. Cilbert, P.M. Petroff, G.J. Dolan, S.J. Pearton, A.C. Gossard, J.H. English, Appl. Phys. Lett. **49**, 1275 (1986)
5. T. Fujui, H. Saito, Appl. Phys. Lett. **50**, 824 (1987)
6. H. Sasaki, Jpn. J. Appl. Phys. **19**, 94 (1980)
7. P.M. Petroff, A.C. Gossard, R.A. Logan, W. Weigmann, Appl. Phys. Lett. **41**, 635 (1982)
8. H. Temkin, G.J. Dolan, M.B. Panish, S.N.G. Chu, Appl. Phys. Lett. **50**, 413 (1988)
9. I. Miller, A. Miller, A. Shahar, U. Koren, P.J. Corvini, Appl. Phys. Lett. **54**, 188 (1989)
10. L.L. Chang, H. Esaki, C.A. Chang, L. Esaki, Phys. Rev. Lett. **38**, 1489 (1977)

11. K. Hess, M.S. Shur, J.J. Drunnond, H. Morkoc, IEEE Trans. Electron. Devices **ED-30**, 07 (1983)
12. G. Bastard, *Wave Mechanics Applied to Semiconductor Hetero-structures, Les Editions de Physique* (Halsted, Les Ulis, New York, 1988)
13. M.J. Kelly, *Low Dimensional Semiconductors: Materials, Physics, Technology, Devices* (Oxford University Press, Oxford, 1995)
14. C. Weisbuch, B. Vinter, *Quantum Semiconductor Structures* (Boston Academic Press, Boston, 1991)
15. N.T. Linch, Festkorperprobleme **23**, 27 (1985)
16. D.R. Sciferes, C. Lindstrom, R.D. Burnham, W. Streifer, T.L. Paoli, Electron. Lett. **19**, 170 (1983)
17. P.M. Solomon, Proc. IEEE **70**, 489 (1982)
18. T.E. Schlesinger, T. Kuech, Appl. Phys. Lett. **49**, 519 (1986)
19. D. Kasemet, C.S. Hong, N.B. Patel, P.D. Dapkus, Appl. Phys. Lett. **41**, 912 (1982)
20. K. Woodbridge, P. Blood, E.D. Pletcher, P.J. Hulyer, Appl. Phys. Lett. **45**, 16 (1984)
21. S. Tarucha, H.O. Okamoto, Appl. Phys. Lett. **45**, 16 (1984)
22. H. Heiblum, D.C. Thomas, C.M. Knoedler, M.I. Nathan, Appl. Phys. Lett. **47**, 1105 (1985)
23. O. Aina, M. Mattingly, F.Y. Juan, P.K. Bhattacharyya, Appl. Phys. Lett. **50**, 43 (1987)
24. I. Suemune, L.A. Coldren, IEEE J. Quant. Electron. **24**, 1178 (1988)
25. D.A.B. Miller, D.S. Chemla, T.C. Damen, J.H. Wood, A.C. Burrus, A.C. Gossard, W. Weigmann, IEEE J. Quant. Electron. **21**, 1462 (1985)
26. J.W. Rowe, J.L. Shay, Phys. Rev. B **3**, 451 (1973)
27. H. Kildal, Phys. Rev. B **10**, 5082 (1974)
28. J. Bodnar, in *Proceedings of the International Conference on Physics of Narrow-gap Semiconductors* (Polish Science Publishers, Warsaw, 1978)
29. G.P. Chuiko, N.N. Chuiko, Sov. Phys. Semicond. **15**, 739 (1981)
30. K.P. Ghatak, S.N. Biswas, Proc. SPIE **1484**, 149 (1991)
31. A. Rogalski, J. Alloys Comp. **371**, 53 (2004)
32. A. Baumgartner, A. Chaggar, A. Patanè, L. Eaves, M. Henini, Appl. Phys. Lett. **92**, 091121 (2008)
33. J. Devenson, R. Teissier, O. Cathabard, A.N. Baranov, Proc. SPIE **6909**, 69090U (2008)
34. B.S. Passmore, J. Wu, M.O. Manasreh, G.J. Salamo, Appl. Phys. Lett. **91**, 233508 (2007)
35. M. Mikhailova, N. Stoyanov, I. Andreev, B. Zhurtanov, S. Kizhaev, E. Kunitsyna, K. Salikhov, Y. Yakovlev, Proc. SPIE **6585**, 658526 (2007)
36. W. Kruppa, J.B. Boos, B.R. Bennett, N.A. Papanicolaou, D. Park, R. Bass, Electron. Lett. **42**, 688 (2006)
37. E.O. Kane, in *Semiconductors and Semimetals*, vol. 1, ed. by R.K. Willardson, A.C. Beer, Academic Press, New York, **75** (1966)
38. B.R. Nag, *Electron Transport in Compound Semiconductors* (Springer, Heidelberg, 1980)
39. A.N. Chakravarti, K.P. Ghatak, A. Dhar, K.K. Ghosh, S. Ghosh, Acta Phys. Polon A **60**, 151 (1981)
40. G.E. Stillman, C.M. Wolfe, J.O. Dimmock, in *Semiconductors and Semimetals*, **12**, ed. by R. K. Willardon, A.C. Beer Academic Press, New York, **169** (1977)
41. D.J. Newson, A. Karobe, Semicond. Sci. Tech. **3**, 786 (1988)
42. E.D. Palik, G.S. Picus, S. Teither, R.E. Wallis, Phys. Rev. **475** (1961)
43. P.Y. Lu, C.H. Wung, C.M. Williams, S.N.G. Chu, C.M. Stiles, Appl. Phys. Lett. **49**, 1372 (1986)
44. N.R. Taskar, I.B. Bhat, K.K. Prat, D. Terry, H. Ehasani, S.K. Ghandhi, J. Vac. Sci. Tech. **7A**, 281 (1989)
45. F. Koch, *Springer Series in Solid States Sciences* (Springer, Germany, 1984)
46. L.R. Tomasetta, H.D. Law, R.C. Eden, I. Reyhimy, K. Nakano, IEEE J. Quant. Electron. **14**, 800 (1978)
47. T. Yamato, K. Sakai, S. Akiba, Y. Suematsu, IEEE J. Quantum Electron. **14**, 95 (1978)
48. T.P. Pearsall, B.I. Miller, R.J. Capik, Appl. Phys. Lett. **28**, 499 (1976)

49. M.A. Washington, R.E. Nahory, M.A. Pollack, E.D. Beeke, Appl. Phys. Lett. **33**, 854 (1978)
50. M.I. Timmons, S.M. Bedair, R.J. Markunas, J.A. Hutchby, in *Proceedings of the 16th IEEE Photovoltaic Specialist Conference* (IEEE, San Diego, California 666, 1982)
51. J.A. Zapien, Y.K. Liu, Y.Y. Shan, H. Tang, C.S. Lee, S.T. Lee, Appl. Phys. Lett. **90**, 213114 (2007)
52. M. Park, Proc. SPIE **2524**, 142 (1995)
53. S.-G. Hur, E.T. Kim, J.H. Lee, G.H. Kim, S.G. Yoon, Electrochem. Solid-State Lett. **11**, H176 (2008)
54. H. Kroemer, Rev. Mod. Phys. **73**, 783 (2001)
55. T. Nguyen Duy, J. Meslage, G. Pichard, J. Crys. Growth **72**, 490 (1985)
56. T. Aramoto, F. Adurodija, Y. Nishiyama, T. Arita, A. Hanafusa, K. Omura, A. Morita, Solar Energy Mater. Solar Cells **75**, 211 (2003)
57. H.B. Barber, J. Electron. Mater. **25**, 1232 (1996)
58. S. Taniguchi, T. Hino, S. Itoh, K. Nakano, N. Nakayama, A. Ishibashi, M. Ikeda, Electron. Lett. **32**, 552 (1996)
59. J.J. Hopfield, J. Appl. Phys. **32**, 2277 (1961)
60. G.P. Agrawal, N.K. Dutta, *Semiconductor Lasers* (Van Nostrand Reinhold, New York, 1993)
61. S. Chatterjee, U. Pal, Opt. Eng. (Bellingham) **32**, 2923 (1993)
62. T.K. Chaudhuri, Int. J. Energy Res. **16**, 481 (1992)
63. J.H. Dughaish, Phys. B **322**, 205 (2002)
64. C. Wood, Rep. Prog. Phys. **51**, 459 (1988)
65. K.F. Hsu, S. Loo, F. Guo, W. Chen, J.S. Dyck, C. Uher, T. Hogan, E.K. Polychroniadis, M. G. Kanatzidis, Science **303**, 818 (2004)
66. J. Androulakis, K.F. Hsu, R. Pcionek, H. Kong, C. Uher, J.J. D'Angelo, A. Downey, T. Hogan, M.G. Kanatzidis, Adv. Mater. **18**, 1170 (2006)
67. P.F.P. Poudeu, J. D'Angelo, A.D. Downey, J.L. Short, T.P. Hogan, M.G. Kanatzidis, Angew. Chem. Int. Ed. **45**, 3835 (2006)
68. P.F. Poudeu, J. D'Angelo, H. Kong, A. Downey, J.L. Short, R. Pcionek, T.P. Hogan, C. Uher, M.G. Kanatzidis, J. Am. Chem. Soc. **128**, 14347 (2006)
69. J.R. Sootsman, R.J. Pcionek, H. Kong, C. Uher, M.G. Kanatzidis, Chem. Mater. **18**, 4993 (2006)
70. A.J. Mountvala, G. Abowitz, J. Am. Ceram. Soc. **48**, 651 (1965)
71. E.I. Rogacheva, I.M. Krivulkin, O.N. Nashchekina, AYu. Sipatov, V.A. Volobuev, M.S. Dresselhaus, Appl. Phys. Lett. **78**, 3238 (2001)
72. H.S. Lee, B. Cheong, T.S. Lee, K.S. Lee, W.M. Kim, J.W. Lee, S.H. Cho, J.Y. Huh, Appl. Phys. Lett. **85**, 2782 (2004)
73. K. Kishimoto, M. Tsukamoto, T. Koyanagi, J. Appl. Phys. **92**, 5331 (2002)
74. E.I. Rogacheva, O.N. Nashchekina, S.N. Grigorov, M.A. Us, M.S. Dresselhaus, S.B. Cronin, Nanotechnology **14**, 53 (2003)
75. E.I. Rogacheva, O.N. Nashchekina, A.V. Meriuts, S.G. Lyubchenko, M.S. Dresselhaus, G. Dresselhaus, Appl. Phys. Lett. **86**, 063103 (2005)
76. E.I. Rogacheva, S.N. Grigorov, O.N. Nashchekina, T.V. Tavrina, S.G. Lyubchenko, AYu. Sipatov, V.V. Volobuev, A.G. Fedorov, M.S. Dresselhaus, Thin Solid Films **493**, 41 (2005)
77. X. Qiu, Y. Lou, A.C.S. Samia, A. Devadoss, J.D. Burgess, S. Dayal, C. Burda, Angew. Chem. Int. Ed. **44**, 5855 (2005)
78. C. Wang, G. Zhang, S. Fan, Y. Li, J. Phys. Chem. Solids **62**, 1957 (2001)
79. B. Poudel, W.Z. Wang, D.Z. Wang, J.Y. Huang, Z.F. Ren, J. Nanosci. Nanotechnol. **6**, 1050 (2006)
80. B. Zhang, J. He, T.M. Tritt, Appl. Phys. Lett. **88**, 043119 (2006)
81. W. Heiss, H. Groiss, E. KaQWmann, G. Hesser, M. Böberl, G. Springholz, F. Schäffler, K. Koike, H. Harada, M. Yano, Appl. Phys. Lett. **88**, 192109 (2006)
82. B.A. Akimov, V.A. Bogoyavlenskiy, L.I. Ryabova, V.N. Vasil'kov, Phys. Rev. B **61**, 16045 (2000)

83. Ya. A. Ugai, A.M. Samoilov, M.K. Sharov, O.B. Yatsenko, B.A. Akimov, Inorg. Mater. **38**, 12 (2002)
84. Ya. A. Ugai, A.M. Samoilov, S.A. Buchnev, Yu. V. Synorov, M.K. Sharov, Inorg. Mater. **38**, 450 (2002)
85. A.M. Samoilov, S.A. Buchnev, YuV Synorov, B.L. Agapov, A.M. Khoviv, Inorg. Mater. **39**, 1132 (2003)
86. A.M. Samoilov, S.A. Buchnev, E.A. Dolgopolova, YuV Synorov, A.M. Khoviv, Inorg. Mater. **40**, 349 (2004)
87. H. Murakami, W. Hattori, R. Aoki, Phys. C **269**, 83 (1996)
88. H. Murakami, W. Hattori, Y. Mizomata, R. Aoki, Phys. C **273**, 41 (1996)
89. H. Murakami, R. Aoki, K. Sakai, Thin Solid Films **27**, 343 (1999)
90. B.A. Volkov, L.I. Ryabova, D.R. Khokhlov, Phys. Usp. **45**, 819 (2002). (and references therein)
91. F. Hüe, M. Hÿtch, H. Bender, F. Houdellier, A. Claverie, Phys. Rev. Lett. **100**, 156602 (2008)
92. S. Banerjee, K.A. Shore, C.J. Mitchell, J.L. Sly, M. Missous, IEE Proc. Circuits Devices Syst. **152**, 497 (2005)
93. M. Razeghi, A. Evans, S. Slivken, J.S. Yu, J.G. Zheng, V.P. Dravid, Proc. SPIE **5840**, 54 (2005)
94. R.A. Stradling, Semicond. Sci. Technol. **6**, C52 (1991)
95. P.K. Weimer, Proc. IEEE **52**, 608 (1964)
96. G. Ribakovs, A.A. Gundjian, IEEE J. Quant. Electron. **QE-14**, 42 (1978)
97. S.K. Dey, J. Vac. Sci. Technol. **10**, 227 (1973)
98. S.J. Lynch, Thin Solid Films **102**, 47 (1983)
99. V.V. Kudzin, V.S. Kulakov, D.R. Pape', S.V. Kulakov, V.V. Molotok, IEEE. Ultrason. Symp. **1**, 749 (1997)
100. F. Hatami, V. Lordi, J.S. Harris, H. Kostial, W.T. Masselink, J. Appl. Phys. **97**, 096106 (2005)
101. B.W. Wessels, J. Electrochem. Soc. **722**, 402 (1975)
102. D.W.L. Tolfree, J. Sci. Instrum. **41**, 788 (1964)
103. P.B. Hart, Proc. IEEE **61**, 880 (1973)
104. M.A. Hines, G.D. Scholes, Adv. Mater. **15**, 1844 (2003)
105. C.A. Wang, R.K. Huang, D.A. Shiau, M.K. Connors, P.G. Murphy, P.W. O'Brien, A.C. Anderson, D.M. DePoy, G. Nichols, M.N. Palmisiano, Appl. Phys. Lett. **83**, 1286 (2003)
106. C.W. Hitchcock, R.J. Gutmann, J.M. Borrego, I.B. Bhat, G.W. Charache, IEEE Trans. Electron. Devices **46**, 2154 (1999)
107. H.J. Goldsmid, R.W. Douglas, Br. J. Appl. Phys. **5**, 386 (1954)
108. F.D. Rosi, B. Abeles, R.V. Jensen, J. Phys. Chem. Sol. **10**, 191 (1959)
109. T.M. Tritt (ed.), *Semiconductors and Semimetals*, vol. **69**, **70** and **71**: Recent Trends in Thermoelectric Materials Research I, II and III (Academic Press, New York, 2000)
110. D.M. Rowe (ed.), *CRC Handbook of Thermoelectrics* (CRC Press, Boca Raton, 1995)
111. D.M. Rowe, C.M. Bhandari, *Modern Thermoelectrics* (Reston Publishing Company, Virginia, 1983)
112. D.M. Rowe (ed.), *Thermoelectrics Handbook: Macro to Nano* (CRC Press, Boca Raton, 2006)
113. H. Choi, M. Chang, M. Jo, S.J. Jung, H. Hwang, Electrochem. Solid-State Lett. **11**, H154 (2008)
114. S. Cova, M. Ghioni, A. Lacaita, C. Samori, F. Zappa, Appl. Opt. **35**, 1956 (1996)
115. H.W.H. Lee, B.R. Taylor, S.M. Kauzlarich, *Nonlinear Optics: Materials, Fundamentals, and Applications* **12**, (Technical Digest, 2000)
116. E. Brundermann, U. Heugen, A. Bergner, R. Schiwon, G.W. Schwaab, S. Ebbinghaus, D.R. Chamberlin, E.E. Haller, M. Havenith, in *29th International Conference on Infrared and MillimeterWaves and 12th International Conference on Terahertz, Electronics*, vol **283** (2004)

117. A.N. Baranov, T.I. Voronina, N.S. Zimogorova, L.M. Kauskaya, Y.P. Yakoviev, Sov. Phys. Semicond. **19**, 1676 (1985)
118. M. Yano, Y. Suzuki, T. Ishii, Y. Matsushima, M. Kimata, Jpn. J. Appl. Phys. **17**, 2091 (1978)
119. F.S. Yuang, Y.K. Su, N.Y. Li, Jpn. J. Appl. Phys. **30**, 207 (1991)
120. F.S. Yuang, Y.K. Su, N.Y. Li, K.J. Gan, J. Appl. Phys. **68**, 6383 (1990)
121. Y.K. Su, S.M. Chen, J. Appl. Phys. **73**, 8349 (1993)
122. S.K. Haywood, A.B. Henriques, N.J. Mason, R.J. Nicholas, P.J. Walker, Semicond. Sci. Technol. **3**, 315 (1988)
123. A.N. Chakravarti, K.P. Ghatak, K.K. Ghosh, S. Ghosh, H.M. Mukherjee, Phys. Stat. Sol. B **108**, 609 (1981)
124. K.P. Ghatak, M. Mondal, Z. fur Physik B **64**, 223 (1986)
125. P.K. Chakraborty, A. Sinha, S. Bhattacharya, K.P. Ghatak, Physica B **390**, 325 (2007)
126. P.K. Chakraborty, K.P. Ghatak, J. Phys. Chem. Solids **62**, 1061 (2001)
127. P.K. Chakraborty, K.P. Ghatak, Phys. Letts. A **288**, 335 (2001)
128. P.K. Chakraborty, K.P. Ghatak, Phys. D, Appl. Phys **32**, 1999 (2001)
129. M. Abramowitz, I.A. Stegun, *Handbook of Mathematical Functions with Formulas, Graphs and Mathematical Tables* (Wiley, New York, 1964)
130. I.S. Gradshteyn, I.M. Ryzhik, *Tables of Integrals, Series and Products* (Academic Press, New York, 1965)
131. V. Heine, Proc. Phys. Soc. **81**, 300 (1963)
132. J.N. Schulman, Y.C. Chang, Phys. Rev. B **24**, 4445 (1981)
133. S. Adachi, J. Appl. Phys. **58**, R11 (1985)
134. K.P. Ghatak, J.P. Banerjee, D. Bhattacharya, Nanotechnology **7**, 110 (1996)
135. S. Bhattacharya, S. Choudhury, K.P Ghatak, Superlatt. and Microstruct. **48**, 257 (2010)
136. K.P. Ghatak, S. Bhattacharya, S. Pahari, D. De, S. Ghosh, M. Mitra, Ann. Phys. **17**, 195 (2008)
137. S. Pahari, S. Bhattacharya, K.P Ghatak, J.Comput. Theor. Nanosci. (Invited Paper), **6**, 2088 (2009)
138. M. Krieehbaum, P. Kocevar, H. Pascher, G. Bauer, IEEE QE **24**, 1727 (1988)
139. M.S. Lundstrom, J. Guo, *Nanoscale Transistors, Device Physics, Modeling and Simulation* (Springer, USA, 2006)
140. R. Saito, G. Dresselhaus, M.S. Dresselhaus, *Physical Properties of Carbon Nanotubes* (Imperial College Press, London, 1998)
141. X. Yang, J. Ni, Phys. Rev. B **72**, 195426 (2005)
142. W. Mintmire, C.T. White, Phys. Rev. Letts. **81**, 2506 (1998)
143. G.L. Bir, G.E. Pikus, *Symmetry and Strain—Induced effects in Semiconductors* (Nauka, Russia, 1972)
144. M. Mondal, K.P. Ghatak, Phys. Stat. Sol. (B) **135**, K21 (1986)
145. C.C. Wu, C.J. Lin, J. Low, Temp. Phys. **57**, 469 (1984)
146. Y. Yamada, J. Phys. Soc. Jpn. **35**, 1600 (1973)
147. A.N. Chakravarti, K.P. Ghatak, K.K. Ghosh, S. Ghosh, H.M. Mukherjee, Phys. Stat. Sol. B **118**, 843 (1983)
148. S. Bhattacharya, D. De, S.M. Adhikari, K.P. Ghatak, Superlatt. and Microstruc. **51**, 203 (2012)
149. K.P. Ghatak, M. Mondal, Z.F. Naturforschung **41A**, 821 (1986)
150. A.N. Chakravarti, A.K. Choudhury, K.P. Ghatak, S. Ghosh, A. Dhar, Appl. Phys. **25**, 105 (1981)
151. P.K. Chakraborty, G.C. Datta, K.P. Ghatak, Phys. Scrip. **68**, 368 (2003)
152. A.N. Chakravarti, K.P. Ghatak, A. Dhar, K.K. Ghosh, S. Ghosh, Appl. Phys. **A26**, 165 (1981)
153. K.P. Ghatak, S. Bhattacharya, S.K. Biswas, A. Dey, A.K. Dasgupta, Phys. Scrip. **75**, 820 (2007)
154. K.P. Ghatak, M. Mondal, ZF Physik B. **B69**, 471 (1988)

155. A.N. Chakravarti, K.P. Ghatak, K.K. Ghosh, S. Ghosh, A. Dhar, ZF Physik B. **47**, 149 (1982)
156. H.A. Lyden, Phys. Rev. **135**, A514 (1964)
157. E.D Palik, G.B. Wright, in *Semiconductors and Semimetals*, ed. by R.K Willardson and A.C Beer **3**, (Academic Press, New York, USA,1967), p. 421
158. M. Mondal, K.P.Ghatak, Phys. Letts. **131 A**, 529 (1988)
159. K.P. Ghatak, B. Mitra, Int. J. Electron. **72**, 541 (1992)
160. B. Mitra, A. Ghoshal, K.P. Ghatak, Nouvo Cimento D **12D**, 891 (1990)
161. K.P. Ghatak, S.N. Biswas, Nonlinear Opt Quantum Opt **4**, 347 (1993)
162. K.P.Ghatak, A. Ghoshal. B. Mitra., Nouvo Cimento **14D**, 903 (1992)
163. K.P. Ghatak, A. Ghoshal, B. Mitra, Nouvo Cimento. **13D**, 867 (1991)
164. B. Mitra, K.P. Ghatak, Solid State Electron. **32**, 177 (1989)
165. M. Mondal, N. Chattapadhyay, K.P. Ghatak, J. Low Temp. Phys. **66**, 131 (1987)
166. P.N. Hai, W.M. Chen, I.A. Buyanova, H.P. Xin, C.W. Tu, Appl. Phys. Lett. **77**, 1843 (2000)
167. D.P. DiVincenzo, E.J. Mele, Phys. Rev. B **29**, 1685 (1984)
168. P. Perlin, E. Litwin-Staszewska, B. Suchanek, W. Knap, J. Camassel, T. Suski, R. Piotrzkowski, I. Grzegory, S. Porowski, E. Kaminska, J.C. Chervin, Appl. Phys. Lett. **68**, 1114 (1996)
169. G.E. Smith, Phys. Rev. Lett. **9**, 487 (1962)
170. D. Schneider, D. Rurup, A. Plichta, H.-U. Grubert, A. Schlachetzki, K. Hansen, Z. Phys. B **95**, 281 (1994)
171. F. Masia, G. Pettinari, A. Polimeni, M. Felici, A. Miriametro, M. Capizzi, A. Lindsay, S.B. Healy, E.P. O'Reilly, A. Cristofoli, G. Bais, M. Piccin, S. Rubini, F. Martelli, A. Franciosi, P.J. Klar, K. Volz, W. Stolz, Phys. Rev. B **73**, 073201 (2006)
172. V.K. Arora, H. Jeafarian, Phys. Rev. B. **13**, 4457 (1976)
173. S.E. Ostapov, V.V. Zhikharevich, V.G. Deibuk, Semicond. Phys. Quan. Electron. Optoelectron. **9**, 29 (2006)
174. M.J. Aubin, L.G. Caron, J.P. Jay Gerin, Phys. Rev. B **15**, 3872 (1977)
175. S.L. Sewall, R.R. Cooney, P. Kambhampati, Appl. Phys. Lett. **94**, 243116 (2009)
176. K. Tanaka, N. Kotera; in *20th International Conference on Indium Phosphide and Related Materials*, (Versailles, France, 2008), pp. 1–4
177. M. Singh, P.R. Wallace, S.D. Jog, J. Erushanov, J. Phys. Chem. Solids **45**, 409 (1984)
178. W. Zawadzki, Adv. Phys. **23**, 435 (1974)
179. K.P. Ghatak, M. Mondal, J. Appl. Phys. **69**, 1666 (1991)
180. K.P. Ghatak, B. Mitra, D.K. Basu, B. Nag, Nonlinear Opt. **17**, 171 (1997)
181. B.R. Nag, A.N. Chakravarti, Solid State Electron. **18**, 109 (1975)
182. B.R. Nag, A.N. Chakravarti, Phys. Stal Sol. (A) **22**, K153 (1974)
183. J.O. Dimmock, in *The Physics of Semimetals and Narrowgap Semiconductors*, ed. by D.L. Carter, R.T. Bates (Pergamon Press, Oxford, 1971)
184. D.G. Seiler, B.D. Bajaj, A.E. Stephens, Phys. Rev. B **16**, 2822 (1977)
185. A.V. Germaneko, G.M. Minkov, Phys. Stat. Sol. (B) **184**, 9 (1994)
186. G.L. Bir, G.E. Pikus, *Symmetry and Strain—Induced effects in Semiconductors* (Nauka, Russia, 1972)
187. J. Bouat, J.C. Thuillier, Surf. Sci. **73**, 528 (1978)
188. G.J. Rees, *Physics of Compounds,* in Proceedings of the 13th International Conference ed. By F.G. Fumi (North Holland Company, 1976), p.1166
189. P.R. Entage, Phys. Rev. **138**, A246 (1965)
190. M. Stordeur, W. Kuhnberger, Phys. Stat. Sol. **69**, 377 (1975)
191. D.R. Lovett, *Semimetals and Narrow-Bandgap Semiconductor* (Pion Limited, UK, 1977)
192. H. Kohler, Phys. Stat. Sol. **74**, 591 (1976)
193. M. Cardona, W. Paul, H. Brooks Helv, Acta Phys. **33**, 329 (1960)
194. A.F. Gibson in *Proceeding of International School of Physics*, ENRICO FEPMI, course XIII, ed. By R.A. Smith (Academic Press, New York, 1963), p.171
195. C.C. Wang, N.W. Ressler, Phys. Rev. **2**, 1827 (1970)

196. P.C. Mathur, S. Jain, Phys. Rev. **19**, 1359 (1979)
197. E.L. Ivchenko, G.E. Pikus Sov, Phys. Semicond. **13**, 579 (1979)
198. M.H. Cohen, Phys. Rev. **121**, 387 (1961)
199. J.W. McClure, K.H. Choi, Solid State Comm. **21**, 1015 (1977)
200. E. Bangert, P. Kastner, Phys. Stat. Sol. (B) **61**, 503 (1974)
201. G.M.T. Foley, P.N. Langenberg, Phys. Rev. B **15B**, 4850 (1977)
202. V.I. Ivanov-Omskii, A.S. Mekhtiev, S.A. Rustambekova, E.N. Ukraintsev, Phys. Stat. Sol. (B) **119**, 159 (1983)
203. D.G. Seiler, W.M. Beeker, L.M. Roth, Phys. Rev. **1**, 764 (1970)
204. H.I. Zhang, Phys. Rev. B **1**, 3450 (1970)
205. U. Rossler, Solid State Commun. **49**, 943 (1984)
206. J. Johnson, D.H. Dickey, Phys. Rev. **1**, 2676 (1970)
207. V.G. Agafonov, P.M. Valov, B.S. Ryvkin, I.D. Yarashetskin, Sov. Phys. Semiconduct. **12**, 1182 (1978)
208. L.A. Vassilev, Phys. State Sol. (B) **121**, 203 (1984)
209. N.S. Averkiev, V.M. Asnin, A.A. Bakun, A.M. Danishevskii, E.L. Ivchenko, G.E. Pikus, A. A. Rogachev, Sov. Phys. Semicond. **18**, 379 (1984)
210. R.W. Cunningham, Phys. Rev. **167**, 761 (1968)
211. A.I. Yekimov, A.A. Onushchenko, A.G. Plyukhin, A.L. Efros, J. Expt. Theor. Phys. **88**, 1490 (1985)
212. B.J. Roman, A.W. Ewald, Phys. Rev. **B5**, 3914 (1972)
213. G.P. Chuiko, Sov. Phys. Semiconduct. **19**, 1381 (1985)

Chapter 2
The EP from Nano Wires (NWs) of Heavily Doped (HD) Non-parabolic Semiconductors

2.1 Introduction

It is well-known that in nano wires (NWs), the restriction of the motion of the carriers along two directions may be viewed as carrier confinement by two infinitely deep 1D rectangular potential wells, along any two orthogonal directions leading to quantization of the wave vectors along the said directions, allowing 1D carrier transport [1–4]. With the help of modern fabricational techniques, such one dimensional quantized structures have been experimentally realized and enjoy an enormous range of important applications in the realm of nanoscience in the quantum regime. They have generated much interest in the analysis of nanostructured devices for investigating their electronic, optical and allied properties [5–8]. Examples of such new applications are based on the different transport properties of ballistic charge carriers which include quantum resistors [9–14], resonant tunneling diodes and band filters [15, 16], quantum switches [17], quantum sensors [18–20], quantum logic gates [21, 22], quantum transistors and sub tuners [23–25], heterojunction FETs [26, 27], high-speed digital networks [28–31], high-frequency microwave circuits [32], optical modulators [33], optical switching systems [34–36], and other devices.

In this chapter in Sects. 2.2.1, 2.2.2, 2.2.3, 2.2.4, 2.2.5, 2.2.6, 2.2.7, 2.2.8, 2.2.9, 2.2.10 and 2.2.11 we have investigated the EP from NWs of HD non-linear optical, III-V, II-VI, stressed Kane type, Te, GaP, $PtSb_2$, Bi_2Te_3, Ge and GaAs respectively. The Sect. 2.3 contains the result and discussions pertaining to this chapter. The Sect. 2.4 presents 24 open research problems.

K.P. Ghatak, *Einstein's Photoemission*, Springer Tracts in Modern Physics 262,
DOI 10.1007/978-3-319-11188-9_2

2.2 Theoretical Background

2.2.1 The EP from Nano Wires of HD Nonlinear Optical Semiconductors

The dispersion relation of the 1D electrons in this case can be written following (1.32) as

$$\frac{\hbar^2(\mathrm{n}_z\pi/\mathrm{d}_z)^2}{2m_{\|}^* T_{21}(\mathrm{E},\eta_g)}+\frac{\hbar^2(\mathrm{n}_y\pi/\mathrm{d}_y)^2}{2m_{\|}^* T_{22}(\mathrm{E},\eta_g)}+\frac{\hbar^2 k_x^2}{2m_{\|}^* T_{21}(\mathrm{E},\eta_g)}=1 \tag{2.1}$$

where, $n_z(=1,2,3,\ldots)$, d_z are the size quantum number and the nano-thickness along the z-direction respectively, $n_y(=1,2,3,\ldots)$ and d_y are the size quantum number and the nano-thickness along the y-direction respectively.

The 1D DOS function per sub-band is given by

$$N_{1D}(E)=\frac{2g_v}{\pi}\frac{\partial k_x}{\partial E} \tag{2.2}$$

The velocity of the emitted electrons along the x-direction can be written as

$$v_x(E)=\frac{1}{\hbar}\frac{\partial E}{\partial k_x} \tag{2.3}$$

Therefore the photocurrent is given by

$$I=\frac{\alpha_o e g_v}{2}\sum_{n_y=1}^{n_{y\max}}\sum_{n_z=1}^{n_{z\max}}\int_{\Delta_1}^{\infty}\left(\frac{2}{\pi}\frac{\partial k_x}{\partial E}\right)\left(\frac{1}{\hbar}\frac{\partial E}{\partial k_x}\right)f(E)\,dE \tag{2.4}$$

where,

$$\Delta_1\equiv E'+W-hv. \tag{2.5}$$

Using (2.4), one can write,

$$I=\frac{\alpha_o e g_v k_B T}{\pi\hbar}\text{Real part of}\sum_{n_y=1}^{n_{y\max}}\sum_{n_z=1}^{n_{z\max}}F_0(\eta_{61HD}), \tag{2.6}$$

where

$$\eta_{61HD}\equiv\left[\frac{E_{F1HDNW}-\left(E'_{1HDNW}+W-hv\right)}{k_B T}\right]$$

E_{F1HDNW} in the Fermi energy in this case, E'_{1HDNW} is the complex sub-band energy which can be expressed in this case as

$$\frac{\hbar^2(\mathrm{n}_z\pi/\mathrm{d}_z)^2}{2m_{\|}^*T_{21}(\mathrm{E}'_{1HDNW},\eta_g)}+\frac{\hbar^2(\mathrm{n}_y\pi/\mathrm{d}_y)^2}{2m_{\|}^*T_{22}(\mathrm{E}'_{1HDNW},\eta_g)}=1 \tag{2.7}$$

The EEM in this case in given by

$$m^*(\mathrm{E}_{F1HDNW},\mathrm{n}_y,\mathrm{n}_z,\eta_g)=\frac{\hbar^2}{2}[\text{Real part of}\frac{\partial}{\partial(\mathrm{E}_{F1HDNW})}[T_{1HDNW}(\mathrm{E},\mathrm{n}_y,\mathrm{n}_z,\eta_g)]^2] \tag{2.8}$$

where

$$T_{1HDNW}(\mathrm{E},\mathrm{n}_y,\mathrm{n}_z,\eta_g)=[[1-\frac{\hbar^2(\mathrm{n}_z\pi/\mathrm{d}_z)^2}{2m_{\|}^*T_{21}(\mathrm{E},\eta_g)}-\frac{\hbar^2(\mathrm{n}_y\pi/\mathrm{d}_y)^2}{2m_{\|}^*T_{22}(\mathrm{E},\eta_g)}]\frac{2m_{\|}^*T_{21}(\mathrm{E},\eta_g)}{\hbar^2}]^{1/2} \tag{2.9}$$

Thus, we observe that the EEM is the function of size quantum numbers in both the directions and the Fermi energy due to the combined influence of the crystal filed splitting constant and the anisotropic spin-orbit splitting constants respectively. Besides it is a function of η_g *due to which the EEM exists in the band gap, which is otherwise impossible.*

Thus, it appears that the evaluation of J_{1D} requires an expression of carrier statistics which can, in turn, be written as

$$n_{1D}=\left(\frac{2g_v}{\pi}\right)\text{Real part of}\sum_{n_y=1}^{n_{y\max}}\sum_{n_z=1}^{n_{z\max}}[T_{1HDNW}(\mathrm{E}_{F1HDNW},\mathrm{n}_y,\mathrm{n}_z,\eta_g)+T_{2HDNW}(\mathrm{E}_{F1HDNW},\mathrm{n}_y,\mathrm{n}_z,\eta_g)] \tag{2.10}$$

where $T_{2HDNW}(\mathrm{E}_{F1HDNW},\mathrm{n}_y,\mathrm{n}_z,\eta_g)=\sum_{r=1}^{s}L(r)[T_{1HDNW}(\mathrm{E}_{F1HDNW},\mathrm{n}_y,\mathrm{n}_z,\eta_g)]$,

In the absence of band-tails, for electron motion along x-direction only, the 1D electron dispersion law in this case can be written following (1.2) as

$$\gamma(E)=f_1(E)k_x^2+f_1(E)(\pi n_y/d_y)^2+f_2(E)(\pi n_z/d_z)^2 \tag{2.11}$$

The sub-band energy (E'_1) are given by the equation

$$\gamma(E'_1)=f_1(E'_1)(\pi n_y/d_y)^2+f_2(E'_1)(\pi n_z/d_z)^2 \tag{2.12}$$

The EP in this case is given by

$$I=\frac{\alpha_o e g_v k_B T}{\pi\hbar}\sum_{n_y=1}^{n_{y\max}}\sum_{n_z=1}^{n_{z\max}}F_0(\eta_{62})$$

where

$$\eta_{62} \equiv \left[\frac{E_{F1d} - \left(E_1' + W - hv\right)}{k_B T}\right], \tag{2.13}$$

and E_{F1d} is the Fermi energy in this case

The electron concentration per unit length can be written as

$$n_{1D} = \left(\frac{2g_v}{\pi}\right) \sum_{n_y=1}^{n_{y\max}} \sum_{n_z=1}^{n_{z\max}} \left[t_1(\mathrm{E}_{F1d}, \mathrm{n_y}, \mathrm{n_z}) + t_2(\mathrm{E}_{F1d}, \mathrm{n_y}, \mathrm{n_z})\right] \tag{2.14}$$

where

$$t_1\left(E_{F1d}, n_y, n_z\right) \equiv \left[\gamma(E_{F1d}) - f_1(E_{F1d})\left(\pi n_y/d_y\right)^2 - f_2(E_{F1d})(\pi n_z/d_z)^2\right]^{1/2} [f_1(E_{F1d})]^{-1/2}$$

and

$$t_2\left(E_{F1d}, n_y, n_z\right) \equiv \sum_{r=1}^{s} L(r)\left[t_1\left(E_{F1d}, n_y, n_z\right)\right].$$

2.2.2 The EP from Nano Wires of HD III-V Semiconductors

(i) Three band model of Kane

The dispersion relation of the 1D electrons in this case can be written following (1.55) as

$$\frac{\hbar^2(\mathrm{n_z}\pi/\mathrm{d_z})^2}{2m_c} + \frac{\hbar^2(\mathrm{n_y}\pi/\mathrm{d_y})^2}{2m_c} + \frac{\hbar^2 k_x^2}{2m_c} = T_{31}(\mathrm{E}, \eta_g) + iT_{31}(\mathrm{E}, \eta_g) \tag{2.15}$$

The EP in this case is given by

$$I = \frac{\alpha_o e g_v k_B T}{\pi\hbar} \text{Real part of} \sum_{n_y=1}^{n_{y\max}} \sum_{n_z=1}^{n_{z\max}} F_0(\eta_{63HD}), \tag{2.16}$$

where

$$\eta_{63HD} \equiv \left[\frac{E_{F1HDNW} - \left(E_{2HDNW}' + W - hv\right)}{k_B T}\right],$$

and E_{2HDNW}' is the sub-band energy in this case which can be expressed as

$$\frac{\hbar^2(n_z\pi/d_z)^2}{2m_c}+\frac{\hbar^2(n_y\pi/d_y)^2}{2m_c}=T_{31}(E'_{2HDNW},\eta_g)+iT_{31}(E'_{2HDNW},\eta_g) \tag{2.17}$$

The EEM in this case is given by

$$m^*(E_{F1HDNW},\eta_g)=m_c[T'_{31}(E_{F1HDNW},\eta_g)] \tag{2.18}$$

Thus, it appears that the evaluation of J_{1D} requires an expression of carrier statistics which can, in turn, be written as

$$n_{1D}=\left(\frac{2g_v}{\pi}\right)\sum_{n_y=1}^{n_{y\max}}\sum_{n_z=1}^{n_{z\max}}[T_{3HDNW}(E_{F1HDNW},n_y,n_z,\eta_g)+T_{4HDNW}(E_{F1HDNW},n_y,n_z,\eta_g)] \tag{2.19}$$

where

$$T_{3HDNW}(E_{F1HDNW},n_y,n_z,\eta_g)=\left[[T_{31}(E_{F1HDNW},\eta_g)+iT_{31}(E_{F1HDNW},\eta_g)\right.$$
$$\left.-\frac{\hbar^2(n_z\pi/d_z)^2}{2m_c}-\frac{\hbar^2(n_y\pi/d_y)^2}{2m_c}]\frac{2m_c}{\hbar^2}\right]^{1/2}$$

where $T_{4HDNW}(E_{F1HDNW},n_y,n_z,\eta_g)=\sum_{r=1}^{s}L(r)[T_{3HDNW}(E_{F1HDNW},n_y,n_z,\eta_g)]$

The one dimensional electron dispersion law is given by

$$\frac{\hbar^2k_x^2}{2m_c}+G_2(n_y,n_z)=I_{11}(E) \tag{2.20}$$

where,

$$G_2(n_y,n_z)\equiv(\hbar^2\pi^2/2m_c)\left[(n_y/d_y)^2+(n_z/d_z)^2\right]$$

The sub band energy E'_2 can be written as

$$G_2(n_y,n_z)=I_{11}(E'_2) \tag{2.21}$$

The EP in this case is given by

$$I=\frac{\alpha_o e g_v k_B T}{\pi\hbar}\sum_{n_y=1}^{n_{y\max}}\sum_{n_z=1}^{n_{z\max}}F_0(\eta_{64}) \tag{2.22}$$

where

$$\eta_{64}\equiv\left[\frac{E_{F1d}-(E'_2+W-h\nu)}{k_BT}\right],$$

The electron statistics in this case can be written as

$$n_{1D} = \frac{2g_v\sqrt{2m_c}}{\pi\hbar}\sum_{n_y=1}^{n_{ymax}}\sum_{n_z=1}^{n_{zmax}}\left[t_3(E_{F1d},n_y,n_z) + t_4(E_{F1d},n_y,n_z)\right] \tag{2.23}$$

where

$$t_3\left(E_{F1d},n_y,n_z\right) \equiv \left[I_{11}(E_{F1d}) - G_2\left(n_y,n_z\right)\right]^{1/2},$$
$$t_4\left(E_{F1d},n_y,n_z\right) \equiv \sum_{r=1}^{S_0} L(r)\left[t_3\left(E_{F1d},n_y,n_z\right)\right].$$

(ii) Two band model of Kane

The dispersion relation of the 1D electrons in this case can be written as

$$\frac{\hbar^2(n_z\pi/d_z)^2}{2m_c} + \frac{\hbar^2(n_y\pi/d_y)^2}{2m_c} + \frac{\hbar^2k_x^2}{2m_c} = \gamma_2(E,\eta_g) \tag{2.24}$$

The EP in this case is given by

$$I = \frac{\alpha_o e g_v k_B T}{\pi\hbar}\sum_{n_y=1}^{n_{ymax}}\sum_{n_z=1}^{n_{zmax}} F_0(\eta_{64HD}), \tag{2.25}$$

where

$$\eta_{64HD} \equiv \left[\frac{E_{F1HDNW} - \left(E'_{3HDNW} + W - h\nu\right)}{k_BT}\right]$$

and E'_{3HDNW} is the sub-band energy in this case which can be expressed as

$$\frac{\hbar^2(n_z\pi/d_z)^2}{2m_c} + \frac{\hbar^2(n_y\pi/d_y)^2}{2m_c} = \gamma_2(E'_{3HDNW},\eta_g) \tag{2.26}$$

The EEM in this case is given by

$$m^*(E_{F1HDNW},\eta_g) = m_c[\gamma'_2(E_{F1HDNW},\eta_g)] \tag{2.27a}$$

Thus, it appears that the evaluation of J_{1D} requires an expression of carrier statistics which can, in turn, be written as

$$n_{1D} = \left(\frac{2g_v}{\pi}\right)\sum_{n_y=1}^{n_{ymax}}\sum_{n_z=1}^{n_{zmax}}\left[T_{7HDNW}(E_{F1HDNW},n_y,n_z,\eta_g) + T_{8HDNW}(E_{F1HDNW},n_y,n_z,\eta_g)\right] \tag{2.27b}$$

where

$$T_{7HDNW}(E_{F1HDNW}, n_y, n_z, \eta_g) = \left[\left[\gamma_2(E_{F1HDNW}, \eta_g) - \frac{\hbar^2 (n_z \pi / d_z)^2}{2m_c} - \frac{\hbar^2 (n_y \pi / d_y)^2}{2m_c}\right] \frac{2m_c}{\hbar^2}\right]^{1/2}$$

and

$$T_{8HDNW}(E_{F1HDNW}, n_y, n_z, \eta_g) = \sum\nolimits_{r=1}^{s} L(r)[T_{7HDNW}(E_{F1HDNW}, n_y, n_z, \eta_g)].$$

The expression of 1D dispersion relation, for NWs of III-V materials whose energy band structures are defined by the two-band model of Kane in the absence of band tailing assumes the form

$$E(1 + \alpha E) = \frac{\hbar^2 k_x^2}{2m_c} + G_2(n_y, n_z) \tag{2.28}$$

In this case, the quantized energy E_3' is given by

$$E_3' = (2\alpha)^{-1}\left[-1 + \sqrt{1 + 4\alpha G_2(n_y, n_z)}\right] \tag{2.29}$$

The EP in this case is given by

$$I = \frac{\alpha_o e g_v k_B T}{\pi \hbar} \sum_{n_y=1}^{n_{y\max}} \sum_{n_z=1}^{n_{z\max}} F_0(\eta_{65}), \tag{2.30}$$

where

$$\eta_{65} \equiv \left[\frac{E_{F1d} - (E_3' + W - hv)}{k_B T}\right],$$

The carrier statistics in the case can be expressed as

$$n_{1D} = \frac{2g_v}{\pi} \frac{\sqrt{2m_c}}{\hbar} \sum_{n_y=1}^{n_{y\max}} \sum_{n_z=1}^{n_{z\max}} [t_5(E_{F1d}, n_y, n_z) + t_6(E_{F1d}, n_y, n_z)] \tag{2.31}$$

where

$$t_5(E_{F1d}, n_y, n_z) \equiv [E_{F1d}(1 + \alpha E_{F1d}) - G_2(n_y, n_z)]^{1/2},$$

$$t_6(E_{F1d}, n_y, n_z) \equiv \sum_{r=1}^{s} L(r)[t_5(E_{F1d}, n_y, n_z)].$$

(iii) Parabolic energy bands

The dispersion relation of the 1D electrons in this case can be written as

$$\frac{\hbar^2(\mathrm{n}_z\pi/\mathrm{d}_z)^2}{2m_c}+\frac{\hbar^2(\mathrm{n}_y\pi/\mathrm{d}_y)^2}{2m_c}+\frac{\hbar^2k_x^2}{2m_c}=\gamma_3(\mathrm{E},\eta_g) \tag{2.32}$$

The EP in this case is given by

$$I=\frac{\alpha_o e g_v k_B T}{\pi\hbar}\sum_{n_y=1}^{n_{y\max}}\sum_{n_z=1}^{n_{z\max}}F_0(\eta_{66HD}), \tag{2.33}$$

where

$$\eta_{66HD}\equiv\left[\frac{E_{F1HDNW}-\left(E'_{5HDNW}+W-hv\right)}{k_BT}\right]$$

and E'_{5HDNW} is the sub-band energy in this case which can be expressed as

$$\frac{\hbar^2(\mathrm{n}_z\pi/\mathrm{d}_z)^2}{2m_c}+\frac{\hbar^2(\mathrm{n}_y\pi/\mathrm{d}_y)^2}{2m_c}=\gamma_3(E'_{5HDNW},\eta_g) \tag{2.34}$$

The EEM in this case is given by

$$m^*(\mathrm{E}_{F1HDNW},\eta_g)=\mathrm{m}_c[\gamma_3'(\mathrm{E}_{F1HDNW},\eta_g)] \tag{2.35a}$$

Thus, it appears that the evaluation of J_{1D} requires an expression of carrier statistics which can, in turn, be written as

$$n_{1D}=\left(\frac{2g_v}{\pi}\right)\sum_{n_y=1}^{n_{y\max}}\sum_{n_z=1}^{n_{z\max}}\left[T_{9HDNW}(\mathrm{E}_{F1HDNW},\mathrm{n}_y,\mathrm{n}_z,\eta_g)+T_{10HDNW}(\mathrm{E}_{F1HDNW},\mathrm{n}_y,\mathrm{n}_z,\eta_g)\right] \tag{2.35b}$$

where

$$T_{9HDNW}(\mathrm{E}_{F1HDNW},\mathrm{n}_y,\mathrm{n}_z,\eta_g)=[[\gamma_3(\mathrm{E}_{F1HDNW},\eta_g)-\frac{\hbar^2(\mathrm{n}_z\pi/\mathrm{d}_z)^2}{2m_c}-\frac{\hbar^2(\mathrm{n}_y\pi/\mathrm{d}_y)^2}{2m_c}]\frac{2m_c}{\hbar^2}]^{1/2}$$

where $\mathrm{T}_{10HDNW}(\mathrm{E}_{F1HDNW},\mathrm{n}_y,\mathrm{n}_z,\eta_g)=\sum_{r=1}^{s}L(r)[\mathrm{T}_{9HDNW}(\mathrm{E}_{F1HDNW},\mathrm{n}_y,\mathrm{n}_z,\eta_g)]$,

The expression of 1D dispersion relation, for NWs of III-V materials whose energy band structures are defined by the two-band model of Kane in the absence of band tailing assumes the form

$$E = \frac{\hbar^2 k_x^2}{2m_c} + G_2\left(n_y, n_z\right) \tag{2.36}$$

In this case, the quantized energy E_7' is given by

$$E_7' = G_2\left(n_y, n_z\right) \tag{2.37}$$

The EP in this case is given by

$$I = \frac{\alpha_o e g_v k_B T}{\pi \hbar} \sum_{n_y=1}^{n_{y\max}} \sum_{n_z=1}^{n_{z\max}} F_0(\eta_{67}), \tag{2.38}$$

where

$$\eta_{67} \equiv \left[\frac{E_{F1d} - \left(E_7' + W - hv\right)}{k_B T}\right]$$

The carrier statistics in the case can be expressed as

$$n_{1D} = \frac{2g_v\sqrt{2m_c \pi k_B T}}{h} \sum_{n_y=1}^{n_{y\max}} \sum_{n_z=1}^{n_{z\max}} F_{\frac{-1}{2}}(\bar{\eta}_{67}), \tag{2.39}$$

where

$$\bar{\eta}_{67} = \left[E_{F1d} - E_7'\right]\left(k_B T\right)^{-1}$$

Converting the summation over quantum numbers to the corresponding integrations in (2.38), the photocurrent density from semiconductors having isotropic parabolic energy bands with non-degenerate electron concentration gets transformed into the well known form as given in the preface. Besides, (2.39) is well-known in the literature.

(iv) The Model of Stillman et al.

The dispersion relation of the 1D electrons in this case can be written as

$$\frac{\hbar^2 (\mathrm{n}_z \pi/\mathrm{d}_z)^2}{2m_c} + \frac{\hbar^2 (\mathrm{n}_y \pi/\mathrm{d}_y)^2}{2m_c} + \frac{\hbar^2 k_x^2}{2m_c} = \theta_4(\mathrm{E}, \eta_g) \tag{2.40}$$

where

$$\theta_4(\mathrm{E}, \eta_g) = \mathrm{I}_{12}(\mathrm{E}, \eta_g)$$

The EP in this case is given by

$$I = \frac{\alpha_o e g_v k_B T}{\pi \hbar} \sum_{n_y=1}^{n_{y\max}} \sum_{n_z=1}^{n_{z\max}} F_0(\eta_{69HD}), \tag{2.41}$$

where

$$\eta_{69HD} \equiv \left[\frac{E_{F1HDNW} - \left(E'_{9HDNW} + W - hv\right)}{k_B T}\right]$$

and E'_{9HDNW} is the sub-band energy in this case which can be expressed as

$$\frac{\hbar^2(\mathrm{n}_z\pi/\mathrm{d}_z)^2}{2m_c} + \frac{\hbar^2(\mathrm{n}_y\pi/\mathrm{d}_y)^2}{2m_c} = \theta_4(E'_{9HDNW}, \eta_g) \tag{2.42}$$

The EEM in this case is given by

$$m^*(\mathrm{E}_{F1HDNW}, \eta_g) = \mathrm{m}_c[\theta'_4(\mathrm{E}_{F1HDNW}, \eta_g)] \tag{2.43a}$$

Thus, it appears that the evaluation of J_{1D} requires an expression of carrier statistics which can, in turn, be written as

$$n_{1D} = \left(\frac{2g_v}{\pi}\right)\sum_{n_y=1}^{n_{y\max}}\sum_{n_z=1}^{n_{z\max}}\left[T_{11HDNW}(\mathrm{E}_{F1HDNW}, \mathrm{n}_y, \mathrm{n}_z, \eta_g) + T_{12HDNW}(\mathrm{E}_{F1HDNW}, \mathrm{n}_y, \mathrm{n}_z, \eta_g)\right] \tag{2.43b}$$

where

$$T_{11HDNW}(\mathrm{E}_{F1HDNW}, \mathrm{n}_y, \mathrm{n}_z, \eta_g) = \left[\left[\theta_4(\mathrm{E}_{F1HDNW}, \eta_g) - \frac{\hbar^2(\mathrm{n}_z\pi/\mathrm{d}_z)^2}{2m_c} - \frac{\hbar^2(\mathrm{n}_y\pi/\mathrm{d}_y)^2}{2m_c}\right]\frac{2m_c}{\hbar^2}\right]^{1/2}$$

where $\mathrm{T}_{12HDNW}(\mathrm{E}_{F1HDNW}, \mathrm{n}_y, \mathrm{n}_z, \eta_g) = \sum_{r=1}^{s} L(r)[\mathrm{T}_{11HDNW}(\mathrm{E}_{F1HDNW}, \mathrm{n}_y, \mathrm{n}_z, \eta_g)]$,

The expression of 1D dispersion relation, for NWs of III-V materials whose energy band structures are defined by the model of Stillman et al. in the absence of band tailing assumes the form

$$I_{12}(E) = \frac{\hbar^2 k_x^2}{2m_c} + G_2(n_y, n_z) \tag{2.44}$$

In this case, the quantized energy E'_9 is given by

$$I_{12}(E'_9) = G_2(n_y, n_z) \tag{2.45}$$

The EP in this case is given by

$$I = \frac{\alpha_o e g_v k_B T}{\pi\hbar}\sum_{n_y=1}^{n_{y\max}}\sum_{n_z=1}^{n_{z\max}} F_0(\eta_{69}),$$

where

$$\eta_{69} \equiv \left[\frac{E_{F1d} - \left(E'_9 + W - h\nu\right)}{k_B T} \right] \quad (2.46)$$

The carrier statistics in the case can be expressed as

$$n_{1D} = \frac{2g_v}{\pi} \frac{\sqrt{2m_c}}{\hbar} \sum_{n_y=1}^{n_{y\max}} \sum_{n_z=1}^{n_{z\max}} \left[P_9(E_{F1d}, n_y, n_z) + Q_9(E_{F1d}, n_y, n_z) \right] \quad (2.47)$$

where

$$P_9\left(E_{F1d}, n_y, n_z\right) \equiv \left[I_{12}(E_{F1d}) - G_2\left(n_y, n_z\right) \right]^{1/2}$$

$$\text{and } Q_9\left(E_{F1d}, n_y, n_z\right) \equiv \sum_{r=1}^{s} L(r) \left[P_9\left(E_{F1d}, n_y, n_z\right) \right].$$

(v) The Model of Palik et al.

The dispersion relation of the 1D electrons in this case can be written as

$$\frac{\hbar^2 (\mathrm{n}_z \pi / \mathrm{d}_z)^2}{2m_c} + \frac{\hbar^2 (\mathrm{n}_y \pi / \mathrm{d}_y)^2}{2m_c} + \frac{\hbar^2 k_x^2}{2m_c} = \theta_5(\mathrm{E}, \eta_g) \quad (2.48)$$

where

$$\theta_5(\mathrm{E}, \eta_g) = I_{13}(\mathrm{E}, \eta_g)$$

The EP in this case is given by

$$I = \frac{\alpha_o e g_v k_B T}{\pi \hbar} \sum_{n_y=1}^{n_{y\max}} \sum_{n_z=1}^{n_{z\max}} F_0(\eta_{610HD}), \quad (2.49)$$

where

$$\eta_{610HD} \equiv \left[\frac{E_{F1HDNW} - \left(E'_{10HDNW} + W - h\nu\right)}{k_B T} \right],$$

and E'_{10HDNW} is the sub-band energy in this case which can be expressed as

$$\frac{\hbar^2 (\mathrm{n}_z \pi / \mathrm{d}_z)^2}{2m_c} + \frac{\hbar^2 (\mathrm{n}_y \pi / \mathrm{d}_y)^2}{2m_c} = \theta_5(E'_{10HDNW}, \eta_g) \quad (2.50)$$

The EEM in this case is given by

$$m^{*}(\mathrm{E}_{F1HDNW}, \eta_g) = \mathrm{m}_c[\theta_5'(\mathrm{E}_{F1HDNW}, \eta_g)] \tag{2.51a}$$

Thus, it appears that the evaluation of J_{1D} requires an expression of carrier statistics which can, in turn, be written as

$$n_{1D} = \left(\frac{2g_v}{\pi}\right) \sum_{n_y=1}^{n_{y\max}} \sum_{n_z=1}^{n_{z\max}} \left[T_{13HDNW}(\mathrm{E}_{F1HDNW}, \mathrm{n}_y, \mathrm{n}_z, \eta_g) + T_{14HDNW}(\mathrm{E}_{F1HDNW}, \mathrm{n}_y, \mathrm{n}_z, \eta_g)\right] \tag{2.51b}$$

where

$$T_{13HDNW}(\mathrm{E}_{F1HDNW}, \mathrm{n}_y, \mathrm{n}_z, \eta_g) = \left[\left[\theta_5(\mathrm{E}_{F1HDNW}, \eta_g) - \frac{\hbar^2(\mathrm{n}_z\pi/\mathrm{d}_z)^2}{2m_c} - \frac{\hbar^2(\mathrm{n}_y\pi/\mathrm{d}_y)^2}{2m_c}\right]\frac{2m_c}{\hbar^2}\right]^{1/2}$$

where $\mathrm{T}_{14HDNW}(\mathrm{E}_{F1HDNW}, \mathrm{n}_y, \mathrm{n}_z, \eta_g) = \sum_{r=1}^{s} L(r)[\mathrm{T}_{13HDNW}(\mathrm{E}_{F1HDNW}, \mathrm{n}_y, \mathrm{n}_z, \eta_g)]$,

The expression of 1D dispersion relation, for NWs of III-V materials whose energy band structures are defined by the model of Palik et al. in the absence of band tailing assumes the form

$$I_{13}(E) = \frac{\hbar^2 k_x^2}{2m_c} + G_2(n_y, n_z) \tag{2.52}$$

In this case, the quantized energy E'_{10} is given by

$$I_{13}(E'_{10}) = G_2(n_y, n_z) \tag{2.53}$$

The EP in this case is given by

$$I = \frac{\alpha_o e g_v k_B T}{\pi\hbar} \sum_{n_y=1}^{n_{y\max}} \sum_{n_z=1}^{n_{z\max}} F_0(\eta_{610}), \tag{2.54}$$

where

$$\eta_{610} \equiv \left[\frac{E_{F1d} - (E'_{10} + W - hv)}{k_B T}\right],$$

The carrier statistics in the case can be expressed as

$$n_{1D} = \frac{2g_v}{\pi}\frac{\sqrt{2m_c}}{\hbar}\sum_{n_y=1}^{n_{y\max}}\sum_{n_z=1}^{n_{z\max}}\left[P_{11}(E_{F1d}, n_y, n_z) + Q_{12}(E_{F1d}, n_y, n_z)\right] \tag{2.55}$$

where

$$P_{11}\left(E_{F1d}, n_y, n_z\right) \equiv \left[I_{13}(E_{F1d}) - G_2\left(n_y, n_z\right)\right]^{1/2}$$

$$\text{and}\, Q_{12}\left(E_{F1d}, n_y, n_z\right) \equiv \sum_{r=1}^{s} L(r)\left[P_{11}\left(E_{F1d}, n_y, n_z\right)\right].$$

2.2.3 The EP from Nano Wires of HD II-VI Semiconductors

The 1D electron dispersion law in NW of HD II-VI semiconductors can be written following (1.141) as

$$\gamma_3(E, \eta_g) = a_0'[(\frac{n_x\pi}{d_x})^2 + (\frac{n_y\pi}{d_y})^2] \pm \bar{\lambda}_0[(\frac{n_x\pi}{d_x})^2 + (\frac{n_y\pi}{d_y})^2]^{1/2} + \frac{\hbar^2 k_z^2}{2m_\parallel^*} \tag{2.56}$$

The EP in this case is given by

$$I = \frac{\alpha_o e g_v k_B T}{2\pi\hbar}\sum_{n_y=1}^{n_{y\max}}\sum_{n_z=1}^{n_{z\max}} F_0(\eta_{613HD}), \tag{2.57}$$

where

$$\eta_{613HD} \equiv \left[\frac{E_{F1HDNW} - \left(E'_{13HDNW} + W - h\nu\right)}{k_B T}\right]$$

and E'_{13HDNW} is the sub-band energy in this case which can be expressed as

$$\gamma_3(E'_{13HDNW}, \eta_g) = a_0'[(\frac{n_x\pi}{d_x})^2 + (\frac{n_y\pi}{d_y})^2] \pm \bar{\lambda}_0[(\frac{n_x\pi}{d_x})^2 + (\frac{n_y\pi}{d_y})^2]^{1/2} \tag{2.58}$$

The EEM in this case is given by

$$m^*(E_{F1HDNW}, \eta_g) = m_\parallel^* \gamma_3'(E_{F1HDNW}, \eta_g) \tag{2.59}$$

Thus, it appears that the evaluation of J_{1D} requires an expression of carrier statistics which can, in turn, be written as

$$n_{1D} = \left(\frac{g_v}{\pi}\right) \sum_{n_x=1}^{n_{x\max}} \sum_{n_y=1}^{n_{y\max}} \left[T_{17HDNW}(\mathrm{E}_{F1HDNW}, \mathrm{n}_x, \mathrm{n}_y, \eta_g) + T_{18HDNW}(\mathrm{E}_{F1HDNW}, \mathrm{n}_x, \mathrm{n}_y, \eta_g)\right] \tag{2.60}$$

where

$$T_{17HDNW}(\mathrm{E}_{F1HDNW}, \mathrm{n}_x, \mathrm{n}_y, \eta_g) = [[\gamma_3(E_{F1HDNW}, \eta_g) - a_0'[(\frac{n_x\pi}{d_x})^2 + (\frac{n_y\pi}{d_y})^2] \mp \bar{\lambda}_0[(\frac{n_x\pi}{d_x})^2 + (\frac{n_y\pi}{d_y})^2]^{1/2}](\frac{2m_{\|}^*}{\hbar^2})]^{1/2}$$

and

$$\mathrm{T}_{18HDNW}(\mathrm{E}_{F1HDNW}, \mathrm{n}_x, \mathrm{n}_y, \eta_g) = \sum_{r=1}^{s} L(r)[\mathrm{T}_{17HDNW}(\mathrm{E}_{F1HDNW}, \mathrm{n}_x, \mathrm{n}_y, \eta_g)],$$

The 1D dispersion relation for NWs of II-VI materials in the absence of band-tails can be written as

$$E = b_0' k_z^2 + G_{3,\pm}(n_x, n_y) \tag{2.61}$$

where

$$G_{3,\pm}(n_x, n_y) \equiv \left[a_0'\left\{\left(\frac{\pi n_x}{d_x}\right)^2 + \left(\frac{\pi n_y}{d_y}\right)^2\right\} \pm \bar{\lambda}_0\left\{\left(\frac{\pi n_x}{d_x}\right)^2 + \left(\frac{\pi n_y}{d_y}\right)^2\right\}^{1/2}\right]$$

The EP photocurrent from NWs of II-VI materials is given by

$$I = \frac{\alpha_0 e g_v k_B T}{2\pi\hbar} \sum_{n_x=1}^{n_{x\max}} \sum_{n_y=1}^{n_{y\max}} \left\{F_0\left\{(k_BT)^{-1}\left[E_{F1d} - \left[G_{3,+}(n_x, n_y) + W - hv\right]\right]\right\} + F_0\left\{(k_BT)^{-1}\left[E_{F1d} - \left[G_{3,-}(n_x, n_y) + W - hv\right]\right]\right\}\right\} \tag{2.62}$$

The 1D electron statistics can be written as

$$n_{1D} = \frac{g_v\sqrt{2m_{\|}^*\pi k_BT}}{h} \sum_{n_y=1}^{n_{y\max}} \sum_{n_z=1}^{n_{z\max}} F_{\frac{-1}{2}}(\eta_{68,\pm}), \eta_{68,\pm} = (k_BT)^{-1}\left[E_{F1d} - \left[G_{3,\pm}(n_x, n_y)\right]\right]. \tag{2.63}$$

2.2.4 The EP from Nano Wires of HD IV-VI Semiconductors

(i) Dimmock Model

The 1D electron dispersion law in NW of HD IV-VI semiconductors can be expressed following (1.174) as

$$\begin{aligned}
&\gamma_2(E,\eta_g)+\alpha\gamma_3(E,\eta_g)\left(\frac{\hbar^2}{2x_4}(\frac{n_x\pi}{d_x})^2+\frac{\hbar^2}{2x_5}(\frac{n_y\pi}{d_y})^2\right)+\alpha\gamma_3(E,\eta_g)\frac{\hbar^2}{2x_6}k_z^2\\
&\quad-\left(1+\alpha\gamma_3(E,\eta_g)\right)\left(\frac{\hbar^2}{2x_1}(\frac{n_x\pi}{d_x})^2+\frac{\hbar^2}{2x_2}(\frac{n_y\pi}{d_y})^2\right)\\
&\quad-\alpha\left(\frac{\hbar^2}{2x_1}(\frac{n_x\pi}{d_x})^2+\frac{\hbar^2}{2x_2}(\frac{n_y\pi}{d_y})^2\right)\left(\frac{\hbar^2}{2x_4}(\frac{n_x\pi}{d_x})^2+\frac{\hbar^2}{2x_5}(\frac{n_y\pi}{d_y})^2\right)\\
&\quad-\alpha\left(\frac{\hbar^2}{2x_1}(\frac{n_x\pi}{d_x})^2+\frac{\hbar^2}{2x_2}(\frac{n_y\pi}{d_y})^2\right)\frac{\hbar^2}{2x_6}k_z^2\\
&\quad-\left(1+\alpha\gamma_3(E,\eta_g)\right)\frac{\hbar^2}{2x_3}k_z^2\\
&\quad-\alpha\frac{\hbar^2}{2x_3}k_z^2\left(\frac{\hbar^2}{2x_4}(\frac{n_x\pi}{d_x})^2+\frac{\hbar^2}{2x_5}(\frac{n_y\pi}{d_y})^2\right)-\alpha\frac{\hbar^4k_z^4}{4x_3x_6}\\
&=\frac{\hbar^2}{2m_1}(\frac{n_x\pi}{d_x})^2+\frac{\hbar^2}{2m_2}(\frac{n_y\pi}{d_y})^2+\frac{\hbar^2}{2m_3}k_z^2
\end{aligned}\tag{2.64}$$

Equation (2.64) can be written as

$$k_z=T_{36}(E,\eta_g,n_x,n_y)\tag{2.65}$$

where

$$\begin{aligned}
T_{36}(E,\eta_g,n_x,n_y)&=[(2C_{22})^{-1}[-B_{HD}(E,\eta_g,n_x,n_y)\\
&\quad+\sqrt{B_{HD}^2(E,\eta_g,n_x,n_y)+4C_{22}A_{HD}(E,\eta_g,n_x,n_y)}]]^{1/2}
\end{aligned}$$

$$\begin{aligned}
C_{22}&=(\alpha\frac{\hbar^4}{4x_3x_6}),\ B_{HD}(E,\eta_g,n_x,n_y)\\
&=[\alpha\left(\frac{\hbar^2}{2x_1}(\frac{n_x\pi}{d_x})^2+\frac{\hbar^2}{2x_2}(\frac{n_y\pi}{d_y})^2\right)\frac{\hbar^2}{2x_6}\\
&\quad+\left(1+\alpha\gamma_3(E,\eta_g)\right)\frac{\hbar^2}{2x_3}-\alpha\gamma_3(E,\eta_g)\frac{\hbar^2}{2x_6}\\
&\quad+\frac{\hbar^2}{2m_3}+\alpha\frac{\hbar^2}{2x_3}\left(\frac{\hbar^2}{2x_4}(\frac{n_x\pi}{d_x})^2+\frac{\hbar^2}{2x_5}(\frac{n_y\pi}{d_y})^2\right)]
\end{aligned}$$

and

$$A_{HD}(E,\eta_g,n_x,n_y) = [-[\frac{\hbar^2}{2m_1}(\frac{n_x\pi}{d_x})^2 + \frac{\hbar^2}{2m_2}(\frac{n_y\pi}{d_y})^2]\gamma_2(E,\eta_g) + \alpha\gamma_3(E,\eta_g)\left(\frac{\hbar^2}{2x_4}(\frac{n_x\pi}{d_x})^2 + \frac{\hbar^2}{2x_5}(\frac{n_y\pi}{d_y})^2\right) - \alpha\left(\frac{\hbar^2}{2x_1}(\frac{n_x\pi}{d_x})^2 + \frac{\hbar^2}{2x_2}(\frac{n_y\pi}{d_y})^2\right) \left(\frac{\hbar^2}{2x_4}(\frac{n_x\pi}{d_x})^2 + \frac{\hbar^2}{2x_5}(\frac{n_y\pi}{d_y})^2\right) - (1+\alpha\gamma_3(E,\eta_g))\left(\frac{\hbar^2}{2x_1}(\frac{n_x\pi}{d_x})^2 + \frac{\hbar^2}{2x_2}(\frac{n_y\pi}{d_y})^2\right)]$$

The EP in this case is given by

$$I = \frac{\alpha_o e g_v k_B T}{\pi\hbar}\sum_{n_x=1}^{n_{x\max}}\sum_{n_y=1}^{n_{y\max}} F_0(\eta_{614HD}), \tag{2.66}$$

where

$$\eta_{614HD} \equiv \left[\frac{E_{F1HDNW} - (E'_{14HDNW} + W - h\nu)}{k_B T}\right], \tag{2.66}$$

and E'_{14HDNW} is the sub-band energy in this case which can be expressed as

$$0 = T_{36}(E'_{14HDNW},\eta_g,n_x,n_y) \tag{2.67}$$

The EEM in this case is given by

$$m^*(E_{F1HDNW},\eta_g,n_x,n_y) = \frac{\hbar^2}{2}\frac{\partial}{\partial E}[T_{36}^2(E_{F1HDNW},\eta_g,n_x,n_y)] \tag{2.68}$$

Thus, it appears that the evaluation of J_{1D} requires an expression of carrier statistics which can, in turn, be written as

$$n_{1D} = \left(\frac{2g_v}{\pi}\right)\sum_{n_x=1}^{n_{x\max}}\sum_{n_y=1}^{n_{y\max}}\left[T_{36HDNW}(E_{F1HDNW},n_x,n_y,\eta_g) + T_{37HDNW}(E_{F1HDNW},n_x,n_y,\eta_g)\right] \tag{2.69}$$

where

$$T_{36HDNW}(\mathrm{E}_{F1HDNW},\mathrm{n}_x,\mathrm{n}_y,\eta_g) = T_{36}(\mathrm{E}_{F1HDNW},\mathrm{n}_x,\mathrm{n}_y,\eta_g)$$

where $\mathrm{T}_{37HDNW}(\mathrm{E}_{F1HDNW},\mathrm{n}_x,\mathrm{n}_y,\eta_g) = \sum_{r=1}^{s} L(r)[\mathrm{T}_{36HDNW}(\mathrm{E}_{F1HDNW},\mathrm{n}_x,\mathrm{n}_y,\eta_g)]$,

The 1D electron dispersion law in NW of IV-VI semiconductors in the absence of band tails can be expressed as

$$
\begin{aligned}
&E(1+\alpha E)+\alpha E\left(\frac{\hbar^2}{2x_4}(\frac{n_x\pi}{d_x})^2+\frac{\hbar^2}{2x_5}(\frac{n_y\pi}{d_y})^2\right)\\
&\quad+\alpha E\frac{\hbar^2}{2x_6}k_z^2-(1+\alpha E)\left(\frac{\hbar^2}{2x_1}(\frac{n_x\pi}{d_x})^2+\frac{\hbar^2}{2x_2}(\frac{n_y\pi}{d_y})^2\right)\\
&\quad-\alpha\left(\frac{\hbar^2}{2x_1}(\frac{n_x\pi}{d_x})^2+\frac{\hbar^2}{2x_2}(\frac{n_y\pi}{d_y})^2\right)\left(\frac{\hbar^2}{2x_4}(\frac{n_x\pi}{d_x})^2+\frac{\hbar^2}{2x_5}(\frac{n_y\pi}{d_y})^2\right)\\
&\quad-\alpha\left(\frac{\hbar^2}{2x_1}(\frac{n_x\pi}{d_x})^2+\frac{\hbar^2}{2x_2}(\frac{n_y\pi}{d_y})^2\right)\frac{\hbar^2}{2x_6}k_z^2-(1+\alpha E)\frac{\hbar^2}{2x_3}k_z^2\\
&\quad-\alpha\frac{\hbar^2}{2x_3}k_z^2\left(\frac{\hbar^2}{2x_4}(\frac{n_x\pi}{d_x})^2+\frac{\hbar^2}{2x_5}(\frac{n_y\pi}{d_y})^2\right)-\alpha\frac{\hbar^4k_z^4}{4x_3x_6}\\
&=\frac{\hbar^2}{2m_1}(\frac{n_x\pi}{d_x})^2+\frac{\hbar^2}{2m_2}(\frac{n_y\pi}{d_y})^2+\frac{\hbar^2}{2m_3}k_z^2
\end{aligned}
\tag{2.70}
$$

Equation (2.70) can be written as

$$k_z=T_{40}(E,n_x,n_y) \tag{2.71}$$

where

$$
\begin{aligned}
T_{40}(E,n_x,n_y)=&[(2C_{22})^{-1}[-B_0(E,n_x,n_y)\\
&+\sqrt{B_0^2(E,n_x,n_y)+4C_{22}A_0(E,n_x,n_y)}\,]]^{1/2}
\end{aligned}
$$

where

$$
\begin{aligned}
B_0(E,n_x,n_y)=&[\alpha\left(\frac{\hbar^2}{2x_1}(\frac{n_x\pi}{d_x})^2+\frac{\hbar^2}{2x_2}(\frac{n_y\pi}{d_y})^2\right)\frac{\hbar^2}{2x_6}+(1+\alpha E)\frac{\hbar^2}{2x_3}\\
&-\alpha E\frac{\hbar^2}{2x_6}+\frac{\hbar^2}{2m_3}+\alpha\frac{\hbar^2}{2x_3}\left(\frac{\hbar^2}{2x_4}(\frac{n_x\pi}{d_x})^2+\frac{\hbar^2}{2x_5}(\frac{n_y\pi}{d_y})^2\right)]
\end{aligned}
$$

and

$$
\begin{aligned}
A_0(E,n_x,n_y)=&[-[\frac{\hbar^2}{2m_1}(\frac{n_x\pi}{d_x})^2+\frac{\hbar^2}{2m_2}(\frac{n_y\pi}{d_y})^2]E(1+\alpha E)\\
&+\alpha E\left(\frac{\hbar^2}{2x_4}(\frac{n_x\pi}{d_x})^2+\frac{\hbar^2}{2x_5}(\frac{n_y\pi}{d_y})^2\right)\\
&-\alpha\left(\frac{\hbar^2}{2x_1}(\frac{n_x\pi}{d_x})^2+\frac{\hbar^2}{2x_2}(\frac{n_y\pi}{d_y})^2\right)\left(\frac{\hbar^2}{2x_4}(\frac{n_x\pi}{d_x})^2\right.\\
&\left.+\frac{\hbar^2}{2x_5}(\frac{n_y\pi}{d_y})^2\right)-(1+\alpha E)\left(\frac{\hbar^2}{2x_1}(\frac{n_x\pi}{d_x})^2+\frac{\hbar^2}{2x_2}(\frac{n_y\pi}{d_y})^2\right)]
\end{aligned}
$$

The EP in this case is given by

$$I = \frac{\alpha_o e g_v k_B T}{\pi \hbar} \sum_{n_x=1}^{n_{x\max}} \sum_{n_y=1}^{n_{y\max}} F_0(\eta_{615}), \tag{2.72}$$

where

$$\eta_{615} \equiv \left[\frac{E_{F1d} - \left(E'_{20} + W - h\nu\right)}{k_B T}\right],$$

and E'_{20} is the sub-band energy in this case which can be expressed as

$$0 = T_{40}(E'_{20}, n_x, n_y) \tag{2.73}$$

The EEM in this case is given by

$$m^*(E_{F1d}, n_x, n_y) = \frac{\hbar^2}{2} \frac{\partial}{\partial E} [T_{40}^2(E_{F1d}, n_x, n_y)] \tag{2.74}$$

Thus, it appears that the evaluation of J_{1D} requires an expression of carrier statistics which can, in turn, be written as

$$n_{1D} = \left(\frac{2g_v}{\pi}\right) \sum_{n_x=1}^{n_{x\max}} \sum_{n_y=1}^{n_{y\max}} \left[T_{40}(\mathrm{E}_{F1d}, \mathrm{n}_x, \mathrm{n}_y) + T_{41}(\mathrm{E}_{F1d}, \mathrm{n}_x, \mathrm{n}_y)\right] \tag{2.75}$$

where $T_{41}(\mathrm{E}_{F1d}, \mathrm{n}_x, \mathrm{n}_y) = \sum_{r=1}^{s} L(r)[T_{40}(\mathrm{E}_{F1d}, \mathrm{n}_x, \mathrm{n}_y)]$.

(ii) Bangert and Kastner Model

Following (1.194d), the 1D dispersion relation in NW of IV-VI semiconductors in accordance with the present model can be written as

$$F_1(E, \eta_g)[(\frac{n_x \pi}{d_x})^2 + (\frac{n_y \pi}{d_y})^2] + F_2(E, \eta_g) k_z^2 = 1 \tag{2.76}$$

The (2.76) can be written as

$$k_z = T_{60}(E, \eta_g, n_x, n_y) \tag{2.77}$$

where

$$T_{60}(E, \eta_g, n_x, n_y) = [[1 - F_1(E, \eta_g)[(\frac{n_x \pi}{d_x})^2 + (\frac{n_y \pi}{d_y})^2]][F_2(E, \eta_g)]^{-1}]^{1/2}$$

The EP in this case is given by

$$I = \frac{\alpha_o e g_v k_B T}{\pi\hbar} \sum_{n_x=1}^{n_{x\max}} \sum_{n_y=1}^{n_{y\max}} F_0(\eta_{615HD}), \tag{2.78}$$

where

$$\eta_{615HD} \equiv \left[\frac{E_{F1HDNW} - (E'_{15HDNW} + W - h\nu)}{k_B T}\right],$$

and E'_{15HDNW} is the sub-band energy in this case which can be expressed as

$$0 = T_{60}(E'_{15HDNW}, \eta_g, n_x, n_y) \tag{2.79}$$

The EEM in this case is given by

$$m^*(E_{F1HDNW}, \eta_g, n_x, n_y) = \frac{\hbar^2}{2}\frac{\partial}{\partial E}[T_{40}^2(E_{F1HDNW}, \eta_g, n_x, n_y)] \tag{2.80}$$

Thus, it appears that the evaluation of J_{1D} requires an expression of carrier statistics which can, in turn, be written as

$$n_{1D} = \left(\frac{2g_v}{\pi}\right) \sum_{n_x=1}^{n_{x\max}} \sum_{n_y=1}^{n_{y\max}} [T_{50HDNW}(E_{F1HDNW}, n_x, n_y, \eta_g) + T_{51HDNW}(E_{F1HDNW}, n_x, n_y, \eta_g)] \tag{2.81}$$

where

$$T_{50HDNW}(\mathrm{E}_{F1HDNW}, \mathrm{n}_x, \mathrm{n}_y, \eta_g) = T_{40}(\mathrm{E}_{F1HDNW}, \mathrm{n}_x, \mathrm{n}_y, \eta_g)$$

and $\mathrm{T}_{51HDNW}(\mathrm{E}_{F1HDNW}, \mathrm{n}_x, \mathrm{n}_y, \eta_g) = \sum_{r=1}^{s} L(r)[\mathrm{T}_{50HDNW}(\mathrm{E}_{F1HDNW}, \mathrm{n}_x, \mathrm{n}_y, \eta_g)]$,

The 1D dispersion relation in the absence of band tailing can be written in this case following (1.194b) as

$$\omega_1(\mathrm{E})[(\frac{\pi n_x}{d_x})^2 + (\frac{\pi n_y}{d_y})^2] + \omega_2(\mathrm{E})k_z^2 = 1 \tag{2.82}$$

The (2.82) can be written as

$$k_z = T_{61}(E, n_x, n_y) \tag{2.83}$$

where

$$T_{61}(E, n_x, n_y) = [[1 - \omega_1(E)[(\frac{n_x\pi}{d_x})^2 + (\frac{n_y\pi}{d_y})^2]][\omega_2(E)]^{-1}]^{1/2}$$

The EP in this case is given by

$$I = \frac{\alpha_o e g_v k_B T}{\pi\hbar} \sum_{n_x=1}^{n_{x\max}} \sum_{n_y=1}^{n_{y\max}} F_0(\eta_{616}), \tag{2.84}$$

where

$$\eta_{616} \equiv \left[\frac{E_{F1d} - (E'_{21} + W - h\nu)}{k_B T}\right],$$

and E'_{21} is the sub-band energy in this case which can be expressed as

$$0 = T_{61}(E'_{21}, n_x, n_y) \tag{2.85}$$

The EEM in this case is given by

$$m^*(E_{F1d}, n_x, n_y) = \frac{\hbar^2}{2} \frac{\partial}{\partial E}[T_{61}^2(E_{F1d}, n_x, n_y)] \tag{2.86}$$

Thus, it appears that the evaluation of J_{1D} requires an expression of carrier statistics which can, in turn, be written as

$$n_{1D} = \left(\frac{2g_v}{\pi}\right) \sum_{n_x=1}^{n_{x\max}} \sum_{n_y=1}^{n_{y\max}} \left[T_{61}(\mathrm{E}_{F1d}, \mathrm{n}_x, \mathrm{n}_y) + T_{62}(\mathrm{E}_{F1d}, \mathrm{n}_x, \mathrm{n}_y)\right] \tag{2.87}$$

where $T_{62}(\mathrm{E}_{F1d}, \mathrm{n}_x, \mathrm{n}_y) = \sum_{r=1}^{s} L(r)[T_{61}(\mathrm{E}_{F1d}, \mathrm{n}_x, \mathrm{n}_y)]$.

2.2.5 The EP from QWs of HD Stressed Kane Type Semiconductors

The 1D dispersion relation in this case can be written following (1.206) as

$$P_{11}(E, \eta_g)(\frac{\pi n_x}{d_x})^2 + Q_{11}(E, \eta_g)(\frac{\pi n_y}{d_y})^2 + S_{11}(E, \eta_g)k_z^2 = 1 \tag{2.88}$$

The (2.88) can be written as

$$k_z = T_{70}(E, \eta_g, n_x, n_y) \tag{2.89}$$

where

$$T_{70}(E,\eta_g,n_x,n_y) = [[1 - P_{11}(E,\eta_g)(\frac{\pi n_x}{d_x})^2 + Q_{11}(E,\eta_g)(\frac{\pi n_y}{d_y})^2][S_{11}(E,\eta_g)]^{-1}]^{1/2}$$

The EP in this case is given by

$$I = \frac{\alpha_o e g_v k_B T}{\pi\hbar}\sum_{n_x=1}^{n_{x\max}}\sum_{n_y=1}^{n_{y\max}} F_0(\eta_{630HD}), \tag{2.90}$$

where

$$\eta_{630HD} \equiv \left[\frac{E_{F1HDNW} - \left(E'_{30HDNW} + W - hv\right)}{k_B T}\right],$$

and E'_{30HDNW} is the sub-band energy in this case which can be expressed as

$$0 = T_{70}(E'_{30HDNW}, \eta_g, n_x, n_y) \tag{2.91}$$

The EEM in this case is given by

$$m^*(E_{F1HDNW},\eta_g,n_x,n_y) = \frac{\hbar^2}{2}\frac{\partial}{\partial E}[T_{70}^2(E_{F1HDNW},\eta_g,n_x,n_y)] \tag{2.92}$$

Thus, it appears that the evaluation of J_{1D} requires an expression of carrier statistics which can, in turn, be written as

$$n_{1D} = \left(\frac{2g_v}{\pi}\right)\sum_{n_x=1}^{n_{x\max}}\sum_{n_y=1}^{n_{y\max}} [T_{70HDNW}(\mathrm{E}_{F1HDNW},\mathrm{n}_x,\mathrm{n}_y,\eta_g) + \mathrm{T}_{71HDNW}(\mathrm{E}_{F1HDNW},\mathrm{n}_x,\mathrm{n}_y,\eta_g)] \tag{2.93}$$

where

$$T_{70HDNW}(\mathrm{E}_{F1HDNW},\mathrm{n}_x,\mathrm{n}_y,\eta_g) = T_{70}(\mathrm{E}_{F1HDNW},\mathrm{n}_x,\mathrm{n}_y,\eta_g)$$

and

$$\mathrm{T}_{71HDNW}(\mathrm{E}_{F1HDNW},\mathrm{n}_x,\mathrm{n}_y,\eta_g) = \sum_{r=1}^{s} L(r)[\mathrm{T}_{70HDNW}(\mathrm{E}_{F1HDNW},\mathrm{n}_x,\mathrm{n}_y,\eta_g)]$$

In the absence of band tailing the 1D dispersion relation in this case assumes the form

$$k_z = t_{70}(E,n_x,n_y) \tag{2.94}$$

where

$$t_{70}(E,n_x,n_y)=[\bar{c}_0(E)[1-(\frac{\pi n_x}{d_x\bar{a}_0(E)})^2-(\frac{\pi n_y}{d_y\bar{b}_0(E)})^2]]^{1/2}$$

The EP in this case is given by

$$I=\frac{\alpha_o e g_v k_B T}{\pi\hbar}\sum_{n_x=1}^{n_{x\max}}\sum_{n_y=1}^{n_{y\max}}F_0(\eta_{642}), \tag{2.95}$$

where

$$\eta_{642}\equiv\left[\frac{E_{F1d}-\left(E'_{42}+W-hv\right)}{k_BT}\right],$$

and E'_{42} is the sub-band energy in this case which can be expressed as

$$0=t_{60}(E'_{42},n_x,n_y) \tag{2.96}$$

The EEM in this case is given by

$$m^*(E_{F1d},n_x,n_y)=\frac{\hbar^2}{2}\frac{\partial}{\partial E}[t_{60}^2(E_{F1d},n_x,n_y)] \tag{2.97}$$

Thus, it appears that the evaluation of J_{1D} requires an expression of carrier statistics which can, in turn, be written as

$$n_{1D}=\left(\frac{2g_v}{\pi}\right)\sum_{n_x=1}^{n_{x\max}}\sum_{n_y=1}^{n_{y\max}}\left[t_{60}(\mathrm{E}_{F1d},\mathrm{n}_x,\mathrm{n}_y)+t_{61}(\mathrm{E}_{F1d},\mathrm{n}_x,\mathrm{n}_y)\right] \tag{2.98}$$

where $t_{61}(\mathrm{E}_{F1d},\mathrm{n}_x,\mathrm{n}_y)=\sum_{r=1}^{s}L(r)[\mathrm{t}_{60}(\mathrm{E}_{F1d},\mathrm{n}_x,\mathrm{n}_y)]$.

2.2.6 The EP from Nano Wires of HD Te

The 1D dispersion relation may be written in this case following (1.235) as

$$k_x=t_{72}(E,n_y,n_z,\eta_g) \tag{2.99}$$

where

$$t_{72}(E,n_y,n_z,\eta_g)=\Big[-(\frac{n_y\pi}{d_y})^2+\psi_{5HD}(E,\eta_g)-\psi_6(\frac{\pi n_z}{d_z})^2 \pm\psi_7[\psi_{8HD}^2(E,\eta_g)-(\frac{\pi n_z}{d_z})^2]^{1/2}\Big]^{1/2}$$

The EP in this case is given by

$$I = \frac{\alpha_o e g_v k_B T}{2\pi\hbar} \sum_{n_y=1}^{n_{y\max}} \sum_{n_z=1}^{n_{z\max}} F_0(\eta_{631HD}) \tag{2.100}$$

where

$$\eta_{631HD} \equiv \left[\frac{E_{F1HDNW} - \left(E'_{31HDNW} + W - h\nu\right)}{k_B T}\right],$$

and E'_{31HDNW} is the sub-band energy in this case which can be expressed as

$$0 = t_{72}(E'_{31HDNW}, \eta_g, n_y, n_z) \tag{2.101}$$

The EEM in this case is given by

$$m^*(E_{F1HDNW}, \eta_g, n_y, n_z) = \frac{\hbar^2}{2}\frac{\partial}{\partial E}[t_{72}^2(E_{F1HDNW}, \eta_g, n_y, n_z)] \tag{2.102}$$

Thus, it appears that the evaluation of J_{1D} requires an expression of carrier statistics which can, in turn, be written as

$$n_{1D} = \left(\frac{2g_v}{\pi}\right) \sum_{n_y=1}^{n_{y\max}} \sum_{n_z=1}^{n_{z\max}} \left[t_{72HDNW}(E_{F1HDNW}, n_y, n_z, \eta_g) + t_{73HDNW}(E_{F1HDNW}, n_y, n_z, \eta_g)\right] \tag{2.103}$$

where

$$t_{72HDNW}(E_{F1HDNW}, n_y, n_z, \eta_g) = t_{72}(E_{F1HDNW}, n_y, n_z, \eta_g)$$

and

$$t_{73HDNW}(E_{F1HDNW}, n_y, n_z, \eta_g) = \sum_{r=1}^{s} L(r)[t_{72HDNW}(E_{F1HDNW}, n_y, n_z, \eta_g)],$$

In the absence of band tailing the 1D dispersion relation in this case assumes the form

$$k_x = H_{70}(E, n_y, n_z) \tag{2.104}$$

where

$$H_{70}(E, n_y, n_z) = [-(\frac{n_y\pi}{d_y})^2 + \psi_5(E) - \psi_6(\frac{\pi n_z}{d_z})^2 \pm \psi_7[\psi_8^2(E) - (\frac{\pi n_z}{d_z})^2]^{1/2}]^{1/2}$$

The EP in this case is given by

$$I = \frac{\alpha_o e g_v k_B T}{2\pi\hbar} \sum_{n_y=1}^{n_{y\max}} \sum_{n_z=1}^{n_{z\max}} F_0(\eta_{644}) \tag{2.105}$$

where

$$\eta_{644} \equiv \left[\frac{E_{F1d} - \left(E'_{44} + W - hv\right)}{k_B T}\right],$$

and E'_{44} is the sub-band energy in this case which can be expressed as

$$0 = H_{70}(E'_{44}, n_y, n_z) \tag{2.106}$$

The EEM in this case is given by

$$m^*(E_{F1d}, n_y, n_z) = \frac{\hbar^2}{2} \frac{\partial}{\partial E} [\mathrm{H}_{70}^2(E_{F1d}, n_y, n_z)] \tag{2.107}$$

Thus, it appears that the evaluation of J_{1D} requires an expression of carrier statistics which can, in turn, be written as

$$n_{1D} = \left(\frac{2g_v}{\pi}\right) \sum_{n_y=1}^{n_{y\max}} \sum_{n_z=1}^{n_{z\max}} \left[H_{70}(\mathrm{E}_{F1d}, \mathrm{n}_y, \mathrm{n}_z) + H_{71}(\mathrm{E}_{F1d}, \mathrm{n}_y, \mathrm{n}_z)\right] \tag{2.108}$$

where $H_{71}(\mathrm{E}_{F1d}, \mathrm{n}_y, \mathrm{n}_z) = \sum_{r=1}^{s} L(r)[\mathrm{H}_{70}(\mathrm{E}_{F1d}, \mathrm{n}_y, \mathrm{n}_z)]$.

2.2.7 The EP from Nano Wires of HD GaP

The 1D dispersion relation may be written in this case following (1.253) as

$$k_x = u_{70}(\mathrm{E}, \mathrm{n}_y, \mathrm{n}_z, \eta_g) \tag{2.109}$$

where

$$u_{70}(\mathrm{E}, \mathrm{n}_y, \mathrm{n}_z, \eta_g) = \left[-\left(\frac{n_y \pi}{d_y}\right)^2 + t_{11}\gamma_3(\mathrm{E}, \eta_g) + t_{21} - t_{31}\left(\frac{n_z \pi}{d_z}\right)^2 - t_{41}\left[\left(\frac{n_z \pi}{d_z}\right)^2 + t_5^2(\mathrm{E}, \eta_g)\right]^{1/2}\right]^{1/2}$$

The EP in this case is given by

$$I = \frac{\alpha_o e g_v k_B T}{\pi\hbar}\sum_{n_y=1}^{n_{y\max}}\sum_{n_z=1}^{n_{z\max}} F_0(\eta_{632HD}), \tag{2.110}$$

where

$$\eta_{632HD} \equiv \left[\frac{E_{F1HDNW} - \left(E'_{32HDNW} + W - h\nu\right)}{k_B T}\right],$$

and E'_{32HDNW} is the sub-band energy in this case which can be expressed as

$$0 = u_{70}(E'_{32HDNW}, \eta_g, n_y, n_z) \tag{2.111}$$

The EEM in this case is given by

$$m^*(E_{F1HDNW}, \eta_g, n_y, n_z) = \frac{\hbar^2}{2}\frac{\partial}{\partial E}[\mathrm{u}_{70}^2(E_{F1HDNW}, \eta_g, n_y, n_z)] \tag{2.112}$$

Thus, it appears that the evaluation of J_{1D} requires an expression of carrier statistics which can, in turn, be written as

$$n_{1D} = \left(\frac{2g_v}{\pi}\right)\sum_{n_y=1}^{n_{y\max}}\sum_{n_z=1}^{n_{z\max}}\left[u_{70HDNW}(\mathrm{E}_{F1HDNW}, \mathrm{n}_y, \mathrm{n}_z, \eta_g) + u_{71HDNW}(\mathrm{E}_{F1HDNW}, \mathrm{n}_y, \mathrm{n}_z, \eta_g)\right] \tag{2.113}$$

where

$$u_{70HDNW}(\mathrm{E}_{F1HDNW}, \mathrm{n}_y, \mathrm{n}_z, \eta_g) = u_{70}(\mathrm{E}_{F1HDNW}, \mathrm{n}_y, \mathrm{n}_z, \eta_g)$$

and

$$u_{71HDNW}(\mathrm{E}_{F1HDNW}, \mathrm{n}_y, \mathrm{n}_z, \eta_g) = \sum_{r=1}^{s} L(r)[\mathrm{u}_{70HDNW}(\mathrm{E}_{F1HDNW}, \mathrm{n}_y, \mathrm{n}_z, \eta_g)],$$

In the absence of band tailing the 1D dispersion relation in this case can be written using (1.260) as

$$k_x = X_{71}(\mathrm{E}, \mathrm{n}_y, \mathrm{n}_z) \tag{2.114}$$

where

$$X_{71}(\mathrm{E}, \mathrm{n}_y, \mathrm{n}_z) = [-(\frac{n_y \pi}{d_y})^2 + t_{42}(E, n_z)]^{1/2}$$

The EP in this case is given by

$$I = \frac{\alpha_o e g_v k_B T}{2\pi\hbar} \sum_{n_y=1}^{n_{y\max}} \sum_{n_z=1}^{n_{z\max}} F_0(\eta_{646}),$$
$$\text{where } \eta_{646} \equiv \left[\frac{E_{F1d} - \left(E'_{46} + W - hv\right)}{k_B T}\right], \tag{2.115}$$

and E'_{46} is the sub-band energy in this case which can be expressed as

$$0 = X_{71}(E'_{46}, n_y, n_z) \tag{2.116}$$

The EEM in this case is given by

$$m^*(E_{F1d}, n_y, n_z) = \frac{\hbar^2}{2} \frac{\partial}{\partial E} [\mathrm{X}_{71}^2(E_{F1d}, n_y, n_z)] \tag{2.117}$$

Thus, it appears that the evaluation of J_{1D} requires an expression of carrier statistics which can, in turn, be written as

$$n_{1D} = \left(\frac{2g_v}{\pi}\right) \sum_{n_y=1}^{n_{y\max}} \sum_{n_z=1}^{n_{z\max}} \left[X_{71}(\mathrm{E}_{F1d}, \mathrm{n}_y, \mathrm{n}_z) + X_{72}(\mathrm{E}_{F1d}, \mathrm{n}_y, \mathrm{n}_z)\right] \tag{2.118}$$

where $X_{72}(\mathrm{E}_{F1d}, \mathrm{n}_y, \mathrm{n}_z) = \sum_{r=1}^{s} L(r)[\mathrm{X}_{71}(\mathrm{E}_{F1d}, \mathrm{n}_y, \mathrm{n}_z)]$.

2.2.8 The EP from Nano Wires of HD $PtSb_2$

The 1D dispersion relation may be written in this case following (1.275) as

$$k_x = V_{70}(\mathrm{E}, \mathrm{n}_y, \mathrm{n}_z, \eta_g) \tag{2.119}$$

where $V_{70}(\mathrm{E}, \mathrm{n}_y, \mathrm{n}_z, \eta_g) = [-(\frac{n_y \pi}{d_y})^2 + A_{60}(\mathrm{E}, \eta_g, \mathrm{n}_y)]^{1/2}$

The EP in this case is given by

$$I = \frac{\alpha_o e g_v k_B T}{\pi\hbar} \sum_{n_y=1}^{n_{y\max}} \sum_{n_z=1}^{n_{z\max}} F_0(\eta_{634HD}), \tag{2.120}$$

where

$$\eta_{634HD} \equiv \left[\frac{E_{F1HDNW} - \left(E'_{34HDNW} + W - hv\right)}{k_B T}\right],$$

and E'_{34HDNW} is the sub-band energy in this case which can be expressed as

$$0 = V_{70}(E'_{34HDNW}, \eta_g, n_y, n_z) \tag{2.121}$$

The EEM in this case is given by

$$m^*(E_{F1HDNW}, \eta_g, n_y, n_z) = \frac{\hbar^2}{2}\frac{\partial}{\partial E}[\mathrm{V}_{70}^2(E_{F1HDNW}, \eta_g, n_y, n_z)] \tag{2.122}$$

Thus, it appears that the evaluation of J_{1D} requires an expression of carrier statistics which can, in turn, be written as

$$n_{1D} = \left(\frac{2g_v}{\pi}\right)\sum_{n_y=1}^{n_{y\max}}\sum_{n_z=1}^{n_{z\max}}[V_{70HDNW}(\mathrm{E}_{F1HDNW}, \mathrm{n}_y, \mathrm{n}_z, \eta_g) + V_{71HDNW}(\mathrm{E}_{F1HDNW}, \mathrm{n}_y, \mathrm{n}_z, \eta_g)] \tag{2.123}$$

where

$$V_{70HDNW}(\mathrm{E}_{F1HDNW}, \mathrm{n}_y, \mathrm{n}_z, \eta_g) = V_{70}(\mathrm{E}_{F1HDNW}, \mathrm{n}_y, \mathrm{n}_z, \eta_g)$$

and

$$V_{71HDNW}(\mathrm{E}_{F1HDNW}, \mathrm{n}_y, \mathrm{n}_z, \eta_g) = \sum_{r=1}^{s} L(r)[V_{70HDNW}(\mathrm{E}_{F1HDNW}, \mathrm{n}_y, \mathrm{n}_z, \eta_g)],$$

In the absence of band tailing the 1D dispersion relation in this case can be written using (1.278) as

$$k_x = D_{71}(\mathrm{E}, \mathrm{n}_y, \mathrm{n}_z) \tag{2.124}$$

where

$$D_{71}(\mathrm{E}, \mathrm{n}_y, \mathrm{n}_z) = [-(\frac{n_y\pi}{d_y})^2 + t_{44}(\mathrm{E}, \mathrm{n}_z)]^{1/2}$$

The EP in this case is given by

$$I = \frac{\alpha_o e g_v k_B T}{2\pi\hbar}\sum_{n_y=1}^{n_{y\max}}\sum_{n_z=1}^{n_{z\max}} F_0(\eta_{648}), \tag{2.125}$$

where

$$\eta_{648} \equiv \left[\frac{E_{F1d} - (E'_{48} + W - hv)}{k_B T}\right],$$

and E'_{48} is the sub-band energy in this case which can be expressed as

$$0 = D_{71}(E'_{48}, n_y, n_z) \tag{2.126}$$

The EEM in this case is given by

$$m^*(E_{F1d}, n_y, n_z) = \frac{\hbar^2}{2} \frac{\partial}{\partial E}[\mathrm{D}_{71}^2(E_{F1d}, n_y, n_z)] \tag{2.127}$$

Thus, it appears that the evaluation of J_{1D} requires an expression of carrier statistics which can, in turn, be written as

$$n_{1D} = \left(\frac{2g_v}{\pi}\right) \sum_{n_y=1}^{n_{y\max}} \sum_{n_z=1}^{n_{z\max}} \left[D_{71}(\mathrm{E}_{F1d}, \mathrm{n}_y, \mathrm{n}_z) + D_{72}(\mathrm{E}_{F1d}, \mathrm{n}_y, \mathrm{n}_z)\right] \tag{2.128a}$$

where $D_{72}(\mathrm{E}_{F1d}, \mathrm{n}_y, \mathrm{n}_z) = \sum_{r=1}^{s} L(r)[\mathrm{D}_{71}(\mathrm{E}_{F1d}, \mathrm{n}_y, \mathrm{n}_z)]$.

2.2.9 The EP from Nano Wires of HD Bi_2Te_3

The dispersion relation in this case can be written following (1.285) as

$$k_x = J_{70}(\mathrm{E}, \mathrm{n}_y, \mathrm{n}_z, \eta_g) \tag{2.128b}$$

where

$$J_{70}(\mathrm{E}, \mathrm{n}_y, \mathrm{n}_z, \eta_g) = [[\gamma_2(\mathrm{E}, \eta_g) - \overline{\omega}_2(\frac{n_y \pi}{d_y})^2 - \overline{\omega}_3(\frac{n_z \pi}{d_z})^2 - 2\overline{\omega}_4 \frac{n_y n_z \pi^2}{d_y d_z}](\overline{\omega}_1)^{-1}]^{1/2}$$

The EP in this case is given by

$$I = \frac{\alpha_o e g_v k_B T}{\pi \hbar} \sum_{n_y=1}^{n_{y\max}} \sum_{n_z=1}^{n_{z\max}} F_0(\eta_{650HD}), \tag{2.129}$$

where

$$\eta_{650HD} \equiv \left[\frac{E_{F1HDNW} - (E'_{50HDNW} + W - hv)}{k_B T}\right],$$

and E'_{50HDNW} is the sub-band energy in this case which can be expressed as

$$0 = J_{70}(E'_{50HDNW}, \eta_g, n_y, n_z) \tag{2.130}$$

The EEM in this case is given by

$$m^*(E_{F1HDNW}, \eta_g, n_y, n_z) = \frac{\hbar^2}{2}\frac{\partial}{\partial E}[J_{70}^2(E_{F1HDNW}, \eta_g, n_y, n_z)] \quad (2.131)$$

Thus, it appears that the evaluation of J_{1D} requires an expression of carrier statistics which can, in turn, be written as

$$n_{1D} = \left(\frac{2g_v}{\pi}\right)\sum_{n_y=1}^{n_{y\max}}\sum_{n_z=1}^{n_{z\max}}\left[J_{70HDNW}(E_{F1HDNW}, n_y, n_z, \eta_g) + J_{71HDNW}(E_{F1HDNW}, n_y, n_z, \eta_g)\right] \quad (2.132)$$

where

$$J_{70HDNW}(\mathrm{E}_{F1HDNW}, \mathrm{n}_y, \mathrm{n}_z, \eta_g) = J_{70}(\mathrm{E}_{F1HDNW}, \mathrm{n}_y, \mathrm{n}_z, \eta_g)$$

and

$$J_{71HDNW}(\mathrm{E}_{F1HDNW}, \mathrm{n}_y, \mathrm{n}_z, \eta_g) = \sum_{r=1}^{s} L(r)[\mathrm{J}_{70HDNW}(\mathrm{E}_{F1HDNW}, \mathrm{n}_y, \mathrm{n}_z, \eta_g)],$$

In the absence of band tailing the 1D dispersion relation in this case can be written using (1.278) as

$$k_x = B_{71}(\mathrm{E}, \mathrm{n}_y, \mathrm{n}_z) \quad (2.133)$$

where

$$B_{71}(\mathrm{E}, \mathrm{n}_y, \mathrm{n}_z) = [[\mathrm{E}(1+\alpha\mathrm{E}) - \overline{\omega}_2(\frac{n_y\pi}{d_y})^2 - \overline{\omega}_3(\frac{n_z\pi}{d_z})^2 - 2\overline{\omega}_4\frac{n_y n_z \pi^2}{d_y d_z}](\overline{\omega}_1)^{-1}]^{1/2}$$

The EP in this case is given by

$$I = \frac{\alpha_o e g_v k_B T}{2\pi\hbar}\sum_{n_y=1}^{n_{y\max}}\sum_{n_z=1}^{n_{z\max}} F_0(\eta_{650}), \quad (2.134)$$

where

$$\eta_{650} \equiv \left[\frac{E_{F1d} - (E'_{50} + W - hv)}{k_B T}\right],$$

and E'_{50} is the sub-band energy in this case which can be expressed as

$$0 = B_{71}(E'_{50}, n_y, n_z) \quad (2.135)$$

The EEM in this case is given by

$$m^*(E_{F1d}, n_y, n_z) = \frac{\hbar^2}{2}\frac{\partial}{\partial E}[\mathrm{B}_{71}^2(E_{F1d}, n_y, n_z)] \tag{2.136}$$

Thus, it appears that the evaluation of J_{1D} requires an expression of carrier statistics which can, in turn, be written as

$$n_{1D} = \left(\frac{2g_v}{\pi}\right)\sum_{n_y=1}^{n_{y\max}}\sum_{n_z=1}^{n_{z\max}}\left[B_{71}(\mathrm{E}_{F1d}, \mathrm{n}_y, \mathrm{n}_z) + B_{72}(\mathrm{E}_{F1d}, \mathrm{n}_y, \mathrm{n}_z)\right] \tag{2.137}$$

where $B_{72}(\mathrm{E}_{F1d}, \mathrm{n}_y, \mathrm{n}_z) = \sum_{r=1}^{s} L(r)[\mathrm{B}_{71}(\mathrm{E}_{F1d}, \mathrm{n}_y, \mathrm{n}_z)]$.

2.2.10 The EP from Nano Wires of HD Ge

(a) Model of Cardona et al.

The dispersion relation in accordance with this model in the present case can be written following (1.306b) as

$$k_x = L_{70}(\mathrm{E}, \mathrm{n}_y, \mathrm{n}_z, \eta_g) \tag{2.138}$$

where

$$L_{70}(\mathrm{E}, \mathrm{n}_y, \mathrm{n}_z, \eta_g) = [[\gamma_2(\mathrm{E}, \eta_g) + \alpha[\frac{\hbar^2}{2m_\parallel^*}(\frac{n_z\pi}{d_z})^2]^2 - (1 + 2\alpha\gamma_3(\mathrm{E}, \eta_g))\frac{\hbar^2}{2m_\parallel^*}(\frac{n_z\pi}{d_z})^2](\frac{2m_\parallel^*}{\hbar^2})]^{1/2}$$

The EP in this case is given by

$$I = \frac{\alpha_o e g_v k_B T}{\pi\hbar}\sum_{n_y=1}^{n_{y\max}}\sum_{n_z=1}^{n_{z\max}} F_0(\eta_{652HD}), \tag{2.139}$$

where

$$\eta_{652HD} \equiv \left[\frac{E_{F1HDNW} - \left(E'_{52HDNW} + W - hv\right)}{k_B T}\right],$$

and E'_{52HDNW} is the sub-band energy in this case which can be expressed as

$$0 = L_{70}(E'_{52HDNW}, \eta_g, n_y, n_z) \tag{2.140}$$

The EEM in this case is given by

$$m^*(E_{F1HDNW}, \eta_g, n_y, n_z) = \frac{\hbar^2}{2} \frac{\partial}{\partial E} [\mathrm{L}_{70}^2(E_{F1HDNW}, \eta_g, n_y, n_z)] \tag{2.141}$$

Thus, it appears that the evaluation of J_{1D} requires an expression of carrier statistics which can, in turn, be written as

$$n_{1D} = \left(\frac{2g_v}{\pi}\right) \sum_{n_y=1}^{n_{ymax}} \sum_{n_z=1}^{n_{zmax}} \left[L_{70HDNW}(\mathrm{E}_{F1HDNW}, \mathrm{n}_y, \mathrm{n}_z, \eta_g) + L_{71HDNW}(\mathrm{E}_{F1HDNW}, \mathrm{n}_y, \mathrm{n}_z, \eta_g)\right] \tag{2.142}$$

where

$$L_{70HDNW}(\mathrm{E}_{F1HDNW}, \mathrm{n}_y, \mathrm{n}_z, \eta_g) = L_{70}(\mathrm{E}_{F1HDNW}, \mathrm{n}_y, \mathrm{n}_z, \eta_g)$$

and

$$L_{71HDNW}(\mathrm{E}_{F1HDNW}, \mathrm{n}_y, \mathrm{n}_z, \eta_g) = \sum_{r=1}^{s} L(r)[L_{70HDNW}(\mathrm{E}_{F1HDNW}, \mathrm{n}_y, \mathrm{n}_z, \eta_g)]$$

In the absence of band tailing the 1D dispersion relation in this case can be written using (1.278) as

$$k_x = B_{77}(\mathrm{E}, \mathrm{n}_y, \mathrm{n}_z) \tag{2.143}$$

where

$$B_{77}(E, n_y, n_z) = [[E(1+\alpha E) + \alpha[\frac{\hbar^2}{2m_\parallel^*}(\frac{n_z\pi}{d_z})^2]^2 - (1+2\alpha E)\frac{\hbar^2}{2m_\parallel^*}(\frac{n_z\pi}{d_z})^2](\frac{2m_\parallel^*}{\hbar^2})]^{1/2}$$

The EP in this case is given by

$$I = \frac{\alpha_o e g_v k_B T}{2\pi\hbar} \sum_{n_y=1}^{n_{ymax}} \sum_{n_z=1}^{n_{zmax}} F_0(\eta_{660}), \tag{2.144}$$

where

$$\eta_{660} \equiv \left[\frac{E_{F1d} - (E'_{60} + W - h\nu)}{k_B T}\right],$$

and E'_{60} is the sub-band energy in this case which can be expressed as

$$0 = B_{77}(E'_{60}, n_y, n_z) \tag{2.145}$$

The EEM in this case is given by

$$m^*(E_{F1d}, n_y, n_z) = \frac{\hbar^2}{2} \frac{\partial}{\partial E}[\mathrm{B}_{77}^2(E_{F1d}, n_y, n_z)] \tag{2.146}$$

Thus, it appears that the evaluation of J_{1D} requires an expression of carrier statistics which can, in turn, be written as

$$n_{1D} = \left(\frac{2g_v}{\pi}\right) \sum_{n_y=1}^{n_{y\max}} \sum_{n_z=1}^{n_{z\max}} \left[B_{77}(\mathrm{E}_{F1d}, \mathrm{n}_y, \mathrm{n}_z) + B_{78}(\mathrm{E}_{F1d}, \mathrm{n}_y, \mathrm{n}_z)\right] \tag{2.147}$$

where $B_{78}(\mathrm{E}_{F1d}, \mathrm{n}_y, \mathrm{n}_z) = \sum_{r=1}^{s} L(r)[\mathrm{B}_{77}(\mathrm{E}_{F1d}, \mathrm{n}_y, \mathrm{n}_z)]$.

(b) Model of Wang et al.

The dispersion relation in accordance with this model in the present case can be written following (1.326) as

$$k_x = \beta_{70}(\mathrm{E}, \mathrm{n}_y, \mathrm{n}_z, \eta_g) \tag{2.148}$$

where

$$\beta_{70}(\mathrm{E}, \mathrm{n}_y, \mathrm{n}_z, \eta_g) = [-(\frac{n_y\pi}{d_y})^2 + \frac{2m_\perp^*}{\hbar^2}[\bar{\alpha}_8 - \bar{\alpha}_9(\frac{\pi n_z}{d_z})^2 - \bar{\alpha}_{10}[(\frac{\pi n_z}{d_z})^4 + \bar{\alpha}_{11}(\frac{\pi n_z}{d_z})^2 + \bar{\alpha}_{12}(E, \eta_g)]^{1/2}]]^{1/2}$$

The EP in this case is given by

$$I = \frac{\alpha_o e g_v k_B T}{\pi\hbar} \sum_{n_y=1}^{n_{y\max}} \sum_{n_z=1}^{n_{z\max}} F_0(\eta_{654HD}) \tag{2.149}$$

where

$$\eta_{654HD} \equiv \left[\frac{E_{F1HDNW} - (E'_{54HDNW} + W - hv)}{k_B T}\right],$$

and E'_{54HDNW} is the sub-band energy in this case which can be expressed as

$$0 = \beta_{70}(E'_{54HDNW}, \eta_g, n_y, n_z) \tag{2.150}$$

The EEM in this case is given by

$$m^{*}(E_{F1HDNW},\eta_g,n_y,n_z)=\frac{\hbar^2}{2}\frac{\partial}{\partial E}[\beta_{70}^{2}(E_{F1HDNW},\eta_g,n_y,n_z)] \quad (2.151)$$

Thus, it appears that the evaluation of J_{1D} requires an expression of carrier statistics which can, in turn, be written as

$$n_{1D}=\left(\frac{2g_v}{\pi}\right)\sum_{n_y=1}^{n_{y\max}}\sum_{n_z=1}^{n_{z\max}}[\beta_{70HDNW}(\mathrm{E}_{F1HDNW},\mathrm{n}_y,\mathrm{n}_z,\eta_g)+\beta_{71HDNW}(\mathrm{E}_{F1HDNW},\mathrm{n}_y,\mathrm{n}_z,\eta_g)] \quad (2.152)$$

where

$$\beta_{70HDNW}(\mathrm{E}_{F1HDNW},\mathrm{n}_y,\mathrm{n}_z,\eta_g)=\beta_{70}(\mathrm{E}_{F1HDNW},\mathrm{n}_y,\mathrm{n}_z,\eta_g)$$

and

$$\beta_{71HDNW}(\mathrm{E}_{F1HDNW},\mathrm{n}_y,\mathrm{n}_z,\eta_g)=\sum_{r=1}^{s}L(r)[\beta_{70HDNW}(\mathrm{E}_{F1HDNW},\mathrm{n}_y,\mathrm{n}_z,\eta_g)]$$

In the absence of band tailing the 1D dispersion relation in this case can be written using (1.278) as

$$k_x=P_{77}(\mathrm{E},\mathrm{n}_y,\mathrm{n}_z) \quad (2.153)$$

where

$$P_{77}(\mathrm{E},\mathrm{n}_y,\mathrm{n}_z)=[[I_1(\mathrm{E},\mathrm{n}_z)-\frac{\hbar^2}{2m_2^*}(\frac{n_y\pi}{d_y})^2](\frac{2m_1^*}{\hbar^2})]^{1/2}$$

The EP in this case is given by

$$I=\frac{\alpha_o e g_v k_B T}{2\pi\hbar}\sum_{n_y=1}^{n_{y\max}}\sum_{n_z=1}^{n_{z\max}}F_0(\eta_{680}), \quad (2.154)$$

where

$$\eta_{680}\equiv\left[\frac{E_{F1d}-(E'_{80}+W-hv)}{k_BT}\right],$$

and E'_{80} is the sub-band energy in this case which can be expressed as

$$0=P_{77}(E'_{80},n_y,n_z) \quad (2.155)$$

The EEM in this case is given by

$$m^*(E_{F1d}, n_y, n_z) = \frac{\hbar^2}{2}\frac{\partial}{\partial E}[P_{77}^2(E_{F1d}, n_y, n_z)] \tag{2.156}$$

Thus, it appears that the evaluation of J_{1D} requires an expression of carrier statistics which can, in turn, be written as

$$n_{1D} = \left(\frac{2g_v}{\pi}\right)\sum_{n_y=1}^{n_{y\max}}\sum_{n_z=1}^{n_{z\max}}\left[P_{77}(\mathrm{E}_{F1d}, \mathrm{n_y}, \mathrm{n_z}) + P_{78}(\mathrm{E}_{F1d}, \mathrm{n_y}, \mathrm{n_z})\right] \tag{2.157}$$

where $P_{78}(\mathrm{E}_{F1d}, \mathrm{n_y}, \mathrm{n_z}) = \sum_{r=1}^{s} L(r)[\mathrm{P}_{77}(\mathrm{E}_{F1d}, \mathrm{n_y}, \mathrm{n_z})]$.

2.2.11 The EP from Nano Wires of HD GaSb

The dispersion relation of the 1D electrons in this case can be written as

$$\frac{\hbar^2(\mathrm{n_z}\pi/\mathrm{d_z})^2}{2m_c} + \frac{\hbar^2(\mathrm{n_y}\pi/\mathrm{d_y})^2}{2m_c} + \frac{\hbar^2 k_x^2}{2m_c} = I_{36}(\mathrm{E}, \eta_g) \tag{2.158}$$

The EP in this case is given by

$$I = \frac{\alpha_o e g_v k_B T}{\pi\hbar}\sum_{n_y=1}^{n_{y\max}}\sum_{n_z=1}^{n_{z\max}} F_0(\eta_{100HD}), \tag{2.159}$$

where

$$\eta_{100HD} \equiv \left[\frac{E_{F1HDNW} - \left(E'_{100HDNW} + W - hv\right)}{k_B T}\right]$$

and $E'_{100HDNW}$ is the sub-band energy in this case which can be expressed as

$$\frac{\hbar^2(\mathrm{n_z}\pi/\mathrm{d_z})^2}{2m_c} + \frac{\hbar^2(\mathrm{n_y}\pi/\mathrm{d_y})^2}{2m_c} = I_{36}(E'_{100HDNW}, \eta_g) \tag{2.160}$$

The EEM in this case is given by

$$m^*(\mathrm{E}_{F1HDNW}, \eta_g) = \mathrm{m}_c[\mathrm{I}'_{36}(\mathrm{E}_{F1HDNW}, \eta_g)] \tag{2.161}$$

Thus, it appears that the evaluation of J_{1D} requires an expression of carrier statistics which can, in turn, be written as

$$n_{1D} = \left(\frac{2g_v}{\pi}\right) \sum_{n_y=1}^{n_{ymax}} \sum_{n_z=1}^{n_{zmax}} \left[R_{7HDNW}(\mathrm{E}_{F1HDNW}, \mathrm{n}_y, \mathrm{n}_z, \eta_g) + R_{8HDNW}(\mathrm{E}_{F1HDNW}, \mathrm{n}_y, \mathrm{n}_z, \eta_g)\right] \tag{2.162}$$

where

$$R_{7HDNW}(\mathrm{E}_{F1HDNW}, \mathrm{n}_y, \mathrm{n}_z, \eta_g) = [[\mathrm{I}_{36}(\mathrm{E}_{F1HDNW}, \eta_g) - \frac{\hbar^2(\mathrm{n}_z\pi/\mathrm{d}_z)^2}{2m_c} - \frac{\hbar^2(\mathrm{n}_y\pi/\mathrm{d}_y)^2}{2m_c}] \frac{2m_c}{\hbar^2}]^{1/2}$$

where $R_{8HDNW}(\mathrm{E}_{F1HDNW}, \mathrm{n}_y, \mathrm{n}_z, \eta_g) = \sum_{r=1}^{s} L(\mathrm{r})[\mathrm{R}_{7HDNW}(\mathrm{E}_{F1HDNW}, \mathrm{n}_y, \mathrm{n}_z, \eta_g)]$

The expression of 1D dispersion relation, for NWs of GaSb whose energy band structures in the absence of band tailing assumes the form

$$I_{36}(E) = \frac{\hbar^2 k_x^2}{2m_c} + G_2(n_y, n_z) \tag{2.163}$$

In this case, the quantized energy E'_{101} is given by

$$I_{36}(E'_{101}) = G_2(n_y, n_z) \tag{2.164}$$

The EP in this case is given by

$$I = \frac{\alpha_o e g_v k_B T}{\pi\hbar} \sum_{n_y=1}^{n_{ymax}} \sum_{n_z=1}^{n_{zmax}} F_0(\eta_{101}), \tag{2.165}$$

where

$$\eta_{101} \equiv \left[\frac{E_{F1d} - (E'_{101} + W - hv)}{k_B T}\right]$$

The carrier statistics in the case can be expressed as

$$n_{1D} = \frac{2g_v}{\pi} \frac{\sqrt{2m_c}}{\hbar} \sum_{n_y=1}^{n_{ymax}} \sum_{n_z=1}^{n_{zmax}} \left[R_{101}(E_{F1d}, n_y, n_z) + R_{102}(E_{F1d}, n_y, n_z)\right] \tag{2.166}$$

where

$$R_{101}\left(E_{F1d}, n_y, n_z\right) \equiv \left[I_{36}(E_{F1d}) - G_2\left(n_y, n_z\right)\right]^{1/2},$$

$$R_{102}\left(E_{F1d}, n_y, n_z\right) \equiv \sum_{r=1}^{s} L(r)\left[R_{101}\left(E_{F1d}, n_y, n_z\right)\right].$$

2.3 Results and Discussion

Using the appropriate equations and taking the energy band constants as given in the Table 1.1, we have plotted the normalized EP from NWs of HD $CdGeAs_2$ (an example of nonlinear optical materials) as a function of d_x as shown in plot (a) of Fig. 2.1, in which the plot (b) corresponds to $\delta = 0$. The plot (c) has been drawn in accordance with the three band model of Kane and the plot (d) refers to the two band model of Kane together with the plot (e) exhibiting the variation in accordance with the parabolic energy bands for the overall assessments of the energy band constants on the EP in this case. The Fig. 2.2 exhibits the plots of the normalized EP from NWs of HD $CdGeAs_2$ as a function of the normalized incident photon energy

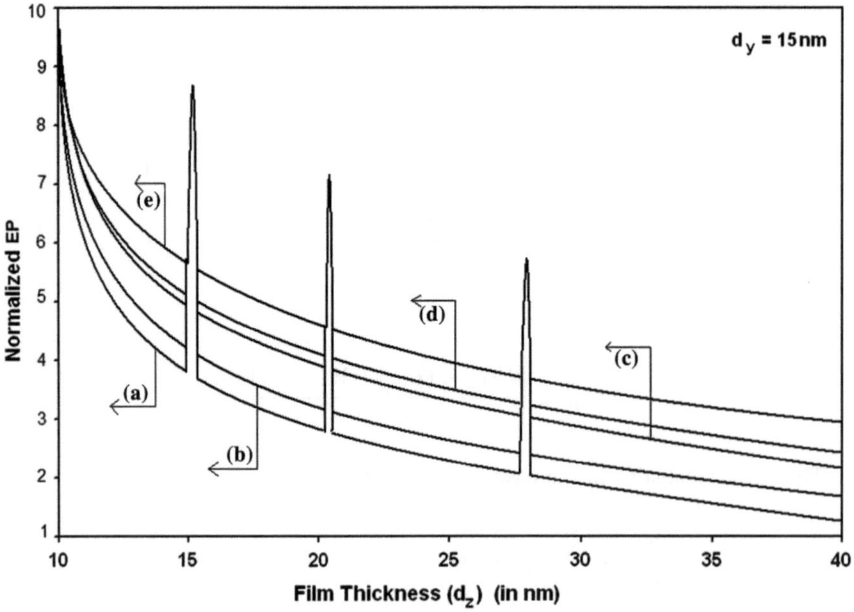

Fig. 2.1 Plot of the normalized EP from NWs of HD $CdGeAs_2$ as a function of d_z in accordance with *a* generalized band model, *b* $\delta = 0$, *c* the three-band model of Kane, *d* the two band model of Kane and *e* the parabolic energy bands

for all cases Figs. 2.1 and 2.3 shows the dependence of the said variable on the normalized electron degeneracy for all cases of Fig. 2.1.

The normalized EP from NWs of HD n-InAs (an example of III-V materials) in accordance with the HD three and two band models of Kane as functions of film thickness, normalized incident photon energy and the normalized electron degeneracy have, respectively, been presented in Figs. 2.4, 2.5 and 2.6. The Figs. 2.7, 2.8 and 2.9 exhibit the variations of normalized EP from NWs of HD n-InSb as functions of film thickness, normalized incident photon energy and the normalized electron degeneracy respectively. The variations of the normalized EP from NWs of HD CdS (an example of II-VI materials) as functions of thickness, normalized incident photon energy and normalized electron degeneracy have respectively been drawn in Figs. 2.10, 2.11 and 2.12, where the plots for $\overline{\lambda}_0 = 0$ have further been drawn for the purpose of assessing the influence of the splitting of the two-spin states by the spin orbit coupling and the crystalline field. The thickness, normalized photon energy and the normalized electron degeneracy dependences of normalized EP from NWs of HD GaP have been shown in Figs. 2.13, 2.14 and 2.15 respectively. The dependence of normalized EP with reference to the aforementioned variables from NWs of HD n-Ge and $PtSb_2$, has been shown in Figs. 2.16, 2.17, 2.18, 2.19, 2.20 and 2.21 in accordance with the HD models of Cardona et al., Wang and Ressler and Emtage respectively. Figures 2.22, 2.23 and 2.24 manifest the variations of the normalized EP from from NWs of HD stressed n-InSb as

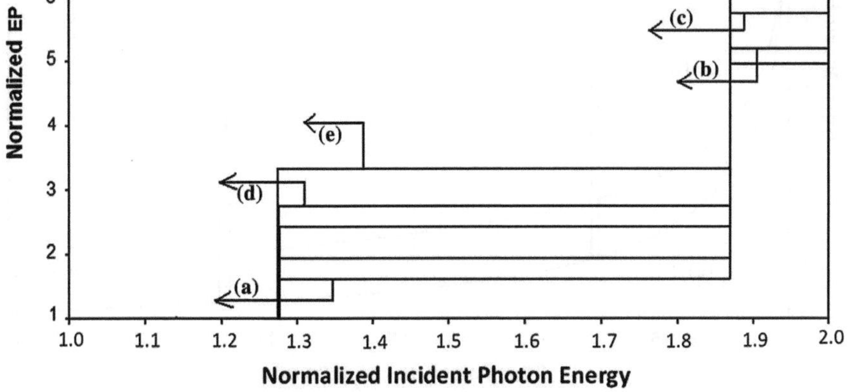

Fig. 2.2 Plot of the normalized EP from NWs of HD $CdGeAs_2$ as a function of normalized incident photon energy for all cases of Fig. 2.1

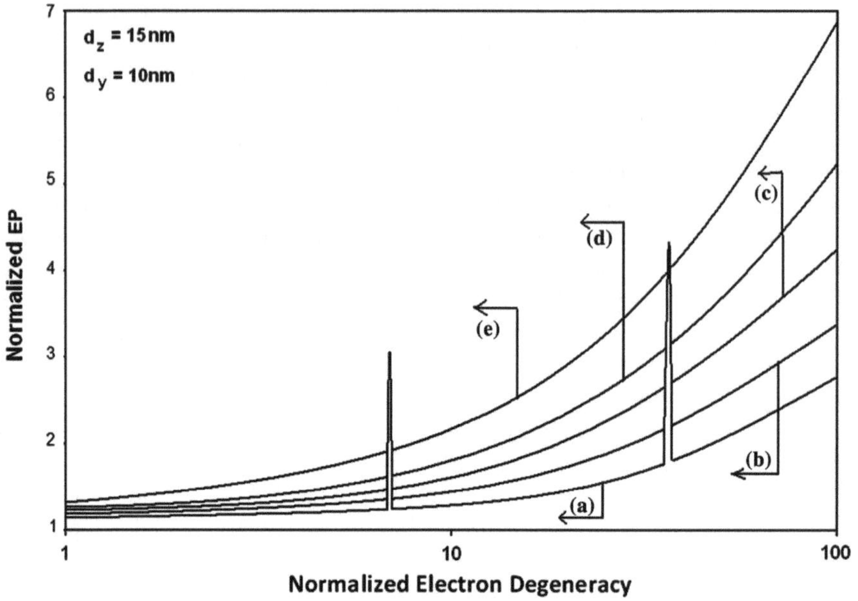

Fig. 2.3 Plot of the normalized EP from NWs of HD $CdGeAs_2$ as a function of normalized electron degeneracy for all cases of Fig. 2.1

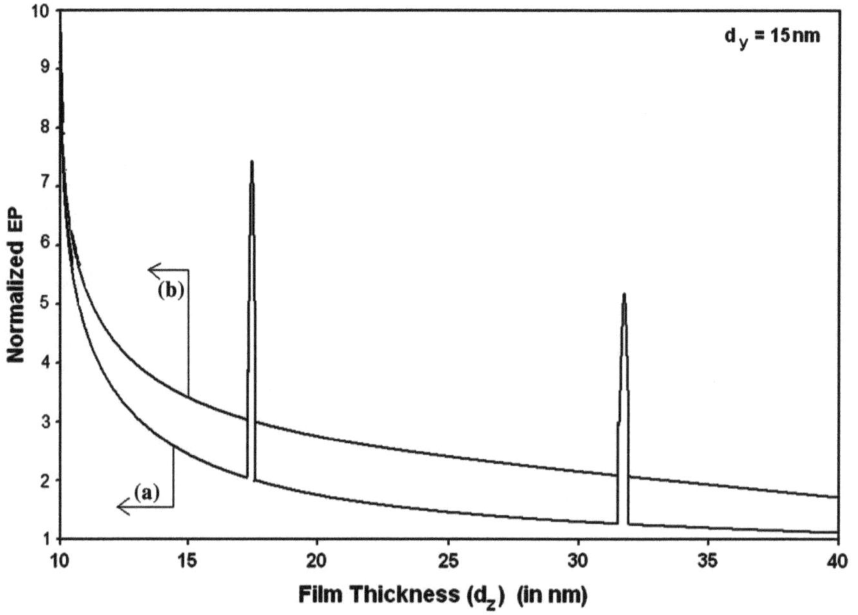

Fig. 2.4 Plot of the normalized EP from NWs of HD n-InAs as a function of d_z in accordance with *a* the three band model of Kane and *b* the two band model of Kane

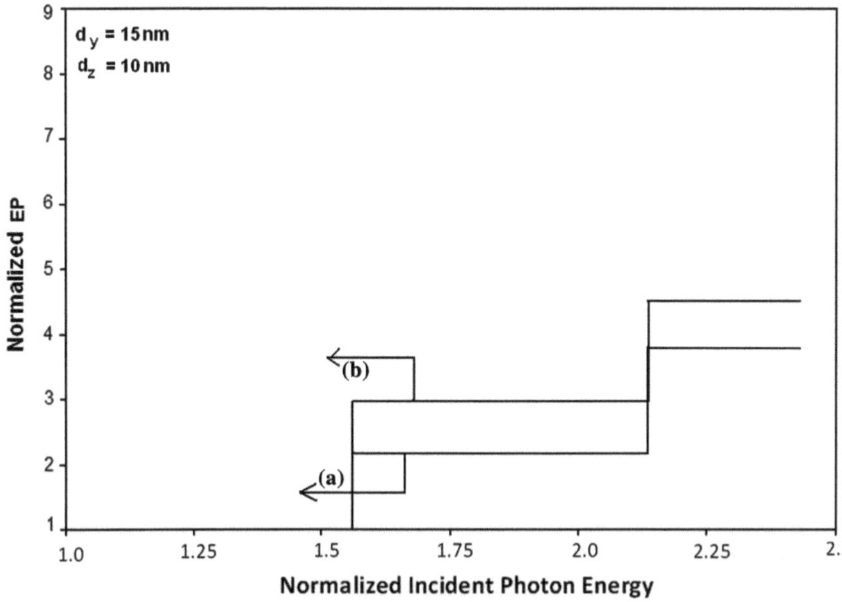

Fig. 2.5 Plot of the normalized EP from NWs of HD n-InAs as a function of normalized incident photon energy in accordance with *a* the three band model of Kane and *b* the two band model of Kane

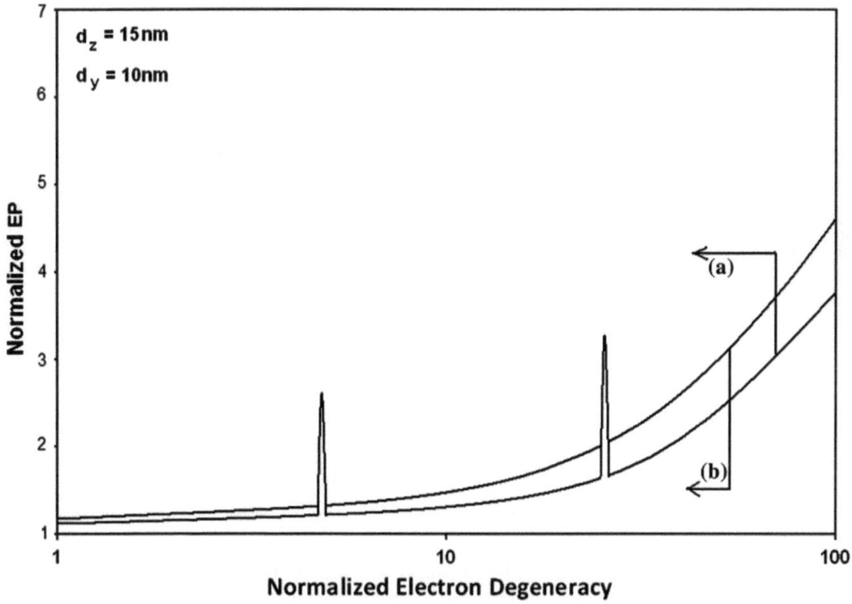

Fig. 2.6 Plot of the normalized EP from NWs of HD n-InAs as a function of normalized electron degeneracy in accordance with *a* the three band model of Kane and *b* the two band model of Kane

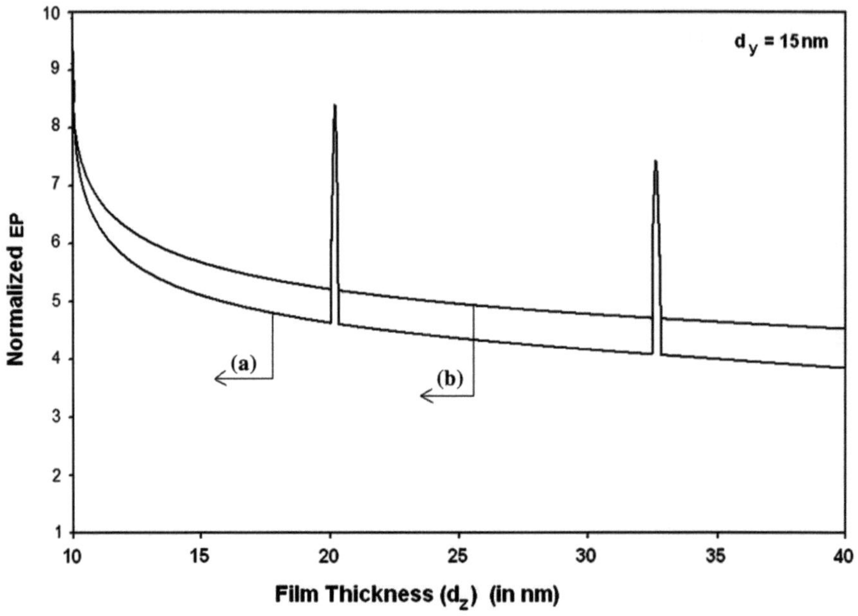

Fig. 2.7 Plot of the normalized EP from NWs of HD n-InSb as a function of d_z in accordance with *a* the three band model of Kane and *b* the two band model of Kane

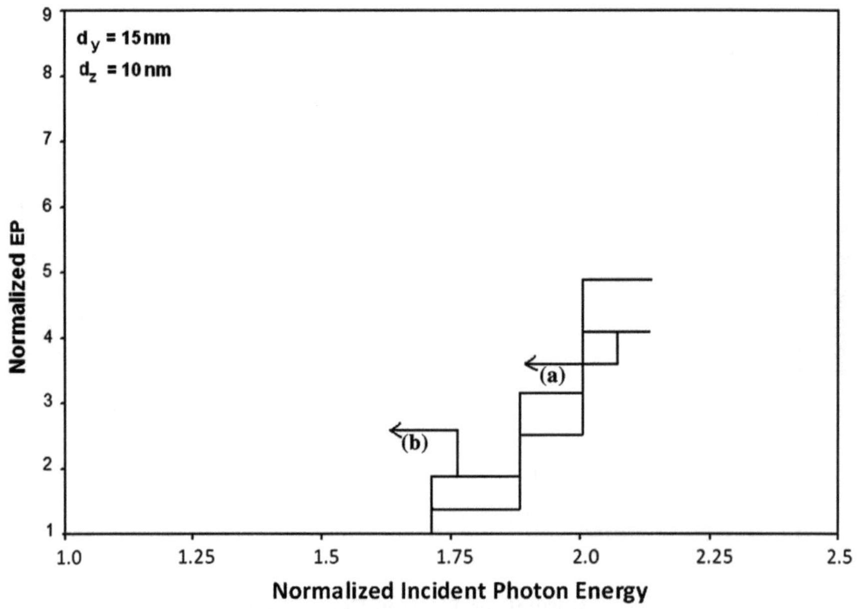

Fig. 2.8 Plot of the normalized EP from NWs of HD n-InSb as a function of normalized incident photon energy in accordance with *a* the three band model of Kane and *b* the two band model of Kane

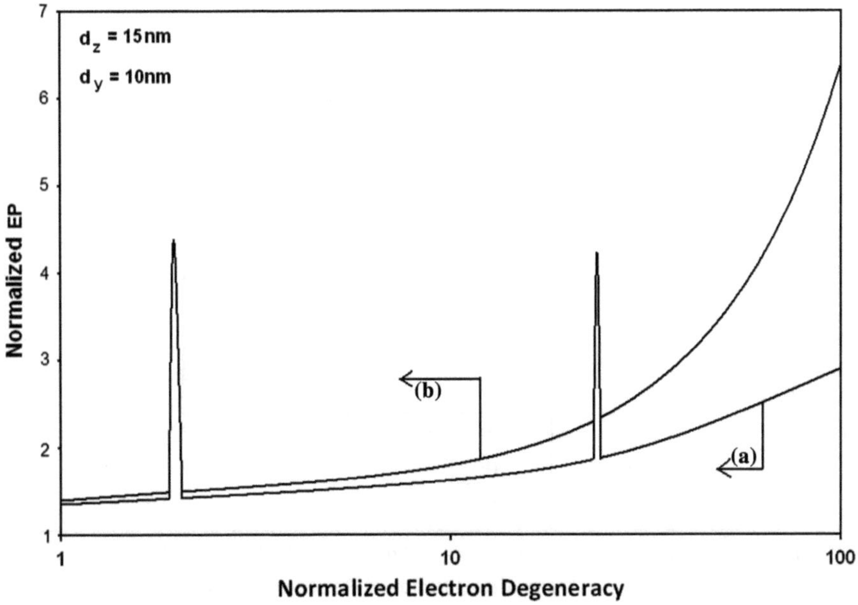

Fig. 2.9 Plot of the normalized EP from NWs of HD n-InSb as a function of normalized electron degeneracy in accordance with *a* the three band model of Kane and *b* the two band model of Kane

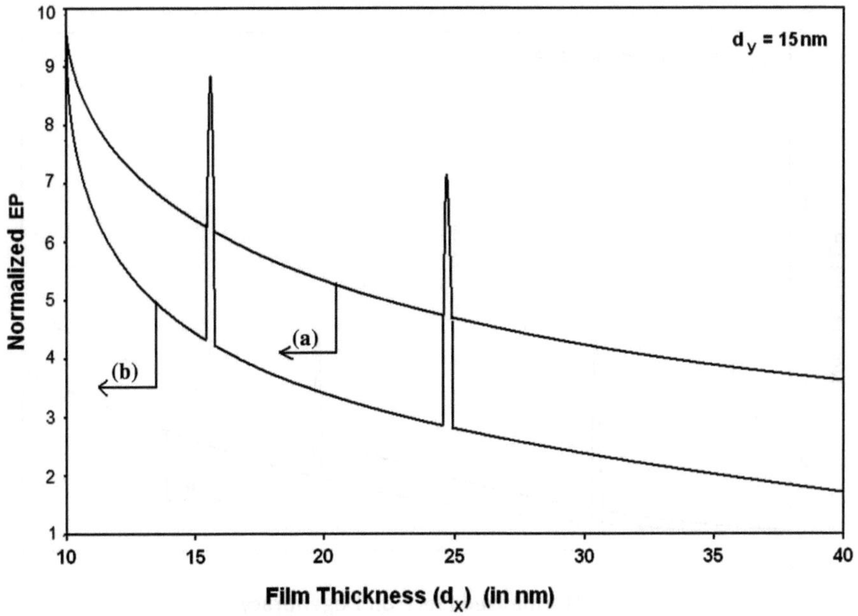

Fig. 2.10 Plot of the normalized EP from NWs of HD CdS as a function of d_z with *a* $\bar{\lambda}_0 \neq 0$ and *b* $\bar{\lambda}_0 = 0$

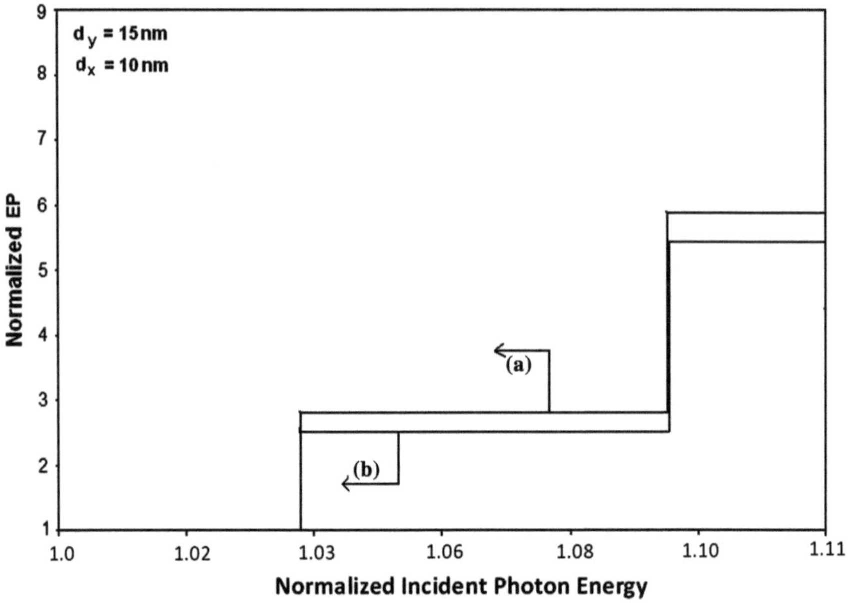

Fig. 2.11 Plot of the normalized EP from NWs of HD CdS as a function of normalized incident photon energy with *a* $\bar{\lambda}_0 \neq 0$ and *b* $\bar{\lambda}_0 = 0$

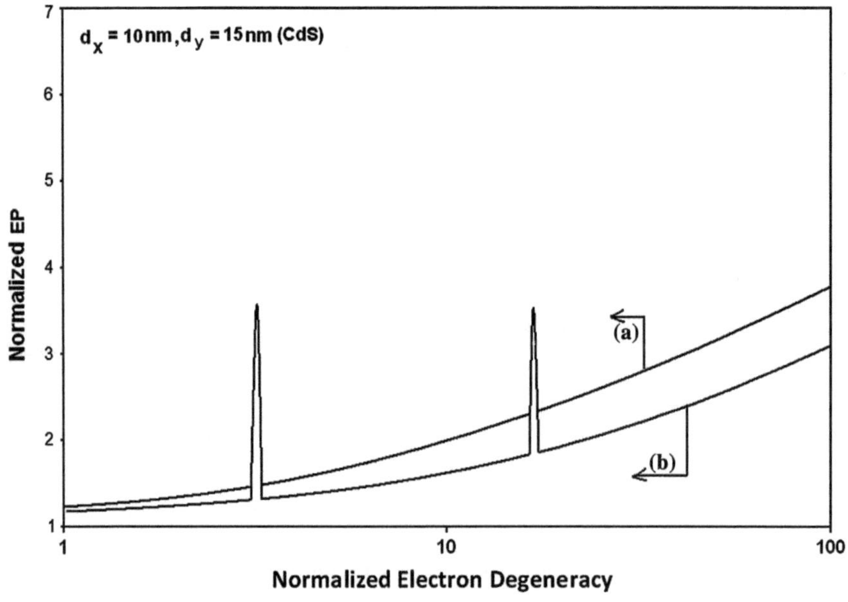

Fig. 2.12 Plot of the normalized EP from NWs of HD CdS as a function of normalized electron degeneracy with *a* $\bar{\lambda}_0 \neq 0$ and *b* $\bar{\lambda}_0 = 0$

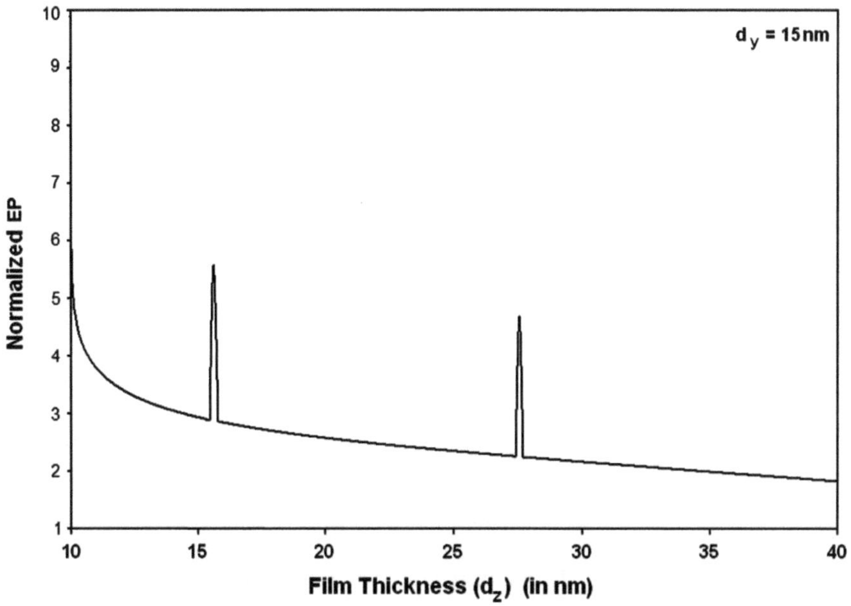

Fig. 2.13 Plot of the normalized EP from NWs of HD n-GaP as a function of d_z

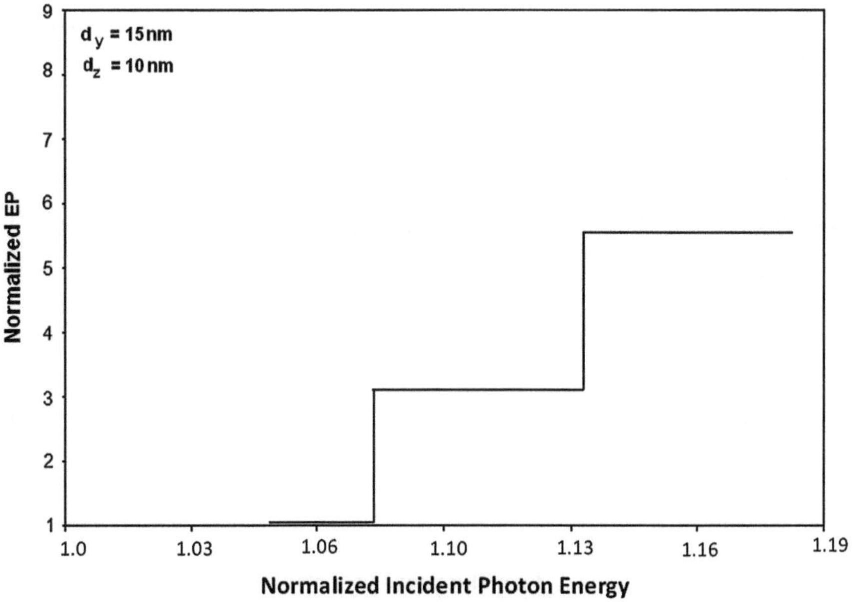

Fig. 2.14 Plot of the normalized EP from NWs of HD n-GaP as a function of normalized incident photon energy

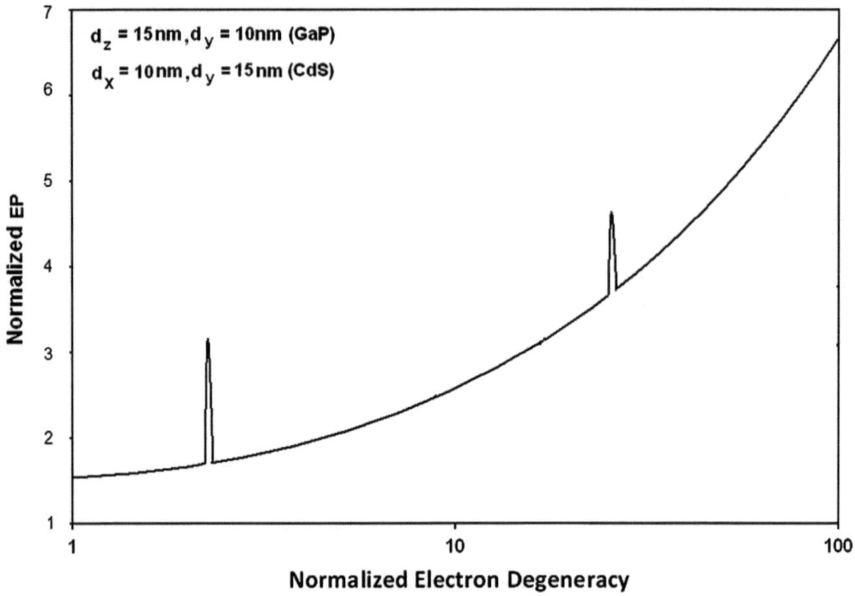

Fig. 2.15 Plot of the normalized EP from NWs of HD n-GaP as a function of normalized electron degeneracy

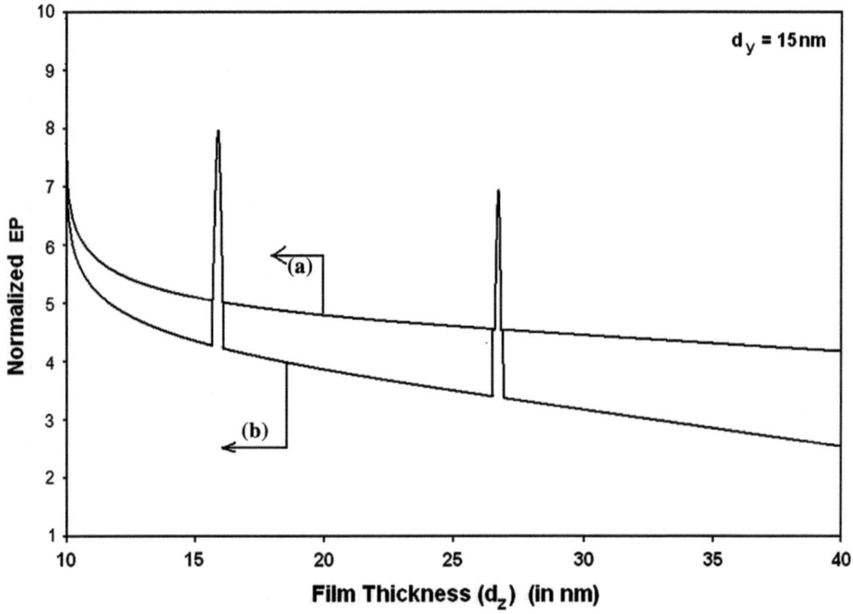

Fig. 2.16 Plot of the normalized EP from NWs of HD n-Ge as a function of thickness in accordance with the models of *a* Cardona et al. and *b* Wang et al.

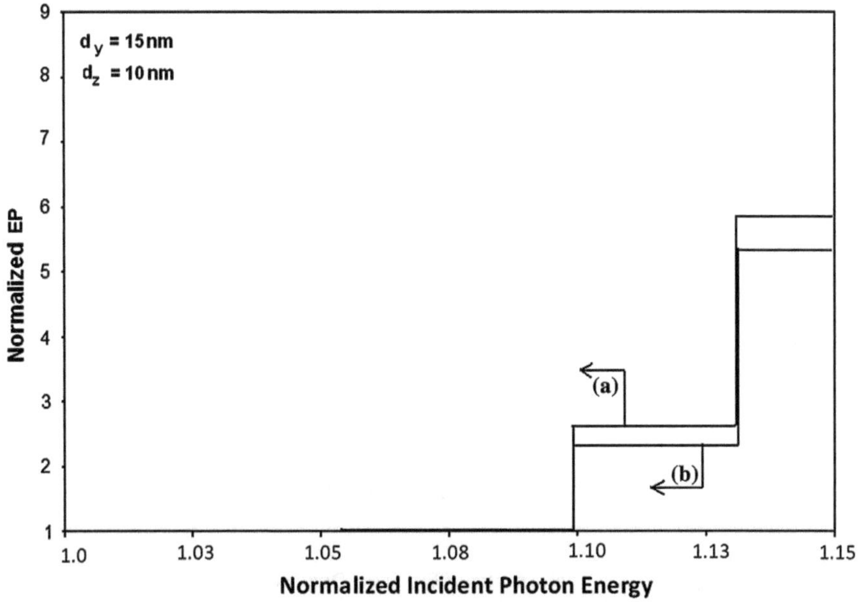

Fig. 2.17 Plot of the normalized EP from NWs of HD n-Ge as a function of normalized incident photon energy for all the cases of Fig. 2.16

functions of the film thickness, normalized incident photon energy and the normalized electron degeneracy respectively.

The Figs. 2.25, 2.26 and 2.27 exhibit the normalized EP as functions of film thickness, normalized incident photon energy and normalized electron degeneracy from NWs of HD PbTe as a function of film thickness in accordance with the models of (a) the Dimmok and (b) the Bangert and Kastner respectively. The plots (c) and (d) exhibit the same for PbSe.

The influence of quantum confinement is immediately apparent from Figs. 2.1, 2.4, 2.7, 2.10, 2.13, 2.16, 2.19, 2.22 and 2.25. Since the EP depends strongly on the thickness of the quantum-confined materials in contrast with the corresponding bulk specimens. The EP decreases with increasing film thickness in an oscillatory way with different numerical magnitudes for HD NWs as compared with HD QWs. It appears from the aforementioned figures that the EP exhibits spikes for particular values of film thickness which, in turn, depends on the particular band structure of the specific material. Moreover, the EP from HD NWs of different compounds can become several orders of magnitude larger than of bulk specimens of the same materials, which is also a direct signature of quantum confinement. This oscillatory dependence will be less and less prominent with increasing film thickness. It appears from Figs. 2.3, 2.6, 2.9, 2.12, 2.15, 2.18, 2.21 and 2.24 that the EP increases with increasing degeneracy and also exhibits spikes for all types of quantum confinement as considered in this chapter. For bulk specimens of the same material, the EP will be found to increase continuously with increasing electron

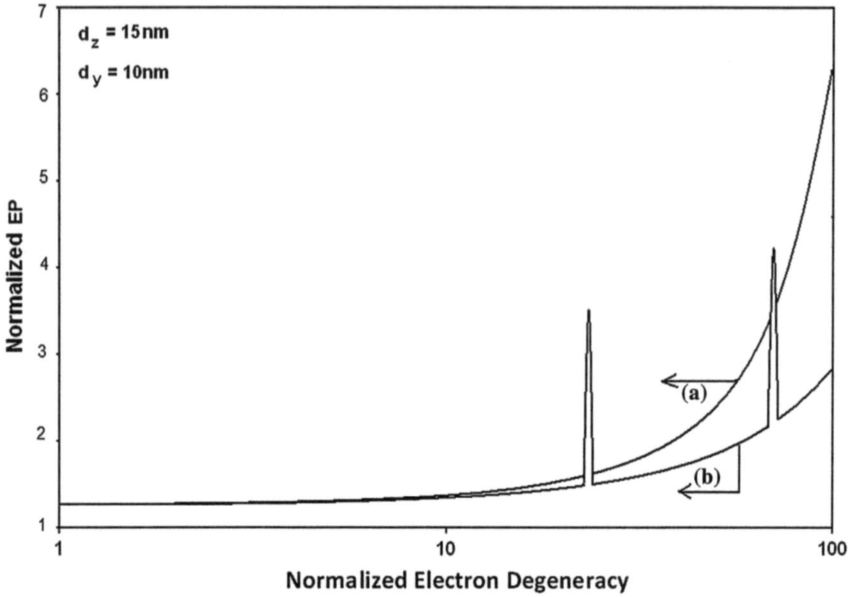

Fig. 2.18 Plot of the normalized EP from NWs of HD n-Ge as a function of normalized electron degeneracy for all the cases of Fig. 2.16

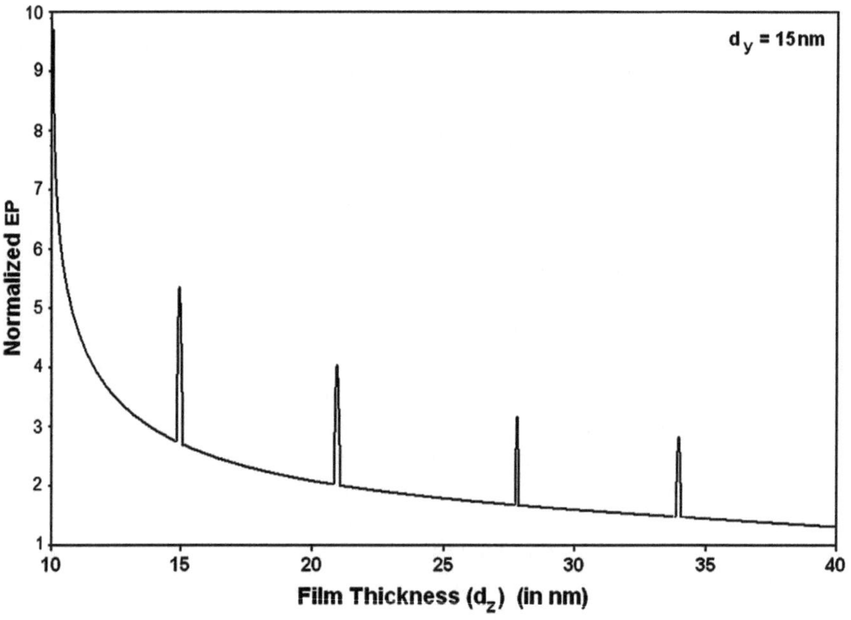

Fig. 2.19 Plot of the normalized EP from NWs of HD n-$PtSb_2$ as a function of film thickness

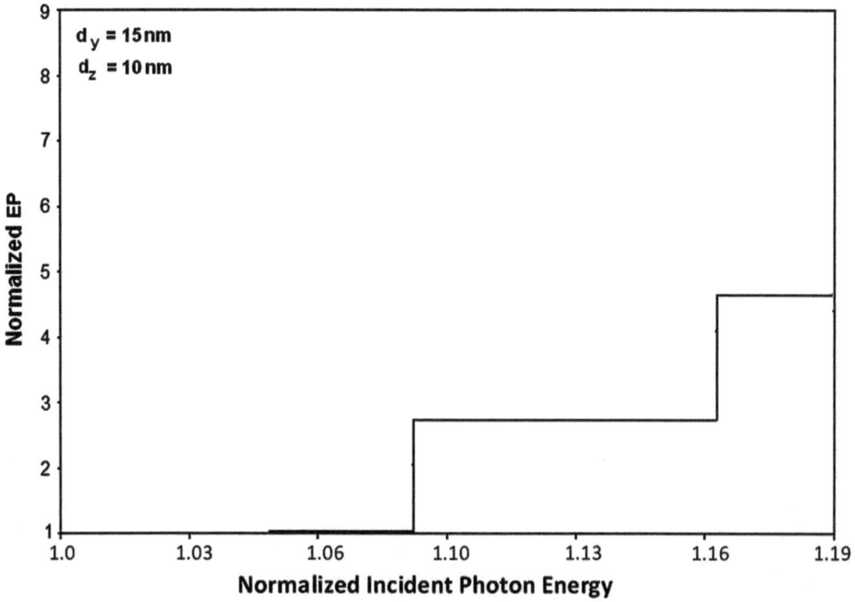

Fig. 2.20 Plot of the normalized EP from NWs of HD n-$PtSb_2$ as a function of normalized incident photon energy

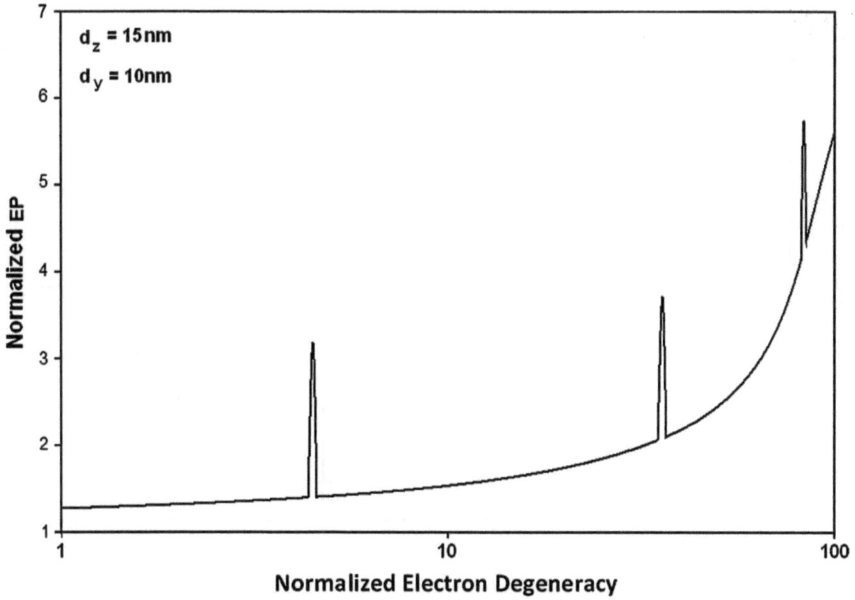

Fig. 2.21 Plot of the normalized EP from NWs of HD n-$PtSb_2$ as a function of normalized electron degeneracy

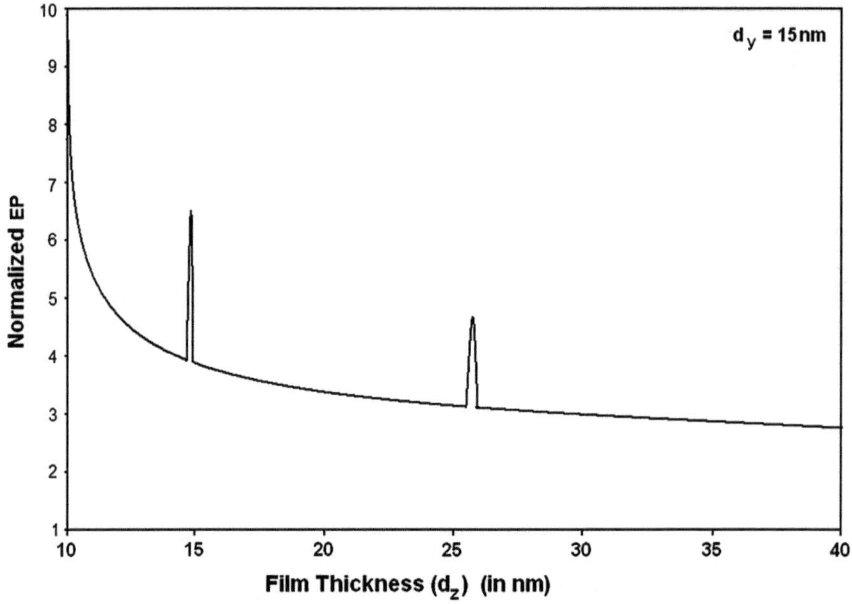

Fig. 2.22 Plot of the normalized EP from NWs of HD stressed n-InSb as a function of film thickness

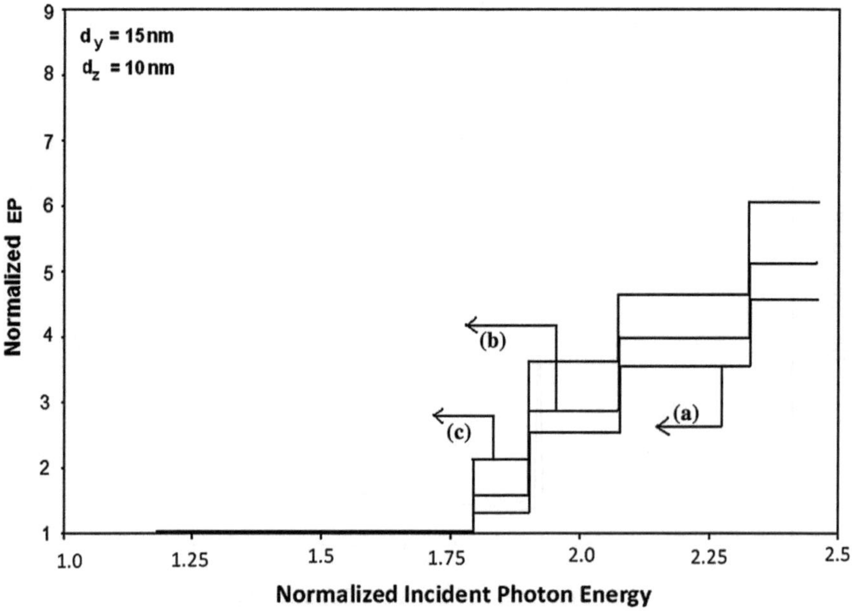

Fig. 2.23 Plot of the normalized EP from NWs of HD stressed n-InSb as a function of normalized incident photon energy

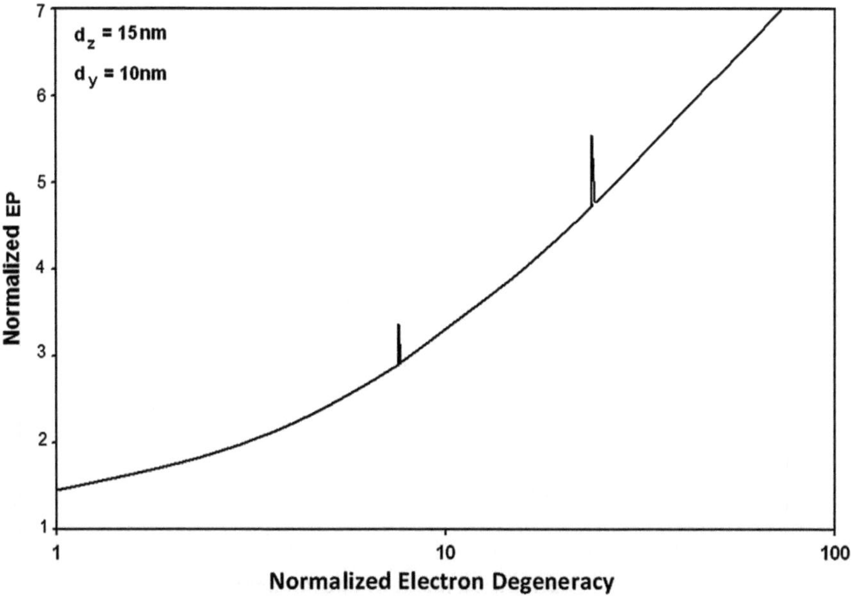

Fig. 2.24 Plot of the normalized EP from NWs of HD stressed n-InSb as a function of normalized electron degeneracy

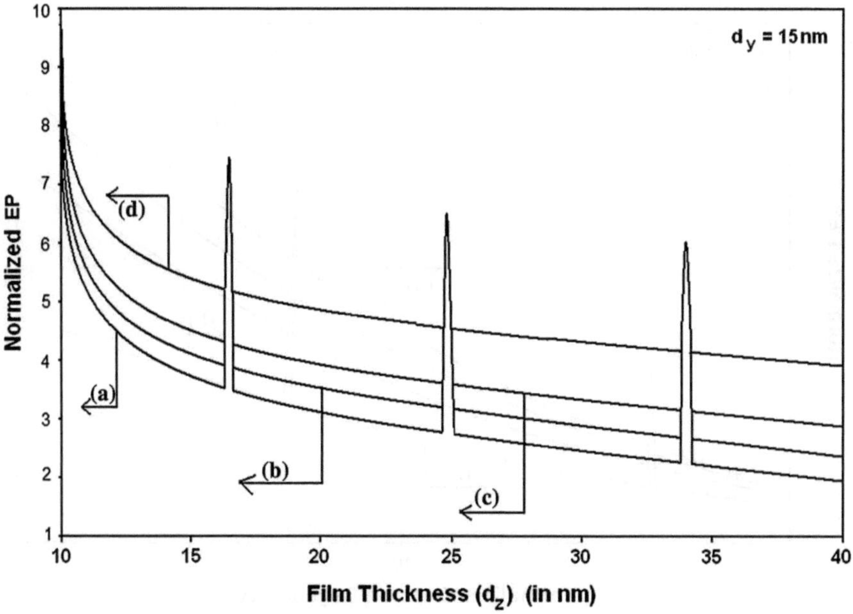

Fig. 2.25 Plot of the normalized EP from NWs of HD PbTe as a function of film thickness in accordance with the models of *a* the Dimmok and *b* the Bangert and Kastner respectively. The plots *c* and *d* exhibit the same for PbSe

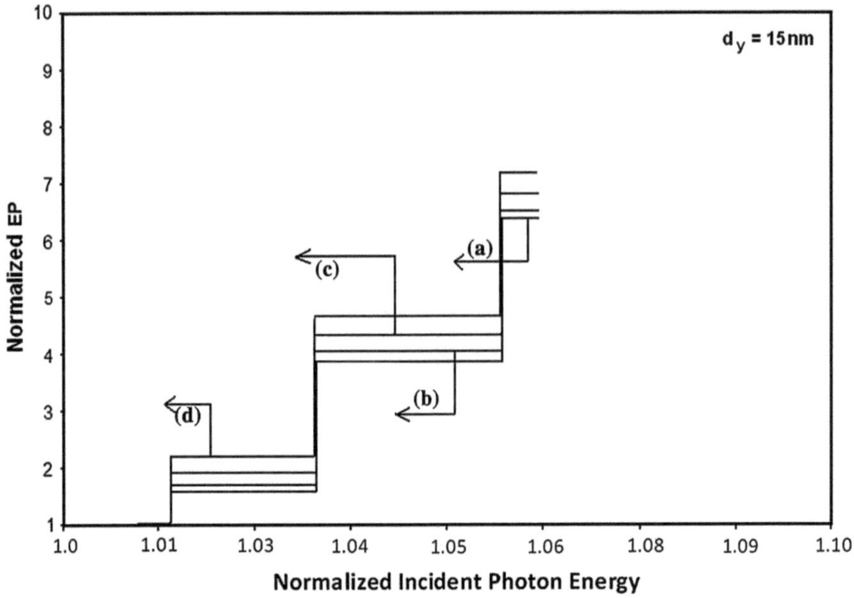

Fig. 2.26 Plot of the normalized EP from QWs of HD PbTe as a function of normalized incident photon energy in accordance with the models of *a* the Dimmok and *b* the Bangert and Kastner respectively. The plots *c* and *d* exhibit the same for PbSe

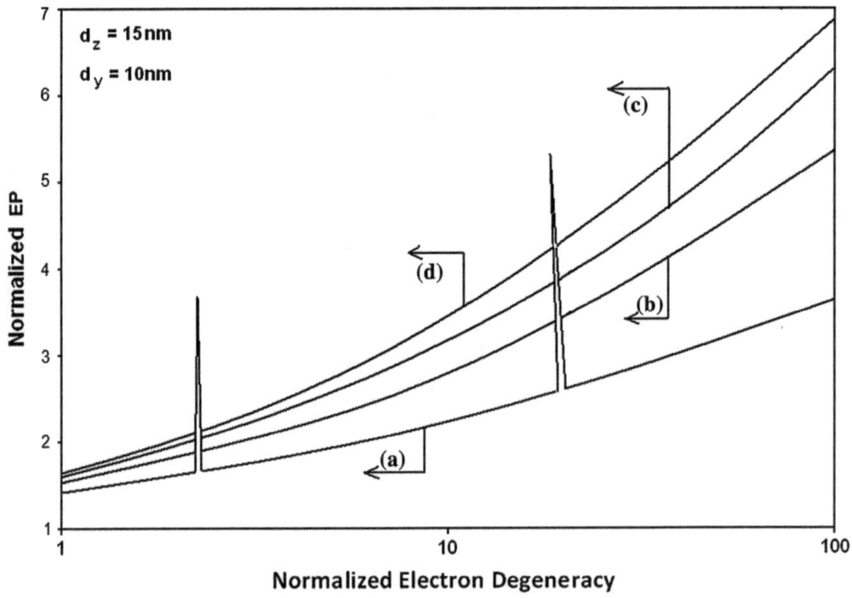

Fig. 2.27 Plot of the normalized EP from QWs of HD PbTe as a function of electron degeneracy in accordance with the models of *a* the Dimmok and *b* the Bangert and Kastner respectively. The plots *c* and *d* exhibit the same for PbSe

degeneracy in a non-oscillatory manner. The Figs. 2.2, 2.5, 2.8, 2.11, 2.14, 2.17, 2.20 and 2.23 illustrate the dependence of the EP from quantum-confined materials on the normalized incident photon energy. The EP increases with increasing photon energy in a step like manner for all the figures. The appearance of the discrete jumps in all the figures is due to the redistribution of the electrons among the quantized energy levels when the size quantum number corresponding to the highest occupied level changes from one fixed value to the others. With varying electron degeneracy, a change is reflected in the EP through the redistribution of the electrons among the size-quantized levels. It may be noted that at the transition zone from one sub band to another, the height of the peaks between any two sub-bands decreases with the increasing in the degree of quantum confinement and is clearly shown in all the curves. It should be noted that although, the EP varies in various manners with all the variables as evident from all the figures, the rates of variations are totally band-structure dependent.

Finally, it may be noted that the basic aim of this chapter is not solely to demonstrate the influence of quantum confinement on the EP from NWs of HD non-linear optical, III-V, II-VI, IV-VI, n-GaP, n-Ge, $PtSb_2$, and stressed compounds respectively but also to formulate the appropriate electron statistics in the most generalized form, since the transport and other phenomena in quantized structures having different band structures and the derivation of the expressions of many important electronic properties are based on the temperature-dependent electron statistics in such materials.

2.4 Open Research Problems

(R.2.1) Investigate the EP for bulk specimens of the HD semiconductors in the presences of Gaussian, exponential, Kane, Halperian, Lax and Bonch-Burevich types of band tails for all systems whose unperturbed carrier energy spectra are defined in R.1.1 in the presence of strain.

(R.2.2) Investigate the EP for NWs of all the HD semiconductors as considered in R.1.2.

(R.2.3) Investigate the EP in the presence of strain for HD bulk specimens of the negative refractive index, organic, magnetic and other advanced optical materials in the presence of an arbitrarily oriented alternating electric field.

(R.2.4) Investigate the EP for the NWs of HD negative refractive index, organic, magnetic and other advanced optical materials in the presence of an arbitrarily oriented alternating electric field.

(R.2.5) Investigate the EP for the multiple NWs of HD materials whose unperturbed carrier energy spectra are defined in R.1.1.

(R.2.6) Investigate the EP for all the appropriate HD low dimensional systems of this chapter in the presence of finite potential wells.

(R.2.7) Investigate the EP for all the appropriate HD low dimensional systems of this chapter in the presence of parabolic potential wells.

(R.2.8) Investigate the EP for all the appropriate HD systems of this chapter forming quantum rings.

(R.2.9) Investigate the EP for all the above appropriate problems in the presence of elliptical Hill and quantum square rings in the presence of strain.

(R.2.10) Investigate the EP for parabolic cylindrical HD low dimensional systems in the presence of an arbitrarily oriented alternating electric field for all the HD materials whose unperturbed carrier energy spectra are defined in R.1.1 in the presence of strain.

(R.2.11) Investigate the EP for HD low dimensional systems of the negative refractive index and other advanced optical materials in the presence of an arbitrarily oriented alternating electric field and non-uniform light waves and in the presence of strain.

(R.2.12) Investigate the EP for triangular HD low dimensional systems of the negative refractive index, organic, magnetic and other advanced optical materials in the presence of an arbitrarily oriented alternating electric field in the presence of strain.

(R.2.13) Investigate the EP for all the problems of (R.1.12) in the presence of arbitrarily oriented magnetic field.

(R.2.14) Investigate the EP for all the problems of (R.1.12) in the presence of alternating electric field.

(R.2.15) Investigate the EP for all the problems of (R.1.12) in the presence of alternating magnetic field.

(R.2.16) Investigate the EP for all the problems of (R.1.12) in the presence of crossed electric field and quantizing magnetic fields.

(R.2.17) Investigate the EP for all the problems of (R.1.12) in the presence of crossed alternating electric field and alternating quantizing magnetic fields.

(R.2.18) Investigate the EP for HD NWs of the negative refractive index, organic and magnetic materials.

(R.2.19) Investigate the EP for HD NWs of the negative refractive index, organic and magnetic materials in the presence of alternating time dependent magnetic field.

(R.2.20) Investigate the EP for HD NWs of the negative refractive index, organic and magnetic materials in the presence of in the presence of crossed alternating electric field and alternating quantizing magnetic fields.

(R.2.21) (a) Investigate the EP for HD NWs of the negative refractive index, organic, magnetic and other advanced optical materials in the presence of an arbitrarily oriented alternating electric field considering many body effects.

(b) Investigate all the appropriate problems of this chapter for a Dirac electron.

(R.2.22) Investigate all the appropriate problems of this chapter by including the many body, image force, broadening and hot carrier effects respectively.

(R.2.23) Investigate all the appropriate problems of this chapter by removing all the mathematical approximations and establishing the respective appropriate uniqueness conditions.

References

1. P. Harrison, *Quantum Wells Wires and Dots* (Wiley, NY, 2002)
2. B.K. Ridley, *Electrons and Phonons in Semiconductors Multilayers* (Cambridge University Press, Cambridge, 1997)
3. G. Bastard, *Wave Mechanics Applied to Semiconductor Heterostructures* (Halsted; Les Ulis Les Editions de Physique, New York, 1988)
4. V.V. Martin, A.A. Kochelap, M.A. Stroscio, *Quantum Heterostructures* (Cambridge University Press, Cambridge, 1999)
5. C.S. Lent, D.J. Kirkner, J. Appl. Phys. **67**, 6353 (1990)
6. F. Sols, M. Macucci, U. Ravaioli, K. Hess, Appl. Phys. Lett. **54**, 350 (1980)
7. C.S. Kim, A.M. Satanin, Y.S. Joe, R.M. Cosby, Phys. Rev. B **60**, 10962 (1999)
8. S. Midgley, J.B. Wang, Phys. Rev. B **64**, 153304 (2001)
9. T. Sugaya, J.P. Bird, M. Ogura, Y. Sugiyama, D.K. Ferry, K.Y. Jang, Appl. Phys. Lett. **80**, 434 (2002)
10. B. Kane, G. Facer, A. Dzurak, N. Lumpkin, R. Clark, L. PfeiKer, K. West, Appl. Phys. Lett. **72**, 3506 (1998)
11. C. Dekker, Phys. Today **52**, 22 (1999)
12. A. Yacoby, H.L. Stormer, N.S. Wingreen, L.N. Pfeiffer, K.W. Baldwin, K.W. West, Phys. Rev. Lett. **77**, 4612 (1996)
13. Y. Hayamizu, M. Yoshita, S. Watanabe, H.A.L. PfeiKer, K. West, Appl. Phys. Lett. **81**, 4937 (2002)
14. S. Frank, P. Poncharal, Z.L. Wang, W.A. Heer, Science **280**, 1744 (1998)
15. I. Kamiya, I.I. Tanaka, K. Tanaka, F. Yamada, Y. Shinozuka, H. Sakaki, Physica E **13**, 131 (2002)
16. A.K. Geim, P.C. Main, N. LaScala, L. Eaves, T.J. Foster, P.H. Beton, J.W. Sakai, F.W. Sheard, M. Henini, G. Hill et al., Phys. Rev. Lett. **72**, 2061 (1994)
17. A.S. Melinkov, V.M. Vinokur, Nature **415**, 60 (2002)
18. K. Schwab, E.A. henriksen, J.M. Worlock, M.L. Roukes, Nature **404**, 974 (2000)
19. L. Kouwenhoven, Nature **403**, 374 (2000)
20. S. Komiyama, O. Astafiev, V. Antonov, H. Hirai, Nature **403**, 405 (2000)
21. E. Paspalakis, Z. Kis, E. Voutsinas, A.F. Terziz, Phys. Rev. B **69**, 155316 (2004)
22. J.H. Jefferson, M. Fearn, D.L.J. Tipton, T.P. Spiller, Phys. Rev. A **66**, 042328 (2002)
23. J. Appenzeller, C. Schroer, T. Schapers, A. Hart, A. Froster, B. Lengeler, H. Luth, Phys. Rev. B **53**, 9959 (1996)
24. J. Appenzeller, C. Schroer, J. Appl. Phys. **87**, 31659 (2002)
25. P. Debray, O.E. Raichev, M. Rahman, R. Akis, W.C. Mitchel, Appl. Phys. Lett. **74**, 768 (1999)
26. P.M. Solomon, Proc. IEEE **70**, 489 (1982)
27. T.E. Schlesinger, T. Kuech, Appl. Phys. Lett. **49**, 519 (1986)
28. D. Kasemet, C.S. Hong, N.B. Patel, P.D. Dapkus, Appl. Phys. Letts. **41**, 912 (1982)

29. K. Woodbridge, P. Blood, E.D. Pletcher, P.J. Hulyer, Appl. Phys. Lett. **45**, 16 (1984)
30. S. Tarucha, H.O. Okamoto, Appl. Phys. Letts. **45**, 16 (1984)
31. H. Heiblum, D.C. Thomas, C.M. Knoedler, M.I. Nathan, Appl. Phys. Letts. **47**, 1105 (1985)
32. O. Aina, M. Mattingly, F.Y. Juan, P.K. Bhattacharyya, Appl. Phys. Letts. **50**, 43 (1987)
33. I. Suemune, L.A. Coldren, IEEE J. Quant. Electronic. **24**, 1178 (1988)
34. D.A.B. Miller, D.S. Chemla, T.C. Damen, J.H. Wood, A.C. Burrus, A.C. Gossard, W. Weigmann, IEEE J. Quant. Electron. **21**, 1462 (1985)
35. J.S. Blakemore, *Semiconductor Statistics* (Dover, New York, 1987)
36. K.P. Ghatak, S. Bhattacharya, S.K. Biswas, A. Dey, A.K. Dasgupta, Phys. Scr. **75**, 820 (2007)

Chapter 3
The EP from Quantum Box of Heavily Doped (HD) Non-parabolic Semiconductors

3.1 Introduction

It is well known that as the dimension of the QWs increases from 1D to 3D, the degree of freedom of the free carriers decreases drastically and the density-of-states function changes from the Heaviside step function in OWs to the Dirac's delta function in Quantum Box (QB) [1, 2].

The QBs can be used for visualizing and tracking molecular processes in cells using standard fluorescence microscopy [3–6]. They display minimal photobleaching [7], thus allowing molecular tracking over prolonged periods and consequently, single molecule can be tracked by using optical fluorescence microscopy [8, 9]. The salient features of quantum dot lasers [10–12] include low threshold currents, higher power, and great stability as compared with the conventional one and the QBs find extensive applications in nano-robotics [13–16], neural networks [17–19] and high density memory or storage media [20]. The QBs are also used in nano-photonics [21] because of their theoretically high quantum yield and have been suggested as implementations of qubits for quantum information processing [22]. The QBs also find applications in diode lasers [23], amplifiers [24, 25], and optical sensors [26, 27]. High-quality QBs are well suited for optical encoding [28, 29] because of their broad excitation profiles and narrow emission spectra. The new generations of QBs have far-reaching potential for the accurate investigations of intracellular processes at the single-molecule level, high-resolution cellular imaging, long-term in vivo observation of cell trafficking, tumor targeting, and diagnostics [30, 31]. The QB nanotechnology is one of the most promising candidates for use in solid-state quantum computation [32, 33]. It may also be noted that the QBs are being used in single electron transistors [34, 35], photovoltaic devices [36, 37], photoelectrics [38], ultrafast all-optical switches and logic gates [39–42], organic dyes [43–45] and in other types of nano devices.

In this chapter in Sects. 3.2.1, 3.2.2, 3.2.3, 3.2.4, 3.2.5, 3.2.6, 3.2.7, 3.2.8, 3.2.9, 3.2.10 and 3.2.11 we have investigated the EP from QBs of HD non-linear optical,

K.P. Ghatak, *Einstein's Photoemission*, Springer Tracts in Modern Physics 262,
DOI 10.1007/978-3-319-11188-9_3

III-V, II-VI, stressed Kane type, Te, GaP, $PtSb_2$, Bi_2Te_3, Ge and GaAs respectively. The Sect. 3.3 contains the result and discussions pertaining to this chapter. The Sect. 3.4 presents 23 open research problems.

3.2 Theoretical Background

3.2.1 The EP from QB of HD Nonlinear Optical Semiconductors

The dispersion relation in this case can be written following (2.1) as

$$\frac{\hbar^2(\mathrm{n}_z\pi/\mathrm{d}_z)^2}{2m_{\|}^* T_{21}(\mathrm{E}_{1QBHD},\eta_g)} + \frac{\hbar^2(\mathrm{n}_y\pi/\mathrm{d}_y)^2}{2m_{\|}^* T_{22}(\mathrm{E}_{1QBHD},\eta_g)} + \frac{\hbar^2(\mathrm{n}_x\pi/\mathrm{d}_x)^2}{2m_{\|}^* T_{21}(\mathrm{E}_{1QBHD},\eta_g)} = 1 \quad (3.1)$$

where E_{1QBHD} is the totally quantized energy in this case.

The total density-of-states function in this case is given by

$$N_{0DT}(E) = \frac{2g_v}{d_x d_y d_z} \sum_{n_x=1}^{n_{x\max}} \sum_{n_y=1}^{n_{y\max}} \sum_{n_z=1}^{n_{z\max}} \delta'(E - \mathrm{E}_{1QBHD}) \quad (3.2)$$

where $\delta'(E - \mathrm{E}_{1QBHD})$ is the Dirac's Delta function.

Using (3.2) and Fermi-Dirac occupation probability factor, the total electron concentration can be written as

$$n_{0D} = \frac{2g_v}{d_x d_y d_z} \text{Real part of} \sum_{n_x=1}^{n_{x\max}} \sum_{n_y=1}^{n_{y\max}} \sum_{n_z=1}^{n_{z\max}} F_{-1}(\eta_{31HD}) \quad (3.3)$$

where $\eta_{31HD} \equiv (k_B T)^{-1}(E_{F0DHD} - \mathrm{E}_{1QBHD})$ and E_{F0DHD} is the Fermi energy in this case.

Therefore the electron concentration per sub-band is given by

$$\Delta n_{0D} = \frac{2g_v}{d_x d_y d_z} \text{Real part of } [F_{-1}(\eta_{31HD})] \quad (3.4)$$

The quantized energy along Z direction the $E_{n_{zHD1}}$ in this case is given by

$$\frac{\hbar^2(\mathrm{n}_z\pi/\mathrm{d}_z)^2}{2m_{\|}^* T_{21}(E_{n_{zHD1}},\eta_g)} = 1 \quad (3.5)$$

The expression of the total photo-emitted current density in this case is

$$J_{0D} = \frac{e\alpha_0 g_v}{2} \text{ Real part of} \sum_{n_x=1}^{n_{x_{\max}}} \sum_{n_y=1}^{n_{y_{\max}}} \sum_{n_{z_{\min}}}^{n_{z_{\max}}} (\Delta n_{0D}) v_z(E_{n_{zHD1}}) \quad (3.6)$$

where $v_z(E_{n_{zHD1}}) = [n_z \pi \hbar][m_{\parallel}^* d_z T_{21}'(E_{n_{zHD1}}, \eta_g)]^{-1}$.

For the purpose of comparison we shall also formulate the EP in the absence of band tails is this case.

Let $E_{n_i} (i = x, y \text{ and } z)$ be the quantized energy levels due to infinitely deep potential well along ith-axis with $n_i (= 1, 2, 3, \ldots)$ as the size quantum numbers. Therefore, from (1.2), one can write

$$\gamma(E_{n_x}) = f_1(E_{n_x}) \left(\frac{\pi n_x}{d_x}\right)^2 \quad (3.7)$$

$$\gamma(E_{n_y}) = f_1(E_{n_y}) \left(\frac{\pi n_y}{d_y}\right)^2 \quad (3.8)$$

$$\gamma(E_{n_z}) = f_2(E_{n_z}) \left(\frac{\pi n_z}{d_z}\right)^2 \quad (3.9)$$

From (1.2), the totally quantized energy (E_{QD1}) can be expressed as

$$\gamma(E_{QD1}) = f_1(E_{QD1}) \left[\left(\frac{\pi n_x}{d_x}\right)^2 + \left(\frac{\pi n_y}{d_y}\right)^2 \right] + f_2(E_{QD1}) \left[\left(\frac{\pi n_z}{d_z}\right)^2 \right] \quad (3.10)$$

The total density-of-states function in this case is given by

$$N_{0DT}(E) = \frac{2g_v}{d_x d_y d_z} \sum_{n_x=1}^{n_{x_{\max}}} \sum_{n_y=1}^{n_{y_{\max}}} \sum_{n_z=1}^{n_{z_{\max}}} \delta'(E - E_{QD1}) \quad (3.11)$$

The total electron concentration in this case can be written as

$$n_{0D} = \frac{2g_v}{d_x d_y d_z} \sum_{n_x=1}^{n_{x_{\max}}} \sum_{n_y=1}^{n_{y_{\max}}} \sum_{n_z=1}^{n_{z_{\max}}} F_{-1}(\eta_{31}) \text{ where } \eta_{31} \equiv (k_B T)^{-1} (E_{F0D} - E_{QD1}) \quad (3.12)$$

Therefore the electron concentration per sub-band is given by

$$\Delta n_{0D} = \left(\frac{2g_v}{d_x d_y d_z}\right) F_{-1}(\eta_{31}) \quad (3.13)$$

The expression of the total photo-emitted current density in this case is

$$J_{0D} = \frac{e\alpha_0 g_v}{2} \sum_{n_x=1}^{n_{x_{\max}}} \sum_{n_y=1}^{n_{y_{\max}}} \sum_{n_{z_{\min}}}^{n_{z_{\max}}} (\Delta n_{0D}) v_z(E_{n_z}) \tag{3.14}$$

where the $n_{z_{\min}}$ should be the nearest integer of the following inequality

$$n_{z_{\min}} \geq \frac{d_z}{\pi} \frac{\sqrt{\gamma(W - hv)}}{\sqrt{f_2(W - hv)}} \tag{3.15}$$

and the velocity of the photo-emitted electrons in the n_zth sub-band can be written as

$$v_z(E_{n_z}) = \frac{1}{\hbar} Q_1(E_{n_z}) \tag{3.16}$$

in which $Q_1(E_{n_z}) \equiv \frac{2f_2^{3/2}(E_{n_z})[\gamma(E_{n_z})]^{1/2}}{[f_2(E_{n_z})\gamma'(E_{n_z}) - \gamma(E_{n_z})f_2'(E_{n_z})]}$

Using the appropriate equations we get

$$J_{0D} = \left(\frac{e\alpha_0 g_v}{\hbar d_x d_y d_z}\right) \sum_{n_x=1}^{n_{x_{\max}}} \sum_{n_y=1}^{n_{y_{\max}}} \sum_{n_{z_{\min}}}^{n_{z_{\max}}} F_{-1}(\eta_{31}) Q_1(E_{n_z}). \tag{3.17}$$

3.2.2 The EP from QB of HD III-V Semiconductors

The dispersion relation of the conduction electrons of III-V semiconductors are described by the models of Kane (both three and two bands) Stillman et al. and Palik et al., respectively. For the purpose of complete and coherent presentation, the EP effect in QBs of HD III-V compounds have also been investigated in accordance with the aforementioned different dispersion relations for relative comparison as follows:

(a) The three band model of Kane
 The dispersion relation in this case can be written following (2.15) as

$$\frac{\hbar^2(\mathrm{n}_z\pi/\mathrm{d}_z)^2}{2m_c} + \frac{\hbar^2(\mathrm{n}_y\pi/\mathrm{d}_y)^2}{2m_c} + \frac{\hbar^2(\mathrm{n}_x\pi/\mathrm{d}_x)^2}{2m_c} = T_{44}(\mathrm{E}_{2QBHD}, \eta_g) \tag{3.18}$$

where $T_{44}(\mathrm{E}_{2QBHD}, \eta_g) = T_{31}(\mathrm{E}_{2QBHD}, \eta_g) + iT_{31}(\mathrm{E}_{2QBHD}, \eta_g)$ and E_{2QBHD} is the totally quantized energy in this case.

The total electron concentration can be written as

$$n_{0D} = \frac{2g_v}{d_x d_y d_z} \text{ Real part of} \sum_{n_x=1}^{n_{x_{\max}}} \sum_{n_y=1}^{n_{y_{\max}}} \sum_{n_z=1}^{n_{z_{\max}}} F_{-1}(\eta_{32HD}) \quad (3.19)$$

where $\eta_{32HD} \equiv (k_B T)^{-1}(E_{F0DHD} - E_{2QBHD})$.
Therefore the electron concentration per sub-band is given by

$$\Delta n_{0D} = \frac{2g_v}{d_x d_y d_z} \text{ Real part of } [F_{-1}(\eta_{32HD})] \quad (3.20)$$

The quantized energy along Z direction the $E_{n_{zHD2}}$ in this case is given by

$$\frac{\hbar^2 (\mathrm{n}_z \pi / \mathrm{d}_z)^2}{2m_c T_{44}(E_{n_{zHD2}}, \eta_g)} = 1 \quad (3.21)$$

The expression of the total photo-emitted current density in this case is

$$J_{0D} = \frac{e\alpha_0 g_v}{2} \text{ Real part of} \sum_{n_x=1}^{n_{x_{\max}}} \sum_{n_y=1}^{n_{y_{\max}}} \sum_{n_{z_{\min}}}^{n_{z_{\max}}} (\Delta n_{0D}) v_{z2}(E_{n_{zHD2}}) \quad (3.22)$$

where $v_{z2}(E_{n_{zHD2}}) = [n_z \pi \hbar][m_c d_z T'_{44}(E_{n_{zHD2}}, \eta_g)]^{-1}$
The quantized energy levels (E_{n_x}, E_{n_y} and E_{n_z} along x, y, and z directions respectively) in the absence of band tails in QBs of III-V semiconductors in accordance with the three band model of Kane can be expressed as

$$I_{11}(E_{n_x}) = \frac{\hbar^2}{2m_c}\left(\frac{\pi n_x}{d_x}\right)^2 \quad (3.23)$$

$$I_{11}(E_{n_y}) = \frac{\hbar^2}{2m_c}\left(\frac{\pi n_y}{d_y}\right)^2 \quad (3.24)$$

$$\text{and } I_{11}(E_{n_z}) = \frac{\hbar^2}{2m_c}\left(\frac{\pi n_z}{d_z}\right)^2 \quad (3.25)$$

The totally quantized energy (E_{QD2}) (changed) in this case assumes the form

$$I_{11}(E_{QD2}) = \frac{\hbar^2 \pi^2}{2m_c}\left[\left(\frac{n_x}{d_x}\right)^2 + \left(\frac{n_y}{d_y}\right)^2 + \left(\frac{n_z}{d_z}\right)^2\right] \quad (3.26)$$

The velocity of the photoelectron in the n_zth quantized level is given by

$$v_z(E_{n_z}) = \sqrt{\frac{2}{m_c}} Q_2(E_{n_z}) \tag{3.27}$$

where, $Q_2(E_{n_z}) \equiv \left[(I_{11}(E_{n_z}))'\right]^{-1}\left(\sqrt{I_{11}(E_{n_z})}\right)$
Thus the photo-emitted current density can be expressed as

$$J_{0D} = \left[\frac{\alpha_0 e g_v}{(d_x d_y d_z)}\sqrt{\frac{2}{m_c}}\right]\sum_{n_x=1}^{n_{x_{\max}}}\sum_{n_y=1}^{n_{y_{\max}}}\sum_{n_{z_{\min}}}^{n_{z_{\max}}} Q_2(E_{n_z})F_{-1}(\eta_{32}) \tag{3.28}$$

in which $n_{z_{\min}}$ is the nearest integer of the following inequality,

$$n_{z_{\min}} \geq \left(\frac{d_z}{\pi}\frac{\sqrt{2m_c}}{\hbar}\left[I_{11}(W - h\nu)\right]^{1/2}\right) \tag{3.29}$$

and $\eta_{32} = (k_B T)^{-1}(E_{F0D} - E_{QD2})$
The electron concentration in this case is given by

$$n_{0D} = \frac{2g_v}{d_x d_y d_z}\sum_{n_x=1}^{n_{x_{\max}}}\sum_{n_y=1}^{n_{y_{\max}}}\sum_{n_z=1}^{n_{z_{\max}}} F_{-1}(\eta_{32}). \tag{3.30}$$

(b) The two band model of Kane
The dispersion relation in this case can be written following (2.24) as

$$\frac{\hbar^2(\mathrm{n}_z\pi/\mathrm{d}_z)^2}{2m_c} + \frac{\hbar^2(\mathrm{n}_y\pi/\mathrm{d}_y)^2}{2m_c} + \frac{\hbar^2(\mathrm{n}_x\pi/\mathrm{d}_x)^2}{2m_c} = \gamma_2(\mathrm{E}_{3QBHD}, \eta_g) \tag{3.31}$$

and E_{3QBHD} is the totally quantized energy in this case.
The total electron concentration can be written as

$$n_{0D} = \frac{2g_v}{d_x d_y d_z}\sum_{n_x=1}^{n_{x_{\max}}}\sum_{n_y=1}^{n_{y_{\max}}}\sum_{n_z=1}^{n_{z_{\max}}} F_{-1}(\eta_{33HD}) \tag{3.32}$$

where $\eta_{33HD} \equiv (k_B T)^{-1}(E_{F0DHD} - \mathrm{E}_{3QBHD})$.
Therefore the electron concentration per sub-band is given by

$$\Delta n_{0D} = \frac{2g_v}{d_x d_y d_z}\left[F_{-1}(\eta_{33HD})\right] \tag{3.33}$$

The quantized energy along Z direction the $E_{n_{zHD3}}$ in this case is given by

$$\frac{\hbar^2(n_z\pi/d_z)^2}{2m_c\gamma_2(E_{n_{zHD3}},\eta_g)}=1 \tag{3.34}$$

The expression of the total photo-emitted current density in this case is

$$J_{0D}=\frac{e\alpha_0 g_v}{2}\sum_{n_x=1}^{n_{x_{\max}}}\sum_{n_y=1}^{n_{y_{\max}}}\sum_{n_{z_{\min}}}^{n_{z_{\max}}}(\Delta n_{0D})v_{z3}(E_{n_{zHD3}}) \tag{3.35}$$

where $v_{z3}(E_{n_{zHD3}})=[n_z\pi\hbar][m_c d_z\gamma_2'(E_{n_{zHD3}},\eta_g)]^{-1}$

For two band model of Kane and the photo-emitted current density in the absence of band tails assumes the form

$$J_{0D}=\frac{\alpha_0 g_v e}{d_x d_y d_z}\left(\sqrt{\frac{2}{m_c}}\right)\sum_{n_x=1}^{n_{x_{\max}}}\sum_{n_y=1}^{n_{y_{\max}}}\sum_{n_{z_{\min}}}^{n_{z_{\max}}}Q_3(E_{n_z})F_{-1}(\eta_{33}) \tag{3.36}$$

where, $n_{z_{\min}}\geq\left(\frac{d_z}{\pi}\frac{\sqrt{2m_c}}{\hbar}[(W-h\nu)[1+\alpha(W-h\nu)]]^{1/2}\right)$, $Q_3(E_{n_z})\equiv\left[\sqrt{E_{n_z}(1+\alpha E_{n_z})}/(1+2\alpha E_{n_z})\right]$ and $\eta_{33}\equiv\frac{1}{k_BT}[E_{F0D}-E_{QD3}]$

The E_{n_z} obeys the equation

$$E_{n_z}(1+\alpha E_{n_z})=\frac{\hbar^2\pi^2}{2m_c}\left(\frac{n_z}{d_z}\right)^2 \tag{3.37}$$

The totally quantized energy (E_{QD3}) in this case is given by

$$E_{QD3}(1+\alpha E_{QD3})=\frac{\hbar^2\pi^2}{2m_c}\left[\left(\frac{n_x}{d_x}\right)^2+\left(\frac{n_y}{d_y}\right)^2+\left(\frac{n_z}{d_z}\right)^2\right] \tag{3.38}$$

The electron concentration in this case is given by

$$n_{0D}=\frac{2g_v}{d_x d_y d_z}\sum_{n_x=1}^{n_{x_{\max}}}\sum_{n_y=1}^{n_{y_{\max}}}\sum_{n_z=1}^{n_{z_{\max}}}F_{-1}(\eta_{33}). \tag{3.39}$$

(c) The parabolic energy bands
The dispersion relation in this case can be written following (2.32) as

$$\frac{\hbar^2(n_z\pi/d_z)^2}{2m_c}+\frac{\hbar^2(n_y\pi/d_y)^2}{2m_c}+\frac{\hbar^2(n_x\pi/d_x)^2}{2m_c}=\gamma_3(E_{4QBHD},\eta_g) \tag{3.40}$$

and E_{4QBHD} is the totally quantized energy in this case.

The total electron concentration can be written as

$$n_{0D} = \frac{2g_v}{d_x d_y d_z} \sum_{n_x=1}^{n_{x_{\max}}} \sum_{n_y=1}^{n_{y_{\max}}} \sum_{n_z=1}^{n_{z_{\max}}} F_{-1}(\eta_{34HD}) \tag{3.41}$$

where $\eta_{34HD} \equiv (k_B T)^{-1}(E_{F0DHD} - \mathrm{E}_{4QBHD})$.
Therefore the electron concentration per sub-band is given by

$$\Delta n_{0D} = \frac{2g_v}{d_x d_y d_z} \left[F_{-1}(\eta_{34HD})\right] \tag{3.42}$$

The quantized energy along Z direction the $E_{n_{zHD4}}$ in this case is given by

$$\frac{\hbar^2 (\mathrm{n}_z \pi/\mathrm{d}_z)^2}{2m_c \gamma_3 (E_{n_{zHD4}}, \eta_g)} = 1 \tag{3.43}$$

The expression of the total photo-emitted current density in this case is

$$J_{0D} = \frac{e\alpha_0 g_v}{2} \sum_{n_x=1}^{n_{x_{\max}}} \sum_{n_y=1}^{n_{y_{\max}}} \sum_{n_{z_{\min}}}^{n_{z_{\max}}} (\Delta n_{0D}) v_{z4}\left(E_{n_{zHD4}}\right) \tag{3.44}$$

where $v_{z4}(E_{n_{zHD4}}) = [n_z \pi \hbar][m_c d_z \gamma_3'(E_{n_{zHD4}}, \eta_g)]^{-1}$
In the absence of band tails the expressions for Δn_{0D}, E_{n_z}, $n_{z_{\min}}$, J_{0D} and total electron concentration (n_{0D}), for QBs of wide-gap materials can, respectively, be written as

$$\Delta n_{0D} = \frac{2g_v}{d_x d_y d_z} F_{-1}(\eta') \tag{3.45}$$

$$E_{n_z} = \frac{\hbar^2}{2m_c}\left(\frac{n_z \pi}{d_z}\right)^2 \tag{3.46}$$

$$n_{z_{\min}} \geq \left(\frac{d_z}{\pi}\right) \frac{\sqrt{2m_c}}{\hbar} \sqrt{(W - h\nu)} \tag{3.47}$$

$$J_{0D} = \frac{\alpha_o e \hbar \pi g_v}{m_c d_x d_y d_z^2} \sum_{n_x=1}^{n_{x\max}} \sum_{n_y=1}^{n_{y\max}} \sum_{n_{z_{\min}}}^{n_{z\max}} [n_z F_{-1}(\eta')] \tag{3.48}$$

$$n_{0D} = \frac{2g_v}{d_x d_y d_z} \sum_{n_x=1}^{n_{x\max}} \sum_{n_y=1}^{n_{y\max}} \sum_{n_z=1}^{n_{z\max}} F_{-1}(\eta') \tag{3.49}$$

where, $\eta' \equiv (k_B T)^{-1}\left[E_{F0D} - \frac{\hbar^2}{2m_c}\left(\frac{n_z \pi}{d_z}\right)^2\right]$.

(d) The Model of Stillman et al.
The dispersion relation of the electrons in this case can be written following (2.40) as

$$\frac{\hbar^2(\mathrm{n}_z\pi/\mathrm{d}_z)^2}{2m_c}+\frac{\hbar^2(\mathrm{n}_y\pi/\mathrm{d}_y)^2}{2m_c}+\frac{\hbar^2(\mathrm{n}_x\pi/\mathrm{d}_x)^2}{2m_c}=\theta_4(\mathrm{E}_{5QBHD},\eta_g) \tag{3.50}$$

and E_{5QBHD} is the totally quantized energy in this case.
The total electron concentration can be written as

$$n_{0D}=\frac{2g_v}{d_xd_yd_z}\sum_{n_x=1}^{n_{x\max}}\sum_{n_y=1}^{n_{y\max}}\sum_{n_z=1}^{n_{z\max}}F_{-1}(\eta_{35HD}) \tag{3.51}$$

where $\eta_{35HD}\equiv(k_BT)^{-1}(E_{F0DHD}-\mathrm{E}_{5QBHD})$.
Therefore the electron concentration per sub-band is given by

$$\Delta n_{0D}=\frac{2g_v}{d_xd_yd_z}\,[F_{-1}(\eta_{35HD})] \tag{3.52}$$

The quantized energy along Z direction the $E_{n_{zHD4}}$ in this case is given by

$$\frac{\hbar^2(\mathrm{n}_z\pi/\mathrm{d}_z)^2}{2m_c\theta_4(E_{n_{zHD5}},\eta_g)}=1 \tag{3.53}$$

The expression of the total photo-emitted current density in this case is

$$J_{0D}=\frac{e\alpha_0g_v}{2}\sum_{n_x=1}^{n_{x\max}}\sum_{n_y=1}^{n_{y\max}}\sum_{n_{z\min}}^{n_{z\max}}(\Delta n_{0D})v_{z5}(E_{n_{zHD5}}) \tag{3.54}$$

where $v_{z5}(E_{n_{zHD5}})=[n_z\pi\hbar][m_cd_z\theta_4'(E_{n_{zHD5}},\eta_g)]^{-1}$
In the absence of band tails, the photoelectric current density in this case can be written as

$$J_{0D}=\frac{\alpha_0g_ve}{d_xd_yd_z}\left(\frac{2}{m_c}\right)^{1/2}\sum_{n_x=1}^{n_{x\max}}\sum_{n_y=1}^{n_{y\max}}\sum_{n_{z\min}}^{n_{z\max}}Q_5(E_{n_z})F_{-1}(\eta_{35}) \tag{3.55}$$

where $n_{z_{\min}}\geq\frac{d_z}{\pi}\frac{\sqrt{2m_c}}{\hbar}[I_{12}(W-hv)]^{1/2}$, $Q_5(E_{n_z})\equiv\left[[I_{12}(E_{n_z})]^{1/2}[I_{12}'(E_{n_z})]^{-1}\right]$
and $\eta_{35}\equiv(k_BT)^{-1}[E_{F0D}-E_{QD5}]$

The E_{n_z} obeys the equation

$$\left(\hbar^2/2m_c\right)\left(\pi n_z/d_z\right)^2 \equiv I_{12}\left(E_{n_z}\right) \tag{3.56}$$

The E_{QD5} in this case can be defined as

$$I_{12}(E_{QD5}) = \frac{\hbar^2\pi^2}{2m_c}\left[\left(\frac{n_x}{d_x}\right)^2 + \left(\frac{n_y}{d_y}\right)^2 + \left(\frac{n_z}{d_z}\right)^2\right] \tag{3.57}$$

The electron concentration in this case is given by

$$n_{0D} = \frac{2g_v}{d_x d_y d_z}\sum_{n_x=1}^{n_{x\max}}\sum_{n_y=1}^{n_{y\max}}\sum_{n_z=1}^{n_{z\max}} F_{-1}(\eta_{35}). \tag{3.58}$$

(e) The model of Palik et al.
The dispersion relation of the electrons in this case can be written following (2.48) as

$$\frac{\hbar^2(\mathrm{n}_z\pi/\mathrm{d}_z)^2}{2m_c} + \frac{\hbar^2(\mathrm{n}_y\pi/\mathrm{d}_y)^2}{2m_c} + \frac{\hbar^2(\mathrm{n}_x\pi/\mathrm{d}_x)^2}{2m_c} = \theta_5(\mathrm{E}_{6QBHD}, \eta_g) \tag{3.59}$$

and E_{6QBHD} is the totally quantized energy in this case.
The total electron concentration can be written as

$$n_{0D} = \frac{2g_v}{d_x d_y d_z}\sum_{n_x=1}^{n_{x\max}}\sum_{n_y=1}^{n_{y\max}}\sum_{n_z=1}^{n_{z\max}} F_{-1}(\eta_{36HD}) \tag{3.60}$$

where $\eta_{36HD} \equiv (k_B T)^{-1}(E_{F0DHD} - E_{6QBHD})$.
Therefore the electron concentration per sub-band is given by

$$\Delta n_{0D} = \frac{2g_v}{d_x d_y d_z}\left[F_{-1}(\eta_{36HD})\right] \tag{3.61}$$

The quantized energy along Z direction the $E_{n_{zHD6}}$ in this case is given by

$$\frac{\hbar^2(\mathrm{n}_z\pi/\mathrm{d}_z)^2}{2m_c\theta_5(E_{n_{zHD6}}, \eta_g)} = 1 \tag{3.62}$$

The expression of the total photo-emitted current density in this case is

$$J_{0D} = \frac{e\alpha_0 g_v}{2}\sum_{n_x=1}^{n_{x\max}}\sum_{n_y=1}^{n_{y\max}}\sum_{n_{z\min}}^{n_{z\max}} (\Delta n_{0D}) v_{z6}\left(E_{n_{zHD6}}\right) \tag{3.63}$$

where $v_{z6}(E_{n_{zHD6}}) = [n_z\pi\hbar][m_c d_z \theta_5'(E_{n_{zHD6}}, \eta_g)]^{-1}$

The photoemission current density in the absence of band tails in this case is given by

$$J_{0D} = \frac{\alpha_0 e g_v}{\hbar d_x d_y d_z}\left(\frac{2}{m_c}\right)^{1/2} \sum_{n_x=1}^{n_{x\max}} \sum_{n_y=1}^{n_{y\max}} \sum_{n_{z\min}}^{n_{z\max}} Q_7(E_{n_z}) F_{-1}(\eta_{37}) \tag{3.64}$$

where, $n_{z_{\min}} \geq \left(\frac{\sqrt{2m_c}}{\hbar}\right)\left(\frac{d_z}{\pi}\right)[I_{13}(W - h\nu)]^{1/2}$, $Q_7(E_{n_z}) \equiv \left[[I_{13}(E_{n_z})]^{1/2}/I'_{13}(E_{n_z})\right]$
and $\eta_{37} \equiv \frac{(E_{F0D} - E_{QD7})}{k_B T}$
The E_{n_z} and E_{QD7} are defined by the following equations:

$$I_{13}(E_{n_z}) = \frac{\hbar^2}{2m_c}\left(\frac{\pi n_z}{d_z}\right)^2 \tag{3.65}$$

$$I_{13}(E_{QD7}) = \frac{\hbar^2}{2m_c}\left[\left(\frac{\pi n_x}{d_x}\right)^2 + \left(\frac{\pi n_y}{d_y}\right)^2 + \left(\frac{\pi n_z}{d_z}\right)^2\right] \tag{3.66}$$

The electron concentration in this case is given by

$$n_{0D} = \frac{2g_v}{d_x d_y d_z} \sum_{n_x=1}^{n_{x\max}} \sum_{n_y=1}^{n_{y\max}} \sum_{n_z=1}^{n_{z\max}} F_{-1}(\eta_{37}). \tag{3.67}$$

3.2.3 The EP from QB of HD II-VI Semiconductors

The 0D electron dispersion law in QB of HD II-VI semiconductors can be written following (2.56) as

$$\gamma_3(\mathrm{E}_{7QBHD}, \eta_g) = a'_0\left[\left(\frac{n_x\pi}{d_x}\right)^2 + \left(\frac{n_y\pi}{d_y}\right)^2\right] \pm \bar{\lambda}_0\left[\left(\frac{n_x\pi}{d_x}\right)^2 + \left(\frac{n_y\pi}{d_y}\right)^2\right]^{1/2} + \frac{\hbar^2(\mathrm{n}_z\pi/\mathrm{d}_z)^2}{2m^*_{\parallel}} \tag{3.68}$$

where E_{7QBHD} is the totally quantized energy in this case.
The total electron concentration can be written as

$$n_{0D} = \frac{g_v}{d_x d_y d_z} \sum_{n_x=1}^{n_{x\max}} \sum_{n_y=1}^{n_{y\max}} \sum_{n_z=1}^{n_{z\max}} F_{-1}(\eta_{37HD}) \tag{3.69}$$

where $\eta_{37HD} \equiv (k_B T)^{-1}(E_{F0DHD} - \mathrm{E}_{7QBHD})$.

Therefore the electron concentration per sub-band is given by

$$\Delta n_{0D} = \frac{g_v}{d_x d_y d_z}\ [F_{-1}(\eta_{37HD})] \tag{3.70}$$

The quantized energy along Z direction the $E_{n_{zHD7}}$ in this case is given by

$$\frac{\hbar^2(\mathrm{n}_z\pi/\mathrm{d}_z)^2}{2m_{\|}^*\gamma_3(E_{n_{zHD7}},\eta_g)} = 1 \tag{3.71}$$

The expression of the total photo-emitted current density in this case is

$$J_{0D} = \frac{e\alpha_0 g_v}{4}\sum_{n_x=1}^{n_{x\max}}\sum_{n_y=1}^{n_{y\max}}\sum_{n_{z\min}}^{n_{z\max}}(\Delta n_{0D})v_{z7}\left(E_{n_{zHD7}}\right) \tag{3.72}$$

where $v_{z7}\left(E_{n_{zHD7}}\right) = [n_z\pi\hbar]\left[m_{\|}^* d_z\gamma_3'(E_{n_{zHD7}},\eta_g)\right]^{-1}$

In the absence of band tails the totally quantized energy $E_{QD10,\pm}$ in this case can be expressed as

$$E_{QD10,\pm} = a_0'\left[\left(\frac{\pi n_x}{d_x}\right)^2 + \left(\frac{\pi n_y}{d_y}\right)^2\right] + \frac{1}{2m_{\|}^*}\left(\frac{\hbar\pi n_z}{d_z}\right)^2 \pm \bar{\lambda}_0\left[\left(\frac{\pi n_x}{d_x}\right)^2 + \left(\frac{\pi n_y}{d_y}\right)^2\right]^{1/2} \tag{3.73}$$

The $v_z\left(E_{n_z}\right)$ and E_{n_z} are given by

$$v_z\left(E_{n_z}\right) = \frac{\hbar}{m_{\|}^*}\left(\frac{\pi n_z}{d_z}\right) \tag{3.74}$$

$$E_{n_z} = \frac{\hbar^2}{2m_{\|}^*}\left(\frac{\pi n_z}{d_z}\right)^2 \tag{3.75}$$

The electron concentration can be written as

$$n_{0D} = \frac{g_v}{d_x d_y d_z}\sum_{n_x=1}^{n_{x\max}}\sum_{n_y=1}^{n_{y\max}}\sum_{n_z=1}^{n_{z\max}}\left[F_{-1}\left(\eta_{40,\pm}\right)\right] \tag{3.76}$$

where, $\eta_{40,\pm} = \frac{1}{k_B T}\left[E_{F0D} - E_{QD10,\pm}\right]$

The photo-emitted current density is given by

$$J_{0D} = \frac{\alpha_0 g_v\hbar\pi e}{2d_x d_y d_z^2 m_{\|}^*}\sum_{n_x=1}^{n_{x\max}}\sum_{n_y=1}^{n_{y\max}}\sum_{n_{z\min}}^{n_{z\max}} n_z F_{-1}\left(\eta_{40,\pm}\right) \tag{3.77}$$

where, $n_{z_{\min}} \geq \frac{d_z}{\pi}\frac{\sqrt{2m_{\|}^*}}{\hbar}(W - h\nu)^{1/2}$.

3.2.4 The EP from QB of HD IV-VI Semiconductors

(a) Dimmock Model
In this case the dispersion relation of the electrons can be written as the following (2.65) as

$$\frac{n_z \pi}{d_z} = T_{36}(\mathrm{E}_{8QBHD}, \eta_g, n_x, n_y) \tag{3.78}$$

where E_{8QBHD} is the totally quantized energy in this case.
The total electron concentration can be written as

$$n_{0D} = \frac{g_v}{d_x d_y d_z} \sum_{n_{x=1}}^{n_{x_{\max}}} \sum_{n_{y=1}}^{n_{y_{\max}}} \sum_{n_{z=1}}^{n_{z_{\max}}} F_{-1}(\eta_{38HD}) \tag{3.79}$$

where $\eta_{38HD} \equiv (k_B T)^{-1}(E_{F0DHD} - \mathrm{E}_{8QBHD})$.
Therefore the electron concentration per sub-band is given by

$$\Delta n_{0D} = \frac{g_v}{d_x d_y d_z} \left[F_{-1}(\eta_{38HD})\right] \tag{3.80}$$

The quantized energy along Z direction the $E_{n_{zHD8}}$ in this case is given by

$$\begin{aligned} &\gamma_2(E_{n_{zHD8}}, \eta_g) + \alpha\gamma_3(E_{n_{zHD8}}, \eta_g)\frac{\hbar^2}{2x_6}\left(\frac{n_z \pi}{d_z}\right)^2 \\ &- [1 + \alpha\gamma_3(E_{n_{zHD8}}, \eta_g)]\frac{\hbar^2}{2x_3}\left(\frac{n_z \pi}{d_z}\right)^2 - \frac{\alpha\hbar^4}{4x_3 x_6}\left(\frac{n_z \pi}{d_z}\right)^4 = \frac{\hbar^2}{2m_3}\left(\frac{n_z \pi}{d_z}\right)^2 \end{aligned} \tag{3.81}$$

The expression of the total photo-emitted current density in this case is

$$J_{0D} = \frac{e\alpha_0 g_v}{4} \sum_{n_{x=1}}^{n_{x_{\max}}} \sum_{n_{y=1}}^{n_{y_{\max}}} \sum_{n_{z_{\min}}}^{n_{z_{\max}}} (\Delta n_{0D}) v_{z8}(E_{n_{zHD8}}) \tag{3.82}$$

where

$$\begin{aligned} v_{z8}(E_{n_{zHD8}}) = &\frac{1}{\hbar}[\gamma_2'(E_{n_{zHD8}}, \eta_g) + \alpha\gamma_3'(E_{n_{zHD8}}, \eta_g)\frac{\hbar^2}{2x_6}\left(\frac{n_z \pi}{d_z}\right)^2 \\ &- \alpha\gamma_3'(E_{n_{zHD8}}, \eta_g)\frac{\hbar^2}{2x_3}\left(\frac{n_z \pi}{d_z}\right)^2]^{-1} \cdot \left[\frac{\hbar^2}{m_3}\left(\frac{n_z \pi}{d_z}\right) + \frac{\alpha\hbar^4}{x_3 x_6}\left(\frac{n_z \pi}{d_z}\right)^3 \right. \\ &\left. + [1 + \alpha\gamma_3(E_{n_{zHD8}}, \eta_g)]\frac{\hbar^2}{x_3}\left(\frac{n_z \pi}{d_z}\right)\right] \end{aligned}$$

In the absence of band tailing, the electron dispersion relation in this case can be written following (2.71) as

$$\frac{n_z\pi}{d_z} = T_{40}(E_{QD11}, n_x, n_y) \tag{3.83}$$

where E_{QD11} is the totally quantized energy in this case.
In this case E_{n_z} and $v_z(E_{n_z})$ are given by

$$\begin{aligned} &E_{n_z}(1+\alpha E_{n_z}) + \alpha E_{n_z}\frac{\hbar^2}{2x_6}\left(\frac{n_z\pi}{d_z}\right)^2 - (1+\alpha E_{n_z})\frac{\hbar^2}{2x_3}\left(\frac{n_z\pi}{d_z}\right)^2 \\ &- \frac{\hbar^4}{4x_3x_6}\left(\frac{n_z\pi}{d_z}\right)^4 = \frac{\hbar^2}{2m_3}\left(\frac{n_z\pi}{d_z}\right)^2 \end{aligned} \tag{3.84}$$

$$v_z(E_{n_z}) = \frac{1}{\hbar}\frac{\left[(1+\alpha E_{n_z})\frac{\hbar^2}{x_3}\left(\frac{n_z\pi}{d_z}\right) + \alpha\frac{\hbar^4}{x_3x_6}\left(\frac{n_z\pi}{d_z}\right)^3\right]}{\left[(1+2\alpha E_{n_z}) + \frac{\alpha\hbar^2}{2x_6}\left(\frac{n_z\pi}{d_z}\right)^2 - \alpha\frac{\hbar^2}{2x_3}\left(\frac{n_z\pi}{d_z}\right)^2\right]} \tag{3.85}$$

The electron concentration per band can be written as

$$\Delta n_{0D} = \frac{g_v}{d_x d_y d_z}[F_{-1}(\eta_{41})] \tag{3.86}$$

where, $\eta_{41} = \frac{1}{k_B T}[E_{F0D} - E_{QD11}]$
The photo emitted current density is given by

$$J_{0D} = \frac{e\alpha_0 g_v}{4}\sum_{n_x=1}^{n_{x_{\max}}}\sum_{n_y=1}^{n_{y_{\max}}}\sum_{n_{z_{\min}}}^{n_{z_{\max}}}(\Delta n_{0D})v_z(E_{n_z}). \tag{3.87}$$

(b) Bangert and Kastner Model
The electron dispersion relation in this case is given by following (2.77) as

$$\frac{n_z\pi}{d_z} = T_{60}(\mathrm{E}_{9QBHD}, \eta_g, n_x, n_y) \tag{3.88}$$

where E_{9QBHD} is the totally quantized energy in this case.
The total electron concentration can be written as

$$n_{0D} = \frac{g_v}{d_x d_y d_z}\sum_{n_x=1}^{n_{x_{\max}}}\sum_{n_y=1}^{n_{y_{\max}}}\sum_{n_z=1}^{n_{z_{\max}}}F_{-1}(\eta_{39HD}) \tag{3.89}$$

where $\eta_{39HD} \equiv (k_B T)^{-1}(E_{F0DHD} - E_{9QBHD})$.

Therefore the electron concentration per sub-band is given by

$$\Delta n_{0D} = \frac{g_v}{d_x d_y d_z} \left[F_{-1}(\eta_{39HD})\right] \tag{3.90}$$

The quantized energy along Z direction the $E_{n_{zHD9}}$ in this case is given by

$$F_2(E_{n_{zHD9}}, \eta_g)\left(\frac{n_z \pi}{d_z}\right)^2 = 1 \tag{3.91}$$

The expression of the total photo-emitted current density in this case is

$$J_{0D} = \frac{e\alpha_0 g_v}{4} \sum_{n_x=1}^{n_{x_{\max}}} \sum_{n_y=1}^{n_{y_{\max}}} \sum_{n_{z_{\min}}}^{n_{z_{\max}}} (\Delta n_{0D}) v_{z9}\left(E_{n_{zHD9}}\right) \tag{3.92}$$

where $v_{z9}\left(E_{n_{zHD9}}\right) = \left[-2\left(\frac{n_z \pi}{d_z}\right) F_2(E_{n_{zHD9}}, \eta_g)\right]\left[\hbar F_2'(E_{n_{zHD9}}, \eta_g)\right]^{-1}$

In the absence of band-tails following (2.83) the dispersion relation is given by

$$\frac{n_z \pi}{d_z} = T_{61}(\mathrm{E}_{12QD}, \eta_g, n_x, n_y) \tag{3.93}$$

where E_{12QD} is the totally quantized energy in this case.
The total electron concentration can be written as

$$n_{0D} = \frac{g_v}{d_x d_y d_z} \sum_{n_x=1}^{n_{x_{\max}}} \sum_{n_y=1}^{n_{y_{\max}}} \sum_{n_z=1}^{n_{z_{\max}}} F_{-1}(\eta_{42}) \tag{3.94}$$

where $\eta_{42} \equiv (k_B T)^{-1}(E_{F0D} - \mathrm{E}_{12QD})$

Therefore the electron concentration per sub-band is given by

$$\Delta n_{0D} = \frac{g_v}{d_x d_y d_z} \left[F_{-1}(\eta_{42})\right] \tag{3.95}$$

The quantized energy along Z direction the E_{n_z} in this case is given by

$$\omega_2(E_{n_z})\left(\frac{n_z \pi}{d_z}\right)^2 = 1$$

The expression of the total photo-emitted current density in this case is

$$J_{0D} = \frac{e\alpha_0 g_v}{4} \sum_{n_x=1}^{n_{x_{\max}}} \sum_{n_y=1}^{n_{y_{\max}}} \sum_{n_{z_{\min}}}^{n_{z_{\max}}} (\Delta n_{0D}) v(E_{n_z}) \tag{3.96}$$

where $v\left(E_{n_z}\right) = \left[-2\left(\frac{n_z \pi}{d_z}\right)\omega_2(E_{n_z}, \eta_g)\right]\left[\hbar \omega_2'(E_{n_z}, \eta_g)\right]^{-1}$.

3.2.5 The EP from QB of HD Stressed Kane Type Semiconductors

The electron dispersion relation in this case is given by following (2.89) as

$$\frac{n_z\pi}{d_z} = T_{70}(\mathrm{E}_{10QBHD}, \eta_g, n_x, n_y) \tag{3.97}$$

where E_{10QBHD} is the totally quantized energy in this case.

The total electron concentration can be written as

$$n_{0D} = \frac{2g_v}{d_x d_y d_z} \sum_{n_x=1}^{n_{x_{\max}}} \sum_{n_y=1}^{n_{y_{\max}}} \sum_{n_z=1}^{n_{z_{\max}}} F_{-1}(\eta_{40HD}) \tag{3.98}$$

where $\eta_{40HD} \equiv (k_B T)^{-1}(E_{F0DHD} - E_{10QBHD})$.

Therefore the electron concentration per sub-band is given by

$$\Delta n_{0D} = \frac{g_v}{d_x d_y d_z} [F_{-1}(\eta_{40HD})] \tag{3.99}$$

The quantized energy along Z direction the $E_{n_{zHD10}}$ in this case is given by

$$S_{11}(E_{n_{zHD10}}, \eta_g)(\frac{n_z\pi}{d_z})^2 = 1 \tag{3.100}$$

The expression of the total photo-emitted current density in this case is

$$J_{0D} = \frac{e\alpha_0 g_v}{2} \sum_{n_x=1}^{n_{x_{\max}}} \sum_{n_y=1}^{n_{y_{\max}}} \sum_{n_{z_{\min}}}^{n_{z_{\max}}} (\Delta n_{0D}) v_{z10}\left(E_{n_{zHD10}}\right) \tag{3.101}$$

where $v_{z10}\left(E_{n_{zHD10}}\right) = [-2(\frac{n_z\pi}{d_z}) S_{11}(E_{n_{zHD10}}, \eta_g)][\hbar S'_{11}(E_{n_{zHD10}}, \eta_g)]^{-1}$

In the absence of band-tails the $v_z\left(E_{n_z}\right)$ for QBs of stressed materials can be written form (1.195) as

$$v_z\left(E_{n_z}\right) = \frac{Q_{21}\left(E_{n_z}\right)}{\hbar} \text{ where } Q_{21}\left(E_{n_z}\right) = \frac{1}{\left[\bar{c}_0\left(E_{n_z}\right)\right]'} \tag{3.102}$$

The E_{n_z} can be expressed through the equation

$$\bar{c}_0\left(E_{n_z}\right) = \frac{n_z\pi}{d_z} \tag{3.103}$$

The totally quantized energy E_{QD23} in this case assumes the form

$$\left(\frac{\pi n_x}{d_x}\right)^2[\bar{a}_0(E_{QD23})]^{-2}+\left(\frac{\pi n_y}{d_y}\right)^2[\bar{b}_0(E_{QD23})]^{-2}+\left(\frac{\pi n_z}{d_z}\right)^2[\bar{c}_0(E_{QD23})]^{-2}=1 \tag{3.104}$$

The electron concentration is given by

$$n_{0D}=\frac{2g_v}{d_xd_yd_z}\sum_{n_x=1}^{n_{x\max}}\sum_{n_y=1}^{n_{y\max}}\sum_{n_z=1}^{n_{z\max}}F_{-1}(\bar{\eta}_{53})\text{ where }\bar{\eta}_{53}=(k_BT)^{-1}(E_{F0D}-E_{QD23}) \tag{3.105}$$

The photo-emitted current density can be written as

$$J_{0D}=\frac{\alpha_0 e g_v}{d_xd_yd_z\hbar}\sum_{n_x=1}^{n_{x\max}}\sum_{n_y=1}^{n_{y\max}}\sum_{n_z=1}^{n_{z\max}}Q_{21}(E_{n_z})F_{-1}(\bar{\eta}_{53}) \tag{3.106}$$

where $n_{z_{\min}}$ satisfies the following inequality

$$n_{z_{\min}}\geq\left(\frac{d_z}{\pi}\right)[\bar{c}_0(W-h\nu)].$$

3.2.6 The EP from QB of HD Te

The 0D dispersion relation may be written in this case following (2.99) as

$$\frac{n_x\pi}{d_x}=t_{72}(\mathrm{E}_{11QBHD},\mathrm{n}_y,\mathrm{n}_z,\eta_g) \tag{3.107}$$

where E_{11QBHD} is the totally quantized energy in this case.

The total electron concentration can be written as

$$n_{0D}=\frac{g_v}{d_xd_yd_z}\sum_{n_x=1}^{n_{x\max}}\sum_{n_y=1}^{n_{y\max}}\sum_{n_z=1}^{n_{z\max}}F_{-1}(\eta_{42HD}) \tag{3.108}$$

where $\eta_{42HD}\equiv(k_BT)^{-1}(E_{F0DHD}-\mathrm{E}_{12QBHD})$.

Therefore the electron concentration per sub-band is given by

$$\Delta n_{0D}=\frac{g_v}{d_xd_yd_z}[F_{-1}(\eta_{42HD})] \tag{3.109}$$

The quantized energy along Z direction the $E_{n_{zHD11}}$ in this case is given by

$$[\psi_6(\frac{n_z\pi}{d_z})^2 - \psi_{5HD}(E_{n_{zHD11}},\eta_g)]^2 = \psi^2_{8HD}(E_{n_{zHD11}},\eta_g) - (\frac{n_z\pi}{d_z})^2 \quad (3.111)$$

The expression of the total photo-emitted current density in this case is

$$J_{0D} = \frac{e\alpha_0 g_v}{4}\sum_{n_x=1}^{n_{x\max}}\sum_{n_y=1}^{n_{y\max}}\sum_{n_{z\min}}^{n_{z\max}}(\Delta n_{0D})v_{z11}(E_{n_{zHD11}}) \quad (3.112)$$

where

$$[v_{z11}(E_{n_{zHD11}})]^{-1} = \frac{\hbar[\psi'_{5HD}(E_{n_{zHD11}},\eta_g) \pm \psi_{8HD}(E_{n_{zHD11}},\eta_g)\psi'_{8HD}(E_{n_{zHD11}},\eta_g)[\psi^2_{8HD}(E_{n_{zHD11}},\eta_g) - (\frac{n_z\pi}{d_z})^2]^{-1/2}]}{[2\psi_6(\frac{n_z\pi}{d_z}) \pm (\frac{n_z\pi}{d_z})[\psi^2_{8HD}(E_{n_{zHD11}},\eta_g) - (\frac{n_z\pi}{d_z})^2]^{-1/2}]}$$

In the absence of doping $v_z\left(E_{n_{z,\pm}}\right)$ in this case is given by

$$v_z\left(E_{n_{z,\pm}}\right) = \frac{1}{\hbar}Q_{12}\left(E_{n_{z,\pm}}\right) \quad (3.113)$$

where $Q_{12}\left(E_{n_{z,\pm}}\right) = \sqrt{2\psi_1(\frac{n_z\pi}{d_z}) \pm \psi_3}$ in which $E_{n_{z,\pm}} = \psi_1\left(\frac{\pi n_z}{d_z}\right)^2 \pm \left(\frac{\pi n_z}{d_z}\right)\psi_3$

The totally quantized energy can be written as

$$E_{QD14,\pm} = \psi_1\left(\frac{\pi n_z}{d_z}\right)^2 + \psi_2\left[\left(\frac{\pi n_x}{d_x}\right)^2 + \left(\frac{\pi n_y}{d_y}\right)^2\right] \pm \left[\psi_3^2\left(\frac{\pi n_z}{d_z}\right)^2 + \psi_4^2\left[\left(\frac{\pi n_x}{d_x}\right)^2 + \left(\frac{\pi n_y}{d_y}\right)^2\right]\right]^{1/2} \quad (3.114)$$

The electron concentration is given by

$$n_{0D} = \frac{g_v}{d_x d_y d_z}\sum_{n_x=1}^{n_{x\max}}\sum_{n_y=1}^{n_{y\max}}\sum_{n_z=1}^{n_{z\max}}F_{-1}(\eta_{44,\pm}) \quad (3.115)$$

where $\eta_{44,\pm} = (E_{F0D} - E_{QD14,\pm})/k_BT$.

The photo-emitted current density is given by

$$J_{0D} = \frac{\alpha_0 e g_v}{2\hbar d_x d_y d_z}\sum_{n_x=1}^{n_{x\max}}\sum_{n_y=1}^{n_{y\max}}\sum_{n_z=1}^{n_{z\max}}Q_{12}\left(E_{n_{z,\pm}}\right)F_{-1}(\eta_{44,\pm}) \quad (3.116)$$

where $n_{z_{\min}}$ can be determined from

$$W - h\nu = \psi_1\left(\frac{\pi n_{z_{\min}}}{d_z}\right)^2 \pm \psi_3 . \frac{\pi}{d_z} . n_{z_{\min}}. \quad (3.117)$$

3.2.7 The EP from QB of HD Gallium Phosphide

The 0D dispersion relation may be written in this case following (2.109) as

$$\frac{n_x \pi}{d_x} = u_{70}(\mathrm{E}_{14QBHD}, \mathrm{n}_y, \mathrm{n}_z, \eta_g) \tag{3.118}$$

where E_{14QBHD} is the totally quantized energy in this case.

The total electron concentration can be written as

$$n_{0D} = \frac{g_v}{d_x d_y d_z} \sum_{n_x=1}^{n_{x_{\max}}} \sum_{n_y=1}^{n_{y_{\max}}} \sum_{n_z=1}^{n_{z_{\max}}} F_{-1}(\eta_{44HD}) \tag{3.119}$$

where $\eta_{44HD} \equiv (k_B T)^{-1}(E_{F0DHD} - \mathrm{E}_{14QBHD})$.

Therefore the electron concentration per sub-band is given by

$$\Delta n_{0D} = \frac{g_v}{d_x d_y d_z} [F_{-1}(\eta_{44HD})] \tag{3.120}$$

The quantized energy along Z direction the $E_{n_{zHD14}}$ in this case is given by

$$t_{11}\gamma_3(E_{n_{zHD14}}, \eta_g) + t_{21} - t_{31}\left(\frac{n_z\pi}{d_z}\right)^2 - t_{41}\left[\left(\frac{n_z\pi}{d_z}\right)^2 + t_5^2(E_{n_{zHD14}}, \eta_g)\right)^{1/2} = 0 \tag{3.121}$$

The expression of the total photo-emitted current density in this case is

$$J_{0D} = \frac{e\alpha_0 g_v}{4} \sum_{n_x=1}^{n_{x_{\max}}} \sum_{n_y=1}^{n_{y_{\max}}} \sum_{n_{z_{\min}}}^{n_{z_{\max}}} (\Delta n_{0D}) v_{z14}(E_{n_{zHD14}}) \tag{3.122}$$

where $[v_{z14}(E_{n_{zHD14}})] = \dfrac{[2t_{31}(\frac{n_z\pi}{d_z}) + t_4(\frac{n_z\pi}{d_z})[(\frac{n_z\pi}{d_z})^2 + t_5^2(E_{n_{zHD14}}, \eta_g)]^{-1/2}]}{\hbar[t_{11}\gamma_3'(E_{n_{zHD14}}, \eta_g) - t_4 t_5(E_{n_{zHD14}}, \eta_g) t_5'(E_{n_{zHD14}}, \eta_g)[(\frac{n_z\pi}{d_z})^2 + t_5^2(E_{n_{zHD14}}, \eta_g)]^{-1/2}]}$

In the absence of doping, the totally quantized energy (E_{QD16}) in this case can be written as

$$E_{QD16} = \frac{\hbar^2}{2m_\perp^*}\left[\left(\frac{\pi n_x}{d_x}\right)^2 + \left(\frac{\pi n_y}{d_y}\right)^2\right] + \frac{\hbar^2}{2m_\parallel^*}\left[\left(\frac{\pi n_x}{d_x}\right)^2 + \left(\frac{\pi n_y}{d_y}\right)^2 + \left(\frac{\pi n_z}{d_z}\right)^2\right]$$
$$- \left[\frac{\hbar^4 k_0^2}{m_\parallel^{*2}}\left[\left(\frac{\pi n_x}{d_x}\right)^2 + \left(\frac{\pi n_y}{d_y}\right)^2 + \left(\frac{\pi n_z}{d_z}\right)^2\right] + |V_G|^2\right]^{1/2} + |V_G| \tag{3.123}$$

The $v_z(E_{n_z})$ and E_{n_z} can be respectively written as

$$v_z(E_{n_z}) = \frac{\hbar}{m_{||}^*}\left(\frac{\pi n_z}{d_z}\right)\bar{Q}_{10}(E_{n_z}) \tag{3.124}$$

where, $\bar{Q}_{10}(E_{n_z}) = \left[1 - \left(\frac{\hbar^2 k_0^2}{m_{||}^*}\right)\left\{|V_G|^2 + \frac{\hbar^4 k_0^2}{\left(m_{||}^*\right)^2}\left(\frac{\pi n_z}{d_z}\right)^2\right\}^{-1/2}\right]$

$$\text{and } E_{n_z} = \frac{\hbar^2}{2m_{||}^*}\left(\frac{\pi n_z}{d_z}\right)^2 - \left[\frac{\hbar^4 k_0^2}{m_{||}^{*2}}\left(\frac{\pi n_z}{d_z}\right)^2 + |V_G|^2\right]^{1/2} + |V_G| \tag{3.125}$$

The electron concentration assumes the form

$$n_{0D} = \frac{2g_v}{d_x d_y d_z}\sum_{n_x=1}^{n_{x\max}}\sum_{n_y=1}^{n_{y\max}}\sum_{n_z=1}^{n_{z\max}}[F_{-1}(\eta_{46})] \tag{3.126}$$

where, $\eta_{46} = \frac{1}{k_B T}[E_{F0D} - E_{QD16}]$

The photo-emitted current density is given by

$$J_{0D} = \frac{\alpha_0 \pi \hbar g_v e}{d_x d_y d_z^2 m_{||}^*}\sum_{n_x=1}^{n_{x\max}}\sum_{n_y=1}^{n_{y\max}}\sum_{n_{z\min}}^{n_{z\max}}\bar{Q}_{10}(E_{n_z})F_{-1}(\eta_{46}) \tag{3.127}$$

where, $n_{z_{\min}}$ is the nearest integer of the following equation

$$(W - h\nu) = \frac{\hbar^2}{2m_{||}^*}\left(\frac{\pi n_{z_{\min}}}{d_z}\right)^2 + |V_G| - \left[|V_G|^2 + \frac{\hbar^4 k_0^2}{\left(m_{||}^*\right)^2}\left(\frac{\pi n_{z_{\min}}}{d_z}\right)^2\right]^{1/2}. \tag{3.128}$$

3.2.8 The EP from QB of HD Platinum Antimonide

The 0D dispersion relation may be written in this case following (2.119) as

$$\frac{n_x \pi}{d_x} = V_{70}(\mathrm{E}_{15QBHD}, \mathrm{n}_y, \mathrm{n}_z, \eta_g) \tag{3.129}$$

where E_{15QBHD} is the totally quantized energy in this case.

The total electron concentration can be written as

$$n_{0D} = \frac{g_v}{d_x d_y d_z} \sum_{n_x=1}^{n_{x\max}} \sum_{n_y=1}^{n_{y\max}} \sum_{n_z=1}^{n_{z\max}} F_{-1}(\eta_{45HD}) \tag{3.130}$$

where $\eta_{45HD} \equiv (k_B T)^{-1}(E_{F0DHD} - \mathrm{E}_{15QBHD})$.

Therefore the electron concentration per sub-band is given by

$$\Delta n_{0D} = \frac{g_v}{d_x d_y d_z} \left[F_{-1}(\eta_{45HD})\right] \tag{3.131}$$

The quantized energy along Z direction the $E_{n_{zHD15}}$ in this case is given by

$$T_{41}\left(\frac{n_z\pi}{d_z}\right)^4 = T_{51}(E_{n_{zHD15}}, \eta_g)\left(\frac{n_z\pi}{d_z}\right)^2 + T_{61}(E_{n_{zHD15}}, \eta_g) \tag{3.132}$$

The expression of the total photo-emitted current density in this case is

$$J_{0D} = \frac{e\alpha_0 g_v}{4} \sum_{n_x=1}^{n_{x\max}} \sum_{n_y=1}^{n_{y\max}} \sum_{n_{z\min}}^{n_{z\max}} (\Delta n_{0D}) v_{z15}(E_{n_{zHD15}}) \tag{3.133}$$

where $\left[v_{z15}(E_{n_{zHD15}})\right] = \frac{\left[4T_{41}\left(\frac{n_z\pi}{d_z}\right)^3 - 2T_{51}(E_{n_{zHD15}}, \eta_g)\left(\frac{n_z\pi}{d_z}\right)\right]}{\hbar\left[T'_{51}(E_{n_{zHD15}}, \eta_g)\left(\frac{n_z\pi}{d_z}\right)^2 + T'_{61}(E_{n_{zHD15}}, \eta_g)\right]}$

In the absence of band tailing the 0D dispersion relation in this case can be written using (2.124) as

$$\frac{n_x\pi}{d_x} = D_{71}(\mathrm{E}_{QD17}, \mathrm{n}_y, \mathrm{n}_z) \tag{3.134}$$

where E_{QD17} is the totally quantized energy in this case

The total electron concentration can be written as

$$n_{0D} = \frac{g_v}{d_x d_y d_z} \sum_{n_x=1}^{n_{x\max}} \sum_{n_y=1}^{n_{y\max}} \sum_{n_z=1}^{n_{z\max}} F_{-1}(\eta_{50}) \tag{3.135}$$

where $\eta_{50} \equiv (k_B T)^{-1}(E_{F0D} - E_{QD17})$

Therefore the electron concentration per sub-band is given by

$$\Delta n_{0D} = \frac{g_v}{d_x d_y d_z} \left[F_{-1}(\eta_{50})\right] \tag{3.136}$$

The quantized energy along Z direction the E_{n_z} in this case is given by

$$[E_{n_z} + \omega_2(\frac{n_z\pi}{d_z})^2][E_{n_z} - \omega_4(\frac{n_z\pi}{d_z})^2 + \delta_0] = I_1(\frac{n_z\pi}{d_z})^4 \tag{3.137}$$

The expression of the total photo-emitted current density in this case is

$$J_{0D} = \frac{e\alpha_0 g_v}{4} \sum_{n_x=1}^{n_{x\max}} \sum_{n_y=1}^{n_{y\max}} \sum_{n_{z\min}}^{n_{z\max}} (\Delta n_{0D}) v_z(E_{n_z}) \tag{3.138}$$

where $v_z(E_{n_z}) = \frac{[4I_1(\frac{n_z\pi}{d_z})^3 + 2(\frac{n_z\pi}{d_z})\omega_4(E_{n_z} + \omega_2(\frac{n_z\pi}{d_z})^2) - 2(\frac{n_z\pi}{d_z})\omega_2(E_{n_z} + \delta_0 - \omega_4(\frac{n_z\pi}{d_z})^2)]}{\hbar[2E_{n_z} + \delta_0 + (\omega_2 - \omega_4)(\frac{n_z\pi}{d_z})^2]}$.

3.2.9 The EP from QB of HD Bismuth Telluride

The dispersion relation in this case can be written following (2.128b) as

$$\frac{n_x\pi}{d_x} = J_{70}(\mathrm{E}_{18QBHD}, \mathrm{n}_y, \mathrm{n}_z, \eta_g) \tag{3.139}$$

where E_{18QBHD} is the totally quantized energy in this case.

The total electron concentration can be written as

$$n_{0D} = \frac{2g_v}{d_x d_y d_z} \sum_{n_x=1}^{n_{x\max}} \sum_{n_y=1}^{n_{y\max}} \sum_{n_z=1}^{n_{z\max}} F_{-1}(\eta_{48HD}) \tag{3.140}$$

where $\eta_{48HD} \equiv (k_B T)^{-1}(E_{F0DHD} - \mathrm{E}_{18QBHD})$.

Therefore the electron concentration per sub-band is given by

$$\Delta n_{0D} = \frac{2g_v}{d_x d_y d_z}[F_{-1}(\eta_{48HD})] \tag{3.141}$$

The quantized energy along Z direction the $E_{n_{zHD20}}$ in this case is given by

$$\gamma_2(E_{n_{zHD20}}, \eta_g) = \bar{\omega}_3(\frac{n_z\pi}{d_z})^2 + 2\bar{\omega}_4(\frac{n_y n_z \pi^2}{d_y d_z}) \tag{3.142}$$

The expression of the total photo-emitted current density in this case is

$$J_{0D} = \frac{e\alpha_0 g_v}{2} \sum_{n_x=1}^{n_{x\max}} \sum_{n_y=1}^{n_{y\max}} \sum_{n_{z\min}}^{n_{z\max}} (\Delta n_{0D}) v_{z20}(E_{n_{zHD20}}) \tag{3.143}$$

where $[v_{z20}(E_{n_{zHD20}})] = \frac{2}{\hbar}[\gamma_2'(E_{n_{zHD20}}, \eta_g)]^{-1}[\omega_3(\frac{n_z\pi}{d_z}) + \omega_4(\frac{n_y\pi}{d_y})]$

The dispersion relation of the conduction electrons in Bi_2Te_3 in the absence of doping can be written following (1.284) as

$$E(1+\alpha E)=\frac{\hbar^2}{2m_0}\left(\alpha_{11}k_x^2+\alpha_{22}k_y^2+\alpha_{33}k_z^2+2\alpha_{23}k_yk_z\right) \tag{3.144}$$

where $\alpha_{11}, \alpha_{22}, \alpha_{33}$ and α_{23} are spectrum constants.

The $v_z(E_{n_z})$ is given by

$$v_z(E_{n_z})=\frac{\pi\hbar\alpha_{33}n_z}{d_zm_0(1+2\alpha E_{n_z})} \tag{3.145}$$

in which E_{n_z} should be determined from the following equation

$$E_{n_z}(1+\alpha E_{n_z})=\frac{\hbar^2}{2m_0}\left(\alpha_{33}\left(\frac{\pi n_z}{d_z}\right)^2\right) \tag{3.146}$$

The totally quantized energy E_{QD33} can be written as

$$E_{QD33}(1+\alpha E_{QD33})=\frac{\hbar^2}{2m_0}\left[\alpha_{11}\left(\frac{n_x\pi}{d_x}\right)^2+\alpha_{22}\left(\frac{n_y\pi}{d_y}\right)^2+\alpha_{33}\left(\frac{n_z\pi}{d_z}\right)^2+2\alpha_{23}\left(\frac{\pi^2n_yn_z}{d_yd_z}\right)\right] \tag{3.147}$$

The hole concentration is given by

$$p_{0D}=\frac{2g_v}{d_xd_yd_z}\sum_{n_x=1}^{n_{x\max}}\sum_{n_y=1}^{n_{y\max}}\sum_{n_z=1}^{n_{z\max}}[1+\exp(\eta_{62})]^{-1} \text{ where } \eta_{62}=\frac{1}{k_BT}[E_{F0D}-E_{QD33}] \tag{3.148}$$

The photo-emitted current density assumes the form

$$J_{0D}=\frac{\pi\alpha_0eg_v\hbar\alpha_{33}}{m_0d_xd_yd_z^2}\sum_{n_x=1}^{n_{x\max}}\sum_{n_y=1}^{n_{y\max}}\sum_{n_{z\min}}^{n_{z\max}}\frac{n_z[1+\exp(\eta_{62})]^{-1}}{(1+2\alpha E_{n_z})} \tag{3.149}$$

where $n_{z_{\min}}$ should be the nearest integer of the following equation

$$(W-h\nu)[1+\alpha(W-h\nu)]=\frac{\hbar^2\alpha_{33}}{2m_0}[\pi n_{z_{\min}}/d_z]^2. \tag{3.150}$$

3.2.10 The EP from QB of HD Germinium

(a) Model of Cardona et al.
The dispersion relation in accordance with this model in the present case can be written following (2.138) as

$$\frac{n_x \pi}{d_x} = L_{70}(\mathrm{E}_{20QBHD}, \mathrm{n}_y, \mathrm{n}_z, \eta_g) \tag{3.151}$$

where E_{20QBHD} is the totally quantized energy in this case.
The total electron concentration can be written as

$$n_{0D} = \frac{2g_v}{d_x d_y d_z} \sum_{n_x=1}^{n_{x\max}} \sum_{n_y=1}^{n_{y\max}} \sum_{n_z=1}^{n_{z\max}} F_{-1}(\eta_{50HD}) \tag{3.152}$$

where $\eta_{50HD} \equiv (k_B T)^{-1}(E_{F0DHD} - \mathrm{E}_{20QBHD})$.
Therefore the electron concentration per sub-band is given by

$$\Delta n_{0D} = \frac{2g_v}{d_x d_y d_z} [F_{-1}(\eta_{50HD})] \tag{3.153}$$

The quantized energy along Z direction the $E_{n_{zHD22}}$ in this case is given by

$$0 = \gamma_2(E_{n_{zHD22}}, \eta_g) + \alpha[\frac{\hbar^2}{2m_{\parallel}^*}(\frac{n_z \pi}{d_z})^2]^2 - [1 + 2\gamma_3(E_{n_{zHD22}}, \eta_g)]\frac{\hbar^2}{2m_{\parallel}^*}(\frac{n_z \pi}{d_z})^2 \tag{3.154}$$

The expression of the total photo-emitted current density in this case is

$$J_{0D} = \frac{e\alpha_0 g_v}{2} \sum_{n_x=1}^{n_{x\max}} \sum_{n_y=1}^{n_{y\max}} \sum_{n_{z\min}}^{n_{z\max}} (\Delta n_{0D}) v_{z22}(E_{n_{zHD22}}) \tag{3.155}$$

where $[v_{z22}(E_{n_{zHD22}})] = \frac{[1+\alpha\gamma_3(E_{n_{zHD22}},\eta_g)\frac{\hbar^2}{m_{\parallel}^*} - \alpha(\frac{n_z\pi}{d_z})^3(\frac{\hbar^2}{m_{\parallel}^*})^2]}{\hbar[\gamma_2'(E_{n_{zHD22}},\eta_g) - \alpha\gamma_3'(E_{n_{zHD22}},\eta_g)(\frac{\hbar^2}{m_{\parallel}^*})(\frac{n_z\pi}{d_z})^2]}$

In the absence of doping the totally quantized energy E_{QD30} in this case can be written as

$$E_{QD30} = -\frac{E_{g_0}}{2} + \frac{\hbar^2}{2m_{\|}^*}\left(\frac{\pi n_z}{d_z}\right)^2 + \left[\frac{E_{g_0}^2}{4} + E_{g_0}\left(\frac{\hbar^2}{2m_{\perp}^*}\right)\left\{\left(\frac{\pi n_x}{d_x}\right)^2 + \left(\frac{\pi n_y}{d_y}\right)^2\right\}\right]^{1/2} \tag{3.156}$$

The $v_z(E_{n_z})$ and E_{n_z} can be respectively written as

$$v_z(E_{n_z}) = \frac{\hbar}{m_{\|}^*}\left(\frac{\pi n_z}{d_z}\right) \tag{3.157}$$

$$E_{n_z} = \frac{\hbar^2}{2m_{\|}^*}\left(\frac{\pi n_z}{d_z}\right)^2 \tag{3.158}$$

The electron concentration assumes the form

$$n_{0D} = \frac{2g_v}{d_x d_y d_z}\sum_{n_x=1}^{n_{x\max}}\sum_{n_y=1}^{n_{y\max}}\sum_{n_z=1}^{n_{z\max}}[F_{-1}(\eta_{42})] \tag{3.159}$$

where, $\eta_{42} = \frac{1}{k_B T}[E_{F0D} - E_{QD30}]$ in which E_{QD30} is determined from (3.156). The EP is given by

$$J_{0D} = \frac{\alpha_0 \pi \hbar g_v e}{d_x d_y d_z^2 m_{\|}^*}\sum_{n_x=1}^{n_{x\max}}\sum_{n_y=1}^{n_{y\max}}\sum_{n_{z_{\min}}}^{n_{z\max}} n_z F_{-1}(\eta_{42}) \tag{3.160}$$

where, $n_{z_{\min}}$ is the nearest integer of the following inequality

$$n_{z_{\min}} \geq \frac{d_z}{\pi}\frac{\sqrt{2m_{\|}^*}}{\hbar}(W - h\nu)^{1/2}. \tag{3.161}$$

(b) Model of Wang and Ressler

The dispersion relation in accordance with this model in the present case can be written following (2.148) as

$$\frac{n_x \pi}{d_x} = \beta_{70}(\mathrm{E}_{24QBHD}, \mathrm{n}_y, \mathrm{n}_z, \eta_g) \tag{3.162}$$

where E_{24QBHD} is the totally quantized energy in this case. The total electron concentration can be written as

$$n_{0D} = \frac{2g_v}{d_x d_y d_z}\sum_{n_x=1}^{n_{x_{\max}}}\sum_{n_y=1}^{n_{y_{\max}}}\sum_{n_z=1}^{n_{z_{\max}}} F_{-1}(\eta_{54HD}) \tag{3.163}$$

where $\eta_{54HD} \equiv (k_B T)^{-1}(E_{F0DHD} - \mathrm{E}_{24QBHD})$.

Therefore the electron concentration per sub-band is given by

$$\Delta n_{0D} = \frac{2g_v}{d_x d_y d_z} \; [F_{-1}(\eta_{54HD})] \tag{3.164}$$

The quantized energy along Z direction the $E_{n_{zHD24}}$ in this case is given by

$$[\bar{\alpha}_8 - \bar{\alpha}_9 (\frac{n_z \pi}{d_z})^2 - \bar{\alpha}_{10} [(\frac{n_z \pi}{d_z})^4 + \bar{\alpha}_{11} (\frac{n_z \pi}{d_z})^2 + \bar{\alpha}_{12} (E_{n_{zHD24}}, \eta_g)]^{1/2}] = 0 \tag{3.165}$$

The expression of the total photo-emitted current density in this case is

$$J_{0D} = \frac{e\alpha_0 g_v}{2} \sum_{n_{x=1}}^{n_{x_{\max}}} \sum_{n_{y=1}}^{n_{y_{\max}}} \sum_{n_{z_{\min}}}^{n_{z_{\max}}} (\Delta n_{0D}) v_{z24} \left(E_{n_{zHD24}}\right) \tag{3.166}$$

where

$$v_{z24} \left(E_{n_{zHD24}}\right) = \frac{1}{\hbar} [\frac{[(2(\frac{n_z \pi}{d_z})\bar{\alpha}_9)\sqrt{(\frac{n_z \pi}{d_z})^4 + \bar{\alpha}_{11}(\frac{n_z \pi}{d_z})^2 \bar{\alpha}_{11}(E_{n_{zHD24}}, \eta_g)} + 2\bar{\alpha}_{10}(\frac{n_z \pi}{d_z})^3 + \bar{\alpha}_{10}\bar{\alpha}_{11}(\frac{n_z \pi}{d_z})]}{[\bar{\alpha}_8 \sqrt{(\frac{n_z \pi}{d_z})^4 + \bar{\alpha}_{11}(\frac{n_z \pi}{d_z})^2 \bar{\alpha}_{11}(E_{n_{zHD24}}, \eta_g)} + \frac{\bar{\alpha}_{10}[\bar{\alpha}_{12}(E_{n_{zHD24}}, \eta_g)]'}{2}]}]$$

In the absence of doping, the totally quantized energy E_{QD40} in this case is given by

$$E_{QD40} = \frac{\hbar^2}{2m_{||}^*}\left(\frac{\pi n_z}{d_z}\right)^2 + \frac{\hbar^2}{2m_{\perp}^*}\left\{\left(\frac{\pi n_x}{d_x}\right)^2 + \left(\frac{\pi n_y}{d_y}\right)^2\right\} - \bar{c}_1\left(\frac{\hbar^2}{2m_{\perp}^*}\right)^2\left\{\left(\frac{\pi n_x}{d_x}\right)^2 + \left(\frac{\pi n_y}{d_y}\right)^2\right\}^2$$
$$- \bar{d}_1\left(\frac{\hbar^2}{2m_{\perp}^*}\right)\left\{\left(\frac{\pi n_x}{d_x}\right)^2 + \left(\frac{\pi n_y}{d_y}\right)^2\right\}\left(\frac{\hbar^2}{2m_{||}^*}\right)\left(\frac{\pi n_z}{d_z}\right)^2 - \bar{e}_1\left(\frac{\hbar^2}{2m_{||}^*}\right)^2\left(\frac{\pi n_z}{d_z}\right)^4 \tag{3.167}$$

The $v_z\left(E_{n_z}\right)$ and E_{n_z} can be respectively written as

$$v_z\left(E_{n_z}\right) = \frac{\hbar}{m_{||}^*}\left(\frac{n_z \pi}{d_z}\right)\bar{Q}_{11}\left(E_{n_z}\right) \tag{3.168}$$

where, $Q_{11}\left(E_{n_z}\right) = \left[1 - \bar{e}_1 \frac{\hbar^2}{\left(m_{||}^*\right)^2}\left(\frac{\pi n_z}{d_z}\right)^2\right]$

$$E_{n_z} = \frac{\hbar^2}{2m_{||}^*}\left(\frac{\pi n_z}{d_z}\right)^2 - \bar{e}_1\left(\frac{\hbar^2}{2m_{||}^*}\right)^2\left(\frac{\pi n_z}{d_z}\right)^4 \tag{3.169}$$

The electron concentration assumes the form

$$n_{0D} = \frac{2g_v}{d_x d_y d_z} \sum_{n_x=1}^{n_{x\max}} \sum_{n_y=1}^{n_{y\max}} \sum_{n_z=1}^{n_{z\max}} [F_{-1}(\eta_{50})] \tag{3.170}$$

where, $\eta_{50} = \frac{1}{k_B T}[E_{F0D} - E_{QD40}]$
The photo-emitted current density is given by

$$J_{0D} = \frac{\alpha_0 \pi \hbar g_v e}{d_x d_y d_z^2 m_{\|}^*} \sum_{n_x=1}^{n_{x\max}} \sum_{n_y=1}^{n_{y\max}} \sum_{n_{z\min}}^{n_{z\max}} \bar{Q}_{11}(E_{n_z}) F_{-1}(\eta_{50}) \tag{3.171}$$

where, $n_{z_{\min}}$ is the nearest integer of the following inequality

$$(W - h\nu) = \frac{\hbar^2}{2m_{\|}^*}\left(\frac{\pi n_{z_{\min}}}{d_z}\right)^2 - \bar{e}_1 \left(\frac{\hbar^2}{2m_{\|}^*}\right)^2 \left(\frac{\pi n_{z_{\min}}}{d_z}\right)^4. \tag{3.172}$$

3.2.11 The EP from QB of HD Gallium Antimonide

The dispersion relation of the 0D electrons in this case can be written following (2.158) as

$$\frac{\hbar^2 (\mathrm{n}_z \pi/\mathrm{d}_z)^2}{2m_c} + \frac{\hbar^2 (\mathrm{n}_y \pi/\mathrm{d}_y)^2}{2m_c} + \frac{\hbar^2 \left(\frac{n_x \pi}{d_x}\right)^2}{2m_c} = I_{36}(\mathrm{E}_{30QBHD}, \eta_g) \tag{3.173}$$

where E_{30QBHD} is the totally quantized energy in this case.
The total electron concentration can be written as

$$n_{0D} = \frac{2g_v}{d_x d_y d_z} \sum_{n_x=1}^{n_{x_{\max}}} \sum_{n_y=1}^{n_{y_{\max}}} \sum_{n_z=1}^{n_{z_{\max}}} F_{-1}(\eta_{60HD}) \tag{3.174}$$

where $\eta_{60HD} \equiv (k_B T)^{-1}(E_{F0DHD} - \mathrm{E}_{30QBHD})$.
Therefore the electron concentration per sub-band is given by

$$\Delta n_{0D} = \frac{2g_v}{d_x d_y d_z}[F_{-1}(\eta_{60HD})] \tag{3.175}$$

The quantized energy along Z direction the $E_{n_{zHD30}}$ in this case is given by

$$\frac{\hbar^2}{2m_c}(\frac{n_z\pi}{d_z})^2 = I_{36}(E_{n_{zHD30}}, \eta_g) \tag{3.176}$$

The expression of the total photo-emitted current density in this case is

$$J_{0D} = \frac{e\alpha_0 g_v}{2}\sum_{n_x=1}^{n_{x_{\max}}}\sum_{n_y=1}^{n_{y_{\max}}}\sum_{n_{z_{\min}}}^{n_{z_{\max}}}(\Delta n_{0D})v_{z30}(E_{n_{zHD30}}) \tag{3.177}$$

where $v_{z30}(E_{n_{zHD30}}) = \frac{\hbar n_z\pi}{d_z m_c I'_{36}(E_{n_{zHD30}},\eta_g)}$

In the absence of band tails, the (1.337) can be written as

$$E = \alpha_9 k^2 + \frac{E_{g1}}{2}\left[\sqrt{1+\alpha_{10}k^2} - 1\right] \tag{3.178}$$

where $\alpha_9 = \frac{\hbar^2}{2m_0}$ and $\alpha_{10} = \left(\frac{(2\hbar^2)}{(E_{g1})}\right)\left[\frac{1}{m_c} - \frac{1}{m_0}\right]$.

From (3.178), we get

$$k^2 = \frac{E}{\alpha_9} + \alpha_{11} - [\alpha_{12}E + \alpha_{13}]^{1/2} \tag{3.179}$$

where $\alpha_{11} = \frac{E_{g1}^2}{8\alpha_9^2}\left[\alpha_{10} + \frac{4\alpha_9}{E_{g1}}\right]$, $\alpha_{12} = \left(\frac{E_{g1}^2}{\alpha_9^3}\right)$, $\alpha_{13} = \frac{E_{g1}^4}{64\alpha_9^4}\left[\alpha_{10}^2 + \frac{10\alpha_9^2}{E_{g1}^2} - \frac{8\alpha_9\alpha_{10}}{E_{g1}}\right]$.

The $v_z(E_{n_z})$ in this case is given by

$$v_z(E_{n_z}) = \frac{1}{\hbar}\bar{Q}_{19}(E_{n_z}) \tag{3.180}$$

where $\bar{Q}_{19}(E_{n_z}) = \left[2\alpha_9\left(\frac{\pi n_z}{d_z}\right) + \frac{E_{g1}}{2}\left[\alpha_{10}\left(\frac{\pi n_z}{d_z}\right)\left(1+\alpha_{10}\left(\frac{\pi n_z}{d_z}\right)^2\right)^{-1/2}\right]\right]$ in which E_{n_z} should be determined from the following equation

$$E_{n_z} = \left(\frac{2n_z\pi}{d_z}\right)\left[\alpha_9 + \frac{\alpha_{10}E_{g1}}{4}\left[1+\alpha_{10}\left(\frac{\pi n_z}{d_z}\right)^2\right]^{-1/2}\right] \tag{3.181}$$

The totally quantized energy E_{QD60} assumes the form

$$E_{QD60} = \alpha_9\left[\left(\frac{\pi n_x}{d_x}\right)^2 + \left(\frac{\pi n_y}{d_y}\right)^2 + \left(\frac{\pi n_z}{d_z}\right)^2\right] + \frac{E_{g1}}{2}\left[\sqrt{1+\alpha_{10}\left[\left(\frac{\pi n_x}{d_x}\right)^2 + \left(\frac{\pi n_y}{d_y}\right)^2 + \left(\frac{\pi n_z}{d_z}\right)^2\right]} - 1\right] \quad (3.182)$$

The electron concentration is given by

$$n_{0D} = \frac{2g_v}{d_x d_y d_z}\sum_{n_x=1}^{n_{x\max}}\sum_{n_y=1}^{n_{y\max}}\sum_{n_{z\min}}^{n_{z\max}} F_{-1}(\eta_{70}) \text{ where } \eta_{70} = (E_{F0D} - E_{QD60})/(k_B T) \quad (3.183)$$

The photo-emitted current density can be written as

$$J_{0D} = \frac{\alpha_0 e g_v}{\hbar d_x d_y d_z}\sum_{n_x=1}^{n_{x\max}}\sum_{n_y=1}^{n_{y\max}}\sum_{n_{z\min}}^{n_{z\max}} \bar{Q}_{19}(E_{n_z})F_{-1}(\eta_{70}) \quad (3.184)$$

where $n_{z_{\min}}$ should be determined from the following equation

$$W - hv = \alpha_9\left(\frac{\pi n_{z_{\min}}}{d_z}\right)^2 + \frac{E_{g1}}{2}\left[-1 + \sqrt{1+\alpha_{10}\left(\frac{\pi n_{z_{\min}}}{d_z}\right)^2}\right]. \quad (3.185)$$

3.3 Results and Discussion

Using the appropriate equations, we have plotted the normalized EP from QBs of HD $CdGeAs_2$ as a function of film thickness as shown in plot (a) of Fig. 3.1, where the plot (b) indicates the case for $\delta = 0$. The plots (c) and (d) represent the photoemission in this case in accordance with three band models of Kane. In Figs. 3.2 and 3.3, the aforementioned variable has been plotted as functions of normalized incident photon energy and normalized electron degeneracy respectively for all cases of Fig. 3.1. In Fig. 3.4, the normalized EP from QBs of HD n-InSb has been plotted as a function of d_z in accordance with the (a) three band model of Kane, (b) two band model of Kane, the models of (c) Stillman et al. and (d) Palik et al. respectively. The plots (e–h) refer to QBs of HD GaAs in accordance with the said models respectively in the same figure. In Figs. 3.5 and 3.6, the normalized EP for all cases of Fig. 3.4 has been plotted as functions of normalized incident photon energy and the normalized electron degeneracy respectively. The Fig. 3.7 exhibits the normalized EP from QBs of HD CdS as a function of d_z as shown in plot (a), in which the plot (b) is valid for $\bar{\lambda}_0 = 0$. The plot (c) in the same figure is valid for

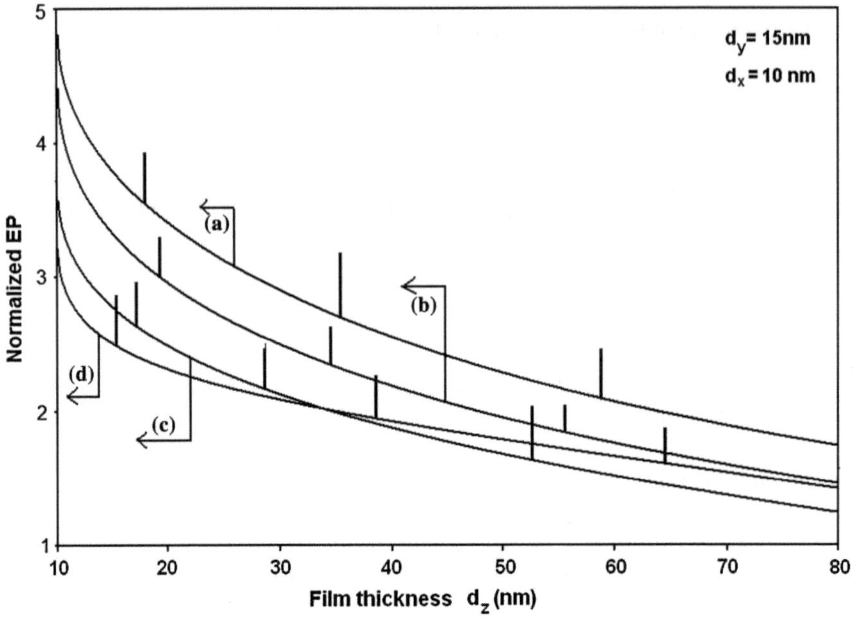

Fig. 3.1 Plot of the normalized EP from QBs of HD $CdGeAs_2$ as a function of d_z in accordance with *a* generalized band model *b* $\delta = 0$, *c* the three-band model of Kane and *d* the two band model of Kane

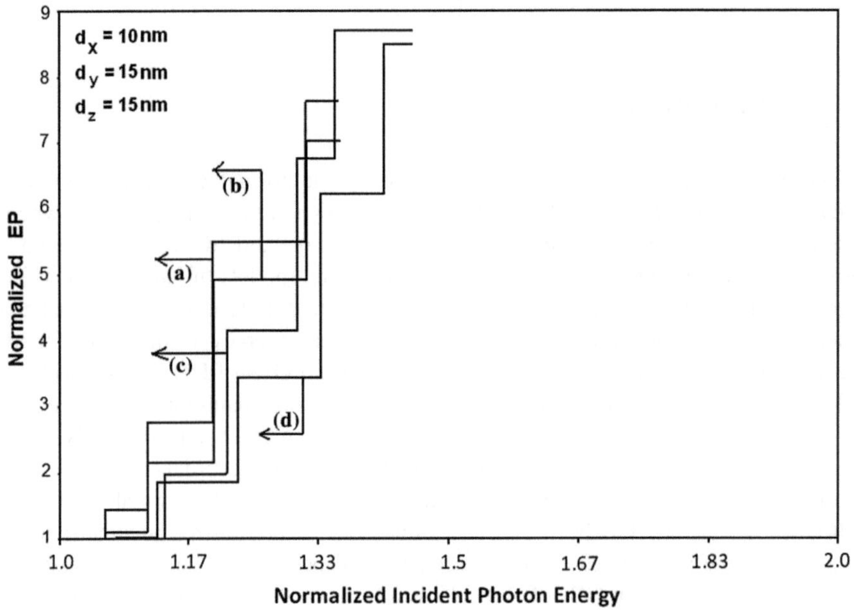

Fig. 3.2 Plot of the normalized EP from QBs of HD $CdGeAs_2$ as a function of normalized incident photon energy for all the cases of Fig. 3.1

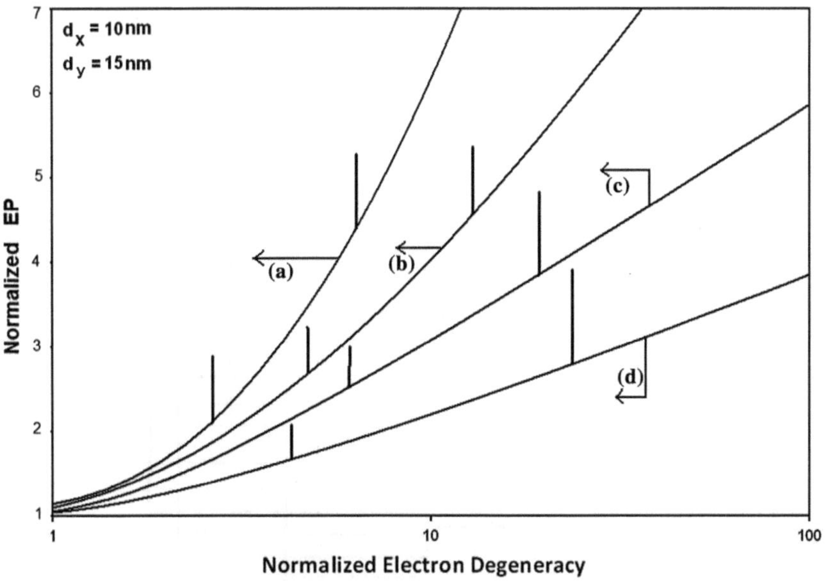

Fig. 3.3 Plot of the normalized EP from QBs of HD $CdGeAs_2$ as a function of normalized electron degeneracy for all the cases of Fig. 3.1

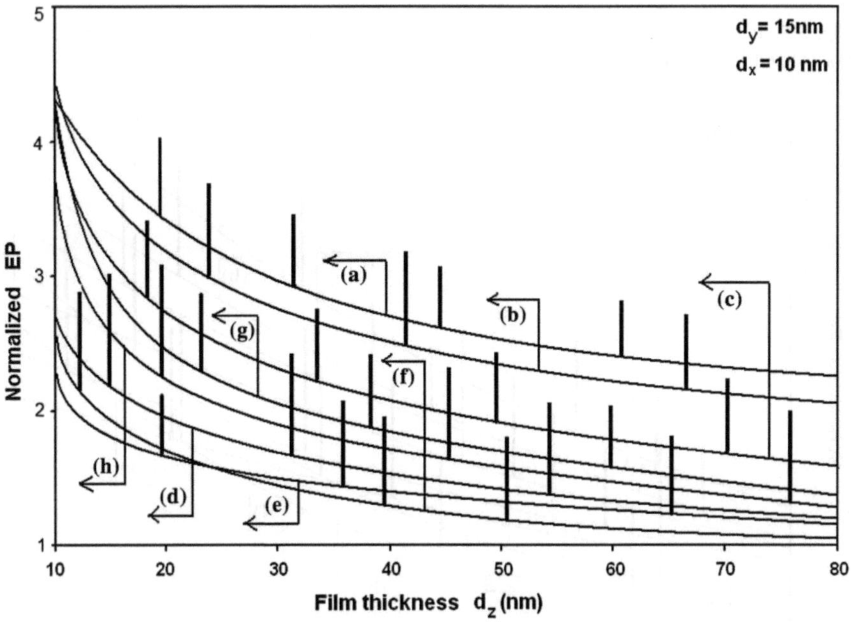

Fig. 3.4 Plot of the normalized EP as a function of d_z from QBs of HD n-InSb in accordance with the *a* three band model of Kane, *b* two band model of Kane, *c* model of Stillman et al. and *d* model of Palik et al. respectively. The plots *e–h* refer to QBs of HD GaAs in accordance with the said models

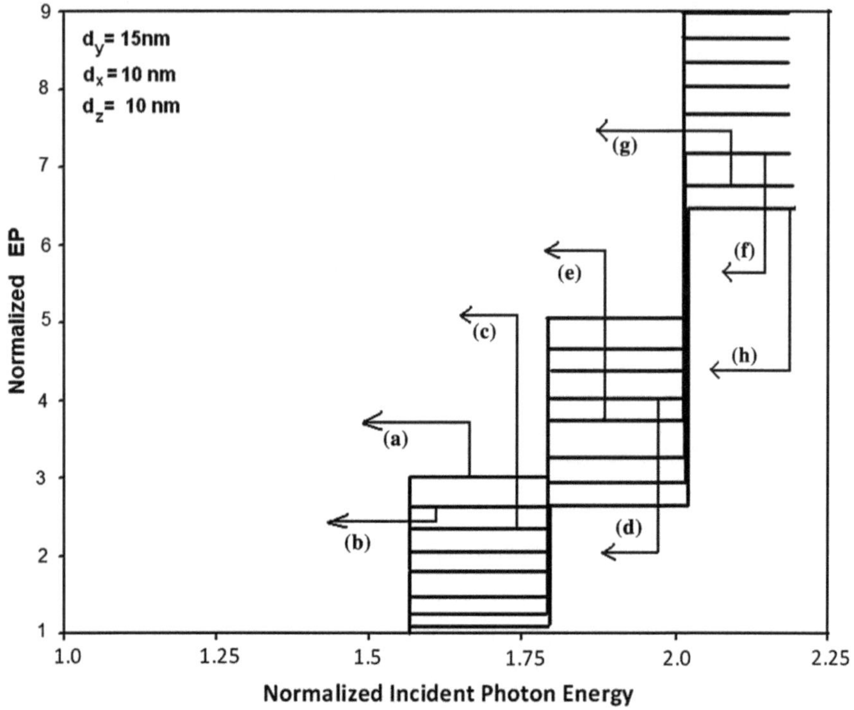

Fig. 3.5 Plot of the normalized EP from QBs of HD n-InSb and HD n-GaAs as a function of normalized incident photon energy for all the cases of Fig. 3.4

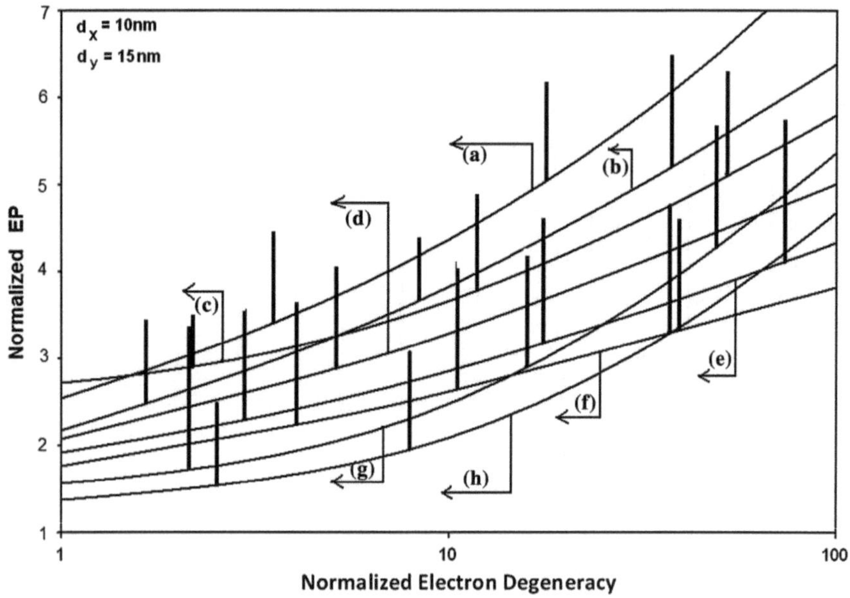

Fig. 3.6 Plot of the normalized EP from QBs of HD n-InSb and HD n-GaAs as a function of normalized electron degeneracy for all the cases of Fig. 3.4

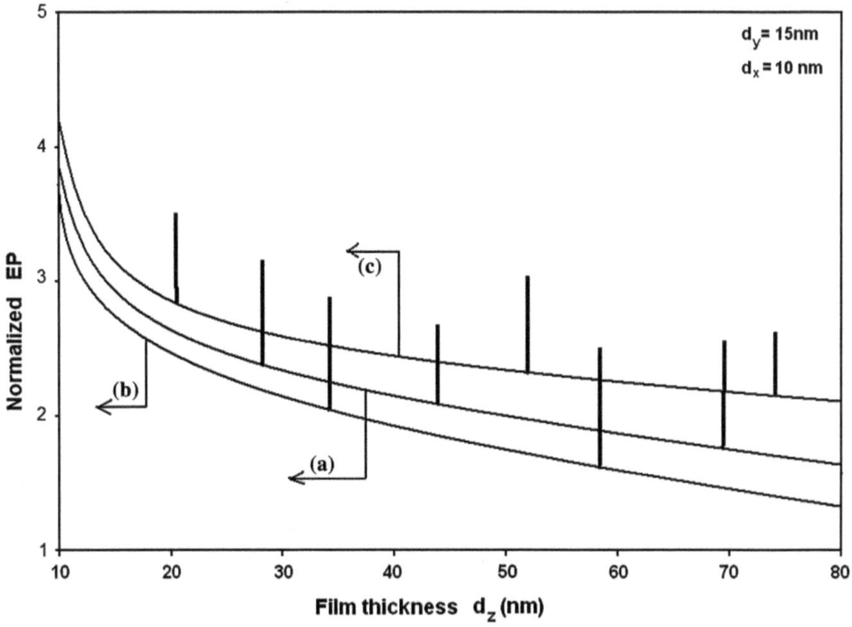

Fig. 3.7 Plot of the normalized EP from QBs of *a* HD CdS with $\bar{\lambda}_0 \neq 0$, *b* $\bar{\lambda}_0 = 0$ respectively and *c* HD GaP as a function of d_z

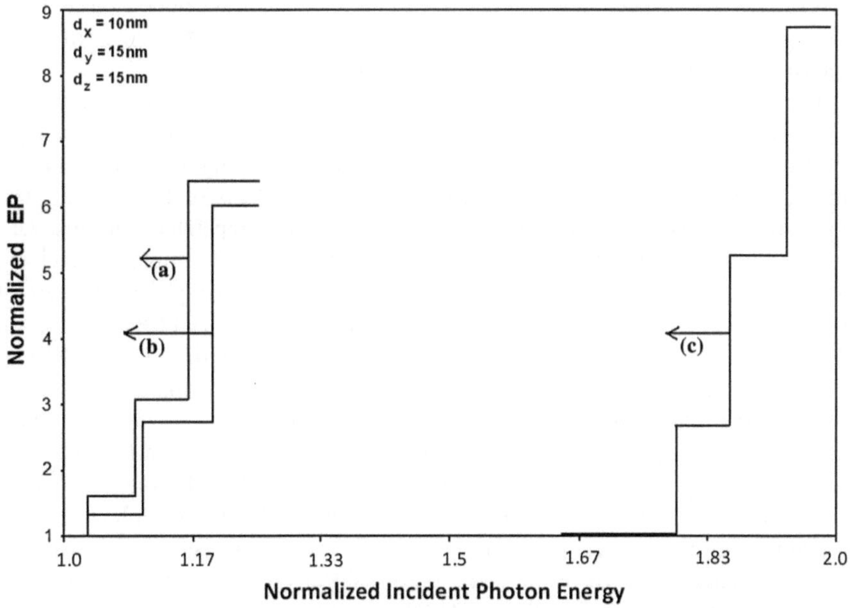

Fig. 3.8 Plot of the normalized EP from QBs of *a* HD CdS with $\bar{\lambda}_0 \neq 0$, *b* $\bar{\lambda}_0 = 0$ respectively and *c* HD GaP as a function of normalized incident photon energy

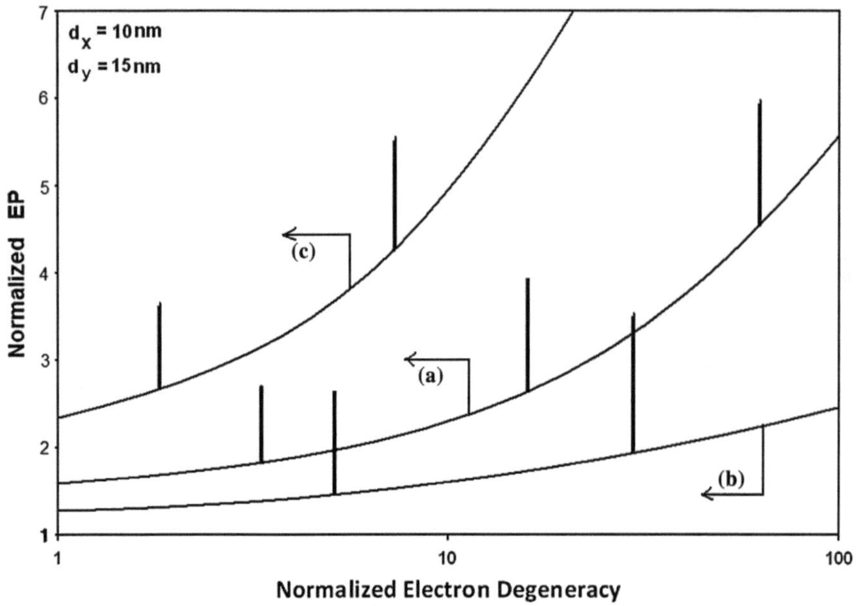

Fig. 3.9 Plot of the normalized EP from QBs of *a* HD CdS with $\bar{\lambda}_0 \neq 0$, *b* $\bar{\lambda}_0 = 0$ respectively and *c* HD GaP as a function of normalized electron degeneracy

QBs of HD GaP respectively. In Figs. 3.8 and 3.9, the aforementioned variable has been plotted as functions of normalized incident photon energy and normalized electron degeneracy respectively for all cases of Fig. 3.7.

The Fig. 3.10 exhibits the plot of the normalized EP as a function of film thickness from QBs of HD n-Ge in accordance with the models of (a) Cardona et al. and (b) Wang et al. respectively. The curve (c) in the same figure refers to the EP from QBs of HD Tellurium in accordance with the models of Bouat et al. The curves (d) and (e) refer to Te for two different values of temperature. In Figs. 3.11 and 3.12, the normalized EP has been plotted as functions of normalized incident photon energy and normalized electron degeneracy respectively for all cases of Fig. 3.10. In Fig. 3.13, we have drawn the plots of the normalized EP as a function of film thickness from HD QBs of (a) n-$PtSb_2$ (b) zerogap materials, taking HgTe as an example and (c) $Pb_{1-x}Ge_xTe$. In Figs. 3.14 and 3.15, the normalized EP in this case has been plotted as functions of normalized incident photon energy and normalized carrier degeneracy respectively for all cases of Fig. 3.13. In Fig. 3.16, we have drawn the plots of the normalized EP as a function of film thickness from QBs of HD GaSb in accordance with the model of Mathur et al. for three different values of temperature respectively. The curve (d) in the same figure refers to the QBs of HD stressed materials, where stressed n-InSb has been considered as an example. In Figs. 3.17 and 3.18, the normalized EP in this case has been plotted as functions of

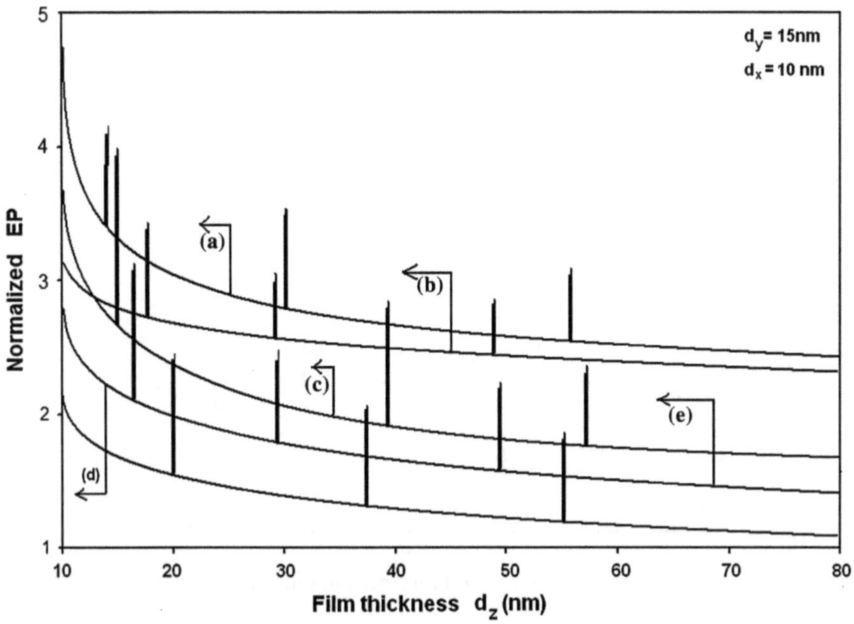

Fig. 3.10 Plot of the normalized EP from HD QBs of *a* n-Ge in accordance with the model of Cardona et al. *b* n-Ge in accordance with the model of Wang et al. as a function of film thickness The *curve c* refers to the same plot for HD QBs of Tellurium in accordance with the model of Bouat et al. The *curves d* and *e* exhibit the same dependence for Te for two different temperatures

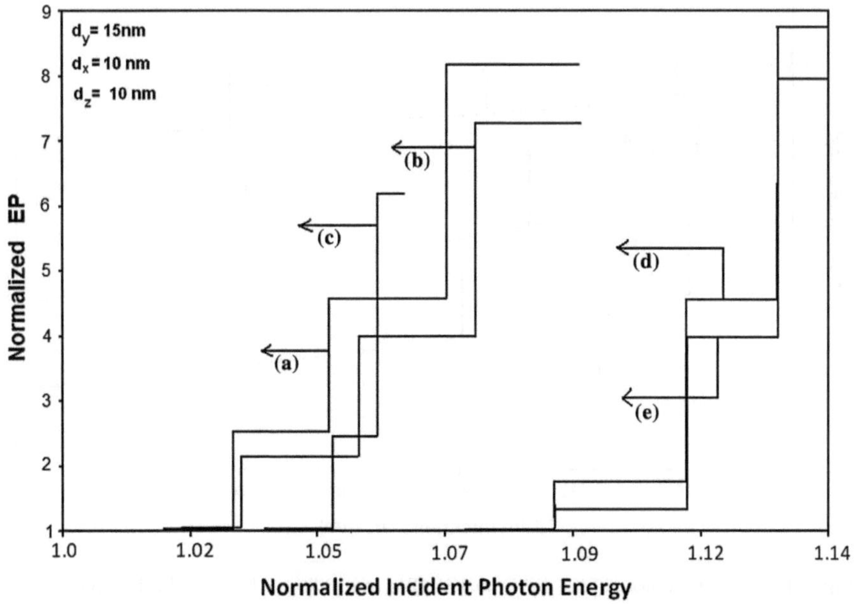

Fig. 3.11 Plot of the normalized EP as a function of normalized incident photon energy for all the cases of Fig. 3.10

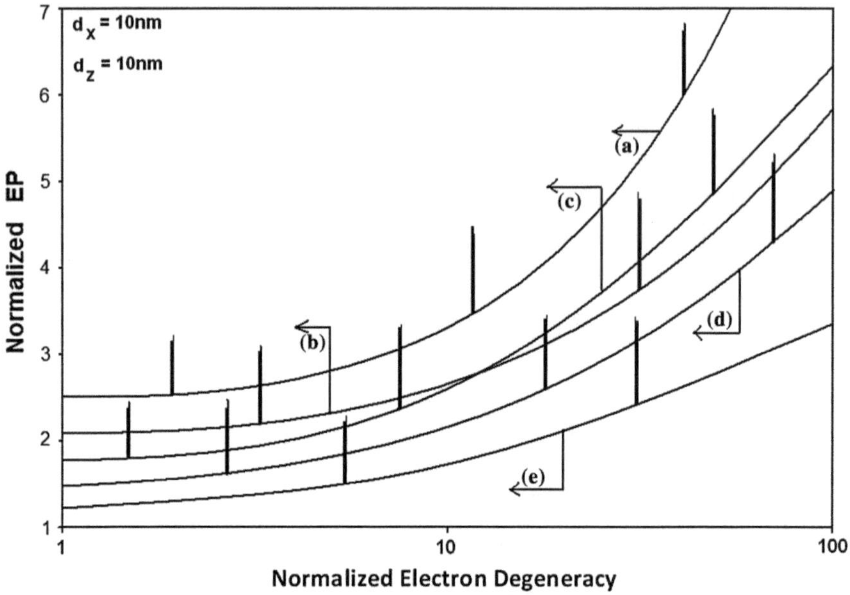

Fig. 3.12 Plot of the normalized EP as a function of normalized electron degeneracy for all the cases of Fig. 3.10

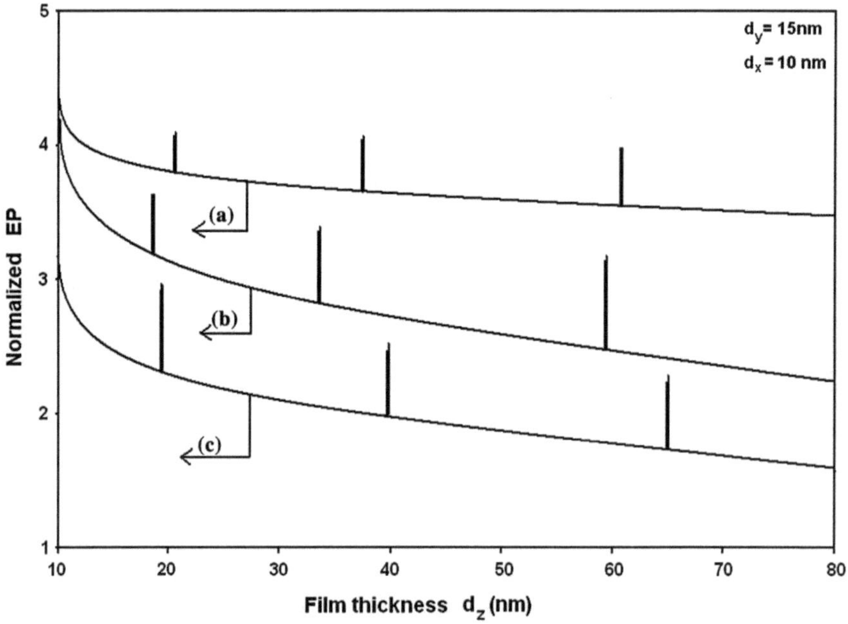

Fig. 3.13 Plot of the normalized EP from HD QBs of *a* $PtSb_2$, *b* HgTe and *c* $Pb_{1-x}Ge_xTe$ as a function of film thickness

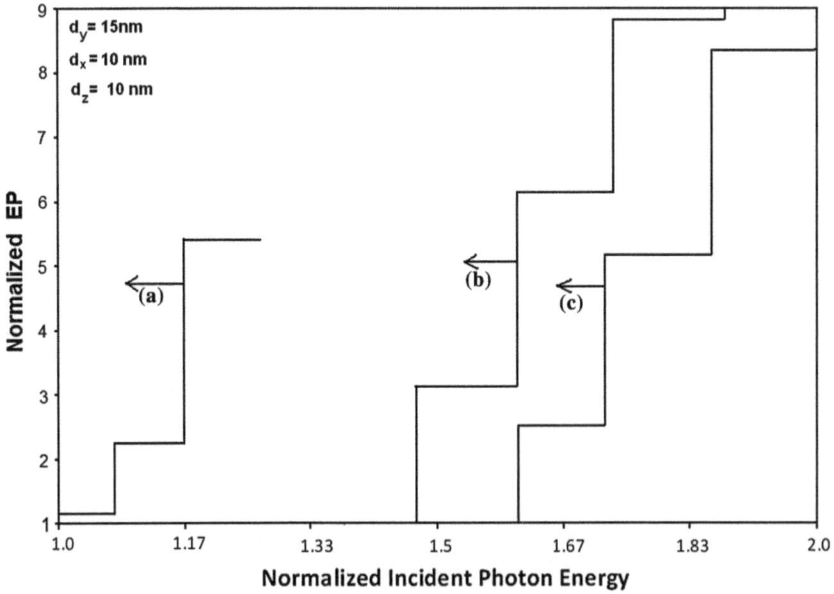

Fig. 3.14 Plot of the normalized EP from HD QBs of *a* $PtSb_2$, *b* HgTe and *c* $Pb_{1-x}Ge_xTe$ as a function of normalized incident photon energy

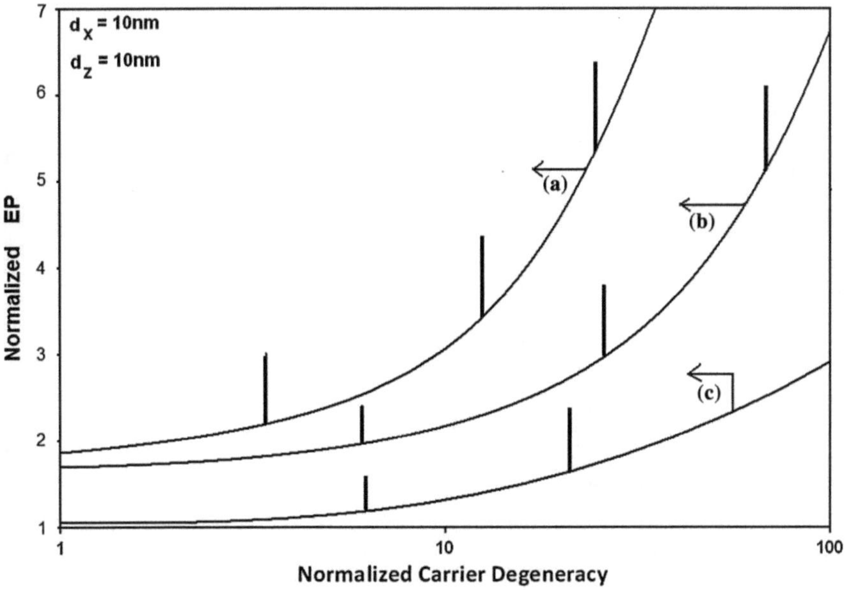

Fig. 3.15 Plot of the normalized EP from HD QBs of *a* $PtSb_2$, *b* HgTe and *c* $Pb_{1-x}Ge_xTe$ as a function of normalized carrier degeneracy

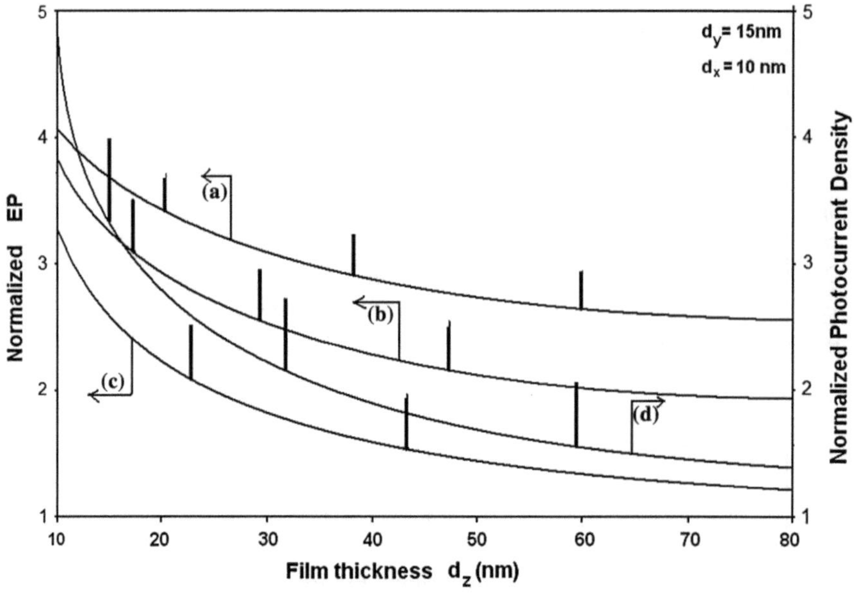

Fig. 3.16 Plot of the normalized EP from HD QBs of GaSb in accordance with the model of Mathur et al. for three different values of temperature as a function of film thickness. The plot *d* refers to QBs of HD stressed InSb

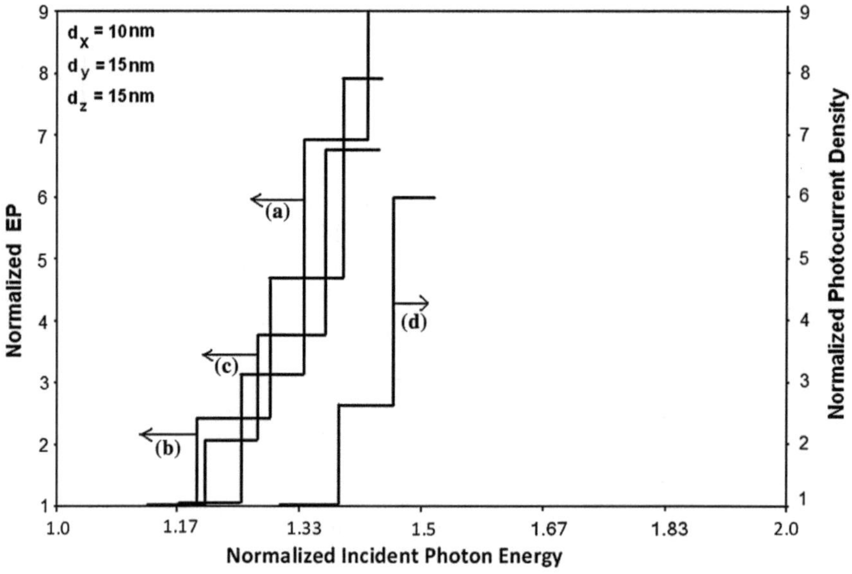

Fig. 3.17 Plot of the normalized EP as a function of normalized incident photon energy for all the cases of Fig. 3.16

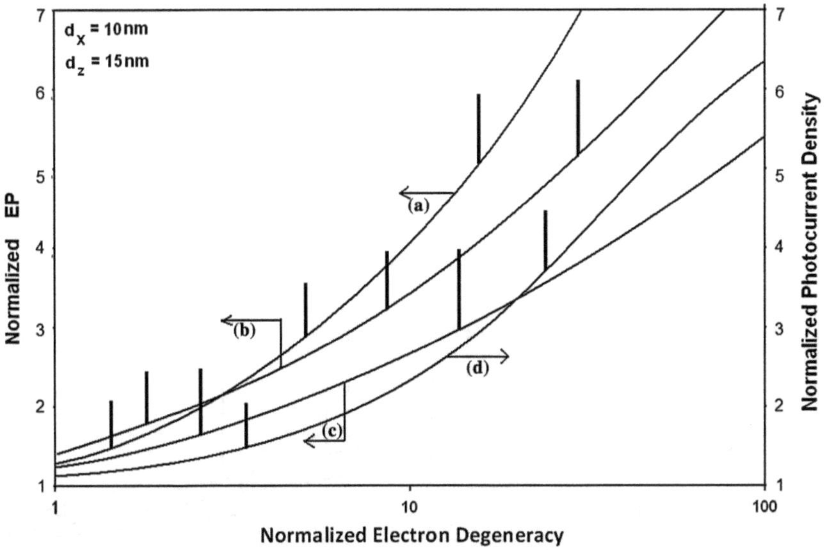

Fig. 3.18 Plot of the normalized EP as a function of normalized electron degeneracy for all the cases of Fig. 3.16

normalized incident photon energy and the normalized electron degeneracy respectively for all cases of Fig. 3.16.

The Fig. 3.19 exhibits the plot of the normalized EP as a function of film thickness from QBs of HD PbSe in accordance with the models of (a) Dimmock, (b) the Cohen, (c) Bangert et al. and (d) the Lax respectively. In Figs. 3.20 and 3.21, the same variable has been plotted as functions of normalized incident photon energy and normalized electron degeneracy respectively for all cases of Fig. 3.19. The Fig. 3.22 depicts the plot of the normalized EP as a function of film thickness from QBs of IV-VI materials taking HD n-PbTe as an example in accordance with the models of (a) Dimmock (b) Cohen, (c) Bangert et al. (d) Lax respectively. Besides, the plot (e) of Fig. 3.22 exhibits the EP from QBs of II-V materials taking HD CdSb as an example. In Figs. 3.23 and 3.24, the normalized EP in this case has been plotted as functions of normalized incident photon energy and normalized carrier degeneracy respectively for all cases of Fig. 3.22.

The Fig. 3.25 shows the plot of the normalized EP as a function of film thickness from QBs of HD Bi_2Te_3 for two different temperatures respectively. In Figs. 3.26 and 3.27, the normalized EP in this case has been plotted as functions of normalized incident photon energy and normalized carrier degeneracy respectively for all cases of Fig. 3.25.

It appears from the Figs. 3.1, 3.4, 3.7, 3.10, 3.13, 3.16, 3.19, 3.22 and 3.25 that the normalized EP increases with decreasing film thickness and exhibits spikes for various values of d_z which are totally band structure dependent. The Figs. 3.2, 3.5, 3.8, 3.11, 3.14, 3.17, 3.20, 3.23 and 3.26 exhibit the step-functional dependence of the normalized EP from the HD QBs of different materials with the incident photon

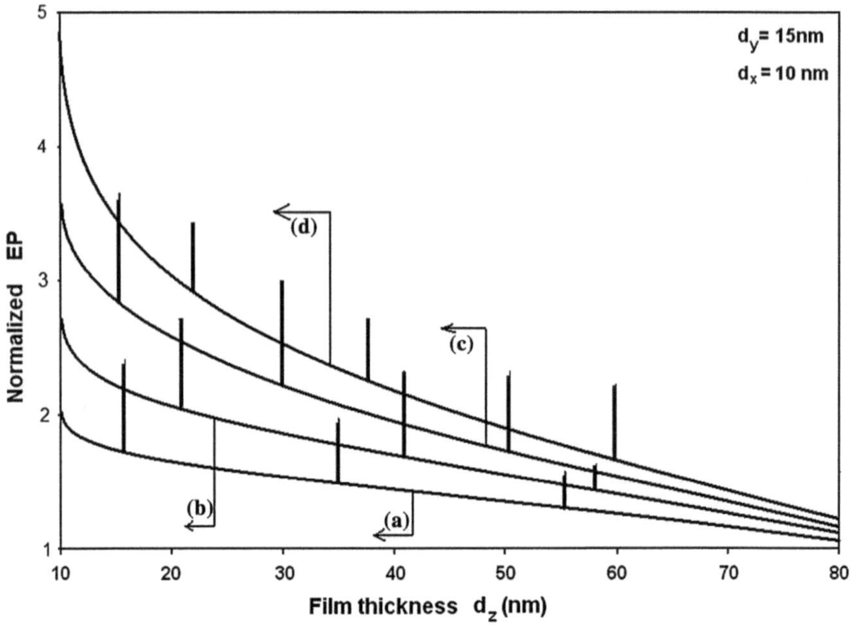

Fig. 3.19 Plot of the normalized EP as a function of film thickness from QBs of HD PbSe in accordance with the models of *a* Dimmock and *b* the Cohen, *c* Bangert et al. and *d* the Lax respectively

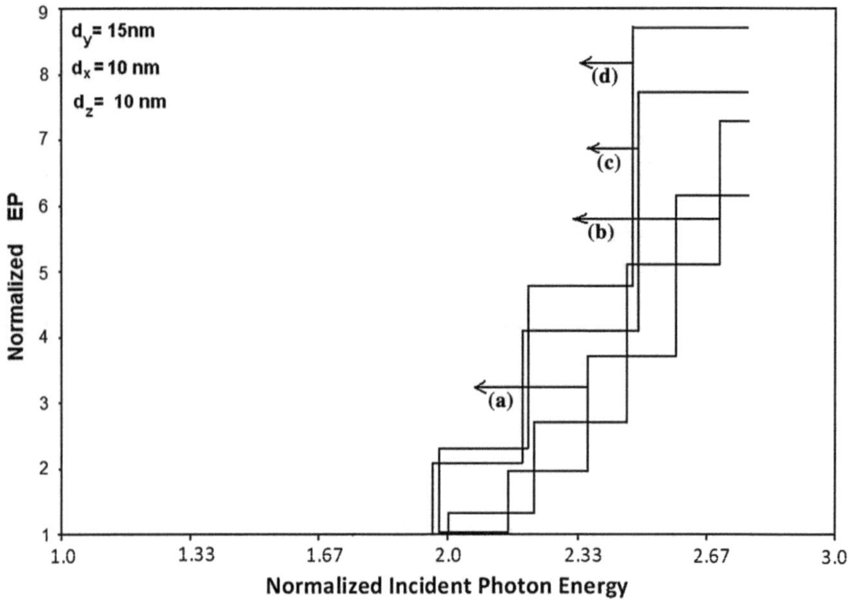

Fig. 3.20 Plot of the normalized EP as a function of normalized incident photon energy for all the cases of Fig. 3.19

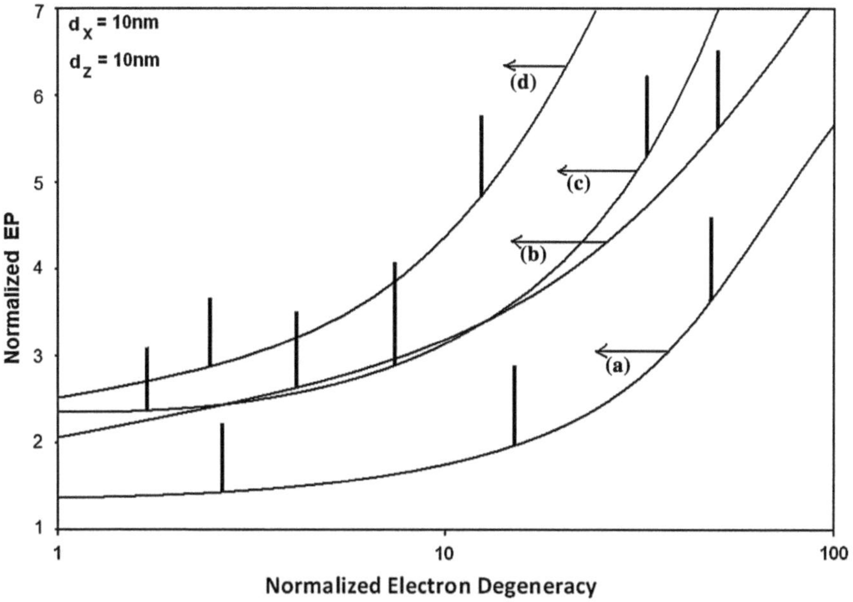

Fig. 3.21 Plot of the normalized EP as a function of normalized electron degeneracy for all the cases of Fig. 3.19

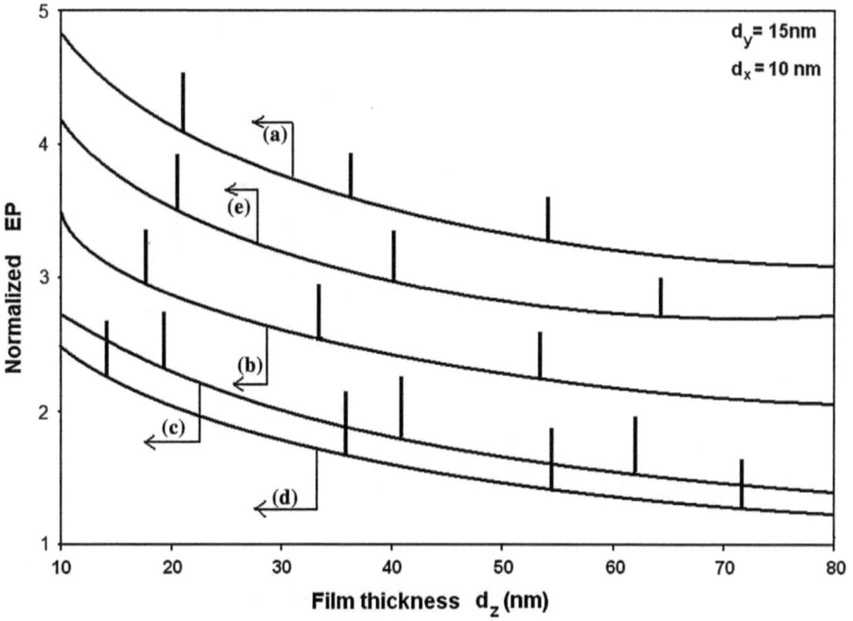

Fig. 3.22 Plot of the normalized EP from QBs of HD PbTe in accordance with the models of *a* Dimmock, *b* Cohen, *c* Bangert et al. and *d* Lax respectively as a function of film thickness. The plot *e* refers to QBs of HD CdSb

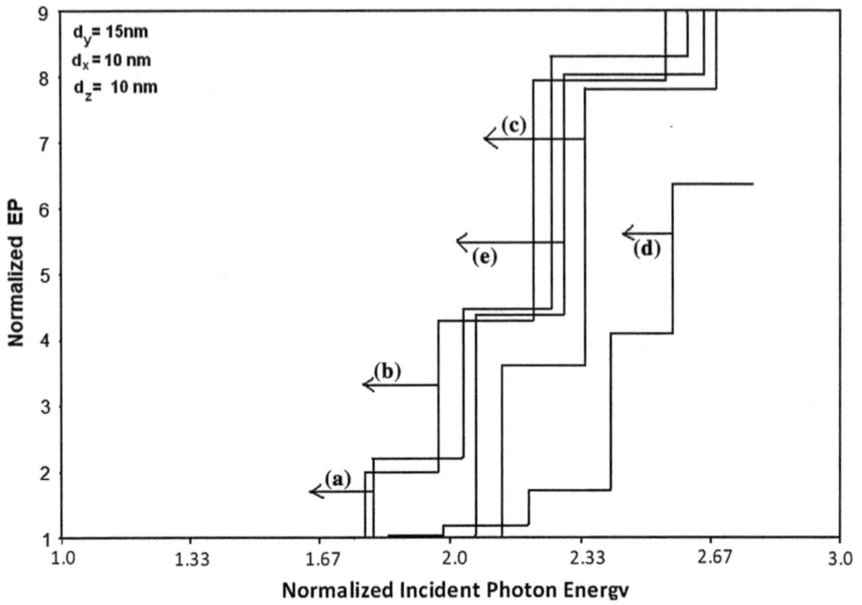

Fig. 3.23 Plot of the normalized EP as a function of normalized incident photon energy for all the cases of Fig. 3.22

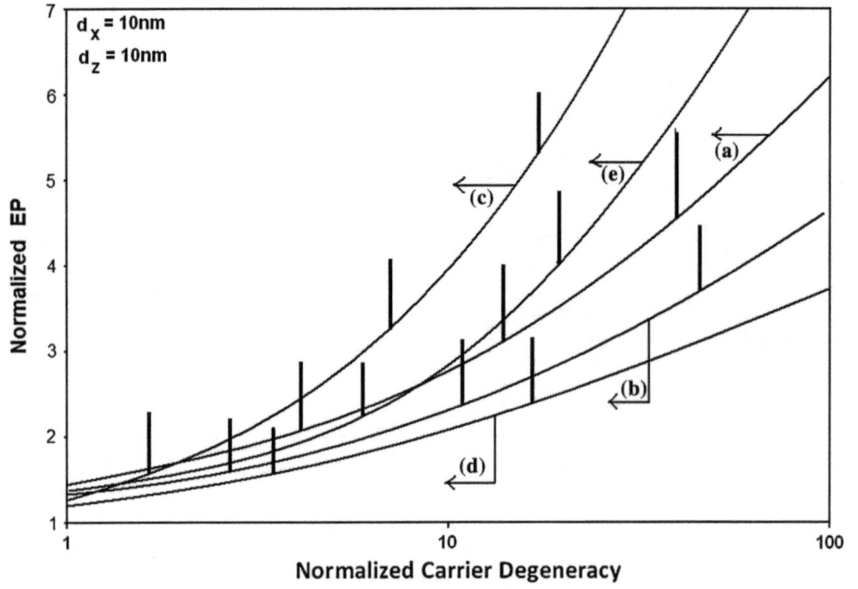

Fig. 3.24 Plot of the normalized EP as a function of normalized carrier degeneracy for all the cases Fig. 3.22

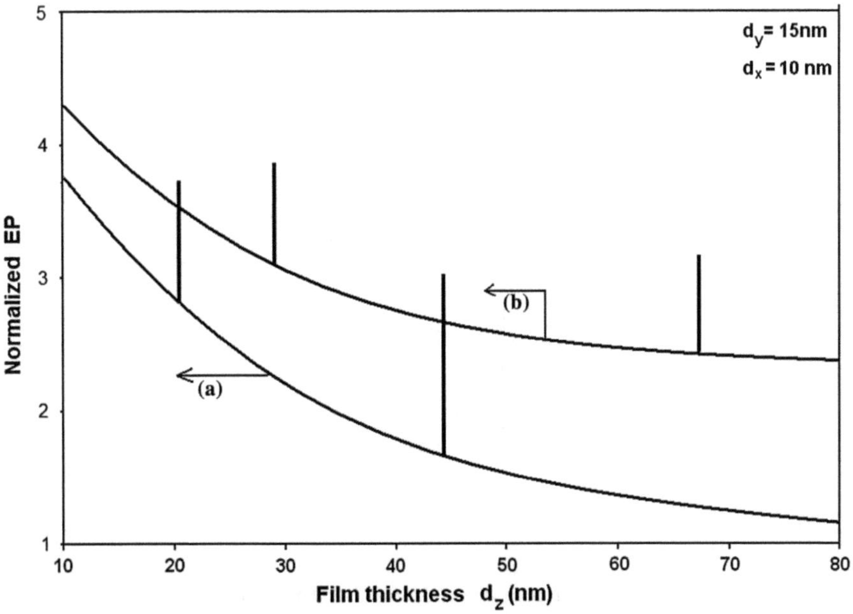

Fig. 3.25 Plot of the normalized EP from QBs of HD Bi_2Te_3 for two different temperatures as a function of film thickness

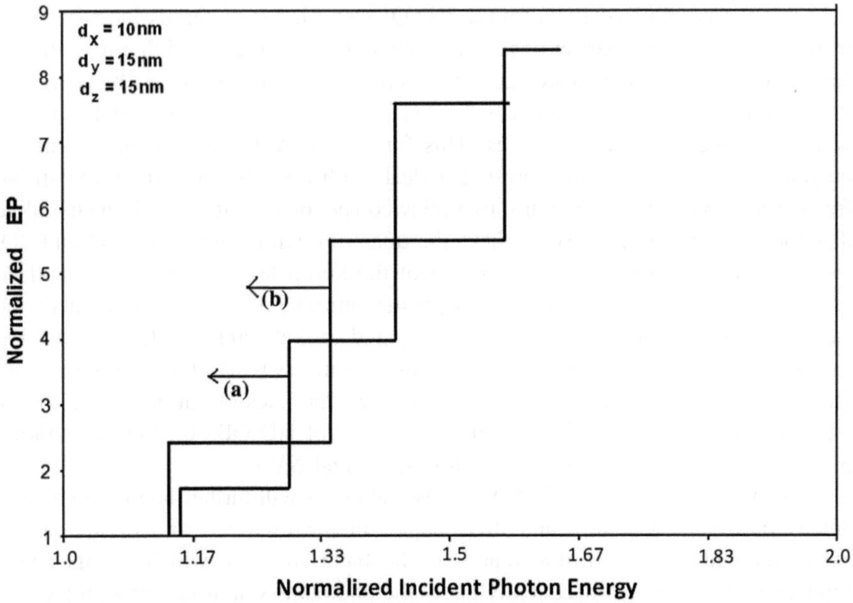

Fig. 3.26 Plot of the normalized EP from QBs of HD Bi_2Te_3 for two different temperatures as a function of normalized incident photon energy

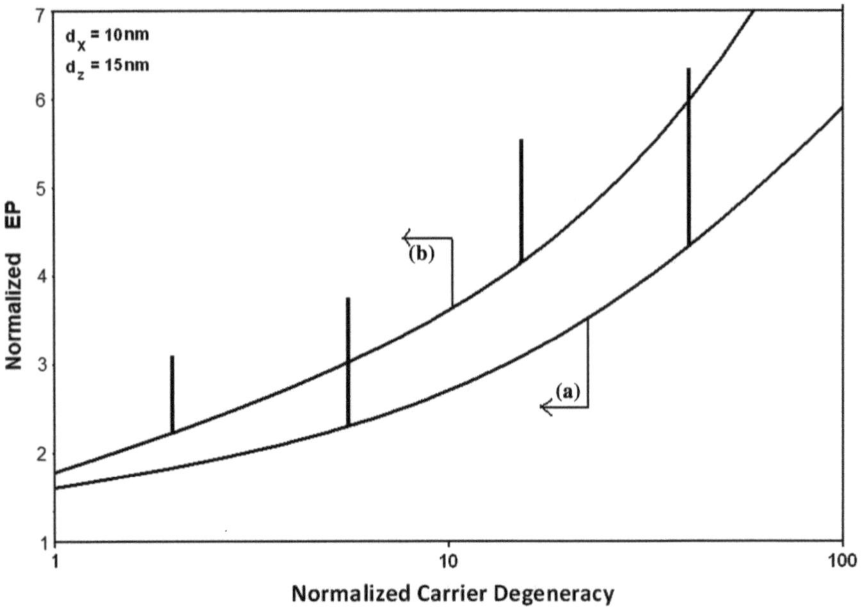

Fig. 3.27 Plot of the normalized EP from QBs of HD Bi_2Te_3 for two different temperatures as a function of normalized carrier degeneracy

energy. It is apparent from the Figs. 3.3, 3.6, 3.9, 3.12, 3.15, 3.18, 3.21, 3.24 and 3.27 that the normalized EP from the HD QBs of various materials increases with increasing normalized carrier degeneracy and exhibits spikes for different values of carrier concentration which are again band structure dependent. It may be noted that the HD QBs lead to the discrete energy levels, somewhat like atomic energy levels, which produce very large changes. This follows from the inherent nature of the quantum confinement of the carrier gas dealt with here. In QBs, there remain no free carrier states in between any two allowed sets of size-quantized levels unlike that found for QWs and NWs where the quantum confinements are 1D and 2D, respectively. Consequently, the crossing of the Fermi level by the size-quantized levels in HD QBs would have much greater impact on the redistribution of the carriers among the allowed levels, as compared to that found for QWs and NWs respectively. Although the EP varies in various manners with all the variables in all the limiting cases as evident from all the figures, the rates of variations are totally band-structure dependent. The quantum signature of HD QBs for the EP is rather prominent as compared to the same from QWs and NWs.

The photoemission from HD QWs, NWs and QBs will further be investigated in details in Chaps. 1 and 2 and this chapter in the presence of external photo-excitation with the realization that it is the band structure which changes in a fundamental way and consequently alters the photoemission together with the fact that in general all the physical properties of all the electronic materials changes radically leading to new physical concepts.

3.4 Open Research Problems

(R.3.1) Investigate the EP for bulk specimens of the HDS under non uniform strain in the presences of Gaussian, exponential, Kane, Halperian, Lax and Bonch-Burevich types of band tails for all systems whose unperturbed carrier energy spectra are defined in R.1.1.

(R.3.2) Investigate the EP for QBs of all the HD semiconductors as considered in R.3.1.

(R.3.3) Investigate the EP in the presence of non uniform strain for HD bulk specimens of the negative refractive index, organic, magnetic and other advanced optical materials in the presence of an arbitrarily oriented alternating electric field.

(R.3.4) Investigate the EP for the QBs of HD negative refractive index, organic, magnetic and other advanced optical materials in the presence of an arbitrarily oriented alternating electric field.

(R.3.5) Investigate the EP for the multiple QBs of HD materials whose unperturbed carrier energy spectra are defined in R.1.1.

(R.3.6) Investigate the EP for all the appropriate HD zero dimensional systems of this chapter in the presence of finite potential wells.

(R.3.7) Investigate the EP for all the appropriate HD zero dimensional systems of this chapter in the presence of parabolic potential wells.

(R.3.8) Investigate the EP for all the appropriate HD totally quantized systems of this chapter forming quantum rings.

(R.3.9) Investigate the EP for all the above appropriate problems in the presence of elliptical Hill and quantum square rings in the presence of strain.

(R.3.10) Investigate the EP for parabolic cylindrical HD low dimensional systems in the presence of an arbitrarily oriented alternating electric field for all the HD materials whose unperturbed carrier energy spectra are defined in R.1.1 in the presence of strain.

(R.3.11) Investigate the EP for HD zero dimensional systems of the negative refractive index and other advanced optical materials in the presence of an arbitrarily oriented alternating electric field and non-uniform light waves and in the presence of strain.

(R.3.12) Investigate the EP for triangular HD zero dimensional systems of the negative refractive index, organic, magnetic and other advanced optical materials in the presence of an arbitrarily oriented alternating electric field in the presence of strain.

(R.3.13) Investigate the EP for all the problems of (R.3.12) in the presence of arbitrarily oriented magnetic field.

(R.3.14) Investigate the EP for all the problems of (R.3.12) in the presence of alternating electric field.

(R.3.15) Investigate the EP for all the problems of (R.3.12) in the presence of alternating magnetic field.

(R.3.16) Investigate the EP for all the problems of (R.3.14) in the presence of crossed electric field and quantizing magnetic fields.

(R.3.17) Investigate the EP for all the problems of (R.1.14) in the presence of crossed alternating electric field and alternating quantizing magnetic fields.

(R.3.18) Investigate the EP for HD QBs of the negative refractive index, organic and magnetic materials.

(R.3.19) Investigate the EP for HD QBs of the negative refractive index, organic and magnetic materials in the presence of alternating time dependent magnetic field.

(R.3.20) Investigate the EP for HD QBs of the negative refractive index, organic and magnetic materials in the presence of in the presence of crossed alternating electric field and alternating quantizing magnetic fields.

(R.3.21) (a) Investigate the EP for HD QBs of the negative refractive index, organic, magnetic and other advanced optical materials in the presence of an arbitrarily oriented alternating electric field considering many body effects.

(b) Investigate all the appropriate problems of this chapter for a Dirac electron.

(R.3.22) Investigate all the appropriate problems of this chapter by including the many body, image force, broadening and hot carrier effects respectively.

(R.3.23) Investigate all the appropriate problems of this chapter by removing all the mathematical approximations and establishing the respective appropriate uniqueness conditions.

References

1. D. Bimberg, M. Grundmann, N.N. Ledentsov, *Quantum Dot Heterostructures* (Wiley, New York, 1999)
2. G. Konstantatos, I. Howard, A. Fischer, S. Howland, J. Clifford, E. Klem, L. Levina, E.H. Sargent, Nature **442**, 180 (2006)
3. J.K. Jaiswal, H. Mattoussi, J.M. Mauro, S.M. Simon, Nat. Biotechnol. **21**, 47 (2003)
4. A. Watson, X. Wu, M. Bruchez, Biotechniques **34**, 296 (2003)
5. J. Nakanishi, Y. Kikuchi, T. Takarada, H. Nakayama, K. Yamaguchi, M. Maeda, J. Am. Chem. Soc. **126**, 16314 (2004)
6. X. Michalet, F.F. Pinaud, L.A. Bentolila, J.M. Tsay, S. Doose, J.J. Li, G. Sundaresan, A.M. Wu, S.S. Gambhir, S. Weiss, Science **307**, 538 (2005)
7. W.G.J.H.M. van Sark, P.L.T.M. Frederix, D.J. Van den Heuvel, H.C.G.A. Bol, J.N.J. van Lingen, de Mello Donegá C. Meijerink A J. Phys. Chem. B **105**, 8281 (2001)
8. E.J. Sánchez, L. Novotny, X.S. Xie, Phys. Rev. Lett. **82**, 4014 (1999)
9. B. Bailey, D.L. Farkas, D.L. Taylor, F. Lanni, Nature **366**, 44 (1993)
10. L.V. Asryan, R.A. Suris, in *Selected Topics in Electronics and Systems*, ed. by E. Borovitskaya, M.S. Shur, vol. 25 (World Scientific, Singapore, 2002)
11. L.V. Asryan, R.A. Suris, Int. J. High Speed Electron. Syst., Special issue on Quantum dot heterostructures—fabrication, application, theory, **12**(1), 111 (2002)
12. L.V. Asryan, S. Luryi, *Future Trends in Microelectronics: The Nano Millennium*, ed. by S. Luryi, J.M. Xu, A. Zaslavsky, Wiley Interscience, New York, (2002) p. 219
13. R.A. Freitas Jr., J. Comput. Theor. Nanosci. **2**, 1 (2005)
14. A. Ferreira, C. Mavroidis, IEEE Robot. Autom. Mag. **13**, 78 (2006)

15. A. Dubey, G. Sharma, C. Mavroidis, S.M. Tomassone, K. Nikitczuk, M.L. Yarmush, J. Comput. Theor. Nanosci. **1**, 18 (2004)
16. C. Mavroidis, A. Dubey, M.L. Yarmush, Ann. Rev. Biomed. Eng. **6**, 363 (2004)
17. Y. Liu, J. A. Starzyk, Z. Zhu, *IEEE Trans. Neural Networks* (2008) [In the press]
18. J.A. Starzyk, H. He, IEEE Trans. Neural Networks **18**(2), 344 (2007)
19. J.A. Starzyk, H. He, IEEE Trans. Circuits Syst. II **54**(2), 176 (2007)
20. E.-S. Hasaneen, E. Heller, R. Bansal, W. Huang, F. Jain, Solid State Electron. **48**, 2055 (2004)
21. T. Kawazoe, S. Tanaka, M. Ohtsu, J. Nanophoton. **2**, 029502 (2008)
22. H.J. Krenner, S. Stufler, M. Sabathil, E.C. Clark, P. Ester, M. Bichler, G. Abstreiter, J.J. Finley, A. Zrenner, New J. Phys. **7**, 184 (2005)
23. A.E. Zhukov, A.R. Kovsh, Quantum Electron. **38**, 409 (2008)
24. M. Sugawara, T. Akiyama, N. Hatori, Y. Nakata, H. Ebe, H. Ishikawa, Meas. Sci. Technol. **13**, 1683 (2002)
25. M. van der Poel, D. Birkedal, J. Hvam, M. Laemmlin, D. Bimberg, in *Conference on Lasers and Electro-Optics (CLEO),* vol. **1**, (2004) p. 16
26. J.M. Costa-Fernandez, Anal. Bioanal. Chem. **384**, 37 (2006)
27. H.S. Djie, C.E. Dimas, D.-N. Wang, B.-S. Ooi, J.C.M. Hwang, G.T. Dang, W.H. Chang, IEEE Sens. J. **7**, 251 (2007)
28. X.-X. Zhu, Y.-C. Cao, X. Jin, J. Yang, X.-F. Hua, H.-Q. Wang, B. Liu, Z. Wang, J.-H. Wang, L. Yang, Y.-D. Zhao, Nanotechnology **19**, 025708 (2008)
29. X. Gao, W.C.W. Chan, S. Nie, J. Biomed. Opt. **7**, 532 (2002)
30. X. Michalet, F.F. Pinaud, L.A. Bentolila, J.M. Tsay, S. Doose, J.J. Li, G. Sundaresan, A.M. Wu, S.S. Gambhir, S. Weiss, Science **307**, 538 (2005)
31. J.K. Jaiswal, E.R. Goldman, H. Mattoussi, S.M. Simon, Nat. Methods **1**, 73 (2004)
32. H. Matsueda, Int. J. Circuit Theo. Appl. **31**, 23 (2003)
33. X. Hu, S. Das Sarma, Phys. Status Solidi B **238**, 360 (2003)
34. G.-L. Chen, D.M.T. Kuo, W.-T. Lai, P.-W. Li, Nanotechnology **18**, 475402 (2007)
35. A.G. Pogosov, M.V. Budantsev, A.A. Shevyrin, A.E. Plotnikov, A.K. Bakarov, A.I. Toropov, JETP Lett. **87**, 150 (2008)
36. K.W. Johnston, A.G. Pattantyus-Abraham, J.P. Clifford, S.H. Myrskog, D.D. MacNeil, L. Levina, E.H. Sargent, Appl. Phys. Letts. **92**, 151115 (2008)
37. K.S. Leschkies, R. Divakar, J. Basu, E.E. Pommer, J.E. Boercker, C.B. Carter, U.R. Kortshagen, D.J. Norris, E.S. Aydil, Nano Lett. **7**, 1793 (2007)
38. I.-S. Liu, H.-H. Lo, C.-T. Chien, Y.-Y. Lin, C.-W. Chen, Y.-F. Chen, W.-F. Su, S.-C. Liou, J. Mater. Chem. **18**, 675 (2008)
39. N. Hitoshi, Y. Sugimoto, K. Nanamoto, N. Ikeda, Y. Tanaka, Y. Nakamura, S. Ohkouchi, Y. Watanabe, K. Inoue, H. Ishikawa, K. Asakawa, Opt. Express **12**, 6606 (2004)
40. N. Yamamoto, T. Matsuno, H. Takai, N. Ohtani, Jpn. J. Appl. Phys. **44**, 4749 (2005)
41. T. Yamada, Y. Kinoshita, S. Kasai, H. Hasegawa, Y. Amemiya, Jpn. J. Appl. Phys. **40**, 4485 (2001)
42. K. Asakawa, Y. Sugimoto, Y. Watanabe, N. Ozaki, A. Mizutani, Y. Takata, Y. Kitagawa, H. Ishikawa, N. Ikeda, K. Awazu, X. Wang, A. Watanabe, S. Nakamura, S. Ohkouchi, K. Inoue, M. Kristensen, O. Sigmund, P.I. Borel, R. Baets, New J. Phys. **8**, 208 (2006)
43. A.R. Clapp, I.L. Medintz, B.R. Fisher, G.P. Anderson, H. Mattoussi, J. Am. Chem. Soc. **127**, 1242 (2005)
44. L. Shi, B. Hernandez, M. Selke, J. Am. Chem. Soc. **128**, 6278 (2006)
45. C. Wu, J. Zheng, C. Huang, J. Lai, S. Li, C. Chen, Y. Zhao, Angew. Chem. Int. Ed. **46**, 5393 (2007)

Chapter 4
The EP from Heavily Doped (HD) Quantized Superlattices

4.1 Introduction

In recent years, modern fabrication techniques have generated altogether a new dimension in the arena of quantum effect devices through the experimental realization of an important artificial structure known as semiconductor superlattice (SL) by growing two similar but different semiconducting compounds in alternate layers with finite thicknesses [1–33]. The materials forming the alternate layers have the same kind of band structure but different energy gaps. The concept of SL was developed for the first time by Keldysh [34–38] SL and was successfully fabricated by Esaki and Tsu [34–38]. The SLs are being extensively used in thermal sensors [39, 40], quantum cascade lasers [41–43], photodetectors [44, 45], light emitting diodes [46–49], multiplication [50], frequency multiplication [51], photocathodes [52, 53], thin film transistor [54], solar cells [55, 56], infrared imaging [57], thermal imaging [58, 59], infrared sensing [60] and also in other microelectronic devices.

The most extensively studied III-V SL is the one consisting of alternate layers of GaAs and $Ga_{1-x}Al_xAs$ owing to the relative easiness of fabrication. The GaAs and $Ga_{1-x}Al_xAs$ layers form the quantum wells and the potential barriers respectively. The III-V SL's are attractive for the realization of high speed electronic and optoelectronic devices [61]. In addition to SLs with usual structure, other types of SLs such as II-VI [62], IV-VI [63] and HgTe/CdTe [64] SL's have also been investigated in the literature. The IV-VI SLs exhibit quite different properties as compared to the III-V SL due to the specific band structure of the constituent materials [65]. The epitaxial growth of II-VI SL is a relatively recent development and the primary motivation for studying the mentioned SLs made of materials with the large band gap is in their potential for optoelectronic operation in the blue [65]. HgTe/CdTe SL's have raised a great deal of attention since 1979, when as a promising new materials for long wavelength infrared detectors and other electro-optical applications [66]. Interest in Hg-based SL's has been further increased as new properties with potential device applications were revealed [66, 67]. These features arise from

K.P. Ghatak, *Einstein's Photoemission*, Springer Tracts in Modern Physics 262,
DOI 10.1007/978-3-319-11188-9_4

the unique zero band gap material HgTe [68] and the direct band gap semiconductor CdTe which can be described by the three band mode of Kane [69]. The combination of the aforementioned materials with specified dispersion relation makes HgTe/CdTe SL very attractive, especially because of the tailoring of the material properties for various applications by varying the energy band constants of the SLs.

We note that all the aforementioned SLs have been proposed with the assumption that the interfaces between the layers are sharply defined, of zero thickness, i.e., devoid of any interface effects. The SL potential distribution may be then considered as a one dimensional array of rectangular potential wells. The aforementioned advanced experimental techniques may produce SLs with physical interfaces between the two materials crystallographically abrupt; adjoining their interface will change at least on an atomic scale. As the potential form changes from a well (barrier) to a barrier (well), an intermediate potential region exists for the electrons [70]. The influence of finite thickness of the interfaces on the electron dispersion law is very important, since; the electron energy spectrum governs the electron transport in SLs. In addition to it, for effective mass SLs, the electronic subbands appear continually in real space [71].

In this chapter, the magneto EP from III-V, II-VI, IV-VI, HgTe/CdTe and strained layer quantum well heavily doped superlattices (QWHDSLs) with graded interfaces has been studied in Sects. 4.2.1–4.2.5. From Sects. 4.2.6–4.2.10, the magneto EP from III-V, II-VI, IV-VI, HgTe/CdTe and strained layer quantum well heavily doped effective mass superlattices respectively has been presented. In Sects. 4.2.11–4.2.20, the same from the quantum dots of the aforementioned heavily doped SLs has been investigated. Section 4.3 contains the result and discussions pertinent to this chapter. The last Sect. 4.4 presents open research problems.

4.2 Theoretical Background

4.2.1 The Magneto EP from III-V Quantum Well HD Superlattices with Graded Interfaces

The electron dispersion law in bulk specimens of the heavily doped constituent materials of III-V SLs whose un-doped energy band structures are defined by three band model of Kane can be expressed as

$$\frac{\hbar^2 k^2}{2m_{cj}^*} = T_{1j}(E, \Delta_j, E_{gj}, \eta_{gj}) + iT_{2j}(E, \Delta_j, E_{gj}, \eta_{gj}) \tag{4.1}$$

where

$$
\begin{aligned}
\text{j} = 1,2,\ T_{ij}(E,\Delta_j,E_{gj},\eta_{gi}) = {} & (2/(1+Erf(E/\eta_{gi}))[(\alpha_j b_j/c_j)\cdot\ \theta_0(E,\eta_{gj}) \\
& + [(\alpha_j c_j + b_j c_j - \alpha_j b_j)/c_j^2]\cdot\gamma_0(E,\eta_{gj}) \\
& + [(1/c_j)(1-(\alpha_j/c_j))(1-(b_j/c_j))\frac{1}{2}[1+Erf(E/\eta_{gj})] \\
& - (1/c_j)(1-(\alpha_j/c_j))(1-b_j/c_j))(2/(c_j\eta_{gj}\sqrt{\pi})\exp(-u_j^2) \\
& [\sum_{p=1}^{\infty}(\exp(-p^2/4)/p)\sinh(pu_j)]],
\end{aligned}
$$

$$b_j \equiv (E_{gj}+\Delta_j)^{-1},\ c_j \equiv (E_{gj}+\frac{2}{3}\Delta_j)^{-1},\ u_j \equiv \frac{1+c_jE}{c_j\eta_{gj}} \text{ and}$$

$$T_{2j}(E,\Delta_j,E_{gj},\eta_{gj}) \equiv \left(\frac{2}{1+Erf(E/\eta_{gj})}\right)\frac{1}{c_j}\left(1-\frac{\alpha_j}{c_j}\right)\left(1-\frac{b_j}{c_j}\right)\frac{\sqrt{\pi}}{c_j\eta_{gj}}\exp(-u_j^2).$$

Therefore, the dispersion law of the electrons of heavily doped quantum well III-V SLs with graded interfaces can be expressed as [71]

$$k_z^2 = G_8 + iH_8 \tag{4.2}$$

where $\mathrm{G}_8 = \left[\frac{C_7^2-D_7^2}{L_0^2} - k_s^2\right], C_7 = \cos^{-1}(\overline{\omega_7}),\ \overline{\omega_7} = (2)^{\frac{-1}{2}}[(1-G_7^2-H_7^2) - \sqrt{(1-G_7^2-H_7^2)^2+4G_7^2}]^{\frac{1}{2}}$

$$a_{20} = \left[\sqrt{\frac{M_{s2}(0,\eta_{g2})}{M_{s1}(0,\eta_{g1})}}+1\right]^2\left[4\left(\frac{M_{s2}(0,\eta_{g2})}{M_{s1}(0,\eta_{g1})}\right)^{1/2}\right]^{-1},$$

$$a_{21} = \left[\sqrt{\frac{M_{s2}(0,\eta_{g2})}{M_{s1}(0,\eta_{g1})}}-1\right]^2\left[4\left(\frac{M_{s2}(0,\eta_{g2})}{M_{s1}(0,\eta_{g1})}\right)^{1/2}\right]^{-1}$$

$$C_{40}(E,k_x,k_y,\eta_{g1}) = [1-\overline{P}_1(E,\eta_{g1})k_x^2 - \overline{Q}_1(E,\eta_{g1})k_y^2]^{1/2}[\overline{S}_1(E,\eta_{g1})]^{-1/2}$$

$$D_{40}(E,k_x,k_y,\eta_{g2}) = [1-\overline{P}_2(E,\eta_{g2})k_x^2 - \overline{Q}_2(E,\eta_{g2})k_y^2]^{1/2}[\overline{S}_2(E,\eta_{g2})]^{-1/2}$$

$$
\begin{aligned}
G_7 = {} & [G_1 + (\rho_5 G_2/2) - (\rho_6 H_2/2) + \Delta_0/2)\{\rho_6 H_2 - \rho_8 H_3 + \rho_9 H_4 - \rho_{10} H_4 \\
& + \rho_{11} H_5 - \rho_{12} H_5 + (1/12)(\rho_{12} G_6 - \rho_{14} H_6)\}], \\
G_1 = {} & [(\cos(\mathrm{h}_1))(\cosh(\mathrm{h}_2))(\cosh(g_1))(\cos(g_2)) \\
& + (\sin(\mathrm{h}_1))(\sinh(\mathrm{h}_2))(\sinh(g_1))(\sin(g_2))], \mathrm{h}_1 = e_1(b_0-\Delta_0), e_1 = 2^{\frac{-1}{2}}(\sqrt{t_1^2+t_2^2}+t_1)^{\frac{1}{2}}
\end{aligned}
$$

$$t_1 = [(2m_{c1}^*/\hbar^2) \cdot T_{11}(E, E_{g1}, \Delta_1, \eta_{g1}) - k_s^2],\ t_2 = [(2m_{c1}^*/\hbar^2)\ T_{21}(E, E_{g1}, \Delta_1, \eta_{g1})],$$
$$\mathrm{h}_2 = e_2(b_0 - \Delta_0), e_2 = 2^{\frac{-1}{2}}(\sqrt{t_1^2 + t_2^2} - t_1)^{\frac{1}{2}}, \mathrm{g}_1 = d_1(a_0 - \Delta_0), \mathrm{d}_1 = 2^{\frac{-1}{2}}(\sqrt{x_1^2 + y_1^2} + x_1)^{\frac{1}{2}},$$
$$x_1 = [-(2m_{c2}^*/\hbar^2) \cdot T_{11}(E - V_0, E_{g2}, \Delta_2, \eta_{g2}) + k_s^2],\ y_1 = [(2m_{c2}^*/\hbar^2)\ T_{22}(E - V_0, E_{g2}, \Delta_2, \eta_{g2})],$$
$$\mathrm{g}_2 = d_2(a_0 - \Delta_0), \mathrm{d}_2 = 2^{\frac{-1}{2}}(\sqrt{x_1^2 + y_2^2} - x_1)^{\frac{1}{2}}, \rho_5 = (\rho_3^2 + \rho_4^2)^{-1}[\rho_1\rho_3 - \rho_2\rho_4],$$
$$\rho_1 = [d_1^2 + e_2^2 - d_2^2 - e_1^2], \rho_3 = [\mathrm{d}_1 e_1 + \mathrm{d}_2 e_2],\ \rho_2 = 2[d_1 d_2 + e_1 e_2], \rho_4 = [d_1 e_2 - e_1 d_2],$$
$$G_2 = [(\sin(\mathrm{h}_1))((\cosh(\mathrm{h}_2))(\sinh(g_1))(\cos(g_2)) + (\cos(\mathrm{h}_1))(\sinh(\mathrm{h}_2))(\cosh(g_1))(\sin(g_2))],$$
$$\rho_6 = (\rho_3^2 + \rho_4^2)^{-1}[\rho_1\rho_4 + \rho_2\rho_3],$$

$$H_2 = [(\sin(\mathrm{h}_1))(\cosh(\mathrm{h}_2))(\sin(g_2))(\cosh(g_1)) - (\cos(\mathrm{h}_1))(\sinh(\mathrm{h}_2))(\sinh(g_1))(\cos(g_2))],$$
$$\rho_7 = [(e_1^2 + e_2^2)^{-1}[e_1(d_1^2 - d_2^2) - 2d_1 d_2 e_2] - 3e_1],$$
$$G_3 = [(\sin(\mathrm{h}_1))(\cosh(\mathrm{h}_2))(\sin(g_1))(\cos(g_2)) + (\cos(\mathrm{h}_1))(\sinh(\mathrm{h}_2))(\sinh(g_1))(\sin(g_2))],$$
$$\rho_8 = [(e_1^2 + e_2^2)^{-1}[e_1(d_1^2 - d_2^2) + 2d_1 d_2 e_2] + 3e_1],$$

$$H_3 = [(\sin(\mathrm{h}_1))(\cosh(\mathrm{h}_2))(\sin(g_2))(\sinh(g_1)) - (\cos(\mathrm{h}_1))(\sinh(\mathrm{h}_2))(\cosh(g_1))(\cos(g_2))],$$
$$\rho_9 = [(d_1^2 + d_2^2)^{-1}[d_1(e_2^2 - e_1^2) + 2e_2 d_2 e_1] + 3d_1],$$
$$G_4 = [(\cos(\mathrm{h}_1))(\cosh(\mathrm{h}_2))(\cos(g_2))(\sin(g_1)) - (\sin(\mathrm{h}_1))(\sinh(\mathrm{h}_2))(\cosh(g_1))(\sin(g_2))],$$
$$\rho_{10} = [-(d_1^2 + d_2^2)^{-1}[d_2(-e_2^2 + e_1^2) + 2e_2 d_2 e_1] + 3d_2],$$

$$H_4 = [(\cos(\mathrm{h}_1))(\cosh(\mathrm{h}_2))(\cosh(g_1))(\sin(g_2)) + (\sin(\mathrm{h}_1))(\sinh(\mathrm{h}_2))(\sinh(g_1))(\cos(g_2))],$$
$$\rho_{11} = 2[d_1^2 + e_2^2 - d_2^2 - e_1^2],$$
$$G_5 = [(\cos(\mathrm{h}_1))(\cosh(\mathrm{h}_2))(\cos(g_2))(\cosh(g_1)) - (\sin(\mathrm{h}_1))(\sinh(\mathrm{h}_2))(\sinh(g_1))(\sin(g_2))],$$
$$\rho_{12} = 4[d_1 d_2 + e_1 e_2],$$
$$H_5 = [(\cos(\mathrm{h}_1))(\cosh(\mathrm{h}_2))(\sinh(g_1))(\sin(g_2)) + (\sin(\mathrm{h}_1))(\sinh(\mathrm{h}_2))(\cosh(g_1))(\cos(g_2))],$$
$$\rho_{13} = [\{5(d_1 e_1^3 - 3e_1 e_2^2 d_1) + 5d_2(e_1^3 - 3e_1^2 e_2)\}(d_1^2 + d_2^2)^{-1} + (e_1^2 + e_2^2)^{-1}\{5(e_1 d_1^3 - 3d_2 e_1^2 d_1) + 5(d_2^3 e_2 - 3d_1^2 d_2 e_2)\} - 34(d_1 e_1 + d_2 e_2)],$$
$$G_6 = [(\sinh(\mathrm{h}_1))(\cosh(\mathrm{h}_2))(\sinh(g_1))(\cos(g_2)) + (\cos(\mathrm{h}_1))(\sinh(\mathrm{h}_2))(\cosh(g_1))(\sin(g_2))],$$
$$\rho_{14} = [\{5(d_1 e_2^3 - 3e_1 e_1^2 d_1) + 5d_2(-e_1^3 + 3e_2^2 e_1)\}(d_1^2 + d_2^2)^{-1} + (e_1^2 + e_2^2)^{-1}\{5(-e_1 d_2^3 + 3d_1^2 d_2 e_1) + 5(-d_1^3 e_2 + 3d_2^2 d_1 e_2)\} + 34(d_1 e_2 + d_2 e_1)],$$

$$H_6 = [(\sin(\mathrm{h}_1))(\cosh(\mathrm{h}_2))(\cosh(g_1))(\sin(g_2)) - (\cos(\mathrm{h}_1))(\sinh(\mathrm{h}_2))(\sinh(g_1))(\cos(g_2))],$$
$$H_7 = [H_1 + (\rho_5 H_2/2) + (\rho_6 G_2/2) + (\Delta_0/2)\{\rho_8 G_3 + \rho_7 H_3 + \rho_{10} G_4 + \rho_9 H_4 + \rho_{12} G_5 + \rho_{11} H_5 + (1/12)(\rho_{14} G_6 + \rho_{13} H_6)\}],$$
$$H_1 = [(\sin(\mathrm{h}_1))(\sinh(\mathrm{h}_2))(\cosh(g_1))(\cos(g_2)) + (\cos(\mathrm{h}_1))(\cosh(\mathrm{h}_2))(\sinh(g_1))(\sin(g_2))],$$
$$D_7 = \sinh^{-1}(\overline{\omega_7}), \mathrm{H}_8 = (2C_7 D_7/L_0^2)$$

The simplified dispersion relation of heavily doped quantum well III-V superlattices with graded interfaces under magnetic quantization can be expressed as

$$k_z^2 = G_{8E,n} + iH_{8E,n} \tag{4.3}$$

where

$$G_{8E,n} = \left[\frac{C_{7E,n}^2 - D_{7E,n}^2}{L_0^2} - \left\{\frac{2eB}{\hbar}\left(n+\frac{1}{2}\right)\right\}\right], C_{7E,n} = \cos^{-1}(\overline{\omega_{7E,n}}),$$

$$\overline{\omega_{7E,n}} = (2)^{\frac{-1}{2}}[(1 - G_{7E,n}^2 - H_{7E,n}^2) - \sqrt{(1 - G_{7E,n}^2 - H_{7E,n}^2)^2 + 4G_{7E,n}^2}]^{\frac{1}{2}}$$

$$\begin{aligned}
G_{7E,n} &= [G_{1E,n} + (\rho_{5E,n}G_{2E,n}/2) - (\rho_{6E,n}H_{2E,n}/2) \\
&\quad + (\Delta_0/2)\{\rho_{6E,n}H_{2E,n} - \rho_{8E,n}H_{3E,n} + \rho_{9E,n}H_{4E,n} - \rho_{10E,n}H_{4E,n} \\
&\quad + \rho_{11E,n}H_{5E,n} - \rho_{12E,n}H_{5E,n} + (1/12)(\rho_{12E,n}G_{6E,n} - \rho_{14E,n}H_{6E,n})\}], \\
G_{1E,n} &= [(\cos(\mathrm{h}_{1E,n}))(\cosh(\mathrm{h}_{2E,n}))(\cosh(g_{1E,n}))(\cos(g_{2E,n})) \\
&\quad + (\sin(\mathrm{h}_{1E,n}))(\sinh(\mathrm{h}_{2E,n}))(\sinh(g_{1E,n}))(\sin(g_{2E,n}))], \\
\mathrm{h}_{1E,n} &= e_{1E,n}(b_0 - \Delta_0), e_{1E,n} = 2^{\frac{-1}{2}}(\sqrt{t_{1E,n}^2 + t_2^2} + t_{1E,n})^{\frac{1}{2}}, \\
t_{1E,n} &= [(2m_{c1}^*/\hbar^2).T_{11}(E, E_{g1}, \Delta_1\eta_{g1}) - \left\{\frac{2eB}{\hbar}\left(n+\frac{1}{2}\right)\right\}], \\
t_2 &= [(2m_{c1}^*/\hbar^2)T_{21}(E, E_{g1}, \Delta_1, \eta_{g1})], \\
\mathrm{h}_{2E,n} &= e_{2E,n}(b_0 - \Delta_0), e_{2E,n} = 2^{\frac{-1}{2}}(\sqrt{t_{1E,n}^2 + t_2^2} - t_{1E,n})^{\frac{1}{2}}, \\
g_{1E,n} &= d_{1E,n}(a_0 - \Delta_0), \mathrm{d}_{1E,n} = 2^{\frac{-1}{2}}(\sqrt{x_{1E,n}^2 + y_1^2} + x_{1E,n})^{\frac{1}{2}}, \\
x_{1E,n} &= [-(2m_{c2}^*/\hbar^2)_{11}(E - V_0, E_{g2}, \Delta_2, \eta_{g2}) + \left\{\frac{2eB}{\hbar}\left(n+\frac{1}{2}\right)\right\}], \\
\mathrm{y}_1 &= [(2m_{c2}^*/\hbar^2)T_{22}(E - V_0, E_{g2}, \Delta_2, \eta_{g2})], \\
g_{2E,n} &= d_{2E,n}(a_0 - \Delta_0),\ d_{2E,n} = 2^{\frac{-1}{2}}(\sqrt{x_{1E,n}^2 + y_1^2} - x_{1E,n})^{\frac{1}{2}}, \\
\rho_{5E,n} &= (\rho_{3E,n}^2 + \rho_{4E,n}^2)^{-1}[\rho_{1E,n}\rho_{3E,n} - \rho_{2E,n}\rho_{4E,n}], \\
\rho_{1E,n} &= [d_{1E,n}^2 + e_{2E,n}^2 - d_{2E,n}^2 - e_{1E,n}^2],\ \rho_{3E,n} = [\mathrm{d}_{1E,n}e_{1E,n} + \mathrm{d}_{2E,n}e_{2E,n}], \\
\rho_{2E,n} &= 2[d_{1E,n}d_{2E,n} + e_{1E,n}e_{2E,n}],\ \rho_{4E,n} = [d_{1E,n}e_{2E,n} - e_{1E,n}d_{2E,n}], \\
G_{2E,n} &= [(\sin(\mathrm{h}_{1E,n}))(\cosh(\mathrm{h}_{2E,n}))(\sinh(g_{1E,n}))(\cos(g_{2E,n})) \\
&\quad + (\cos(\mathrm{h}_{1E,n}))(\sinh(\mathrm{h}_{2E,n}))(\cosh(g_{1E,n}))(\sin(g_{2E,n}))], \\
\rho_{6E,n} &= (\rho_{3E,n}^2 + \rho_{4E,n}^2)^{-1}[\rho_{1E,n}\rho_{4E,n} + \rho_{2E,n}\rho_{3E,n}],
\end{aligned}$$

$$H_{2E,n} = [(\sin(h_{1E,n}))(\cosh(h_{2E,n}))(\sin(g_{2E,n}))(\cos(g_{1E,n})) - (\cos(h_{1E,n}))(\sinh(h_{2E,n}))(\sinh(g_{1E,n}))(\cos(g_{2E,n}))],$$

$$\rho_{7E,n} = (e_{1E,n}^2 + e_{2E,n}^2)^{-1}[e_{1E,n}(d_{1E,n}^2 - d_{2E,n}^2) - 2d_{1E,n}d_{2E,n}e_{2E,n}] - 3e_{1E,n}],$$

$$G_{3E,n} = [(\sin(h_{1E,n}))(\cosh(h_{2E,n}))(\cosh(g_{1E,n}))(\cos(g_{2E,n})) + (\cos(h_{1E,n}))(\sinh(h_{2E,n}))(\sinh(g_{1E,n}))(\sin(g_{2E,n}))],$$

$$\rho_{8E,n} = (e_{1E,n}^2 + e_{2E,n}^2)^{-1}[e_{2E,n}(d_{1E,n}^2 - d_{2E,n}^2) + 2d_{1E,n}d_{2E,n}e_{1E,n}] + 3e_{2E,n}],$$

$$H_{3E,n} = [(\sin(h_{1E,n}))(\cosh(h_{2E,n}))(\sin(g_{2E,n}))(\sinh(g_{1E,n})) - (\cos(h_{1E,n}))(\sinh(h_{2E,n}))(\cosh(g_{1E,n}))(\cos(g_{2E,n}))],$$

$$\rho_{9E,n} = (d_{1E,n}^2 + d_{2E,n}^2)^{-1}[d_{1E,n}(e_{2E,n}^2 - e_{1E,n}^2) + 2e_{2E,n}d_{2E,n}e_{1E,n}] + 3d_{1E,n}],$$

$$G_{4E,n} = [(\cos(h_{1E,n}))(\cosh(h_{2E,n}))(\cos(g_{2E,n}))(\sinh(g_{1E,n})) - (\sin(h_{1E,n}))(\sinh(h_{2E,n}))(\cosh(g_{1E,n}))(\sin(g_{2E,n}))],$$

$$\rho_{10E,n} = (-d_{1E,n}^2 + d_{2E,n}^2)^{-1}[d_{2E,n}(-e_{2E,n}^2 + e_{1E,n}^2) + 2e_{2E,n}d_{2E,n}e_{1E,n}] + 3d_{2E,n}],$$

$$H_{4E,n} = [(\cos(h_{1E,n}))(\cosh(h_{2E,n}))(\cosh(g_{1E,n}))(\sin(g_{2E,n})) + (\sin(h_{1E,n}))(\sinh(h_{2E,n}))(\sinh(g_{1E,n}))(\cos(g_{2E,n}))],$$

$$\rho_{11E,n} = 2[d_{1E,n}^2 + e_{2E,n}^2 - d_{2E,n}^2 - e_{1E,n}^2],$$

$$G_{5E,n} = [(\cos(h_{1E,n}))(\cosh(h_{2E,n}))(\cos(g_{2E,n}))(\cosh(g_{1E,n})) - (\sin(h_{1E,n}))(\sinh(h_{2E,n}))(\sinh(g_{1E,n}))(\sin(g_{2E,n}))],$$

$$\rho_{12E,n} = 4[d_{1E,n}d_{2E,n} + e_{1E,n}e_{2E,n}],$$

$$H_{5E,n} = [(\cos(h_{1E,n}))(\cosh(h_{2E,n}))(\sinh(g_{1E,n}))(\sin(g_{2E,n})) + (\sin(h_{1E,n}))(\sinh(h_{2E,n}))(\cosh(g_1))(\cos(g_2))],$$

$$\begin{aligned}\rho_{13E,n} = [\{&5(d_{1E,n}e_{1E,n}^3 - 3e_{1E,n}e_{2E,n}^2d_{1E,n}) \\ &+ 5d_{2E,n}(e_{1E,n}^2 - 3e_{1E,n}^2e_{2E,n})\}(d_{1E,n}^2 + d_{2E,n}^2)^{-1} \\ &+ (e_{1E,n}^2 + e_{2E,n}^2)^{-1}\{5(e_{1E,n}d_{1E,n}^3 - 3d_{2E,n}e_{1E,n}^2d_{1E,n}) \\ &+ 5(d_{2E,n}^3e_{2E,n} - 3d_{1E,n}^2d_{2E,n}e_{2E,n})\} - 34(d_{1E,n}e_{1E,n} + d_{2E,n}e_{2E,n})],\end{aligned}$$

$$G_{6E,n} = [(\sin(h_{1E,n}))(\cosh(h_{2E,n}))(\sinh(g_{1E,n}))(\cos(g_{2E,n})) + (\cos(h_{1E,n}))(\sinh(h_{2E,n}))(\cosh(g_1))(\sin(g_2))],$$

$$\begin{aligned}\rho_{14E,n} = [\{&5(d_{1E,n}e_{2E,n}^3 - 3e_{2E,n}e_{1E,n}^2d_{1E,n}) \\ &+ 5d_{2E,n}(-e_{1E,n}^3 + 3e_{2E,n}^2e_{1E,n})\}(d_{1E,n}^2 + d_{2E,n}^2)^{-1}, \\ &+ (e_{1E,n}^2 + e_{2E,n}^2)^{-1}\{5(-e_{1E,n}d_{2E,n}^3 + 3d_{1E,n}^2d_{2E,n}e_{1E,n}) \\ &+ 5(-d_{1E,n}^3e_{2E,n} + 3d_{2E,n}^2d_{1E,n}e_{2E,n})\} \\ &+ 34(d_{1E,n}e_{2E,n} - d_{2E,n}e_{1E,n})],\end{aligned}$$

$$H_{6E,n} = [(\sin(\mathrm{h}_{1E,n}))(\cosh(\mathrm{h}_{2E,n}))(\cosh(g_{1E,n}))(\sin(g_{2E,n})) - (\cos(\mathrm{h}_{1E,n}))(\sinh(\mathrm{h}_{2E,n}))(\sinh(g_{1E,n}))(\cos(g_{2E,n}))],$$

$$H_{7E,n} = [H_{1E,n} + (\rho_{5E,n}H_{2E,n}/2) + (\rho_{6E,n}G_{2E,n}/2) + (\Delta_0/2)\{\rho_{8E,n}G_{3E,n} + \rho_{7E,n}H_{3E,n} + \rho_{10E,n}G_{4E,n} + \rho_{9E,n}H_{4E,n} + \rho_{12E,n}G_{5E,n} + \rho_{11E,n}H_{5E,n} + (1/12)(\rho_{14E,n}G_{6E,n} + \rho_{13E,n}H_{6E,n})\}],$$

$$H_{1E,n} = [(\sin(\mathrm{h}_{1E,n}))(\sinh(\mathrm{h}_{2E,n}))(\cosh(g_{1E,n}))(\cos(g_{2E,n})) + (\cos(\mathrm{h}_{1E,n}))(\cosh(\mathrm{h}_{2E,n}))(\sinh(g_{1E,n}))(\sin(g_{2E,n}))],$$

$$D_{7E,n} = \sinh^{-1}(\overline{\omega_{7E,n}}),\ \mathrm{H}_{8E,n} = (2C_{7E,n}D_{7E,n}/L_0^2)$$

The dispersion relation in heavily doped quantum well III-V superlattices under magnetic quantization assumes the form

$$\left(\frac{n_z\pi}{d_z}\right)^2 = G_{8E_{41,n}} + iH_{8E_{41,n}} \tag{4.4}$$

where $E_{41,n}$ is the totally quantized energy in this case.

The electron concentration in this case is given by

$$n_{0SL} = \frac{eBg_v}{\pi\hbar}\ \text{Real part of} \sum_{n=0}^{n_{\max}}\sum_{n_z=1}^{n_{z\max}} F_{-1}(\eta_{41}) \tag{4.5}$$

where $\eta_{41} = (k_BT)^{-1}(E_{FSL} - E_{41,n})$ and E_{FSL} is the Fermi energy in this case.

The EP in this case(J_{SL}) is given by

$$J_{SL} = \frac{\alpha_0 e^2 Bg_v}{2\pi\hbar d_z}\ \text{Real part of} \sum_{n=0}^{n_{\max}}\sum_{n_{z\min}}^{n_{z\max}} F_{-1}(\eta_{41})v_{z1}(E_{nzSL1}) \tag{4.6a}$$

where

$$v_{z1}(\mathrm{E}_{nzSL1}) = \frac{2\pi n_z}{\hbar \mathrm{d}_z(\mathrm{G}_8' + \mathrm{i}\mathrm{H}_8')\big|_{k_x=0,k_y=0,k_z=\frac{n_z\pi}{d_z}\ and\ E=\mathrm{E}_{nzSL1}}} \tag{4.6b}$$

E_{nzSL1} is the quantized energy along z direction and is obtained by substituting $k_x = 0, k_y = 0, k_z = \frac{n_z\pi}{d_z}\ and\ E = E_{nzSL1}$ in (4.2) and the primes denote the differentiation with respect to energy.

4.2.2 The Magneto EP from II-VI Quantum Well HD Superlattices with Graded Interfaces

The electron energy spectra of the heavily doped constituent materials of II-VI SLs are given by

$$\gamma_3(E,\eta_{g1}) = \frac{\hbar^2 k_s^2}{2m^*_{\perp,1}} + \frac{\hbar^2 k_z^2}{2m^*_{\parallel,1}} \pm C_0 k_s \quad (4.7)$$

and

$$\frac{\hbar^2 k^2}{2m^*_{c2}} = T_{12}(E,\Delta_2,E_{g2},\eta_{g2}) + iT_{22}(E,\Delta_2,E_{g2},\eta_{g2}) \quad (4.8)$$

where $m^*_{\perp,1}$ and $m^*_{\parallel,1}$ are the transverse and longitudinal effective electron masses respectively at the edge of the conduction band for the first material. The energy-wave vector dispersion relation of the conduction electrons in heavily doped quantum well II-VI SLs with graded interfaces can be expressed as

$$k_z^2 = G_{19} + iH_{19} \quad (4.9)$$

where

$$\mathrm{G}_{19} = \left[\frac{C_{18}^2 - D_{18}^2}{L_0^2} - k_s^2\right],$$

$$C_{18} = \cos^{-1}(\omega_{18}), \omega_{18} = (2)^{\frac{-1}{2}}[(1 - G_{18}^2 - H_{18}^2) - \sqrt{(1 - G_{18}^2 - H_{18}^2)^2 + 4G_{18}^2}]^{\frac{1}{2}},$$

$$G_{18} = \frac{1}{2}[G_{11} + G_{12} + \Delta_0(G_{13} + G_{14}) + \Delta_0(G_{15} + G_{16})],$$

$$G_{11} = 2(\cos(g_1))(\cos(g_2))(\cos\gamma_{11}(E,k_s))$$

$$\gamma_{11}(E,k_s) = k_{21}(E,k_s)(b_0 - \Delta), k_{21}(E,k_s) = \{[\gamma_3(E,\eta_{g1}) - \frac{\hbar^2 k_s^2}{2m^*_{\perp,1}} \pm C_0 k_s]\frac{2m^*_{\parallel,1}}{\hbar^2}\}^{1/2},$$

$$G_{12} = ([\Omega_1(E,k_s)(\sinh g_1)(\cos g_2) - \Omega_2(E,k_s)(\sin g_2)(\cosh g_1)](\sin\gamma_{11}(E,k_s)))$$

$$\Omega_1(E,k_s) = [\frac{d_1}{k_{21}(E,k_s)} - \frac{k_{21}(E,k_s)d_1}{d_1^2 + d_2^2}] \text{ and } \Omega_2(E,k_s) = [\frac{d}{k_{21}(E,k_s)} + \frac{k_{21}(E,k_s)d_2}{d_1^2 + d_2^2}]$$

$$G_{13} = ([\Omega_3(E,k_s)(\cosh g_1)(\cos g_2) - \Omega_4(E,k_s)(\sinh g_1)(\sin g_2)](\sin\gamma_{11}(E,k_s)))$$

$$\Omega_3(E,k_s) = [\frac{d_1^2 - d_2^1}{k_{21}(E,k_s)} - 3k_{21}(E,k_s)], \Omega_4(E,k_s) = [\frac{2d_1 d_2}{k_{21}(E,k_s)}]$$

$$G_{14} = ([\Omega_5(E,k_s)(\sinh g_1)(\cos g_2) - \Omega_6(E,k_s)(\sin g_1)(\cosh g_2)](\cos\gamma_{11}(E,k_s))).$$

$$\Omega_5(E,k_5) = [3d_1 - \frac{d_1}{d_1^2+d_2^2}k_{21}^2(E,k_s)], \Omega_6(E,k_s) = [3d_2 + \frac{d_2}{d_1^2+d_2^2}k_{21}^2(E,k_s)]$$

$$G_{15} = ([\Omega_9(E,k_s)(\cosh g_1)(\cos g_2) - \Omega_{10}(E,k_s)(\sinh g_1)(\sin g_2)](\cos\gamma_{11}(E,k_s)))$$

$$\Omega_9(E,k_5) = [2d_1^2 - 2d_2^2 - k_{21}^2(E,k_s)], \Omega_{10}(E,k_s) = [2d_1d_2]$$

$$G_{16} = ([\Omega_7(E,k_s)(\sinh g_1)(\cos g_2) - \Omega_8(E,k_s)(\sin g_1)(\cosh g_2)](\sin\gamma_{11}(E,k_s)/12)),$$

$$\Omega_7(E,k_s) = [\frac{5d_1}{d_1^2+d_2^2}k_{21}^3(E,k_s) + \frac{5(d_1^3-3d_2^2d_1)}{k_{21}(E,k_s)} - 34k_{21}(E,k_s)d_1],$$

$$\Omega_8(E,k_s) = [\frac{5d_2}{d_1^2+d_2^2}k_{21}^3(E,k_s) + \frac{5(d_2^3-3d_2^2d_1}{k_{21}(E,k_s)} + 34k_{21}(E,k_s)d_2],$$

$$H_{18} = \frac{1}{2}[H_{11} + H_{12} + \Delta_0(H_{13}+H_{14}) + \Delta_0(H_{15}+H_{16})],$$

$$H_{11} = 2(\sinh\ g_1\ \sin\ g_2\ \cos\gamma_{11}(E,k_s)),$$

$$H_{12} = ([\Omega_2(E,k_s)(\sinh g_1)(\cos g_2) + \Omega_1(E,k_s)(\sin g_2)(\cosh g_1)](\sin\gamma_{11}(E,k_s))),$$

$$H_{13} = ([\Omega_4(E,k_s)(\cosh g_1)(\cos g_2) + \Omega_3(E,k_s)(\sinh g_1)(\sin g_2)](\sin\gamma_{11}(E,k_s))),$$

$$H_{14} = ([\Omega_6(E,k_s)(\sinh g_1)(\cos g_2) + \Omega_5(E,k_s)(\sin g_1)(\cosh g_2)](\cos\gamma_{11}(E,k_s))),$$

$$H_{15} = ([\Omega_{10}(E,k_s)(\cosh g_1)(\cos g_2) + \Omega_9(E,k_s)(\sinh g_1)(\sin g_2)](\cos\gamma_{11}(E,k_s))),$$

$$H_{16} = ([\Omega_8(E,k_s)(\sinh g_1)(\cos g_2) + \Omega_7(E,k_s)(\sin g_1)(\cosh g_2)](\sin\gamma_{11}(E,k_s)/12)),$$

$$\mathrm{H}_{19} = \left[\frac{2C_{18}D_{18}}{L_0^2}\right] \text{ and } D_{18} = \sinh^{-1}(\omega_{18})$$

The simplified dispersion relation in heavily doped quantum well II-VI superlattices with graded interfaces under magnetic quantization can be expressed as

$$k_z^2 = G_{19E,n} + iH_{19E,n} \tag{4.10}$$

$$\text{where } \mathrm{G}_{19E,n} = \left[\frac{C_{18E,n}^2 - D_{18E,n}^2}{L_0^2} - \left(\frac{2eB}{\hbar}\left(n+\frac{1}{2}\right)\right)\right],$$

$$C_{18E,n} = \cos^{-1}(\omega_{18E,n}), \omega_{18E,n} = (2)^{\frac{-1}{2}}[(1 - G_{18E,n}^2 - H_{18E,n}^2) - \sqrt{(1 - G_{18E,n}^2 - H_{18E,n}^2)^2 + 4G_{180D}^2}]^{\frac{1}{2}},$$

$$G_{18E,n} = \frac{1}{2}[G_{11E,n} + G_{12E,n} + \Delta_0(G_{13E,n} + G_{14E,n}) + \Delta_0(G_{15E,n} + G_{16E,n})],$$

$$G_{11E,n} = 2(\cos(g_{1E,n}))(\cos(g_{2E,n}))(\cos\gamma_{11}(E,n)),\ \gamma_{11}(E,n) = k_{21}(E,n)(b_0 - \Delta_0),$$

$$k_{21}(E,n) = \{[\gamma_3(E,\eta_{g1}) - \frac{\hbar^2}{2m_{\perp,1}^*}\left\{\frac{2eB}{\hbar}\left(n+\frac{1}{2}\right)\right\} \pm C_0\left\{\frac{2eB}{\hbar}\left(n+\frac{1}{2}\right)\right\}^{1/2}]\frac{2m_{\|,1}^*}{\hbar^2}\}^{1/2},$$

$$G_{12E,n} = ([\Omega_1(E,n)(\sinh g_{1E,n})(\cos g_{2E,n}) - \Omega_2(E,n)(\sin g_{2E,n})(\cosh g_{1E,n})](\sin \gamma_{11}(E,n)))$$

$$\Omega_1(E,n) = [\frac{d_{1E,n}}{k_{21}(E,n)} - \frac{k_{21}(E,n)d_{1E,n}}{d_{1E,n}^2 + d_{2E,n}^2}],$$

$$\Omega_2(E,n) = [\frac{d_{2E,n}}{k_{21}(E,n)} + \frac{k_{21}(E,n)d_{2E,n}}{d_{1E,n}^2 + d_{2E,n}^2}],$$

$$G_{13E,n} = ([\Omega_3(E,n)(\cosh g_{1E,n})(\cos g_{2E,n}) - \Omega_4(E,n)(\sinh g_{1E,n})(\sin g_{2E,n})](\sin \gamma_{11}(E,n)))$$

$$\Omega_3(E,n) = [\frac{d_{1E,n}^2 - d_{2E,n}^1}{k_{21}(E,n)} - 3k_{21}(E,n)],\ \Omega_4(E,n) = [\frac{2d_{1E,n}d_{2E,n}}{k_{21}(E,n)}]$$

$$G_{14E,n} = ([\Omega_5(E,n)(\sinh g_{1E,n})(\cos g_{2E,n}) - \Omega_6(E,n)(\sin g_{1E,n})(\cosh g_{2E,n})](\cos \gamma_{11}(E,n))).$$

$$\Omega_5(E,n) = [3d_{1E,n} - \frac{d_{1E,n}}{d_{1E,n}^2 + d_{2E,n}^2}k_{21}^2(E,n)],$$

$$\Omega_6(E,n) = [3d_{2E,n} + \frac{d_{2E,n}}{d_{1E,n}^2 + d_{2E,n}^2}k_{21}^2(E,n)]$$

$$G_{15E,n} = ([\Omega_9(E,n)(\cosh g_{1E,n})(\cos g_{2E,n}) - \Omega_{10}(E,n)(\sinh g_{1E,n})(\sin g_{2E,n})](\cos \gamma_{11}(E,n)))$$

$$\Omega_9(E,n) = [2d_{1E,n}^2 - 2d_{2E,n}^2 - k_{21}^2(E,n)],\ \Omega_{10}(E,n) = [2d_{1E,n}d_{2E,n}]$$

$$G_{16E,n} = ([\Omega_7(E,n)(\sinh g_{1E,n})(\cos g_{2E,n}) - \Omega_8(E,n)(\sin g_{1E,n})(\cosh g_{2E,n})](\sin \gamma_{11}(E,n)/12)),$$

$$\Omega_7(E,n) = [\frac{5d_{1E,n}}{d_{1E,n}^2 + d_{2E,n}^2}k_{21}^3(E,n) + \frac{5(d_{1E,n}^3 - 3d_{2E,n}^2 d_{1E,n})}{k_{21}(E,n)} - 34k_{21}(E,n)d_{1E,n}],$$

$$\Omega_8(E,n) = [\frac{5d_{2E,n}}{d_{1E,n}^2 + d_{2E,n}^2}k_{21}^3(E,n) + \frac{5(d_{2E,n}^3 - 3d_{2E,n}^2 d_{1E,n})}{k_{21}(E,n)} + 34k_{21}(E,n)d_{2E,n}]$$

$$H_{18E,n} = \frac{1}{2}[H_{11E,n} + H_{12E,n} + \Delta_0(H_{13E,n} + H_{14E,n}) + \Delta_0(H_{15E,n} + H_{16E,n})],$$

$$H_{11E,n} = 2(\sinh\ g_{1E,n})(\sin\ g_{2E,n})(\cos\gamma_{11}(E,n)),$$
$$H_{12E,n} = ([\Omega_2(E,n)(\sinh g_{1E,n})(\cos g_{2E,n}) + \Omega_1(E,n)(\sin g_{2E,n})(\cosh g_{1E,n})](\sin\gamma_{11}(E,n))),$$
$$H_{13E,n} = ([\Omega_4(E,n)(\cosh g_{1E,n})(\cos g_{2E,n}) + \Omega_3(E,n)(\sinh g_{1E,n})(\sin g_{2E,n})](\sin\gamma_{11}(E,n))),$$
$$H_{14E,n} = ([\Omega_6(E,n)(\sinh g_{1E,n})(\cos g_{2E,n}) + \Omega_5(E,n)(\sin g_{1E,n})(\cosh g_{2E,n})](\cos\gamma_{11}(E,n))),$$
$$H_{15E,n} = ([\Omega_{10}(E,n)(\cosh g_{1E,n})(\cos g_{2E,n}) + \Omega_9(E,n)(\sinh g_{1E,n})(\sin g_{2E,n})](\cos\gamma_{11}(E,n))),$$
$$H_{16E,n} = ([\Omega_8(E,n)(\sinh g_{1E,n})(\cos g_{2E,n}) + \Omega_7(E,n)(\sin g_{1E,n})(\cosh g_{2E,n})](\sin\gamma_{11}(E,n)/12)),$$
$$\mathrm{H}_{19E,n} = \left[\frac{2C_{18E,n}D_{18E,n}}{L_0^2}\right] \text{ and } D_{18E,n} = \sinh^{-1}(\omega_{18E,n})$$

The dispersion relation in quantum well heavily doped II-VI superlattices under magnetic quantization assumes the form

$$\left(\frac{n_z\pi}{d_z}\right)^2 = G_{19E_{42,n}} + iH_{19E_{42,n}} \tag{4.11}$$

where $E_{42,n}$ is the totally quantized energy in this case.

The electron concentration in this case is given by

$$n_{0SL} = \frac{eBg_v}{\pi\hbar} \text{ Real part of } \sum_{n=0}^{n_{\max}}\sum_{n_z=1}^{n_{z\max}} F_{-1}(\eta_{42}) \tag{4.12}$$

where $\eta_{42} = (\mathrm{k}_B\mathrm{T})^{-1}(\mathrm{E}_{FSL} - \mathrm{E}_{42,n})$ and E_{FSL} is the Fermi energy in this case.

The EP in this case(J_{SL}) is given by

$$J_{SL} = \frac{\alpha_0 e^2 Bg_v}{2\pi\hbar d_z} \text{ Real part of } \sum_{n=0}^{n_{\max}}\sum_{n_{z\min}}^{n_{z\max}} F_{-1}(\eta_{42})v_{z2}(E_{nzSL2}) \tag{4.13a}$$

where

$$v_{z2}(\mathrm{E}_{nzSL2}) = \frac{2\pi n_z}{\hbar \mathrm{d}_z(\mathrm{G}'_{19} + \mathrm{i}\mathrm{H}'_{19})\big|_{k_x=0,k_y=0,k_z=\frac{n_z\pi}{d_z} \text{ and } E=\mathrm{E}_{nzSL2}}} \tag{4.13b}$$

E_{nzSL2} is the quantized energy along z direction and is obtained by substituting $k_x = 0, k_y = 0, k_z = \frac{n_z\pi}{d_z}$ and $E = \mathrm{E}_{nzSL2}$ in (4.9) and the primes denote the differentiation with respect to energy.

4.2.3 The Magneto EP from IV-VI Quantum Well HD Superlattices with Graded Interfaces

The E-k dispersion relation of the conduction electrons of the heavily doped constituent materials of the IV-VI SLs can be expressed as

$$k_z^2 = [2\bar{p}_{9,i}]^{-1}[-\bar{q}_{9,i}(E,k_s,\eta_{gi}) + [[\bar{q}_{9,i}(E,k_s,\eta_{gi})]^2 + 4\bar{p}_{9,i}\bar{R}_{9,i}(E,k_s,\eta_{gi})]^{\frac{1}{2}}] \quad (4.14)$$

where, $\bar{p}_{9,i} = (\alpha_i\hbar^4)/(4m_{l,i}^- m_{li}^+)$, $\quad$ i = 1, 2, $\bar{q}_{9,i}(E,k_s,\eta_{gi}) = [(\hbar^2/2)((1/m_{li}^*) + (1/m_{li}^-)) + \alpha_i(\hbar^4/4)$ $k_s^2((1/m_{li}^+ m_{ti}^-) + (1/m_{ti}^+ m_{li}^-)) - \alpha_i\gamma_3(E,\eta_{gi})((1/m_{li}^+) - (1/m_{li}^-]$ and

$$\bar{R}_{9,i}(E,k_s,\eta_{gi}) = [\gamma_2(E,\eta_{gi}) + \gamma_3(E,\eta_{gi})[(\hbar^2/2)\alpha_i k_s^2((1/m_{ti}^*) - (1/m_{ti}^-))] - [(\hbar^2/2)k_s^2((1/m_{ti}^*) + (1/m_{ti}^-))] - \alpha_i(\hbar^6/4)k_s^4((1/m_{ti}^+ m_{ti}^-))]$$

The electron dispersion law in heavily doped quantum well IV-VI SLs with graded interfaces can be expressed as

$$\cos(L_0 k) = \frac{1}{2}\Phi_2(E,k_s) \quad (4.15)$$

where

$$\begin{aligned}\Phi_2(E,k_s) \equiv {} & [2\cosh\{\beta_2(E,k_s)\}\cos\{\gamma_2(E,k_s)\} + \varepsilon_2(E,k_s)\sinh\{\beta_2(E,k_s)\}\sin\{\gamma_{22}(E,k_s)\} \\ & + \Delta_0\left[\left(\frac{\{K_{112}(E,k_s)\}^2}{K_{212}(E,k_s)} - 3K_{212}(E,k_s)\right)\cosh\{\beta_2(E,k_s)\}\sin\{\gamma_{22}(E,k_s)\}\right] \\ & + \left(3K_{112}(E,k_s) - \frac{\{K_{212}(E,k_s)\}^2}{K_{112}(E,k_s)}\right)\sinh\{\beta_2(E,k_s)\}\cos\{\gamma_{22}(E,k_s)\}] \\ & + \Delta_0\left[2\left(\{K_{112}(E,k_s)\}^2 - \{K_{212}(E,k_s)\}^2\right)\cosh\{\beta_2(E,k_s)\}\cos\{\gamma_{22}(E,k_s)\}\right. \\ & \left. + \frac{1}{12}\left[\frac{5\{K_{112}(E,k_s)\}^3}{K_{212}(E,k_s)} + \frac{5\{K_{212}(E,k_s)\}^3}{K_{112}(E,k_s)} - 34K_{212}(E,k_s)K_{112}(E,k_s)\right]\sin\{\beta_2(E,k_s)\}\sin\{\gamma_{22}(E,k_s)\}\right]\Bigg],\end{aligned}$$

$$\begin{aligned}\beta_2(E,k_s) &\equiv K_{112}(E,k_s)[a_0 - \Delta_0], \\ k_{112}^2(E,k_s) &= [2\bar{p}_{9,2}]^{-1}[-\bar{q}_{9,2}(E-V_0,k_s,\eta_{g2}) - [[\bar{q}_{9,2}(E-V_0,k_s,\eta_{g2})]^2 \\ &\quad + 4\bar{p}_{9,2}\bar{R}_{9,2}(E-V_0,k_s,\eta_{g2})]^{\frac{1}{2}}], \\ \gamma_{22}(E,k_s) &\equiv K_{212}(E,k_s)[b_0 - \Delta_0], \\ k_{212}^2(E,k_s) &= [2\bar{p}_{9,1}]^{-1}[-\bar{q}_{9,1}(E,k_s,\eta_{g1}) + [[\bar{q}_{9,1}(E,k_s,\eta_{g1})]^2 \\ &\quad + 4\bar{p}_{9,1}\bar{R}_{9,1}(E,k_s,\eta_{g1})]^{\frac{1}{2}}], \text{ and} \\ \varepsilon_2(E,k_s) &\equiv \left[\frac{K_{112}(E,k_s)}{K_{212}(E,k_s)} - \frac{K_{212}(E,k_s)}{K_{112}(E,k_s)}\right].\end{aligned}$$

The simplified dispersion relation in heavily doped quantum well IV-VI superlattices with graded interfaces under magnetic quantization can be expressed as

$$k_z^2 = \frac{1}{L_0^2}\left[\cos^{-1}\left\{\frac{1}{2}\Phi_2(E,n)\right\}\right]^2 - \frac{2eB}{\hbar}\left(n+\frac{1}{2}\right) \tag{4.16}$$

where

$$\begin{aligned}\Phi_2(E,n) \equiv \Big[&2\cosh\{\beta_2(E,n)\}\cos\{\gamma_2(E,n)\} + \varepsilon_2(E,n)\sinh\{\beta_2(E,n)\}\sin\{\gamma_{22}(E,n)\}\\ &+ \Delta_0[((\{K_{112}(E,n)\}^2/K_{212}(E,n)) - 3K_{212}(E,n))\cosh\{\beta_2(E,n)\}\sin\{\gamma_{22}(E,n)\}\\ &+ \left(3K_{112}(E,n) - \frac{\{K_{212}(E,n)\}^2}{K_{112}(E,n)}\right)\sinh\{\beta_2(E,n)\}\cos\{\gamma_{22}(E,n)\}]\\ &+ \Delta_0\Big[2\left(\{K_{112}(E,n)\}^2 - \{K_{212}(E,n)\}^2\right)\cosh\{\beta_2(E,n)\}\cos\{\gamma_{22}(E,n)\}\\ &+ \frac{1}{12}\left[\frac{5\{K_{112}(E,n)\}^3}{K_{212}(E,n)} + \frac{5\{K_{212}(E,n)\}^3}{K_{112}(E,n)} - 34K_{212}(E,n)K_{112}(E,n)\right]\sinh\{\beta_2(E,n)\}\sin\{\gamma_{22}(E,n)\}\Big]\Big],\end{aligned}$$

$$\beta_2(E,n) \equiv K_{112}(E,n)[a_0 - \Delta_0],$$

$$\begin{aligned}k_{112}^2(E,n) = [2\bar{p}_{9,2n}]^{-1}[-\bar{q}_{9,2n}(E - V_0,\eta_{g2}) - [[\bar{q}_{9,2n}(E - V_0,\eta_{g2})]^2\\ + 4\bar{p}_{9,2n}\bar{R}_{9,2n}(E - V_0,\eta_{g2})]^{\frac{1}{2}}],\end{aligned}$$

$$\begin{aligned}\bar{q}_{9,2n}(E - V_0,\eta_{g2}) = [(\hbar^2/2)((1/m_{t2}^*) + (1/m_{t2}^-)) + \alpha_2(\hbar^4/4)\frac{2eB}{\hbar}\left(n+\frac{1}{2}\right)((1/m_{t2}^+ m_{t2}^-\\ + (1/m_{t2}^+ m_{t2}^-)) - \alpha_2\gamma_3(E - V_0,\eta_{g2})((1/m_{t2}^+) - (1/m_{t2}^-)],\end{aligned}$$

$$\begin{aligned}\bar{R}_{9,2n}(E - V_0,\eta_{g2}) = [\gamma_2(E - V_0,\eta_{g2}) + \gamma_3(E - V_0,\eta_{g2})[\hbar^2/2)\alpha_2\frac{2eB}{\hbar}\left(n+\frac{1}{2}\right)((1/m_{t2}^*)\\ - (1/m_{t2}^-))] - [(\hbar^2/2)k_{s0}^2((1/m_{t2}^*) + (1/m_{t2}^-))]\\ - \alpha_2(\hbar^6/4)\left[\frac{2eB}{\hbar}\left(n+\frac{1}{2}\right)\right]^2((1/m_{t2}^* m_{t2}^-))],\end{aligned}$$

$$\begin{aligned}\gamma_2(E,n) = K_{212}(E,n)[b_0 - \Delta_0], k_{212}^2(E,n) = [2\bar{p}_{9,1n}]^{-1}[-\bar{q}_{9,1n}(E,\eta_{g1})\\ + [[\bar{q}_{9,1n}(E,\eta_{g1})]^2 + 4\bar{p}_{9,1n}\bar{R}_{9,1n}(E,\eta_{g1})]^{\frac{1}{2}}]\end{aligned}$$

$$\bar{q}_{9,1n}(E,\eta_{g1}) = [(\hbar^2/2)((1/m_{l1}^*)+(1/m_{l1}^-))+\alpha_1(\hbar^4/4)\frac{2eB}{\hbar}\left(n+\frac{1}{2}\right)((1/m_{l1}^+m_{t1}^-)$$
$$+(1/m_{t1}^+m_{l1}^-))-\alpha_1\gamma_3(E,\eta_{g1})((1/m_{l1}^+)-(1/m_{l1}^-)],$$
$$\bar{R}_{9,1}(E,\eta_{g1}) = [\gamma_2(E,\eta_{g1})+\gamma_3(E,\eta_{g1})[\hbar^2/2)\alpha_1(2eB/\hbar)(n+\frac{1}{2})((1/m_{t1}^*)$$
$$-(1/m_{t1}^-))]-[\hbar^2/2)k_{s0}^2((1/m_{t1}^*)+(1/m_{t1}^-))]$$
$$-\alpha_1(\hbar^6/4)((2eB/\hbar)(n+\frac{1}{2}))^2((1/m_{t1}^+m_{t1}^-))] \text{ and}$$

$$\varepsilon_2(E,n) \equiv \left[\frac{K_{112}(E,n)}{K_{212}(E,n)}-\frac{K_{212}(E,n)}{K_{112}(E,n)}\right].$$

The dispersion relation in quantum well heavily doped quantum well IV-VI superlattices under magnetic quantization assumes the form

$$\left(\frac{\pi n_z}{d_z}\right)^2=\frac{1}{L_0^2}\left[\cos^{-1}\left\{\frac{1}{2}\varphi_2(E_{43,n},n)\right\}\right]^2-\frac{2|e|B}{\hbar}\left(n+\frac{1}{2}\right) \tag{4.17a}$$

where $E_{43,n}$ is the totally quantized energy in this case.

The electron concentration in this case is given by

$$n_{0SL}=\frac{eBg_v}{\pi\hbar}\sum_{n=0}^{n_{\max}}\sum_{n_z=1}^{n_{z\max}}F_{-1}(\eta_{43}) \tag{4.17b}$$

where $\eta_{43}=(k_BT)^{-1}(E_{FSL}-E_{43,n})$ and E_{FSL} is the Fermi energy in this case.

The EP in this case(J_{SL}) is given by

$$J_{SL}=\frac{\alpha_0e^2Bg_v}{2\pi\hbar d_z}\sum_{n=0}^{n_{\max}}\sum_{n_{z\min}}^{n_{z\max}}F_{-1}(\eta_{43})v_{z3}(E_{nzSL3}) \tag{4.17c}$$

where

$$v_{z3}(\mathrm{E}_{nzSL3})=\frac{2L_0\sin(\frac{L_0n_z\pi}{d_z})}{\hbar\varphi_2'(\mathrm{E}_{nzSL3},0)} \tag{4.17d}$$

E_{nzSL3} is the quantized energy along z direction and is obtained from the equation

$$\cos[(\frac{L_0n_z\pi}{d_z})]=\frac{1}{2}\varphi_2(\mathrm{E}_{nzSL3},0). \tag{4.17e}$$

4.2.4 The Magneto EP from HgTe/CdTe Quantum Well HD Superlattices with Graded Interfaces

The electron energy spectra of the constituent materials of HgTe/CdTe SLs are given by

$$k^2 = \left[\frac{B_{01}^2 + 4A_1E - B_{01}\sqrt{B_{01}^2 + 4A_1E}}{2A_1^2}\right] \tag{4.18}$$

$$\text{and } \frac{\hbar^2 k^2}{2m_{c2}^*} = T_{12}(E, \Delta_2, E_{g2}, \eta_{g2}) + iT_{22}(E, \Delta_2, E_{g2}, \eta_{g2}) \tag{4.19}$$

where $B_{01} = (3|e|^2/128\varepsilon_{sc1})$, $A_1 = (\hbar^2/2m_{c1}^*)$. ε_{sc1} is the semiconductor permittivity of the first material. The energy-wave vector dispersion relation of the conduction electrons in heavily doped quantum well HgTe/CdTe SLs with graded interfaces can be expressed as

$$k_z^2 = G_{192} + iH_{192} \tag{4.20}$$

where

$$G_{192} = [((C_{182}^2 - D_{182}^2)/L_0^2) - k_s^2],$$

$$\begin{aligned} C_{182} &= \cos^{-1}(\omega_{182}), \omega_{182} = (2)^{\frac{-1}{2}}[(1 - G_{182}^2 - H_{182}^2) \\ &\quad - \sqrt{(1 - G_{182}^2 - H_{182}^2)^2 + 4G_{182}^2}]^{\frac{1}{2}}, \\ G_{182} &= \frac{1}{2}[G_{112} + G_{122} + \Delta_0(G_{132} + G_{142}) + \Delta_0(G_{152} + G_{162})], \\ G_{112} &= 2(\cos(g_{12}))(\cos(g_{22}))(\cos\gamma_8(E, k_s)) \end{aligned}$$

$$\gamma_8(E, k_s) = k_8(E, k_s)(b_0 - \Delta_0),\ k_8(E, k_s) = \left[\frac{B_{01}^2 + 4A_1E - B_{01}\sqrt{B_{01}^2 + 4A_1E}}{2A_1^2} - k_s^2\right]^{1/2},$$

$$G_{122} = ([\Omega_{12}(E, k_s)(\sinh g_{12})(\cos g_{22}) - \Omega_{22}(E, k_s)(\sin g_{22})(\cosh g_{12})](\sin \gamma_8(E, k_s)))$$

$$\Omega_{12}(E, k_s) = [\frac{d_{12}}{k_8(E, k_s)} - \frac{k_8(E, k_s)d_{12}}{d_{12}^2 + d_{22}^2}],\ \Omega_{22}(E, k_s) = [\frac{d_{22}}{k_8(E, k_s)} + \frac{k_8(E, k_s)d_{22}}{d_{12}^2 + d_{22}^2}],$$

$$G_{132} = ([\Omega_{32}(E, k_s)(\cosh g_{12})(\cos g_{22}) - \Omega_{42}(E, k_s)(\sin g_{12})(\sin g_{22})](\sin \gamma_8(E, k_s))),$$

$$\Omega_{32}(E, k_s) = [\frac{d_{12}^2 - d_{22}^2}{k_8(E, k_s)} - 3k_8(E, k_s)],\ \Omega_{42}(E, k_s) = [\frac{2d_{12}d_{22}}{k_8(E, k_s)}],$$

$$G_{142} = ([\Omega_{52}(E, k_s)(\sinh g_{12})(\cos g_{22}) - \Omega_{62}(E, k_s)(\sin g_{12})(\cosh g_{22})](\cos \gamma_8(E, k_s))),$$

$$\Omega_{52}(E, k_s) = [3d_{12} - \frac{d_{12}}{d_{12}^2 + d_{22}^2}k_8^2(E, k_s)],\ \Omega_{62}(E, k_s) = [3d_{22} + \frac{d_{22}}{d_{12}^2 + d_{22}^2}k_8^2(E, k_s)],$$

$$G_{152} = ([\Omega_{92}(E,k_s)(\cosh\ g_{12})(\cos\ g_{22}) - \Omega_{102}(E,k_s)(\sinh\ g_{12})(\sin\ g_{22})](\cos\ \gamma_8(E,k_s))),$$
$$\Omega_{92}(E,k_s) = [2d_{12}^2 - 2d_{22}^2 - k_8^2(E,k_s)],\ \Omega_{102}(E,k_s) = [2d_{12}d_{22}],$$
$$G_{162} = ([\Omega_{72}(E,k_s)(\sinh\ g_{12})(\cos\ g_{22}) - \Omega_{82}(E,k_s)(\sin\ g_{12})(\cosh\ g_{22})](\sin\ \gamma_8(E,k_s/12))),$$

$$\Omega_{72}(E,k_s) = [\frac{5d_{12}}{d_{12}^2 + d_{22}^2} k_8^2(E,k_s) + \frac{5(d_{12}^3 - 3_{22}^2 d_{12})}{k_8(E,k_s)} - 34k_8(E,k_s)d_{12}],$$
$$\Omega_{82}(E,k_s) = [\frac{5d_{12}}{d_{12}^2 + d_{22}^2} k_8^2(E,k_s) + \frac{5(d_{22}^3 - 3_{22}^2 d_{12})}{k_8(E,k_s)} + 34k_8(E,k_s)d_{22}],$$

$$H_{182} = \frac{1}{2}[H_{112} + H_{122} + \Delta_0(H_{132} + H_{142}) + \Delta_0(H_{152} + H_{162})],$$
$$H_{112} = 2(\sinh\ g_{12}\ \sin\ g_{22}\ \cos\ \gamma_8(E,k_s)),$$
$$H_{122} = ([\Omega_{22}(E,k_s)(\sinh\ g_{12})(\cos\ g_{22}) + \Omega_{12}(E,k_s)(\sin\ g_{22})(\cosh\ g_{12})](\sin\ \gamma_8(E,k_s))),$$
$$H_{132} = ([\Omega_{42}(E,k_s)(\cosh\ g_{12})(\cos\ g_{22}) + \Omega_{32}(E,k_s)(\sinh\ g_{12})(\sin\ g_{22})](\sin\ \gamma_8(E,k_s))),$$
$$H_{142} = ([\Omega_{62}(E,k_s)(\sinh\ g_{12})(\cos\ g_{22}) + \Omega_{52}(E,k_s)(\sin\ g_{12})(\cosh\ g_{22})](\cos\ \gamma_8(E,k_s))),$$
$$H_{152} = ([\Omega_{102}(E,k_s)(\cosh\ g_{12})(\cos\ g_{22}) + \Omega_{92}(E,k_s)(\sinh\ g_{12})(\sin\ g_{22})](\cos\ \gamma_8(E,k_s))),$$
$$H_{162} = ([\Omega_{82}(E,k_s)(\sinh\ g_{12})(\cos\ g_{22}) + \Omega_{72}(E,k_s)(\sin\ g_{12})(\cosh\ g_{22})](\sin\ \gamma_8(E,k_s/12)),$$
$$\mathrm{H}_{192} = [((2\mathrm{C}_{182}D_{182})/L_0^2)] \text{ and } \mathrm{D}_{182} = \sinh^{-1}(\omega_{182})$$

The simplified dispersion relation in heavily doped quantum well HgTe/CdTe superlatices with graded interfaces under magnetic quantization can be expressed as

$$(k_z)^2 = G_{192E,n} + iH_{192E,n} \tag{4.21}$$

where

$$G_{192E,n} = \left[\frac{C_{182E,n}^2 - D_{182E,n}^2}{L_0^2} - (2eB/\hbar)(n + (1/2))\right],$$
$$C_{1820D} = \cos^{-1}(\omega_{182E,n}), \omega_{182E,n} = (2)^{\frac{-1}{2}}[(1 - G_{182E,n}^2 - H_{182E,n}^2) - \sqrt{(1 - G_{182E,n}^2 - H_{182E,n}^2)^2 + 4G_{182E,n}^2}]^{\frac{1}{2}},$$
$$G_{182E,n} = \frac{1}{2}[G_{112E,n} + G_{122E,n} + \Delta_0(G_{132E,n} + G_{142E,n}) + \Delta_0(G_{152E,n} + G_{162E,n})],$$

$$G_{112E,n} = 2(\cos(g_{12}))(\cos(g_{22}))(\cos\gamma_8(E,n)),\ \gamma_8(E,n) = k_8(E,n)(b_0 - \Delta_0),$$
$$k_8(E,n) = \left[\frac{B_{01}^2 + 4A_1E - B_{01}\sqrt{B_{01}^2 + 4A_1E}}{2A_1^2} - (2eB/\hbar)(n + (1/2))\right]^{1/2},$$

$$G_{120D} = ([\Omega_{12}(E,n)(\sinh\ g_{12E,n})(\cos\ g_{22E,n}) - \Omega_{22}(E,n)(\sin\ g_{22E,n})(\cosh g_{12E,n})](\sin\ \gamma_8(E,k_s))),$$

$$\Omega_{12}(E,n) = [\frac{d_{12E,n}}{k_8(E,k_s)} - \frac{k_8(E,k_s)d_{12E,n}}{d_{12E,n}^2 + d_{22E,n}^2}],\ \Omega_{22}(E,n) = [\frac{d_{22E,n}}{k_8(E,n)} + \frac{k_8(E,n)d_{22E,n}}{d_{12E,n}^2 + d_{22E,n}^2}],$$

$$G_{1320D} = ([\Omega_{32}(E,n)(\cosh\ g_{12E,n})(\cos\ g_{22E,n}) - \Omega_{42}(E,n)(\sinh\ g_{12E,n})(\sin g_{22})](\sin\ \gamma_8(E,n))),$$

$$\Omega_{32}(E,n) = [\frac{d_{12E,n}^2 - d_{2E,n}^2}{k_8(E,n)} - 3k_8(E,n)],\ \Omega_{42}(E,n) = [\frac{2d_{12E,n}d_{22E,n}}{k_8(E,n)}],$$

$$G_{1420D} = ([\Omega_{52}(E,n)(\sinh\ g_{12E,n})(\cos\ g_{22E,n}) - \Omega_{62}(E,n)(\sin\ g_{12E,n})(\cosh g_{22E,n})](\cos\ \gamma_8(E,n))),$$

$$\Omega_{52}(E,n) = [3d_{12E,n} - \frac{d_{12E,n}}{d_{12E,n}^2 + d_{22E,n}^2}k_8^2(E,n)],$$

$$\Omega_{62}(E,n) = [3d_{22E,n} + \frac{d_{22E,n}}{d_{12E,n}^2 + d_{22E,n}^2}k_8^2(E,n)],$$

$$G_{1520D} = ([\Omega_{92}(E,n)(\cosh\ g_{12E,n})(\cos\ g_{22E,n}) - \Omega_{102}(E,n)(\sinh\ g_{12E,n})(\sin\ g_{22E,n})](\cos\ \gamma_8(E,n))),$$

$$\Omega_{92}(E,n) = [2d_{12E,n}^2 - 2d_{22E,n}^2 - k_8^2(E,n)],\ \Omega_{102}(E,n) = [2d_{12E,n}d_{22E,n}],$$

$$G_{162E,n} = ([\Omega_{72}(E,n)(\sinh\ g_{12E,n})(\cos\ g_{22E,n}) - \Omega_{82}(E,n)(\sin\ g_{12E,n})(\cosh\ g_{22E,n})](\sin\ \gamma_{80D}(E,n/12)),$$

$$\Omega_{72}(E,n) = [\frac{5d_{12E,n}}{d_{12E,n}^2 + d_{22E,n}^2}k_8^2(E,n) + \frac{5(d_{12E,n}^3 - 3d_{22E,n}^2 d_{12E,n})}{k_8(E,n)} - 34k_8(E,n)d_{12E,n}],$$

$$\Omega_{82}(E,n) = [\frac{5d_{22E,n}}{d_{12E,n}^2 + d_{22E,n}^2}k_8^2(E,n) + \frac{5(d_{22E,n}^3 - 3d_{22E,n}^2 d_{12E,n})}{k_8(E,n)} + 34k_8(E,n)d_{22E,n}],$$

$$H_{182E,n} = \frac{1}{2}[H_{112E,n} + H_{122E,n} + \Delta_0(H_{132E,n} + H_{142E,n}) + \Delta_0(H_{152E,n} + H_{162E,n})]$$

$$H_{112E,n} = 2(\sinh\ g_{12E,n})(\sin\ g_{22E,n})(\cos\ \gamma_8(E,n))),$$

$$H_{1220D} = ([\Omega_{22}(E,n)(\sinh\ g_{12E,n})(\cos\ g_{22E,n}) + \Omega_{12}(E,n)(\sin\ g_{22E,n})(\cosh\ g_{12E,n})](\sin\ \gamma_8(E,n))),$$

$$H_{132E,n} = ([\Omega_{42}(E,n)(\cosh\ g_{12E,n})(\cos\ g_{22E,n}) + \Omega_{32}(E,n)(\sinh\ g_{12E,n})(\sin\ g_{22E,n})](\sin\ \gamma_8(E,n))),$$

$$H_{142E,n} = ([\Omega_{62}(E,n)(\sinh\ g_{12E,n})(\cos\ g_{22E,n}) + \Omega_{52}(E,n)(\sin\ g_{12E,n})(\cosh\ g_{22E,n})](\cos\ \gamma_8(E,n))),$$

$$H_{1520D} = ([\Omega_{102}(E,n)(\cosh\ g_{12E,n})(\cos\ g_{22E,n}) + \Omega_{92}(E,n)(\sinh\ g_{12E,n})(\sin\ g_{22E,n})](\cos\ \gamma_8(E,n))),$$

$$H_{162E,n} = ([\Omega_{82}(E,n)(\sinh\ g_{12E,n})(\cos\ g_{22E,n}) + \Omega_{72}(E,n)(\sin\ g_{12E,n})(\cosh\ g_{22E,n})](\sin\ \gamma_8(E,n/12)),$$

$H_{192E,n} = [((2C_{182E,n}D_{182E,n})/L_0^2)]$ and $D_{182E,n} = \sinh^{-1}(\omega_{182E,n})$

The dispersion relation in quantum well heavily doped HgTe/CdTe superllatices under magnetic quantization assumes the form

$$\left(\frac{n_z\pi}{d_z}\right)^2 = G_{192E_{44,n}} + iH_{192E_{44,n}} \quad (4.22a)$$

where $E_{44,n}$ is the totally quantized energy in this case.

The electron concentration in this case is given by

$$n_{0SL} = \frac{eBg_v}{\pi\hbar} \text{ Real part of } \sum_{n=0}^{n_{\max}} \sum_{n_z=1}^{n_{z\max}} F_{-1}(\eta_{44}) \quad (4.22b)$$

where $\eta_{44} = (k_B T)^{-1}(E_{FSL} - E_{44,n})$ and E_{FSL} is the Fermi energy in this case.

The EP in this case(J_{SL}) is given by

$$J_{SL} = \frac{\alpha_0 e^2 B g_v}{2\pi\hbar d_z} \text{ Real part of } \sum_{n=0}^{n_{\max}} \sum_{n_{z\min}}^{n_{z\max}} F_{-1}(\eta_{44}) v_{z4}(E_{nzSL4}) \quad (4.22c)$$

where

$$v_{z4}(E_{nzSL4}) = \frac{2\pi n_z}{\hbar d_z (G'_{192} + iH'_{192})\big|_{k_x=0, k_y=0, k_z=\frac{n_z\pi}{d_z} \text{ and } E=E_{nzSL4}}} \quad (4.22d)$$

E_{nzSL4} is the quantized energy along z direction and is obtained by substituting $k_x = 0, k_y = 0, k_z = \frac{n_z\pi}{d_z}$ *and* $E = E_{nzSL4}$ in (4.20) and the primes denote the differentiation with respect to energy.

4.2.5 The Magneto EP from Strained Layer Quantum Well HD Superlattices with Graded Interfaces

The dispersion relation of the conduction electrons of the constituent materials of the strained layer super lattices can be expressed as

$$[E - T_{1i}]k_x^2 + [E - T_{2i}]k_y^2 + [E - T_{3i}]k_z^2 = q_i E^3 - R_i E^3 + V_i E + \zeta_i \quad (4.23)$$

where

$$T_{1i} = \overline{\theta}_i, \overline{\theta}_i = \left[E_{gi} - C_{1i}^c \varepsilon_i - (a_i + C_{1i}^c)\varepsilon_i + \frac{3}{2} b_i \varepsilon_{xxi} - \frac{b_i \varepsilon_i}{2} + \frac{\sqrt{3} d_i \varepsilon_{xyi}}{2}\right],$$

$$T_{2i} = \omega_i, \omega_i = \left[E_{gi} - C_{1i}^c \varepsilon_i - (a_i + C_{1i}^c)\varepsilon_i + \frac{3}{2} b_i \varepsilon_{xxi} - \frac{b_i \varepsilon_i}{2} - \frac{\sqrt{3} d_i \varepsilon_{xyi}}{2}\right],$$

$$T_{3i} = \delta_i, \delta_i = \left[E_{gi} - C_{1i}^c \varepsilon_i + (a_i + C_{1i}^c)\varepsilon_i + \frac{3}{2} b_i \varepsilon_{zzi} - \frac{b_i \varepsilon_i}{2}\right]$$

$$R_i = q_i\left[2A_i + C_{1i}^c \varepsilon_i\right], \mathrm{q}_i = \frac{3}{2B_{2i}^2}, \ \mathrm{A}_i = E_{gi} - C_{1i}^c \varepsilon_i,$$

$$V_i = q_i\left[A_i^2 - \frac{2C_{2i}^2 \varepsilon_{xyi}}{3} + 2A_i C_{1i}^c \varepsilon_i\right], \varsigma_i = q_i\left[\frac{2C_{2i}^2 \varepsilon_{xyi}}{3} - C_{1i}^c \varepsilon_i A_i^2\right]$$

Therefore the electron energy spectrum in HD stressed materials can be written as

$$\overline{P_i}(E, \eta_{gi}) k_x^2 + \overline{Q_i}(E, \eta_{gi}) k_y^2 + \overline{S_i}(E, \eta_{gi}) k_z^2 = 1 \tag{4.24}$$

where

$$\overline{P_i}(E, \eta_{gi}) = \frac{\left[\gamma_0(E, \eta_{gi}) - I_0 T_{1i}\right]}{\overline{\Delta_i}(E, \eta_{gi})},$$

$$\overline{\Delta_i}(E, \eta_{gi}) = \left[\frac{-q_i \eta_{gi}^3}{2\sqrt{\pi}} \exp\left(\frac{-E^2}{\eta_{gi}^2}\right)\left[1 + \frac{E^2}{\eta_{gi}^2}\right] - R_i \theta_0(E, \eta_{gi}) + V_i \gamma_0(E, \eta_{gi}) + \frac{\varsigma_i}{2}\left[1 + Erf\left(\frac{E}{\eta_{gi}}\right)\right]\right],$$

$$I_0 = \frac{1}{2}[1 + Erf(E/\eta_{gi})], \overline{Q_i}(E, \eta_{gi}) = \frac{\left[\gamma_0(E, \eta_{gi}) - I_0 T_{2i}\right]}{\overline{\Delta_i}(E, \eta_{gi})} \text{ and } \overline{S_i}(E, \eta_{gi}) = \frac{\left[\gamma_0(E, \eta_{gi}) - I_0 T_{3i}\right]}{\overline{\Delta_i}(E, \eta_{gi})}$$

The energy-wave vector dispersion relation of the conduction electrons in heavily doped strained layer quantum well SLs with graded interfaces can be expressed as

$$\cos(L_0 k) = \frac{1}{2}\overline{\Phi}_6(E, k_s) \tag{4.25}$$

where

$$\begin{aligned}
\overline{\Phi_6}(E, k_s) = {} & \left[2\cosh\left[T_4(E, \eta_{g2})\right]\cos\left[T_5(E, \eta_{g1})\right]\right] + \left[T_6(E, k_s)\right]\sinh\left[T_4(E, \eta_{g2})\right]\sin\left[T_5(E, \eta_{g1})\right] \\
& + \Delta_0\left[\left(\frac{k_0^2(E, \eta_{g2})}{k_0'(E, \eta_{g1})} - 3k_0'(E, \eta_{g1})\right)\cosh\left[T_4(E, \eta_{g2})\right]\sin\left[T_5(E, \eta_{g1})\right]\right. \\
& \left. + \left(3k_0(E, \eta_{g2}) - \frac{k'^2(E, \eta_{g1})}{k_0(E, \eta_{g2})}\right)\sinh\left[T_4(E, \eta_{g2})\right]\cos\left[T_5(E, \eta_{g1})\right]\right] \\
& + \Delta_0\left[2\left(k_0^2(E, \eta_{g2}) - k'^2(E, \eta_{g1})\right)\cosh\left[T_4(E, \eta_{g2})\right]\cos\left[T_5(E, \eta_{g1})\right]\right] \\
& + \frac{1}{12}\left(\frac{5k_0^3(E, \eta_{g2})}{k_0'(E, \eta_{g1})} + \frac{5k'^3(E, \eta_{g1})}{k_0(E, \eta_{g2})} - 34k_0(E, \eta_{g2})k'(E, \eta_{g1})\right)\sinh\left[T_5(E, \eta_{g2})\right]\sin\left[T_5(E, \eta_{g1})\right]
\end{aligned}$$

$$
\begin{aligned}
\left[T_4(E,\eta_{g2})\right] &= k_0(E,\eta_{g2})[a_0-\Delta_0],\\
k_0(E,\eta_{g2}) &= \left[\overline{S_2}(E-V_0,\eta_{g2})\right]^{-1/2}\left[\overline{P_2}(E-V_0,\eta_{g2})k_x^2+\overline{Q_2}(E-V_0,\eta_{g2})k_y^2-1\right]^{1/2},\\
[T_5(E,\eta_{g1})] &= k'(E,\eta_{g1})[b_0-\Delta_0],\\
k'(E,\eta_{g1}) &= \left[\overline{S_1}(E,\eta_{g1})\right]^{-1/2}\left[1-\overline{P_1}(E,\eta_{g1})k_x^2-\overline{Q_1}(E,\eta_{g1})k_y^2\right]^{1/2} \text{ and}\\
T_6(E,k_s) &= \left[\frac{k_0(E,\eta_{g2})}{k'(E,\eta_{g1})}-\frac{k'(E,\eta_{g1})}{k_0(E,\eta_{g2})}\right]
\end{aligned}
$$

Therefore the dispersion relation of the conduction electrons in heavily doped strained layer quantum well QEPs with graded interfaces can be expressed as

$$\cos(L_0k)=\frac{1}{2}\overline{\varphi_6}(\mathrm{E,n}) \tag{4.26}$$

where

$$
\begin{aligned}
\overline{\phi_6}(E,n) = {}& \left[2\cosh\left[T_4(E,n,\eta_{g2})\right]\cos\left[T_5(E,n,\eta_{g1})\right]\right]+[T_6(E,n)]\sinh[T_4(E,n,\eta_{g2})]\sin[T_5(E,n,\eta_{g1})]\\
&+\Delta_0\left[\left(\frac{k_0^2(E,n,\eta_{g2})}{k_0'(E,n,\eta_{g1})}-3k_0'(E,n,\eta_{g1})\right)\cosh\left[T_4(E,n,\eta_{g2})\right]\sin\left[T_5(E,n,\eta_{g1})\right]\right.\\
&\left.+\left(3k_0(E,n,\eta_{g2})-\frac{k_0'^2(E,n,\eta_{g1})}{k_0(E,n,\eta_{g2})}\right)\sinh\left[T_4(E,n,\eta_{g2})\right]\cos\left[T_5(E,n,\eta_{g1})\right]\right]\\
&+\Delta_0\left[2(k_0^2(E,n,\eta_{g2})-k_{0D}'^2(k,n,\eta_{g1}))\cosh\left[T_4(E,n,\eta_{g2})\right]\cos\left[T_5(E,n,\eta_{g1})\right]\right]\\
&\left.+\frac{1}{12}\left(\frac{5k_0^3(E,n,\eta_{g2})}{k_0'(k,n,\eta_{g1}}+\frac{5k_0'^3(E,n,\eta_{g1})}{k_0(E,n,\eta_{g2})}-34k_0(E,n,\eta_{g2})k_0'(E,n,\eta_{g1})\right)\sinh\left[T_4(E,n,\eta_{g2})\right]\sin[T_5(E,n,\eta_{g1})]\right]
\end{aligned}
$$

$$
\begin{aligned}
\left[T_4(E,n,\eta_{g2})\right] &= k_0(E,n,\eta_{g2})[a_0-\Delta_0],\\
k_0(E,n,\eta_{g2}) &= \left[\overline{S_2}(E-V_0,\eta_{g2})\right]^{-1/2}\cdot\left[\left[(n+1/2)\hbar eB/(\sqrt{\rho_1(E)\rho_2(E)})\right]-1\right]^{1/2},\\
&\rho_1(E)=\hbar^2/(2\overline{P_2}(E-V_0,\eta_{g2})),\ \rho_2(E)=\hbar^2/(2\overline{Q_2}(E-V_0,\eta_{g2}))\\
[T_5(E,n,\eta_{g1})] &= k_0'(E,n,\eta_{g1})[b_0-\Delta_0],\\
k_0'(E,n,\eta_{g1}) &= \left[\overline{S_1}(E,n,\eta_{g1})\right]^{1/2}\left[1-\left[(n+1/2)\hbar eB/(\sqrt{\rho_3(E)\rho_4(E)})\right]\right]^{1/2}\\
&\rho_3(E)=\hbar^2/(2\overline{P_1}(E,\eta_{g1})),\ \rho_4(E)=\hbar^2/(2\overline{Q_1}(E,\eta_{g1}))\\
T_6(E,n) &= \left[\frac{k_0(E,n,\eta_{g2})}{k_0'(E,n,\eta_{g1})}-\frac{k_0'(E,n,\eta_{g1})}{k_0(E,n,\eta_{g2})}\right]
\end{aligned}
$$

The energy-wave vector dispersion relation of the conduction electrons in heavily doped quantum well strained layer SLs with graded interfaces can be expressed as

$$\left(\frac{\pi n_z}{d_z}\right)^2 = \frac{1}{L_0^2}\left[\cos^{-1}\left\{\frac{1}{2}\overline{\varphi}_6(E_{47,n}, n)\right\}\right]^2 - \frac{2|e|B}{\hbar}\left(n + \frac{1}{2}\right) \qquad (4.27)$$

where $E_{47,n}$ is the totally quantized energy in this case.

The electron concentration in this case is given by

$$n_{0SL} = \frac{eBg_v}{\pi\hbar}\sum_{n=0}^{n_{\max}}\sum_{n_z=1}^{n_{z\max}} F_{-1}(\eta_{47}) \qquad (4.28a)$$

where $\eta_{47} = (k_B T)^{-1}(E_{FSL} - E_{47,n})$ and E_{FSL} is the Fermi energy in this case.

The EP in this case(J_{SL}) is given by

$$J_{SL} = \frac{\alpha_0 e^2 Bg_v}{2\pi\hbar d_z}\sum_{n=0}^{n_{\max}}\sum_{n_{z\min}}^{n_{z\max}} F_{-1}(\eta_{47}) v_{z7}(E_{nzSL7}) \qquad (4.28b)$$

where

$$v_{z7}(E_{nzSL7}) = \frac{2L_0 \sin(\frac{L_0 n_z \pi}{d_z})}{\hbar[\overline{\varphi}_6(E_{nzSL7}, 0)]'} \qquad (4.28c)$$

E_{nzSL7} is the quantized energy along z direction and is obtained from the equation

$$\cos[(\frac{L_0 n_z \pi}{d_z})] = \frac{1}{2}\overline{\varphi}_6(E_{nzSL7}, 0). \qquad (4.28d)$$

4.2.6 The Magneto EP from III-V Quantum Well HD Effective Mass Super Lattices

Following Sasaki [71], the electron dispersion law in III-V heavily doped effective mass super-lattices (EMSLs) can be written as

$$k_x^2 = \left[\frac{1}{L_0^2}\{\cos^{-1}(f_{21}(E, k_y, k_z))\}^2 - k_\perp^2\right] \qquad (4.29)$$

in which

$$f_{21}(E,k_y,k_z) = a_1 \cos\left[a_0 C_{21}(E,k_\perp,\eta_{g1}) + b_0 D_{21}(E,k_\perp,\eta_{g2})\right] - a_2 \cos\left[a_0 C_{21}(E,k_\perp,\eta_{g1}) - b_0 D_{21}(E,k_\perp,\eta_{g2})\right], k_\perp^2 = k_y^2 + k_z^2,$$

$$a_1 = \left[\sqrt{\frac{M_2(0,\eta_{g2})}{M_1(0,\eta_{g1})}} + 1\right]^2 \left[4\left(\frac{M_2(0,\eta_{g2})}{M_1(0,\eta_{g1})}\right)^{1/2}\right]^{-1},$$

$$a_2 = \left[\sqrt{\frac{M_2(0,\eta_{g2})}{M_1(0,\eta_{g1})}} - 1\right]^2 \left[4\left(\frac{M_2(0,\eta_{g2})}{M_1(0,\eta_{g1})}\right)^{1/2}\right]^{-1},$$

$$M_i(0,\eta_{gi}) = m_{ci}^*\left[\frac{-2}{\sqrt{\pi}}T(0,\eta_{gi}) + 2\left[\frac{\alpha_i b_i}{c_i}\frac{\eta_{gi}}{\sqrt{\pi}} + \frac{1}{2}\left(\frac{\alpha_i c_i + c_i b_i - \alpha_i b_i}{c_i^2}\right) + \frac{1}{\sqrt{\pi}c_i}\left(1-\frac{\alpha_i}{c_i}\right)\left(1-\frac{b_i}{c_i}\right) - \frac{1}{c_i}\left(1-\frac{\alpha_i}{c_i}\right)\left(1-\frac{b_i}{c_i}\right)\frac{2}{c_i\eta_{gi}\sqrt{\pi}}\left\{\frac{-2}{c_i\eta_{gi}^2}\exp\left(\frac{-1}{c_i^2\eta_{gi}^2}\right)\left(\sum_{p=1}^{\alpha}\left(\exp\left(\frac{-p^2}{4}\right)\right)\frac{1}{p}\sinh\left(\frac{p}{c_i\eta_{gi}}\right)\right) + \exp\left(\frac{-1}{c_i^2\eta_{gi}^2}\right)\left(\sum_{p=1}^{\alpha}\exp\left(\frac{-p^2}{4}\right)\frac{1}{\eta_{gi}}\cosh\left(\frac{p}{c_i\eta_{gi}}\right)\right)\right\}\right]\right],$$

$$\mathrm{T}(0,\eta_{gi}) = 2\left[\frac{\alpha_i b_i}{c_i}\frac{\eta_{gi}^2}{4} + \left(\frac{\alpha_i c_i + b_i c_i - \alpha_i b_i}{c_i^2}\right)\frac{\eta_{gi}}{2\sqrt{\pi}} + \frac{1}{2c_i}\left(1-\frac{\alpha_i}{c_i}\right)\left(1-\frac{b_i}{c_i}\right) - \frac{1}{c_i}\left(1-\frac{\alpha_i}{c_i}\right)\left(1-\frac{b_i}{c_i}\right)\frac{2}{c_i\eta_{gi}\sqrt{\pi}}\exp\left(\frac{-1}{c_i^2\eta_{gi}^2}\right)\sum_{p=1}^{\alpha}\frac{\exp(-p^2/4)}{p}\sinh\left(\frac{p}{c_i\eta_{gi}}\right)\right],$$

$$C_{21}(E,k_\perp,\eta_{g1}) = e_1 + ie_2, D_{21}(E,k_\perp,\eta_{g2}) = e_3 + ie_4,$$

$$e_1 = [((\sqrt{t_1^2+t_2^2}+t_1)/2)]^{\frac{1}{2}}, e_2 = [((\sqrt{t_1^2+t_2^2}-t_1)/2)]^{\frac{1}{2}},$$

$$t_1 = \left[\frac{2m_{c1}^*}{\hbar^2}T_{11}(E,\Delta_1,\eta_{g1},E_{g1}) - k_\perp^2\right], \ \mathrm{t}_2 = \frac{2m_{c1}^*}{\hbar^2}T_{21}(E,\Delta_1,\eta_{g1},E_{g1}),$$

$$e_3 = \left[\frac{\sqrt{t_3^2+t_4^2}+t_3}{2}\right]^{1/2}, e_4 = \left[\frac{\sqrt{t_3^2+t_4^2}-t_3}{2}\right]^{1/2}$$

$$t_3 = \left[\frac{2m_{c2}^*}{\hbar^2}T_{12}(E,\Delta_2,\eta_{g2},E_{g2}) - k_\perp^2\right], t_4 = \frac{2m_{c2}^*}{\hbar^2}T_{22}(E,\Delta_2,\eta_{g2},E_{g2}),$$

Therefore (4.29) can be expressed as

$$k_x^2 = \delta_7 + i\delta_8 \tag{4.30}$$

where

$$\delta_7 = \left[\frac{1}{L_0^2}\left(\delta_5^2 - \delta_6^2\right) - k_\perp^2\right], \delta_5 = \cos^{-1} p_5,$$

$$\mathrm{p}_5 = \left[\frac{1 - \delta_3^2 - \delta_4^2 - \sqrt{\left(1 - \delta_3^2 - \delta_4^2\right)^2 + 4\delta_4^2}}{2}\right]^{1/2},$$

$$\delta_3 = (a_1 \ \cos\Delta_1 \cosh\Delta_2 - a_2 \cos\Delta_3 \cosh\Delta_4),$$
$$\delta_4 = (a_1 \ \sin\Delta_1 \sinh\Delta_2 - a_2 \sin\Delta_3 \sinh\Delta_4),$$
$$\Delta_1 = (a_0 e_1 + b_0 e_3), \Delta_2 = (a_0 e_2 + b_0 e_4), \Delta_3 = (a_0 e_1 - b_0 e_3), \Delta_4 = (a_0 e_2 - b_0 e_4),$$
$$\delta_6 = \sinh^{-1} \mathrm{p}_5 \text{ and } \delta_8 = [2\delta_5\delta_6/L_0^2]$$

Therefore the magneto electron dispersion relation in this case assumes the form

$$(k_x)^2 = \delta_{7E,n} + i\delta_{8E,n} \tag{4.31}$$

where

$$\delta_{7E,n} = \left[\frac{1}{L_0^2}\left(\delta_{5E,n}^2 - \delta_{6E,n}^2\right) - \left\{\frac{2eB}{\hbar}\left(n + \frac{1}{2}\right)\right\}\right], \delta_{5E,n} = \cos^{-1} p_{5E,n},$$

$$\mathrm{p}_{5E,n} = \left[\frac{1 - \delta_{3E,n}^2 - \delta_{4E,n}^2 - \sqrt{(1 - \delta_{3E,n}^2 - \delta_{4E,n}^2)^2 + 4\delta_{4E,n}^2}}{2}\right]^{1/2},$$

$$\delta_{3E,n} = (a_1 \cos\Delta_{1E,n} \cosh\Delta_{2E,n} - a_2 \cos\Delta_{3E,n} \cosh\Delta_{4E,n}),$$
$$\delta_{4E,n} = (a_1 \sin\Delta_{1E,n} \sinh\Delta_{2E,n} - a_2 \sin\Delta_{3E,n} \sinh\Delta_{4E,n}),$$

$$\Delta_{1E,n} = (a_0 e_{1E,n} + b_0 e_{3E,n}),\ \Delta_{2E,n} = (a_0 e_{2E,n} + b_0 e_{4E,n}),\ \Delta_{3E,n} = (a_0 e_{1E,n} - b_0 e_{3E,n}),$$
$$\Delta_{4E,n} = (a_0 e_{2E,n} - b_0 e_{4E,n}),$$
$$\delta_{6E,n} = \sinh^{-1} \mathrm{p}_{5E,n} \text{ and } \delta_{8E,n} = [2\delta_{5E,n}\delta_{6E,n}/L_0^2],$$

$$e_{1E,n} = [((\sqrt{t_{1E,n}^2 + t_2^2} + t_{1E,n})/2]^{\frac{1}{2}},\ e_{2E,n} = [((\sqrt{t_{1E,n}^2 + t_2^2} - t_{1E,n})/2]^{\frac{1}{2}},$$

$$e_{3E,n} = \left[\frac{\sqrt{t_{3E,n}^2 + t_4^2} + t_{3E,n}}{2}\right]^{1/2}, e_{4E,n} = \left[\frac{\sqrt{t_{3E,n}^2 + t_4^2} - t_{3E,n}}{2}\right]^{1/2},$$

$$t_{1E,n} = \left[\frac{2m_{c1}^*}{\hbar^2} T_{11}(E, \Delta_1, \eta_{g1}, E_g) - \frac{2eB}{\hbar}\left(n + \frac{1}{2}\right)\right], t_{3E,n} = \left[\frac{2m_{c2}^*}{\hbar^2} T_{12}(E, \Delta_2, \eta_{g2}, E_{g2}) - \frac{2eB}{\hbar}\left(n + \frac{1}{2}\right)\right]$$

The dispersion relation in quantum well heavily doped III-V superlattices under magnetic quantization assumes the form

$$\left(\frac{n_x\pi}{d_x}\right)^2 = \delta_{7A1,n} + i\delta_{8A1,n} \tag{4.32}$$

where $A1$ is the totally quantized energy in this case.

The electron concentration in this case is given by

$$n_{0SL} = \frac{eBg_v}{\pi\hbar} \text{ Real part of } \sum_{n=0}^{n_{\max}} \sum_{n_z=1}^{n_{z\max}} F_{-1}(\eta_{45}) \tag{4.33a}$$

where $\eta_{45} = (k_B T)^{-1}(E_{FSL} - A1)$ and E_{FSL} is the Fermi energy in this case.

The EP in this case(J_{SL}) is given by

$$J_{SL} = \frac{\alpha_0 e^2 Bg_v}{2\pi\hbar d_x} \text{ Real part of } \sum_{n=0}^{n_{\max}} \sum_{n_{x\min}}^{n_{x\max}} F_{-1}(\eta_{45}) v_{x5}(E_{nzSL5}) \tag{4.33b}$$

where

$$v_{x5}(E_{nzSL5}) = \frac{2\pi n_x}{\hbar d_x(\delta_7' + i\delta_8')\big|_{k_y=0,k_z=0,k_x=\frac{n_x\pi}{d_x} \text{ and } E=E_{nzSL5}}} \tag{4.33c}$$

E_{nzSL5} is the quantized energy along x direction and is obtained by substituting $k_y = 0, k_z = 0, k_x = \frac{n_x\pi}{d_x}$ and $E = E_{nzSL5}$ in (4.30) and the primes denote the differentiation with respect to energy.

4.2.7 The Magneto EP from II-VI Quantum Well HD Effective Mass Super Lattices

Following Sasaki [70], the electron dispersion law in heavily doped II-VI EMSLs can be written as

$$k_z^2 = \Delta_{13} + i\Delta_{14}, \tag{4.34}$$

where

$$\Delta_{13} = \left[\frac{1}{L_0^2}(\Delta_{11}^2 - \Delta_{12}^2) - k_s^2\right]$$

$$\Delta_{11} = \cos^{-1} p_6, \mathrm{p}_6 = \left[\frac{1 - \Delta_9^2 - \Delta_{10}^2 - \sqrt{(1 - \Delta_9^2 - \Delta_{10}^2)^2 + 4\Delta_{10}^2}}{2}\right]^{1/2},$$

$$\Delta_9 = (\overline{a_1} \cos \Delta_6 \cosh \Delta_7 - \overline{a_2} \cos \Delta_8 \cosh \Delta_7),$$
$$\Delta_{10} = (\overline{a_1} \sin \Delta_6 \sinh \Delta_7 + \overline{a_2} \sin \Delta_8 \sinh \Delta_7),$$

$$\Delta_6 = \left[a_0 C_{22}(E, k_s, \eta_{g1}) + b_0 e_3\right], \Delta_7 = b_0 e_4, \Delta_8 = \left[a_0 C_{22}(E, K_s, \eta_{g1}) - b_0 e_3\right],$$

$$C_{22}(E, k_s, \eta_{g1}) = \left[\frac{2m_{\|,1}^*}{\hbar^2}\left\{\gamma_3(E, \eta_{g1}) - \frac{\hbar^2 k_s^2}{2m_{\perp,1}^*} \mp C_0 k_s\right\}\right]^{1/2},$$

$$\overline{a_1} = \left[\sqrt{\frac{M_2(0, \eta_{g2})}{\overline{M_1}(0, \eta_{g1})}} + 1\right]^2 \left[4\left(\frac{M_2(0, \eta_{g2})}{\overline{M_1}(0, \eta_{g1})}\right)^{1/2}\right]^{-1}, \overline{M_1}(0, \eta_{g1}) = m_{c1}^*\left(1 - \frac{2}{\pi}\right),$$

$$\overline{a_2} = \left[\sqrt{\frac{M_2(0, \eta_{g2})}{\overline{M_1}(0, \eta_{g1})}} - 1\right]^2 \left[4\left(\frac{M_2(0, \eta_{g2})}{\overline{M_1}(0, \eta_{g1})}\right)^{1/2}\right]^{-1}$$

$$\Delta_{12} = \cos^{-1} p_6, \Delta_{14} = \frac{2\Delta_{11}\Delta_{12}}{L_0^2}$$

The electron dispersion law in heavily doped II-VI QEP can be written as

$$(k_z)^2 = \Delta_{13\,E,n} + i\Delta_{14\,E,n}, \tag{4.35}$$

where

$$\Delta_{13E,n} = \left[\frac{1}{L_0^2}(\Delta_{11E,n}^2 - \Delta_{12E,n}^2) - \left\{\frac{2eB}{\hbar}(n + \frac{1}{2})\right\}\right]$$

$$\Delta_{11E,n} = \cos^{-1} p_{6E,n}, \mathrm{p}_{6E,n} = \left[\frac{1 - \Delta_{9E,n}^2 - \Delta_{10E,n}^2 - \sqrt{\left(1 - \Delta_{9E,n}^2 - \Delta_{10E,n}^2\right)^2 + 4\Delta_{10E,n}^2}}{2}\right]^{\frac{1}{2}},$$

$$\Delta_{9E,n} = \left(\overline{a_1}\cos\Delta_{6E,n}\cosh\Delta_{7E,n} - \overline{a_2}\cos\Delta_{8E,n}\cosh\Delta_{7E,n}\right),$$
$$\Delta_{10E,n} = \left(\overline{a_1}\sin\Delta_{6E,n}\sinh\Delta_{7E,n} + \overline{a_2}\sin\Delta_{8E,n}\sinh\Delta_{7E,n}\right),$$
$$\Delta_{6E,n} = \left[a_0 C_{22E,n}\left(E_{E,n},\eta_{g1}\right) + b_0 e_{3E,n}\right],$$
$$\Delta_{7E,n} = b_0 e_{4E,n},\ \Delta_{8E,n} = \left[a_0 C_{22E,n}\left(E_{E,n},\eta_{g1}\right) - b_0 e_{3E,n}\right],$$
$$C_{22E,n}(E_{E,n}\eta_{g1}) = \left[\frac{2m^*_{\|,1}}{\hbar^2}\left\{\gamma_3(E_{E,n},\eta_{g1}) - \frac{\hbar^2}{2m^*_{\perp,1}}\left\{\frac{2eB}{\hbar}(n+\frac{1}{2})\right\} \mp C_0\left[\left\{\frac{2eB}{\hbar}(n+\frac{1}{2})\right\}\right]^{1/2}\right\}\right]^{1/2},$$
$$\Delta_{12E,n} = \cos^{-1} p_{6E,n},\ \Delta_{14E,n} = \frac{2\Delta_{11E,n}\Delta_{12E,n}}{L_0^2},$$

The dispersion relation in quantum well heavily doped III-V superlattices under magnetic quantization assumes the form

$$\left(\frac{n_z\pi}{d_z}\right)^2 = \Delta_{13,A2,n} + i\Delta_{14,A2,n} \tag{4.36}$$

where $A2$ is the totally quantized energy in this case.

The electron concentration in this case is given by

$$n_{0SL} = \frac{eBg_v}{\pi\hbar}\ \text{Real part of} \sum_{n=0}^{n_{\max}}\sum_{n_z=1}^{n_{z\max}} F_{-1}(\eta_{46}) \tag{4.37a}$$

where $\eta_{46} = (k_B T)^{-1}(E_{FSL} - A2)$ and E_{FSL} is the Fermi energy in this case.

The EP in this case(J_{SL}) is given by

$$J_{SL} = \frac{\alpha_0 e^2 B g_v}{2\pi\hbar d_z}\ \text{Real part of} \sum_{n=0}^{n_{\max}}\sum_{n_{z\min}}^{n_{z\max}} F_{-1}(\eta_{46}) v_{z6}(E_{nzSL6}) \tag{4.37b}$$

where

$$v_{z6}(E_{nzSL6}) = \frac{2\pi n_z}{\hbar d_z(\Delta'_{13} + i\Delta'_{14})\big|_{k_x=0,k_y=0,k_z=\frac{n_z\pi}{d_z}\ and\ E=E_{nzSL6}}} \tag{4.37c}$$

E_{nzSL6} is the quantized energy along x direction and is obtained by substituting $k_x = 0, k_y = 0, k_z = \frac{n_z\pi}{d_z}$ and $E = E_{nzSL6}$ in (4.34) and the primes denote the differentiation with respect to energy.

4.2.8 The Magneto EP from IV-VI Quantum Well HD Effective Mass Super Lattices

Following Sasaki [70], the electron dispersion law in IV-VI, EMSLs can be written as

$$k_z^2 = \left[\frac{1}{L_0^2}\{\cos^{-1}(f_{23}(E,k_x,k_y))\}^2 - k_s^2\right] \tag{4.38}$$

where

$$\begin{aligned} f_{23}(E,k_x,k_y) &= a_3\cos\left[a_0C_{23}(E,k_x,k_y,\eta_{g1}) + b_0D_{23}(E,k_x,k_y,\eta_{g1})\right] \\ &\quad - a_4\cos\left[a_0C_{23}(E,k_x,k_y,\eta_{g2}) - b_0D_{23}(E,k_x,k_y,\eta_{g2})\right], \end{aligned}$$

$$a_3 = \left[\sqrt{\frac{M_3(0,\eta_{g2})}{M_3(0,\eta_{g1})}} + 1\right]^2\left[4\left(\frac{M_3(0,\eta_{g2})}{M_3(0,\eta_{g1})}\right)^{1/2}\right]^{-1},$$

$$a_4 = \left[\sqrt{\frac{M_3(0,\eta_{g2})}{M_3(0,\eta_{g1})}} - 1\right]^2\left[4\left(\frac{M_3(0,\eta_{g2})}{M_3(0,\eta_{g1})}\right)^{1/2}\right]^{-1},$$

$$\begin{aligned} M_3(0,\eta_{gi}) &= (4\overline{p_{9,i}})^{-1}\left[\left\{\alpha_i\left(1-\frac{2}{\pi}\right)\left(\frac{1}{m_{l,i}^+}-\frac{1}{m_{l,i}^-}\right)\right\} + \left[[\overline{q_{9,i}}(0,\eta_{gi})]^2\right.\right. \\ &\quad \left.+ (4\overline{p_{9,i}})\overline{R_{9,i}}(0,\eta_{gi})]^{-1/2}\left[\alpha_i\left(1-\frac{2}{\pi}\right)\left(\frac{1}{m_{l,i}^+}-\frac{1}{m_{l,i}^-}\right)\overline{q_{9,i}}(0,\eta_{gi}) + 2\overline{p_{9,i}}\left(1-\frac{2}{\pi}+\frac{\alpha_i\eta_{gi}}{\sqrt{\pi}}\right)\right]\right], \end{aligned}$$

$$\overline{p_{9,i}} = \frac{\alpha_i\hbar^4}{4m_{l,i}^+m_{l,i}^-}, \overline{q_{9,i}}(0,\eta_{gi}) = \left[\frac{\hbar^2}{2}\left(\frac{1}{m_{l,i}^+}+\frac{1}{m_{l,i}^-}\right) - \frac{\alpha_i\eta_{gi}}{\sqrt{\pi}}\left(\frac{1}{m_{l,i}^+}-\frac{1}{m_{l,i}^-}\right)\right], \overline{R_{9,i}}(0,\eta_{gi}) = \left[\frac{\eta_{gi}}{\sqrt{\pi}}+\frac{\alpha_i\eta_{gi}^2}{2}\right],$$

$$\begin{aligned} C_{23}(E,k_x,k_y,\eta_{g1}) &= \left[[2\overline{p_{9,i}}].^{-1}\left[-\overline{q_{9,i}}(E,k_x,k_y,\eta_{g1}) + \left[\{\overline{q_{9,i}}(E,k_x,k_y,\eta_{g1})\}^2\right.\right.\right. \\ &\quad \left.\left.\left.+ (4\overline{p_{9,i}})\overline{R_{9,i}}(E,k_x,k_y,\eta_{g1})\right]^{1/2}\right]\right]^{1/2}, \end{aligned}$$

$$\begin{aligned} D_{23}(E,k_x,k_y,\eta_{g2}) &= \left[[2\overline{p_{9,2}}]^{-1}\left[-\overline{q_{9,2}}(E,k_x,k_y,\eta_{g1}) + \left[\{\overline{q_{9,2}}(E,k_x,k_y,\eta_{g1})\}^2\right.\right.\right. \\ &\quad \left.\left.\left.+ (4\overline{p_{9,2}})\overline{R_{9,2}}(E,k_x,k_y,\eta_{g1})\right]^{1/2}\right]\right]^{1/2}, \end{aligned}$$

$$\overline{q_{9,i}}(E,k_x,k_y,\eta_{gi}) = \left[\frac{\hbar^2}{2}\left(\frac{1}{m_{l,i}^{+}}+\frac{1}{m_{l,i}^{-}}\right)+\alpha_i\frac{\hbar^4}{4}k_s^2\left(\frac{1}{m_{l,i}^{+}m_{t,i}^{-}}+\frac{1}{m_{t,i}^{+}m_{l,i}^{-}}\right)\right.$$
$$\left.-\alpha_i\gamma_3(E,\eta_{gi})\left(\frac{1}{m_{l,i}^{+}}-\frac{1}{m_{l,i}^{-}}\right)\right],$$
$$\overline{R_{9,i}}(E,k_x,k_y,\eta_{gi}) = \left[\gamma_2(E,\eta_{gi})+\gamma_3(E,\eta_{gi})\alpha_i\frac{\hbar^2}{2}k_s^2\left(\frac{1}{m_{t,i}^{+}}-\frac{1}{m_{t,i}^{-}}\right)\right.$$
$$\left.-\frac{\hbar^2}{2}k_s^2\left(\frac{1}{m_{t,i}^{*}}-\frac{1}{m_{t,i}^{-}}\right)-\frac{\alpha\hbar^2}{4}\frac{k_s^4}{m_{t,i}^{-}m_{t,i}^{+}}\right], a_5 = \left[\sqrt{\frac{m_2^*}{m_1^*}}+1\right]^2\left[4\left(\frac{m_2^*}{m_1^*}\right)^{1/2}\right]^{-1}$$

Therefore the electron dispersion law in heavily doped IV-VI, EMSLs under magnetic quantization can be written as

$$(k_z)^2 = [[1/L_0^2]\{cos^{-1}(f_{23}(E,n))\}^2 - (\frac{2eB}{\hbar}(n+\frac{1}{2}))] \tag{4.39}$$

where
$$f_{23}(E,n) = a_3\cos\left[a_0C_{23E,n}(E,n,\eta_{g1})+b_0D_{23E,n}(E,n,\eta_{g1})\right]$$
$$-a_4\cos\left[a_0C_{23E,n}(E,n,\eta_{g2})-b_0D_{23E,n}(E,n,\eta_{g2})\right],$$

$$C_{23}(E,n,\eta_{g1}) = \left[\left[2\overline{p_{9,1}}\right]^{-1}\left[-\overline{q_{9,1}}(E,n,\eta_{g1})+\left[\{\overline{q_{9,1}}(E,n,\eta_{g1})\}^2\right.\right.\right.$$
$$\left.\left.\left.+\left(4\overline{p_{9,1}}\right)\overline{R_{9,1}}(E,n,\eta_{g1})\right]^{1/2}\right]\right]^{1/2},$$
$$D_{23}(E,n,\eta_{g2}) = \left[\left[2\overline{p_{9,2}}\right]^{-1}\left[-\overline{q_{9,2}}(E,n,\eta_{g2})+\left[\{\overline{q_{9,2}}(E,n,\eta_{g2})\}^2\right.\right.\right.$$
$$\left.\left.\left.+\left(4\overline{p_{9,2}}\right)\overline{R_{9,2}}(E,n,\eta_{g2})\right]^{1/2}\right]\right]^{1/2},$$

$$\overline{q_{9,i}}(E,n,\eta_{gi}) = \left[\frac{\hbar^2}{2}\left(\frac{1}{m_{l,i}^{*}}+\frac{1}{m_{l,i}^{-}}\right)+\alpha_i\frac{\hbar^4}{4}(\frac{2eB}{\hbar}(n+\frac{1}{2}))\left(\frac{1}{m_{l,i}^{+}m_{t,i}^{-}}+\frac{1}{m_{t,i}^{+}m_{l,i}^{-}}\right)\right.$$
$$\left.-\alpha_i\gamma_3(E,\eta_{gi})\left(\frac{1}{m_{l,i}^{+}}-\frac{1}{m_{l,i}^{-}}\right)\right],$$
$$\overline{R_{9,i}}(E,n,\eta_{gi}) = \left[\gamma_2(E,\eta_{gi})+\gamma_3(E,\eta_{gi})\alpha_i\frac{\hbar^2}{2}(\frac{2eB}{\hbar}(n+\frac{1}{2}))\left(\frac{1}{m_{t,i}^{+}}-\frac{1}{m_{t,i}^{-}}\right)\right.$$
$$\left.-\frac{\hbar^2}{2}(\frac{2eB}{\hbar}(n+\frac{1}{2}))\left(\frac{1}{m_{t,i}^{*}}-\frac{1}{m_{t,i}^{-}}\right)-\frac{\alpha\hbar^6}{4}\frac{(\frac{2eB}{\hbar}(n+\frac{1}{2}))^2}{m_{t,i}^{-}m_{t,i}^{+}}\right],$$

The energy-wave vector dispersion relation of the conduction electrons in heavily doped quantum well strained layer SLs with graded interfaces can be expressed as

$$\left(\frac{\pi n_z}{d_z}\right)^2 = \frac{1}{L_0^2}\left[\cos^{-1}\left\{\frac{1}{2}f_{23}(A3,n)\right\}\right]^2 - \frac{2|e|B}{\hbar}\left(n+\frac{1}{2}\right) \tag{4.40}$$

where $A3$ is the totally quantized energy in this case.

The electron concentration in this case is given by

$$n_{0SL} = \frac{eBg_v}{\pi\hbar}\sum_{n=0}^{n_{\max}}\sum_{n_z=1}^{n_{z\max}} F_{-1}(\eta_{48}) \tag{4.41a}$$

where $\eta_{47} = (k_B T)^{-1}(E_{FSL} - E_{47,n})$ and E_{FSL} is the fermi energy in this case.

The EP in this case(J_{SL}) is given by

$$J_{SL} = \frac{\alpha_0 e^2 Bg_v}{2\pi\hbar d_z}\sum_{n=0}^{n_{\max}}\sum_{n_{z\min}}^{n_{z\max}} F_{-1}(\eta_{48}) v_{z8}(E_{nzSL8}) \tag{4.41b}$$

where

$$v_{z8}(E_{nzSL8}) = \frac{2L_0 \sin(\frac{L_0 n_z \pi}{d_z})}{\hbar[f_{23}(E_{nzSL8},0)]'} \tag{4.41c}$$

E_{nzSL8} is the quantized energy along z direction and is obtained from the equation

$$\cos[(\frac{L_0 n_z \pi}{d_z})] = \frac{1}{2}f_{23}(E_{nzSL8},0). \tag{4.41d}$$

4.2.9 The Magneto EP from HgTe/CdTe Quantum Well HD Effective Mass Super Lattices

Following Sasaki [71], the electron dispersion law in heavily doped HgTe/CdTe EMSLs can be written as

$$k_z^2 = \Delta_{13H} + i\Delta_{14H} \tag{4.42}$$

where $\Delta_{13H} = \left[\frac{1}{L_0^2}(\Delta_{11H}^2 - \Delta_{12H}^2) - k_s^2\right]$

$$\Delta_{11H} = \cos^{-1} p_{6H}, p_{6H} = \left[\frac{1 - \Delta_{9H}^2 - \Delta_{10H}^2 - \sqrt{(1 - \Delta_{9H}^2 - \Delta_{10H}^2)^2 + 4\Delta_{10H}^2}}{2}\right]^{1/2},$$

$$\Delta_{9H} = (\overline{a_{1H}} \cos \Delta_{5H} \cosh \Delta_{6H} - \overline{a_{2H}} \cos \Delta_{7H} \cosh \Delta_{6H}),$$
$$\Delta_{10H} = (\overline{a_{1H}} \sin \Delta_{5H} \sinh \Delta_{6H} + \overline{a_{2H}} \sin \Delta_{7H} \sinh \Delta_{6H}),$$

$$\Delta_{5H} = [a_0 C_{22H}(E, k_s, \eta_{g1}) + b_0 e_3], \Delta_{6H} = b_0 e_4, \Delta_{7H} = [a_0 C_{22H}(E, K_s, \eta_{g1}) - b_0 e_3],$$
$$C_{22H}(E, k_s, \eta_{g1}) = \left[\frac{B_{01}^2 + 2A_1 E - B_{01}(B_{01}^2 + 4A_1 E)}{2A_1^2} - k_s^2\right]^{1/2},$$

$$\overline{a_{1H}} = \left[\sqrt{\frac{M_2(0, \eta_{g2})}{m_{c1}^*}} + 1\right]^2 \left[4\left(\frac{M_2(0, \eta_{g2})}{m_{c1}^*}\right)^{1/2}\right]^{-1},$$
$$\overline{a_{2H}} = \left[\sqrt{\frac{M_2(0, \eta_{g2})}{m_{c1}^*}} - 1\right]^2 \left[4\left(\frac{M_2(0, \eta_{g2})}{m_{c1}^*}\right)^{1/2}\right]^{-1}$$
$$\Delta_{12H} = \cos^{-1} p_{6H}, \Delta_{14H} = \frac{2\Delta_{11H}\Delta_{12H}}{L_0^2}$$

The electron dispersion law in heavily doped HgTe/CdTe EMSLs under magnetic quantization can be written as

$$(k_z)^2 = \Delta_{13HE,n} + i\Delta_{14HE,n} \quad (4.43)$$

where $\Delta_{13HE,n} = [(1/L_0^2)(\Delta_{11HE,n}^2 - \Delta_{12HE,n}^2) - \frac{2eB}{\hbar}(n + \frac{1}{2})]$

$$\Delta_{11HE,n} = \cos^{-1} p_{6HE,n}, p_{6HE,n} = [((1 - \Delta_{9HE,n}^2 - \Delta_{10HE,n}^2 - \sqrt{(1 - \Delta_{9HE,n}^2 - \Delta_{10HE,n}^2)^2 + 4\Delta_{10HE,n}^2})/2)]^{\frac{1}{2}}$$

$$\Delta_{9HE,n} = (\overline{a_{1H}} \cos \Delta_{5HE,n} \cosh \Delta_{6HE,n} - \overline{a_{2H}} \cos \Delta_{7HE,n} \cosh \Delta_{6HE,n}),$$
$$\Delta_{10HE,n} = (\overline{a_{1H}} \sin \Delta_{5HE,n} \sinh \Delta_{6HE,n} + \overline{a_{2H}} \sin \Delta_{7HE,n} \sinh \Delta_{6HE,n}),$$
$$\Delta_{5HE,n} = [a_0 C_{22HE,n}(E_{E,n}, \eta_{gi}) + b_0 e_3], \Delta_{6HE,n} = b_0 e_4,$$
$$\Delta_{7HE,n} = [a_0 C_{22HE,n}(E_{E,n} eta_{g1}) - b_0 e_3],$$
$$C_{22HE,n}(E_{E,n}, \eta_{g1}) = \left[\frac{B_{01}^2 + 2A_1 E_{E,n} - B_{01}(B_{01}^2 + 4A_1 E_{E,n})}{2A_1^2} - [\frac{2eB}{\hbar}(n + \frac{1}{2})]\right]^{1/2},$$
$$\Delta_{12HE,n} = \cos^{-1} p_{6HE,n}, \Delta_{14HE,n} = \frac{2\Delta_{11HE,n}\Delta_{12HE,n}}{L_0^2}$$

The dispersion relation in quantum well heavily doped III-V superlattices under magnetic quantization assumes the form

$$\left(\frac{n_z\pi}{d_z}\right)^2 = \Delta_{13,A4,n} + i\Delta_{14,A4,n} \tag{4.44}$$

where $A4$ is the totally quantized energy in this case.

The electron concentration in this case is given by

$$n_{0SL} = \frac{eBg_v}{\pi\hbar} \text{ Real part of } \sum_{n=0}^{n_{\max}} \sum_{n_z=1}^{n_{z\max}} F_{-1}(\eta_{48}) \tag{4.45a}$$

where $\eta_{48} = (\mathrm{k}_B\mathrm{T})^{-1}(\mathrm{E}_{FSL} - A4)$

The EP in this case(J_{SL}) is given by

$$J_{SL} = \frac{\alpha_0 e^2 Bg_v}{2\pi\hbar d_z} \text{ Real part of } \sum_{n=0}^{n_{\max}} \sum_{n_{z\min}}^{n_{z\max}} F_{-1}(\eta_{48}) v_{z81}(E_{nzSL81}) \tag{4.45b}$$

where

$$v_{z81}(\mathrm{E}_{nzSL81}) = \frac{2\pi n_z}{\hbar d_z(\Delta'_{13} + \mathrm{i}\Delta'_{14})|_{k_x=0,k_y=0,k_z=\frac{n_z\pi}{d_z} \text{ and } E=\mathrm{E}_{nzSL81}}} \tag{4.45c}$$

E_{nzSL81} is the quantized energy along x direction and is obtained by substituting $k_x = 0, k_y = 0, k_z = \frac{n_z\pi}{d_z}$ and $E = E_{nzSL81}$ in (4.34) and the primes denote the differentiation with respect to energy.

4.2.10 The Magneto EP from Strained Layer Quantum Well HD Effective Mass Super Lattices

The dispersion relation of the constituent materials of heavily doped III-V super lattices can be written as

$$\overline{P}_i(\mathrm{E}, \eta_{gi})k_x^2 + \overline{Q}_i(\mathrm{E}, \eta_{gi})\mathrm{k}_y^2 + \overline{S}_i(\mathrm{E}, \eta_{gi})\mathrm{k}_z^2 = 1 \tag{4.46}$$

where $\overline{P}_i(\mathrm{E}, \eta_{gi}) = (\gamma_0(\mathrm{E}, \eta_{gi}) - \mathrm{I}_0\mathrm{T}_{1i})(\overline{\Delta}_i(\mathrm{E}, \eta_{gi}))^{-1}$, $\mathrm{I}_0 = (1/2)[1 + Erf(\mathrm{E}/\eta_{gi})]$,

$$T_{1i} = [E_{gi} - C_{1i}^c\varepsilon_i - (a_i + C_{1i}^c)\varepsilon_i + (3/2)b_i\varepsilon_{xxi} - (b_i\varepsilon_i/2) + (\sqrt{3}d_i\varepsilon_{xyi}/2)],$$

$$\bar{\Delta}_i(E,\eta_{gi}) = [(-q_i\eta_{gi}^3/2\sqrt{\pi})\exp(-(E^2/\eta_{gi}^2))[1+(E^2/\eta_{gi}^2)] - R_i\theta_0(E,\eta_{gi}) + V_i\gamma_0(E,\eta_{gi}) + (\zeta_i/2)[1+Erf(E/\eta_{gi})]], q_i = (3/2B_{2i}^2), R_i = q_i[2A_i + C_{1i}^c\varepsilon_i], A_i = E_{gi} - C_{1i}^c\varepsilon_i,$$

$$V_i = q_i[A_i^2 - (2C_{2i}^2\varepsilon_{xyi}/3) + 2A_iC_{1i}^c\varepsilon_i], \varsigma_i = q_i[2C_{2i}^2\varepsilon_{xyi}/3) - C_{1i}^c\varepsilon_iA_i^2],$$
$$\bar{Q}_i(E,\eta_{gi}) = (\gamma_0(E,\eta_{gi}) - I_0T_{2i})(\bar{\Delta}_i(E,\eta_{gi}))^{-1}, T_{2i} = [E_{gi} - C_{1i}^c\varepsilon_i - (a_i + C_{1i}^c)\varepsilon_i + (3/2)b_i\varepsilon_{xxi} - (b_i\varepsilon_i/2) - (\sqrt{3}d_i\varepsilon_{xyi}/2)],\ \bar{S}_i(E,\eta_{gi}) = \gamma_0(E,\eta_{gi}) - I_0T_{3i})(\bar{\Delta}_i(E,\eta_{gi}))^{-1},$$
$$T_{3i} = [E_{gi} - C_{1i}^c\varepsilon_i + (a_i + C_{1i}^c)\varepsilon_i + (3/2)b_i\varepsilon_{zzi} - (b_i\varepsilon_i/2)],$$

The electron energy spectrum in heavily doped strained layer effective mass super-lattices can be written as

$$(k_z^2) = \left[\frac{1}{L_0^2}\{\cos^{-1}(f_{40}(E,k_x,k_y))\}^2 - k_s^2\right] \tag{4.47}$$

where $f_{40}(E,k_x,k_y) = a_{20}\cos\left[a_0C_{40}(E,k_x,k_y,\eta_{g1}) + b_0D_{40}(E,k_x,k_y,\eta_{g1})\right] - a_{21}\cos\left[a_0C_{40}(E,k_x,k_y,\eta_{g2}) - b_0D_{40}(E,k_x,k_y,\eta_{g2})\right]$,

$$a_{20} = \left[\sqrt{\frac{M_{s2}(0,\eta_{g2})}{M_{s1}(0,\eta_{g1})}} + 1\right]^2\left[4\left(\frac{M_{s2}(0,\eta_{g2})}{M_{s1}(0,\eta_{g1})}\right)^{1/2}\right]^{-1},$$

$$M_{si}(0,\eta_{gi}) = (\hbar/2)\rho_i(\eta_{gi})$$

$$\rho(\eta_{gi}) = [(\eta_{gi}/2\sqrt{\pi}) - (T_{3i}/2)]^{-2} \times [\{(\eta_{gi}/2\sqrt{\pi}) - (T_{3i}/2)\}\{(V_i/2) - (R_i\eta_{gi}/\sqrt{\pi}) + (\varsigma_i + \eta_{gi}\sqrt{\pi})\} - ((1/2) - (T_{3i}/\eta_{gi}\sqrt{\pi}))\{(\varsigma_i/2) + (V_i\eta_{gi}/2\sqrt{\pi}) - (R_i\eta_{gi}^2/4) - (q_i\eta_{gi}^3/2\sqrt{\pi})\}]$$

$$a_{20} = \left[\sqrt{\frac{M_{s2}(0,\eta_{g2})}{M_{s1}(0,\eta_{g1})}} + 1\right]^2\left[4\left(\frac{M_{s2}(0,\eta_{g2})}{M_{s1}(0,\eta_{g1})}\right)^{1/2}\right]^{-1},$$
$$a_{21} = \left[\sqrt{\frac{M_{s2}(0,\eta_{g2})}{M_{s1}(0,\eta_{g1})}} - 1\right]^2\left[4\left(\frac{M_{s2}(0,\eta_{g2})}{M_{s1}(0,\eta_{g1})}\right)^{1/2}\right]^{-1}$$
$$C_{40}(E,k_x,k_y,\eta_{g1}) = [1 - \bar{P}_1(E,\eta_{g1})k_x^2 - \bar{Q}_1(E,\eta_{g1})k_y^2]^{1/2}[\bar{S}_1(E,\eta_{g1})]^{-1/2}$$
$$D_{40}(E,k_x,k_y,\eta_{g2}) = [1 - \bar{P}_2(E,\eta_{g2})k_x^2 - \bar{Q}_1(E,\eta_{g2})k_y^2]^{1/2}[\bar{S}_2(E,\eta_{g2})]^{-1/2}$$

Therefore, the electron dispersion law in heavily doped strained layer effective mass quantum dot super-lattices can be expressed as

$$(k_z)^2 = [\frac{1}{L_0^2}\{\cos^{-1}(\mathrm{f}_{40}(\mathrm{E},\mathrm{n}))\}^2 - (\frac{2eB}{\hbar}(\mathrm{n}+\frac{1}{2}))] \tag{4.48}$$

where
$$\begin{aligned} f_{40}(E,n) &= a_{20}\cos\left[a_0 C_{40}(E,n,\eta_{g1}) + b_0 D_{40}(E,n,\eta_{g1})\right] \\ &\quad - a_{21}\cos\left[a_0 C_{40}(E,n,\eta_{g2}) - b_0 D_{40}(E,n,\eta_{g2})\right], \end{aligned}$$

$$C_{40}(E,n,\eta_{g1}) = [1 - \frac{\hbar eB}{\phi_{50}(E,\eta_{g1})}(n+\frac{1}{2})]^{1/2}[\bar{S}_1(E,\eta_{g1})]^{-1/2},$$

$$\phi_{50}(E,\eta_{g1}) = \sqrt{\psi_{50}(E,\eta_{g1})\psi_{51}(E,\eta_{g1})},$$

$$\psi_{50}(E,\eta_{g1}) = \frac{\hbar^2}{2\bar{P}_1(E,\eta_{g1})}, \psi_{51}(E,\eta_{g1}) = \frac{\hbar^2}{2\bar{Q}_1(E,\eta_{g1})}$$

$$D_{40}(E,n,\eta_{g2}) = [1 - \frac{\hbar eB}{\phi_{501}(E,\eta_{g2})}(n+\frac{1}{2})]^{1/2}[\bar{S}_2(E,\eta_{g2})]^{-1/2}$$

$$\phi_{501}(E,\eta_{g2}) = \sqrt{\psi_{501}(E,\eta_{g2})\psi_{511}(E,\eta_{g2})}$$

$$\psi_{501}(E,\eta_{g2}) = \frac{\hbar^2}{2\bar{P}_2(E,\eta_{g2})}, \psi_{511}(E,\eta_{g2}) = \frac{\hbar^2}{2\bar{Q}_2(E,\eta_{g2})}$$

The energy-wave vector dispersion relation of the conduction electrons in heavily doped quantum well strained layer SLs with graded interfaces can be expressed as

$$\left(\frac{\pi n_z}{d_z}\right)^2 = \frac{1}{L_0^2}\left[\cos^{-1}\left\{\frac{1}{2}f_{40}(A8,n)\right\}\right]^2 - \frac{2|e|B}{\hbar}\left(n+\frac{1}{2}\right) \tag{4.49}$$

where $A8$ is the totally quantized energy in this case.

The electron concentration in this case is given by

$$n_{0SL} = \frac{eBg_v}{\pi\hbar}\sum_{n=0}^{n_{\max}}\sum_{n_z=1}^{n_{z\max}} F_{-1}(\eta_{50}) \tag{4.50a}$$

where $\eta_{50} = (k_B T)^{-1}(E_{FSL} - A8)$

The EP in this case(J_{SL}) is given by

$$J_{SL} = \frac{\alpha_0 e^2 B g_v}{2\pi\hbar d_z}\sum_{n=0}^{n_{\max}}\sum_{n_{z\min}}^{n_{z\max}} F_{-1}(\eta_{50})v_{z50}(\mathrm{E}_{nzSL50}) \tag{4.50b}$$

where $\nu_{z8}(\mathrm{E}_{nzSL50}) = \frac{2L_0 \sin(\frac{L_0 n_z \pi}{d_z})}{\hbar[f_{40}(E_{nzSL50},0)]'}$, E_{nzSL50} is the quantized energy along z direction and is obtained from the equation

$$\cos[(\frac{L_0 n_z \pi}{d_z})] = \frac{1}{2} f_{40}(\mathrm{E}_{nzSL50}, 0). \tag{4.50c}$$

4.2.11 The EP from III-V Quantum Dot HD Superlattices with Graded Interfaces

The electron concentration in this case is given by

$$n_{0QDSL} = \frac{2g_v}{d_x d_y d_z} \text{ Real part of } \sum_{n_x=1}^{n_{x\max}} \sum_{n_y=1}^{n_{y\max}} \sum_{n_z=1}^{n_{z\max}} F_{-1}(\eta_{100}) \tag{4.51}$$

where $\eta_{100} \equiv (k_B T)^{-1}(E_{FQDHDSL} - \bar{e}_{11})$ *and* E_{FDHDSL} *is the Fermi energy in this case and and* $\bar{e}_{11}$ *is the totally quantized energy which can be obtained by substituting* $k_x = \frac{n_x \pi}{d_x}, k_y = \frac{n_y \pi}{d_y}, k_z = \frac{n_z \pi}{d_z}$ *and* $E = \bar{e}_{11}$ in (4.2).

The electron density per sub-band assumes the form

$$n_{0QDSL} = \frac{2g_v}{d_x d_y d_z} \text{ Real part of } [F_{-1}(\eta_{100})] \tag{4.52}$$

The EP in this case(J_{SL}) is given by

$$J_{QDSL} = \frac{\alpha_0 e g_v}{d_x d_y d_z} \text{ Real part of } \sum_{n_x=1}^{n_{x\max}} \sum_{n_y=1}^{n_{y\max}} \sum_{n_{z\min}}^{n_{z\max}} F_{-1}(\eta_{100}) \nu_{z1}(E_{nzSL1}). \tag{4.53}$$

4.2.12 The EP from II-VI Quantum Dot HD Superlattices with Graded Interfaces

The electron concentration in this case is given by

$$n_{0QDSL} = \frac{2g_v}{d_x d_y d_z} \text{ Real part of } \sum_{n_x=1}^{n_{x\max}} \sum_{n_y=1}^{n_{y\max}} \sum_{n_z=1}^{n_{z\max}} F_{-1}(\eta_{101}) \tag{4.54}$$

where $\eta_{101} \equiv (k_B T)^{-1}(E_{FQDHDSL} - \bar{e}_{21})$ *and* $\bar{e}_{21}$ *is the totally quantized energy which can be obtained by substituting* $k_x = \frac{n_x \pi}{d_x}, k_y = \frac{n_y \pi}{d_y}, k_z = \frac{n_z \pi}{d_z}$ *and* $E = \bar{e}_{21}$ in (4.9).

The electron density per sub-band assumes the form

$$n_{0QDSL} = \frac{2g_v}{d_x d_y d_z} \text{ Real part of } [F_{-1}(\eta_{101})] \tag{4.55}$$

The EP in this case(J_{SL}) is given by

$$J_{QDSL} = \frac{\alpha_0 e g_v}{d_x d_y d_z} \text{ Real part of } \sum_{n_x=1}^{n_{xmax}} \sum_{n_y=1}^{n_{ymax}} \sum_{n_{zmin}}^{n_{zmax}} F_{-1}(\eta_{101}) v_{z2}(\mathrm{E}_{nzSL2}) \tag{4.56}$$

4.2.13 The EP from IV-VI Quantum Dot HD Superlattices with Graded Interfaces

The electron concentration in this case is given by

$$n_{0QDSL} = \frac{2g_v}{d_x d_y d_z} \text{ Real part of } \sum_{n_x=1}^{n_{x_{max}}} \sum_{n_y=1}^{n_{y_{max}}} \sum_{n_z=1}^{n_{z_{max}}} F_{-1}(\eta_{102}) \tag{4.57}$$

where $\eta_{102} \equiv (k_B T)^{-1}(E_{FQDHDSL} - \bar{e}_{31})$ *and* $\bar{e}_{31}$ *is the totally quantized energy which can be obtained by substituting* $k_x = \frac{n_x \pi}{d_x}, k_y = \frac{n_y \pi}{d_y}, k_z = \frac{n_z \pi}{d_z}$ *and* $E = \bar{e}_{31}$ in (4.15).

The electron density per sub-band assumes the form

$$n_{0QDSL} = \frac{2g_v}{d_x d_y d_z} \text{ Real part of } [F_{-1}(\eta_{102})] \tag{4.58}$$

The EP in this case(J_{SL}) is given by

$$J_{QDSL} = \frac{\alpha_0 e g_v}{d_x d_y d_z} \text{ Real part of } \sum_{n_x=1}^{n_{xmax}} \sum_{n_y=1}^{n_{ymax}} \sum_{n_{zmin}}^{n_{zmax}} F_{-1}(\eta_{102}) v_{z3}(\mathrm{E}_{nzSL3}) \tag{4.59}$$

4.2.14 The EP from HgTe/CdTe Quantum Dot HD Superlattices with Graded Interfaces

The electron concentration in this case is given by

$$n_{0QDSL} = \frac{2g_v}{d_x d_y d_z} \text{ Real part of } \sum_{n_x=1}^{n_{x_{\max}}} \sum_{n_y=1}^{n_{y_{\max}}} \sum_{n_z=1}^{n_{z_{\max}}} F_{-1}(\eta_{104}) \tag{4.60}$$

where $\eta_{104} \equiv (k_B T)^{-1}(E_{FQDHDSL} - \bar{e}_{41})$ *and* $\bar{e}_{31}$ *is the totally quantized energy which can be obtained by substituting* $k_x = \frac{n_x \pi}{d_x}, k_y = \frac{n_y \pi}{d_y}$, $k_z = \frac{n_z \pi}{d_z}$ *and* $E = \bar{e}_{41}$ in (4.20).

The electron density per sub-band assumes the form

$$n_{0QDSL} = \frac{2g_v}{d_x d_y d_z} \text{ Real part of } [F_{-1}(\eta_{104})] \tag{4.61}$$

The EP in this case(J_{SL}) is given by

$$J_{QDSL} = \frac{\alpha_0 e g_v}{d_x d_y d_z} \text{ Real part of } \sum_{n_x=1}^{n_{x\max}} \sum_{n_y=1}^{n_{y\max}} \sum_{n_{z\min}}^{n_{z\max}} F_{-1}(\eta_{104}) v_{z4}(\mathrm{E}_{nzSL4}). \tag{4.62}$$

4.2.15 The EP from Strained Layer Quantum Dot HD Superlattices with Graded Interfaces

The electron concentration in this case is given by

$$n_{0QDSL} = \frac{2g_v}{d_x d_y d_z} \text{ Real part of } \sum_{n_x=1}^{n_{x_{\max}}} \sum_{n_y=1}^{n_{y_{\max}}} \sum_{n_z=1}^{n_{z_{\max}}} F_{-1}(\eta_{105}) \tag{4.63}$$

where $\eta_{105} \equiv (k_B T)^{-1}(E_{FQDHDSL} - \bar{e}_{51})$ *and* $\bar{e}_{51}$ *is the totally quantized energy which can be obtained by substituting* $k_x = \frac{n_x \pi}{d_x}, k_y = \frac{n_y \pi}{d_y}$, $k_z = \frac{n_z \pi}{d_z}$ *and* $E = \bar{e}_{51}$ in (4.25).

The electron density per sub-band assumes the form

$$n_{0QDSL} = \frac{2g_v}{d_x d_y d_z} \text{ Real part of } [F_{-1}(\eta_{105})] \tag{4.64}$$

The EP in this case(J_{SL}) is given by

$$J_{QDSL} = \frac{\alpha_0 e g_v}{d_x d_y d_z} \text{ Real part of } \sum_{n_x=1}^{n_{xmax}} \sum_{n_y=1}^{n_{ymax}} \sum_{n_{zmin}}^{n_{zmax}} F_{-1}(\eta_{105}) v_{z7}(\mathrm{E}_{nzSL7}). \quad (4.65)$$

4.2.16 The EP from III-V Quantum Dot HD Effective Mass Superlattices

The electron concentration in this case is given by

$$n_{0QDSL} = \frac{2g_v}{d_x d_y d_z} \text{ Real part of } \sum_{n_x=1}^{n_{x_{max}}} \sum_{n_y=1}^{n_{y_{max}}} \sum_{n_z=1}^{n_{z_{max}}} F_{-1}(\eta_{106}) \quad (4.66)$$

where $\eta_{106} \equiv (k_B T)^{-1}(E_{FQDHDSL} - \bar{e}_{61})$ *and* $\bar{e}_{61}$ *is the totally quantized energy which can be obtained by substituting* $k_x = \frac{n_x \pi}{d_x}, k_y = \frac{n_y \pi}{d_y}$, $k_z = \frac{n_z \pi}{d_z}$ *and* $E = \bar{e}_{61}$ in (4.30).

The electron density per sub-band assumes the form

$$n_{0QDSL} = \frac{2g_v}{d_x d_y d_z} \text{ Real part of } [F_{-1}(\eta_{106})] \quad (4.67)$$

The EP in this case(J_{SL}) is given by

$$J_{QDSL} = \frac{\alpha_0 e g_v}{d_x d_y d_z} \text{ Real part of } \sum_{n_x=1}^{n_{xmax}} \sum_{n_y=1}^{n_{ymax}} \sum_{n_{zmin}}^{n_{zmax}} F_{-1}(\eta_{106}) v_{z5}(E_{nzSL5}). \quad (4.68)$$

4.2.17 The EP from Heavily Doped Effective Mass Quantum Dot II-VI Super-Lattices

The electron concentration in this case is given by

$$n_{0QDSL} = \frac{2g_v}{d_x d_y d_z} \text{ Real part of } \sum_{n_x=1}^{n_{x_{max}}} \sum_{n_y=1}^{n_{y_{max}}} \sum_{n_z=1}^{n_{z_{max}}} F_{-1}(\eta_{107}) \quad (4.69)$$

where $\eta_{107} \equiv (k_B T)^{-1}(E_{FQDHDSL} - \bar{e}_{71})$ and $\bar{e}_{71}$ *is the totally quantized energy which can be obtained by substituting*
$k_x = \frac{n_x \pi}{d_x}, k_y = \frac{n_y \pi}{d_y}$, $k_z = \frac{n_z \pi}{d_z}$ *and* $E = \bar{e}_{71}$ in (4.34).

The electron density per sub-band assumes the form

$$n_{0QDSL} = \frac{2g_v}{d_x d_y d_z} \text{ Real part of } [F_{-1}(\eta_{107})] \tag{4.70}$$

The EP in this case(J_{SL}) is given by

$$J_{QDSL} = \frac{\alpha_0 e g_v}{d_x d_y d_z} \text{ Real part of } \sum_{n_x=1}^{n_{xmax}} \sum_{n_y=1}^{n_{ymax}} \sum_{n_{zmin}}^{n_{zmax}} F_{-1}(\eta_{107}) v_{z6}(\mathrm{E}_{nzSL6}). \tag{4.71}$$

4.2.18 The EP from Heavily Doped Effective Mass Quantum Dot IV-VI Super-Lattices

The electron concentration in this case is given by

$$n_{0QDSL} = \frac{2g_v}{d_x d_y d_z} \text{ Real part of } \sum_{n_x=1}^{n_{x_{max}}} \sum_{n_y=1}^{n_{y_{max}}} \sum_{n_z=1}^{n_{z_{max}}} F_{-1}(\eta_{108}) \tag{4.72}$$

where $\eta_{108} \equiv (k_B T)^{-1}(E_{FQDHDSL} - \bar{e}_{81})$ *and* $\bar{e}_{81}$ *is the totally quantized energy which can be obtained by substituting* $k_x = \frac{n_x \pi}{d_x}, k_y = \frac{n_y \pi}{d_y}$, $k_z = \frac{n_z \pi}{d_z}$ *and* $E = \bar{e}_{81}$ in (4.38).

The electron density per sub-band assumes the form

$$n_{0QDSL} = \frac{2g_v}{d_x d_y d_z} \text{ Real part of } [F_{-1}(\eta_{108})] \tag{4.73}$$

The EP in this case(J_{SL}) is given by

$$J_{QDSL} = \frac{\alpha_0 e g_v}{d_x d_y d_z} \text{ Real part of } \sum_{n_x=1}^{n_{xmax}} \sum_{n_y=1}^{n_{ymax}} \sum_{n_{zmin}}^{n_{zmax}} F_{-1}(\eta_{108}) v_{z8}(E_{nzSL8}). \tag{4.74}$$

4.2.19 The EP from Heavily Doped Effective Mass HgTe/CdTe Quantum Dot Super-Lattices

The electron concentration in this case is given by

$$n_{0QDSL} = \frac{2g_v}{d_x d_y d_z} \text{ Real part of } \sum_{n_x=1}^{n_{x_{max}}} \sum_{n_y=1}^{n_{y_{max}}} \sum_{n_z=1}^{n_{z_{max}}} F_{-1}(\eta_{109}) \tag{4.75}$$

where $\eta_{109} \equiv (k_B T)^{-1}(E_{FQDHDSL} - \bar{e}_{91})$ *and* $\bar{e}_{91}$ *is the totally quantized energy which can be obtained by substituting* $k_x = \frac{n_x \pi}{d_x}, k_y = \frac{n_y \pi}{d_y}, k_z = \frac{n_z \pi}{d_z}$ *and* $E = \bar{e}_{91}$ in (4.42).

The electron density per sub-band assumes the form

$$n_{0QDSL} = \frac{2g_v}{d_x d_y d_z} \text{ Real part of } [F_{-1}(\eta_{109})] \tag{4.76}$$

The EP in this case(J_{SL}) is given by

$$J_{QDSL} = \frac{\alpha_0 e g_v}{d_x d_y d_z} \text{ Real part of } \sum_{n_x=1}^{n_{xmax}} \sum_{n_y=1}^{n_{ymax}} \sum_{n_{zmin}}^{n_{zmax}} F_{-1}(\eta_{109}) v_{z81}(\mathrm{E}_{nzSL81}). \tag{4.77}$$

4.2.20 The EP from Heavily Doped Strained Layer Effective Mass Quantum Dot Super-Lattices

The electron concentration in this case is given by

$$n_{0QDSL} = \frac{2g_v}{d_x d_y d_z} \sum_{n_x=1}^{n_{x_{max}}} \sum_{n_y=1}^{n_{y_{max}}} \sum_{n_z=1}^{n_{z_{max}}} F_{-1}(\eta_{110}) \tag{4.78}$$

where $\eta_{110} \equiv (k_B T)^{-1}(E_{FQDHDSL} - \bar{e}_{110})$ *and* $\bar{e}_{110}$ *is the totally quantized energy which can be obtained by substituting* $k_x = \frac{n_x \pi}{d_x}, k_y = \frac{n_y \pi}{d_y}, k_z = \frac{n_z \pi}{d_z}$ *and* $E = \bar{e}_{110}$ in (4.42).

The electron density per sub-band assumes the form

$$n_{0QDSL} = \frac{2g_v}{d_x d_y d_z} [F_{-1}(\eta_{110})] \tag{4.79}$$

The EP in this case(J_{SL}) is given by

$$J_{QDSL} = \frac{\alpha_0 e g_v}{d_x d_y d_z} \sum_{n_x=1}^{n_{xmax}} \sum_{n_y=1}^{n_{ymax}} \sum_{n_{zmin}}^{n_{zmax}} F_{-1}(\eta_{110}) v_{z50}(E_{nzSL50}). \tag{4.80}$$

4.3 Results and Discussion

Using the appropriate equations in Figs. 4.1, 4.2, 4.3 and 4.4, the normalized EP from QW HD III-V SLs (taking $GaAs/Ga_{1-x}Al_xAs$ and $In_xGa_{1-x}As/InP$ QW HD SLs) with graded interfaces under the magnetic quantization has been plotted as functions of the inverse quantizing magnetic field, normalized electron degeneracy, film thickness and the normalized incident photon energy respectively. It appears from Fig. 4.1 that the EP in this case oscillates with inverse quantizing magnetic field due to SdH effect. The Fig. 4.2 exhibits the fact that the EP increases with increasing carrier degeneracy in an oscillatory way and the nature of oscillations is different as compared with Fig. 4.1. From Fig. 4.3, it can be inferred that the EP oscillates with film thickness and for certain values of film thickness the EP exhibits very large values. From Fig. 4.4, it appears that EP increases with increasing photon energy in quantum steps. The plot of the normalized magneto EP from II-VI HD QWSLs (taking CdS/ZnSe QW HD SL as an example) with graded interfaces as a function of inverse quantizing magnetic field has been shown in the curve (b) of Fig. 4.5 where the plot (a) has been drawn with $\bar{\lambda}_0 = 0$ for the purpose of assessing the splitting of the two spin states by the spin orbit coupling and the crystalline field on the magneto EP in this case. The plot (c) of Fig. 4.5 has been drawn for HgTe/CdTe QW HD SL whereas the plot (d) is valid for IV-VI QW HD SL (using PbSe/PbTe as an example). The Figs. 4.6, 4.7 and 4.8 demonstrate the plots of the EP as functions of normalized electron degeneracy, film thickness and normalized incident photon energy respectively for all the cases of Fig. 4.5. The plot of the normalized magneto EP for III-V QW effective mass HD SLs (taking HD $GaAs/Ga_{1-x}Al_xAs$ as an example) as a

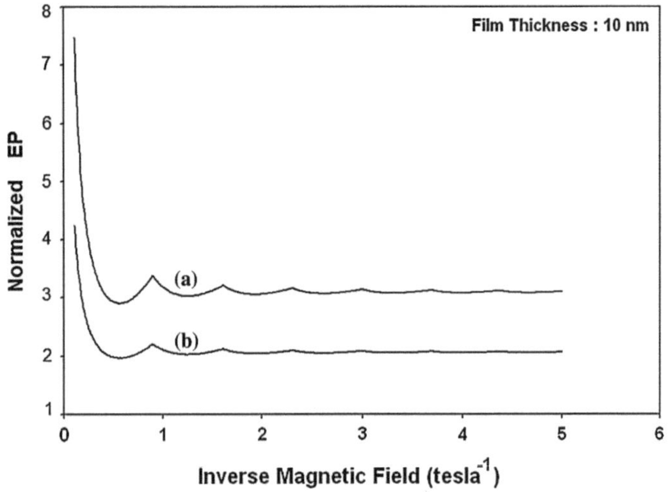

Fig. 4.1 Plot of the normalized EP from **a** $GaAs/Ga_{1-x}Al_xAs$ and **b** $In_xGa_{1-x}As/InP$ QWHD superlattices with graded interfaces as a function of inverse quantizing magnetic field

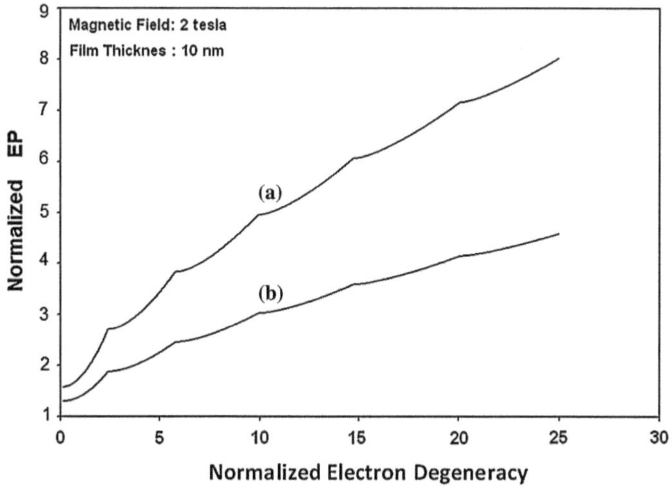

Fig. 4.2 Plot of the normalized EP from **a** $GaAs/Ga_{1-x}Al_xAs$ and **b** $In_xGa_{1-x}As/InP$ QWHD superlattices with graded interfaces under quantizing magnetic field as a function of normalized electron degeneracy

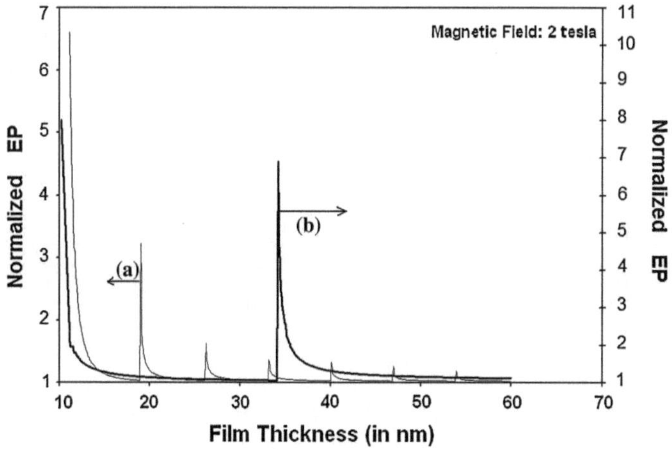

Fig. 4.3 Plot of the normalized EP from **a** $GaAs/Ga_{1-x}Al_xAs$ and **b** $In_xGa_{1-x}As/InP$ QWHD superlattices with graded interfaces under quantizing magnetic field as a function of film thickness

function of inverse quantizing magnetic field has been shown in the curve (a) of Fig. 4.9. The plots (b), (c) and (d) in the same figure have been drawn for II-VI QW effective mass HD SL (taking CdS/ZnSe HD SL as an example, IV-VI QW effective mass HD SL (taking PbSe/PbTe HD SL as an example) and HgTe/CdTe QW HD

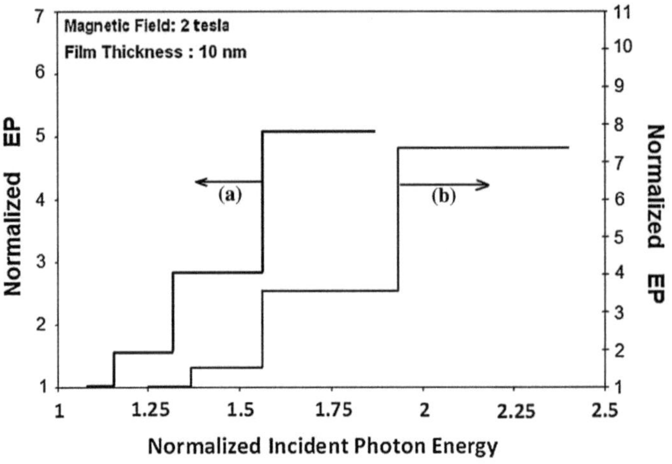

Fig. 4.4 Plot of the normalized EP from **a** $GaAs/Ga_{1-x}Al_xAs$ and **b** $In_xGa_{1-x}As/InP$ QWHD superlattices with graded interfaces under quantizing magnetic field as a function of normalized incident photon energy

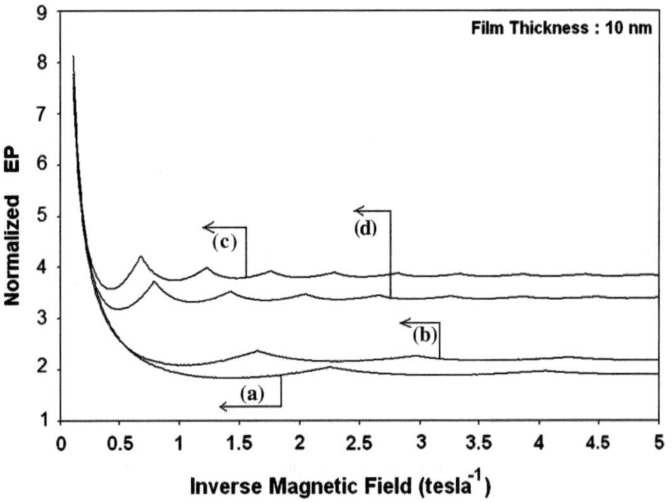

Fig. 4.5 Plot of the normalized EP from **a** CdS/ZnSe with $\bar{\lambda}_o = 0$, **b** CdS/ZnSe with $\bar{\lambda}_o \neq 0$ **c** HgTe/CdTe and **d** PbSe/PbTe QWHD superlattices with graded interfaces as a function of inverse magnetic field

effective mass SL respectively. The plots for normalized EP as functions of normalized electron degeneracy, film thickness and normalized incident photon energy for all the cases of Fig. 4.9 has respectively been drawn in the Figs. 4.10, 4.11 and 4.12. The plot (a) of Fig. 4.13 exhibits the variation of the normalized EP as a

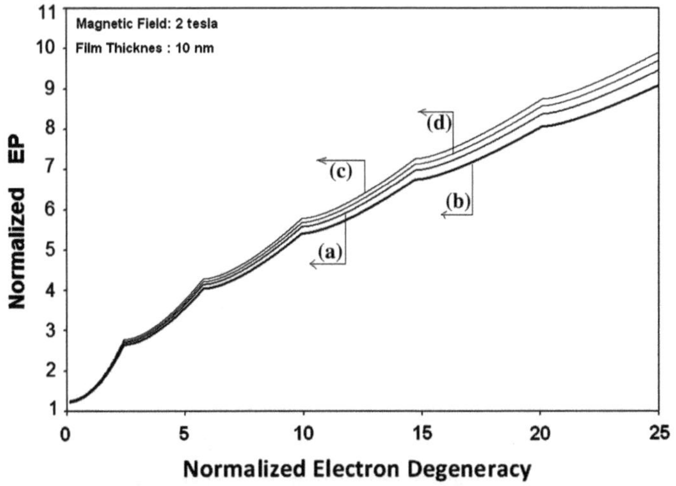

Fig. 4.6 Plot of the normalized EP from **a** CdS/ZnSe with $\bar{\lambda}_o = 0$, **b** CdS/ZnSe with $\bar{\lambda}_o \neq 0$ **c** HgTe/CdTe and **d** PbSe/PbTe QWHD superlattices with graded interfaces under quantizing magnetic field as a function of normalized electron degeneracy

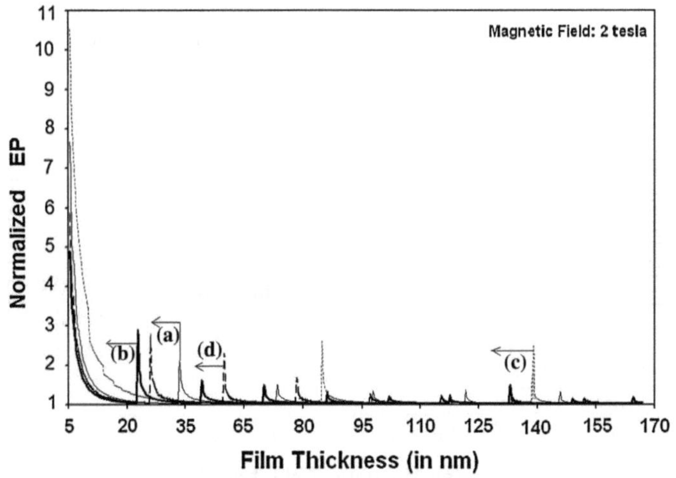

Fig. 4.7 Plot of the normalized EP from **a** CdS/ZnSe with $\bar{\lambda}_o = 0$, **b** CdS/ZnSe with $\bar{\lambda}_o \neq 0$ **c** HgTe/CdTe and **d** PbSe/PbTe QWHD superlattices with graded interfaces under quantizing magnetic fieldas a function of film thickness

function of film thickness for QB HD SLs with graded interfaces of (a) HgTe/CdTe, (b) HgTe/$Hg_{1-x}Cd_xTe$ (an example of III-V QB HD SLs, (c) CdS/ZnSe (an example of II-VI QB HD SLs with $\bar{\lambda}_0 \neq 0$) and (d) PbSe/PbTe (an example of IV-VI QB HD SLs respectively. The Figs. 4.14 and 4.15 demonstrate the plots for normalized EP as

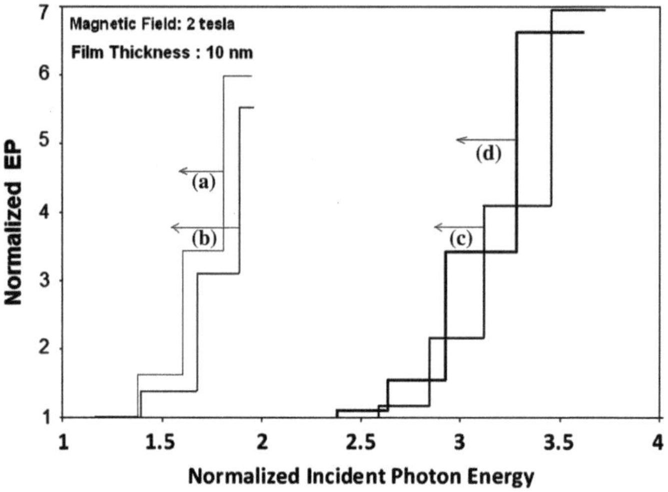

Fig. 4.8 Plot of the normalized EP from **a** CdS/ZnSe with $\bar{\lambda}_o = 0$, **b** CdS/ZnSe with $\bar{\lambda}_o \neq 0$ **c** HgTe/CdTe and **d** PbSe/PbTe QWHD superlattices with graded interfaces under quantizing magnetic field as a function of normalized incident photon energy

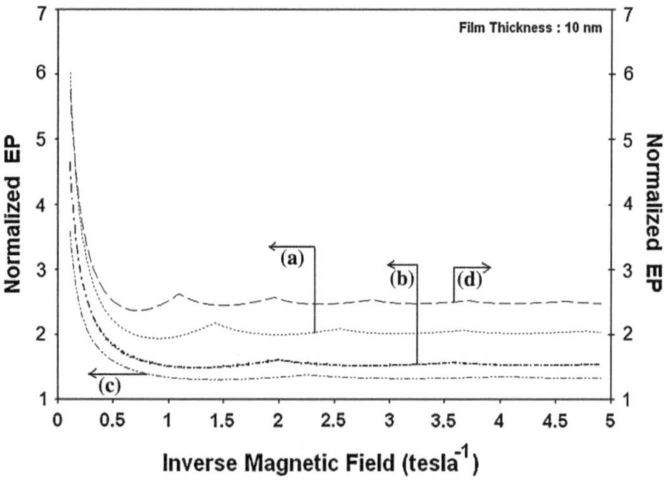

Fig. 4.9 Plot of the normalized EP from **a** $GaAs/Ga_{1-x}Al_xAs$, **b** CdS/ZnSe with $\bar{\lambda}_o \neq 0$, (c) PbSe/PbTe and (d) HgTe/CdTe QW effective mass HD superlattices as a function of inverse quantizing magnetic field

functions of normalized electron degeneracy and the normalized incident photon energy respectively for all the cases of Fig. 4.13. The plot (a) of Fig. 4.16 exhibits the variation of the normalized EP as a function of film thickness for QB effective HD SLs of (a) HgTe/CdTe, (b) $HgTe/Hg_{1-x}Cd_xTe$ (an example of III-V HD SLs, (c)

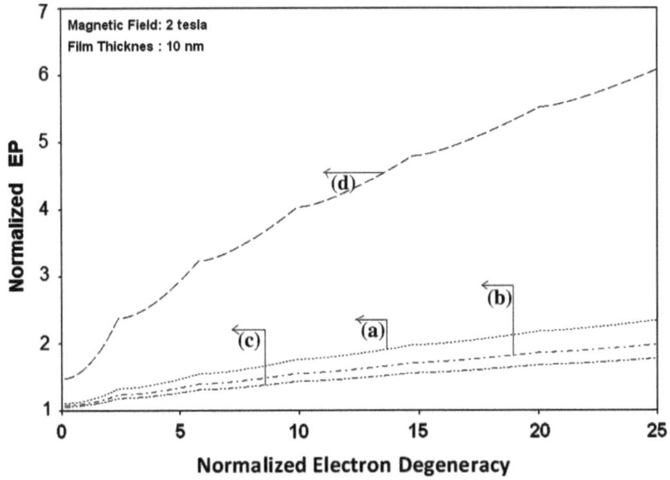

Fig. 4.10 Plot of the normalized EP from **a** $GaAs/Ga_{1-x}Al_xAs$, **b** CdS/ZnSe with $\bar{\lambda}_o \neq 0$, **c** PbSe/PbTe and **d** HgTe/CdTe QW effective mass HD superlattices under quantizing magnetic field as a function of normalized electron degeneracy

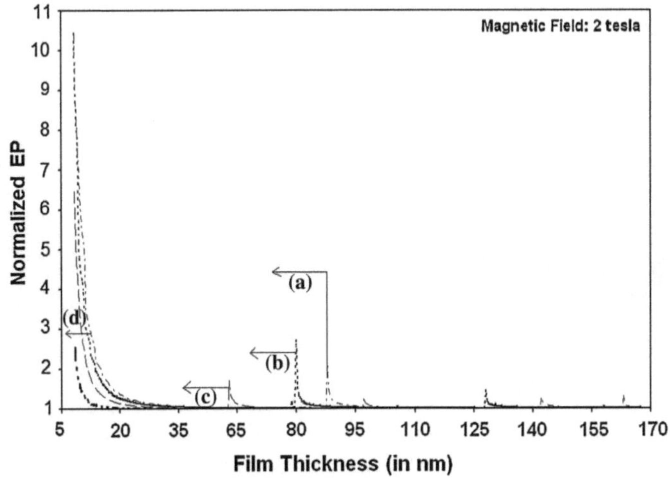

Fig. 4.11 Plot of the normalized EP from **a** $GaAs/Ga_{1-x}Al_xAs$, **b**CdS/ZnSe with $\bar{\lambda}_o \neq 0$, **c** PbSe/PbTe and **d** HgTe/CdTe QW effective mass HD superlattices under quantizing magnetic field as a function of film thickness

CdS/ZnSe (an example of II-VI QB HD SLs with $\bar{\lambda}_0 \neq 0$) and (d) PbSe/PbTe (an example of IV-VI QB HD SLs respectively.

The Figs. 4.17 and 4.18 exhibit the plots for EP as functions of normalized electron degeneracy and the normalized incident photon energy respectively for all

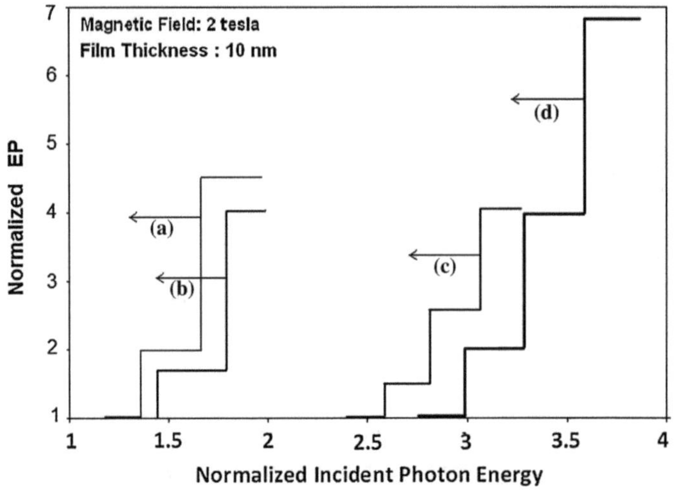

Fig. 4.12 Plot of the normalized EP from **a** GaAs/$Ga_{1-x}Al_xAs$, **b** CdS/ZnSe with $\bar{\lambda}_o \neq 0$, **c** PbSe/PbTe and **d** HgTe/CdTe QW effective mass HD superlattices under quantizing magnetic field as a function of normalized incident photon energy

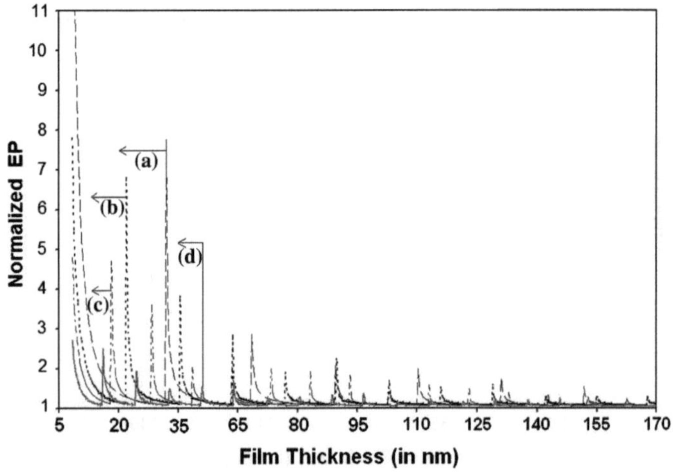

Fig. 4.13 Plot of the normalized EP from **a** HgTe/CdTe **b** HgTe/$Hg_{1-x}Cd_xTe$, **c** CdS/ZnSe with $\bar{\lambda}_o \neq 0$ and **d** PbSe/PbTe QB HD superlattices with graded interfaces as a function of film thickness

the cases of Fig. 4.16. The nature of variation of the plots in the different types of HD SLs under different physical conditions as shown from Figs. 4.5, 4.6, 4.7, 4.8, 4.9, 4.10, 4.11, 4.12, 4.13, 4.14, 4.15, 4.16, 4.17 and 4.18 have already been discussed in describing the plots of Figs. 4.1, 4.2, 4.3 and 4.4. Finally, it is logical

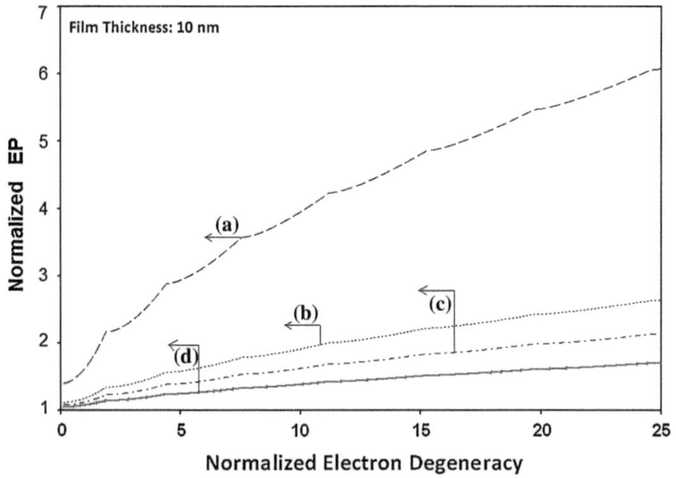

Fig. 4.14 Plot of the normalized EP from **a** HgTe/CdTe **b** HgTe/$Hg_{1-x}Cd_xTe$, **c** CdS/ZnSe with $\bar{\lambda}_o \neq 0$ and **d** PbSe/PbTe QB HD superlattices with graded interfaces as a function of normalized electron degeneracy

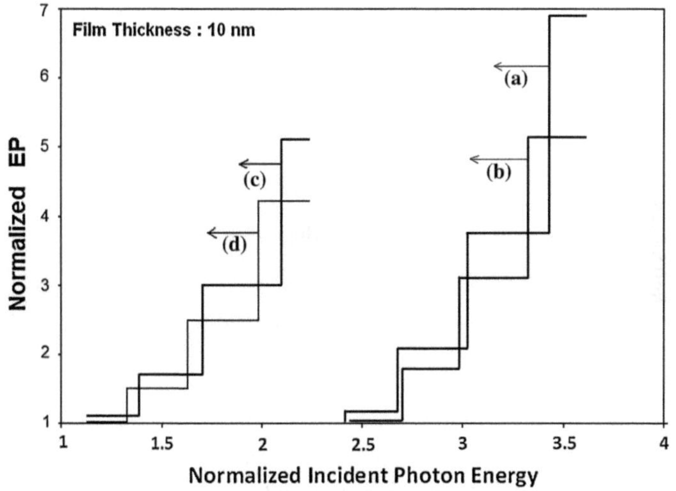

Fig. 4.15 Plot of the normalized EP from **a** HgTe/CdTe **b** HgTe/$Hg_{1-x}Cd_xTe$, **c** CdS/ZnSe with $\bar{\lambda}_o \neq 0$ and **d** PbSe/PbTe QB HD superlattices with graded interfaces as a function of normalized incident photon energy

to conclude that the numerical values of the EP are totally different in all cases which exhibit the signature of the respective band structure of HD SL under different physical conditions and the rates of variation are again totally energy spectrum dependent.

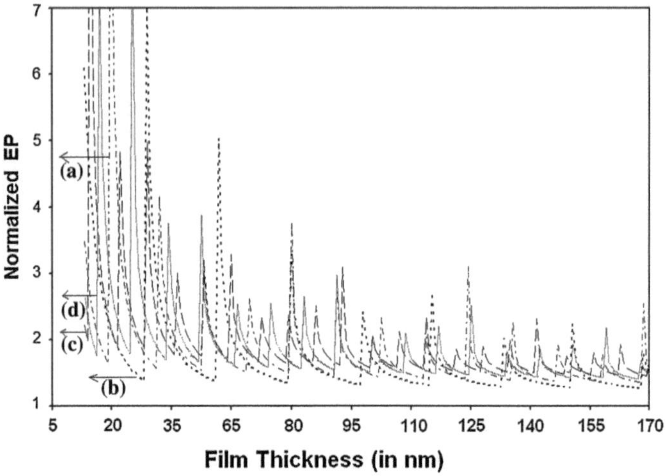

Fig. 4.16 Plot of the normalized EP from **a** HgTe/CdTe **b** $HgTe/Hg_{1-x}Cd_xTe$, **c** CdS/ZnSe with $\bar{\lambda}_o \neq 0$ and **d** PbSe/PbTe QB HD superlattices as a function of film thickness

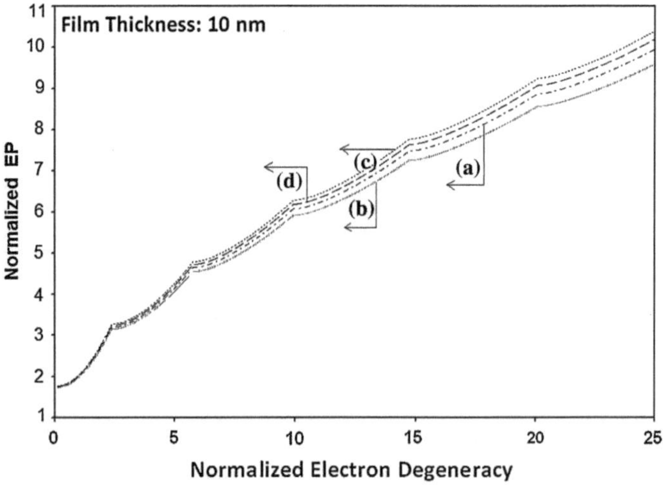

Fig. 4.17 Plot of the normalized EP from **a** HgTe/CdTe **b** $HgTe/Hg_{1-x}Cd_xTe$, **c** CdS/ZnSe with $\bar{\lambda}_o \neq 0$ and **d** PbSe/PbTe QW effective mass HD superlattices as a function of normalized electron degeneracy

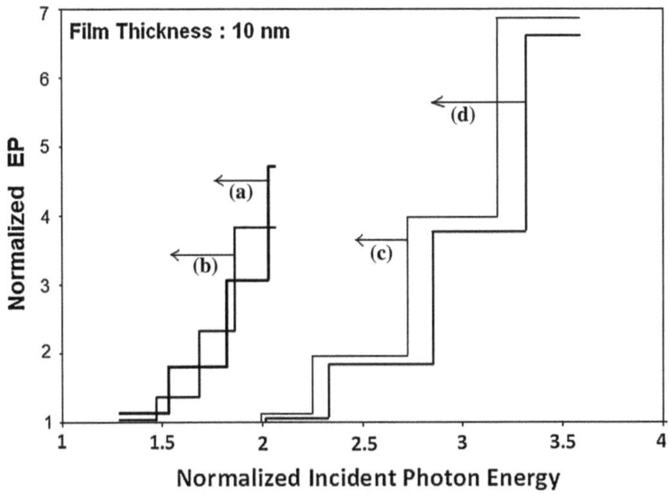

Fig. 4.18 Plot of the normalized EP from **a** HgTe/CdTe **b** $HgTe/Hg_{1-x}Cd_xTe$, **c** CdS/ZnSe with $\bar{\lambda}_o \neq 0$ and **d** PbSe/PbTe QW effective mass HD superlattices as a function of normalized incident photon energy

4.4 Open Research Problem

(R.4.1) Investigate the influence of arbitrarily oriented alternating quantizing magnetic field and strain on the EP for all types of HD super-lattices whose carrier energy spectra are described in this book.

References

1. S. Mukherjee, S.N. Mitra, P.K. Bose, A.R. Ghatak, A. Neoigi, J.P. Banerjee, A. Sinha, M. Pal, S. Bhattacharya, K.P. Ghatak, J. Comput. Theor. Nanosci. **4**, 550 (2007)
2. N.G. Anderson, W.D. Laidig, R.M. Kolbas, Y.C. Lo, J. Appl. Phys. **60**, 2361 (1986)
3. N. Paitya, K.P. Ghatak, J. Adv. Phys. **1**, 161 (2012)
4. N. Paitya, S. Bhattacharya, D. De, K.P. Ghatak, Adv. Sci. Engg. Medi. **4**, 96 (2012)
5. S. Bhattacharya, D. Dkdhjskj, S.M. Adhikari, K.P. Ghatak, Superlatt. Microst. **51**, 203 (2012)
6. D. De, S. Bhattacharya, S.M. Adhikari, A. Kumar, P.K. Bose, K.P. Ghatak, Beilstein J. Nanotech. **2**, 339 (2012)
7. D. De, A. Kumar, S.M. Adhikari, S. Pahari, N. Islam, P. Banerjee, S.K. Biswas, S. Bhattacharya, K.P. Ghatak, Superlatt. Microstruct. **47**, 377 (2010)
8. S. Pahari, S. Bhattacharya, S. Roy, A. Saha, D. De, K. P. Ghatak, Superlatt. Microstruct. **46**, (760) (2009)
9. S. Pahari, S. Bhattacharya, K. P. Ghatak J. Comput. Theory. Nanosci. **6**, (2088) (2009)
10. S.K. Biswas, A.R. Ghatak, A. Neogi, A. Sharma, S. Bhattacharya, K.P. Ghatak, Phys. E: Low-dimen. Sys. and Nanostruct. **36**, 163 (2007)

11. L.J. Singh, S. Choudhury, D. Baruah, S.K. Biswas, S. Pahari, K.P. Ghatak, Phys. B: Conden. Matter **368**, 188 (2005)
12. S. Chowdhary, L.J. Singh, K.P. Ghatak, Phys. B: Conden. Matter **365**, 5 (2005)
13. L.J. Singh, S. Choudhary, A. Mallik, K.P. Ghatak, J. Comput. Theo. Nanosci. **2**, 287 (2005)
14. K.P. Ghatak, J. Mukhopadhyay, J.P. Banerjee, SPIE Proc. Ser. **4746**, 1292 (2002)
15. K.P. Ghatak, S. Dutta, D.K. Basu, B. Nag, Il Nuovo Cimento. D **20**, 227 (1998)
16. K.P. Ghatak, D.K. Basu, B. Nag, J. Phys. Chem. Solids. **58**, 133 (1997)
17. K.P. Ghatak, B. De, Mat. Resc. Soc. Proc. **300**, 513 (1993)
18. K.P. Ghatak, B. Mitra, Il Nuovo Cimento. D **15**, 97 (1993)
19. K.P. Ghatak, Inter. Soci.Opt. and Photon. Proc. Soc. Photo Opt. Instrum. Engg. **1626**, 115 (1992)
20. K.P. Ghatak, A. Ghoshal, Phys. Stat. Sol. (b) **170**, K27 (1992)
21. K.P. Ghatak, S. Bhattacharya, S.N. Biswas, Proc. Soc. Photo opt. instrum. Engg. **836**, 72 (1988)
22. K.P. Ghatak, A. Ghoshal, S.N. Biswas, M. Mondal, Proc. Soc. Photo Opt. Instrum. Engg. **1308**, 356 (1990)
23. K.P. Ghatak, B. De, Proc. Wide Bandgap Semiconductors Symposium, Material Research Society (377) (1992)
24. K.P. Ghatak, B. De, Defect Engg. Semi. Growth, Processing and Device Technoogy Symposium, Material Research Society **262**, 911 (1992)
25. S.N. Biswas, K.P. Ghatak, Int. J. Electron. Theor. Exp. **70**, 125 (1991)
26. B. Mitra, K.P. Ghatak, Phys. Lett. A **146**, 357 (1990)
27. B. Mitra, K.P. Ghatak, Phys. Lett. A **142**, 401 (1989)
28. K.P. Ghatak, B. Mitra, A. Ghoshal, Phy. Stat. Sol. (b) **154**, K121 (1989)
29. B. Mitra, K.P. Ghatak, Phys. Stat. Sol. (b). **149**, K117 (1988)
30. K.P. Ghatak, S.N. Biswas, Proc. Soc. Photo Opt. Instrum. Eng. **792**, 239 (1987)
31. S. Bhattacharyya, K. P. Ghatak, S. Biswas, OE/Fibers' 87, Int. Soc. Opt. Photon. (73) (1987)
32. M. Mondal, K.P. Ghatak, Czech. J. Phys. B. **36**, 1389 (1986)
33. K.P. Ghatak, A.N. Chakravarti, Phys. Stat. Sol. (b). **117**, 707 (1983)
34. L.V. Keldysh, Sov. Phys. Solid State **4**, 1658 (1962)
35. L. Esaki, R. Tsu, IBM J. Res. Develop. **14**, 61 (1970)
36. G. Bastard, *Wave mechanics applied to heterostructures* (Editions de Physique, Les Ulis, France, 1990)
37. E.L. Ivchenko, G. Pikus, *Superlattices and other heterostructures*, (Springer-Berlin, 1995)
38. R. Tsu, *Superlattices to nanoelectronics* (Elsevier, The Netherlands, 2005)
39. P. Fürjes, Cs. Dücs, M. Ádám, J. Zettner, I. Bársony. Superlattices Microstruct. **35**, 455 (2004)
40. T. Borca-Tasciuc, D. Achimov, W.L. Liu, G. Chen, H.W. Ren, C.H. Lin, S.S. Pei, Microscale Thermophys. Eng. **5**, 225 (2001)
41. B.S. Williams, Nat. Photonics **1**, 517 (2007)
42. A. Kosterev, G. Wysocki, Y. Bakhirkin, S. So, R. Lewicki, F. Tittel, R.F. Curl, Appl. Phys. B **90**, 165 (2008)
43. M.A. Belkin, F. Capasso, F. Xie, A. Belyanin, M. Fischer, A. Wittmann, J. Faist, Appl. Phys. Lett. **92**, 201101 (2008)
44. G.J. Brown, F. Szmulowicz, R. Linville, A. Saxler, K. Mahalingam, C.-H. Lin, C.H. Kuo, W.Y. Hwang, IEEE Photonics Technol. Letts. **12**, 684 (2000)
45. H.J. Haugan, G.J. Brown, L. Grazulis, K. Mahalingam, D.H. Tomich, Physics E: Low-Dimensional Syst. Nanostruct. **20**, 527 (2004)
46. S.A. Nikishin, V.V. Kuryatkov, A. Chandolu, B.A. Borisov, G.D. Kipshidze, I. Ahmad, M. Holtz, H. Temkin, Jpn. J. Appl. Phys. **42**, L1362 (2003)
47. Y.K. Su, H.C. Wang, C.L. Lin, W.B. Chen, S.M. Chen, Jpn. J. Appl. Phys. **42**, L751 (2003)
48. C.H. Liu, Y.K. Su, L.W. Wu, S.J. Chang, R.W. Chuang, Semicond. Sci. Technol. **18**, 545 (2003)
49. S.B. Che, I. Nomura, A. Kikuchi, K. Shimomura, K. Kishino, Phys. Stat. Sol. (b) **229**, 1001 (2002)

50. C.P. Endres, F. Lewen, T.F. Giesen, S. SchlEEMer, D.G. Paveliev, Y.I. Koschurinov, V.M. Ustinov, A.E. Zhucov, Rev. Sci. Instrum. **78**, 043106 (2007)
51. F. Klappenberger, K.F. Renk, P. Renk, B. Rieder, Y.I. Koshurinov, D.G. Pavelev, V. Ustinov, A. Zhukov, N. Maleev, A. Vasilyev, Appl. Phys. Letts. **84**, 3924 (2004)
52. X. Jin, Y. Maeda, T. Saka, M. Tanioku, S. Fuchi, T. Ujihara, Y. Takeda, N. Yamamoto, Y. Nakagawa, A. Mano, S. Okumi, M. Yamamoto, T. Nakanishi, H. Horinaka, T. Kato, T. Yasue, T. Koshikawa, J. Crystal Growth **310**, 5039 (2008)
53. X. Jin, N. Yamamoto, Y. Nakagawa, A. Mano, T. Kato, M. Tanioku, T. Ujihara, Y. Takeda, S. Okumi, M. Yamamoto, T. Nakanishi, T. Saka, H. Horinaka, T. Kato, T. Yasue, T. Koshikawa, Appl. Phys. Express **1**, 045002 (2008)
54. B.H. Lee, K.H. Lee, S. Im, M.M. Sung, Org. Electron. **9**, 1146 (2008)
55. P.H. Wu, Y.K. Su, I.L. Chen, C.H. Chiou, J.T. Hsu, W.R. Chen, Jpn. J. Appl. Phys. **45**, L647 (2006)
56. A.C. Varonides, Renew. Energy **33**, 273 (2008)
57. M. Walther, G. Weimann, Phys. Stat. Sol. (b) **203**, 3545 (2006)
58. R. Rehm, M. Walther, J. Schmitz, J. Fleißner, F. Fuchs, J. Ziegler, W. Cabanski, Opto-Electroni. Rev. **14**, 19 (2006)
59. R. Rehm, M. Walther, J. Scmitz, J. Fleissner, J. Ziegler, W. Cabanski, R. Breiter, Electron. Letts. **42**, 577 (2006)
60. G.J. Brown, F. Szmulowicz, H. Haugan, K. Mahalingam, S. Houston, Microelectron. J. **36**, 256 (2005)
61. K.V. Vaidyanathan, R.A. Jullens, C.L. Anderson, H.L. Dunlap, Solid State Electron. **26**, 717 (1983)
62. B.A. Wilson, IEEE J. Quant. Electron. **24**, 1763 (1988)
63. M. Krichbaum, P. Kocevar, H. Pascher, G. Bauer, IEEE J. Quant. Electron. **24**, 717 (1988)
64. J.N. Schulman, T.C. McGill, Appl. Phys. Letts. **34**, 663 (1979)
65. H. Kinoshita, T. Sakashita, H. Fajiyasu, J. Appl. Phys. **52**, 2869 (1981)
66. L. Ghenin, R.G. Mani, J.R. Anderson, J.T. Cheung, Phys. Rev. B **39**, 1419 (1989)
67. C.A. Hoffman, J.R. Mayer, F.J. Bartoli, J.W. Han, J.W. Cook, J.F. Schetzina, J.M. Schubman, Phys. Rev. B. **39**, 5208 (1989)
68. V.A. Yakovlev, Sov. Phys. Semicon. **13**, 692 (1979)
69. E.O. Kane, J. Phys. Chem. Solids **1**, 249 (1957)
70. H.X. Jiang, J.Y. Lin, J. Appl. Phys. **61**, 624 (1987)
71. H. Sasaki, Phys. Rev. B **30**, 7016 (1984)

Part II
The EP from HD III-V Semiconductors and Their Quantized Counter Parts

Chapter 5
The EP from HD Kane Type Semiconductors

5.1 Introduction

With the advent of nano-photonics, there has been a considerable interest in studying the optical processes in semiconductors and their nanostructures [1]. It appears from the literature, that the investigations have been carried out on the assumption that the carrier energy spectra are invariant quantities in the presence of intense light waves, which is not fundamentally true. The physical properties of semiconductors in the presence of light waves which change the basic dispersion relation have relatively less investigated in the literature [2–12]. In this appendix we shall study the EP in HD III-V, ternary and quaternary semiconductors on the basis of newly formulated electron dispersion law under external photo excitation.

In Sect. 5.2.1 of the theoretical background Sect. 5.2, we have formulated the dispersion relation of the conduction electrons of HD III-V, ternary and quaternary materials in the presence of light waves whose unperturbed electron energy spectrum is described by the three-band model of Kane in the absence of band tailing. In Sect. 5.2.2, we have studied the EP for all the aforementioned cases. The Sect. 5.3 contains the results and discussion for this chapter and the Sect. 5.4 contains the open research problems.

5.2 Theoretical Background

5.2.1 The Formulation of the Electron Dispersion Law in the Presence of Light Waves in HD III-V, Ternary and Quaternary Semiconductors

The Hamiltonian $(\hat{H})$ of an electron in the presence of light wave characterized by the vector potential $\vec{A}$ can be written following [11] as

K.P. Ghatak, *Einstein's Photoemission*, Springer Tracts in Modern Physics 262,
DOI 10.1007/978-3-319-11188-9_5

$$\hat{H} = \left[\left| \left(\hat{p} + |e|\vec{A} \right) \right|^2 \Big/ 2m \right] + V(\vec{r}) \tag{5.1}$$

in which, $\hat{p}$ is the momentum operator, $V(\vec{r})$ is the crystal potential and m is the free electron mass.

(5.1) can be expressed as

$$\hat{H} = \hat{H}_0 + \hat{H}' \tag{5.2}$$

where

$$\hat{H}_0 = \frac{\hat{p}^2}{2m} + V(\vec{r})$$

and

$$\hat{H}' = \frac{|e|}{2m}\vec{A} \cdot \hat{p} \tag{5.3}$$

The perturbed Hamiltonian $\hat{H}'$ can be written as

$$\hat{H}' = \left(\frac{-i\hbar|e|}{2m} \right) \left(\vec{A} \cdot \nabla \right) \tag{5.4}$$

where $i = \sqrt{-1}$ and $\hat{p} = -i\hbar\nabla$

The vector potential $(\vec{A})$ of the monochromatic light of plane wave can be expressed as

$$\vec{A} = A_0 \vec{\varepsilon}_s \cos(\vec{s}_0 \cdot \vec{r} - \omega t) \tag{5.5}$$

where A_0 is the amplitude of the light wave, $\vec{\varepsilon}_s$ is the polarization vector, $\vec{s}_0$ is the momentum vector of the incident photon, $\vec{r}$ is the position vector, ω is the angular frequency of light wave and t is the time scale. The matrix element of $\hat{H}'_{nl}$ between initial state, $\psi_l(\vec{q}, \vec{r})$ and final state $\psi_n\left(\vec{k}, \vec{r}\right)$ in different bands can be written as

$$\hat{H}'_{nl} = \frac{|e|}{2m} \left\langle n\vec{k} \middle| \vec{A} \cdot \hat{p} \middle| l\vec{q} \right\rangle \tag{5.6}$$

Using (5.4) and (5.5), we can re-write (5.6) as

$$\hat{H}'_{nl} = \left(\frac{-i\hbar|e|A_0}{4m} \right) \vec{\varepsilon}_s \cdot \left[\left\{ \left\langle n\vec{k} \middle| e^{(i\vec{s}_0 \cdot \vec{r})} \nabla \middle| l\vec{q} \right\rangle e^{-i\omega t} \right\} + \left\{ \left\langle n\vec{k} \middle| e^{(i\vec{s}_0 \cdot \vec{r})} \nabla \middle| l\vec{q} \right\rangle e^{i\omega t} \right\} \right] \tag{5.7}$$

The first matrix element of (5.7) can be written as

$$\left\langle n\vec{k}\middle|e^{(i\vec{s}_0\cdot\vec{r})}\nabla\middle|l\vec{q}\right\rangle = \int e^{\left(i\left[\vec{q}+\vec{s}_0-\vec{k}\right]\cdot\vec{r}\right)} i\vec{q}u_n^*\left(\vec{k},\vec{r}\right)u_l(\vec{q},\vec{r})d^3r + \int e^{\left(i\left[\vec{q}+\vec{s}_0-\vec{k}\right]\cdot\vec{r}\right)} u_n^*\left(\vec{k},\vec{r}\right)\nabla u_l(\vec{q},\vec{r})d^3r \tag{5.8}$$

The functions $u_n^*u_l$ and $u_n^*\nabla u_l$ are periodic. The integral over all space can be separated into a sum over unit cells times an integral over a single unit cell. It is assumed that the wave length of the electromagnetic wave is sufficiently large so that if $\vec{k}$ and $\vec{q}$ are within the Brillouin zone, $\left(\vec{q}+\vec{s}_0-\vec{k}\right)$ is not a reciprocal lattice vector.

Therefore, we can write (5.8) as

$$\begin{aligned}\left\langle n\vec{k}\middle|e^{(i\vec{s}_0\cdot\vec{r})}\nabla\middle|l\vec{q}\right\rangle &= \left[\frac{(2\pi)^3}{\Omega}\right]\left\{i\vec{q}\delta\left(\vec{q}+\vec{s}_0-\vec{k}\right)\delta_{nl} + \delta\left(\vec{q}+\vec{s}_0-\vec{k}\right)\right.\\ &\quad\left.\int_{cell} u_n^*\left(\vec{k},\vec{r}\right)\nabla u_l(\vec{q},\vec{r})d^3r\right\}\\ &= \left[\frac{(2\pi)^3}{\Omega}\right]\left\{\delta\left(\vec{q}+\vec{s}_0-\vec{k}\right)\int_{cell} u_n^*\left(\vec{k},\vec{r}\right)\nabla u_l(\vec{q},\vec{r})d^3r\right\}\end{aligned} \tag{5.9}$$

where Ω is the volume of the unit cell and $\int u_n^*(\vec{k},\vec{r})u_l(\vec{q},\vec{r})d^3r = \delta(\vec{q}-\vec{k})\delta_{nl} = 0$, since $n \neq l$.

The delta function expresses the conservation of wave vector in the absorption of light wave and $\vec{s}_0$ is small compared to the dimension of a typical Brillouin zone and we set $\vec{q} = \vec{k}$.

From (5.8) and (5.9), we can write,

$$\hat{H}'_{nl} = \frac{|e|A_0}{2m}\vec{\varepsilon}_s \cdot \hat{p}_{nl}(\vec{k})\delta(\vec{q}-\vec{k})\cos(\omega t) \tag{5.10}$$

where

$$\hat{p}_{nl}(\vec{k}) = -i\hbar\int u_n^*\nabla u_l d^3r = \int u_n^*(\vec{k},\vec{r})\hat{p}u_l(\vec{k},\vec{r})d^3r$$

Therefore, we can write

$$\hat{H}'_{nl} = \frac{|e|A_0}{2m}\vec{\varepsilon} \cdot \hat{p}_{nl}(\vec{k}) \tag{5.11}$$

where $\vec{\varepsilon} = \vec{\varepsilon}_s\cos\omega t$.

When a photon interacts with a semiconductor, the carriers (i.e., electrons) are generated in the bands which are followed by the inter-band transitions. For example, when the carriers are generated in the valence band, the carriers then make inter-band transition to the conduction band. The transition of the electrons within the same band i.e., $\hat{H}'_{nn} = \left\langle n\vec{k}\middle|\hat{H}'\middle|n\vec{k}\right\rangle$ is neglected. Because, in such a case, i.e., when the carriers are generated within the same bands by photons, are lost by recombination within the aforementioned band resulting zero carriers.

Therefore,

$$\left\langle n\vec{k}\middle|\hat{H}'\middle|n\vec{k}\right\rangle = 0 \tag{5.12}$$

With $n = c$ stands for conduction band and $l = v$ stand for valance band, the energy equation for the conduction electron can approximately be written as

$$I_{11}(E) = \left(\frac{\hbar^2 k^2}{2m_c}\right) + \frac{\left(\frac{|e|A_0}{2m}\right)^2 \left\langle \left|\vec{\varepsilon}\cdot\hat{p}_{cv}(\vec{k})\right|^2 \right\rangle_{av}}{E_c(\vec{k}) - E_v(\vec{k})} \tag{5.13}$$

where $I_{11}(E) \equiv E(aE+1)(bE+1)/(cE+1), a \equiv 1/E_{g_0}$, E_{g_0} is the un-perturbed band-gap, $b \equiv 1/\left(E_{g_0} + \Delta\right)$, $c \equiv 1/\left(E_{g_0} + 2\Delta/3\right)$, and $\left\langle \left|\vec{\varepsilon}\cdot\hat{p}_{cv}(\vec{k})\right|^2 \right\rangle_{av}$ represents the average of the square of the optical matrix element (OME).

For the three-band model of Kane, we can write,

$$\xi_{1k} = E_c(\vec{k}) - E_v(\vec{k}) = (E_{g_0}^2 + E_{g_0}\hbar^2 k^2/m_r)^{1/2} \tag{5.14}$$

where m_r is the reduced mass and is given by $m_r^{-1} = (m_c)^{-1} + m_v^{-1}$, and m_v is the effective mass of the heavy hole at the top of the valance band in the absence of any field.

The doubly degenerate wave functions $u_1\left(\vec{k},\vec{r}\right)$ and $u_2\left(\vec{k},\vec{r}\right)$ can be expressed as [13, 14]

$$u_1\left(\vec{k},\vec{r}\right) = a_{k+}[(is)\downarrow'] + b_{k+}\left[\frac{X' - iY'}{\sqrt{2}}\uparrow'\right] + c_{k+}[Z'\downarrow] \tag{5.15}$$

and

$$u_2\left(\vec{k},\vec{r}\right) = a_{k-}[(is)\uparrow'] - b_{k-}\left[\frac{X' + iY'}{\sqrt{2}}\downarrow'\right] + c_{k-}[Z'\uparrow'] \tag{5.16}$$

s is the s-type atomic orbital in both unprimed and primed coordinates, $\downarrow'$ indicates the spin down function in the primed coordinates,

$$a_{k\pm} \equiv \beta\left[E_{g_0} - (\gamma_{0k\pm})^2 (E_{g_0} - \delta')^{1/2} (E_{g_0} + \delta')^{1/2}\right],$$
$$\beta \equiv \left[(6(E_{g_0} + 2\Delta/3)(E_{g_0} + \Delta))/\chi\right]^{1/2},$$
$$\chi \equiv (6E_{g_0}^2 + 9E_{g_0}\Delta + 4\Delta^2),\ \gamma_{0k\pm} \equiv \left[\frac{(\xi_{1k} \mp E_{g_0})}{2(\xi_{1k} + \delta')}\right]^{1/2},$$

$\xi_{1k} \equiv E_c(\vec{k}) - E_v(\vec{k}) = E_{g_0}\left[1 + 2\left(1 + \frac{m_c}{m_v}\right)\frac{I_{11}(E)}{E_{g_0}}\right]^{1/2} \delta' \equiv (E_{g_0}^2 \Delta)(\chi)^{-1}$, X', X', and Z' are the p-type atomic orbitals in the primed coordinates, $\uparrow'$ indicates the spin-up function in the primed coordinates,

$$b_{k\pm} \equiv \rho\gamma_{0k\pm},\ \rho(4\Delta^2/3\chi)^{1/2},\ c_{k\pm} \equiv t\gamma_{0k\pm} \text{ and } t \equiv \left[6(E_{g_0} + 2\Delta/3)^2/\chi\right]^{1/2}.$$

We can, therefore, write the expression for the optical matrix element (OME) as OME

$$\text{OME} = \hat{p}_{cv}(\vec{k}) = \langle u_1(\vec{k},\vec{r})|\hat{p}|u_2(\vec{k},\vec{r})\rangle \tag{5.17}$$

Since the photon vector has no interaction in the same band for the study of inter-band optical transition, we can therefore write

$$\langle S|\hat{p}|S\rangle = \langle X|\hat{p}|X\rangle = \langle Y|\hat{p}|Y\rangle = \langle Z|\hat{p}|Z\rangle = 0$$

and

$$\langle X|\hat{p}|Y\rangle = \langle Y|\hat{p}|Z\rangle = \langle Z|\hat{p}|X\rangle = 0$$

There are finite interactions between the conduction band (CB) and the valance band (VB) and we can obtain

$$\langle S|\hat{P}|X\rangle = \hat{i}\cdot\hat{P} = \hat{i}\cdot\hat{P}_x$$
$$\langle S|\hat{P}|Y\rangle = \hat{j}\cdot\hat{P} = \hat{j}\cdot\hat{P}_y$$
$$\langle S|\hat{P}|Z\rangle = \hat{k}\cdot\hat{P} = \hat{k}\cdot\hat{P}_z$$

where $\hat{i}$, $\hat{j}$ and $\hat{k}$ are the unit vectors along x, y and z axes respectively.

It is well known [14] that

$$\begin{bmatrix}\uparrow' \\ \downarrow'\end{bmatrix} = \begin{bmatrix} e^{-i\phi/2}\cos(\theta/2) & e^{i\phi/2}\sin(\theta/2) \\ -e^{-i\phi/2}\sin(\theta/2) & e^{i\phi/2}\cos(\theta/2)\end{bmatrix}\begin{bmatrix}\uparrow \\ \downarrow\end{bmatrix}$$

and

$$\begin{bmatrix} X' \\ Y' \\ Z' \end{bmatrix} = \begin{bmatrix} \cos\theta\cos\phi & \cos\theta\sin\phi & -\sin\theta \\ -\sin\phi & \cos\phi & 0 \\ \sin\theta\cos\phi & \sin\theta\sin\phi & \cos\theta \end{bmatrix} \begin{bmatrix} X \\ Y \\ Z \end{bmatrix}$$

Besides, the spin vector can be written as $\vec{S} = \frac{\hbar}{2}\vec{\sigma}$ where $\sigma_x = \begin{bmatrix} 0 & 1 \\ 1 & 0 \end{bmatrix}, \sigma_y \begin{bmatrix} 0 & -i \\ i & 0 \end{bmatrix}$ and $\sigma_z = \begin{bmatrix} 1 & 0 \\ 0 & -1 \end{bmatrix}$.

From above, we can write

$$\hat{p}_{CV}(\vec{k}) = \left\langle u_1(\vec{k},\vec{r})\middle|\hat{P}\middle|u_2(\vec{k},\vec{r})\right\rangle$$
$$= \left\langle \left\{ a_{k_+}[(iS)\downarrow'] + b_{k_+}\left[\left(\frac{X'-iY'}{\sqrt{2}}\right)\uparrow'\right] + c_{k_+}[Z'\downarrow'] \right\} \middle|\hat{P}\middle| \{a_{k_-}[(iS)\uparrow'] \right.$$
$$\left. -b_{k_-}\left[\left(\frac{X'+iY'}{\sqrt{2}}\right)\downarrow' + c_{k_-}[Z'\uparrow']\right]\right\}\right\rangle.$$

Using above relations, we get

$$\begin{aligned}\hat{p}_{CV}(\vec{k}) &= \left\langle u_1(\vec{k},\vec{r})\middle|\hat{P}\middle|u_2(\vec{k},\vec{r})\right\rangle \\ &= \frac{b_{k_+}a_{k_-}}{\sqrt{2}}\left\{\langle (X'-iY')|\hat{P}|iS\rangle\langle\uparrow' \mid \uparrow'\rangle\right\} + c_{k_+}a_{k_-}\left\{\langle Z'|\hat{P}|iS\rangle\langle\downarrow' \mid \uparrow'\rangle\right\} \\ &\quad - \frac{a_{k_+}b_{k_-}}{\sqrt{2}}\left\{\langle iS|\hat{P}|(X'+iY')\rangle\langle\downarrow' \mid \downarrow'\rangle\right\} + a_{k_+}c_{k_-}\left\{\langle iS|\hat{P}|Z'\rangle\langle\downarrow' \mid \uparrow'\rangle\right\}\end{aligned} \tag{5.18}$$

From (5.18), we can write

$$\begin{aligned}\langle (X'-iY')|\hat{P}|iS\rangle &= \langle (X')|\hat{P}|iS\rangle - \langle (iY')|\hat{P}|iS\rangle \\ &= i\int u^*_{X'}\hat{P}S - \int -iu^*_{Y'}\hat{P}iu_X = i\langle X'|\hat{P}|S\rangle - \langle Y'|\hat{P}|S\rangle\end{aligned}$$

From the above relations, for X', Y' and Z', we get

$$|X'\rangle = \cos\theta\cos\phi|X\rangle + \cos\theta\sin\phi|Y\rangle - \sin\theta|Z\rangle$$

Thus,

$$\langle X'|\hat{P}|S\rangle = \cos\theta\cos\phi\langle X|\hat{P}|S\rangle + \cos\theta\sin\phi\langle Y|\hat{P}|S\rangle - \sin\theta\langle Z|\hat{P}|S\rangle = \hat{P}\hat{r}_1$$

where,

$$\hat{r}_1 = \hat{i}\cos\theta\cos\phi + \hat{j}\cos\theta\sin\phi - \hat{k}\sin\theta$$

$$|Y'\rangle = -\sin\phi|X\rangle + \cos\phi|Y\rangle + 0|Z\rangle$$

Thus,

$$\langle Y'|\hat{P}|S\rangle = -\sin\phi\langle X|\hat{P}|S\rangle + \cos\phi\langle Y|\hat{P}|S\rangle + 0\langle Z|\hat{P}|S\rangle = \hat{P}\hat{r}_2$$

where

$$\hat{r}_2 = -\hat{i}\sin\phi + \hat{j}\cos\phi$$

so that

$$\langle (X' - iY')|\hat{P}|S\rangle = \hat{P}(i\hat{r}_1 - \hat{r}_2)$$

Thus,

$$\frac{a_{k_-}b_{k_+}}{\sqrt{2}}\langle (X' - iY')|\hat{P}|S\rangle\langle\uparrow' \mid \uparrow'\rangle = \frac{a_{k_-}b_{k_+}}{\sqrt{2}}\hat{P}(i\hat{r}_1 - \hat{r}_2)\langle\uparrow' \mid \uparrow'\rangle \tag{5.19}$$

Now since,

$$\langle iS|\hat{P}|(X' + iY')\rangle = i\langle S|\hat{P}|X'\rangle - \langle S|\hat{P}|Y'\rangle = \hat{P}(i\hat{r}_1 - \hat{r}_2)$$

We can write,

$$-\left[\frac{a_{k_+}b_{k_-}}{\sqrt{2}}\{\langle iS|\hat{P}|(X' + iY')\rangle\langle\downarrow' \mid \downarrow'\rangle\}\right] = -\left[\frac{a_{k_+}b_{k_-}}{\sqrt{2}}\hat{P}(i\hat{r}_1 - \hat{r}_2)\langle\downarrow' \mid \downarrow'\rangle\right] \tag{5.20}$$

Similarly, we get

$$|Z'\rangle = \sin\theta\cos\phi|X\rangle + \sin\theta\sin\phi|Y\rangle + \cos\theta|Z\rangle$$

So that,

$$\langle Z'|\hat{P}|iS\rangle = i\langle Z'|\hat{P}|S\rangle = i\hat{P}\{\sin\theta\cos\phi\hat{i} + \sin\theta\sin\phi\hat{j} + \cos\theta\hat{k}\} = i\hat{P}\hat{r}_3$$

where,

$$\hat{r}_3 = \hat{i}\sin\theta\cos\phi + \hat{j}\sin\theta\sin\phi + \hat{k}\cos\theta$$

Thus,

$$c_{k_+} a_{k_-} \langle Z'|\hat{P}|iS\rangle \langle \downarrow' \mid \uparrow'\rangle = c_{k_+} a_{k_-} i\hat{P}\hat{r}_3 \langle \downarrow' \mid \uparrow'\rangle \tag{5.21}$$

Similarly, we can write,

$$c_{k_-} a_{k_+} \langle iS|\hat{P}|Z'\rangle \langle \downarrow' \mid \uparrow'\rangle = c_{k_-} a_{k_+} i\hat{P}\hat{r}_3 \langle \downarrow' \mid \uparrow'\rangle \tag{5.22}$$

Therefore, we obtain

$$\begin{aligned} &\frac{a_{k_-} b_{k_+}}{\sqrt{2}} \{\langle (X' - iY')|\hat{P}|S\rangle \langle \uparrow' \mid \uparrow'\rangle\} - \frac{a_{k_+} b_{k_-}}{\sqrt{2}} \{\langle iS|\hat{P}|(X' + iY')\rangle \langle \downarrow' \mid \downarrow'\rangle\} \\ &= \frac{\hat{P}}{\sqrt{2}} (-a_{k_+} b_{k_-} \langle \downarrow' \mid \downarrow'\rangle + a_{k_-} b_{k_+} \langle \uparrow' \mid \uparrow'\rangle)(i\hat{r}_1 - \hat{r}_2) \end{aligned} \tag{5.23}$$

Also, we can write,

$$\begin{aligned} &c_{k_+} a_{k_-} \langle Z'|\hat{P}|iS\rangle \langle \downarrow' \mid \uparrow'\rangle + c_{k_-} a_{k_+} \langle iS|\hat{P}|Z'\rangle \langle \downarrow' \mid \uparrow'\rangle \\ &\quad = i\hat{P}(c_{k_+} a_{k_-} + c_{k_-} a_{k_+})\hat{r}_3 [\langle \downarrow' \mid \downarrow'\rangle] \end{aligned} \tag{5.24}$$

Combining (5.23) and (5.24), we find

$$\begin{aligned} \hat{p}_{CV}(\vec{k}) = &\frac{\hat{P}}{\sqrt{2}} (i\hat{r}_1 - \hat{r}_2)\{(b_{k_+} a_{k_-})\langle \uparrow' \mid \uparrow'\rangle - (b_{k_-} a_{k_+})\langle \downarrow' \mid \downarrow'\rangle\} \\ &+ i\hat{P}\hat{r}_3 (c_{k_+} a_{k_-} - c_{k_-} a_{k_+})\langle \downarrow' \mid \uparrow'\rangle \end{aligned} \tag{5.25}$$

From the above relations, we obtain,

$$\left.\begin{aligned} \uparrow' &= e^{-i\phi/2} \cos(\theta/2) \uparrow + e^{i\phi/2} \sin(\theta/2) \\ \downarrow' &= -e^{-i\phi/2} \sin(\theta/2) \uparrow + e^{i\phi/2} \cos(\theta/2) \end{aligned}\right\} \tag{5.26}$$

Therefore,

$$\begin{aligned} \langle \downarrow' \mid \uparrow'\rangle_x = &- \sin(\theta/2)\cos(\theta/2)\langle \uparrow \mid \uparrow\rangle_x + e^{-i\phi} \cos^2(\theta/2)\langle \downarrow \mid \uparrow\rangle_x \\ &- e^{i\phi} \sin^2(\theta/2)\langle \uparrow \mid \downarrow\rangle_x + \sin(\theta/2)\cos(\theta/2)\langle \downarrow \mid \downarrow\rangle_x \end{aligned} \tag{5.27}$$

But we know from above that

$$\langle \uparrow \mid \uparrow\rangle_x = 0, \langle \downarrow \mid \uparrow\rangle = \frac{1}{2}, \langle \downarrow \mid \uparrow\rangle_x = \frac{1}{2}$$

and $\langle \downarrow \mid \downarrow\rangle_x = 0$

Thus, from (5.27), we get

$$
\begin{aligned}
\langle\downarrow' \mid \uparrow'\rangle_x &= \frac{1}{2}[e^{-i\phi}\cos^2(\theta/2) - e^{i\phi}\sin^2(\theta/2)] \\
&= \frac{1}{2}\left[(\cos\phi - i\sin\phi)\cos^2(\theta/2) - (\cos\phi + i\sin\phi)\sin^2(\theta/2)\right] \\
&= \frac{1}{2}[\cos\phi\cos\theta - i\sin\phi]
\end{aligned}
\tag{5.28}
$$

Similarly, we obtain

$$\langle\uparrow' \mid \uparrow'\rangle_y = \frac{1}{2}[i\cos\phi + \sin\phi\cos\theta]$$

and $\langle\downarrow' \mid \uparrow'\rangle_z = \frac{1}{2}[-\sin\theta]$

Therefore,

$$
\begin{aligned}
\langle\downarrow' \mid \uparrow'\rangle &= \hat{i}\langle\downarrow' \mid \uparrow'\rangle_x + \hat{j}\langle\downarrow' \mid \uparrow'\rangle_y + \hat{k}\langle\downarrow' \mid \uparrow'\rangle_z \\
&= \frac{1}{2}\{(\cos\theta\cos\phi - i\sin\phi)\hat{i} + (i\cos\phi + \sin\phi\cos\theta)\hat{j} - \sin\theta\hat{k}\} \\
&= \frac{1}{2}\left[\{(\cos\theta\cos\phi)\hat{i} + (\sin\phi\cos\theta)\hat{j} - \sin\theta\hat{k}\} + i\{-\hat{i}\sin\phi + \hat{j}\cos\phi\}\right] \\
&= \frac{1}{2}[\hat{r}_1 + i\hat{r}_2] = -\frac{1}{2}i[i\hat{r}_1 - \hat{r}_2]
\end{aligned}
$$

Similarly, we can write

$$\langle\uparrow' \mid \uparrow'\rangle = \frac{1}{2}\left[\hat{i}\sin\theta\cos\phi + \hat{j}\sin\theta\sin\phi + \hat{k}\cos\theta\right] = \frac{1}{2}\hat{r}_3 \text{ and } \langle\downarrow' \mid \downarrow'\rangle = -\frac{1}{2}\hat{r}_3$$

Using the above results and following (5.25) we can write

$$
\begin{aligned}
\hat{p}_{CV}(\vec{k}) &= \frac{\hat{P}}{\sqrt{2}}(i\hat{r}_1 - \hat{r}_2)\left\{\left(a_{k_-}b_{k_+}\right)\langle\uparrow' \mid \uparrow'\rangle - \left(b_{k_-}a_{k_+}\right)\langle\downarrow' \mid \downarrow'\rangle\right\} \\
&\quad + i\hat{P}\hat{r}_3\left\{\left(c_{k_+}a_{k_-} - c_{k_-}a_{k_+}\right)\langle\downarrow' \mid \uparrow'\rangle\right\} \\
&= \frac{\hat{P}}{2}\hat{r}_3(i\hat{r}_1 - \hat{r}_2)\left\{\left(\frac{a_{k_-}b_{k_+}}{\sqrt{2}} + \frac{b_{k_-}a_{k_+}}{\sqrt{2}}\right)\right\} \\
&\quad + \frac{\hat{P}}{2}\hat{r}_3(i\hat{r}_1 - \hat{r}_2)\left\{\left(c_{k_+}a_{k_-} + c_{k_-}a_{k_+}\right)\right\}
\end{aligned}
$$

Thus,

$$\hat{p}_{CV}(\vec{k}) = \frac{\hat{P}}{2}\hat{r}_3(i\hat{r}_1 - i\hat{r}_2)\left\{a_{k_+}\left(\frac{b_{k_-}}{\sqrt{2}} + c_{k_-}\right) + a_{k_-}\left(\frac{b_{k_+}}{\sqrt{2}} + c_{k_+}\right)\right\} \tag{5.29}$$

We can write that,

$$|\hat{r}_1| = |\hat{r}_2| = |\hat{r}_3| = 1, \text{ also}, \hat{P}\hat{r}_3 = \hat{P}_x \sin\theta\cos\phi\hat{i} + \hat{P}_y \sin\theta\sin\phi\hat{j} + \hat{P}_z\cos\theta\hat{k}$$

where,

$$\begin{aligned} \hat{P} &= \langle S|\hat{P}|X\rangle = \langle S|\hat{P}|Y\rangle = \langle S|\hat{P}|Z\rangle, \\ \langle S|\hat{P}|X\rangle &= \int u_C^*(0,\vec{r})\hat{P}u_{VX}(0,\vec{r})d^3r = \hat{P}_{CVX}(0) \\ \text{and } \langle S|\hat{P}|Z\rangle &= \hat{P}_{CVZ}(0) \end{aligned}$$

Thus, $\hat{P} = \hat{P}_{CVX}(0) = \hat{P}_{CVY}(0) = \hat{P}_{CVZ}(0) = \hat{P}_{CV}(0)$ where $\hat{P}_{CV}(0) \equiv \int u_C^*(0,\vec{r})$ $\hat{P}u_V(0,\vec{r})d^3r \equiv \hat{P}$.

For a plane polarized light wave, we have the polarization vector $\vec{\varepsilon}_s = \hat{k}$, when the light wave vector is traveling along the z-axis. Therefore, for a plane polarized light-wave, we have considered $\vec{\varepsilon}_s = \hat{k}$.

Then, from (5.29) we get

$$\left(\vec{\varepsilon}\cdot\hat{p}_{CV}\left(\vec{k}\right)\right) = \vec{k}\cdot\frac{\hat{P}}{2}\hat{r}_3(i\hat{r}_1 - \hat{r}_2)\left[A(\vec{k}) + B(\vec{k})\right]\cos\omega t \tag{5.30}$$

and

$$\left.\begin{aligned} A(\vec{k}) &= a_{k_-}\left(\frac{b_{k_+}}{\sqrt{2}} + c_{k_+}\right) \\ B(\vec{k}) &= a_{k_+}\left(\frac{b_{k_-}}{\sqrt{2}} + c_{k_-}\right) \end{aligned}\right\} \tag{5.31}$$

Thus,

$$\begin{aligned} \left|\vec{\varepsilon}\cdot\hat{p}_{cv}(\vec{k})\right|^2 &= \left|\hat{k}\cdot\frac{\hat{P}}{2}\hat{r}_3\right|^2 |i\hat{r}_1 - \hat{r}_2|^2\left[A(\vec{k}) + B(\vec{k})\right]^2\cos^2\omega t \\ &= \frac{1}{4}\left|\hat{P}_z\cos\theta\right|^2\left[A(\vec{k}) + B(\vec{k})\right]^2\cos^2\omega t \end{aligned} \tag{5.32}$$

So, the average value of $\left|\vec{\varepsilon}\cdot\hat{p}_{cv}(\vec{k})\right|^2$ for a plane polarized light-wave is given by

$$\left\langle \left| \vec{\varepsilon} \cdot \hat{p}_{cv}(\vec{k}) \right|^2 \right\rangle_{av} = \frac{2}{4} \left| \hat{P}_z \right|^2 [A(\vec{k}) + B(\vec{k})]^2 \left(\int_0^{2\pi} d\phi \int_0^{\pi} \cos^2 \theta \sin \theta d\theta \right) \left(\frac{1}{2} \right)$$
$$= \frac{2\pi}{3} \left| \hat{P}_z \right|^2 \left[A(\vec{k}) + B(\vec{k}) \right]^2 \tag{5.33}$$

where $\left| \hat{P}_z \right|^2 = \left(\frac{1}{2}\right) \left| \vec{k} \cdot \hat{p}_{cv}(0) \right|^2$ and

$$\left| \vec{k} \cdot \hat{p}_{cv}(0) \right|^2 = \frac{m^2}{4m_r} \frac{E_{g_0}(E_{g_0} + \Delta)}{\left(E_{g_0} + \frac{2}{3}\Delta\right)} \tag{5.34}$$

We shall express $A(\vec{k})$ and $B(\vec{k})$ in terms of constants of the energy spectra in the following way:

Substituting $a_{k_\pm}$, $b_{k_\pm}$, $c_{k_\pm}$ and $\gamma_{0k_\pm}$ in $A(\vec{k})$ and $B(\vec{k})$ in (5.31) we get

$$A(\vec{k}) = \beta \left(t + \frac{\rho}{\sqrt{2}} \right) \left\{ \left(\frac{E_{g0}}{E_{g_0} + \delta'} \right) \gamma_{0k_+}^2 - \gamma_{0k_+}^2 \gamma_{0k_-}^2 \left(\frac{E_{g0} - \delta'}{E_{g_0} + \delta'} \right) \right\}^{1/2} \tag{5.35}$$

$$B(\vec{k}) = \beta \left(t + \frac{\rho}{\sqrt{2}} \right) \left\{ \left(\frac{E_{g0}}{E_{g_0} + \delta'} \right) \gamma_{0k_-}^2 - \gamma_{0k_+}^2 \gamma_{0k_-}^2 \left(\frac{E_{g0} - \delta'}{E_{g_0} + \delta'} \right) \right\}^{1/2} \tag{5.36}$$

in which, $\gamma_{0_{k\pm}}^2 \equiv \frac{\xi_{1k} - E_{g0}}{2(\xi_{1k} + \delta')} \equiv \frac{1}{2} \left[1 - \left(\frac{E_{g0} + \delta'}{\xi_{1k} + \delta'} \right) \right]$ and $\gamma_{0k_-}^2 \equiv \frac{\xi_{1k} + E_{g0}}{2(\xi_{1k} + \delta')} \equiv \frac{1}{2} \left[1 + \left(\frac{E_{g0} - \delta'}{\xi_{1k} + \delta'} \right) \right]$

Substituting $x \equiv \xi_{1k} + \delta'$ in $\gamma_{0_{k\pm}}^2$, we can write,

$$A(\vec{k}) = \beta \left(t + \frac{\rho}{\sqrt{2}} \right) \left\{ \left(\frac{E_{g0}}{E_{g_0} + \delta'} \right) \frac{1}{2} \left(1 - \frac{E_{g0} + \delta'}{x} \right) - \frac{1}{4} \left(\frac{E_{g_0} - \delta'}{E_{g_0} + \delta'} \right) \right.$$
$$\left. \left(1 - \frac{E_{g0} + \delta'}{x} \right) \left(1 + \frac{E_{g_0} - \delta'}{x} \right) \right\}^{1/2}$$

Thus,

$$A(\vec{k}) = \frac{\beta}{2} \left(t + \frac{\rho}{\sqrt{2}} \right) \left\{ 1 - \frac{2a_0}{x} + \frac{a_1}{x^2} \right\}^{1/2}$$

where

$$a_0 \equiv (E_{g_0}^2 + \delta'^2)(E_{g_0} + \delta')^{-1}$$

and $a_1 \equiv \left(E_{g_0} - \delta'\right)^2$.

After tedious algebra, one can show that

$$A(\vec{k}) = \frac{\beta}{2}\left(t + \frac{\rho}{\sqrt{2}}\right)(E_{g_0} - \delta')\left[\frac{1}{\xi_{1k} + \delta'} - \frac{1}{E_{g_0} + \delta'}\right]^{1/2}\left[\frac{1}{\xi_{1k} + \delta'} - \frac{(E_{g_0} + \delta')}{(E_{g_0} - \delta')^2}\right]^{1/2} \tag{5.37}$$

Similarly, from (5.36), we can write,

$$B(\vec{k}) = \beta\left(t + \frac{\rho}{\sqrt{2}}\right)\left\{\left(\frac{E_{g_0}}{E_{g_0} + \delta'}\right)\frac{1}{2}\left(1 + \frac{E_{g_0} - \delta'}{x}\right) - \frac{1}{4}\left(\frac{E_{g_0} - \delta'}{E_{g_0} + \delta'}\right)\right.$$
$$\left.\left(1 - \frac{E_{g_0} + \delta'}{x}\right)\left(1 + \frac{E_{g_0} - \delta'}{x}\right)\right\}^{1/2}$$

So that, finally we get,

$$B(\vec{k}) = \frac{\beta}{2}\left(t + \frac{\rho}{\sqrt{2}}\right)\left(1 + \frac{E_{g_0} - \delta'}{\xi_{1k} + \delta'}\right) \tag{5.38}$$

Using (5.33), (5.34), (5.37) and (5.38), we can write

$$\left(\frac{|e|A_0}{2m}\right)^2 \frac{\left\langle\left|\vec{\varepsilon}\cdot\hat{p}_{cv}(\vec{k})\right|^2\right\rangle_{av}}{E_c(\vec{k}) - E_v(\vec{k})} = \left(\frac{|e|A_0}{2m}\right)^2 \frac{2\pi}{3}\left|\vec{k}\cdot\hat{p}_{cv}(0)\right|^2 \frac{\beta^2}{4}\left(t + \frac{\rho}{\sqrt{2}}\right)^2$$
$$\frac{1}{\xi_{1k}}\left\{\left(1 + \frac{E_{g_0} - \delta'}{\xi_{1k} + \delta'}\right) + (E_{g_0} - \delta')\right.$$
$$\left.\left[\frac{1}{\xi_{1k} + \delta'} - \frac{1}{E_{g_0} + \delta'}\right]^{1/2}\left[\frac{1}{\xi_{1k} + \delta'} - \frac{E_{g_0} + \delta'}{(E_{g_0} - \delta')^2}\right]^{1/2}\right\}^2 \tag{5.39}$$

Following Nag [12], it can be shown that

$$A_0^2 = \frac{I\lambda^2}{2\pi^2 c^3\sqrt{\varepsilon_{sc}\varepsilon_0}} \tag{5.40}$$

where, I is the light intensity of wavelength λ, ε_0 is the permittivity of free space and c is the velocity of light. Thus, the simplified electron energy spectrum in III-V, ternary and quaternary materials in the presence of light waves can approximately be written as

$$\frac{\hbar^2 k^2}{2m_c} = \beta_0(E,\lambda) \tag{5.41}$$

where $\beta_0(E,\lambda) \equiv [I_{11}(E) - \theta_0(E,\lambda)]$,

$$\theta_0(E,\lambda) \equiv \frac{|e|^2}{96 m_r \pi c^3} \frac{I\lambda^2}{\sqrt{\varepsilon_{sc}\varepsilon_0}} \frac{E_{g_0}(E_{g_0}+\Delta)}{(E_{g_0}+\frac{2}{3}D)} \frac{\beta^2}{4}\left(t+\frac{\rho}{\sqrt{2}}\right)^2 \frac{1}{\phi_0(E)}$$
$$\left\{\left(1+\frac{E_{g_0}-\delta'}{\phi_0(E)+\delta'}\right)+(E_{g_0}-\delta')\left[\frac{1}{\phi_0(E)+\delta'}-\frac{1}{E_{g_0}+\delta'}\right]^{1/2}\right.$$
$$\left.\left[\frac{1}{\phi_0(E)+\delta'}-\frac{E_{g_0}+\delta'}{(E_{g_0}-\delta')^2}\right]^{1/2}\right\}^2$$

and $\phi_0(E) \equiv E_{g_0}\left(1+2\left(1+\frac{m_c}{m_v}\right)\frac{I_{11}(E)}{E_{g_0}}\right)^{1/2}$

Thus, under the limiting condition $\vec{k} \to 0$, from (5.41), we observe that $E \neq 0$ and is positive. Therefore, in the presence of external light waves, the energy of the electron does not tend to zero when $\vec{k} \to 0$, where as for the un-perturbed three band model of Kane, $I_{11}(E) = \left[\hbar^2 k^2/(2m_c)\right]$ in which $E \to 0$ for $\vec{k} \to 0$. As the conduction band is taken as the reference level of energy, therefore the lowest positive value of E for $\vec{k} \to 0$ provides the increased band gap (ΔE_g) of the semiconductor due to photon excitation. The values of the increased band gap can be obtained by computer iteration processes for various values of I and λ respectively.

Special Cases:

(1) For the two-band model of Kane, we have $\Delta \to 0$. Under this condition, $I_{11}(E) \to E(1+aE) = \frac{\hbar^2 k^2}{2m_c}$. Since, $\beta \to 1$, $t \to 1$, $\rho \to 0$, $\delta' \to 0$ for $\Delta \to 0$, from (5.41), we can write the energy spectrum of III-V, ternary and quaternary materials in the presence of external photo-excitation whose unperturbed conduction electrons obey the two band model of Kane as

$$\frac{\hbar^2 k^2}{2m_c} = \tau_0(E,\lambda) \tag{5.42}$$

where $\tau_0(E,\lambda) \equiv E(1+aE) - B_0(E,\lambda)$,

$$B_0(E,\lambda) \equiv \frac{|e|^2 I\lambda^2 E_{g_0}}{384\pi c^3 m_r\sqrt{\varepsilon_{sc}\varepsilon_0}}\frac{1}{\phi_1(E)}$$

$$\left\{\left(1+\frac{E_{g_0}}{\phi_1(E)}\right)+E_{g_0}\left[\frac{1}{\phi_1(E)}-\frac{1}{E_{g_0}}\right]\right\}^2,$$

$$\phi_1(E) \equiv E_{g_0}\left\{1+\frac{2m_c}{m_r}aE(1+aE)\right\}^{1/2}.$$

(2) For relatively wide band gap semiconductors, one can write, $a \to 0$, $b \to 0$, $c \to 0$ and $I_{11}(E) \to E$.

Thus, from (5.42), we get,

$$\frac{\hbar^2 k^2}{2m_c} = \rho_0(E,\lambda) \tag{5.43}$$

where $\rho_0(E,\lambda) \equiv E - \frac{|e|^2 I\lambda^2}{96\pi c^3 m_r\sqrt{\varepsilon_{sc}\varepsilon_0}}\left[1+\left(\frac{2m_c}{m_r}\right)aE\right]^{-3/2}$.

5.2.2 *The EP in the Presence of Light Waves in HD III-V, Ternary and Quaternary Semiconductors*

The (5.40), (5.41) and (5.42) can approximately be written as

$$\frac{\hbar^2 k^2}{2m_c^*} = U_\lambda I_{11}(E) - \mathrm{P}_\lambda \tag{5.44}$$

$$\frac{\hbar^2 k^2}{2m_c^*} = t_{1\lambda}E + t_{2\lambda}E^2 - \delta_\lambda \tag{5.45}$$

and

$$\frac{\hbar^2 k^2}{2m_c^*} = t_{1\lambda}E - \delta_\lambda \tag{5.46a}$$

where $U_\lambda = (1+\theta_\lambda), \theta_\lambda = \frac{C_0}{A}(t_\lambda + \frac{BJ_\lambda}{A})$, $C_0 = \left[\frac{|e|^2 I\lambda^2 E_{g_0}(E_{g_0}+\Delta)}{96 m_r \pi c^3\sqrt{\varepsilon_{sc}\varepsilon_0}(E_{g_0}+\frac{2}{3}\Delta)}\frac{\beta^2}{4}(1+\frac{\rho}{\sqrt{2}})^2\right]$

$$B = [1 + \frac{m*}{m_V}], G_\lambda = [\frac{2B}{(A+\delta')^3} - \frac{BC_\lambda}{(A+\delta')}]$$
$$C_\lambda = [(E_{g_0} + \delta')^{-1} + (E_{g_0} + \delta')(E_{g_0} - \delta')^{-2}](A+\delta')^{-1}$$
$$P_\lambda = \frac{C_0}{A} J_\lambda, J_\lambda = (D_\lambda + 2(E_{g_0} - \delta')\sqrt{f_\lambda}),$$
$$D_\lambda = (1 + \frac{2(E_{g_0} - \delta')}{(A+\delta')}, f_\lambda = [\frac{1}{(A+\delta')^2} + \frac{1}{(E_{g_0} - \delta')^2} - C_\lambda],$$
$$t_{1\lambda} = (1 + \frac{3m_c}{m_r}\alpha\delta_\lambda), \alpha = \frac{1}{E_{g_0}}, \delta_\lambda = \frac{|e|^2 I\lambda^2}{96 m_r \pi c^3 \sqrt{\varepsilon_{sc}\varepsilon_0}} \text{ and } t_{2\lambda} = \alpha t_{1\lambda}$$

Under the condition of heavy doping, following the methods as developed in Chap. 1, the HD dispersion relations in this case in the presence of light waves can be written as

$$\frac{\hbar^2 k^2}{2m_c} = T_1(E, \eta_g, \lambda) \tag{5.46b}$$

$$\frac{\hbar^2 k^2}{2m_c} = T_2(E, \eta_g, \lambda) \tag{5.47}$$

$$\frac{\hbar^2 k^2}{2m_c} = T_3(E, \eta_g, \lambda) \tag{5.48}$$

where

$$T_1(E, \eta_g, \lambda) = [U_\lambda[T_{31}(E, \eta_g) + iT_{32}(E, \eta_g)] - P_\lambda]$$
$$T_2(E, \eta_g, \lambda) = [t_{1\lambda}\gamma_3(E, \eta_g) + (t_{2\lambda})2\theta_0(E, \eta_g)[1 + \text{Erf}(E/\eta_g)]^{-1} - \delta_\lambda]$$

and

$$T_3(E, \eta_g, \lambda) = [t_{1\lambda}\gamma_3(E, \eta_g) - \delta_\lambda]$$

The EEM can be expressed in this case by using (5.46a, 5.46b), (5.47) and (5.48) as

$$m*(E_{FHDL}, \eta_g, \lambda) = m_c \text{ Real part of } [T_1(E_{FHDL}, \eta_g, \lambda)]' \tag{5.49}$$

$$m*(E_{FHDL}, \eta_g, \lambda) = m_c[T_2(E_{FHDL}, \eta_g, \lambda)]' \tag{5.50}$$

$$m*(E_{FHDL}, \eta_g, \lambda) = m_c[T_3(E_{FHDL}, \eta_g, \lambda)]' \tag{5.51}$$

The electron concentration is given by

$$n_0 = \frac{g_v}{3\pi^2}\left(\frac{2m_c}{\hbar^2}\right)^{\frac{3}{2}} \text{Real part of}[\mathrm{T}_1(\mathrm{E}_{FHDL}, \eta_g, \lambda)]^{\frac{3}{2}} \tag{5.52}$$

$$n_0 = \frac{g_v}{3\pi^2}\left(\frac{2m_c}{\hbar^2}\right)^{\frac{3}{2}} [\mathrm{T}_2(\mathrm{E}_{FHDL}, \eta_g, \lambda)]^{\frac{3}{2}} \tag{5.53}$$

$$n_0 = \frac{g_v}{3\pi^2}\left(\frac{2m_c}{\hbar^2}\right)^{\frac{3}{2}} [\mathrm{T}_3(\mathrm{E}_{FHDL}, \eta_g, \lambda)]^{\frac{3}{2}} \tag{5.54}$$

where E_{FHDL} is the Fermi energy in HD III-V semiconductors in the presence of light waves as measured from the age of the unperturbed conduction band in the vertically upward direction.

The velocity along z direction and the density of states function in this case for HD optoelectronic Kane type materials under intense light waves whose conduction electrons in the absence of perturbation obey the three band model of Kane can respectively be written as

$$v_z(\mathrm{E}_1') = \sqrt{\frac{2}{m_c}}\frac{[T_1(E_1', \eta_g, \lambda)]^{1/2}}{T_1'(\mathrm{E}_1', \eta_g, \lambda)} \tag{5.55}$$

$$N(\mathrm{E}_1') = 4\pi g_v(\frac{2m_c}{\hbar^2})^{3/2}\sqrt{T_1(\mathrm{E}_1', \eta_g, \lambda)}[T_1'(\mathrm{E}_1', \eta_g, \lambda)] \tag{5.56}$$

where $\mathrm{E}_1' = E - E_{01HD}, E_{01HD} = \xi_1 + W - hv$, ξ_1 *is the root of the equation*

$$T_1(\xi_1, \eta_g, \lambda) = 0 \tag{5.57}$$

The EP in this case is given by

$$J_{LHD} = \frac{4\pi\alpha_0 e m_c g_v}{h^3} \text{Real part of} \int_{E_{01HD}}^{\alpha} T_1(\mathrm{E}_1', \eta_g, \lambda) f(E) d\mathrm{E}_1' \tag{5.58}$$

Similarly the EP for perturbed two band model of Kane and that of parabolic energy bands can respectively be expressed as

$$J_{LHD} = \frac{4\pi\alpha_0 e m_c g_v}{h^3} \int_{E_{02HD}}^{\alpha} T_2(\mathrm{E}_2', \eta_g, \lambda) f(E) d\mathrm{E}_2' \tag{5.59}$$

and

$$J_{LHD} = \frac{4\pi\alpha_0 e m_c g_v}{h^3} \int\limits_{E_{03HD}}^{\alpha} T_3(E_3', \eta_g, \lambda) f(E) dE_3' \tag{5.60}$$

where $E_2' = E - E_{02HD}, E_{02HD} = \xi_2 + W - h\nu$, ξ_2 *is the root of the equation*

$$T_2(\xi_2, \eta_g, \lambda) = 0 \tag{5.61}$$

and where $E_3' = E - E_{03HD}, E_{03HD} = \xi_3 + W - h\nu$, ξ_3 *is the root of the equation*

$$T_3(\xi_3, \eta_g, \lambda) = 0. \tag{5.62}$$

5.3 Results and Discussion

Using the appropriate equations, the normalized EP from HD n-$Hg_{1-x}Cd_xTe$ has been plotted as functions of normalized I_0 (for a given wavelength and considering red light for which λ is about 640 nm), λ (assuming I_0 = 10 nWm^{-2}) and the normalized electron degeneracy at $T = 4.2$ K in accordance with the perturbed three and two band models of Kane and that of perturbed parabolic energy bands in Figs. 5.1, 5.2 and 5.3 respectively. The Figs. 5.4, 5.5 and 5.6 exhibit all the aforementioned cases for HD n-$In_{1-x}Ga_xAs_yP_{1-y}$ lattice matched to InP respectively.

It appears that the J increases with the increasing electron degeneracy in accordance with all the band models. The combined influence of the energy band constants on the EP from ternary and quaternary materials can easily be assessed from all the figures. It appears that the EP decreases with increasing light intensity for all the materials and also decreases as the wavelength shifts from violet to red.

The influence of light is immediately apparent from all the plots, since the EP depends strongly on the light intensity for all types of perturbed band models, which is in direct contrast with that for the bulk specimens of the said compounds whose formulations depend on the general idea that the band structure is an invariant quantity in the presence of external photo-excitation together with the fact that the physics of EP is being converted mathematically by using the lower limit of integration as E_0 as often used in the literature. The dependence of J_L on light intensity and wavelength reflects the direct signature of the light wave on the band structure dependent physical properties of electronic materials in general in the presence of external photo-excitation and the photon assisted transport for the corresponding HD optoelectronic semiconductor devices. Although J_L tends to decrease with the increasing intensity and the wavelength but the rate of increase is totally band structure dependent.

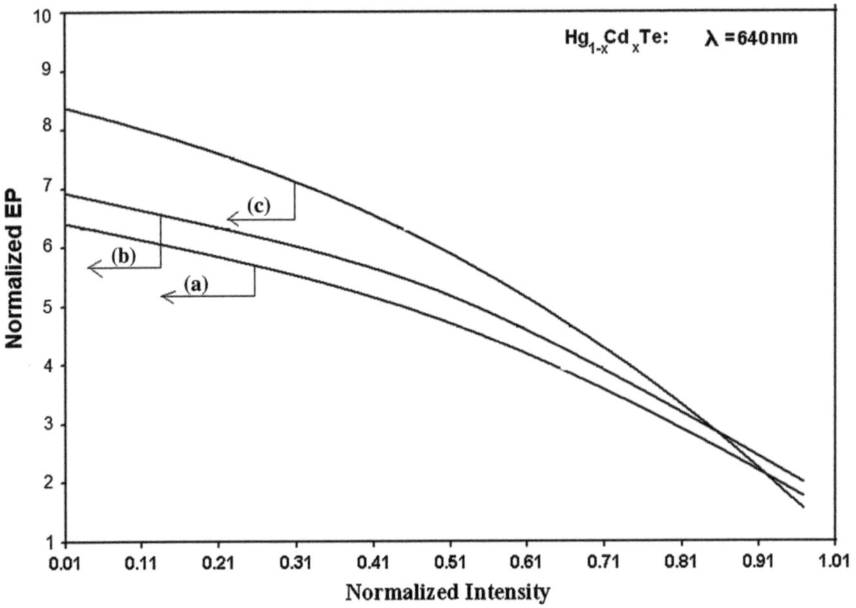

Fig. 5.1 Plot of the normalized EP from HD n-$Hg_{1-x}Cd_xTe$ as a function of normalized light intensity in which the curves (*a*), (*b*) and (*c*) represent the perturbed three and two band models of Kane together with parabolic energy bands respectively

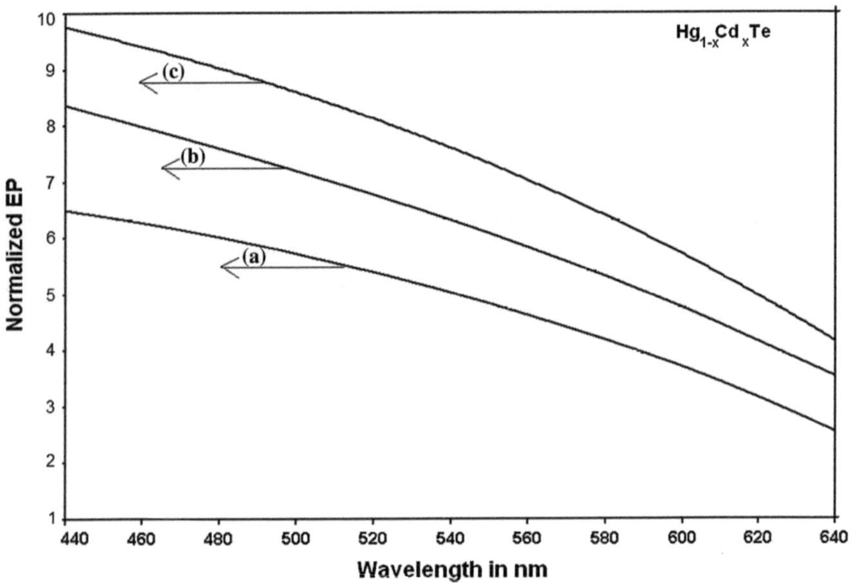

Fig. 5.2 Plot of the normalized EP from HD n-$Hg_{1-x}Cd_xTe$ as a function of wavelength for all cases of Fig. 5.1

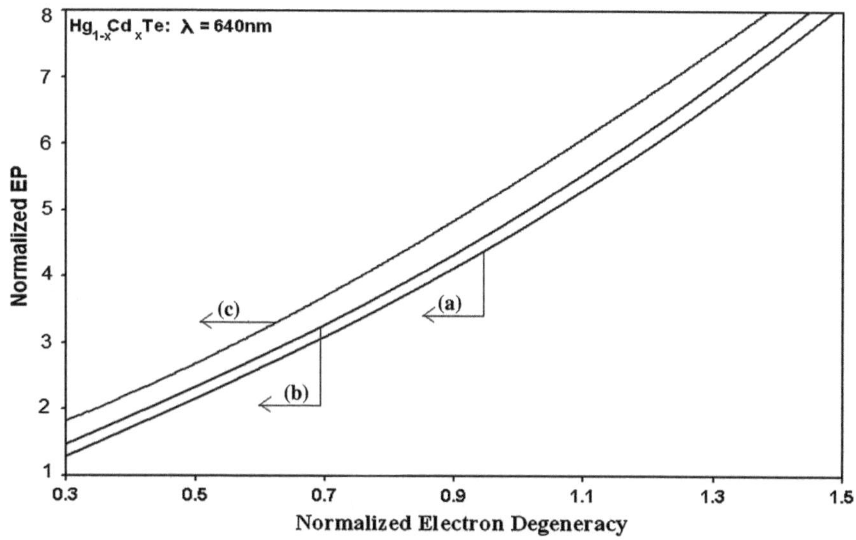

Fig. 5.3 Plot of the normalized EP from HD n-$Hg_{1-x}Cd_xTe$ as a function of normalized electron degeneracy for all cases of Fig. 5.1

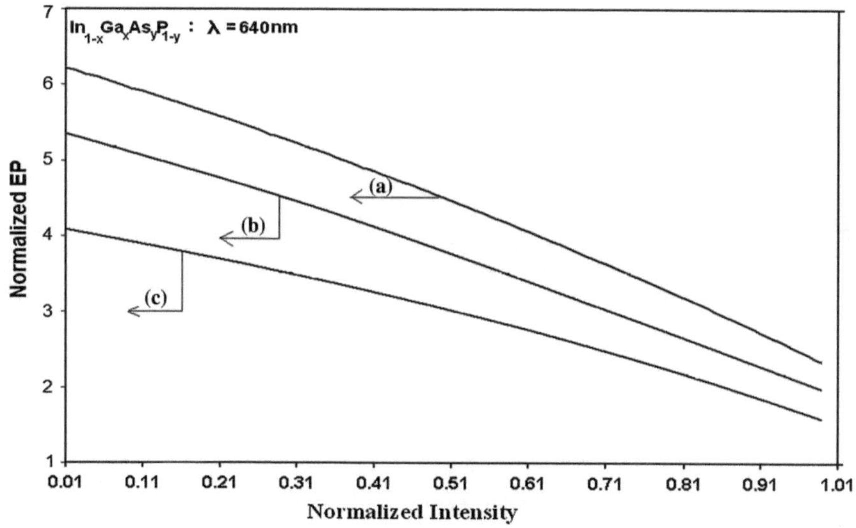

Fig. 5.4 Plot of the normalized EP from HD $In_{1-x}Ga_xAs_yP_{1-y}$ lattice matched to InP as a function of normalized light intensity in which the curves (*a*), (*b*) and (*c*) represent the perturbed three and two band models of Kane together with parabolic energy bands respectively

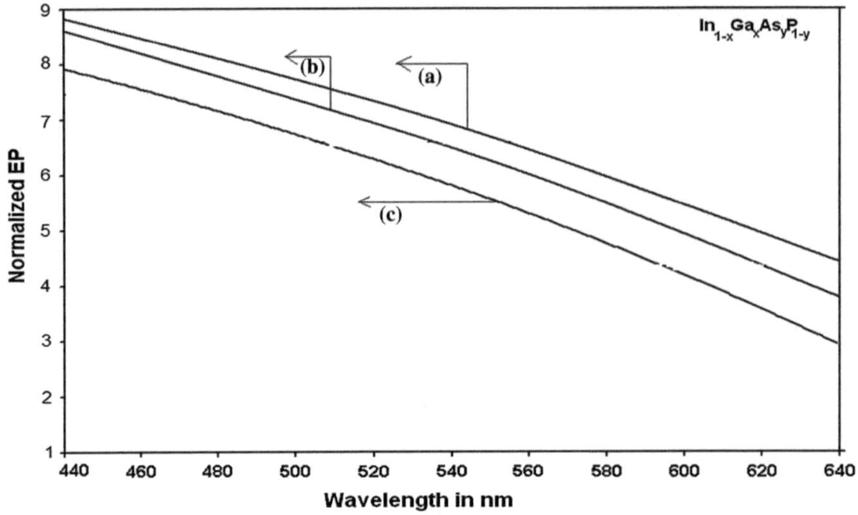

Fig. 5.5 Plot of the normalized EP from HD $In_{1-x}Ga_xAs_yP_{1-y}$ lattice matched to InP as a function of wavelength for all cases of Fig. 5.4

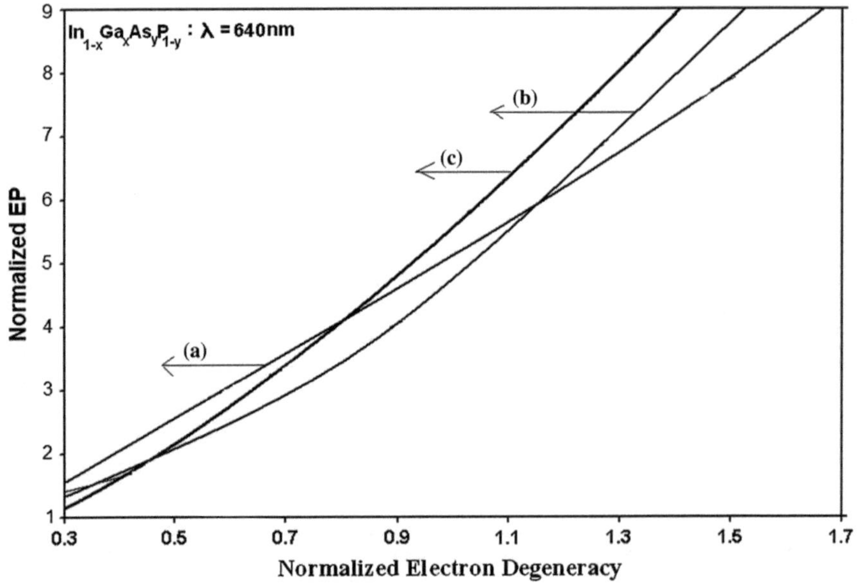

Fig. 5.6 Plot of the normalized EP from HD $In_{1-x}Ga_xAs_yP_{1-y}$ lattice matched to InP as a function of normalized electron degeneracy for all cases of Fig. 5.4

It is worth remarking that our basic Eq. (5.46b) covers various materials having different energy band structures. Under certain limiting conditions, all the results of the EP for different materials having various band structures lead to the well-known expression of the same for wide-gap materials having simplified parabolic energy bands. This indirect test not only exhibits the mathematical compatibility of the formulation but also shows the fact that the presented simple analysis is a more generalized one, since well known result can be obtained under certain limiting conditions of the generalized expressions.

It is worth remarking that the influence of an external photo-excitation is to change radically the original band structure of the material. Because of this change, the photon field causes to increase the band gap of semiconductors. We have proposed two experiments for the measurement of band gap of HD semiconductors under photo-excitation in Chap. 9 in this context.

We have not considered other types of optoelectronic and III-V materials and other external variables for the purpose of concise presentation. Besides, the influence of energy band models and the various band constants on the EP for different materials can also be studied from all the figures of this chapter. The numerical results presented in this chapter would be different for other materials but the nature of variation would be unaltered. The theoretical results as given here would be useful in analyzing various other experimental data related to this phenomenon. Finally, it appears that this theory can be used to investigate the thermoelectric power, the Debye screening length, the magnetic susceptibilities, the Burstien Moss shift, plasma frequency, the Hall coefficient, the specific heat and other different transport coefficients of modern HD optoelectronic devices operated in the presence of light waves.

5.4 Open Research Problems

(R.5.1) Investigate the EP in the presence of intense external light waves for all the HD materials whose respective dispersion relations of the carriers in the absence of any field are given in R.1.1 of Chap. 1.

(R.5.2) Investigate the EP for the heavily–doped semiconductors in the presences of Gaussian, exponential, Kane, Halperian, Lax and Bonch-Burevich types of band tails [16] for all systems whose unperturbed carrier energy spectra are defined in (R.1.1) in the presence of external light waves.

(R.5.3) Investigate the EP in the presence of external light waves for bulk specimens of the HD negative refractive index, organic, magnetic and other advanced optical materials in the presence of an arbitrarily oriented alternating electric field.

(R.5.4) Investigate all the appropriate HD problems of this chapter for a Dirac electron.

(R.5.5) Investigate all the appropriate problems of this chapter by including the many body, broadening and hot carrier effects respectively.

(R.5.6) Investigate all the appropriate problems of this chapter by removing all the mathematical approximations and establishing the respective appropriate uniqueness conditions.

References

1. P.K. Basu, *Theory of Optical Process in Semiconductors, Bulk and Microstructures* (Oxford University Press, Oxford, 1997)
2. K.P. Ghatak, S. Bhattacharya, S. Bhowmik, R. Benedictus, S. Chowdhury, J. Appl. Phys. **103**, 094314 (2008)
3. K.P. Ghatak, S. Bhattacharya, J. Appl. Phys. **102**, 073704 (2007)
4. K.P. Ghatak, S. Bhattacharya, S.K. Biswas, A. De, A.K. Dasgupta, Phys. Scr. **75**, 820 (2007)
5. P.K. Bose, N. Paitya, S. Bhattacharya, D. De, S. Saha, K.M. Chatterjee, S. Pahari, K.P. Ghatak, Quantum Matter **1**, 89 (2012)
6. K.P. Ghatak, S. Bhattacharya, A. Mondal, S. Debbarma, P. Ghorai, A. Bhattacharjee, Quantum Matter **2**, 25 (2013)
7. S. Bhattacharya, D. De, S. Ghosh, J.P. Bannerje, M. Mitra, B. Nag, S. Saha, S.K. Bishwas, M. Paul, J. Comput. Theo. Nanosci. **7**, 1066 (2010)
8. K.P. Ghatak, S. Bhattacharya, S. Pahari, S.N. Mitra, P.K. Bose, J. Phys. Chem. Solids **70**, 122 (2009)
9. S. Bhattacharya, D. De, R. Sarkar, S. Pahari, A. De, A.K. Dasgupta, S.N. Biswas, K.P. Ghatak, J. Comput. Theo. Nanosci. **5**, 1345 (2008)
10. S. Mukherjee, D. De, D. Mukherjee, S. Bhattacharya, A. Sinha, K.P. Ghatak, Phys. B **393**, 347 (2007)
11. K. Seeger, *Semiconductor Physics*, 7th edn. (Springer, Heidelberg, 2006)
12. B.R. Nag, *Physics of Quantum Well Devices* (Kluwer Academic Publishers, The Netherlands, 2000)
13. R.K. Pathria, *Statistical Mechanics*, 2nd edn. (Butterworth-Heinmann, Oxford, 1996)
14. S. Bhatttacharya, K.P. Ghatak, *Fowler-Nordheim Field Emission*, vol. 170, Springer Series in Solid-State Sciences (Springer, Germany, 2012)

Chapter 6
The EP from HD Kane Type Materials Under Magnetic Quantization

6.1 Introduction

In this chapter, the EP under magnetic quantization in HD Kane type materials has been investigated in the presence of external photo-excitation whose conduction electrons obeys the energy wave-vector dispersion relations as given by (5.46b), (5.47) and (5.48) respectively. The Sect. 6.2 contains the theoretical background. The dependence of the magneto EP from HD n-$Hg_{1-x}Cd_xTe$ and n-$In_{1-x}Ga_xAs_yP_{1-y}$ lattice matched to InP on the inverse quantizing magnetic field, the carrier concentration, the intensity of light and the wavelength has been discussed in Sect. 6.3. The Sect. 6.4 presents open research problems pertinent to this chapter.

6.2 Theoretical Background

Using (5.46b), the magneto-dispersion law, in the absence of spin, for HD III-V, ternary and quaternary semiconductors, in the presence of photo-excitation, whose unperturbed conduction electrons obey the three band model of Kane, is given by [1]

$$T_1(E,\eta_g) = \left(n+\frac{1}{2}\right)\hbar\,\omega_0 + \frac{\hbar^2 k_z^2}{2m_c} \tag{6.1}$$

Using (6.1), the DOS function in the present case can be expressed as

$$D_B(E,\eta_g,\lambda) = \frac{g_v|e|\sqrt{2m_c}}{2\pi^2\hbar^2}\sum_{n=0}^{n_{\max}}\left[\{T_1(E,\eta_g,\lambda)\}'\left\{T_1(E,\eta_g,\lambda)-\left(n+\frac{1}{2}\right)\hbar\omega_0\right\}^{-1/2} H(E-E_{n_{l1}})\right] \tag{6.2}$$

K.P. Ghatak, *Einstein's Photoemission*, Springer Tracts in Modern Physics 262,
DOI 10.1007/978-3-319-11188-9_6

where $E_{n_{l1}}$ is the Landau sub-band energies in this case and is given as

$$T_1\left(E_{n_{l1}}, \eta_g, \lambda\right) = \left(n + \frac{1}{2}\right)\hbar\omega_0 \tag{6.3}$$

The EEM in this case assumes the form

$$m^*\left(E_{FHDLB}, \eta_g, \lambda\right) = m_c \text{ Real part of } \left\{T_1\left(E_{FHDLB}, \eta_g, \lambda\right)\right\}' \tag{6.4}$$

where E_{FHDIB} is the Fermi energy under quantizing magnetic field in the presence of light waves as measured from the edge of the conduction band in the vertically upward direction in the absence of any quantization. Combining (6.2) with the Fermi-Dirac occupation probability factor and using the generalized Sommerfeld's lemma [1], the electron concentration can be written as

$$n_0 = \frac{g_v|e|\sqrt{2m_c}}{\pi^2\hbar^2}\sum_{n=0}^{n_{\max}}\left[M_{13}\left(E_{FHDLB}, \eta_g, \lambda\right) + N_{13}\left(E_{FHDLB}, \eta_g, \lambda\right)\right] \tag{6.5}$$

where $M_{13}\left(E_{FHDLB}, \eta_g, \lambda\right) \equiv \text{Real part of } \left[T_1\left(E_{FHDLB}, \eta_g, \lambda\right) - \left(n + \frac{1}{2}\right)\hbar\omega_0\right]^{1/2}$, and $N_{13}\left(E_{FHDLB}, \eta_g, \lambda\right) \equiv \sum_{r=1}^{s} L(r) M_{13}(E_{FHDLB}, \eta_g, \lambda)$.

The EP in this case is given by

$$J = \frac{\alpha_0 e^2 B k_B T}{2\pi^2\hbar^2} \text{ Real part of } \sum_{n=0}^{n_{\max}} F_0(\eta_{61HDLB}) \tag{6.6}$$

where $\eta_{61HDLB} = (k_B T)^{-1}[E_{FHDLB} - (E_{n11} + W - h\nu)]$.

(i) Using (5.47), the magneto-dispersion law, in the absence of spin, for HD III-V, ternary and quaternary semiconductors, in the presence of photo-excitation, whose unperturbed conduction electrons obey the two band model of Kane, is given by

$$T_2\left(E, \eta_g\right) = \left(n + \frac{1}{2}\right)\hbar\omega_0 + \frac{\hbar^2 k_z^2}{2m_c} \tag{6.7}$$

Using (11.62), the DOS function in the present case can be expressed as

$$D_B\left(E, \eta_g, \lambda\right) = \frac{g_v|e|\sqrt{2m_c}}{2\pi^2\hbar^2}\sum_{n=0}^{n_{\max}}\left[\left\{T_2(E, \eta_g, \lambda)\right\}'\left\{T_2\left(E, \eta_g, \lambda\right) - \left(n + \frac{1}{2}\right)\hbar\omega_0\right\}^{-1/2} H(E - E_{n_{l2}})\right] \tag{6.8}$$

where $E_{n_{l2}}$ is the Landau sub-band energies in this case and is given as

$$T_2\left(E_{n_{l2}}, \eta_g, \lambda\right) = \left(n + \frac{1}{2}\right)\hbar\omega_0 \tag{6.9}$$

The EEM in this case assumes the form

$$m^*\left(E_{FHDIB}, \eta_g, \lambda\right) = m_c\left\{T_2\left(E_{FHDIB}, \eta_g, \lambda\right)\right\}' \tag{6.10}$$

where E_{FHDLB} is the Fermi energy under quantizing magnetic field in the presence of light waves as measured from the edge of the conduction band in the vertically upward direction in the absence of any quantization. Combining (6.8) with the Fermi-Dirac occupation probability factor and using the generalized Sommerfeld's lemma, the electron concentration can be written as

$$n_0 = \frac{g_v|e|\sqrt{2m_c}}{\pi^2\hbar^2}\sum_{n=0}^{n_{\max}}\left[M_{23}\left(E_{FHDLB}, \eta_g, \lambda\right) + N_{23}\left(E_{FHDLB}, \eta_g, \lambda\right)\right] \tag{6.11}$$

where $M_{23}\left(E_{FHDLB}, \eta_g, \lambda\right) \equiv \left[T_2\left(E_{FHDLB}, \eta_g, \lambda\right) - \left(n + \frac{1}{2}\right)\hbar\omega_0\right]^{1/2}$ and $N_{23}\left(E_{FHDLB}, \eta_g, \lambda\right) \equiv \sum_{r=1}^{s} L(r)M_{23}\left(E_{FHDLB}, \eta_g, \lambda\right)$.
The EP in this case is given by

$$J = \frac{\alpha_0 e^2 B k_B T}{2\pi^2\hbar^2}\sum_{n=0}^{n_{\max}} F_0\left(\eta_{62HDLB}\right) \tag{6.12}$$

where $\eta_{62HDLB} = (k_BT)^{-1}\left[E_{FHDLB} - (E_{n12} + W - h\nu)\right]$.

(ii) Using (5.48), the magneto-dispersion law, in the absence of spin, for HD III-V, ternary and quaternary semiconductors, in the presence of photo-excitation, whose unperturbed conduction electrons obey the parabolic energy bands, is given by

$$T_3\left(E, \eta_g\right) = \left(n + \frac{1}{2}\right)\hbar\omega_0 + \frac{\hbar^2 k_z^2}{2m_c} \tag{6.13}$$

Using (6.13), the DOS function in the present case can be expressed as

$$D_B\left(E, \eta_g, \lambda\right) = \frac{g_v|e|\sqrt{2m_c}}{2\pi^2\hbar^2}\sum_{n=0}^{n_{\max}}\left[\left\{T_3(E, \eta_g, \lambda)\right\}'\left\{T_3(E, \eta_g, \lambda) - \left(n + \frac{1}{2}\right)\hbar\omega_0\right\}^{-1/2} H\left(E - E_{n_{l2}}\right)\right] \tag{6.14}$$

where $E_{n_{l2}}$ is the Landau sub-band energies in this case and is given as

$$T_3\left(E_{n_{l2}}, \eta_g, \lambda\right) = \left(n + \frac{1}{2}\right)\hbar\omega_0 \tag{6.15}$$

The EEM in this case assumes the form

$$m^*\left(E_{FHDIB}, \eta_g, \lambda\right) = m_c\left\{T_3\left(E_{FHDIB}, \eta_g, \lambda\right)\right\}' \tag{6.16}$$

where E_{FHDLB} is the Fermi energy under quantizing magnetic field in the presence of light waves as measured from the edge of the conduction band in the vertically upward direction in the absence of any quantization. Combining (6.14) with the Fermi-Dirac occupation probability factor and using the generalized Sommerfeld's lemma, the electron concentration can be written as

$$n_0 = \frac{g_v|e|\sqrt{2m_c}}{\pi^2\hbar^2}\sum_{n=0}^{n_{\max}}\left[M_{33}\left(E_{FHDLB}, \eta_g, \lambda\right) + N_{33}\left(E_{FHDLB}, \eta_g, \lambda\right)\right] \tag{6.17}$$

where $M_{33}\left(E_{FHDLB}, \eta_g, \lambda\right) \equiv \left[T_3\left(E_{FHDLB}, \eta_g, \lambda\right) - \left(n + \frac{1}{2}\right)\hbar\omega_0\right]^{1/2}$ and $N_{33}\left(E_{FHDLB}, \eta_g, \lambda\right) \equiv \sum_{r=1}^{s} L(r) M_{33}\left(E_{FHDLB}, \eta_g, \lambda\right)$.

The EP in this case is given by

$$J = \frac{\alpha_0 e^2 B k_B T}{2\pi^2\hbar^2}\sum_{n=0}^{n_{\max}} F_0\left(\eta_{63HDLB}\right) \tag{6.18}$$

where $\eta_{63HDLB} = (k_B T)^{-1}\left[E_{FHDLB} - \left(E_{n13} + W - h\nu\right)\right]$.

6.3 Results and Discussion

Using the appropriate equations we have plotted the normalized magneto EP from HD n-$Hg_{1-x}Cd_xTe$ versus inverse quantizing magnetic field in accordance with the perturbed three and two band models of Kane and that of perturbed parabolic energy bands as shown in Fig. 6.1. The Figs. 6.2, 6.3 and 6.4 exhibit the variation of the aforementioned quantity from HD n-$Hg_{1-x}Cd_xTe$ as functions of the normalized electron degeneracy, the normalized intensity of light and wavelength at $T = 4.2$ K respectively. The Figs. 6.5, 6.6, 6.7 and 6.8 represent the said variations of EP under magnetic quantization from HD n-$In_{1-x}Ga_xAs_yP_{1-y}$ lattice matched to InP.

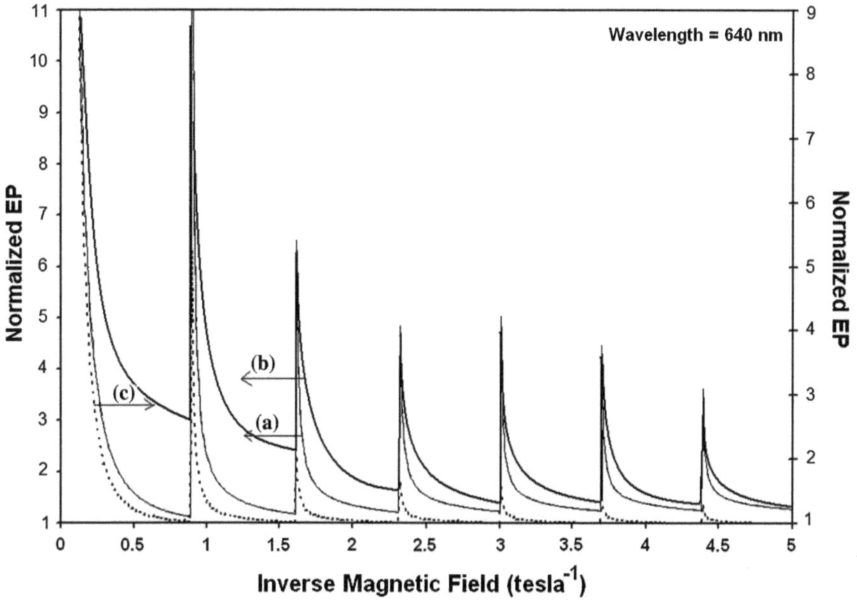

Fig. 6.1 Plot of the normalized EP as a function of inverse magnetic field from HD n-$Hg_{1-x}Cd_xTe$ in which the curves (*a*), (*b*) and (*c*) represent the perturbed three and two band models of Kane together with parabolic energy bands respectively

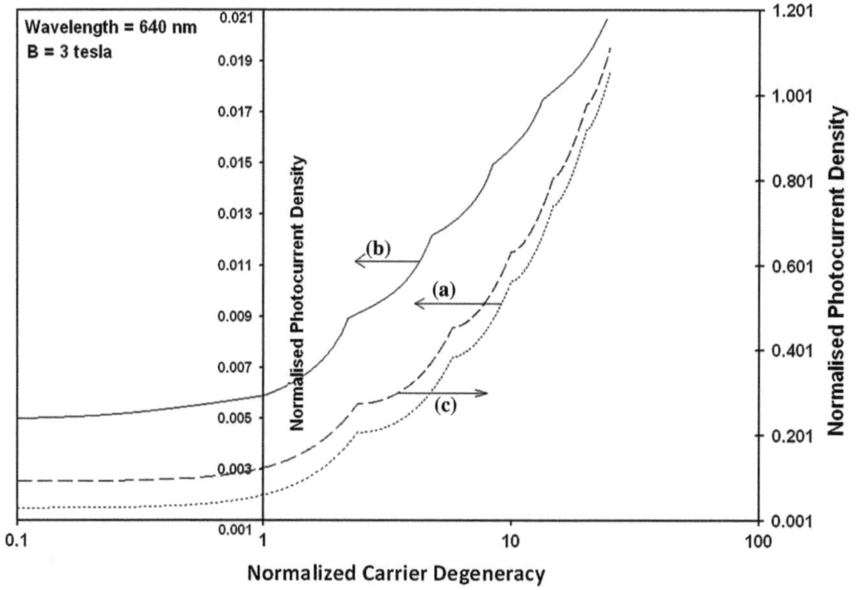

Fig. 6.2 Plot of the normalized EP as a function of normalized carrier degeneracy from HD n-$Hg_{1-x}Cd_xTe$ in which the curves (*a*), (*b*) and (*c*) represent the perturbed three and two band models of Kane together with parabolic energy bands respectively

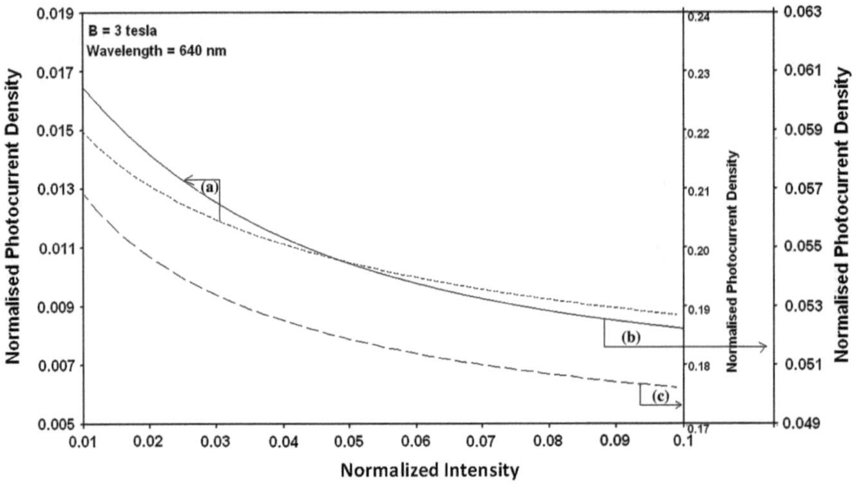

Fig. 6.3 Plot of the normalized EP as a function of normalized light intensity from HD n-$Hg_{1-x}Cd_xTe$ in which the curves (*a*), (*b*) and (*c*) represent the perturbed three and two band models of Kane together with parabolic energy bands respectively

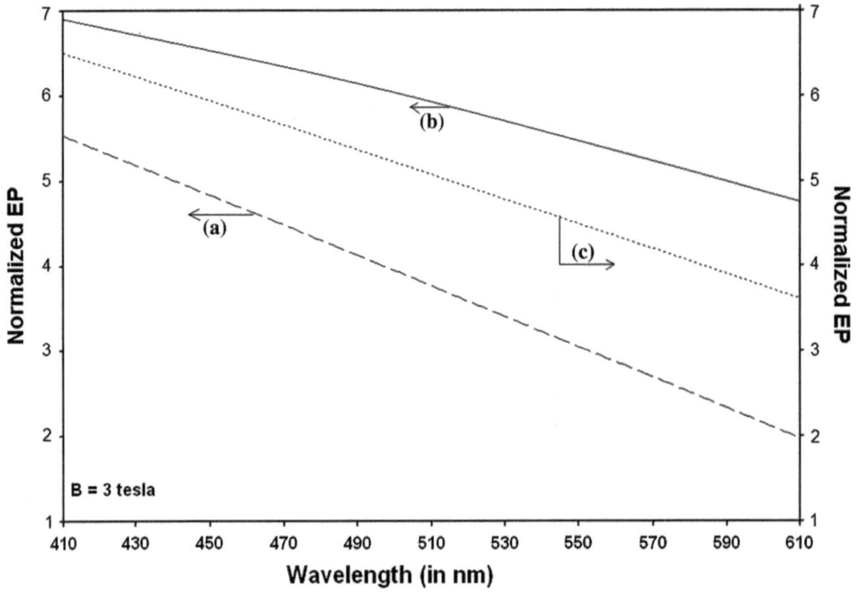

Fig. 6.4 Plot of the normalized EP as a function of wavelength from HD n-$Hg_{1-x}Cd_xTe$ in which the curves (*a*), (*b*) and (*c*) represent the perturbed three and two band models of Kane together with parabolic energy bands respectively

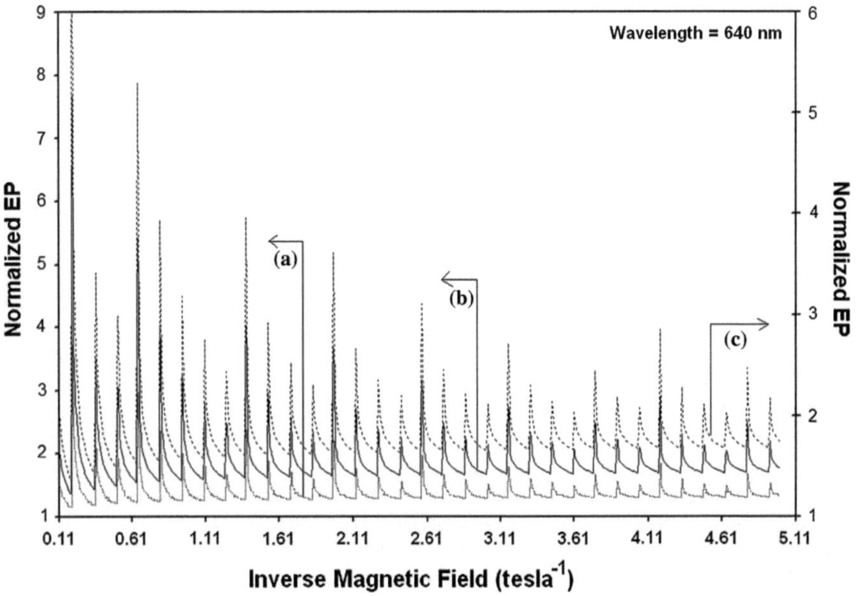

Fig. 6.5 Plot of the normalized EP as a function of inverse magnetic field from HD $In_{1-x}Ga_xAs_yP_{1-y}$ lattice matched to InP in which the curves (*a*), (*b*) and (*c*) represent the perturbed three and two band models of Kane together with parabolic energy bands respectively

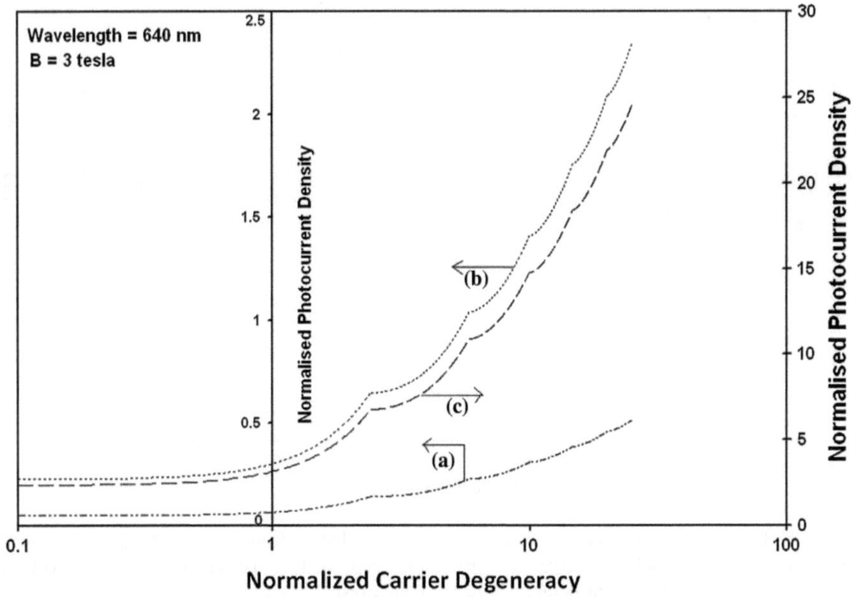

Fig. 6.6 Plot of the normalized EP as a function of normalized carrier degeneracy from HD $In_{1-x}Ga_xAs_yP_{1-y}$ lattice matched to InP in which the curves (*a*), (*b*) and (*c*) represent the perturbed three and two band models of Kane together with parabolic energy bands respectively

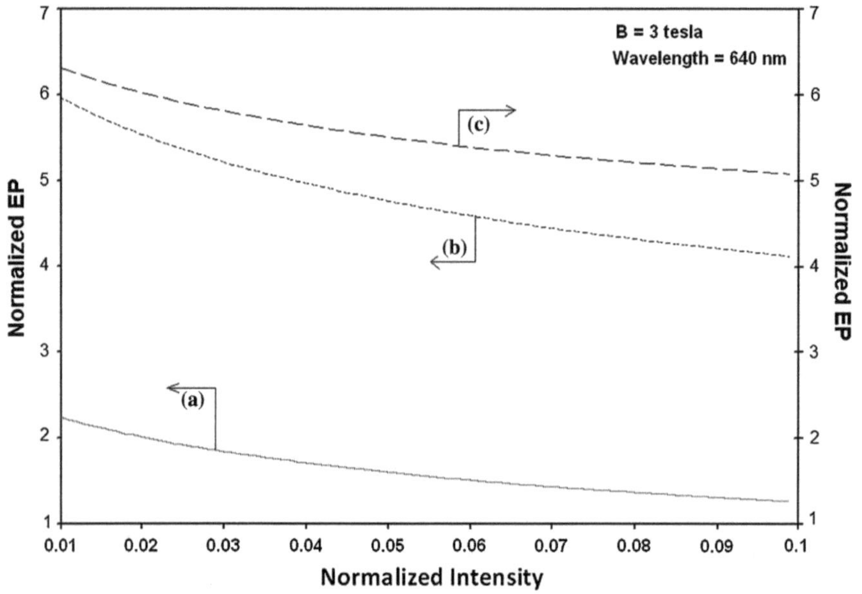

Fig. 6.7 Plot of the normalized EP as a function of normalized light intensity from HD $In_{1-x}Ga_xAs_yP_{1-y}$ lattice matched to InP in which the curves (*a*), (*b*) and (*c*) represent the perturbed three and two band models of Kane together with parabolic energy bands respectively

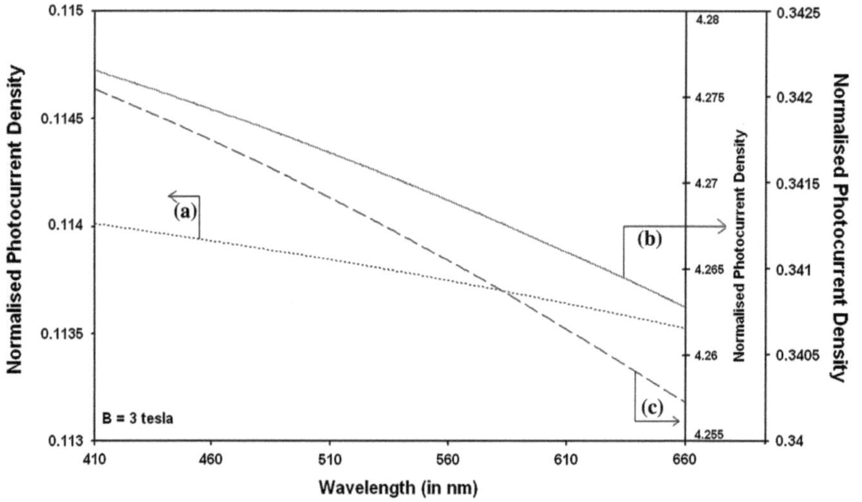

Fig. 6.8 Plot of the normalized EP as a function of wavelength from HD $In_{1-x}Ga_xAs_yP_{1-y}$ lattice matched to InP in which the curves (*a*), (*b*) and (*c*) represent the perturbed three and two band models of Kane together with parabolic energy bands respectively

It appears from Figs. 6.1 and 6.5 that the EP under magnetic quantization oscillates with inverse quantizing magnetic field and the numerical values are different in various cases which is the direct signature of the band structure. It may be noted the origin of the oscillation is the same as that of SdH oscillations. From Figs. 6.2 and 6.6, we observe that the EP oscillates with electron degeneracy, although the nature of oscillation is different. Figures 6.3 and 6.7 exhibits the fact that the normalized magneto EP decreases with increasing intensity and the slopes directly reflects the influence of energy band constants.

The Figs. 6.4 and 6.8 reflect the fact that the magneto EP decreases with increasing wavelength. Finally, we note that the form of the expression of the said physical quantity in this case as given by (6.10)–(6.12) is generalized where the Landau energy and the Fermi energy under magnetic field are the two band structure dependent quantities.

6.4 Open Research Problems

Investigate the following open research problems in the presence of external photo-excitation which changes the band structure in a fundamental way together with the proper inclusion of the electron spin, the variation of work function and the broadening of Landau levels respectively for appropriate problems.

(R.6.1) Investigate the multi-photon EP from all the HD materials whose unperturbed dispersion relations are given in (R.1.1) of Chap. 1 in the presence of arbitrarily oriented photo-excitation and quantizing magnetic field respectively.

(R.6.2) Investigate the multi-photon EP from all the HD materials whose unperturbed dispersion relations are given in (R.1.1) of Chap. 1 in the presence of an arbitrarily oriented non-quantizing non-uniform electric field, photo-excitation and quantizing magnetic field respectively.

(R.6.3) Investigate the multi-photon EP from all the HD materials whose unperturbed dispersion relations are given in (R.1.1) of Chap. 1 in the presence of an arbitrarily oriented non-quantizing alternating electric field, photo-excitation and quantizing magnetic field respectively.

(R.6.4) Investigate the multi-photon EP from all the HD materials whose unperturbed dispersion relations are given in (R.1.1) of Chap. 1 in the presence of an arbitrarily oriented non-quantizing alternating electric field, photo-excitation and quantizing alternating magnetic field respectively.

(R.6.5) Investigate the multi-photon EP from all the HD materials whose unperturbed dispersion relations are given in (R.1.1) of Chap. 1 in the presence of an arbitrarily oriented photo-excitation and crossed electric and quantizing magnetic fields respectively.

(R.6.6) Investigate the multi-photon EP for arbitrarily oriented photo-excitation and quantizing magnetic field from the heavily-doped materials in the presence of Gaussian, exponential, Kane, Halperin, Lax and Bonch-Bruevich types of band for all materials whose unperturbed carrier energy spectra are defined in Chap. 1.

(R.6.7) Investigate the multi-photon EP for arbitrarily oriented photo-excitation and quantizing alternating magnetic field for all the cases of R.6.6.

(R.6.8) Investigate the multi-photon EP for arbitrarily oriented photo-excitation and non-quantizing alternating electric field and quantizing magnetic field for all the cases of R.6.6.

(R.6.9) Investigate the multi-photon EP for arbitrarily oriented photo-excitation and non-uniform alternating electric field and quantizing magnetic field for all the cases of R.6.6.

(R.6.10) Investigate the multi-photon EP for arbitrarily oriented photo-excitation and crossed electric and quantizing magnetic fields for all the cases of R.6.6.

(R.6.11) Investigate the multi-photon EP from HD negative refractive index, organic, magnetic, heavily doped, disordered and other advanced optical materials in the presence of arbitrary oriented photo-excitation and quantizing magnetic field.

(R.6.12) Investigate the multi-photon EP in the presence of arbitrary oriented photo-excitation, quantizing magnetic field and alternating non-quantizing electric field for all the problems of R.6.11.

(R.6.13) Investigate the multi-photon EP in the presence of arbitrary oriented photo-excitation, quantizing magnetic field and non-quantizing non-uniform electric field for all the problems of R.6.11.

(R.6.14) Investigate the multi-photon EP in the presence of arbitrary oriented photo-excitation, alternating quantizing magnetic field and crossed alternating non-quantizing electric field for all the problems of R.6.11.

(R.6.15) Investigate all the problems from R.6.1 to R.6.14 by removing all the mathematical approximations and establishing the respective appropriate uniqueness conditions.

Reference

1. S. Bhattacharya, K.P. Ghatak, *Effective Electron Mass in Low-Dimensional Semiconductors*, vol. 167, Springer Series in Materials Science (Springer, Germany, 2013)

Chapter 7
The EP from QWs, NWs and QBs of HD Optoelectronic Materials

7.1 Introduction

In this chapter, in Sects. 7.2.1, 7.2.2 and 7.2.3 of theoretical background, the EP from QWs, NWs and QBs of HD optoelectronic materials has been studied, whose bulk conduction electrons are defined by the dispersion relations as given by (5.46b), (5.47) and (5.48) respectively. In Sect. 7.2, the EP from the afore-mentioned HD quantum confined materials has been investigated with respect to various external variables and Sect. 7.3 includes the result and discussions. The Sect. 7.4 presents open research problems pertinent to this chapter.

7.2 Theoretical Background

7.2.1 The EP from HD QWs of Optoelectronic Materials

The dispersion relation of the 2D electrons in QWs of HD optoelectronic materials, the conduction electrons of whose bulk samples are defined by the dispersion relations as given by (5.46b), (5.47) and (5.48) can, respectively, be expressed following (7.1) as

$$k_x^2 + k_y^2 = \frac{2m_c T_1(E, \eta_g, \lambda)}{\hbar^2} - \left(\frac{\pi n_{z71}}{d_z}\right)^2 \tag{7.1}$$

$$k_x^2 + k_y^2 = \frac{2m_c T_2(E, \eta_g, \lambda)}{\hbar^2} - \left(\frac{\pi n_{z72}}{d_z}\right)^2 \tag{7.2}$$

K.P. Ghatak, *Einstein's Photoemission*, Springer Tracts in Modern Physics 262,
DOI 10.1007/978-3-319-11188-9_7

$$k_x^2 + k_y^2 = \frac{2m_c T_3(E, \eta_g, \lambda)}{\hbar^2} - \left(\frac{\pi n_{z73}}{d_z}\right)^2 \tag{7.3}$$

where $n_{z7J}(J = 1, 2, 3)$ is the size quantum number.

The 2D electron statistics assumes the form

$$n_{2DL} = \left(\frac{m_c g_v}{\pi \hbar^2}\right) \text{Real part of} \sum_{n_{z71}=1}^{n_{z71\max}} [\phi_{71}(E_{F2DL}, n_{z71}) + \phi_{72}(E_{F2DL}, n_{z71})] \tag{7.4}$$

$$n_{2DL} = \left(\frac{m_c g_v}{\pi \hbar^2}\right) \sum_{n_{z72}=1}^{n_{z72\max}} [\phi_{73}(E_{F2DL}, n_{z72}) + \phi_{74}(E_{F2DL}, n_{z72})] \tag{7.5}$$

$$n_{2DL} = \left(\frac{m_c g_v}{\pi \hbar^2}\right) \sum_{n_{z73}=1}^{n_{z73\max}} [\phi_{75}(E_{F2DL}, n_{z73}) + \phi_{76}(E_{F2DL}, n_{z73})] \tag{7.6}$$

where E_{F2DL} is the Fermi energy in HDQWs in the presence of lightwaves as measured from the edge of the conduction band in the vertically upward direction in the absence of any quantization,

$$\phi_{71}(E_{2DF}, n_{z71}) = \left[\frac{2m_c}{\hbar^2} T_1(E_{F2DL}, \eta_g, \lambda) - \left(\frac{\pi n_{z71}}{d_z}\right)^2\right],$$

$$\phi_{72}(E_{F2DL}, n_{z71}) = \sum_{r=1}^{s_0} L(r)[\phi_{71}(E_{F2DL}, n_{z71})],$$

$$\phi_{73}(E_{F2DL}, n_{z72}) = \left[\frac{2m_c}{\hbar^2} T_2(E_{F2DL}, \eta_g, \lambda) - \left(\frac{\pi n_{z72}}{d_z}\right)^2\right],$$

$$\phi_{74}(E_{F2DL}, n_{z72}) = \sum_{r=1}^{s_0} L(r)[\phi_{73}(E_{F2DL}, n_{z72})],$$

$$\phi_{75}(E_{F2DL}, n_{z73}) = \left[\frac{2m_c}{\hbar^2} T_3(E_{F2DL}, \eta_g, \lambda) - \left(\frac{\pi n_{z73}}{d_z}\right)^2\right], \text{ and}$$

$$\phi_{76}(E_{F2DL}, n_{z73}) = \sum_{r=1}^{s_0} L(r)[\phi_{75}(E_{F2DL}, n_{z73})]$$

The velocity of the electron in the n_{z71}th, n_{z72}th and n_{z73}th sub bands for the 2D electron energy spectra as given by (7.1–7.3) can respectively be written as

$$\upsilon_z(E_{n_{z71}}) = \left(\frac{m_c}{2}\right)^{-1/2} \left[\frac{\sqrt{T_1(E_{n_{z71}}, \eta_g, \lambda)}}{T_1'(E_{n_{z71}}, \eta_g, \lambda)}\right] \tag{7.7}$$

$$\upsilon_z(E_{n_{z72}}) = \left(\frac{m_c}{2}\right)^{-1/2}\left[\frac{\sqrt{T_2(E_{n_{z72}},\eta_g,\lambda)}}{T_2'(E_{n_{z72}},\eta_g,\lambda)}\right] \tag{7.8}$$

$$\upsilon_z(E_{n_{z73}}) = \left(\frac{m_c}{2}\right)^{-1/2}\left[\frac{\sqrt{T_3(E_{n_{z73}},\eta_g,\lambda)}}{T_3'(E_{n_{z73}},\eta_g,\lambda)}\right] \tag{7.9}$$

where the sub-band energies $E_{n_{z71}}, E_{n_{z72}}$ and $E_{n_{z73}}$ are respectively defined through the following equations

$$T_1(E_{n_{z71}},\eta_g,\lambda) = \frac{\hbar^2}{2m_c}\left(\frac{\pi n_{z71}}{d_z}\right)^2 \tag{7.10}$$

$$T_2(E_{n_{z72}},\eta_g,\lambda) = \left(\frac{\hbar^2}{2m_c}\right)\left(\frac{\pi n_{z72}}{d_z}\right)^2 \tag{7.11}$$

and

$$T_3(E_{n_{z73}},\eta_g,\lambda) = \left(\frac{\hbar^2}{2m_c}\right)\left(\frac{\pi n_{z73}}{d_z}\right)^2 \tag{7.12}$$

The respective expressions of the photoemission are given by

$$J_{2DL} = \frac{\alpha_0 g_v e}{\pi\hbar^2 d_z}\left(\frac{m_c}{2}\right)^{-1/2} \text{Real part of} \sum_{n_{z71_{\min}}}^{n_{z71\max}}\left[\frac{\sqrt{T_1(E_{n_{z71}},\eta_g,\lambda)}}{T_1'(E_{n_{z71}},\eta_g,\lambda)}\right][\phi_{71}(E_{F2DL},n_{z71}) + \phi_{72}(E_{F2DL},n_{z71})] \tag{7.13}$$

where $n_{z71_{\min}} \geq \left(\frac{d_z}{\pi}\right)\left(\frac{\sqrt{2m_c}}{\hbar}\right)[T_1(W - h\upsilon,\eta_g,\lambda)]^{1/2}$.

$$J_{2DL} = \frac{\alpha_0 g_v e}{\pi\hbar^2 d_z}\left(\frac{m_c}{2}\right)^{-1/2} \sum_{n_{z72_{\min}}}^{n_{z72\max}}\left[\frac{\sqrt{T_2(E_{n_{z72}},\eta_g,\lambda)}}{T_2'(E_{n_{z72}},\eta_g,\lambda)}\right][\phi_{73}(E_{F2DL},n_{z72}) + \phi_{74}(E_{F2DL},n_{z72})] \tag{7.14}$$

where $n_{z72_{\min}} \geq \left(\frac{d_z}{\pi}\right)\left(\frac{\sqrt{2m_c}}{\hbar}\right)\sqrt{T_2(W - h\upsilon,\eta_g,\lambda)}$ and

$$J_{2DL} = \frac{\alpha_0 g_v e}{\pi \hbar^2 d_z} \left(\frac{m_c}{2}\right)^{-1/2} \sum_{n_{z73_{\min}}}^{n_{z73_{\max}}} \left[\frac{\sqrt{T_3(E_{n_{z73}}, \eta_g, \lambda)}}{T_3'(E_{n_{z73}}, \eta_g, \lambda)} \right] [\phi_{75}(E_{F2DL}, n_{z73}) + \phi_{72}(E_{F2DL}, n_{z73})] \tag{7.15}$$

where $n_{z73_{\min}} \geq \left(\frac{d_z}{\pi}\right)\left(\frac{\sqrt{2m_c}}{\hbar}\right)\left[T_3(W - h\upsilon, \eta_g, \lambda)\right]^{1/2}$.

7.2.2 The EP from HD NWs of Optoelectronic Materials

The dispersion relations of the 1D electrons in NWs of HD optoelectronic materials in the presence of light waves can be expressed from (7.1–7.3) as

$$k_y^2 = \frac{2m_c T_1(E, \eta_g, \lambda)}{\hbar^2} - \left(\frac{\pi n_{z71}}{d_z}\right)^2 - \left(\frac{\pi n_{x71}}{d_x}\right)^2 \tag{7.16}$$

$$k_y^2 = \frac{2m_c T_2(E, \eta_g, \lambda)}{\hbar^2} - \left(\frac{\pi n_{z72}}{d_z}\right)^2 - \left(\frac{\pi n_{x72}}{d_x}\right)^2 \tag{7.17}$$

$$k_y^2 = \frac{2m_c T_3(E, \eta_g, \lambda)}{\hbar^2} - \left(\frac{\pi n_{z73}}{d_z}\right)^2 - \left(\frac{\pi n_{x73}}{d_x}\right)^2 \tag{7.18}$$

where $n_{x7J} (J = 1, 2, 3)$ is the size quantum number.

The electron concentration per unit length are respectively given by

$$n_{1DL} = \frac{2g_v\sqrt{2m_c}}{\pi\hbar} \text{Real part of} \sum_{n_{x71}=1}^{n_{x71_{\max}}} \sum_{n_{z71}=1}^{n_{z71_{\max}}} [\phi_{77}(E_{F1DL}, n_{x71}, n_{z71}) + \phi_{78}(E_{F1DL}, n_{x71}, n_{z71})] \tag{7.19}$$

$$n_{1DL} = \frac{2g_v\sqrt{2m_c}}{\pi\hbar} \sum_{n_{x72}=1}^{n_{x72_{\max}}} \sum_{n_{z72}=1}^{n_{z72_{\max}}} [\phi_{79}(E_{F1DL}, n_{x72}, n_{z72}) + \phi_{80}(E_{F1DL}, n_{x72}, n_{z72})] \tag{7.20}$$

$$n_{1DL} = \frac{2g_v\sqrt{2m_c}}{\pi\hbar} \sum_{n_{x73}=1}^{n_{x73_{\max}}} \sum_{n_{z73}=1}^{n_{z73_{\max}}} [\phi_{81}(E_{F1DL}, n_{x73}, n_{z73}) + \phi_{82}(E_{F1DL}, n_{x73}, n_{z73})] \tag{7.21}$$

where E_{F1DL} is the Fermi energy in NWs in the presence of light waves as measured from the edge of the conduction band in the vertically upward direction in the absence of any quantization,

$$\phi_{77}(E_{F1DL}, n_{x71}, n_{z71}) = \left[T_1(E_{F1DL}, \eta_g, \lambda) - G_{71}(n_{x71}, n_{z71})\right]^{1/2},$$

$$G_{7i}(n_{x7i}, n_{z7i}) = \frac{\hbar^2}{2m_c}\left[\left(\frac{\pi n_{x7i}}{d_x}\right)^2 + \left(\frac{\pi n_{z7i}}{d_z}\right)^2\right],$$

$$\phi_{78}(E_{F1DL}, n_{x71}, n_{z71}) = \sum_{r=1}^{s_0} L(r)[\phi_{77}(E_{F1DL}, n_{x71}, n_{z71})],$$

$$\phi_{79}(E_{F1DL}, n_{x72}, n_{z72}) = \left[T_2(E_{F1DL}, \eta_g, \lambda) - G_{72}(n_{x72}, n_{z72})\right]^{1/2},$$

$$\phi_{80}(E_{F1DL}, n_{x72}, n_{z72}) = \sum_{r=1}^{s} L(r)[\phi_{79}(E_{F1DL}, n_{x72}, n_{z72})],$$

$$\phi_{81}(E_{F1DL}, n_{x73}, n_{z73}) = [\rho_0(E_{F1DL}, \lambda) - G_{73}(n_{x73}, n_{z73})]^{1/2} \text{ and}$$

$$\phi_{82}(E_{F1DL}, n_{x73}, n_{z73}) = \sum_{r=1}^{s} L(r)[\phi_{81}(E_{F1DL}, n_{x73}, n_{z73})].$$

The generalized expression of photo current in this case is given by

$$I_L = \frac{\alpha_0 e g_v k_B T}{\pi\hbar} \sum_{n_{x7i}=1}^{n_{x7i\max}} \sum_{n_{z7i}=1}^{n_{z7i\max}} F_0(\eta_{7i}) \tag{7.22a}$$

where, $\eta_{7i} = \frac{E_{F1DL} - \left(E'_{7i} + W - h\upsilon\right)}{k_B T}$ and E'_{7i} are the sub-band energies in this case and are defined through the following equations

$$\left.\begin{aligned} T_1\left(E'_{71}, \eta_g, \lambda\right) &= G_{71}(n_{x71}, n_{z71}) \\ T_2\left(E'_{72}, \eta_g, \lambda\right) &= G_{72}(n_{x72}, n_{z72}) \\ T_3\left(E'_{73}, \eta_g, \lambda\right) &= G_{73}(n_{x73}, n_{z73}) \end{aligned}\right\} \tag{7.22b}$$

Real Part of the (7.22a and 7.22b) should be used for computing the EP from NWs of HD optoelectronic materials whose unperturbed energy band structures are defined by the three-band model of Kane [1].

7.2.3 The EP from QB of HD Optoelectronic Materials

The dispersion relations of the electrons in QBs of HD optoelectronic materials in the presence of light waves can respectively be expressed from (7.16–7.18) as

$$\frac{2m_c T_1(E_{Q1}, \eta_g, \lambda)}{\hbar^2} = H_{71}(n_{x71}, n_{y71}, n_{z71}) \tag{7.23}$$

$$\frac{2m_c T_2(E_{Q2}, \eta_g, \lambda)}{\hbar^2} = H_{72}(n_{x72}, n_{y72}, n_{z72}) \tag{7.24}$$

$$\frac{2m_c T_3(E_{Q3}, \eta_g, \lambda)}{\hbar^2} = H_{73}(n_{x73}, n_{y73}, n_{z73}) \tag{7.25}$$

where E_{Qi} is the totally quantized energy and $H_{7i}(n_{x7i}, n_{y7i}, n_{z7i}) = \left(\frac{\pi n_{x7i}}{d_x}\right)^2 + \left(\frac{\pi n_{y7i}}{d_y}\right)^2 + \left(\frac{\pi n_{z7i}}{d_z}\right)^2$.

The electron concentration can, in general, be written as

$$n_{0DL} = \left(\frac{2g_v}{d_x d_y d_z}\right) \sum_{n_{x7i}=1}^{n_{x7i\max}} \sum_{n_{y7i}=1}^{n_{y7i\max}} \sum_{n_{z7i}=1}^{n_{z7i\max}} F_{-1}(\eta_{7i0D}) \tag{7.26}$$

where $\eta_{7i0D} = \frac{E_{F0DL} - E_{Qi}}{k_B T}$ and E_{F0DL} is the Fermi energy in QBs of HD optoelectronic materials in the presence of lightwaves as measured from the edge of the conduction band in the vertically upward direction in the absence of any quantization. Real Part of the (7.26) should be used for computing the carrier density from QBs of HD optoelectronic materials whose unperturbed energy band structures are defined by the three-band model of Kane.

The photo-emitted current densities in this case are given by the following equations

$$J_{0DL} = \frac{(\alpha_0 e g_v)}{d_x d_y d_z}\left(\frac{m_c}{2}\right)^{-1/2} \text{Real part of} \sum_{n_{x71}=1}^{n_{x71\max}} \sum_{n_{y71}=1}^{n_{y71\max}} \sum_{n_{z71\min}}^{n_{z71\max}} \left[\frac{\sqrt{T_1(E_{n_{z71}}, \eta_g, \lambda)}}{T_1'(E_{n_{z71}}, \eta_g, \lambda)}\right] F_{-1}(\eta_{710D}) \tag{7.27}$$

$$J_{0DL} = \frac{(\alpha_0 e g_v)}{d_x d_y d_z}\left(\frac{m_c}{2}\right)^{-1/2} \sum_{n_{x72}=1}^{n_{x72\max}} \sum_{n_{y72}=1}^{n_{y72\max}} \sum_{n_{z72\min}}^{n_{z72\max}} \left[\frac{\sqrt{T_2(E_{n_{z72}}, \eta_g, \lambda)}}{T_2'(E_{n_{z72}}, \eta_g, \lambda)}\right] F_{-1}(\eta_{720D}) \tag{7.28}$$

and

$$J_{0DL} = \frac{(\alpha_0 e g_v)}{d_x d_y d_z}\left(\frac{m_c}{2}\right)^{-1/2} \sum_{n_{x73}=1}^{n_{x73\max}} \sum_{n_{y73}=1}^{n_{y73\max}} \sum_{n_{z73\min}}^{n_{z73\max}} \left[\frac{\sqrt{T_3(E_{n_{z73}}, \eta_g, \lambda)}}{T_3'(E_{n_{z73}}, \eta_g, \lambda)}\right] F_{-1}(\eta_{730D}) \tag{7.29}$$

7.3 Results and Discussion

Using the numerical values of the energy band constants the normalized EP has been plotted from QWs of HD n-$Hg_{1-x}Cd_xTe$, under external photo-excitation whose band structure follows the perturbed HD three [using (7.13) and (7.4)] and HD two [using (7.14) and (7.5)] band models of Kane and that of the perturbed HD parabolic [using (7.15) and (7.6)] energy bands as shown by curves (a–c) of Fig. 7.1 as functions of film thickness. The plots of the Figs. 7.2, 7.3 and 7.4 exhibit the dependence of the normalized EP on the normalized electron degeneracy, normalized intensity and wavelength respectively for all cases of Fig. 7.1. The variations of the normalized EP from QWs of HD

n-$In_{1-x}Ga_xAs_yP_{1-y}$ lattice matched to InP as functions of film thickness, normalized carrier degeneracy, normalized incident light intensity and wavelength respectively have been drawn in Figs. 7.5, 7.6, 7.7 and 7.8 for all cases of Fig. 7.1. The dependences of the normalized EP from NWs of HD n-$Hg_{1-x}Cd_xTe$ with respect to film thickness, normalized carrier degeneracy, normalized light intensity and wavelength have been drawn in Figs. 7.9, 7.10, 7.11 and 7.12 in accordance with perturbed HD three [using (7.22a) and (7.19)] and HD two [using (7.22a) and (7.20)] band models of Kane together with HD parabolic [using (7.22a) and (7.21)] energy bands as shown by curves (a–c) respectively. The variations of normalized EP from NWs of HD n-$In_{1-x}Ga_xAs_yP_{1-y}$ lattice matched to InP, have been drawn in Figs. 7.13, 7.14, 7.13 and 7.16 as functions of film thickness, normalized carrier degeneracy, normalized incident light intensity and wavelengths respectively.

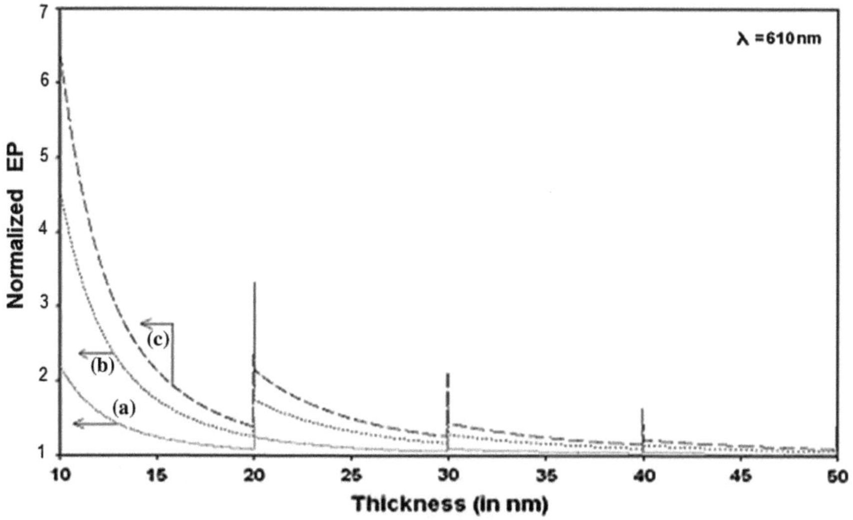

Fig. 7.1 Plot of the normalized EP from QWs of HD n-$Hg_{1-x}Cd_xTe$ as a function of film thickness in which the curves *a–c* represent the perturbed HD three and two band models of Kane together with HD parabolic energy bands respectively

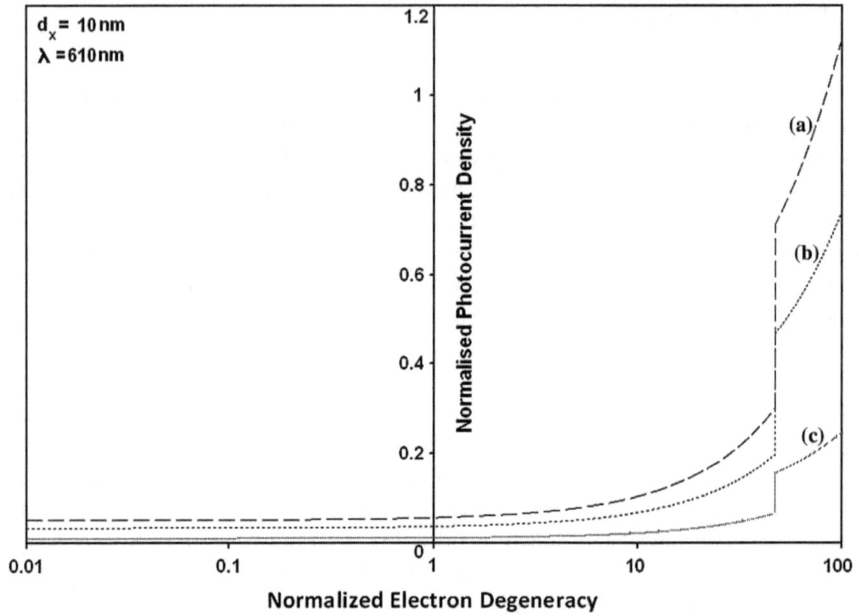

Fig. 7.2 Plot of the normalized EP from QWs of HD n-$Hg_{1-x}Cd_xTe$ as a function of normalized electron degeneracy for all cases of Fig. 7.1

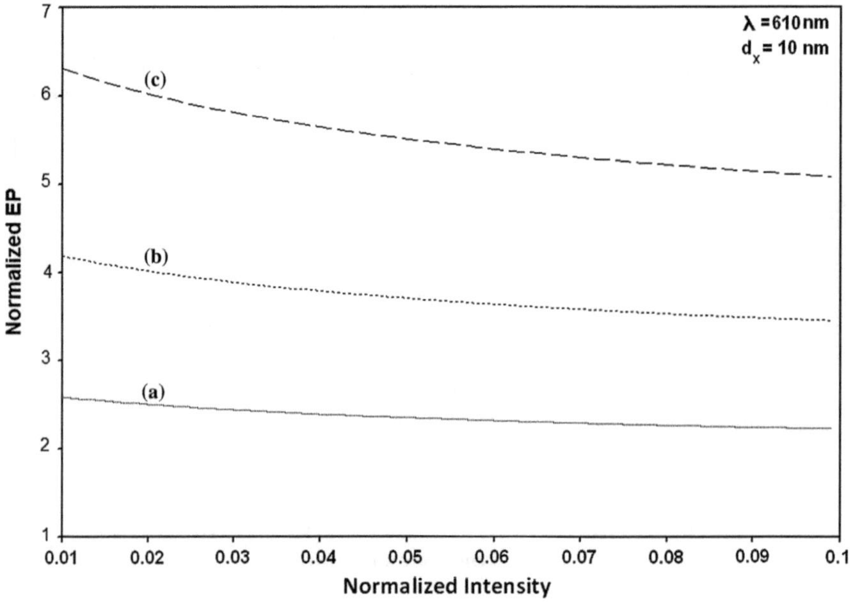

Fig. 7.3 Plot of the normalized EP from QWs of HD n-$Hg_{1-x}Cd_xTe$ as a function of normalized light intensity for all cases of Fig. 7.1

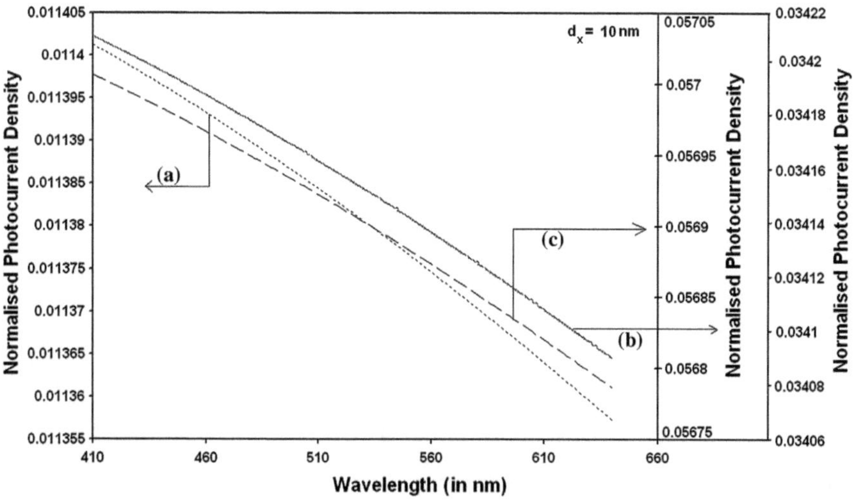

Fig. 7.4 Plot of the normalized EP from QWs of HD n-$Hg_{1-x}Cd_xTe$ as a function of light wavelength for all cases of Fig. 7.1

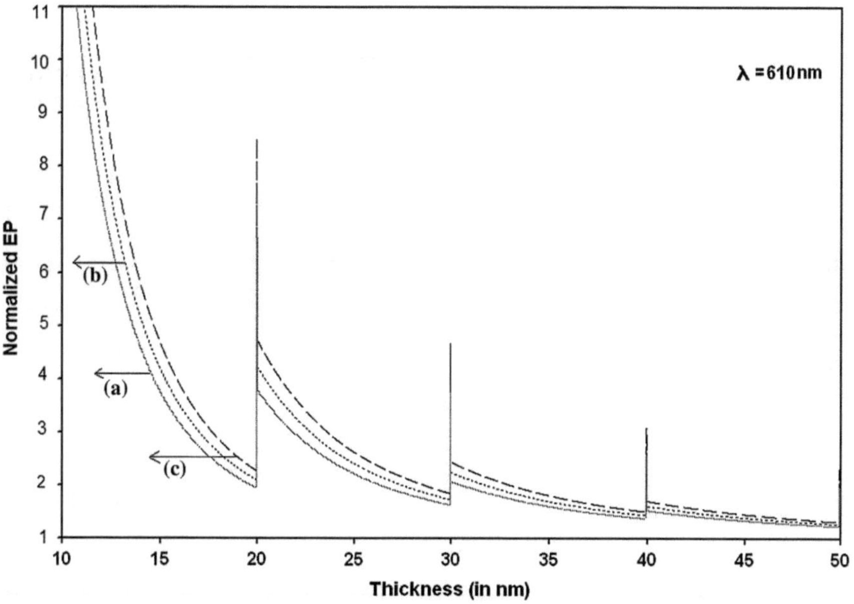

Fig. 7.5 Plot of the normalized EP from QWs of HD n-$In_{1-x}Ga_xAs_yP_{1-y}$ lattice matched to InP as a function of film thickness in which the curves *a–c* represent the perturbed HD three and two band models of Kane together with HD parabolic energy bands respectively

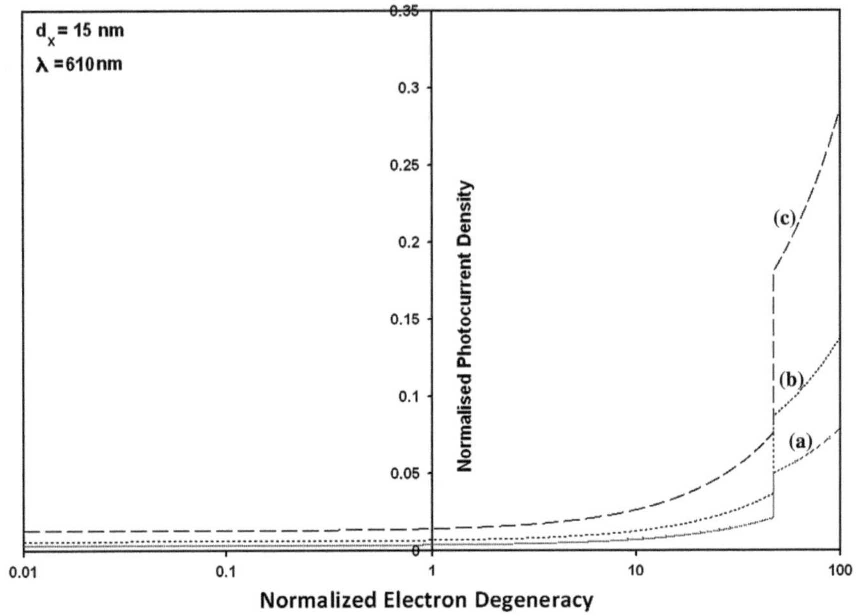

Fig. 7.6 Plot of the normalized EP from QWs of HD n-$In_{1-x}Ga_xAs_yP_{1-y}$ lattice matched to InP as a function of normalized electron degeneracy for all cases of Fig. 7.5

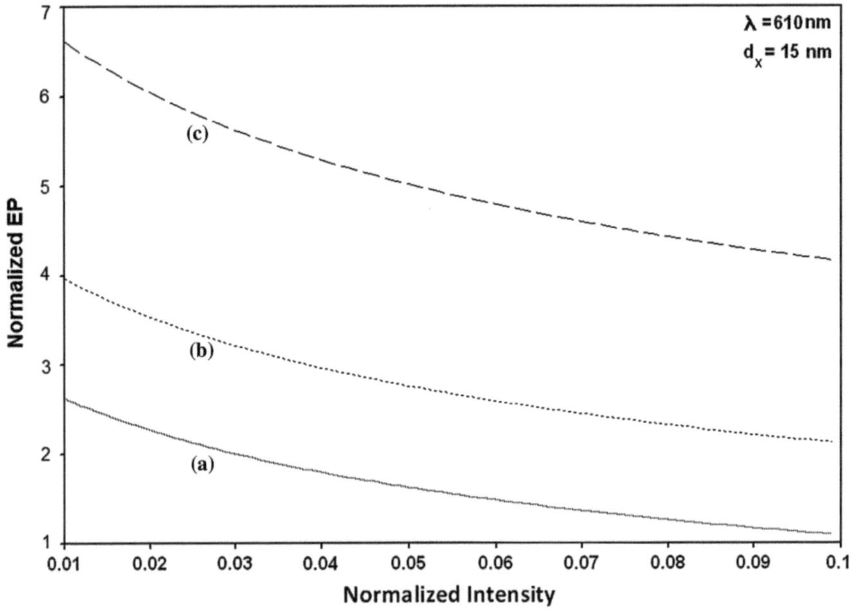

Fig. 7.7 Plot of the normalized from QWs of HD n-$In_{1-x}Ga_xAs_yP_{1-y}$ lattice matched to InP as a function of normalized light intensity for all cases of Fig. 7.5

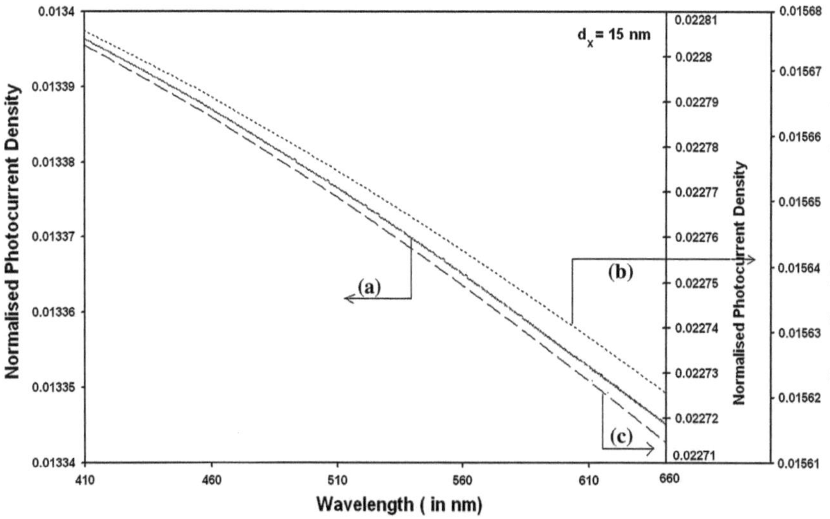

Fig. 7.8 Plot of the normalized EP from QWs of HD n-$In_{1-x}Ga_xAs_yP_{1-y}$ lattice matched to InP as a function of light wavelength for all cases of Fig. 7.5

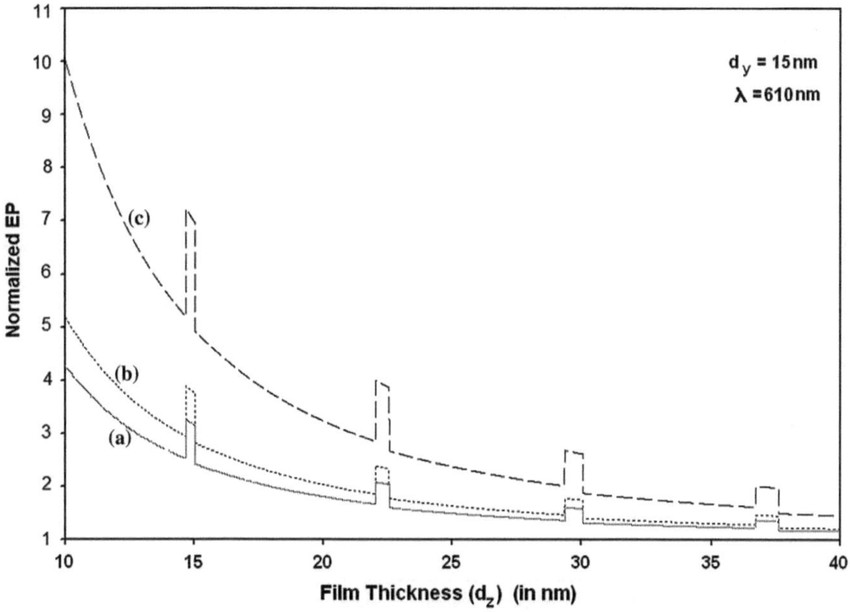

Fig. 7.9 Plot of the normalized EP from NWs of HD n-$Hg_{1-x}Cd_xTe$ as a function of film thickness in which the curves *a–c* represents the perturbed HD three and two band models of Kane together with HD parabolic energy bands respectively

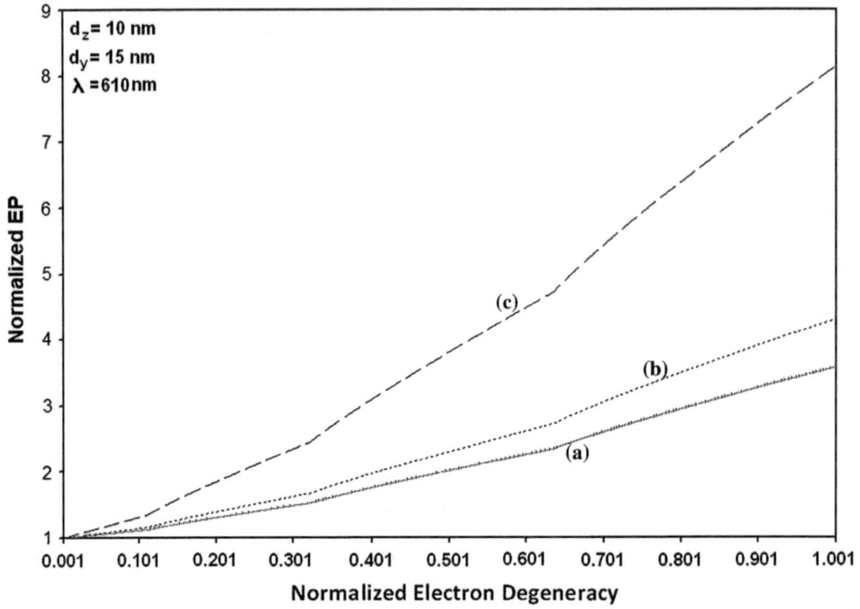

Fig. 7.10 Plot of the normalized EP from NWs of HD n-$Hg_{1-x}Cd_xTe$ as a function of normalized electron degeneracy for all cases of Fig. 7.9

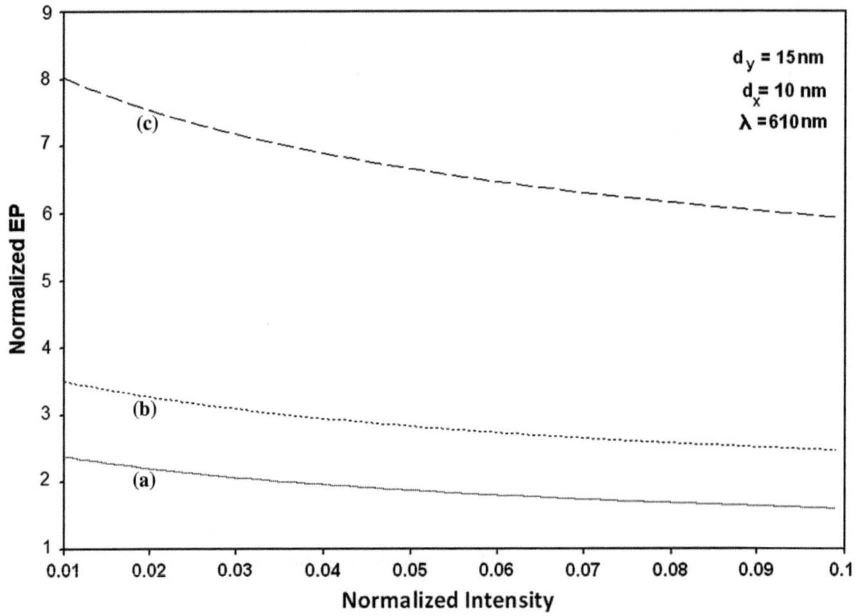

Fig. 7.11 Plot of the normalized EP from NWs of HD n-$Hg_{1-x}Cd_xTe$ as a function of normalized light intensity for all cases of Fig. 7.9

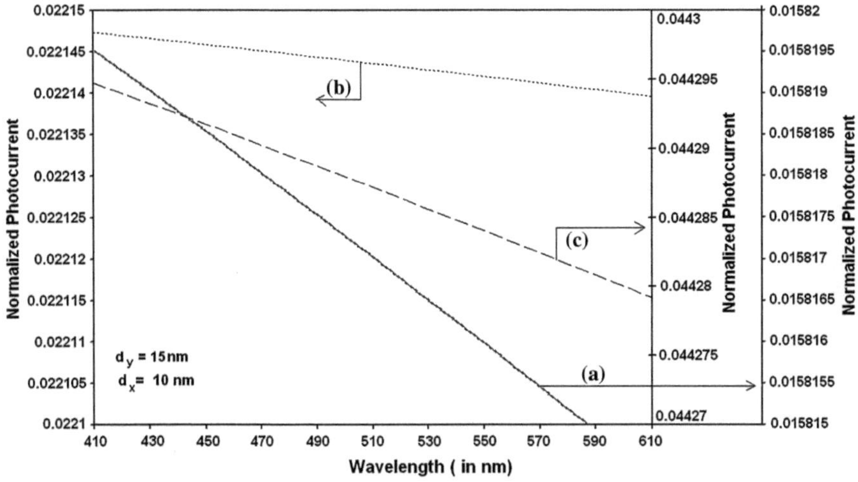

Fig. 7.12 Plot of the normalized EP from NWs of HD n-$Hg_{1-x}Cd_xTe$ as a function of light wavelength for all cases of Fig. 7.9

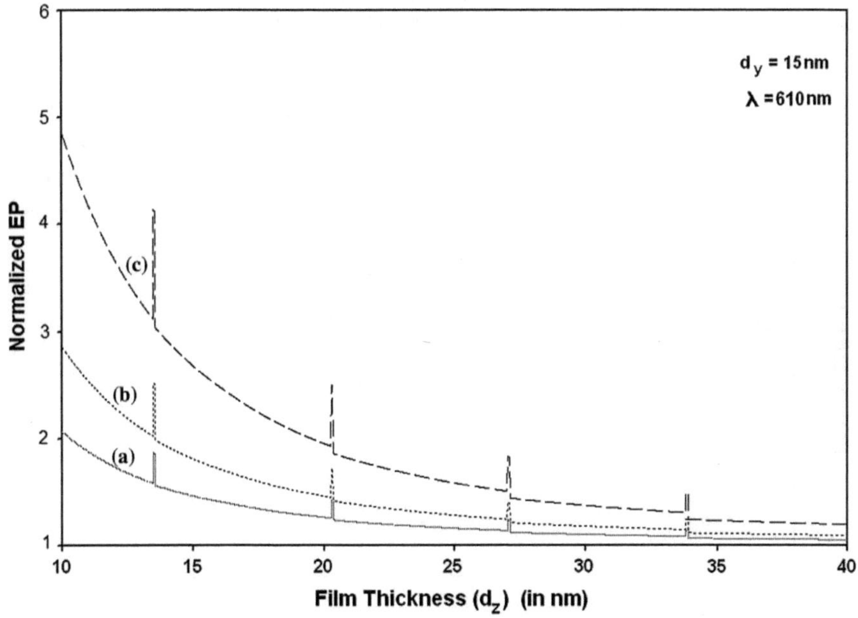

Fig. 7.13 Plot of the normalized EP from NWs of HD n-$In_{1-x}Ga_xAs_yP_{1-y}$ lattice matched to InP as a function of film thickness in which the curves *a–c* represent the perturbed HD three and two band models of Kane together with HD parabolic energy bands respectively

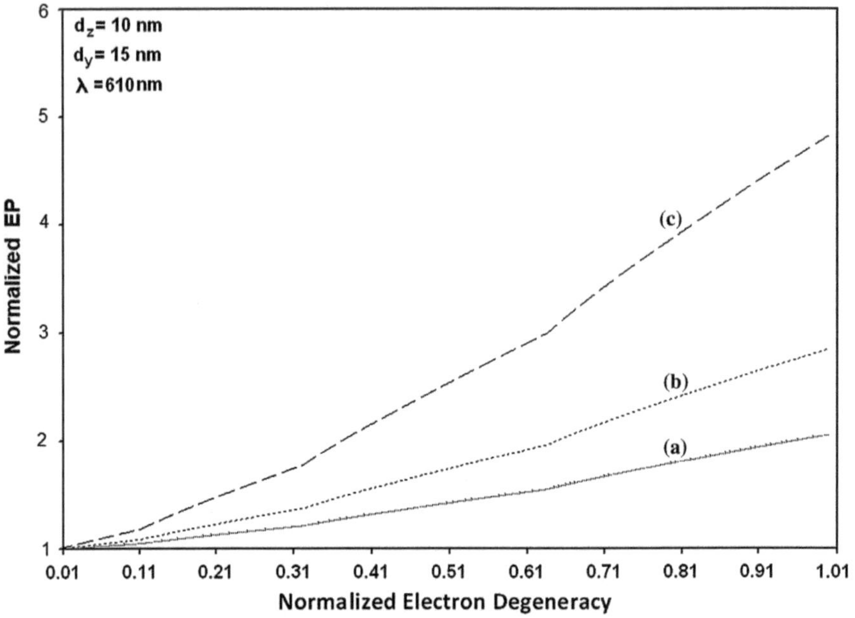

Fig. 7.14 Plot of the normalized EP from NWs of HD n-$In_{1-x}Ga_xAs_yP_{1-y}$ lattice matched to InP as a function of normalized electron degeneracy for all cases of Fig. 7.13

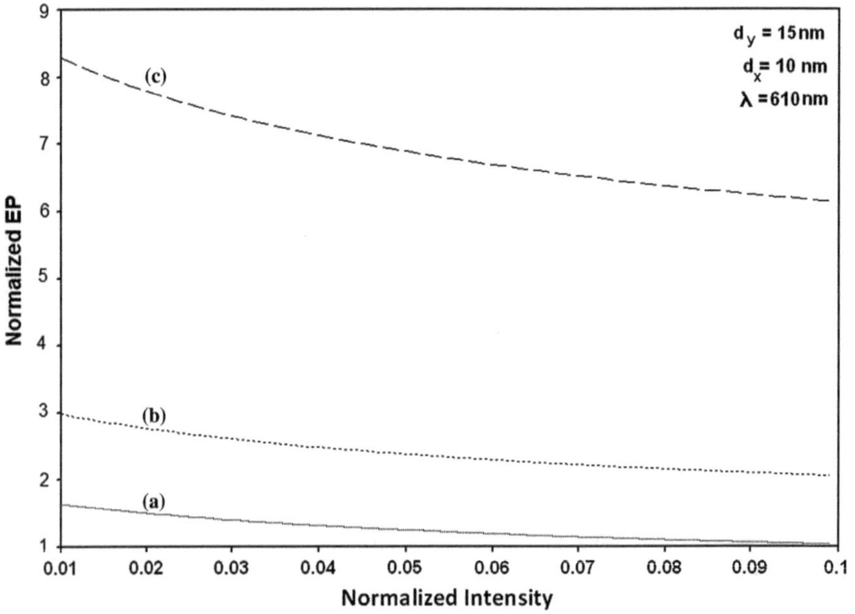

Fig. 7.15 Plot of the normalized EP from NWs of HD n-$In_{1-x}Ga_xAs_yP_{1-y}$ lattice matched to InP as a function of normalized light intensity for all cases of Fig. 7.13

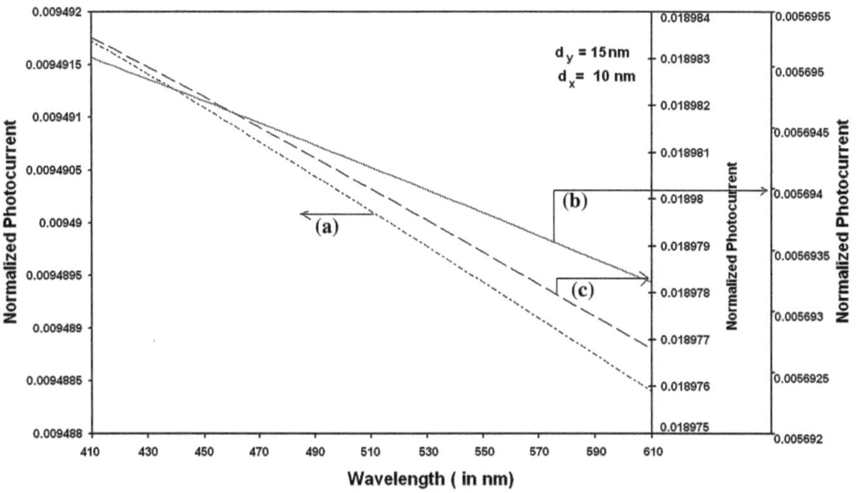

Fig. 7.16 Plot of the normalized EP from NWs of HD n-$In_{1-x}Ga_xAs_yP_{1-y}$ lattice matched to InP as a function of light wavelength for all cases of Fig. 7.13

The dependences of the normalized EP from QBs of HD n-$Hg_{1-x}Cd_xTe$ on the film thickness, normalized carrier degeneracy, normalized light intensity and wavelength have been drawn in Figs. 7.17, 7.18, 7.19 and 7.20 in accordance with perturbed HD three [using (7.27) and (7.26)] and HD two [using (7.28) and (7.26)] band models of Kane together with HD parabolic [using (7.29) and (7.26)] energy bands as shown by curves (a–c) respectively. The variations of normalized EP from QBs of HD n-$In_{1-x}Ga_xAs_yP_{1-y}$ lattice matched to InP, have been drawn in Figs. 7.21, 7.22, 7.23 and 7.24 as functions of film thickness, normalized carrier degeneracy, normalized incident light intensity and wavelengths respectively for all the cases of Fig. 7.17. From Figs. 7.1 and 7.5, it appears that EP from QWs of HD optoelectronic materials decreases with increasing film thickness in oscillatory manners. From Figs. 7.9 and 7.13, it appears that the EP from NWs of HD optoelectronic materials increases with decreasing film thickness exhibiting trapezoidal variation for a very small thickness bandwidth for the whole range of thicknesses considered. The widths of the trapezoids depend on the energy band constants of n-$Hg_{1-x}Cd_xTe$ and n-$In_{1-x}Ga_xAs_yP_{1-y}$ lattice matched to InP respectively.

From Figs. 7.17 and 7.21, we observe that the EP from HD QBs of optoelectronic materials decreases with increasing film thickness exhibiting prominent trapezoidal variation for relatively large thickness bandwidth. These three types of variations are the special signatures of 1D confinement in HD QWs, 2D confinement in HD NWs and 3D confinement in HD QBs of optoelectronic materials respectively in the presence of light. From Figs. 7.2 and 7.6, it appears that the normalized EP from HD QWs increases with increasing carrier degeneracy and for relatively large values of the same variable; it exhibits quantum jumps for all types of band models when the size quantum number changes from one fixed value to

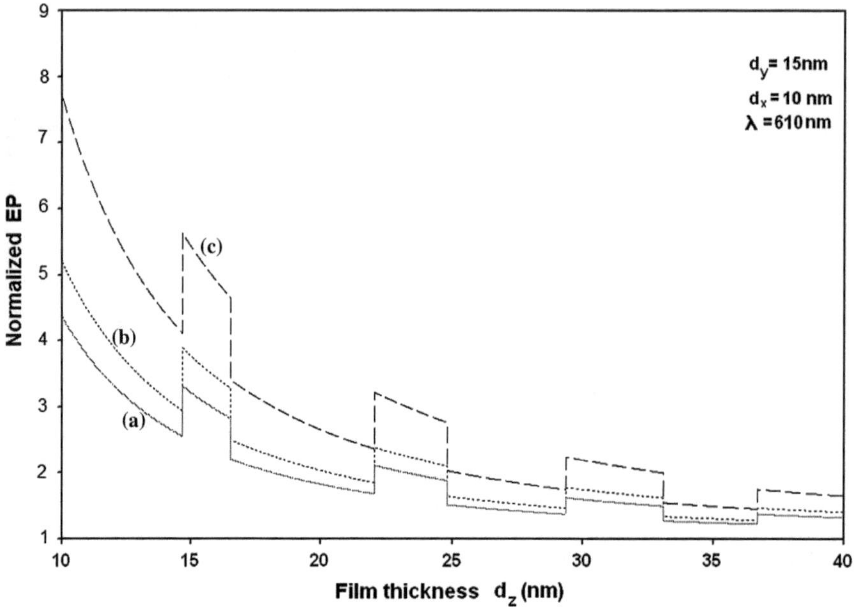

Fig. 7.17 Plot of the normalized EP from HD QBs of n-$Hg_{1-x}Cd_xTe$ as a function of film thickness in which the curves *a*–*c* represent the perturbed HD three and two band models of Kane together with HD parabolic energy bands respectively

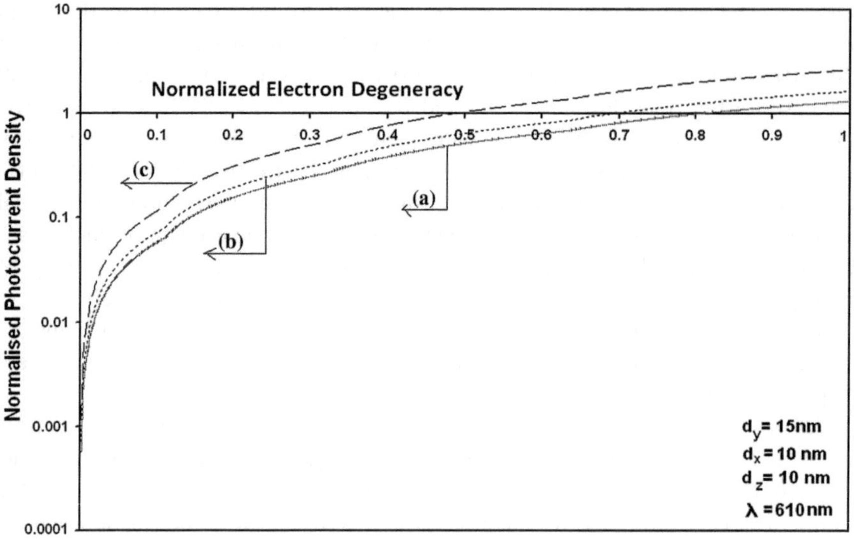

Fig. 7.18 Plot of the normalized EP from HD QBs of n-$Hg_{1-x}Cd_xTe$ as a function of normalized electron degeneracy for all cases of Fig. 7.17

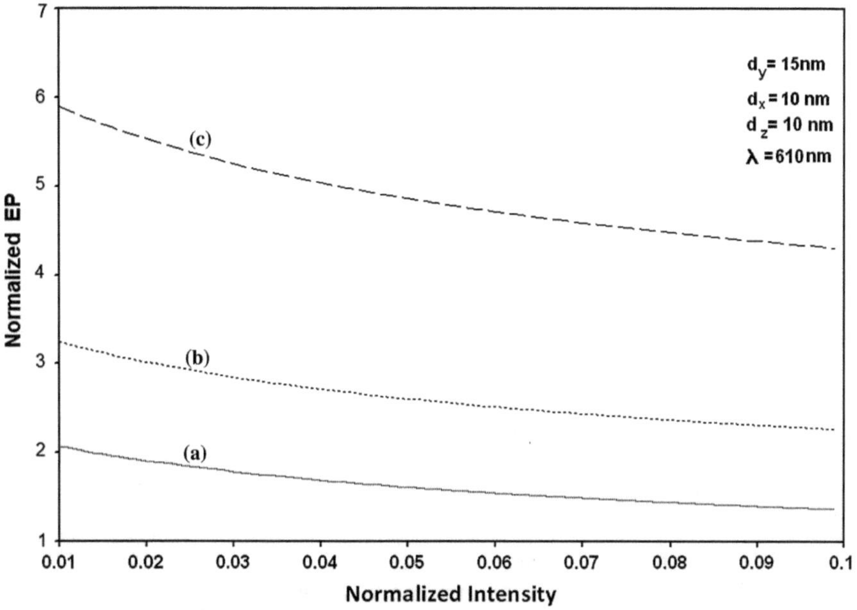

Fig. 7.19 Plot of the normalized EP from HD QBs of n-$Hg_{1-x}Cd_xTe$ as a function of normalized light intensity for all cases of Fig. 7.17

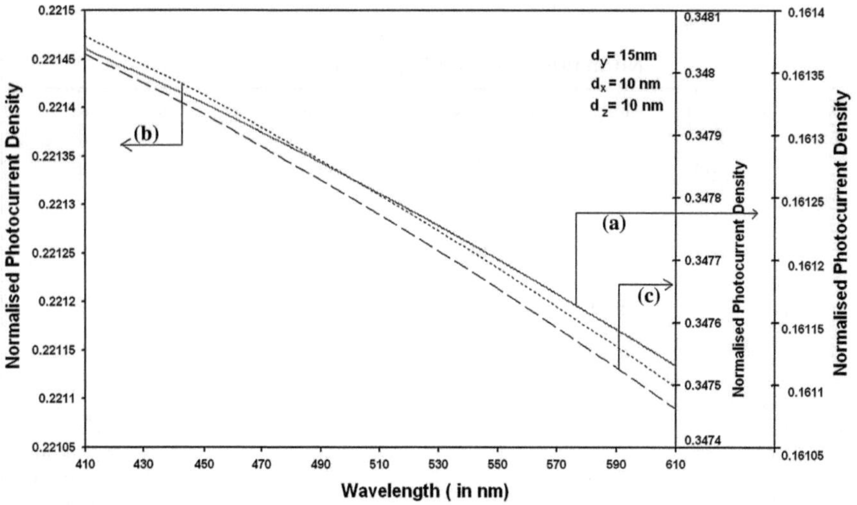

Fig. 7.20 Plot of the normalized EP from HD QBs of n-$Hg_{1-x}Cd_xTe$ as a function of light wavelength for all cases of Fig. 7.17

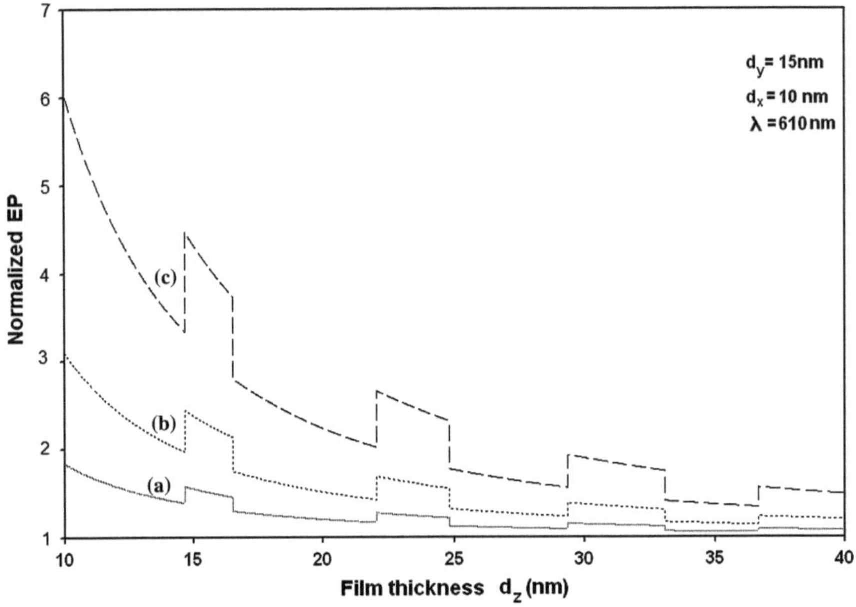

Fig. 7.21 Plot of the normalized EP from HD QBs of n-$In_{1-x}Ga_xAs_yP_{1-y}$ lattice matched to InP as a function of film thickness in which the curves *a–c* represent the perturbed HD three and two band models of Kane together with HD parabolic energy bands respectively

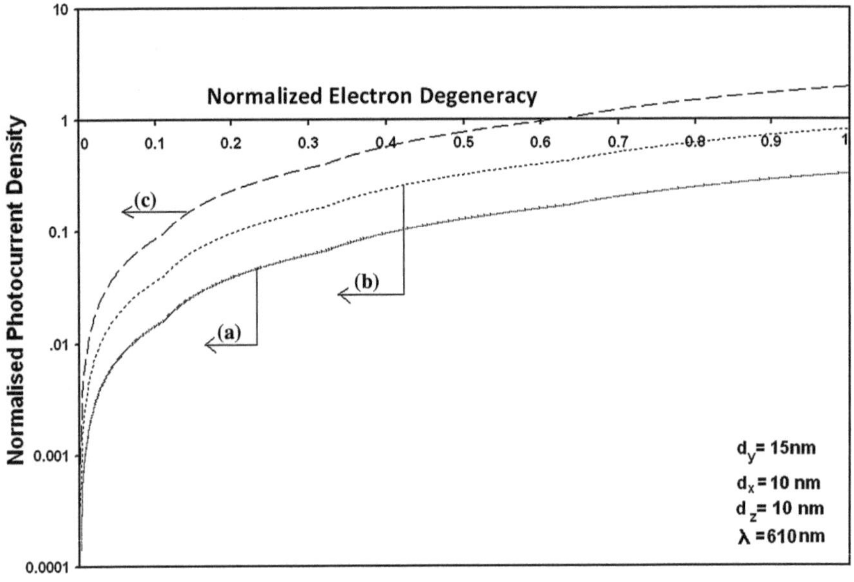

Fig. 7.22 Plot of the normalized EP from HD QBs of n-$In_{1-x}Ga_xAs_yP_{1-y}$ lattice matched to InP as a function of normalized electron degeneracy for all cases of Fig. 7.21

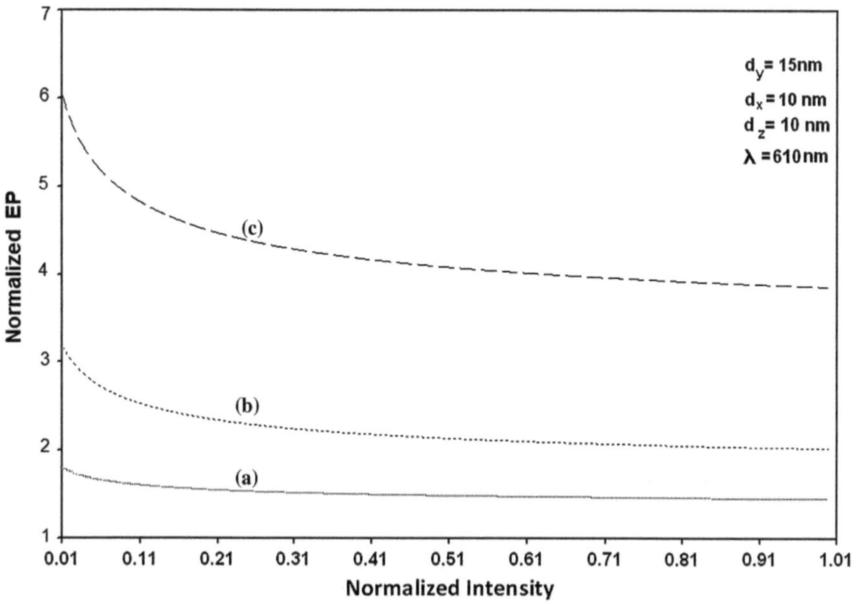

Fig. 7.23 Plot of the normalized EP from HD QBs of n-$In_{1-x}Ga_xAs_yP_{1-y}$ lattice matched to InP as a function of normalized light intensity for all cases of Fig. 7.21

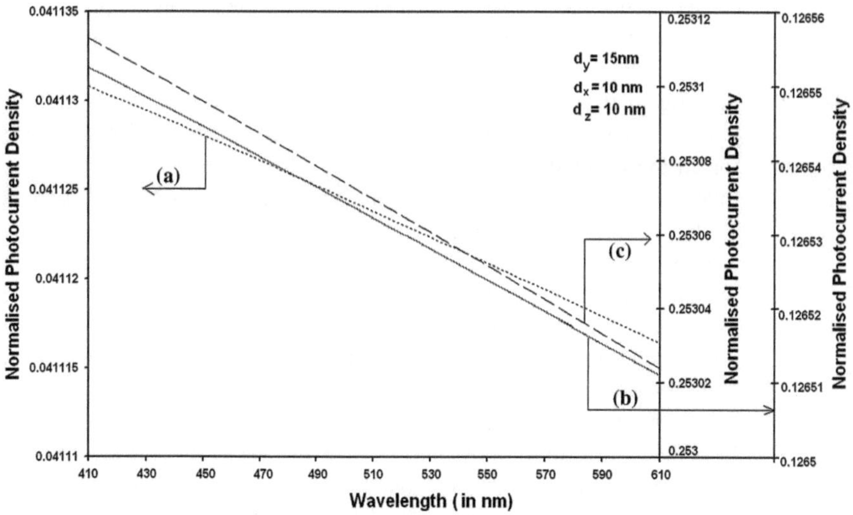

Fig. 7.24 Plot of the normalized EP from HD QBs of n-$In_{1-x}Ga_xAs_yP_{1-y}$ lattice matched to InP as a function of light wavelength for all cases of Fig. 7.21

another. The Figs. 7.10 and 7.14 show respectively that the normalized EP in HD NWs of optoelectronic materials increases with increasing normalized electron degeneracy.

The Figs. 7.18 and 7.22 demonstrate that the EP from HD QBs of optoelectronic materials increases with increasing electron degeneracy again in a different oscillatory manner. From Figs. 7.3, 7.7, 7.11, 7.15, 7.19 and 7.23, it appears that the EP increases with decreasing intensity for all types of quantum confinement. From Figs. 7.4, 7.8, 7.12, 7.16, 7.20 and 7.24, we can conclude that the normalized EP decreases with increasing wavelength for HD QWs, NWs and QBs of optoelectronic materials. Finally, it is apparent from all the figures that the EP from quantum confined HD ternary materials is larger as compared with the quantum confined HD quaternary compounds for all types of quantum confinement.

7.4 Open Research Problems

Investigate the following open research problems in the presence of external photo-excitation which changes the band structure in a fundamental way together with the proper inclusion of the electron spin, the variation of work function and the broadening of Landau levels respectively for appropriate problems.

(R.7.1) Investigate the multi-photon EP from all the quantum confined HD materials (i.e., HD multiple QWs, NWs and QBs) whose unperturbed carrier energy spectra are defined in (R.1.1) of Chap. 1 in the presence of arbitrary oriented photo-excitation and quantizing magnetic field respectively.

(R.7.2) Investigate the multi-photon EP in the presence of arbitrary oriented photo-excitation and alternating quantizing magnetic field respectively for all the problems of R.7.1.

(R.7.3) Investigate the multi-photon EP in the presence of arbitrary oriented photo-excitation, alternating quantizing magnetic field and an additional arbitrary oriented non-quantizing non-uniform electric field respectively for all the problems of R.7.1.

(R.7.4) Investigate the multi-photon EP in the presence of arbitrary oriented photo-excitation, alternating quantizing magnetic field and additional arbitrary oriented non-quantizing alternating electric field respectively for all the problems of R.7.1.

(R.7.5) Investigate the multi-photon EP in the presence of arbitrary oriented photo-excitation, and crossed quantizing magnetic and electric fields respectively for all the problems of R.7.1.

(R.7.6) Investigate the multi-photon EP for arbitrarily oriented photo-excitation and quantizing magnetic field from the entire quantum confined heavily-doped materials in the presence of exponential, Kane, Halperin, Lax and

Bonch-Bruevich types of band tails for all materials whose unperturbed carrier energy spectra are defined in (R.1.1) of Chap. 1.

(R.7.7) Investigate the multi-photon EP for arbitrarily oriented photo-excitation and alternating quantizing magnetic field for all the cases of R.7.6.

(R.7.8) Investigate the multi-photon EP in the presence of arbitrarily oriented photo-excitation, alternating quantizing magnetic field and an additional arbitrarily oriented non-quantizing non-uniform electric field for all the cases of R.7.6.

(R.7.9) Investigate the multi-photon EP in the presence of arbitrary oriented photo-excitation, alternating quantizing magnetic field and additional arbitrary oriented non-quantizing alternating electric field respectively for all the cases of R.7.6.

(R.7.10) Investigate the multi-photon EP in the presence of arbitrary oriented photo-excitation, and crossed quantizing magnetic and electric fields respectively for all the cases of R.7.6.

(R.7.11) Investigate the multi-photon EP for all the appropriate problems from R.7.1 to R.7.10 in the presence of finite potential wells.

(R.7.12) Investigate the multi-photon EP for all the appropriate HD problems from R.7.1 to R.7.10 in the presence of parabolic potential wellsPotential well.

(R.7.13) Investigate the multi-photon EP for all the above appropriate HD problems for quantum rings.

(R.7.14) Investigate the multi-photon EP for all the above appropriate HD problems in the presence of elliptical Hill and quantum square rings respectively.

(R.7.15) Investigate the multi-photon EP from HD nanotubes in the presence of arbitrary photo-excitation.

(R.7.16) Investigate the multi-photon EP from HD nanotubes in the presence of arbitrary photo-excitation and non-quantizing alternating electric field.

(R.7.17) Investigate the multi-photon EP from HD nanotubes in the presence of arbitrary photo-excitation and non-quantizing alternating magnetic field.

(R.7.18) Investigate the multi-photon EP from HD nanotubes in the presence of arbitrary photo-excitation and crossed electric and quantizing magnetic fields.

(R.7.19) Investigate the multi-photon EP from HD semiconductor nanotubes in the presence of arbitrary photo-excitation for all the materials whose unperturbed carrier dispersion laws are defined in (R.1.1) of Chap. 1.

(R.7.20) Investigate the multi-photon EP from HD semiconductor nanotubes in the presence of non-quantizing alternating electric field and arbitrary photo-excitation for all the materials whose unperturbed carrier dispersion laws are defined in (R.1.1) of Chap. 1.

(R.7.21) Investigate the multi-photon EP from HD semiconductor nanotubes in the presence of non-quantizing alternating magnetic field and arbitrary photo-excitation for all the materials whose unperturbed carrier dispersion laws are defined in (R.1.1) of Chap. 1.

(R.7.22) Investigate the multi-photon EP from HD semiconductor nanotubes in the presence of arbitrary photo-excitation and non-uniform electric field for all the materials whose unperturbed carrier dispersion laws are defined in (R.1.1) of Chap. 1.

(R.7.23) Investigate the multi-photon EP from HD semiconductor nanotubes in the presence of arbitrary photo-excitation and alternating quantizing magnetic fields for all the materials whose unperturbed carrier dispersion laws are defined in (R.1.1) of Chap. 1.

(R.7.24) Investigate the multi-photon EP from HD semiconductor nanotubes in the presence of arbitrary photo-excitation and crossed electric and quantizing magnetic fields for all the materials whose unperturbed carrier dispersion laws are defined in (R.1.1) of Chap. 1.

(R.7.25) Investigate the multi-photon EP in the presence of arbitrary photo-excitation for all the appropriate HD nipi structures of the materials whose unperturbed carrier energy spectra are defined in (R.1.1) of Chap. 1.

(R.7.26) Investigate the multi-photon EP in the presence of arbitrary photo-excitation for all the appropriate HD nipi structures of the materials whose unperturbed carrier energy spectra are defined in (R.1.1) of Chap. 1 in the presence of an arbitrarily oriented non-quantizing non-uniform additional electric field.

(R.7.27) Investigate the multi-photon EP for all the appropriate HD nipi structures of the materials whose unperturbed carrier energy spectra are defined in (R.1.1) of Chap. 1 in the presence of an arbitrarily oriented photo-excitation and non-quantizing alternating additional magnetic field.

(R.7.28) Investigate the multi-photon EP for all the appropriate HD nipi structures of the materials whose unperturbed carrier energy spectra are defined in (R.1.1) of Chap. 1 in the presence of an arbitrarily oriented photo-excitation and quantizing alternating additional magnetic field.

(R.7.29) Investigate the multi-photon EP for all the appropriate HD nipi structures of the materials whose unperturbed carrier energy spectra are defined in (R.1.1) of Chap. 1 in the presence of an arbitrarily oriented photo-excitation and crossed electric and quantizing magnetic fields.

(R.7.30) Investigate the multi-photon EP from HD nipi structures for all the appropriate cases of all the above problems.

(R.7.31) Investigate the multi-photon EP in the presence of arbitrary photo-excitation for the appropriate accumulation layers of all the materials whose unperturbed carrier energy spectra are defined in (R.1.1) of Chap. 1.

(R.7.32) Investigate the multi-photon EP in the presence of arbitrary photo-excitation for the appropriate accumulation layers of all the materials whose unperturbed carrier energy spectra are defined in (R.1.1) of

Chap. 1 in the presence of an arbitrarily oriented non-quantizing non-uniform additional electric field.

(R.7.33) Investigate the multi-photon EP for the appropriate accumulation layers of all the materials whose unperturbed carrier energy spectra are defined in (R.1.1) of Chap. 1 in the presence of an arbitrarily oriented photo-excitation and non-quantizing alternating additional magnetic field.

(R.7.34) Investigate the multi-photon EPfor the appropriate accumulation layers of all the materials whose unperturbed carrier energy spectra are defined in (R.1.1) of Chap. 1 in the presence of an arbitrarily oriented photo-excitation and quantizing alternating additional magnetic field.

(R.7.35) Investigate the multi-photon EP for the appropriate accumulation layers of all the materials whose unperturbed carrier energy spectra are defined in (R.1.1) of Chap. 1 in the presence of an arbitrarily oriented photo-excitation and crossed electric and quantizing magnetic fields by considering electron spin and broadening of Landau levels.

(R.7.37) Investigate the multi-photon EP in the presence of arbitrary photo-excitation from wedge shaped and cylindrical HD QBs of all the materials whose unperturbed carrier energy spectra are defined in (R.1.1) of Chap. 1.

(R.7.38) Investigate the multi-photon EP in the presence of arbitrary photo-excitation from wedge shaped and cylindrical HD QBs of all the materials whose unperturbed carrier energy spectra are defined in (R.1.1) of Chap. 1 in the presence of an arbitrarily oriented non-quantizing non-uniform additional electric field.

(R.7.39) Investigate the multi-photon EP from wedge shaped and cylindrical HD QBs of all the materials whose unperturbed carrier energy spectra are defined in (R.1.1) of Chap. 1 in the presence of an arbitrarily oriented photo-excitation and non-quantizing alternating additional magnetic field.

(R.7.40) Investigate the multi-photon EP from wedge shaped and cylindrical HD QBs of all the materials whose unperturbed carrier energy spectra are defined in (R.1.1) of Chap. 1 in the presence of an arbitrarily oriented photo-excitation and quantizing alternating additional magnetic field.

(R.7.41) Investigate the multi-photon EP from wedge shaped and cylindrical HD QBs of all the materials whose unperturbed carrier energy spectra are defined in (R.1.1) of Chap. 1 in the presence of an arbitrarily oriented photo-excitation and crossed electric and quantizing magnetic fields.

(R.7.42) Investigate the multi-photon EP from wedge shaped and cylindrical HD QBs for all the appropriate cases of the above problems.

(R.7.43) Investigate all the problems from R.7.1 to R.7.42 by removing all the mathematical approximations and establishing the respective appropriate uniqueness conditions.

Reference

1. K. P. Ghatak, S. Bhattacharya, D. De, *Einstein Relation in Compound Semiconductors and their nanostructures*. Springer Series in Materials Science, vol. 116 (Springer, Germany, 2008) (and the references cited therein)

Chapter 8
The EP from HD Effective Mass Super Lattices of Optoelectronic Materials

8.1 Introduction

In Chap. 4, the photoemission has been studied from SLs having various band structures assuming that the band structures of the constituent materials are invariant quantities in the presence of external photo-excitation. In this chapter, this assumption has been removed and in Sect. 8.2.1, an attempt is made to study the magneto EP from HD effective mass QWSL of optoelectronic materials. In Sect. 8.2.2, the photoemission from HD effective mass NW SLs of optoelectronic materials has been investigated and in Sect. 8.2.3, the EP from HD effective mass QB SLs of optoelectronic materials has been studied. The Sect. 8.2.4 explores the magneto EP from HD effective mass SLs of optoelectronic materials. The Sects. 8.3 and 8.4 contain respectively the result and discussions and open research problems pertinent to this chapter.

8.2 Theoretical Background

8.2.1 The Magneto EP from HD QWs Effective Mass Super Lattices

The electron dispersion law in III-V effective mass super lattices can be written as [1]

$$k_x^2 = [\frac{1}{L_0^2}[\cos^{-1}\{f_{HD1}(E, \lambda, \eta_g, k_y, k_z)\}]^2 - k_\perp^2] \tag{8.1}$$

K.P. Ghatak, *Einstein's Photoemission*, Springer Tracts in Modern Physics 262,
DOI 10.1007/978-3-319-11188-9_8

where

$$f_{HD1}(E,\lambda,\eta_g,k_y,k_z) = [[\bar{a}_{1HD}\cos[a_0\overline{C}_{1HD}(E,\eta_{g1},\lambda,k_\perp) + b_0\overline{D}_{1HD}(E,\eta_{g2},\lambda,k_\perp)]] - [\bar{a}_{2HD}\cos[a_0\overline{C}_{1HD}(E,\eta_{g1},\lambda,k_\perp) - b_0\overline{D}_{1HD}(E,\eta_{g2},\lambda,k_\perp)]]],$$

$$\bar{a}_{1HD} = [\sqrt{\frac{m_{c2}T_1'(0,\lambda,\eta_{g2})}{m_{c1}T_1'(0,\lambda,\eta_{g1})}} + 1]^2 \cdot [4\sqrt{\frac{m_{c2}T_1'(0,\lambda,\eta_{g2})}{m_{c1}T_1'(0,\lambda,\eta_{g1})}}]^{-1},$$

$$a_{2HD} = [\sqrt{\frac{m_{c2}T_1'(0,\lambda,\eta_{g2})}{m_{c1}T_1'(0,\lambda,\eta_{g1})}} - 1]^2 \cdot [4\sqrt{\frac{m_{c2}T_1'(0,\lambda,\eta_{g2})}{m_{c1}T_1'(0,\lambda,\eta_{g1})}}]^{-1},$$

$$\bar{C}_{1HD}(E,\eta_{g1},\lambda,k_\perp) = [\frac{2m_{c1}}{\hbar^2}T_1(E,\eta_{g1},\lambda) - k_\perp^2]^{1/2} \quad \text{and} \quad \bar{D}_{1HD}(E,\eta_{g1},\lambda,k_\perp) = [\frac{2m_{c2}}{\hbar^2}T_1(E,\eta_{g2},\lambda) - k_\perp^2]^{1/2}$$

In the presence of a quantizing magneticfield B along k_x direction, the magneto electron energy spectrum can be written as

$$k_x^2 = \overline{\omega}_{HD}(E,\eta_g,\lambda,n) \tag{8.2}$$

where $\bar{\omega}_{HD}(E,\eta_g,\lambda,n) = [\frac{1}{L_0^2}[\cos^{-1}\{f_{HD1}(E,\lambda,\eta_g,n)\}]^2 - \frac{2eB}{\hbar}(n+\frac{1}{2})]$,

$$f_{HD1}(E,\lambda,\eta_g,n) = [[\bar{a}_{1HD}\cos[a_0\overline{C}_{1HD}(E,\eta_{g1},\lambda,n) + b_0\overline{D}_{1HD}(E,\eta_{g2},\lambda,n)]] - [\bar{a}_{2HD}\cos[a_0\overline{C}_{1HD}(E,\eta_{g1},\lambda,n) - b_0\overline{D}_{1HD}(E,\eta_{g2},\lambda,n)]]],$$

$$\bar{C}_{1HD}(E,\eta_{g1},\lambda,n) = [\frac{2m_{c1}}{\hbar^2}T_1(E,\eta_{g1},\lambda) - \frac{2eB}{\hbar}(n+\frac{1}{2})]^{1/2} \text{ and}$$

$$\bar{D}_{1HD}(E,\eta_{g1},\lambda,n) = [\frac{2m_{c2}}{\hbar^2}T_1(E,\eta_{g2},\lambda) - \frac{2eB}{\hbar}(n+\frac{1}{2})]^{1/2}$$

The total energy e_{TQ1} in this case can be expressed as

$$(\frac{n_x\pi}{d_x})^2 = \overline{\omega}_{HD}(e_{TQ1},\eta_g,\lambda,n) \tag{8.3}$$

The Z part of the energy E_{ZQ1} in this case can be written as

$$(\frac{n_x\pi}{d_x})L_0 = \cos^{-1}[f_{HD1}(E_{ZQ1},\lambda,\eta_g,0)] \tag{8.4}$$

where

$$[f_{HD1}(E_{ZQ1},\lambda,\eta_g,0)] = [[\bar{a}_{1HD}\cos[a_0\bar{C}_{1HD}(E_{ZQ1},\eta_{g1},\lambda,0)+b_0\bar{D}_{1HD}(E_{ZQ1},\eta_{g2},\lambda,0)]] - [\bar{a}_{2HD}\cos[a_0\bar{C}_{1HD}(E_{ZQ1},\eta_{g1},\lambda,0)-b_0\bar{D}_{1HD}(E_{ZQ1},\eta_{g2},\lambda,0)]]],$$

$$\bar{C}_{1HD}(E_{ZQ1},\eta_{g1},\lambda,0) = \left[\frac{2m_{c1}}{\hbar^2}T_1(E_{ZQ1},\eta_{g1},\lambda)\right]^{1/2} \quad \text{and} \quad \bar{D}_{1HD}(E_{ZQ1},\eta_{g1},\lambda,0) = [\frac{2m_{c2}}{\hbar^2}T_1(E_{ZQ1},\eta_{g2},\lambda)]^{1/2}$$

The electron concentration is given by

$$n_0 = \frac{g_v eB}{\pi\hbar}\text{ Real part of }\sum_{n=0}^{n_{\max}}\sum_{n=1}^{n_{\max}}F_{-1}(\eta_{8SL1}) \tag{8.5}$$

where $\eta_{8SL1} = (k_BT)^{-1}[E_{FF} - E_{TQ1}]$ and E_{FF} is the Fermi energy in this case.

The EP can be written as

$$J = \frac{g_v e^2 B\alpha_0}{hd_x}\text{ Real part of }\sum_{n=0}^{n_{\max}}\sum_{n_{x_{\min}}}^{n_{x\max}}F_{-1}(\eta_{8SL1})v_z(E_{ZQ1}) \tag{8.6}$$

where $v_z(E_{ZQ1}) = \frac{L_0\sqrt{1-f_{HD1}^2(E_{ZQ1},\lambda,\eta_g,0)}}{\hbar f'_{HD1}(E_{ZQ1},\lambda,\eta_g,0)}$.

8.2.2 The EP from HD NW Effective Mass Super Lattices

The dispersion relation in this case is given by

$$k_x^2 = [\frac{1}{L_0^2}[\cos^{-1}\{f_{HD1}(E,\lambda,\eta_g,n_y,n_z)\}]^2 - G_{881}] \tag{8.7}$$

where

$$f_{HD1}(E,\lambda,\eta_g,n_y,n_z) = [[\bar{a}_{1HD}\cos[a_0\overline{C}_{1HD}(E,\eta_{g1},\lambda,n_y,n_z)+b_0\overline{D}_{1HD}(E,\eta_{g2},\lambda,n_y,n_z)]] - [\bar{a}_{2HD}\cos[a_0\overline{C}_{1HD}(E,\eta_{g1},\lambda,n_y,n_z)-b_0\overline{D}_{1HD}(E,\eta_{g2},\lambda,n_y,n_z)]]],$$

$$\bar{C}_{1HD}(E,\eta_{g1},\lambda,n_y,n_z) = [\frac{2m_{c1}}{\hbar^2}T_1(E,\eta_{g1},\lambda) - G_{881}]^{1/2},\ G_{881} = [(\frac{n_y\pi}{d_y})^2 + (\frac{n_z\pi}{d_z})^2] \text{ and}$$

$$\bar{D}_{1HD}(E,\eta_{g1},\lambda,n_y,n_z) = [\frac{2m_{c2}}{\hbar^2}T_1(E,\eta_{g2},\lambda) - G_{881}]^{1/2}$$

The sub-band energy E_{831} is given by

$$0 = [\frac{1}{L_0^2}[\cos^{-1}\{f_{HD1}(E_{831}, \lambda, \eta_g, n_y, n_z)\}]^2 - G_{881}] \tag{8.8}$$

The (8.7) can be written as

$$k_x = [\Delta(E, \lambda, \eta_g, n_y, n_z)] \tag{8.9}$$

where $\Delta(E, \lambda, \eta_g, n_y, n_z) = [\frac{1}{L_0^2}[\cos^{-1}\{f_{HD1}(E, \lambda, \eta_g, n_y, n_z)\}]^2 - G_{881}]^{1/2}$

The electron concentration is given by

$$n_0 = \frac{2g_v}{\pi} \text{Real part of} \sum_{n_y=1}^{n_{y\max}} \sum_{n_z=1}^{n_{z\max}} [\Delta(E_{F81}, \lambda, \eta_g, n_y, n_z) + \Delta_1(E_{F81}, \lambda, \eta_g, n_y, n_z)] \tag{8.10}$$

where $\Delta_1(E_{F81}, \lambda, \eta_g, n_y, n_z) = \sum_{r=1}^{s} L(r)[\Delta(E_{F81}, \lambda, \eta_g, n_y, n_z)]$ and E_{F81} is the Fermi energy in this case

The EP can be written as

$$I_{1LHD} = (\frac{\alpha_0 g_v e k_B T}{\pi\hbar}) \text{Real part of} \sum_{n_y=1}^{n_{y\max}} \sum_{n_z=1}^{n_{z\max}} F_0(\Upsilon_{SLHD1}) \tag{8.11}$$

where $\Upsilon_{SLHD1} = [E_{F81} - (E_{831} + W - h\nu)](k_B T)^{-1}$.

8.2.3 The EP from HD QB Effective Mass Super Lattices

The totally quantized energy E_{TQSL88} in this case is given by

$$(\frac{n_x \pi}{d_x})^2 = [\frac{1}{L_0^2}[\cos^{-1}\{f_{HD1}(E_{TQSL88}, \lambda, \eta_g, n_y, n_z)\}]^2 - G_{881}] \tag{8.12}$$

The electron concentration in this case is given by

$$N_{0L} = \frac{2g_v}{d_x d_y d_z} \text{Real part of} \sum_{n_x=1}^{n_{x\max}} \sum_{n_y=1}^{n_{y\max}} \sum_{n_z=1}^{n_{z\max}} F_{-1}(\eta_{32HD}) \tag{8.13}$$

where $\eta_{32HD} = (k_B T)^{-1}[E_{FQDSLEMHD} - E_{TQSL88}]$ and $E_{FQDSLEMHD}$ is the Fermi energy in this case.

The EP in this case can be expressed as

$$J = \frac{\alpha_0 e g_v}{d_x d_y d_z} \text{ Real part of } \sum_{n_{x\min}}^{n_{x\max}} \sum_{n_y=1}^{n_{y\max}} \sum_{n_z=1}^{n_{z\max}} F_{-1}(\eta_{32HD}) v_Z(E_{ZQ1}). \tag{8.14}$$

8.2.4 *The Magneto EP from HD Effective Mass Super Lattices*

The (8.2) can be written as

$$\Delta_3(E, \eta_g, \lambda, n) = k_x \tag{8.15}$$

where

$$\Delta_3(E, \eta_g, \lambda, n) = [\overline{\omega}_{HD}(E, \eta_g, \lambda, n)]^{1/2}$$

The Landau sub-band energy E_{33HD} in this case can be expressed as

$$\Delta_3(E_{33HD}, \eta_g, \lambda, n) = 0 \tag{8.16}$$

The electron concentration is given by

$$n_0 = \frac{g_v eB}{\pi^2 \hbar} \text{ Real part of } \sum_{n=0}^{n_{y\max}} [\Delta_3(E_{FBSLEMHD}, \eta_g, \lambda, n) + \Delta_4(E_{FBSLEMHD}, \eta_g, \lambda, n)] \tag{8.17}$$

where $\Delta_4(E_{FBSLEMHD}, \eta_g, \lambda, n) = \sum_{r=1}^{s} L(r)[\Delta_3(E_{FBSLEMHD}, \eta_g, \lambda, n)]$ and $E_{FBSLEMHD}$ is the Fermi energy in this case.

The EP assumes the form

$$J_{ML} = \left(\frac{\alpha_0 e^2 B k_B T}{2\pi^2 \hbar^2}\right) \text{ Real part of } \sum_{n=0}^{n_{\max}} F_0(\eta_{701HD}) \tag{8.18}$$

where $\eta_{701HD} = (k_B T)^{-1}[E_{FBSLEMHD} - (E_{33HD} + W - h\nu)]$.

8.3 Results and Discussion

Using the appropriate equations the normalized EP from QW $HgTe/Hg_{1-x}Cd_xTe$ HD effective mass SL has been plotted as a function of inverse quantizing magnetic field as shown in plot (a) of Fig. 8.1 whose constituent materials obey the perturbed HD three band model of Kane in the presence of external photo-excitation. The curves (b)

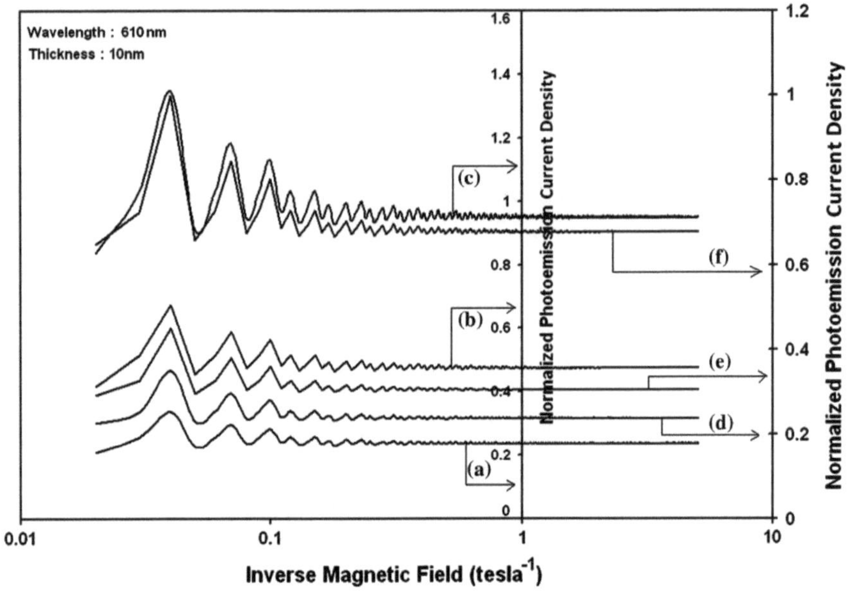

Fig. 8.1 Plot of the normalized EP from HD QW effective mass superlattices of $HgTe/Hg_{1-x}Cd_xTe$ as a function of inverse magnetic field in which the curves *a*, *b* and *c* represent the perturbed HD three and two band models of Kane together with HD parabolic energy bands respectively. The curves *d*, *e* and *f* exhibit the corresponding plots of HD $In_xGa_{1-x}As/InP$SL

and (c) of the same figure have been drawn for perturbed HD two band model of Kane and that of perturbed HD parabolic energy bands respectively. The curves (d), (e) and (f) in the same figure exhibit the corresponding plots of QW $In_xGa_{1-x}As/InP$ effective mass HD SL. The Figs. 8.2, 8.3, 8.4 and 8.5 show the variations of the normalized EP from the said HD SLs as functions of normalized electron degeneracy, normalized intensity, wavelength and thickness respectively for all the cases of Fig. 8.1. Using the appropriate equations, the normalized EP from NW HD effective mass $HgTe/Hg_{1-x}Cd_xTe$ SL as a function of film thickness has been depicted in plot (a) of Fig. 8.6 whose constituent materials obey the perturbed HD three band model of Kane in the presence of external light waves. The curves (b) and (c) of the same figure have been drawn for perturbed HD two band model of Kane and perturbed HD parabolic energy bands respectively. The curves (d), (e) and (f) in the same figure exhibit the corresponding plots of $In_xGa_{1-x}As/InP$ NW HD effective mass SL. The Figs. 8.7 8.8, 8.9 and 8.10 exhibit the plots of the normalized EP as functions of normalized carrier concentration, normalized intensity, wavelength and normalized incident photon energy respectively for all the cases of Fig. 8.6.

Using appropriate equations, the normalized EP from $HgTe/Hg_{1-x}Cd_xTe$ and $In_xGa_{1-x}As/InP$ effective mass QB HD SLs respectively has been plotted for all types of band models as a function of film thickness as shown in Fig. 8.11. Figures 8.12, 8.13, 8.14 and 8.15 exhibit the plots of normalized EP from the said QB HD SLs as functions of normalized electron degeneracy, normalized intensity,

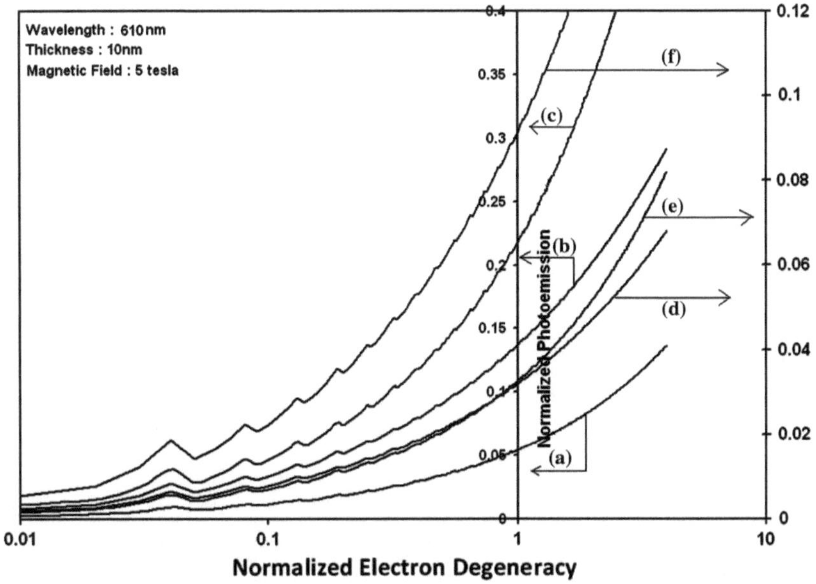

Fig. 8.2 Plot of the normalized EP from QW HD effective mass superlattices of $HgTe/Hg_{1-x}Cd_xTe$ and $In_xGa_{1-x}As/InP$ as a function of normalized electron degeneracy for all cases of Fig. 8.1

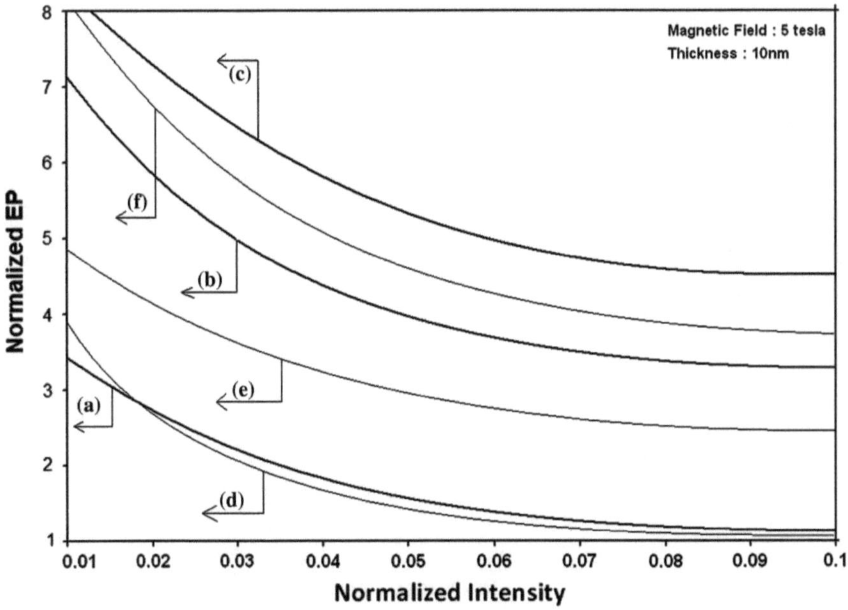

Fig. 8.3 Plot of the normalized EP from QW HD effective mass superlattices of $HgTe/Hg_{1-x}Cd_xTe$ and $In_xGa_{1-x}As/InP$ as a function of normalized light intensity for all cases of Fig. 8.1

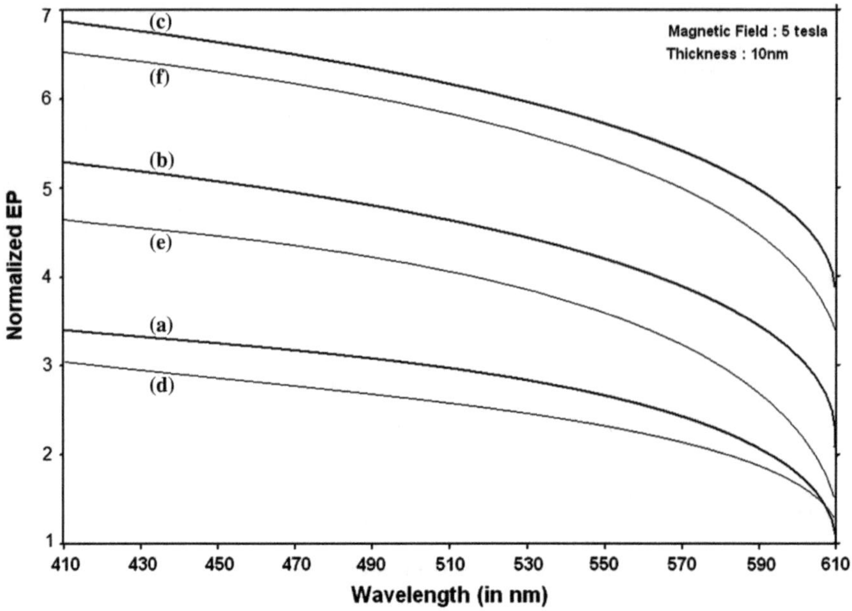

Fig. 8.4 Plot of the normalized EP from QW HD effective mass superlattices of $HgTe/Hg_{1-x}Cd_xTe$ and $In_xGa_{1-x}As/InP$ as a function of light wavelength for all cases of Fig. 8.1

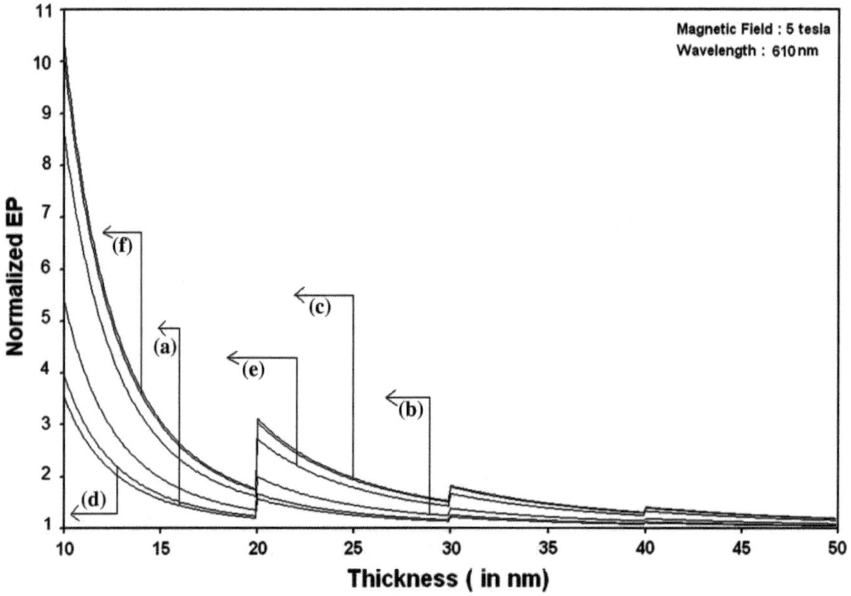

Fig. 8.5 Plot of the normalized EP from QW HD effective mass superlattices of HD $HgTe/Hg_{1-x}Cd_xTe$ as a function of film thickness in which the curves *a*, *b* and *c* represent the perturbed three and two band models of Kane together with HD parabolic energy bands respectively. The curves *d*, *e* and *f* exhibit the corresponding plots of $In_xGa_{1-x}As/InP$ HDSL

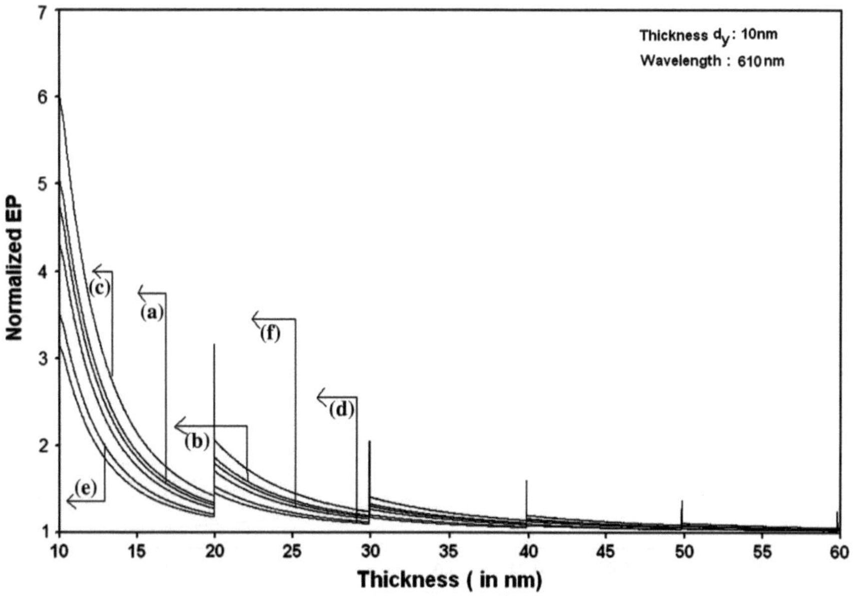

Fig. 8.6 Plot of the normalized EP from NW HD effective mass superlattices of $HgTe/Hg_{1-x}Cd_xTe$ and $In_xGa_{1-x}As/InP$ as function of film thickness for all cases of Fig. 8.5

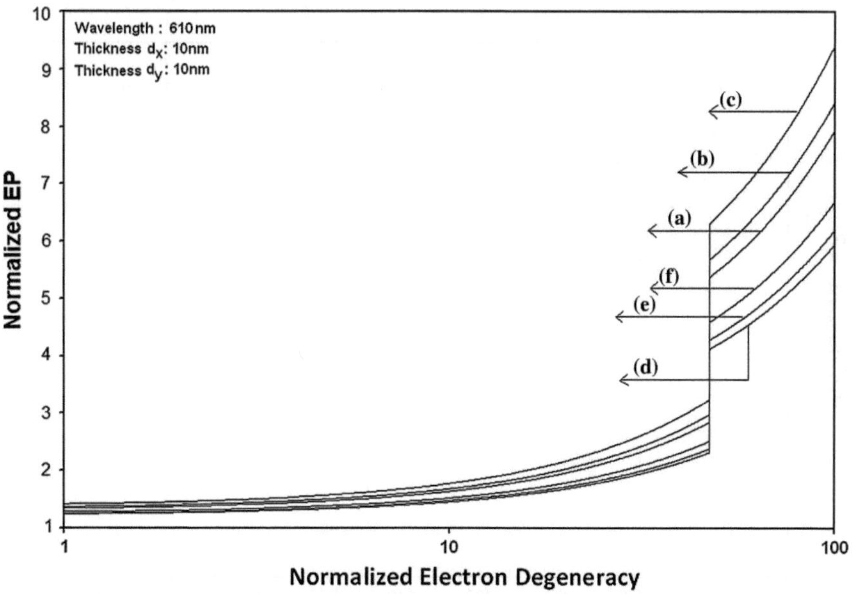

Fig. 8.7 Plot of the normalized EP from NW HD effective mass superlattices of $HgTe/Hg_{1-x}Cd_xTe$ and $In_xGa_{1-x}As/InP$ as function of normalized electron degeneracy for all cases of Fig. 8.5

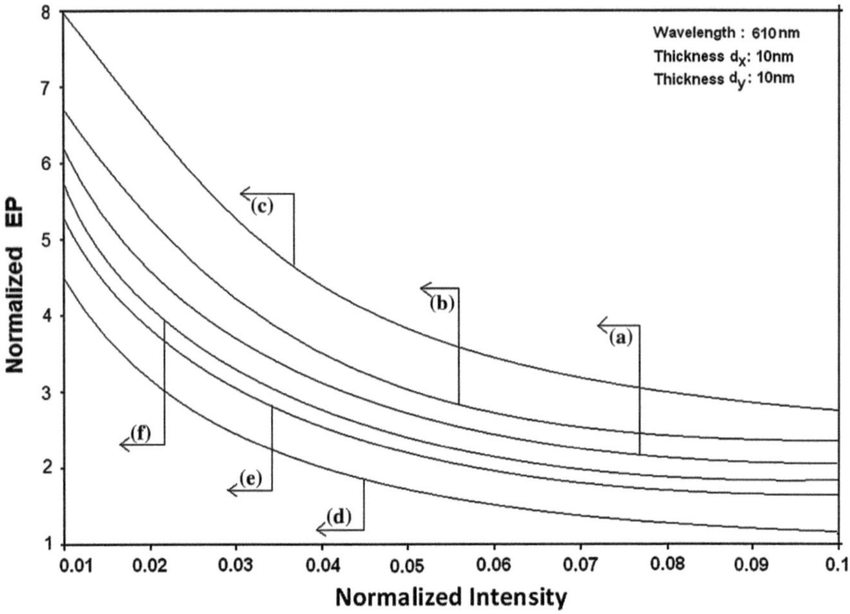

Fig. 8.8 Plot of the normalized EP from NW HD effective mass superlattices of $HgTe/Hg_{1-x}Cd_xTe$ and $In_xGa_{1-x}As/InP$ as function of normalized light intensity for all cases of Fig. 8.5

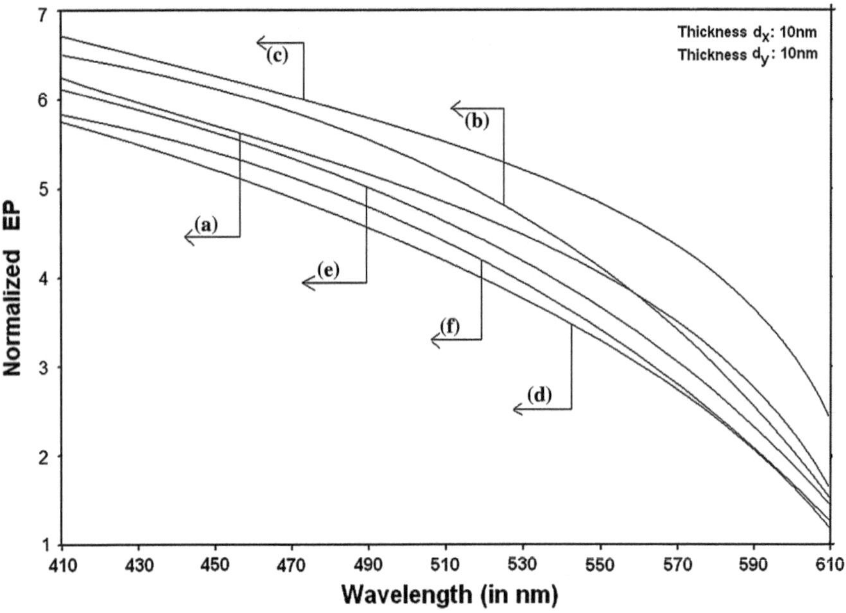

Fig. 8.9 Plot of the normalized EP from NW HD effective mass superlattices of $HgTe/Hg_{1-x}Cd_xTe$ and $In_xGa_{1-x}As/InP$ as function of light wavelength for all cases of Fig. 8.5

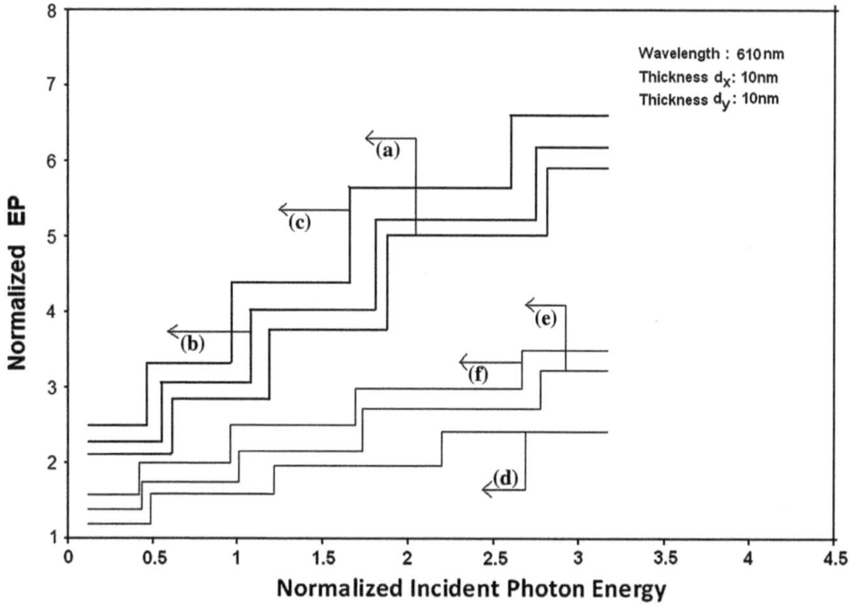

Fig. 8.10 Plot of the normalized EP as function of normalized incident photon energy from NW HD effective mass superlattices of $HgTe/Hg_{1-x}Cd_xTe$ and $In_xGa_{1-x}As/InP$ for all cases of Fig. 8.5

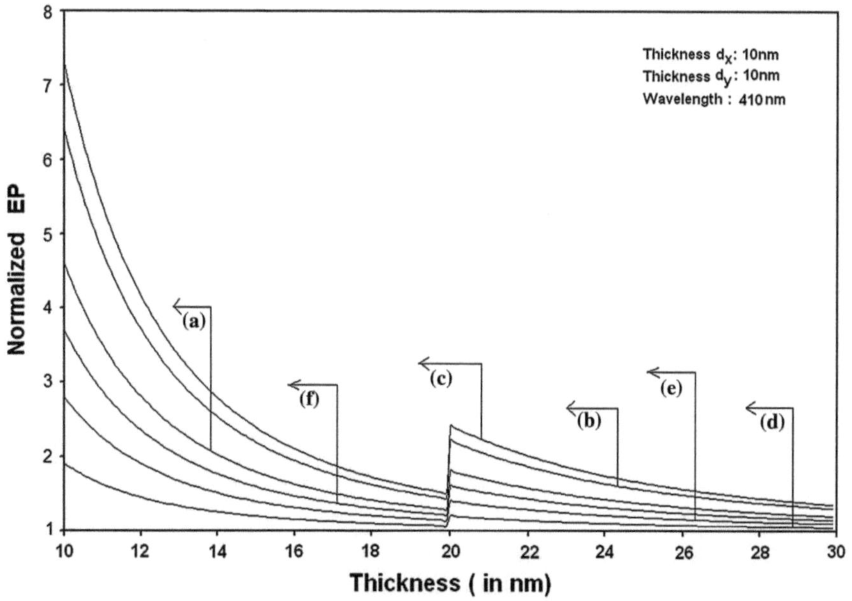

Fig. 8.11 Plot of the normalized EP from QB HD effective mass superlattices of $HgTe/Hg_{1-x}Cd_xTe$ as function of film thickness in which the curves *a*, *b* and *c* represent the perturbed HD three and two band models of Kane together with HD parabolic energy bands respectively. The curves *d*, *e* and *f* exhibit the corresponding plots of $In_xGa_{1-x}As/InP$

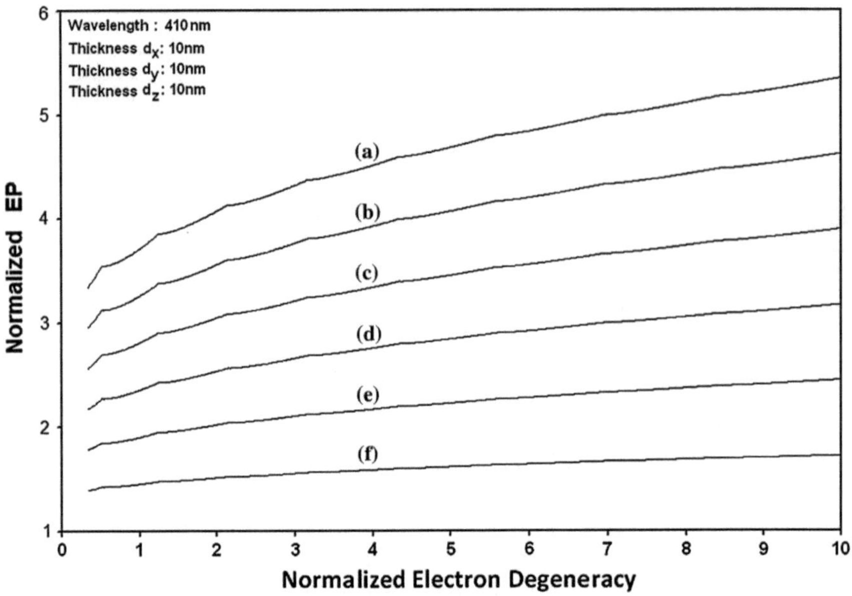

Fig. 8.12 Plot of the normalized EP from QB HD effective mass superlattices of $HgTe/Hg_{1-x}Cd_xTe$ and $In_xGa_{1-x}As/InP$ as function of normalized electron degeneracy for all cases of Fig. 8.11

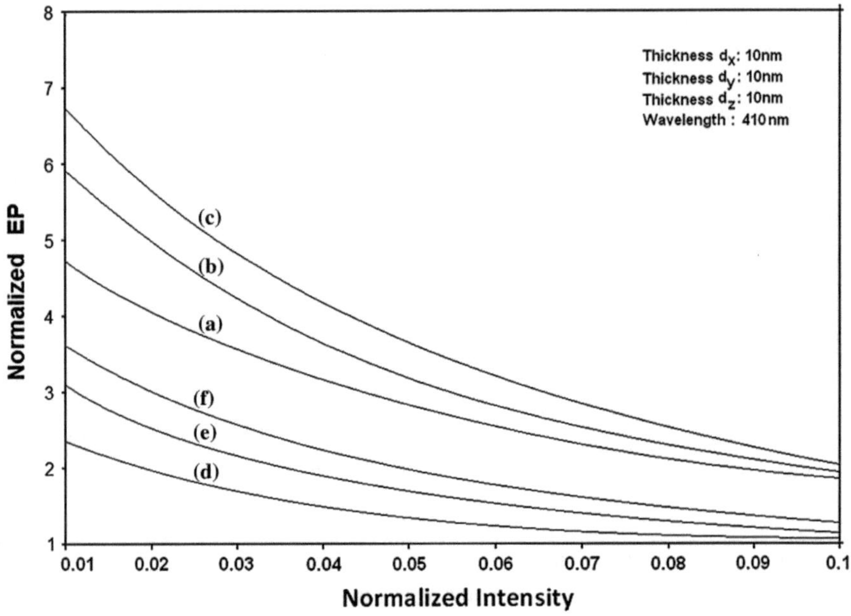

Fig. 8.13 Plot of the normalized EP from QB HD effective mass superlattices of $HgTe/Hg_{1-x}CdxTe$ and $In_xGa_{1-x}As/InP$ as function of normalized light intensity for all cases of Fig. 8.11

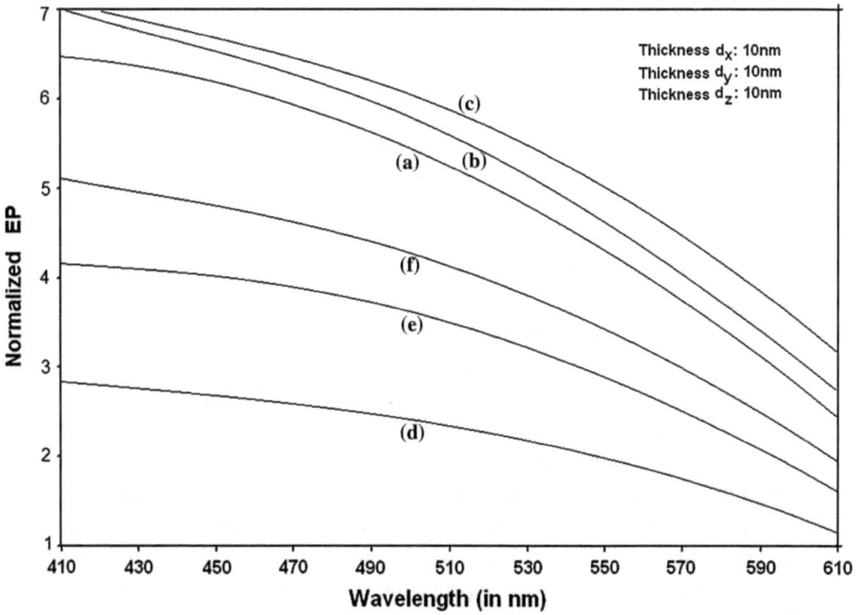

Fig. 8.14 Plot of the normalized EP from QB HD effective mass superlattices of $HgTe/Hg_{1-x}Cd_xTe$ and $In_xGa_{1-x}As/InP$ as function of light wavelength for all cases of Fig. 8.11

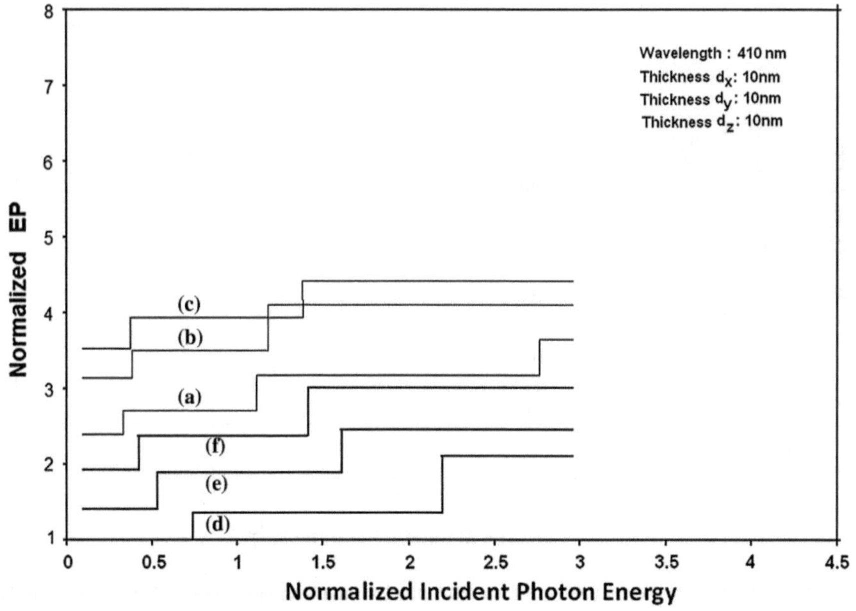

Fig. 8.15 Plot of the normalized EP from QB HD effective mass superlattices of $HgTe/Hg_{1-x}Cd_xTe$ and $In_xGa_{1-x}As/InP$ as function of normalized incident photon energy for all cases of Fig. 8.11

wavelength and normalized incident photon energy respectively for all cases of Fig. 8.11. Using appropriate equations, the normalized EP from effective mass $HgTe/Hg_{1-x}Cd_xTe$ HD SL under magnetic quantization has been plotted as a function of quantizing inverse magnetic field as shown in plot (a) of Fig. 8.16 whose constituent HD materials obey the perturbed HD three band model of Kane in the presence of external photo-excitation. The curves (b) and (c) of the same figure have been drawn for perturbed HD two band model of Kane and perturbed HD parabolic energy bands respectively. The curves (d), (e) and (f) in the same figure exhibit the corresponding plots of $In_xGa_{1-x}As/InP$ HD SL. Figures 8.17, 8.18, 8.19 and 8.20 exhibit the said variation in this case as functions of normalized electron degeneracy, normalized intensity, wavelength and normalized incident photon energy respectively for all the cases of Fig. 8.16.

It appears from Fig. 8.1 that the normalized EP from QW effective mass $HgTe/Hg_{1-x}Cd_xTe$ and $In_xGa_{1-x}As/InP$ HD SLs oscillate with the inverse quantizing magnetic field due to SdH effect where the oscillatory amplitudes and the numerical values are determined by the respective energy band constants. From Fig. 8.2, it appears that the EP increases with increasing carrier concentration in an oscillatory way. The Figs. 8.3 and 8.4 show that the EP decreases with increasing intensity and wavelength in different manners. From Fig. 8.5, it appears that the normalized EP from QW effective mass $HgTe/Hg_{1-x}Cd_xTe$ and $In_xGa_{1-x}As/InP$ HD SLs decreases with increasing film

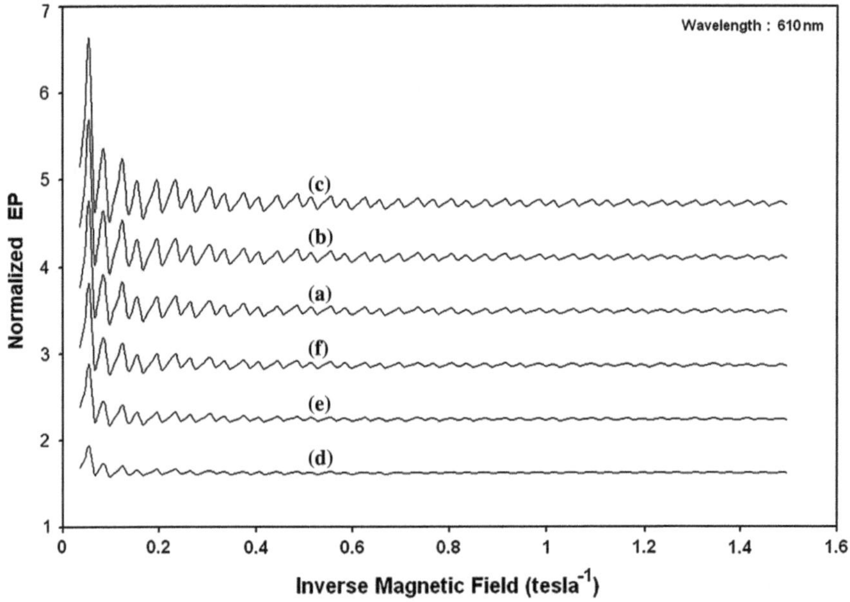

Fig. 8.16 Plot of the normalized EP from HD effective mass superlattices of $HgTe/Hg_{1-x}Cd_xTe$ as function of inverse magnetic field and in which the curves *a*, *b* and *c* represent the perturbed three and two band models of Kane together with parabolic energy bands respectively. The curves *d*, *e* and *f* exhibit the corresponding plots of $In_xGa_{1-x}As/InP$

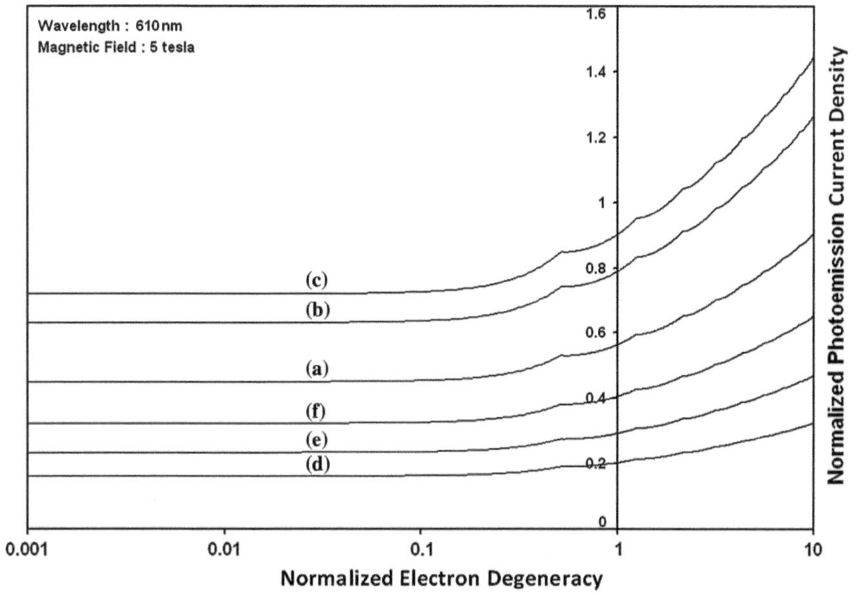

Fig. 8.17 Plot of the normalized magneto EP from HD effective mass superlattices of $HgTe/Hg_{1-x}Cd_xTe$ and $In_xGa_{1-x}As/InP$ as function of normalized electron degeneracy for all cases of Fig. 8.16

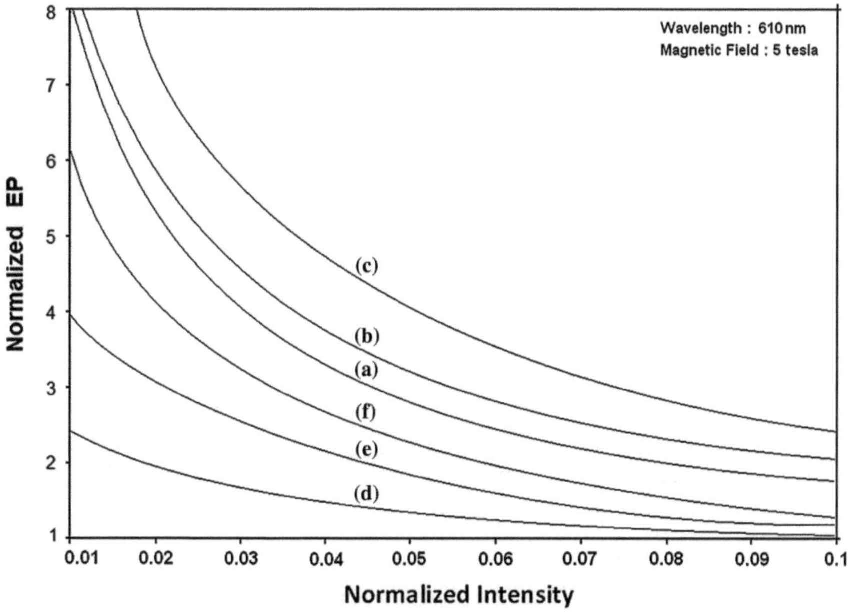

Fig. 8.18 Plot of the normalized magneto EP from HD effective mass superlattices of $HgTe/Hg_{1-x}Cd_xTe$ and $In_xGa_{1-x}As/InP$ as function of normalized light intensity for all cases of Fig. 8.16

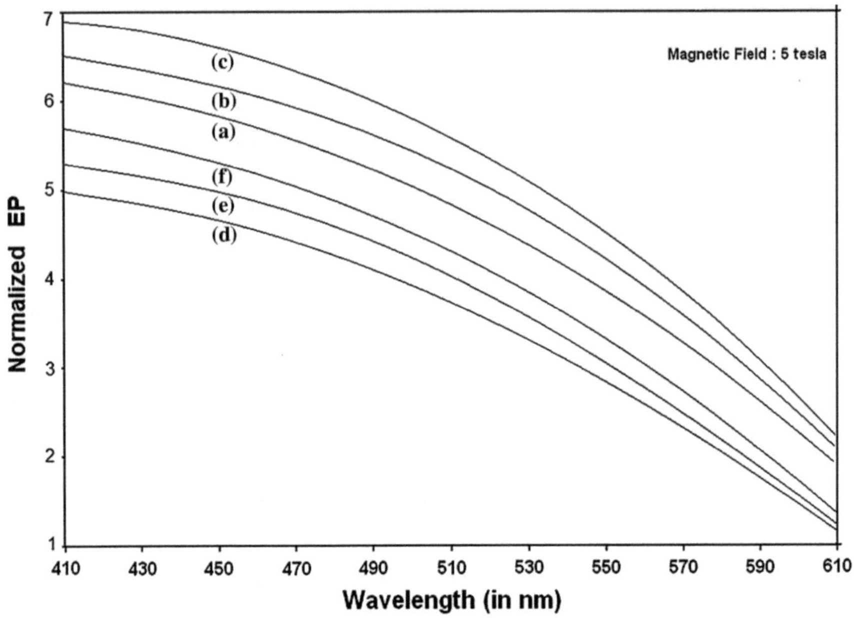

Fig. 8.19 Plot of the normalized magneto EP from HD effective mass superlattices of $HgTe/Hg_{1-x}Cd_xTe$ and $In_xGa_{1-x}As/InP$ as function of light wavelength for all cases of Fig. 8.16

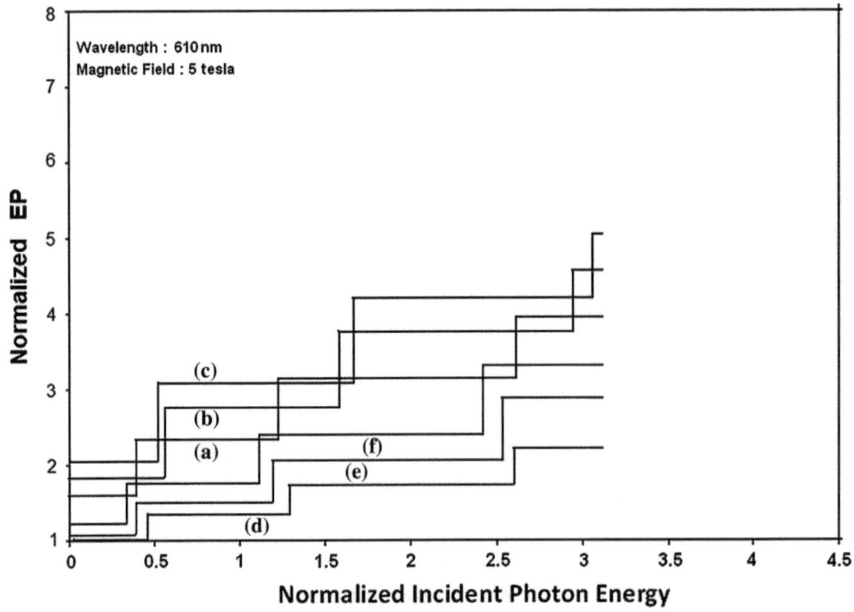

Fig. 8.20 Plot of the normalized magneto EP from HD effective mass superlattices of $HgTe/Hg_{1-x}Cd_xTe$ and $In_xGa_{1-x}As/InP$ as function of normalized incident photon energy for all cases of Fig. 8.16

thickness in oscillatory manner with different numerical values as specified by the energy band constants of the aforementioned HD SLs. From Fig. 8.6, it appears that the normalized EP from NW effective mass $HgTe/Hg_{1-x}Cd_xTe$ and $In_xGa_{1-x}As/InP$ HD SLs increases with decreasing thickness and exhibit large oscillations. From Fig. 8.7, it appears that normalized EP for the said system increases with increasing carrier concentration, exhibiting a quantum jump for a particular value of the said variable for all the models of both the HD SLs. From Figs. 8.8 and 8.9, it can be inferred that the normalized EP in this case increases with decreasing intensity and wavelength in different manners. From Fig. 8.10, it has been observed that the normalized EP from NW effective mass $HgTe/Hg_{1-x}Cd_xTe$ and $In_xGa_{1-x}As/InP$ HD SLs increases with increasing normalized incident photon energy and exhibits quantum steps for specific values of the said variable.

From Fig. 8.11, it appears that EP from QB effective mass $HgTe/Hg_{1-x}Cd_xTe$ and $In_xGa_{1-x}As/InP$ HD SLs exhibit the same type of variations as given in Figs. 8.5 and 8.6 respectively although the physics of QB effective mass HD SLs is completely different as compared with the magneto QW effective mass HD SLs and NW effective mass HD SLs respectively. The different physical phenomena in the former one as compared with the latter two cases yield different numerical values of EP and different thicknesses for exhibiting quantum jump respectively. From Figs. 8.12, 8.13 and 8.14, it appears that EP from QB effective mass $HgTe/Hg_{1-x}Cd_xTe$ and $In_xGa_{1-x}As/InP$ HD SLs increases with increasing carrier concentration, decreasing intensity and decreasing wavelength respectively in various manners. Figure 8.15 demonstrates the fact that the EP from QB effective mass $HgTe/Hg_{1-x}Cd_xTe$ and $In_xGa_{1-x}As/InP$ HD SLs exhibit quantum steps with increasing photon energy for both the cases.

Figure 8.16 exhibits the fact that the normalized EP current density from effective mass $HgTe/Hg_{1-x}Cd_xTe$ and $In_xGa_{1-x}As/InP$ HD SLs oscillates with inverse quantizing magnetic field. Figure 8.17 exhibits the fact that the EP in this case increases with increasing carrier concentration. Figures 8.18 and 8.19 demonstrate that EP decreases with increasing intensity and wavelength in different manners. Finally, from Fig. 8.20, it can be inferred that EP exhibits step functional dependence with increasing photon energy for both the HD SLs in this case with different numerical magnitudes.

8.4 Open Research Problems

Investigate the following open research problems in the presence of external photo-excitation which changes the band structure in a fundamental way together with the proper inclusion of the electron spin, the variation of work function and the broadening of Landau levels respectively for appropriate problems.

(R.8.1) Investigate the EP from quantum confined HD III-V, II-VI, IV-VI, HgTe/CdTe effective mass superlattices together with short period, strained layer, random, Fibonacci, polytype and sawtooth superlattices in the presence of arbitrarily oriented photo-excitation and strain.

(R.8.2) Investigate the multi-photon EP in the presence of arbitrarily oriented photo-excitation and quantizing magnetic field respectively for all the cases of R.8.1.

(R.8.3) Investigate the multi-photon EP in the presence of arbitrarily oriented photo-excitation and non-quantizing non-uniform electric field respectively for all the cases of R.8.1.

(R.8.4) Investigate the multi-photon EP in the presence of arbitrarily oriented photo-excitation and non-quantizing alternating electric field respectively for all the cases of R.8.1.

(R.8.5) Investigate the multi-photon EP in the presence of arbitrarily oriented photo-excitation and crossed electric and quantizing magnetic fields respectively for all the cases of R.8.1.

(R.8.6) Investigate the multi-photon EP from heavily doped quantum confined superlattices for all the problems of R.8.1.

(R.8.7) Investigate the multi-photon EP in the presence of arbitrarily oriented photo-excitation and quantizing magnetic field respectively for all the cases of R.8.1.

(R.8.8) Investigate the multi-photon EP in the presence of arbitrarily oriented photo-excitation and non-quantizing non-uniform electric field respectively for all the cases of R.8.1.

(R.8.9) Investigate the multi-photon EP in the presence of arbitrarily oriented photo-excitation and non-quantizing alternating electric field respectively for all the cases of R.8.1.

(R.8.10) Investigate the multi-photon EP in the presence of arbitrarily oriented photo-excitation and crossed electric and quantizing magnetic fields respectively for all the cases of R.8.1.

(R.8.11) Investigate the EP from quantum confined HD III-V, II-VI, IV-VI, HgTe/CdTe superlatticesQuantum confined HD III-V, II-VI, IV-VI, HgTe/CdTe superlattices with graded interfaces together with short period, strained layer, random, Fibonacci, polytype and sawtooth superlattices in this context in the presence of arbitrarily oriented photo-excitation.

(R.8.12) Investigate the multi-photon EP from heavily doped quantum confined superlattices for all the problems of R.8.11 in the presence non-uniform strain.

(R.8.13) Investigate the multi-photon EP in the presence of arbitrarily oriented photo-excitation and quantizing magnetic field respectively for all the cases of R.8.11.

(R.8.14) Investigate the multi-photon EP in the presence of arbitrarily oriented photo-excitation and non-quantizing non-uniform electric field respectively for all the cases of R.8.11.

(R.8.15) Investigate the multi-photon EP in the presence of arbitrarily oriented photo-excitation and non-quantizing alternating electric field respectively for all the cases of R.8.11.

(R.8.16) Investigate the multi-photon EP in the presence of arbitrarily oriented photo-excitation and crossed electric and quantizing magnetic fields respectively for all the cases of R.8.11.

(R.8.17) Investigate all the problems from R.8.1 to R.8.16 by removing all the mathematical approximations and establishing the respective appropriate uniqueness conditions.

Reference

1. K.P. Ghatak, S. Bhattacharya, D. De, *Einstein Relation in Compound Semiconductors and their Nanostructures*. Springer series in materials science, vol. 116 (Springer, Germany, 2008)

Chapter 9
Few Related Applications and Brief Review of Experimental Results

9.1 Introduction

In this book we have discussed many aspects of EP based on the dispersion relations of different technologically important HDS and their nanostructures. In this chapter we discuss regarding few applications of the content of this book in the Sect. 9.2. We shall also present the brief review of the experimental results in Sect. 9.3. The Sect. 9.4 contains single deep open research problem.

9.2 Different Related Applications

The content of this book finds six applications in the field of materials science and related disciplines in general.

I. *Carrier contribution to the elastic constants*: The knowledge of the carrier contribution to the elastic constants is important in studying the mechanical properties of the materials and has been investigated in the literature [1–23]. The electronic contribution to the second and third order elastic constants for HDS can be written as [1–23]

$$\Delta C_{44} = -\frac{G_0^2}{9} \text{ Real part of } \frac{\partial n_0}{\partial (E_{F_h} - \zeta_i)}, \tag{9.1}$$

and

$$\Delta C_{456} = \frac{G_0^3}{27} \text{ Real part of } \frac{\partial^2 n_0}{\partial (E_{F_h} - \zeta_i)^2}, \tag{9.2}$$

where G_0 is the deformation potential constant.

K.P. Ghatak, *Einstein's Photoemission*, Springer Tracts in Modern Physics 262,
DOI 10.1007/978-3-319-11188-9_9

It is well-known that the thermoelectric power of the carriers in HDS in the presence of a classically large magnetic field is independent of scattering mechanisms and is determined only by their energy band spectra [24–54]. The magnitude of the thermoelectric power G can be written as [24–54]

$$G = \frac{1}{|e|Tn_0} \int_{-\infty}^{\infty} (E - E_{F_h})R(E)\left[-\frac{\partial f_0}{\partial E}\right] dE \tag{9.3}$$

where $R(E)$ is the total number of states. The (9.3) can be written under the condition of carrier degeneracy [24–54] as

$$G = \left(\frac{\pi^2 k_B^2 T}{3|e|n_0}\right) \text{Real part of } \left(\frac{\partial n_0}{\partial(E_{F_h} - \zeta_i)}\right) \tag{9.4}$$

Thus, using (9.1), (9.2) and (9.4), we can write

$$\Delta C_{44} = \left[-n_0 G_0^2 |e| G / \left(3\pi^2 k_B^2 T\right)\right] \tag{9.5}$$

and

$$\Delta C_{456} = \left(n_0 |e| G_0^3 G^2 / (3\pi^4 k_B^3 T)\right)\left(1 + \frac{n_0}{G}\frac{\partial G}{\partial n_0}\right). \tag{9.6}$$

Thus, again the experimental graph of G versus n_0 allows us to determine the electronic contribution to the elastic constants for materials having arbitrary spectra.

II. *Measurement of Band-gap of HDS in the presence of Light Waves*: With the advent of nano-photonics, there has been considerable interest in studying the optical processes in semiconductors and their nanostructures in the presence of intense light waves [55–63]. *It appears from the literature, that the investigations in the presence of external intense photo-excitation have been carried out on the assumption that the carrier energy spectra are invariant quantities under strong external light waves, which is not fundamentally true.* The physical properties of semiconductors in the presence of strong light waves which alter the basic dispersion relations have relatively been much less investigated in [64] as compared with the cases of other external fields needed for the characterization of the low dimensional semiconductors.
With the radical change in the dispersion relation, it is evident that the band gap will also change and in this section we study the normalized incremental band gap (ΔE_g) of HDS as functions of incident light intensity and the wave length respectively in the presence of strong light excitation.
Using the (5.46b)–(5.48), the normalized incremental band gap (ΔE_g) has been plotted as a function of normalized I_0 (for a given wavelength and considering red light for which $\lambda = 660$ nm) at T = 4.2 K in Figs. 9.1 and 9.2 for HD

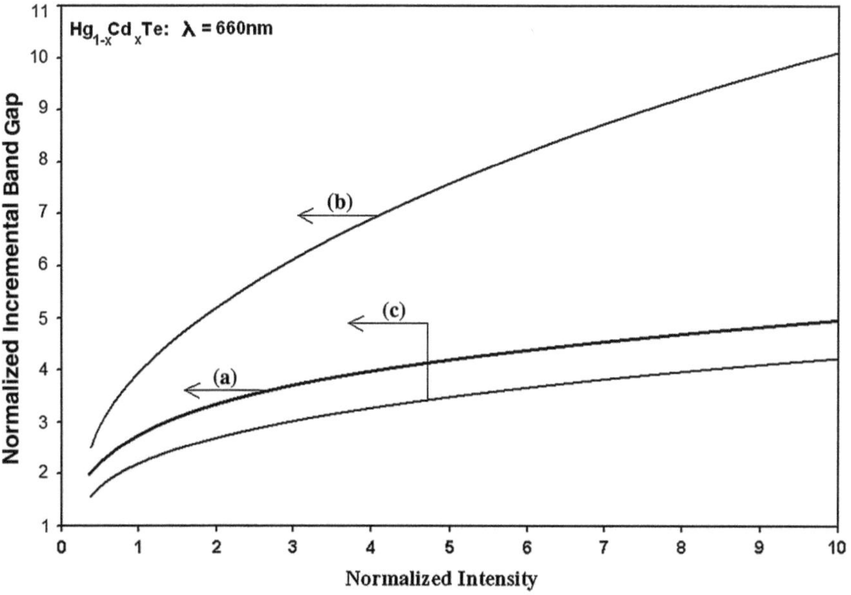

Fig. 9.1 Plots of the normalized incremental band gap (ΔE_g) for HD n-$Hg_{1-x}Cd_xTe$ as a function of normalized light intensity in which the *curves a* and *b* represent the perturbed HD three and two band models of Kane respectively. The *curve c* represents the same variation in HD n-$Hg_{1-x}Cd_xTe$ in accordance with the perturbed parabolic energy bands

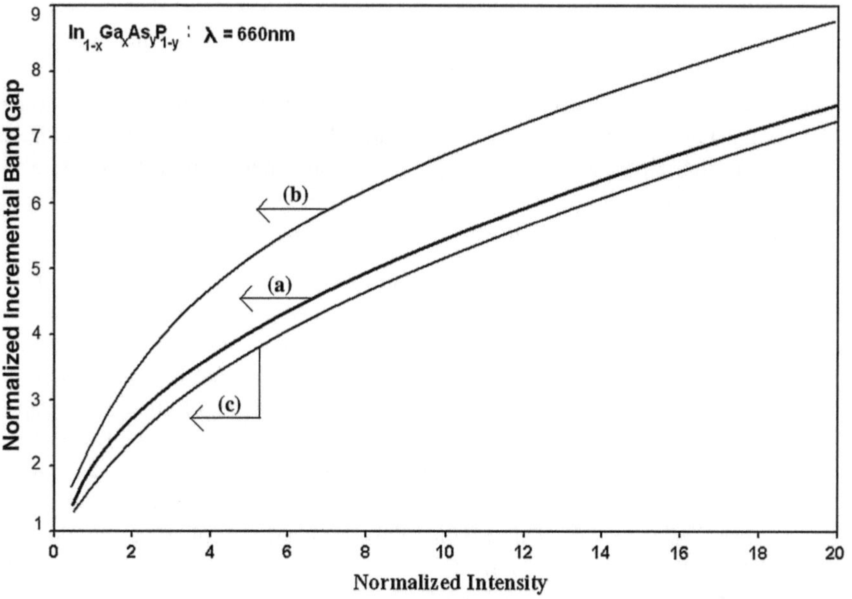

Fig. 9.2 Plots of the normalized incremental band gap (ΔE_g) for HD $In_{1-x}Ga_xAs_yP_{1-y}$ lattice matched to InP as a function of normalized light intensity for all cases of Fig. 9.1

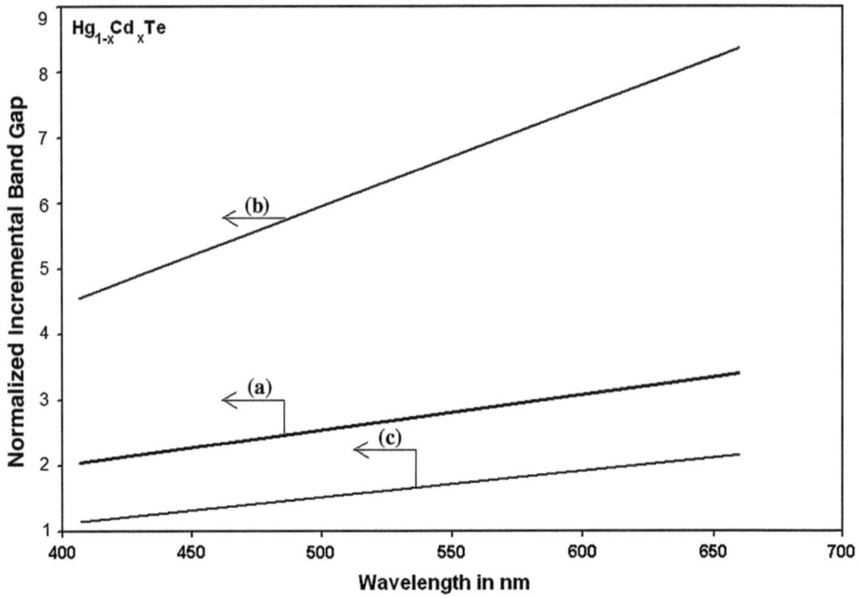

Fig. 9.3 Plots of the normalized incremental band gap (ΔE_g) for HD $Hg_{1-x}Cd_xTe$ as a function of wavelength for all cases of Fig. 9.1

n-$Hg_{1-x}Cd_xTe$ and n-$In_{1-x}Ga_xAs_yP_{1-y}$ lattice matched to InP in accordance with the perturbed three and two band models of Kane and that of perturbed parabolic energy bands respectively. In Figs. 9.3 and 9.4, the normalized incremental band gap has been plotted for the aforementioned optoelectronic compounds as a function of λ. It is worth remarking that the influence of an external photo-excitation is to change radically the original band structure of the material. Because of this change, the photon field causes to increase the band gap of semiconductors. We propose the following two experiments for the measurement of band gap of semiconductors under photo-excitation.

(A) A white light with colour filter is allowed to fall on a semiconductor and the optical absorption coefficient $(\bar{\alpha}_0)$ is being measured experimentally. For different colours of light, $\bar{\alpha}_0$ is measured and $\bar{\alpha}_0$ versus $\hbar\omega$ (the incident photon energy) is plotted and we extrapolate the curve such that $\bar{\alpha}_0 \rightarrow 0$ at a particular value $\hbar\omega_1$. This $\hbar\omega_1$ is the unperturbed band gap of the semiconductor. During this process, we vary the wavelength with fixed I_0. From our present study, we have observed that the band gap of the semiconductor increases for various values of λ when I_0 is fixed (from Figs. 9.3 and 9.4). This implies that the band gap of the semiconductor measured (i.e. $\hbar\omega_1 = E_g$) is not the unperturbed band gap E_{g_0} but the perturbed band gap E_g; where $E_g = E_{g_0} + \Delta E_g$, ΔE_g is the increased band gap at $\hbar\omega_1$. Conventionally, we consider this E_g as the unperturbed band

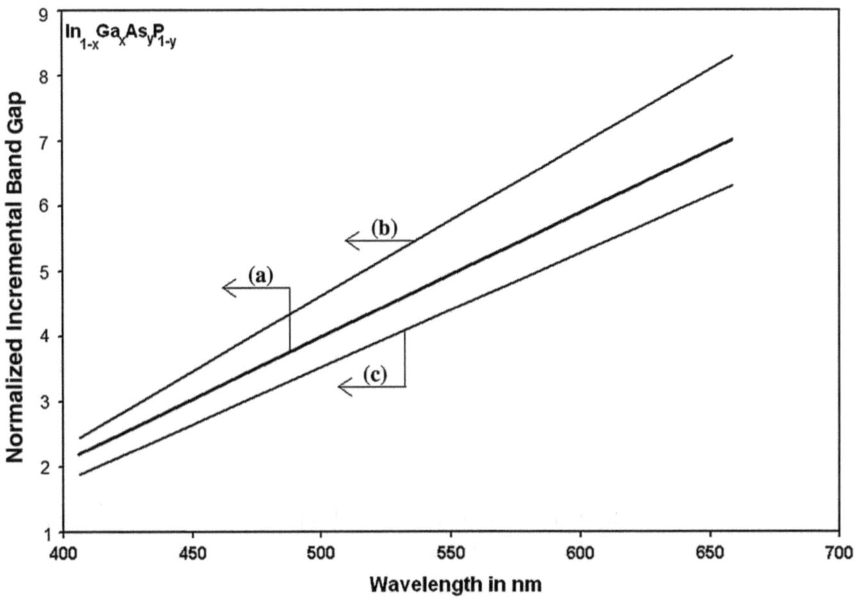

Fig. 9.4 Plots of the normalized incremental band gap (ΔE_g) for HD In $_{1-x}Gs_xAs_yP_{1-y}$ lattice matched to InP as a function of wavelength for all cases of Fig. 9.1

gap of the semiconductor and this particular concept needs modification. Furthermore, if we vary I_0 for a monochromatic light (when λ is fixed) the band gap of the semiconductor will also change consequently (Figs. 9.1 and 9.2). Consequently, the absorption coefficient will change with the intensity of light [64]. For the overall understanding, the detailed theoretical and experimental investigations are needed in this context for various materials having different band structures.

(B) The conventional idea for the measurement of the band gap of the semiconductors is the fact that the minimum photon energy hv (v is the frequency of the monochromatic light) should be equal to the band gap E_{g_0} (unperturbed) of the semiconductor, i.e.,

$$hv = E_{g_0} \tag{9.7}$$

In this case, λ is fixed for a given monochromatic light and the semiconductor is exposed to a light of wavelength λ. Also the intensity of the light is fixed. From Figs. 9.3 and 9.4, we observe that the band

$$hv = E_g \tag{9.8}$$

Furthermore, if we vary the intensity of light (Figs. 9.1 and 9.2) for the study of photoemission, the minimum photon energy should be

$$h\nu_1 = E_{g_1} \tag{9.9}$$

gap of the semiconductor is not E_{g_0} (for a minimum value of $h\nu$) but E_g, the perturbed band gap. Thus, we can rewrite the above equality as where E_{g_1} is the perturbed band gap of the semiconductor due to various intensity of light when ν and ν_1 are different.
Thus, we arrive at the following conclusions:

(a) Under different intensity of light, keeping λ fixed, the condition of band gap measurement is given by

$$h\nu_1 = E_{g_1} = E_{g_0} + \Delta E_{g_1} \tag{9.10}$$

(b) Under different colour of light, keeping the intensity fixed, the condition of band gap measurement assumes the form

$$h\nu = E_g = E_{g_0} + \Delta E_g \tag{9.11}$$

and not the conventional result as given by (9.7).

III. *Diffusion Coefficient of the Minority Carriers*: This particular coefficient in quantum confined lasers can be expressed as

$$D_i/D_0 = dE_{Fi}/dE_F \tag{9.12}$$

where D_i and D_0 are the diffusion coefficients of the minority carriers both in the presence and absence of quantum confinements and E_{Fi} and E_F are the Fermi energies in the respective HD cases. It appears then that, the formulation of the above ratio requires a relation between E_{Fi} and E_F, which, in turn, is determined by the appropriate carrier statistics. Thus, our present study plays an important role in determining the diffusion coefficients of the minority carriers of HD quantum-confined lasers with materials having arbitrary band structures. Therefore in the investigation of the optical excitation of the HD optoelectronic materials which lead to the study of the ambipolar diffusion coefficients the present results contribute significantly.

IV. *Nonlinear Optical Response*: The nonlinear response from the optical excitation of the free carriers is given by [65]

$$Z_0 = \frac{-e^2}{\omega^2\hbar^2}\int_0^\infty \left(k_x\frac{\partial k_x}{\partial E}\right)^{-1} f_0 N(E)dE \tag{9.13}$$

where ω is the optical angular frequency, $N(E)$ is the DOS function. From the various E-k relations of different HD materials under different physical conditions, we can formulate the expression of $N(E)$ and from band structure we can derive the term $\left(k_x \frac{\partial k_x}{\partial E}\right)$ and thus by using the DOS function as formulated, we can study the Z_0 for all types of materials as considered in this monograph.

V. *Third Order Nonlinear Optical Susceptibility*: This particular susceptibility can be written as [66]

$$\chi_{NP}(\omega_1, \omega_2, \omega_3) = \frac{n_0 e^4 \langle \varepsilon^4 \rangle}{24\omega_1\omega_2\omega_3(\omega_1 + \omega_2 + \omega_3)\hbar^4} \tag{9.14}$$

where $n_0\langle\varepsilon^4\rangle = \int_0^\infty \frac{\partial^4 E}{\partial k_z^4} N(E) f_0 \, dE$ and the other notations are defined in [70]. The term $\left(\frac{\partial^4 E}{\partial k_z^4}\right)$ can be formulated by using the dispersion relations of different HD materials as given in appropriate sections of this monograph. Thus one can investigate the $\chi_{NP}(\omega_1, \omega_2, \omega_3)$ for all materials as considered in this monograph.

VI. *Generalized Raman Gain*: The generalized Raman gain in optoelectronic materials can be expressed as [67]

$$R_G = \bar{I}\left(\frac{16\pi^2 c^2}{\hbar\omega\rho g \omega_s^2 n_s n_p}\right)\left(\frac{\Gamma_\rho}{\Gamma}\right)\left(\left(\frac{e^2}{mc^2}\right)^2 m^2 R^2\right) \tag{9.15}$$

where, $\bar{I} = \sum_{n.t_z} [f_0(n, k_z \uparrow) - f_0(n, k_z \downarrow)]$, $f_0(n, k_z \uparrow)$ is the Fermi factor for spin-up Landau levels, $f_0(n, k_z \downarrow)$ is the Fermi factor for spin down Landau levels, n is the Landau quantum number and the other notations are defined in [71]. It appears then the formulation of R_G is determined by the appropriate derivation of the magneto-dispersion relations. By using the different appropriate formulas as formulated in various HD materials in different chapters of this monograph R_G can, in general, be investigated.

9.3 Brief Review of Experimental Results

The experimental aspects of the EP is very wide and even the condensed presentation of which in a chapter highlighting the major points only permanently enjoys the domain of impossibility theorems. Still for the purpose of coherent presentation we embark on a difficult and deep work.

Houdré et al. [68] have presented the **first experimental evidence** of tunnelling and transport of electrons from quantum states in a GaAs/GaAlAs super-lattice or a quantum well to a GaAs surface activated to negative electron affinity. The photocurrent versus light excitation energy showed definite structures which appeared exactly at the calculated energies of the allowed optical transitions between the quantized levels of the valence and conduction bands. The 300- and 30-K results for the super-lattice were successfully compared to luminescence experiments, and could lead to the production of highly polarized electron beams. Lockwood [69] in an important review article has pointed out that amongst a number of diverse approaches to engineering efficient light emission in silicon nano-structures, one system that has received considerable attention has been Si/SiO_2 quantum wells. Engineering such structures has not been easy, because to observe the desired quantum confinement effects, the quantum well thickness has to be less than 5 nm. Nevertheless, such ultra thin structures have now been produced by a variety of techniques. The SiO_2 layers are amorphous, but the silicon layers can range from amorphous through nano-crystalline to single-crystal form. The fundamental band gap of the quantum wells has been measured primarily by optical techniques and strong confinement effects have been observed. A detailed comparison is made between theoretical and experimental determinations of the band gap in Si/SiO_2 quantum wells.

Confinement of electrons in small structures such as a thin film results in discrete quantum well states. Such states can be probed by angle-resolved photoemission and Chiang and Chiang [70] have reviewed the basic physics and applications of quantum well spectroscopy. The energies and lifetime widths of quantum well states in a film depend on the film thickness, the dynamics of electron motion in the film, and the confinement potential. A detailed study allows a determination of the bulk band structure of the film material, the lifetime broadening of the quasi-particle, and the interfacial reflectivity and phase shift, as will be demonstrated with simple examples. Quantification of the photoemission results can be achieved through a simple phase analysis based on the Bohr-Sommerfeld quantization rule. Explicit forms of wave functions can also be constructed for additional information regarding the spatial distribution of the electronic states. From such studies, a detailed understanding of the behaviour of simple quantum wells including the effects of lattice mismatch can be developed, which provides a useful basis for investigating the properties of multi-layers.

Low-temperature optical transmission spectra of several In_xGa_{1-x} As/GaAs strained multiple quantum wells (MQWs) with different well widths and In mole fractions have been measured by Ji et al. [71] the excitonic transitions up to 3C-3H are observed. The notation n c-m H (L) is used to indicate the transitions related to the nth conduction and mth valence heavy (light) hole sub bands. Step like structures corresponding to band-to-band transitions are also observed, which are identified as 1C-1L transitions. The calculated transition energies, taking into account both the strain and the quantum well effects, are in good agreement with the measured values. In these calculations the lattice mismatch between the GaAs buffer and the InGaAs/GaAs MQW is taken into account and the valence-band offset Qv is chosen as an adjustable parameter. By fitting the experimental results to our calculations, we conclude that the light holes are in GaAs barrier region (type II

MQW) and the valence-band offset Qv is determined to be 0.30. A possible system in which the transition from type I to type II for light holes might be observed is also discussed.

Excitonic Photoluminescence (PL) line widths in AlGaAs grown by molecular beam epitaxy (MBE). The line widths of excitonic transitions were measured by Reynolds et al. [72] in $Al_xGa_{1-x}As$, grown by MBE as a function of alloy composition x for values of $x \lesssim 0.43$ using high resolution PL spectroscopy at liquid helium temperature. The values of the line widths thus measured are compared with the results of several theoretical calculations in which the dominant broadening mechanism is assumed to be the statistical potential fluctuations caused by the components of the alloy. An increase in the line width as a function of x is observed which is in essential agreement with the prediction of the various theoretical calculations. The line widths of the excitonic transitions in Al_xGa_{1-x} As observed in the present work are the narrowest ever reported in the literature, for example $\sigma = 2.1$ meV for $x = 0.36$, thus indicating very high quality material.

Martin et al. [73] have measured the valence-band discontinuity at a wurtzite GaN/AIN (0001) hetero junction by means of x-ray photoemission spectroscopy. The method first measures the core level binding energies with respect to the valence-band maximum in both GaN and A1N bulk films. The precise location of the valence band maximum is determined by aligning prominent features in the valence band spectrum with calculated densities of states. Subsequent measurements of separations between Ga and Al core levels for thin over layers of GaN film grown on A1N and vice versa yield a valence band discontinuity of $\Delta Ev = 0.8 + -0.3$ eV in the standard Type I hetero junction alignment.

Electronic defects in n-type GaN were characterized by Götz et al. [74] with the help of photoemission capacitance transient spectroscopy. Conventional deep level transient spectroscopy is of limited use in semiconductors with wide band gaps (e.g., 3.4 eV for GaN at 300 K) because it utilizes thermal energy for charge emission which restricts the accessible range of band gap energies to within ~0.9 eV of either band edge, for practical measurement conditions. For electron photoemission to the conduction band, four deep levels were detected at optical threshold energies of approximately 0.87, 0.97, 1.25, and 1.45 eV. It is suggested that the above photo-detected deep levels may participate in the 2.2 eV defect luminescence transitions, which are also demonstrated for our material.

Benjamin et al. [75] present results of UV photoemission measurements of the surface and interface properties of hetero epitaxial AlGaN on 6H-SiC. Previous results have demonstrated a negative electron affinity of AlN on 6H-SiC. In this study $Al_xGa_{1-x}N$ alloy films were grown by OMVPE and doped with silicon. The analytical techniques included UPS, Auger electron spectroscopy, and LEED. All analysis took place in an integrated UHV transfer system which included the analysis techniques, a surface processing chamber and a gas source MBE. The OMVPE alloy samples were transported in air to the surface characterization system while the AlN and GaN investigations were prepared in situ. The surface electronic states were characterized by surface normal UV photoemission to determine whether the electron affinity was positive or negative. Two aspects of the photoemission distinguish a surface that exhibits a NEA: (1) the spectrum exhibits a sharp peak in the low kinetic energy region, and (2) the width of the spectrum is $h\nu - E_g$. The in situ prepared AlN samples exhibited the characteristics of a NEA while the GaN and $Al_{0.13}Ga_{0.87}N$ samples did not. The $Al_{0.55}Ga_{0.45}N$ sample shows a low positive electron affinity. Annealing of the sample to >400 °C resulted in the disappearance of the sharp emission features, and this effect was related to contaminant effects on the surface. The results suggest the potential of nitride based cold cathode electron emitters.

The polarization of photo emitted electrons from thin GaAs layers grown by MBE has been measured by Maruyama et al. [76]. Polarization as high as 49 % was observed for a 0.2 μm-thick GaAs sample at excitation photon wavelengths longer than 750 nm. The maximum polarization is dependent on the thickness of the GaAs layer, decreasing to about 41 % for a 0.9 μm-thick GaAs sample.

The base-collector junction of GaAs/AlGaAs single hetero junction bipolar transistors has been observed to emit light at avalanche break down by Chen et al. [77]. The spectral distribution curve exhibits broad peaks at 2.03 and 1.43 eV, with the intensities dependent upon the reverse current. These observations suggest that electrons, excited to the upper conduction band by the field, lose their energy by impact ionizing electron-hole pairs and producing the 2.03 eV light, which corresponds to the threshold energy for electron impact ionization. The band-edge emission is the result of direct-gap free-carrier recombination and self-absorption of the high energy transition.

Angle-resolved transmission of s-polarized light in triple-film hetero-opals has been investigated by Khunsin et al. [78] in the spectral range including high-order photonic band gaps, and compared to the transmission of its constituent single-film opals. The interfaces do not destroy the predominantly ballistic light propagation over the studied frequency and angular ranges, but hetero structuring leads to a smoothed angular distribution of intensity of the transmitted light and to the reconstruction of the transmission minima dispersion. The interface transmission function has been extracted by comparing the transmission of the hetero-opal and its components in order to demonstrate the difference. This deviation from the superposition principle was provisionally assigned to light refraction and reflection at the photonic crystal interfaces and to the mismatch between mode group velocities in hetero-opal components.

A 1.7-fold enhancement in the spontaneous emission intensity of dye chromophore loaded in a printable polymer is achieved by Reboud et al. [79] by coupling the dye emission to surface plasmons of metallic nano particles. The nano composite material, embossed into arrays of wires by nano imprint lithography process, shows good imprint properties. The results prove the potential of the prepared luminescent functional materials for micro- and nanofabrication and suggest the use of nano composite materials in prospective nano plasmonic applications.

A hetero junction between two 3-dimensional photonic crystals has been realized by Romanov et al. [80] by interfacing two opal films of different lattice constants. The interface-related transmission minimum has been observed in the frequency range between two directional lowest-order band gaps of the hetero-opal constituents. The interface transmission minimum has been modelled numerically and tentatively explained by formation of the standing wave across the photonic hetero-crystal due to matching of group velocities of optical modes in both parts at this frequency.

The results of optical phenomena investigations in QD and quantum well structures under inter band optical pumping are presented by Aleshkin et al. [81]. Inter-band and intra-band light absorption in nanostructures with QDs has been studied experimentally and theoretically. PL and inter-band light absorption in stepped quantum wells have been investigated including PL studies under pico-second optical pumping. Experimental results have been compared with results of calculation of energy spectrum and transition probabilities. It is shown that inversion of population exists between the third and second excited levels of stepped quantum well.

The PL spectra of samples with ultrathin InGaN layers embedded in AlGaN and GaN matrices are studied experimentally by Usov et al. [82] in the temperature range of 80–300 K. It is shown that the temperature dependences can be understood in the context of Eliseev's model and that, in the active region of the structures under study, the dispersion σ of the exciton-localization energy depends on the average In content in InGaN-alloy layers. Furthermore, the Urbach energy E U, which characterizes the localization energy of excitons in the tails of the density-of-states, was determined from an analysis of the shape of the low-energy slope of the spectrum. It is shown that σ and E U, quantities representing the scale of the exciton-localization effects, vary linearly with the PL-peak wavelength in the range from the ultraviolet to the green region of the spectrum.

Nano wires have been formed by the infiltration of CdTe nano crystals into nano tubes of chrysotile asbestos ($Mg_3Si_2O_5(OH)_4$). PL of a regular array of these templated nano-wires was studied by Bardosova et al. [83] under the excitation of the light from a xenon lamp at different wavelengths. Strong interactions of nano-crystals with structural defects of the template were observed. No dependence of the PL spectra upon polarisation of the laser beam was observed and no shift of the PL band was detected in the light polarised along and across nano wires, thus indicating the weakness of the interaction between nano-crystals in the nano tubes.

Upon deposition of silicon onto the (1 1 0) surface of a silver crystal, Leandri et al. [84] have grown massively parallel one-dimensional Si nano wires. They are imaged in scanning tunnelling microscopy as straight, high aspect ratio, nano-structures, all with the same characteristic width of 16 Å, perfectly aligned along the atomic troughs of the bare surface. Low energy electron diffraction confirms the massively parallel assembly of these self-organized nano wires. Photoemission reveals striking quantized states dispersing only along the length of the nano wires, and extremely sharp, two-components, Si 2p core levels. This demonstrates that in the large ensemble each individual nano wire is a well-defined quantum object comprising only two distinct silicon atomic environments. They also suggest that this self-assembled array of highly perfect Si nano wires provides a simple, atomically precise, novel template that may impact a wide range of applications.

The coverage dependent electronic structure of Cu and Co on vicinal W(110) surfaces has been investigated by Zilkens [85] with angle resolved photoelectron spectroscopy. To prepare the quasi-one-dimensional Cu and Co systems, the method of step edge decoration of the vicinal W(110) surfaces has been used. The vicinal surfaces with step edges in (110), (100) and (111) direction has been investigated using LEED. From the characteristic spot splitting a terrace width of 11 atom rows was determined. The band structures of the flat and the vicinal surfaces have indicated that the step edges have no bearing on the bulk band structure at k parallel = 0. But the surface band structure shows a different dispersion and different energy positions of surface states. An analysis of the W 4fτ/2 core level spectra has resulted in an additional contribution of the step edges in the spectra of the vicinal surfaces with a surface core level shift between 120 and 150 meV. A Cu and Co coverage dependent investigation of the core levels shows that there is no Co induced surface reconstruction and up to 0.15 monolayer no Cu induced surface reconstruction. In the range of 0.15–0.3 monolayer Cu the surface peak shifts to higher binding energies. This is probably a result of a surface reconstruction of the W substrate In the core level spectra with Co coverage the intensity of the surface peak decreases linear with Co coverage and the intensity of a new contribution, the interface structure, increases with Co coverage. With Co respectively Cu coverage the contribution of the step edge shifts to lower respectively higher binding energies. This can be attributed to a charge transfer between the adsorbate and the substrate in different directions.

Spherical Si nano-crystallites with Ge core ($\sim$20 nm in average dot diameter) have been prepared by Darma [86] by controlling selective growth conditions of low-pressure chemical vapour deposition (LPCVD) on ultrathin SiO_2 using alternately pure SiH_4 and 5 % GeH_4 diluted with He. XPS results confirm the highly selective growth of Ge on the pre-grown Si dots and subsequently complete coverage by Si selective growth on Ge/Si dots. Compositional mixing and the crystallinity of Si dots with Ge core as a function of annealing temperature in the range of 550–800 °C has been evaluated by XPS analysis and confirms the diffusion of Ge atoms from Ge core towards the Si clad accompanied by formation of GeO_x at the Si clad surface. The first sub-band energy at the valence band of Si dot with Ge core has been measured as an energy shift at the top of the valence band density of state using XPS. The systematic shift of the valence band maximum towards higher binding energy with progressive deposition in the dot formation indicate the charging effect of dots and SiO_2 layer by photoemission during measurements.

Rowella et al. [87] have obtained Ge nano-crystals from the dewetting process during thermal annealing of an amorphous Ge layer deposited by MBE on a thin SiO_2 layer on Si(001). The Ge nano-crystals were then capped with a thin layer of amorphous Si. The mean nano-crystal size—2.5 to 60 nm—depends on the initial Ge layer thickness. Low-temperature PL measurements were performed to investigate quantum confinement effects on the Ge nano-crystal energy gap and defect states. For the present range of particle sizes, the nano-particle PL emission appeared as a wide near-infrared band near 900 meV although a weak confined band was also observed for the smallest nano-particles. Further thermal annealing of the samples increased the inter band recombination by nearly two orders of magnitude.

Mala et al. [88] have obtained an intense PL peaking near 0.9 eV is emitted by a single $Si_{1-x}Ge_x$ nano-meter-thick layer (NL) with x ≈ 8 % incorporated into Si/ $Si_{0.6}Ge_{0.4}$ cluster multi layers (CMs). The SiGe NL PL does not saturate in output intensity with up to 50 mJ/cm^2 of excitation energy density, and it has nearly a 1,000 times shorter lifetime compared to CM PL, which peaks at ~0.8 eV. These dramatic differences in observed PL properties are attributed to different compositions and structures of the Si/SiGe NL and CM hetero-interfaces.

Carlsson et al. [89] have observed a strong oscillatory photon energy dependence for the intensity of the photoemission peaks due to quantum well states in Na overlayers on Cu(111). The measurements are made at low photon energies $h\upsilon < 8$ eV with Na films, which are between four and eight atomic layers thick. The intensity oscillations are ascribed to the interference between the contributions to the outgoing wave associated with the two tails of a quantum-well state.

Woodruff et al. [90] have studied normal-emission photoemission spectra from the quantum well state in a single monolayer of Ag on V(100) have been studied as a function of photon energy. By comparing the measured binding energy and calculated electron momentum perpendicular to the surface in this state with the unoccupied s-p bands of Ag and V, they have showed that the peak intensity corresponds to the condition expected for a direct transition in V(100). This result is consistent with the fact that the state is in many ways similar to an intrinsic Shockley surface state of a clean surface. PL spectra obtained at 6 K with excitation at 405 nm exhibits the sharp drop at low energy near 700 meV is due to the cut-off in the instrumental response. A strong low-energy PL doublet is seen, with peaks near 780 and 820 meV, together with a much weaker peak at 872 meV. The ratio of intensities of the strong and weak peaks is the same in both samples. The intensities of all three PL peaks decrease with increasing temperature up to 25 K, but the weak peak decreases in intensity faster than that of the strong peaks. The weak peak at 872 meV is most likely the dipole-allowed direct-gap transition expected at 0.863 eV in the superstructure. The small difference in energy between theory and experiment could be the result of a difference in strain within the layer in the sample compared with the ideal (perfect) modeled structure or from assumptions in parameter values in the model. The strong peaks at 820 and 780 meV are assigned to the no-phonon and transverse-optic-phonon emission lines, respectively, of the $Si_{0.4}Ge_{0.6}$ buffer layer. The $\sim$40 meV separation between the two strong peaks is characteristic of the phonon energies in SiGe alloys. The energies of the peaks, however, are much lower than that expected for a bulk $Si_{0.4}Ge_{0.6}$ alloy ($\sim$0.97 eV). The energies and general appearance of these peaks is reminiscent of what has been obtained from PL studies of SiGe nanostructures imbedded in Si. It is therefore likely that this PL arises predominately at the $Si_{0.4}Ge_{0.6}$/superstructure interface where there is type-II band alignment. In conclusion, they have obtained experimental evidence of the predicted direct-gap optically-allowed transition in a special super cell comprised of a number of ultrathin layers of Si and Ge.

The II-VI semiconductor nano-particles have recently attracted a lot of attention due to the possibility of their application in various devices. Tiwari et al. [91] have used chemical method in synthesis of CdS nano-particles and thiophenol was used as capping agent. X-ray diffraction studies of both samples were done. The dc conductivity of CdS increases at a lower rate or is approximately constant up to 500 K and thereafter the conductivity increases at a rapid rate. Beyond T_c it is seen that the portion of the σ_{dc} versus 1,000/T is almost a straight line showing an Arrhenius behaviour. The dielectric constant of nano-particles of CdS is found to be larger than the corresponding values of CdS crystals. It is clearly observed that at lower wavelengths nano-crystalline samples show a blue-shift. The three peaks of sample (S2), A, B and C can be ascribed to the transition from Cd-O complex donor formed by adsorbed oxygen to the valance band, Cd—excess acceptor and the surface states, respectively.

Wilson et al. [92] have analyzed the dynamics and spectroscopy of silicon nano-crystals that emit at visible wavelengths. Size-selective precipitation and size-exclusion chromatography cleanly separate the silicon nano-crystals from larger crystallites and aggregates and provide direct evidence for quantum confinement in luminescence. Measured quantum yields are as high as 50 % at low temperature, principally as a result of efficient oxide passivation. Despite a 0.9—electron-volt shift of the band gap to higher energy, the nano-crystals behave fundamentally as indirect gap materials with low oscillator strength.

Maillard et al. [93] have shown silver nano crystals, self-organized in compact hexagonal networks, on gold and graphite exhibit anisotropic optical properties. From polarized electron photoemission spectroscopy, a two-photon mechanism was demonstrated and an enhancement due to the surface plasmon resonance (SPR) of the nano crystal film was observed. Two SPR peaks appear, due to dipolar interactions and induced by the self-organization of silver nano crystals. This property was used to probe the substrate effect on the plasmon resonance. Its damping was related to particle–substrate interactions.

Xiong et al. [94] have **presented the first photoelectron spectroscopy measurements of QDs** in the gas phase. By coupling a nano-particle aerosol source to a femto-second velocity map imaging photoelectron spectrometer, we apply robust gas-phase photoelectron spectroscopy techniques to colloidal QDs, which typically must be studied in a liquid solvent or while bound to a surface. Working with a flowing aerosol of QDs offers the additional advantages of providing fresh nano-particles for each laser shot and removing perturbations from bonding with a surface or interactions with the solvent. In this work, they have performed a two-photon photo-ionization experiment to show that the photoelectron yield per exciton depends on the physical size of the QD, increasing for smaller dots. Next, using effective mass modelling they have shown that the extent to which the electron wave function of the exciton extends from the QD, the so-called "evanescent electron wave function", increases as the size of the QD decreases. This group show that the photoelectron yield is dominated by the evanescent electron density due to quantum confinement effects, the difference in the density of states inside and outside of the QDs, and the angle-dependent transmission probability of electrons through the surface of the QD. Therefore, the photoelectron yield directly reflects the fraction of evanescent electron wave function that extends outside of the QDs. The work of this group shows that gas-phase photoelectron spectroscopy is a robust and general probe of the electronic structure of QDs, enabling **the first direct measurements of the evanescent exciton wave function.**

Konchenko et al. [95] have studied quantum-confinement effect in the valence band of germanium nano-dots and measured by means of photoemission spectroscopy. The spherical dots of 3–10 nm in diameter were prepared on a 0.3-nm-thick SiO_2 film on Si(111) substrate. Dot-size dependence of the band edge matched the ones expected from the spherical QD model and calculated from the semi-empirical simulation. Colvin et al. [96] have reported **the first application of valence-band photoemission to a quantum-dot system**. Photoemission spectra of Cds QDs, ranging in size from 12 to 35 Å radius, were obtained using photon energies of 20–70 eV. The spectra are qualitatively similar to those obtained for bulk Cds, but show a shift in the valence-band maximum with size.

The measurements of the optical and structural characteristics of various porous Si samples have been correlated by Lockwood et al. [97]. Although the PL peak wavelength shows no correlation with the Si nano-particle size, the optical absorption edge exhibits a **strong inverse correlation that is in excellent agreement with theoretical predictions for the optical gap in Si spheres or QDs**. This constitutes direct evidence for quantum confinement effects in porous Si. The possibility induction of light emission from silicon, an indirect band-gap material in which radiative transitions are unlikely, raises several interesting and technologically important possibilities, especially the fabrication of a truly integrated opto-electronic microchip. Laser diode structures on GaAs substrates with an active region employing laterally associated InAs, QDs obtained by Zhukov et al. [98] by low-temperature MBE exhibit electroluminescence at a wavelength of 1.55–1.6 μm in a temperature range from 20 to 260 K.

Structural and optical properties of thin InGaAsN insertions in GaAs, grown by MBE using an RF nitrogen plasma source, have been investigated by Volovik et al. [99]. Nitrogen incorporation into InGaAs results in a remarkable broadening of the luminescence spectrum as compared with that of InGaAs layer with the same indium content. Correspondingly, a pronounced corrugation of the upper interface and the formation of well defined nanodomains are revealed in cross-sectional and plan-view transmission electron microscope (TEM) images, respectively. Raising the indium concentration in InGaAsN ($N < 1$ %) to 35 % results in the formation of well defined separated three-dimensional (3D) islands. The size of the nanodomains proves that the InGaAsN insertions in GaAs should be regarded as QD structures even in the case of relatively small indium concentrations (25 %) and layer thicknesses (7 nm), which are below the values required for a 2D-3D transition to occur in InGaAs/GaAs growth. Dislocation loops have been found in TEM images of the structures emitting at 1.3 μm. They are expected to be responsible for the degradation of the luminescence intensity of such structures in agreement with the case of long-wavelength InGaAs-GaAs QDs.

QDs formed on GaAs(100) substrates by InAs, deposition followed by (Al,Ga) As or (In,Ga,Al)As overgrowth demonstrate a PL peak that is red shifted (up to 1.3 μm) compared to PL emission of GaAs-covered QDs have been demonstrated by Tsatsul'nikov et al. [100]. The result is attributed to redistribution of InAs molecules in the system in favour of the QDs, stimulated by Al atoms in the cap layer. The deposition of a 1 nm thick AlAs cover layer on top of the InAs–GaAs QDs results in replacement of InAs molecules of the wetting layer by AlAs molecules, leading to a significant increase in the heights of the InAs QDs, as follows from transmission electron microscopy. This effect is directly confirmed by transmission electron microscopy indicating a transition to a Volmer–Weber-like QD arrangement.

Structural and optical properties of InAs QDs overgrown by thin (In, Ga, AI) As layers were investigated by Tsatsul'nikov et al. [101]. Adding In as well as Al during overgrowth of the QDs results in an increase in QD size and a change in QD shape due to reduced of in diffusion from the QDs during overgrowth and transport of in atoms from the wetting layer. This leads to a red shift of the emission and allows to realise 1.3\im emission using QDs in the initial stage of formation.

A two-dimensional photonic crystal with hexagonal lattice of air-holes is patterned into an active planar waveguide containing InAs/InGaAs QDs by Blokhin et al. [102]. Variable-angle reflectivity spectroscopy is used to map out the photonic band structure. Fano-type resonances observed in the measured reflectivity spectra in TE (TM) polarizations along the Γ–K (Γ–M) lattice direction are attributed to resonance coupling of the optically active photonic bands to external light. Angle-resolved PL measurements are shown to trace the band structure of the leaky mode. The revealed three-fold emission intensity enhancement of photonic crystals is ascribed to both Purcell and Bragg scattering effects.

Krestnikov et al. [103] have reported on resonant PL of InGaN inclusions in a GaN matrix. The structures were grown on sapphire substrates using metal-organic chemical vapour deposition. Non resonant pulsed excitation results in a broad PL peak, while resonant excitation into the non resonant PL intensity maximum results in an evolution of a sharp resonant PL peak, having a spectral shape defined by the excitation laser pulse and a radiative decay time close to that revealed for PL under non resonant excitation. Observation of a resonantly excited narrow PL line gives clear proof of the QD nature of luminescence in InGaN–GaN samples. PL decay demonstrates strongly non exponential behaviour evidencing coexistence of QDs having similar ground-state transition energy, but very different electron-hole wave-function overlap.

Borchert et al. [104] have colloidally prepared CdS/HgS/CdS QD, quantum well nano-crystals and CdS/HgS/CdS/HgS/CdS double quantum well nano-crystals stabilized with polyphosphates have been investigated by photoelectron spectroscopy with tuneable synchrotron radiation. High-resolution spectra reveal in addition to a bulk species also a surface environment for Cd, whereas different S species could not be resolved. Although not expected for the ideal structure model, Hg occurs in two distinct environments. The total amount of Hg giving evidence for sample in-homogeneities is of the order of 10 %. Furthermore, to some extent, the XPS experiments allowed characterization of the onion like structure of the QD quantum well nano-crystals. A method for layer thickness determination previously developed for core shell nano-crystals is extended to the more complex case of QD quantum well structures.

Benemanskaya et al. [105] have prepared an array of non-overgrown InAs/GaAs QDs which has been decorated with adsorbed metal atoms in situ in ultrahigh vacuum. Their electron and photoemission properties have been studied. The radical modification of the spectra of the threshold emission from the QDs with increasing cesium coating has been found. Two photoemission channels have been established; they are characterized by considerably different intensities, spectral locations, and widths of the selective bands. It has been shown that the decoration of the QDs makes it possible to control the electronic structure and quantum yield of photoemission, the nature of which is related to the excitation of the electronic states of the GaAs substrate and InAs/GaAs QDs.

Giorgio et al. [106] have discussed the application of photoemission-based microscopy techniques to the compositional characterization of semiconductor QDs and rings. The experimental technique is discussed in detail by them in the said paper. Using this technique, self-assembled III-V and Ge/Si QDs have been studied. These results have also been discussed, both for randomly nucleated and site-controlled QDs. Rowell et al. [107] have grown For Ge nano-dots approximately 20 nm in diameter by annealing a thin amorphous Ge layer deposited by MBE on a mesoporous TiO_2 layer on Si(001), PL was observed as a wide near-infrared band near 800 meV. Using a tight binding theoretical model, the energy-dependent PL spectrum was transformed into a dependence on dot size. The average dot size determined the peak energy of the PL band and its shape depended on the size distribution, including band gap enlargement due to quantum confinement. Combining the dot sample PL with an established dependence of emission efficiency on dot diameter, it was possible to derive a dot size distribution and compare it with results obtained independently from atomic force microscopy.

Heyderman et al. [108] have investigated the periodic square arrays of anti dots in 10 nm-thick cobalt films with anti dot periods, p, ranging from 2 μm down to 200 nm and various ratios of anti dot size to anti dot separation, w/d. For $p = 2$ μm, the extent of modification of the thin film magnetic domain structure increases with increasing w/d, forming domains pinned diagonally between anti dots for $w/d \geq 0.2$ and resulting in a two-dimensional periodic checked domain contrast commensurate with the anti dot lattice for $w/d \geq 0.9$. As p is decreased while maintaining $d \approx w$, they observed a dramatic change in the magnetic domain configuration at $p = 400$ nm resulting in chains of magnetic domains running parallel to the intrinsic hard axis and with lengths corresponding to a multiple of the anti dot period.

Lockwood et al. [109] have grown the super-lattices of Si/SiO_2 at room temperature with atomic layer precision using state of the art MBE and ultraviolet ozone treatment. PL was observed at wavelengths across the visible range for Si layer thicknesses $1 < d < 3$ nm. The fitted peak emission energy $E(eV) = 1.60 + 0.72\, d^{-2}$ is in accordance with effective mass theory for quantum confinement by the wide-gap SiO_2 barriers and also with the bulk amorphous Si band gap. Measurements of the conduction and valence band shifts by x-ray techniques correlate with E(d), confirming the role of quantum confinement and indicating a direct band-to-band recombination mechanism.

The many interesting and unique physical properties of nano-crystalline-Si/amorphous-SiO_2 superlattices (SLs) stem from their vertical periodicity and nearly defect-free, atomically flat, and chemically abrupt nano-crystalline-Si/SiO_2 interfaces. By combining a less than 5 % variation in the initial as-grown amorphous-Si layer thickness with control over the Si nano crystal shape and crystallographic orientation Lockwood and Tsybeskov [110] have produced via an appropriate annealing process, systems of nearly identical Si nano-crystals having remarkably different shapes (spheres, ovoids, bricks, etc.). Such details governing the fabrication of nano-crystalline-Si/amorphous-SiO_2 SLs have dramatic effects on their structural and optical-Raman scattering and PL-properties. The reliable fabrication of Si-based nanostructures with control over the nano-crystal size, shape, and crystallographic orientation is an important first step in their applications in Si photonics.

GaAs/AlAs SLs grown simultaneously on GaAs substrates with the (311)A and (311)B orientations have been studied by Lyubas et al. [111] by PL and high-resolution transmission electron microscopy with a Fourier analysis of images. A periodic interface corrugation is observed for (311)B SLs. A comparison of the structure of (311)A and (311)B SLs indicates that the corrugation occurs in both cases and its period along the $[1\bar{1}1]$ direction is equal to 3.2 nm. The corrugation is less pronounced in (311)B SLs, wherein it exhibits an additional modulation (long-wavelength disorder) with the characteristic lateral size exceeding 10 nm. The vertical correlation of regions rich in GaAs and AlAs, which is well observed in (311)A SLs, is weak in (311)B SLs due to the occurrence of long-wavelength disorder. The optical properties of (311)B SLs are similar to those of (100) ones and differ radically from those of (311)A SLs. As distinct from (311)B, strong PL polarization anisotropy is observed for (311)A SLs. It is shown that it is the interface corrugation rather than the crystallographic (311) surface orientation that determines the optical properties of (311)A corrugated SLs with thin GaAs and AlAs layers.

Lyubas et al. [112] have determined PL properties of type II GaAs/AlAs SLs grown on (311) surface by its polarity. The 3.2 nm lateral periodicity is revealed, using a high resolution transmission electron microscopy and Fourier transform images of SLs, however, it is rather illegible because of not clear corrugating and due to the presence of long-wavelength (>10 nm) disorder. PL spectra of GaAs/AlAs SLs grown on (311)A surface are strongly polarized in relation to the direction of interface corrugation unlike the SLs grown on (311)B surface, where corrugation was weak.

Optical phenomena in the mid-infrared range connected with inter level and inter sub-band charge-carrier transitions in QD and quantum well (QW) hetero-structures under optical and electrical pumping were investigated by Vorobjev et al. [113]. Spectra of inter band PL are also presented. The existence of a meta stable level in funnel-shaped QWs is experimentally confirmed. The inter sub-band transition dynamics in asymmetrical pairs of tunnel-coupled QWs was studied by means of pump-and-probe time-resolved spectroscopy. The measurement of photoemission spectra of $SrTiO_3/LaTiO_3$ super-lattices with a topmost $SrTiO_3$ layer of variable thickness has been reported in [114, 115]. Finite coherent spectral weight with a clear Fermi cut-off was observed at chemically abrupt $SrTiO_3/LaTiO_3$ interfaces, indicating that an "electronic reconstruction" occurs at the interface between the Mott insulator $LaTiO_3$ and the band insulator $SrTiO_3$. For $SrTiO_3/LaTiO_3$ interfaces annealed at high temperatures ($\sim$1,000 °C), which leads to Sr/La atomic inter-diffusion and hence to the formation of $La_{1-x}Sr_xTiO_3$ like material, the intensity of the incoherent part was found to be dramatically reduced whereas the coherent part with a sharp Fermi cut-off is enhanced due to the spread of charge. These important experimental features are well reproduced by layer dynamical-mean-field-theory calculation.

Photonic devices are becoming increasingly important in information and communication technologies. But attempts to integrate photonics with silicon-based microelectronics are hampered by the fact that silicon has an indirect band gap, which prevents efficient electron-photon energy conversion. Light-emitting silicon-based materials have been made using band-structure engineering of SiGe and SiC alloys and Si/Ge SLs, and by exploiting quantum-confinement effects in nanoscale particles and crystallites [116]. The discovery [117, 118] that silicon can be etched electrochemically into a highly porous form that emits light with a high quantum yield has opened up the latter approach to intensive study [119–125]. Lu et al. [126] have reported the fabrication, by MBE technique, of well-defined SLs of silicon and SiO_2, which emit visible light through PL. The said group have shown that this light emission can be explained in terms of quantum confinement of electrons in the two-dimensional silicon layers and these superlattice structures are robust and compatible with standard silicon technology.

Lao et al. [127] have used internal photoemission spectroscopy to determine the conduction band offset of a type-II InAs/GaSb super-lattice (T2SL) pBp photodetector to be eV at 78 K, confirming its unipolar operation. It is also found that phonon-assisted hole transport through the B-region disables its two-colour detection mode around 140 K. In addition, photoemission yield shows a reduction at about an energy of longitudinal-optical phonon above the threshold, confirming carrier-phonon scattering degradation on the photo-response. These results may indicate a pathway for optimizing T2SL detectors in addition to current efforts in material growth, processing, substrate preparation, and device passivation 0.004 (±0.004). The natural considerations that constrain silicon from emitting light efficiently are examined, as are several engineered solutions to this limitation by Iyer and Xie [128]. These include intrinsic and alloy-induced luminescence; radioactively active impurities; quantum-confined structures, including zone folding and the recent developments in porous silicon; and a hybrid approach, the integration of direct band-gap materials onto silicon.

Lin et al. [129] have proposed a new approach for the fabrication of n-type porous silicon layer. A hole-rich p-layer is arranged underneath the n-layer, and the np-junction is under forward biased condition in the etching process. Therefore sufficient holes can drift straight-upward and pass across the np-junction from p-region to n-region to participate in electrochemical reaction during the etching process with an unfailing supply. Illumination is an optional hole-supplier in this approach, so the problem of illumination-depth limitation can be overcome. Strong visible PL emissions are demonstrated on the hole-poor n-type porous layer at about 650 nm.

Lockwood et al. [130] have reported the optical properties of porous GaAs formed electrochemically on n- and p-type GaAs in HCl electrolyte. The porous structure comprises GaAs crystallites ranging in size from micrometers to nanometers and under certain chemical conditions other transparent crystallites of As_2O_3 and Ga_2O_3 form. PL measurements at 295 K reveal an "infrared" PL at $\sim$840 nm and a "green" PL at $\sim$540 nm, which could easily be seen by the naked eye in some samples. The infrared and green PL peak wavelength and intensity varied from sample to sample consistent with an assignment to quantum confinement effects in GaAs micro- and nano-crystallites, respectively. The many and diverse approaches to materials science problems have greatly enhanced the ability in recent times to engineer the physical properties of semiconductors. Silicon, of all semiconductors, underpins nearly all microelectronics today and will continue to do so for some time to come. However, in optoelectronics and, more recently, in photonics, the severe disadvantage of an indirect band gap has limited the application of elemental silicon.

Electroluminescent devices have been developed recently that are based on new materials such as porous silicon and semiconducting polymers. By taking advantage of developments in the preparation and characterization of direct-gap semiconductor nano-crystals, and of electroluminescent polymers, Colvin et al. [131] have constructed a hybrid organic/inorganic electroluminescent device. Light emission arises from the recombination of holes injected into a layer of semiconducting p-paraphenylene vinylene (PPV) with electrons injected into a multilayer film of cadmium selenide nano-crystals. Close matching of the emitting layer of nano-crystals with the work function of the metal contact leads to an operating voltage of only 4 V. At low voltages emission from the CdSe layer occurs. Because of the quantum size effect the colour of this emission can be varied from red to yellow by changing the nano-crystal size. At higher voltages green emission from the polymer layer predominates. Thus this device has a degree of voltage tunability of colour.

Rogach et al. [132] have reported that the colloidal semiconductor nano crystals are promising luminophores for creating a new generation of electroluminescence devices. Research on semiconductor nano crystal based light-emitting diodes (LEDs) has made remarkable advances in just one decade: the external quantum efficiency has improved by over two orders of magnitude and highly saturated colour emission is now the norm. Although the device efficiencies are still more than an order of magnitude lower than those of the purely organic LEDs there are potential advantages associated with nano crystal-based devices, such as a spectrally pure emission colour, which will certainly merit future research. Further developments of nano crystal-based LEDs will be improving material stability, understanding and controlling chemical and physical phenomena at the interfaces, and optimizing charge injection and charge transport.

Highly luminescent (CdSe)ZnS nano crystals, with band edge emission in the red region of the visible spectrum, were successfully synthesized by Reboud et al. [133] and incorporated in a resist, namely mr-NIL 6000. The nano composite material was imprinted by using conventional nano imprint lithography (NIL) process. We report on the fabrication and characterization of nano imprinted photonic crystals in this new functional material. Experiments showed good imprint properties of the NC/polymer based material and that the surface nano structuration improves the light extraction efficiency by over 2 compared to a nano imprinted unpatterned surface.

Bertsch et al. [134] have calculated the two-photon ionization of clusters for photon energies near the surface plasmon resonance. The plasmon was described in a schematic jellium-RPA model assuming a separable residual interaction between electrons determined so that the plasmon energy is reproduced. The ionization rate of a double plasmon excitation was calculated perturbatively. In Na + 93 clusters they found an ionization rate of the order of at most 0.05–0.10 fs^{-1}. This rate was used to determine the ionization probability in an external field in terms of the number of absorbed photon pairs and the duration of the field. They discussed the dependence of the results on the choice of their empirical separable force. The number of emitted electrons per pair of absorbed photons was found to be small, in the range 10^{-5}–10^{-3} **H**.

Fiorini et al. [135] have presented a brief survey of quantum effects in amorphous superstructures, with a particular attention to optical properties. The determination of interface properties by superstructure investigation is discussed, with a special emphasis on the amorphous silicon/amorphous silicon-carbon system. In the review Lockwood [136] has considered a number of diverse approaches to engineering efficient light emission in silicon nanostructures. These different approaches are placed in context and their prospects are assessed for applications in silicon-based photonics.

Yoshimatsu et al. [137] have investigated changes in the electronic structures of digitally controlled $La_{0.6}Sr_{0.4}MnO_3$ (LSMO) layers sandwiched between $SrTiO_3$ as a function of LSMO layer thickness in terms of hard x-ray photo-emission-spectroscopy (HX-PES). The HX-PES spectra show the evolution of Mn 3d derived states near the Fermi level and the occurrence of metal-insulator transition at 8ML. The detailed analysis for the thickness dependent HX-PES spectra reveals the existence of the less conducting and nonmagnetic transition layer with a film thickness of about 4ML in the interface region owing to significant interaction through the interface.

The PL spectroscopy has been used by Rowell et al. [138] to study the incorporation of C in several samples consisting of strained $Si_{1-x-y}Ge_xC_y$ epi-layers lattice matched to Si(001). To obtain the total C concentration, these samples were characterized by both SIMS and Auger emission spectroscopy, and X-ray diffraction data was analyzed to obtain the substitutional C concentration. The difference of the total and substitutional C concentrations, i.e., the non-substitutional carbon fraction, was found to be directly correlated with specific spectral lines in both the room-temperature Raman and low-temperature PL spectra. Shimoda et al. [139]

studied the effects of structure and morphology on lithium storage in single-wall carbon nano tube (SWNT) bundles by electrochemistry and nuclear magnetic resonance techniques. SWNTs were chemically etched to variable lengths and were intercalated with Li. The reversible Li storage capacity increased from LiC(6) in close-end SWNTs to LiC(3) after etching, which was twice the value observed in intercalated graphite. All the nano tubes became metallic upon intercalation of Li, with the density of states at the Fermi level increasing with increasing Li concentration. The enhanced capacity is attributed to Li diffusion into the interior of the SWNTs through the opened ends and sidewall defects.

In this monograph, we have studied the EP from quantum confined HD nonlinear optical, III-V, II-VI, GaP, Ge, $PtSb_2$, stressed materials, GaSb, IV-VI, Tellurium, II-V, Bi_2Te_3, III-V, II-VI, IV-VI and HgTe/CdTe quantum wire superlattices HD with graded interfaces, III-V, II-VI, IV-VI and HgTe/CdTe HD effective mass superlattices under magnetic quantization, quantum confined effective mass HD superlattices and HD superlattices of optoelectronic materials under intense electric field and light waves with graded interfaces on the basis of appropriate carrier energy spectra. Finally it may be noted that although we have considered the EP from a plethora of quantized materials having different band structures theoretically, the detailed experimental works are still needed for an in-depth study of the EP from such low-dimensional HD systems as functions of externally controllable quantities which, in turn, will add new physical phenomenon in the regime of the electron motion in HD nano structured materials and related topics.

9.4 Open Research Problem

(R.9.1) Investigate experimentally the EP for all the HD systems as discussed in this monograph in the presence of arbitrarily oriented strain.

References

1. A.K. Sreedhar, S.C. Gupta, Phys. Rev. B **5**, 3160 (1972)
2. R.W. Keyes, IBM. J. Res. Dev. **5**, 266 (1961)
3. R.W. Keyes, Solid State Phys. **20**, 37 (1967)
4. S. Bhattacharya, S. Chowdhury, K.P. Ghatak, J. Comput. Theor. Nanosci. **3**, 423 (2006)
5. S. Choudhury, L.J. Singh, K.P. Ghatak, Phys. B **365**, 5 (2005)
6. L.J. Singh, S. Choudhary, A. Mallik, K.P. Ghatak, J. Compu. Theor. Nanosci. **2**, 287 (2005)
7. K.P. Ghatak, J.Y. Siddiqui, B. Nag, Phys. Lett. A **282**, 428 (2001)
8. K.P. Ghatak, J.P. Banerjee, B. Nag, J. Appl. Phys. **83**, 1420 (1998)
9. B. Nag, K.P. Ghatak, Nonlinear Opt. **19**, 1 (1998)
10. K.P. Ghatak, B. Nag, Nanostruc. Mater. **10**, 923 (1998)
11. B. Nag, K.P. Ghatak, J. Phys. Chem. Solids **58**, 427 (1997)
12. K.P. Ghatak, D.K. Basu, B. Nag, J. Phys. Chem. Solids **58**, 133 (1997)

13. K.P. Ghatak, J.P. Banerjee, B. Goswami, B. Nag, Nonlin. Opt. Quant. Opt. **16**, 241 (1996)
14. K.P. Ghatak, J.P. Banerjee, D. Bhattacharyya, B. Nag, Nanotechnology **7**, 110 (1996)
15. K.P. Ghatak, J.P. Banerjee, M. Mitra, B. Nag, Nonlinear Opt. **17**, 193 (1996)
16. B. Nag, K.P. Ghatak, Phys. Scr. **54**, 657 (1996)
17. K.P. Ghatak, B. Mitra, Phys. Scr. **46**, 182 (1992)
18. K.P. Ghatak, Int. J. Electron. **71**, 239 (1991)
19. K.P. Ghatak, B.De, S.N. Biswas, M. Mondal, Mechanical behavior of materials and structures in microelectronics, in *MRS Symposium Proceedings, Spring Meeting*, vol. 2216 (1991), p. 191
20. K.P. Ghatak, B. De, in *MRS Symposium Proceedings*, vol. 226 (1991), p. 191
21. K.P. Ghatak, B. Nag, G. Majumdar, in *Proceedings of MRS*, vol. 379 (1995), p. 109
22. K.P. Ghatak, B. Nag, G. Majumdar, in *Proceedings of MRS*, vol. 379 (1995), p. 109
23. D. Baruah, S. Choudhury, K.M. Singh, K.P. Ghatak, J. Phys. Conf. Ser. **61**, 80 (2007)
24. E.A. Arushanov, A.F. Knyazev, A.N. Natepov, S.I. Radautsan, Sov. Phys. Semi. **15**, 828 (1981)
25. S.P. Zelenim, A.S. Kondratev, A.E. Kuchma, Sov. Phys. Semicond. **16**, 355 (1982)
26. F.M. Peeters, P. Vasilopoulos, Phys. Rev. B **46**, 4667 (1992)
27. W. Zawadzki, in *Two-Dimensional Systems, Heterostructures and Superlattices*, ed. by G. Bauer, F. Kuchar, H. Heinrich. Springer Series in Solid State Sciences, vol. 53 (Springer, Heidelberg, 1984)
28. B.M. Askerov, N.F. Gashimzcede, M.M. Panakhov, Sov. Phys. Solid State **29**, 465 (1987)
29. G.P. Chuiko, Sov. Phys. Semicond. **19**, 1279 (1985)
30. K.P. Ghatak, S. Bhattacharya, S. Bhowmik, R. Benedictus, S. Choudhury, J. Appl. Phys. **103**, 034303 (2008)
31. K.P. Ghatak, S.N. Biswas, J. Appl. Phys. **70**, 299 (1991)
32. K.P. Ghatak, S.N. Biswas, J. Low Temp. Phys. **78**, 219 (1990)
33. K.P. Ghatak, M. Mondal, J. Appl. Phys. **65**, 3480 (1989)
34. K.P. Ghatak, B. Nag, Nanostruct. Mater. **5**, 769 (1995)
35. K.P. Ghatak, M. Mondal, Phys. Status Solidi B **185**, K5 (1994)
36. K.P. Ghatak, B. Mitra, Il Nuovo Cimento D **15**, 97 (1993)
37. K.P. Ghatak, S.N. Biswas, Phys. Status Solidi B **140**, K107 (1987)
38. K.P. Ghatak, Il Nuovo Cimento D **13**, 1321 (1991)
39. K.P. Ghatak, A. Ghoshal, Phys. Status Solidi B **170**, K27 (1992)
40. K.P. Ghatak, B. De, B. Nag, P.K. Chakraborty, Mol. Cryst. Liq. Cryst. Sci. Technol. Sect. B Nonlinear Opt. **16**, 221 (1996)
41. K.P. Ghatak, M. Mitra, B. Goswami, B. Nag, Mol. Cryst. Liq. Cryst. Sci. Technol. Sect. B Nonlinear Opt. **16**, 167 (1996)
42. K.P. Ghatak, D. K. Basu, D. Basu, B. Nag, Nuovo Cimento della Societa Italiana di Fisica D-Condensed Matter. Atomic Mol. Chem. Phys. Biophys. **18**, 947 (1996)
43. B. Mitra, K.P. Ghatak, Phys. Lett. A **141**, 81 (1989)
44. S.K. Biswas, A.R. Ghatak, A. Neogi, A. Sharma, S. Bhattacharya, K.P. Ghatak, Physica E **36**, 163 (2007)
45. M. Mondal, A. Ghoshal, K.P. Ghatak, Il Nuovo Cimento D **14**, 63 (1992)
46. K.P. Ghatak, M. Mondal Phys. Status Solidi B **135**, 819 (1986)
47. L.J. Singh, S. Choudhury, D. Baruah, S.K. Biswas, S. Pahari, K.P. Ghatak, Phys. B **368**, 188 (2005)
48. K.P. Ghatak, B. De, M. Mondal, S. N. Biswas, in *Proceedings of Materials Research Society Symposium*, vol. 198 (1990), p. 327
49. K.P. Ghatak, in *Proceedings of SPIE—The International Society for Optical Engineering*, vol. 1584 (1992), p. 435
50. K.P. Ghatak, B. De, in *Proceedings of Materials Research Society Symposium*, vol. 234 (1991), p. 55, 59
51. K.P. Ghatak, B. De, M. Mondal, S.N. Biswas, in *Proceedings of Materials Research Society Symposium*, vol. 184 (1990), p. 261

52. K.P. Ghatak, Influence of band structure on some quantum processes in tetragonal semiconductors, D. Eng. Thesis, Jadavpur University, Kolkata, India, 1991
53. K.P. Ghatak, S.N. Biswas, in *Proceedings of Materials Research Society Symposium*, vol. 216 (1990), p. 465
54. T.M. Tritt, M. Kanatzidis, G. Mahan, H.B. Lyon Jr. (eds.), New Materials for Small Scale Thermoelectric Refrigeration and Power Generation Applications, in *Proceedings of the 1998 Materials Research Society*, vol. 545 (1998)
55. K.P. Ghatak, S. Bhattacharya, S. Pahari, D. De, S. Ghosh, M. Mitra, Ann. Phys. **17**, 5 (2008)
56. P.K. Bose, N. Paitya, S. Bhattacharya, D. De, S. Saha, K.M. Chatterjee, S. Pahari, K.P. Ghatak, Quantum Matter. **1**, 89 (2012)
57. S. Bhattarchya, D. De, S. Ghosh, P. Banerjee, S. Saha, M. Mitra, B. Nag, M. Pal, S.K. Biswas, K.P. Ghatak, J. Comput. Theor. Nanosci. **7**, 1066 (2010)
58. K.P. Ghatak, S. Bhattacharya, S. Pahari, S. N. Mitra, P.K. Bose, D. De, J. Phys. Chem. Solids. **70**, 122 (2009)
59. S. Bhattacharya, N.C. Paul, D. De, K.P. Ghatak, Phys. B Condens. Matter **403**, 4139 (2008)
60. K.P. Ghatak, S. Bhattacharya, D. De, R. Sarkar, S. Pahari, A. Dey, A.K. Dasgupta, J. Comput. Theor. Nanosci. **5**, 1345 (2008)
61. K.P. Ghatak, S. Bhattacharya, S. Pahari, D. De, S. Ghosh, M. Mitra, Ann. Phys. **17**, 195 (2008)
62. K.P. Ghatak, S. Bhattacharya, J. Appl. Phys. **102**, 073704 (2007)
63. S. Mukherjee, D. De, D.J. Mukherjee, S. Bhattacharya, A. Sinha, K.P. Ghatak, Phys. B Condens. Matter **393**, 347 (2007)
64. P.K. Chakraborty, S. Bhattacharya, K.P. Ghatak, J. Appl. Phys. **98**, 053517 (2005)
65. A.S. Filipchenko, I.G. Lang, D.N. Nasledov, S.T. Pavlov, L.N. Radaikine, Phys. Status Solidi B **66**, 417 (1974)
66. M. Wegener, *Extreme Nonlinear Optics* (Springer, Germany, 2005)
67. B.S. Wherreff, W. Wolland, C.R. Pidgeon, R.B. Dennis, S.D. Smith, in *Proceedings of the 12th International Conference of the Physics of the Semiconductors*, ed. by M.H. Pilkahn, R. G. Tenbner (Staffgard, 1978), p.793
68. R. Houdré, C. Hermann, G. Lampel, P.M. Frijlink, A.C. Gossard, Phys. Rev. Lett. **55**, 734 (1985)
69. D.J. Lockwood, ECS Trans. **50**, 39 (2013)
70. T.C. Chiang, T.C. Chiang, Surf. Sci. Rep. **39**, 181 (2000)
71. G. Ji, D. Huang, U.K. Reddy, T.S. Henderson, R. Houdré, H. Morkoç, J. Appl. Phys. **62**, 3366 (1987)
72. D.C. Reynolds, K.K. Bajaj, C.W. Litton, P.W. Yu, J. Klem, C.K. Peng, H. Morkoç, J. Singh, Appl. Phys. Lett. **48**, 727 (1986)
73. G. Martin, S. Strite, A. Botchkarev, A. Agarwal, A. Rockett, W.R.L. Lambrecht, B. Segall, H. Morkoç, J. Elec. Mat. **24**, 225 (1995)
74. W. Götz, N.M. Johnson, R.A. Street, H. Amano, I. Akasaki, Appl. Phys. Lett. **66**, 1340 (1995)
75. M.C. Benjamin, M.D. Bremser, T.W. Weeks Jr, S.W. King, R.F. Davis, R.J. Nemanich, Appl. Surf. Sci. **104–105**, 455–460 (1996)
76. T. Maruyama, R. Prepost, E.L. Garwin, C.K. Sinclair, B. Dunham, S. Kalem, Appl. Phys. Lett. **55**, 1686 (1989)
77. J. Chen, G.B. Gao, D. Huang, J.I. Chyi, M.S. Ünlü, H. Morkoç, Appl. Phys. Lett. **55**, 374 (1989)
78. W. Khunsin, S.G. Romanov, C.M. Sotomayor Torres, J. Ye, R. Zentel, J. Appl. Phys. **104**, 013527 (2008)
79. V. Reboud, N. Kehagias, M. Striccoli, T. Placido, A. Panniello, M.L. Curri, M. Zelsmann, F. Reuther, G. Gruetzner, C.M.S. Torres, J. Vac. Sci. Technol. B **25**, 2642 (2007)
80. S.G. Romanov, D.N. Chigrin, C.M.S. Torres, N. Gaponik, A. Eychmueller, A.L. Rogach, M. Egen, R. Zentel, in *Proceedings of SPIE, Photonic Crystal Materials and Nanostructures*, vol. 5450 (2004), p. 44

81. V.Y. Aleshkin, D.M. Gaponova, D.G. Revin, L.E.Vorobjev, S.N. Danilov, V.Y. Panevin, N. K. Fedosov, D.A. Firsov, V.A. Shalygin, A. D. Andreev, A.E. Zhukov, N.N. Ledentsov, V. M. Ustinov, G.E. Cirlin, V.A. Egorov, F. Fossard, F.H. Julien, E. Towe, D. Pal, S.R. Schmidt, A. Seilmeier, in *Proceedings of SPIE 10th International Symposium on Nanostructures: Physics and Technology*, vol. 5023 (2002), p. 209
82. S.O. Usov, A.F. Tsatsul'nikov, V.V. Lundin, A.V. Sakharov, E.E. Zavarin, N.N. Ledentsov, Semiconductors **42**, 188 (2008)
83. M. Bardosova, S.G. Romanova, C.M.S. Torres, N. Gaponik, A. Eychmueller, Y.A. Kumzerov, Phys. E **37**, 218 (2007)
84. C. Leandri, G.L. Lay, B. Aufray, C. Girardeaux, J.Avila, M.E. Dávila, M.C. Asensio, D.C. Ottaviani, A. Cricenti, Surf. Sci. **574**, L9 (2005)
85. C. Zilkens, The International Nuclear Information System **33**(41), (2002). Ref. No. 33048164
86. Y. Darma, ITB, J. Sci. **40 A**, 88 (2008)
87. N.L. Rowella, D.J. Lockwooda, A. Karmousb, P.D. Szkutnikb, I. Berbezierb, A. Rondab, Superlat. Microstruc. **44**, 305 (2008)
88. S.A. Mala, L. Tsybeskov, D.J. Lockwood, X. Wu, J.-M. Baribeau, Appl. Phys. Lett. **103**, 033103 (2013)
89. A. Carlsson, D. Claesson, S.A. Lindgren, L. Wallden, Phys. Rev. B **52**, 11144 (1995)
90. D.P. Woodruff, M. Milun, P. Pervan, J. Phys. Condens. Matter **11**, L105 (1999)
91. S. Tiwari, S. Tiwari, Crys. Res. Tech. **41**, 78 (2006)
92. W.L. Wilson, P.F. Szajowski, L.E. Brus, Science **262**, 1242 (1993)
93. M. Maillard, P. Monchicourt, M.P. Pileni, Chem. Phys. Lett. **380**, 704 (2003)
94. W. Xiong, D.D. Hickstein, K.J. Schnitzenbaumer, J.L. Ellis, B.B. Palm, K.E. Keister, C. Ding, L.M. Avila, G. Dukovic, J.L. Jimenez, M.M. Murnane, H.C. Kapteyn, Nano Lett. **13**, 2924 (2013)
95. A. Konchenko, Y. Nakayama, I. Matsuda, S. Hasegawa, Y. Nakamura, M. Ichikawa, Phys. Rev. B **73**, 113311 (2006)
96. V.L. Colvin, A.P. Alivisatos, J.G. Tobin, Phys. Rev. Lett. **66**, 2786 (1991)
97. D.J. Lockwood, A. Wang, B. Bryskiewicz, Solid State Commun. **89**, 587 (1994)
98. A.E. Zhukov, B.V. Volovik, S.S. Mikhrin, N.A. Maleev, A.F. Tsatsul'nikov, E.V. Nikitina, I.N. Kayander, V.M. Ustinov, N.N. Ledentsov, Tech. Phys. Lett. **27**, 734 (2001)
99. B.V. Volovik, A.R. Kovsh, W. Passenberg, H. Kuenzel, N. Grote, N.A. Cherkashin, Y.G. Musikhin, N.N. Ledentsov, D. Bimberg, V.M. Ustinov, Semicond. Sci. Technol. **16**, 186 (2001)
100. A.F. Tsatsul'nikov, A.R. Kovsh, A.E. Zhukov, Y.M. Shernyakov, Y.G. Musikhin, V.M. Ustinov, N.A. Bert, P.S. Kop'ev, Zh.I. Alferov, A.M. Mintairov, J.L. Merz, N.N. Ledentsov, D.Bimberg, J. Appl. Phys. **88**, 6272 (2000)
101. A.F. Tsatsul'nikov, D.A. Bedarev, N.A. Bert, A.R. Kovsh, P.S. Kop'ev, N.A. Maleev, Y.G. Musikhin, M.V. Maximov, V.M. Ustinov, B.V. Volovik, A.E. Zhukov, Z.I. Alferov, A.A. Suvorova, P. Werner, in *Proceedings of the 26th International Symposium on Compound Semiconductors*, Berlin, Germany, vol. 22 (1999), p. 261
102. S.A. Blokhin, M.V. Maximov, O.A. Usov, A.V. Nashchekin, E.M. Arakcheeva, E.M. Tanklevskaya, S.A. Gurevich, S.G. Konnikov, A.E. Zhukov, N.N. Ledentsov, V.M. Ustinov, Int. J. Nanosci. **06**, 197 (2007)
103. I.L. Krestnikov, N.N. Ledentsov, A. Hoffmann, D. Bimberg, A.V. Sakharov, W.V. Lundin, A.F. Tsatsul'nikov, A.S. Usikov, Z.I. Alferov, Y.G. Musikhin, D. Gerthsen, Phys. Rev. B **66**, 155310 (2002)
104. H. Borchert, D. Dorfs, C. McGinley, S. Adam, T. Moller, H. Weller, A. Eychmuller, J. Phys. Chem. B **107**, 7486 (2003)
105. G.V. Benemanskaya, M.N. Lapushkin, V.P. Evtikhiev, A.S. Shkol'nik, Pis'ma v Zhurnal Eksperimental'noi i Teoreticheskoi Fiziki **96**, 363 (2012)
106. B. Giorgio, H. Stefan, S. Lucia, J. Nanoelectron. Optoelectron. **6**, 20 (2011)

107. N.L Rowell, D.J Lockwood, G Amiard, L. Favre, A. Ronda, I. Berbezier, M. Faustini, D. Grosso, J. Nanosci. Nanotechnol. **11**, 9190 (2011)
108. L.J. Heyderman, F. Nolting, C. Quitmann, Appl. Phys. Lett. **83**, 1797 (2003)
109. D.J. Lockwood, Z.H. Lu, J.-M. Baribeau, Phys. Rev. Lett. **76**, 539 (1996)
110. D.J. Lockwood, L. Tsybeskov, J. Nanophoton. **2**, 022501 (2008)
111. G.A. Lyubas, N.N. Ledentsov, D. Litvinov, D. Gerthsen, I.P. Soshnikov, V.M. Ustinov, J. Exp. Theo. Phys. Lett. **75**, 179 (2002)
112. G.A. Lyubas, B.R. Semyagin, V.V. Bolotov, N.N. Ledentsov, I.P. Soshnikov, V.M. Ustinov, D. Litvinov, D. Gerthsen, Fizika i Tekhnika Poluprovodnikov **36**, 959 (2002)
113. L.E. Vorobjev, S.N. Danilov, A.V. Gluhovskoy, V.L. Zerova, E.A. Zibik, V.Y. Panevin, D. A. Firsov, V.A. Shalygin, A.D. Andreev, B.V. Volovik, A.E. Zhukov, N.N. Ledentsov, D.A. Livshits, V.M. Ustinov, Y.M. Shernyakov, A.F. Tsatsulnikov, A. Weber, M. Grundmann, S. R. Schmidt, A. Seilmeier, E. Towe, D. Pal, Nanotechnology **12**, 462 (2001)
114. B. Abeles, T. Tiedje, Phys. Rev. Lett. **51**, 2003 (1983)
115. W.L. Wilson, P.F. Szajowski, L.E. Brus, Science **262**, 1242 (1993)
116. M. Takizawa, H. Wadati, K. Tanaka, M. Hashimoto, T. Yoshida, A. Fujimori, A. Chikamtsu, H. Kumigashira, M. Oshima, K. Shibuya, T. Mihara, T. Ohnishi, M. Lippmaa, M. Kawasaki, H. Koinuma, S. Okamoto, A.J. Millis, Phys. Rev. Lett. **97**, 057601 (2006)
117. L.T. Canham, Appl. Phys. Lett. **57**, 1046 (1990)
118. A.G. Cullis, L.T. Canham, Nature **353**, 335 (1991)
119. V. Lehmann, U. Gösele, Appl. Phys. Lett. **58**, 856 (1991)
120. A. Halimaoui, C. Oules, G. Bomchil, A. Bsiesy, F. Gaspard, R. Herino, M. Ligeon, F. Muller, Appl. Phys. Lett. **59**, 304 (1991)
121. V.P. Koch, T. Muschik, A. Kux, B.K. Meyer, F. Koch, V. Lehmann, Appl. Phys. Lett. **61**, 943 (1992)
122. J.P. Proot, C. Delerue, G. Allan, Appl. Phys. Lett. **61**, 1948 (1992)
123. C.G. Van de Walle, J.E. Northrup, Phys. Rev. Lett. **70**, 1116 (1993)
124. T. van Buuren, T. Tiedje, J.R. Dahn, B.M. Way, Appl. Phys. Lett. **63**, 2911 (1993)
125. M.S. Hybertsen, Phys. Rev. Lett. **72**, 1514 (1994)
126. Z.H. Lu, D.J. Lockwood, J.M. Baribeau, Nature **378**, 258 (1995)
127. Y.F. Lao, P.K.D.D.P. Pitigala, A.G. Unil Perera, E. Plis, S.S. Krishna, S.P. Wijewarnasuriya, Appl. Phys. Lett. **103**, 181110 (2013)
128. S.S. Iyer, Y.-H. Xie, Science **260**, 40 (1993)
129. J.C. Lin, W.L Chen, W.C Tsai, Opt. Exp. **14**, 9764 (2006)
130. D.J. Lockwood, P. Schmuki, H.J. Labbé, J.W. Fraser, Phys. E **4**, 102 (1999)
131. V.L. Colvin, M.C. Schlamp, A.P. Alivisatos, Nature **370**, 354 (1994)
132. A.L. Rogach, N. Gaponik, J.M. Lupton, C. Bertoni, D.E. Gallardo, S. Dunn, N.L. Pira, M. Paderi, P. Repetto, S.G. Romanov, C.O. Dwyer, C.M.S. Torres, A. Eychmüller, Angew. Chem. **47**, 6538 (2008)
133. V. Reboud, N. Kehagias, M. Zelsmann, M. Striccoli, M. Tamborra, M.L. Curri, A. Agostiano, M. Fink, F. Reuther, G. Gruetzner, C.M.S. Torres, J. Nanosci. Nanotech. **8**, 535 (2008)
134. G.F. Bertsch, N. Van Giai, N. Vinh, Mau Phys. Rev. A **61**, 033202 (2000)
135. P. Fiorini, M. de Seta, F. Evangelisti, Hel. Phys. Acta **62**, 070645 (1989)
136. D.J. Lockwood, J. Mat. Sci. Mat. Elec. **20**, 235 (2009)
137. K. Yoshimatsu, K. Horiba, H. Kumigashira, E. Ikenaga, M. Oshima, Appl. Phys. Lett. **94**, 071901 (2009)
138. N.L. Rowell, D.J. Lockwood, J.-M. Baribeau, Thin Solid Films **517**, 128 (2008)
139. H. Shimoda, B. Gao, X.P. Tang, A. Kleinhammes, L. Fleming, Y. Wu, O. Zhou, Phys. Rev. Lett. **88**, 015502 (2002)

Chapter 10
Conclusion and Future Research

This monograph deals with the EP in various types of low dimensional HD materials. The intense photo excitation, quantization and strong electric field alter profoundly the basic band structures, which, in turn, generate pinpointed knowledge regarding EP in various HDS and their nanostructures having different carrier energy spectra. The in-depth experimental investigations covering the whole spectrum of solid state and allied science in general, are extremely important to uncover the underlying physics and the related mathematics. We have presented the simplified expressions of EP for few HD quantized structures together with the fact that our investigations are based on the simplified *k.p* formalism of solid-state science without incorporating the advanced field theoretic techniques. In spite of such constraints, the role of band structure behind the curtain, which generates, in turn, new concepts are truly amazing and discussed throughout the text.

We present the last set of few open research problems in this pin pointed topic of research of modern physics.

(R.10.1) Investigate the multi-photon EP from multiple HD QWs, NWs and QBs of negative refractive index, organic, magnetic, heavily doped, disordered and other advanced optical materials in the presence of arbitrarily oriented photo-excitation.

(R.10.2) Investigate multi-photon EP from cylindrical and wedge shaped HD QBs of all the appropriate problems of R.10.1 in the presence of arbitrarily photo-excitation.

(R.10.3) Investigate the multi-photon EP in the presence of an arbitrarily oriented photo-excitation and quantizing magnetic field for all the appropriate cases of R.10.1.

(R.10.4) Investigate the multi-photon EP in the presence of arbitrarily oriented photo-excitation and alternating non-quantizing electric field for all the appropriate cases of R.10.2.

(R.10.5) Investigate the multi-photon EP in the presence of an arbitrarily oriented photo-excitation and non-uniform non-quantizing electric field for all the appropriate cases of R.10.2.

K.P. Ghatak, *Einstein's Photoemission*, Springer Tracts in Modern Physics 262,
DOI 10.1007/978-3-319-11188-9_10

(R.10.6) Investigate the multi-photon EP in the presence of an arbitrarily oriented photo-excitation and crossed electric and quantizing magnetic fields for all the appropriate cases of R.10.1.

(R.10.7) Investigate the multi-photon EP in the presence of an arbitrarily oriented photo-excitation and quantizing magnetic field for all the appropriate cases of R.10.1.

(R.10.8) Investigate the multi-photon EP in the presence of arbitrarily oriented photo-excitation and alternating non-quantizing electric field for all the appropriate cases of R.10.1.

(R.10.9) Investigate the multi-photon EP in the presence of an arbitrarily oriented photo-excitation and non-uniform non-quantizing electric field for all the cases of R.10.1.

(R.10.10) Investigate the multi-photon EP in the presence of an arbitrarily oriented photo-excitation and crossed electric and quantizing magnetic fields for all the appropriate cases of R.10.1.

(R.10.11) Investigate the multi-photon EP from all the systems and the open research problems of Chaps. 1–5 in the presence of many body effects.

(R.10.12) Investigate the multi-photon EP from quantum confined HD III-V, II-VI, IV-VI, HgTe/CdTe superlattices with graded interfaces together with short period, strained layer, random, Fibonacci, polytype and sawtooth superlattices in this context in the presence of arbitrarily oriented photo-excitation.

(R.10.13)
(a) Investigate the static photoelectric effect for all the appropriate problems of this monograph.
(b) Investigate all the appropriate problems for open HDQB of different materials as discussed in this monograph.

(R.10.14) Investigate the EP in the presence of a quantizing magnetic field under exponential, Kane, Halperin, Lax and Bonch-Bruevich band tails [1] for all the problems of this monograph of all the HD materials whose unperturbed carrier energy spectra are defined in Chap. 1 by including spin and broadening effects.

(R.10.15) Investigate all the appropriate problems after proper modifications introducing new theoretical formalisms for the problems as defined in (R.10.14) for HD negative refractive index, macro molecular, nitride and organic materials.

(R.10.16) Investigate all the appropriate problems of this monograph for all types of HD quantum confined p-InSb, p-CuCl and semiconductors having diamond structure valence bands whose dispersion relations of the carriers in bulk materials are given by Cunningham [2], Yekimov et al. [3] and Roman and Ewald [4] respectively.

(R.10.17) Investigate the influence of defect traps and surface states separately on the EP of the HD materials for all the appropriate problems of all the chapters after proper modifications.

(R.10.18) Investigate the EP of the HD materials under the condition of non-equilibrium of the carrier states for all the appropriate problems of this monograph.

(R.10.19) Investigate the EP for all the appropriate problems of this monograph for the corresponding HD p-type semiconductors and their nanostructures.

(R.10.20) Investigate the EP for all the appropriate problems of this monograph for all types of HD semiconductors and their nanostructures under mixed conduction in the presence of strain.

(R.10.21) Investigate the EP for all the appropriate problems of this monograph for all types of HD semiconductors and their nanostructures in the presence of hot electron effects.

(R.10.22) Investigate the EP for all the appropriate problems of this monograph for all types of HD semiconductors and their nanostructures for nonlinear charge transport.

(R.10.23) Investigate the EP for all the appropriate problems of this monograph for all types of HD semiconductors and their nanostructures in the presence of strain in an arbitrary direction.

(R.10.24) Investigate all the appropriate problems of this monograph for strongly correlated electronic HD systems in the presence of strain.

(R.10.25) Investigate all the appropriate problems of this chapter in the presence of arbitrarily oriented photon field and strain.

(R.10.26) Investigate all the appropriate problems of this monograph for all types of HD nanotubes in the presence of strain.

(R.10.27) Investigate all the appropriate problems of this monograph for various types of pentatellurides in the presence of strain.

(R.10.28) Investigate all the appropriate problems of this monograph for HD Bi_2Te_3–Sb_2Te_3 super-lattices in the presence of strain.

(R.10.29) Investigate the influence of temperature-dependent energy band constants for all the appropriate problems of this monograph.

(R.10.30) Investigate the influence of the localization of carriers on the EP in HDS for all the appropriate problems of this monograph.

(R.10.31) Investigate EP for HD p-type SiGe under different appropriate physical conditions as discussed in this monograph in the presence of strain.

(R.10.32) Investigate EP for different metallic alloys under different appropriate physical conditions as discussed in this monograph in the presence of strain.

(R.10.33) Investigate EP for different intermetallic compounds under different appropriate physical conditions as discussed in this monograph in the presence of strain.

(R.10.34) Investigate EP for HD GaN under different appropriate physical conditions as discussed in this monograph in the presence of strain.

(R.10.35) Investigate EP for different disordered HD conductors under different appropriate physical conditions as discussed in this monograph in the presence of strain.

(R.10.36) Investigate EP for various semimetals under different appropriate physical conditions as discussed in this monograph in the presence of strain.

(R.10.37) Investigate all the appropriate problems of this monograph for HD $Bi_2Te_{3-x}Se_x$ and $Bi_{2-x}Sb_xTe_3$ respectively in the presence of strain.

(R.10.38) Investigate all the appropriate problems of this monograph for all types of skutterudites in the presence of strain.

(R.10.39) Investigate all the appropriate problems of this monograph in the presence of crossed electric and quantizing magnetic fields.

(R.10.40) Investigate all the appropriate problems of this monograph in the presence of crossed alternating electric and quantizing magnetic fields.

(R.10.41) Investigate all the appropriate problems of this monograph in the presence of crossed electric and alternating quantizing magnetic fields.

(R.10.42) Investigate all the appropriate problems of this monograph in the presence of alternating crossed electric and alternating quantizing magnetic fields.

(R.10.43) Investigate all the appropriate problems of this monograph in the presence of arbitrarily oriented pulsed electric and quantizing magnetic fields.

(R.10.44) Investigate all the appropriate problems of this monograph in the presence of arbitrarily oriented alternating electric and quantizing magnetic fields.

(R.10.45) Investigate all the appropriate problems of this monograph in the presence of crossed in homogeneous electric and alternating quantizing magnetic fields.

(R.10.46) Investigate all the appropriate problems of this monograph in the presence of arbitrarily oriented electric and alternating quantizing magnetic fields under strain.

(R.10.47) Investigate all the appropriate problems of this monograph in the presence of arbitrarily oriented electric and alternating quantizing magnetic fields under light waves.

(R.10.48) Investigate all the appropriate problems of this monograph in the presence of arbitrarily oriented pulsed electric and alternating quantizing magnetic fields under light waves.

(R.10.49) Investigate all the appropriate problems of this monograph in the presence of arbitrarily oriented inhomogeneous electric and pulsed quantizing magnetic fields in the presence of strain and light waves.

(R.10.50) (a) *Investigate the EP for all types of HD materials of this monograph in the presence of many body effects strain and arbitrarily oriented light waves respectively.*

(b) *Investigate all the appropriate problems of this chapter for the Dirac electron.*
(c) *Investigate all the problems of this monograph by removing all the physical and mathematical approximations and establishing the respective appropriate uniqueness conditions.*

The formulation of EP for all types of HD materials and their quantum confined counter parts considering the influence of all the bands created due to all types of quantizations after removing all the assumptions and establishing the respective appropriate uniqueness conditions is, in general, an extremely difficult problem. 300 open research problems have been presented in this monograph and we hope that the readers will not only solve them but also will generate new concepts, both theoretical and experimental. Incidentally, we can easily infer how little is presented and how much more is yet to be investigated in this exciting topic which is the signature of coexistence of new physics, advanced mathematics combined with the inner fire for performing creative researches in this context from the young scientists since like Kikoin [5] we firmly believe that *A young scientist is no good if his teacher learns nothing from him and gives his teacher nothing to be proud of.* In the mean time our research interest has been shifted and we are leaving this particular beautiful topic with the hope that (R.10.50) alone is sufficient to draw the attention of the researchers from diverse fields and our readers are surely in tune with the fact that *Exposition, criticism, appreciation is the work for second-rate minds* [6].

References

1. B.R. Nag, *Electron Transport in Compound Semiconductors*, vol. 11, Springer Series in Solid State Sciences (Springer, Berlin, 1980)
2. R.W. Cunningham, Phys. Rev. **167**, 761 (1968)
3. A.I. Yekimov, A.A. Onushchenko, A.G. Plyukhin, Al.L. Efros, J. Expt. Theor. Phys. **88**, 1490 (1985)
4. B.J. Roman, A.W. Ewald, Phys. Rev. B **5**, 3914 (1972)
5. I.K. Kikoin, *Science for Everyone: Encounters with Physicists and Physics* (Mir Publishers, Russia, 1989), p. 154
6. G.H. Hardy, *A Mathematician's Apology* (Cambridge University Press, Cambridge, 1990), p. 61

Chapter 11
Appendix A: The EP from HDS Under Magnetic Quantization

11.1 Introduction

It is well known that the band structure of semiconductors can be dramatically changed by applying the external fields. The effects of the quantizing magnetic field on the band structure of compound semiconductors are more striking and can be observed easily in experiments [1–67]. Under magnetic quantization, the motion of the electron parallel to the magnetic field remains unaltered while the area of the wave vector space perpendicular to the direction of the magnetic field gets quantized in accordance with the Landau's rule of area quantization in the wave-vector space [39–67]. The energy levels of the carriers in a magnetic field (with the component of the wave-vector parallel to the direction of magnetic field be equated with zero) are termed as the Landau levels and the quantized energies are known as the Landau sub-bands. It is important to note that the same conclusion may be arrived either by solving the single-particle time independent Schrödinger differential equation in the presence of a quantizing magnetic field or by using the operator method. The quantizing magnetic field tends to remove the degeneracy and increases the band gap. A semiconductor, placed in a magnetic field B, can absorb radiative energy with the frequency $(\omega_0 = (|e|B/m_c))$. This phenomenon is known as cyclotron or diamagnetic resonance. The effect of energy quantization is experimentally noticeable when the separation between any two consecutive Landau levels is greater than k_BT. A number of interesting transport phenomena originate from the change in the basic band structure of the semiconductor in the presence of quantizing magnetic field. These have been widely been investigated and also served as diagnostic tools for characterizing the different materials having various band structures [68–72]. The discreteness in the Landau levels leads to a whole crop of magneto-oscillatory phenomena, important among which are

K.P. Ghatak, *Einstein's Photoemission*, Springer Tracts in Modern Physics 262,
DOI 10.1007/978-3-319-11188-9_11

(i) Shubnikov-de Haas oscillations in magneto-resistance; (ii) De Haas-van Alphen oscillations in magnetic susceptibility; (iii) magneto-phonon oscillations in thermoelectric power, etc.

In this chapter in Sect. 11.2.1, of the theoretical background, the EP has been investigated in HD non linear optical semiconductors in the presence of a quantizing magnetic field. Section 11.2.2 contains the results for HD III-V, ternary and quaternary compounds in accordance with the three and the two band models of Kane. In the same section the EP in accordance with the models of Stillman et al. and Palik et al. have also been studied for the purpose of relative comparison. Section 11.2.3 contains the study of the EP for HD II-VI semiconductors under magnetic quantization. In Sect. 11.2.4, the EP in HD IV-VI materials has been discussed in accordance with the models of Cohen, Lax, Dimmock, Bangert and Kastner and Foley and Landenberg respectively. In Sect. 11.2.5, the magneto-EP for the stressed HD Kane type semiconductors has been investigated. In Sect. 11.2.6, the EP in HD Te has been studied under magnetic quantization. In Sect. 11.2.7, the magneto-EP in n-GaP has been studied. In Sect. 11.2.8, the EP in HD $PtSb_2$, has been explored under magnetic quantization. In Sect. 11.2.9, the magneto-EP in HD Bi_2Te_3 has been studied. In Sect. 11.2.10, the EP in HD Ge has been studied under magnetic quantization in accordance with the models of Cardona et al. and Wang and Ressler respectively. In Sects. 11.2.11 and 11.2.12, the magneto-EP in HD n-GaSb and II-V compounds has respectively been studied. In Sect. 11.2.13 the magneto EP in HD $Pb_{1-x}Ge_xTe$ has been discussed. The last Sect. 11.3 contains 52 open research problems for this Appendix.

11.2 Theoretical Background

11.2.1 The EP from HD Nonlinear Optical Semiconductors Under Magnetic Quantization

The dispersion relation under magnetic quantization in non-linear optical materials can be written as

$$\gamma(E) = \frac{\hbar^2 k_s^2}{2m_\perp^*} f_3(E) + \frac{\hbar^2 k_z^2}{2m_\parallel^*} f_4(E) \pm \frac{eB\hbar E_g}{6}\left[\frac{(E_{g_0}+\Delta_\perp)}{(E_{g_0}+\frac{2}{3}\Delta_\perp)}\right]\left[E + E_{g_0} + \delta + \frac{\Delta_\parallel^2 - \Delta_\perp^2}{3\Delta_\parallel}\right] \tag{11.1}$$

where $f_3(E) = \frac{E_g(E_g+\Delta_\perp)}{(E_g+\frac{2}{3}\Delta_\perp)}[(E+E_g)(E+E_g+\frac{2}{3}\Delta_\parallel) + \delta(E+E_g+\frac{1}{3}\Delta_\parallel) + \frac{1}{9}(\Delta_\parallel^2 - \Delta_\perp^2)]$
and $f_4(E) = \frac{E_g(E_g+\Delta_\parallel)}{(E_g+\frac{2}{3}\Delta_\parallel)}[(E+E_g)(E+E_g+\frac{2}{3}\Delta_\parallel)]$

The (11.1) can be expressed as

$$\frac{\hbar^2 k_z^2}{2m_{\parallel}^*} + \left(\frac{b_{\parallel} c_{\perp}}{b_{\perp} c_{\parallel}}\right)\left(\frac{\hbar^2 k_s^2}{2m_{\perp}^*}\right) = \{[\frac{ab_{\parallel}}{c_{\parallel}} E^2 + \frac{(ac_{\parallel} + b_{\parallel} c_{\parallel} - ab_{\parallel})}{c_{\parallel}^2} E$$
$$+ \frac{1}{c_{\parallel}}(1 - \frac{a}{c_{\parallel}})(1 - \frac{b_{\parallel}}{c_{\parallel}}) - \frac{1}{c_{\parallel}}(1 - \frac{a}{c_{\parallel}})(1 - \frac{b_{\parallel}}{c_{\parallel}})\frac{1}{(c_{\parallel} E + 1)}]$$
$$+ \frac{ab_{\parallel}}{c_{\parallel}}[\delta E + \frac{2}{9}(\Delta_{\parallel}^2 - \Delta_{\perp}^2)] - \frac{2}{9}\frac{ab_{\parallel}}{c_{\parallel}}\frac{(\Delta_{\parallel}^2 - \Delta_{\perp}^2)}{(c_{\parallel} E + 1)}\}$$
$$- \left(\frac{\hbar^2 k_s^2}{2m_{\perp}^*}\right)\{\left(\frac{b_{\parallel} c_{\perp}}{b_{\perp} c_{\parallel}}\right)[\left(\frac{\delta}{2} + \frac{\Delta_{\parallel}^2 - \Delta_{\perp}^2}{6\Delta_{\parallel}}\right)\frac{a}{(aE + 1)}$$
$$+ \left(\frac{\delta}{2} - \frac{\Delta_{\parallel}^2 - \Delta_{\perp}^2}{6\Delta_{\parallel}}\right)\frac{c_{\parallel}}{(c_{\parallel} E + 1)}]\}$$
$$\pm e_1 [\frac{\rho_1}{E + E_g} + \frac{\rho_2}{E + E_g + \frac{2}{3}\Delta_{\parallel}}] \tag{11.2}$$

where $e_1 = [\frac{eB\hbar(E_g + \frac{2}{3}\Delta_{\parallel})}{6(E_g + \Delta_{\parallel})}\frac{(E_g + \Delta_{\perp})}{(E_g + \frac{2}{3}\Delta_{\perp})}]$, $\rho_1 = \frac{(-E_g + G_1)}{(\frac{2}{3}\Delta_{\parallel})}$, $G_1 = [E_{g_0} + \delta + \frac{\Delta_{\parallel}^2 - \Delta_{\perp}^2}{3\Delta_{\parallel}}]$ and $\rho_2 = \frac{3}{2\Delta_{\parallel}}[\frac{\Delta_{\parallel}}{3} + \frac{\Delta_{\perp}^2}{3\Delta_{\parallel}} - \delta]$

Therefore, the dispersion relation of the conduction electrons in HD non-linear optical semiconductors in the presence of a quantizing magnetic field B can be written following the methods as developed in Chap. 1 as

$$\frac{\hbar^2 k_z^2}{2m_{\parallel}^*} = U_{1,\pm}(E, n, \eta_g) + iU_{2,\pm}(E, n, \eta_g) \tag{11.3a}$$

where

$$U_{1,\pm}(E, n, \eta_g) = [\frac{-eB}{m_{\perp}^*}(n + \frac{1}{2})(\frac{b_{\parallel} c_{\perp}}{c_{\parallel} b_{\perp}}) + [\frac{1 + Erf(E/\eta_g)}{2}]^{-1}[\{[\frac{ab_{\parallel}}{c_{\parallel}}\theta_0(E, \eta_g)$$
$$+ \frac{ac_{\parallel} + b_{\parallel} c_{\parallel} - ab_{\parallel}}{c_{\parallel}^2}\gamma_0(E, \eta_g) + \frac{1}{c_{\parallel}}(1 - \frac{a}{c_{\parallel}})(1 - \frac{b_{\parallel}}{c_{\parallel}})[\frac{1 + Erf(E/\eta_g)}{2}]$$
$$- \frac{1}{c_{\parallel}}(1 - \frac{a}{c_{\parallel}})(1 - \frac{b_{\parallel}}{c_{\parallel}})c(\beta_1, E, \eta_g) + \frac{ab_{\parallel}}{c_{\parallel}}[\delta\gamma_0(E, \eta_g)$$
$$+ \frac{2}{9}(\Delta_{\parallel}^2 - \Delta_{\perp}^2)[\frac{1 + Erf(E/\eta_g)}{2}]] - \frac{2}{9}\frac{ab_{\parallel}}{c_{\parallel}}(\Delta_{\parallel}^2 - \Delta_{\perp}^2)c(\beta_1, E, \eta_g)\}$$
$$- \frac{\hbar eB}{m_{\perp}^*}(n + \frac{1}{2})\{(\frac{b_{\parallel} c_{\perp}}{c_{\parallel} b_{\perp}})[(\frac{\delta}{2} + \frac{\Delta_{\parallel}^2 - \Delta_{\perp}^2}{6\Delta_{\parallel}})ac(\beta_2, E, \eta_g)$$
$$+ (\frac{\delta}{2} - \frac{\Delta_{\parallel}^2 - \Delta_{\perp}^2}{6\Delta_{\parallel}})c_{\parallel} c(\beta_1, E, \eta_g)]\} \pm \frac{\rho_1 e_1}{E_g}c(\beta_2, E, \eta_g) \pm \frac{\rho_2 e_1}{(E_g + \frac{2}{3}\Delta_{\parallel})}c(\beta_3, E, \eta_g)],$$

$$U_{2,\pm}(E,n,\eta_g) = [\frac{1+Erf(E/\eta_g)}{2}]^{-1}[\frac{1}{c_{\|}}(1-\frac{a}{c_{\|}})(1-\frac{b_{\|}}{c_{\|}})D(\beta_1,E,\eta_g)$$
$$+\frac{2}{9}\frac{ab_{\|}}{c_{\|}}(\Delta_{\|}^2-\Delta_{\perp}^2)D(\beta_1,E,\eta_g)+\frac{\hbar eB}{m_{\perp}^*}(n+\frac{1}{2})$$
$$\{(\frac{b_{\|}c_{\perp}}{c_{\|}b_{\perp}})[(\frac{\delta}{2}+\frac{\Delta_{\|}^2-\Delta_{\perp}^2}{6\Delta_{\|}})aD(\beta_2,E,\eta_g)$$
$$+(\frac{\delta}{2}-\frac{\Delta_{\|}^2-\Delta_{\perp}^2}{6\Delta_{\|}})c_{\|}D(\beta_1,E,\eta_g)]\}\mp[\frac{\rho_1 e_1}{E_g}D(\beta_2,E,\eta_g)$$
$$\pm\frac{\rho_2 e_1}{(E_g+\frac{2}{3}\Delta_{\|})}D(\beta_3,E,\eta_g)]]$$

$$\beta_1 = c_{\|},\beta_2 = \frac{1}{E_g} = a,\ \beta_3 = \frac{1}{E_g+\frac{2}{3}\Delta_{\|}},$$

$$C(\beta_i,E,\eta_g) = [\frac{2}{\beta_i\eta_g\sqrt{\pi}}]\exp(-u_i^2)\times[\sum_{p=1}^{\infty}\{\frac{\exp(\frac{-p^2}{4})}{p}\}\sinh(pu_i)],u_i = \frac{1+\beta_i E}{\beta_i\eta_g},$$

and $D(\beta_i,E,\eta_g) = [\frac{\sqrt{\pi}}{\beta_i\eta_g}\exp(-u_i^2)]$

The EEM at the Fermi Level can be written from (11.3a) as

$$m_{\pm}^*(E_{FBHD},n,\eta_g) = m_{\|}^* U_{1,\pm}'(E_{FBHD},n,\eta_g) \tag{11.3b}$$

where E_{FBHD} is the Fermi energy in this case.

Therefore the double valued EEM in this case is a function of Fermi energy, magnetic field, quantum number and the scattering potential together with the fact that the EEM exists in the band gap which is the general characteristics of HD materials.

The complex density-of-states function under magnetic quantization is given by

$$\begin{aligned} N_B(E) &= N_{BR}(E)+iN_{BI}(E) \\ &= \frac{eB}{2\pi^2\hbar^2}\sqrt{2m_{11}^*}\sum_{n=0}^{n_{\max}}[\frac{x'}{2\sqrt{x}}+\frac{iy'}{2\sqrt{y}}] \end{aligned} \tag{11.4a}$$

where

$$x = \frac{\sqrt{(U_{1,\pm}(E,n,\eta_g))^2+(U_{2,\pm}(E,n,\eta_g))^2}+(U_{1,\pm}(E,n,\eta_g))}{2}$$
$$y = \frac{\sqrt{(U_{1,\pm}(E,n,\eta_g))^2+(U_{2,\pm}(E,n,\eta_g))^2}-(U_{1,\pm}(E,n,\eta_g))}{2}$$

and x' and y' are the differentiations of x and y with respect to energy E.

Therefore, from (11.4a) we can write

$$N_{BR}(E) = \frac{eB\sqrt{2m_{\|}^{*}}}{4\pi^2\hbar^2}\sum_{n=0}^{n_{\max}}\frac{x'}{\sqrt{x}} \tag{11.4b}$$

and

$$N_{BI}(E) = \frac{eB\sqrt{2m_{\|}^{*}}}{4\pi^2\hbar^2}\sum_{n=0}^{n_{\max}}\frac{y'}{\sqrt{y}} \tag{11.4c}$$

The electron concentration under the condition of extreme carrier degeneracy is given by

$$n_0 = \frac{g_v eB}{2\pi^2\hbar^2}\frac{\sqrt{2m_{\|}^{*}}}{\sqrt{2}}\sum_{n=0}^{n_{\max}}[\sqrt{(U_{1,\pm}(\mathrm{E}_{FBHD}, n, \eta_g))^2 + (U_{2,\pm}(\mathrm{E}_{FBHD}, n, \eta_g))^2} + (U_{1,\pm}(\mathrm{E}_{FBHD}, n, \eta_g))]^{1/2} \tag{11.5}$$

The magneto EP in this case is given by

$$J_B = \frac{\alpha_0 e^2 g_v B k_B T}{2\pi^2\hbar^2}\text{Real part of}\sum_{n=0}^{n_{\max}} F_0(\eta_{1A}) \tag{11.6}$$

where

$$\eta_{1A} = (k_B T)^{-1}[\mathrm{E}_{FHBD} - (\mathrm{E}_{L1} + W - h\upsilon)]$$

where E_{L1} is the complex Landau sub-band energy which can be obtained from (11.3a) by substituting $\mathrm{k}_z = 0$ and $E = E_{L1}$.

11.2.2 The EP from HD Kane Type III-V, Ternary and Quaternary Semiconductors Under Magnetic Quantization

(a) The electron energy spectrum in III-V semiconductors under magnetic quantizaion is given by

$$\frac{E(E+E_g)(E+E_g+\Delta)(E_g+\frac{2}{3}\Delta)}{E_g(E_g+\Delta)(E+E_g+\frac{2}{3}\Delta)} = (n+\frac{1}{2})\hbar\omega_0 + \frac{\hbar^2 k_z^2}{2m_c} \pm \frac{eB\hbar\Delta}{6m_c(E+E_g+\frac{2}{3}\Delta)} \tag{11.7}$$

The (11.7) can be written as

$$[\frac{ab}{c}E^2] + (\frac{ac+bc-ab}{c^2})E + \frac{1}{c}(1-\frac{a}{c})(1-\frac{b}{c}) - \frac{1}{c}(1-\frac{a}{c})(1-\frac{b}{c})\frac{1}{(1+cE)}$$
$$= (n+\frac{1}{2})\hbar\omega_0 + \frac{\hbar^2k_z^2}{2m_c} \pm \frac{eB\hbar\Delta}{6m_c(1+cE)(E_g+\frac{2}{3}\Delta)}$$

where $a = \frac{1}{E_g}$, $b = \frac{1}{E_g+\Delta}$ and $c = \frac{1}{E_g+\frac{2}{3}\Delta}$

Therefore

$$\frac{ab}{c}I(5) + (\frac{ab+bc-ab}{c^2})I(4) + [\frac{1}{c}(1-\frac{a}{c})(1-\frac{b}{c}) - (n+\frac{1}{2})\hbar\omega_0]I(1)$$
$$- g_{\pm}[G(C,E,\eta_g) - iH(C,E,\eta_g)] = \frac{\hbar^2k_z^2}{2m_c}I(1) \tag{11.8}$$

where $g_{\pm} = [\frac{1}{c}(1-\frac{a}{c})(1-\frac{b}{c}) \pm \frac{eB\hbar\Delta}{6m_c(E_g+\frac{2}{3}\Delta)}]$,

$$G(C,E,\eta_g) = [\frac{2}{C\eta_g\sqrt{\pi}}]\exp(-u^2) \times [\sum_{p=1}^{\infty}\{\frac{\exp(\frac{-p^2}{4})}{p}\}\sinh(pu)],\ u = \frac{1+CE}{C\eta_g},$$

$$H(C,E,\eta_g) = [\frac{\sqrt{\pi}}{C\eta_g}\exp(-u^2)]$$

Therefore

$$\frac{\hbar^2k_z^2}{2m_c} = [\frac{ab}{c}]\theta_0(E,\eta_g)[\frac{1+Erf(E/\eta_g)}{2}]^{-1} + (\frac{ac+bc-ab}{c^2})\gamma_0(E,\eta_g)[\frac{1+Erf(E/\eta_g)}{2}]^{-1}$$
$$+ [\frac{1}{c}(1-\frac{a}{c})(1-\frac{b}{c}) - (n+\frac{1}{2})\hbar\omega_0] - g_{\pm}[\frac{1+Erf(E/\eta_g)}{2}]^{-1}[G(C,E,\eta_g) - iH(C,E,\eta_g)] \tag{11.9}$$

Therefore the dispersion relation is given by

$$\frac{\hbar^2k_z^2}{2m_c} = U_{3,\pm}(E,n,\eta_g) + iU_{4,\pm}(E,\eta_g) \tag{11.10a}$$

where

$$U_{3,\pm}(E,n,\eta_g) = [\frac{ab}{c}\theta_0(E,\eta_g) \times [\frac{1+Erf(E/\eta_g)}{2}]^{-1} + (\frac{ac+bc-ab}{c^2})\gamma_0(E,\eta_g)$$
$$\times [\frac{1+Erf(E/\eta_g)}{2}]^{-1} + \frac{1}{c}(1-\frac{a}{c})(1-\frac{b}{c})$$
$$- (n+\frac{1}{2})\hbar\omega_0 + g_{\pm}[\frac{1+Erf(E/\eta_g)}{2}]^{-1}G(C,E,\eta_g)]$$

and $U_{4,\pm}(E,\eta_g) = g_{\pm}[\frac{1+Erf(E/\eta_g)}{2}]^{-1}H(C,E,\eta_g)$

The complex Landau energy E_{nHD1} in this case can be obtained by substituting $k_z = 0$ and $E = E_{nHD}$ in (11.10a)

The EEM at the Fermi Level can be written from (11.10a) as

$$m^*_\pm(E_{FBHD}, n, \eta_g) = m^*_\| U'_{3,\pm}(E_{FBHD}, n, \eta_g) \tag{11.10b}$$

Thus the EEM is a function of Fermi energy, Landau quantum number and scattering potential together with the fact it is double valued due to spin.

The complex density of states function under magnetic quantization is given by

$$\begin{aligned} N_B(E) &= N_{BRI}(E) + iN_{BI1}(E) \\ &= \frac{eB}{2\pi^2\hbar^2}\sqrt{2m_c}\sum_{n=0}^{n_{\max}}\left[\frac{x'_1}{2\sqrt{x_1}} + \frac{iy'_1}{2\sqrt{y_1}}\right] \end{aligned} \tag{11.11a}$$

where

$$x_1 = \frac{\sqrt{(U_{3,\pm}(E,n,\eta_g))^2 + (U_{4,\pm}(E,n,\eta_g))^2} + (U_{3,\pm}(E,n,\eta_g))}{2},$$

$$y_1 = \frac{\sqrt{(U_{3,\pm}(E,n,\eta_g))^2 + (U_{4,\pm}(E,n,\eta_g))^2} - (U_{3,\pm}(E,n,\eta_g))}{2}$$

and x'_1 and y'_1 are the differentiations of x and y with respect to energy E.

From (11.11a) we can write

$$N_{BR1}(E) = \frac{eB\sqrt{2m_c}}{4\pi^2\hbar^2}\sum_{n=0}^{n_{\max}}\frac{x'_1}{\sqrt{x_1}} \tag{11.11b}$$

$$\text{and } N_{BI1}(E) = \frac{eB\sqrt{2m_c}}{4\pi^2\hbar^2}\sum_{n=0}^{n_{\max}}\frac{y'_1}{\sqrt{y_1}} \tag{11.11c}$$

The electron concentration under the condition of extreme degeneracy is given by

$$\begin{aligned} n_0 = \frac{g_v eB}{2\pi^2\hbar^2}\sqrt{m_c}\sum_{n=0}^{n_{\max}}\Big[\sqrt{(U_{3,\pm}(\mathrm{E}_{FBHD},n,\eta_g))^2 + (U_{4,\pm}(\mathrm{E}_{FBHD},n,\eta_g))^2} \\ + (U_{3,\pm}(\mathrm{E}_{FBHD},n,\eta_g))\Big]^{1/2} \end{aligned} \tag{11.12a}$$

The magneto EP in this case is given by

$$J_B = \frac{\alpha_0 e^2 g_v B k_B T}{2\pi^2\hbar^2}\text{Real part of}\sum_{n=0}^{n_{\max}} F_0(\eta_{2A}) \tag{11.12b}$$

where

$$\eta_{2A} = (k_B T)^{-1}[\mathrm{E}_{FHBD} - (\mathrm{E}_{L2} + W - h\upsilon)]$$

where E_{L2} is the complex Landau sub-band energy which can be obtained from (11.10a) by substituting $\mathrm{k}_z = 0$ and $E = E_{L2}$

(b) Two band model of Kane

The magneto-dispersion law in this case is given by

$$\frac{\hbar^2 k^2}{2m_c} = \gamma_2(E, \eta_g) - (n + \frac{1}{2})\hbar\omega_0 \mp \frac{1}{2} g^* \mu_0 B \tag{11.13a}$$

where g^* is the magnitude of the effective g factor at the edge of the conduction band and μ_0 is the Bohr magnetron.

The EEM at the Fermi Level can be written from (11.13a) as

$$m^*(E_{FBHD}, \eta_g) = m_c \gamma_2'(E_{FBHD}, \eta_g) \tag{11.13b}$$

Thus EEM is independent of quantum number.

The electron concentration under the condition of extreme degeneracy is given by

$$n_0 = \frac{g_v eB}{\pi^2 \hbar^2} \sqrt{m_c} \sum_{n=0}^{n_{\max}} (U_{5,\pm}(\mathrm{E}_{FBHD}, n, \eta_g))^{\frac{1}{2}} \tag{11.14a}$$

where

$$U_{5,\pm}(\mathrm{E}_{FBHD}, n, \eta_g) = \gamma_2(\mathrm{E}_{FBHD}, \eta_g) - (n + \frac{1}{2})\hbar\omega_0 \mp \frac{1}{2} g^* \mu_0 B$$

The magneto EP in this case is given by

$$J_B = \frac{\alpha_0 e^2 g_v B k_B T}{2\pi^2 \hbar^2} \sum_{n=0}^{n_{\max}} F_0(\eta_{3A}) \tag{11.14b}$$

where

$$\eta_{3A} = (k_B T)^{-1}[\mathrm{E}_{FHBD} - (\mathrm{E}_{L3} + W - h\upsilon)]$$

where E_{L3} is the Landau sub-band energy which can be obtained from (11.13a) by substituting $\mathrm{k}_z = 0$ and $E = E_{L3}$

(c) Parabolic Energy Bands

The magneto-dispersion law in this case is given by

$$\frac{\hbar^2 k^2}{2m_c} = \gamma_3(E, \eta_g) - (n + \frac{1}{2})\hbar\omega_0 \mp \frac{1}{2} g^* \mu_0 B \tag{11.15a}$$

The EEM at the Fermi Level can be written from (11.15a) as

$$m^*(E_{FBHD}, \eta_g) = m_c \gamma_3'(E_{FBHD}, \eta_g) \tag{11.15b}$$

Thus the EEM in heavily doped parabolic energy bands is a function of Fermi energy and scattering potential whereas in the absence of band-tails the same mass is a constant quantity invariant of any variables.

The electron concentration under the condition of extreme degeneracy is given by

$$n_0 = \frac{g_v eB}{\pi^2 \hbar^2} \sqrt{m_c} \sum_{n=0}^{n_{\max}} (U_{6,\pm}(\mathrm{E}_{FBHD}, n, \eta_g))^{\frac{1}{2}} \tag{11.16a}$$

where $U_{6,\pm}(\mathrm{E}_{FBHD}, n, \eta_g) = \gamma_3(\mathrm{E}_{FBHD}, \eta_g) - (n + \frac{1}{2})\hbar\omega_0 \mp \frac{1}{2} g^* \mu_0 B$

The magneto EP in this case is given by

$$J_B = \frac{\alpha_0 e^2 g_v B k_B T}{2\pi^2 \hbar^2} \sum_{n=0}^{n_{\max}} F_0(\eta_{4A}) \tag{11.16b}$$

where $\eta_{4A} = (k_B T)^{-1} [\mathrm{E}_{FHBD} - (\mathrm{E}_{L4} + W - h\upsilon)]$ where E_{L4} is the Landau sub-band energy which can be obtained from (11.15a) by substituting $\mathrm{k}_z = 0$ and $E = E_{L4}$

(d) The model of Stillman et al.

The (1.107) under the condition of band tailing assumes the form

$$k^2 = \frac{[\bar{t}_{11} - \sqrt{(\bar{t}_{11})^2 - 4\bar{t}_{12}\gamma_3(E, \eta_g)}]}{2\bar{t}_{12}} \tag{11.17}$$

Therefore the magneto dispersion law is given by

$$k_z^2 = U_7(E, n, \eta_g) \tag{11.18a}$$

where $U_7(E,n,\eta_g) = [\frac{[\bar{t}_{11} - \sqrt{(\bar{t}_{11})^2 - 4\bar{t}_{12}\gamma_3(E,\eta_g)}]}{2\bar{t}_{12}} - \frac{2eB}{\hbar}(n+\frac{1}{2})]$

The EEM at the Fermi Level can be written from (11.18a) as

$$m^*(E_{FBHD},\eta_g) = \frac{\hbar^2}{2} U_7'(E_{FBHD},n,\eta_g) \tag{11.18b}$$

The electron concentration under the condition of extreme degeneracy is given by

$$n_0 = \frac{g_v eB}{\pi^2\hbar} \sum_{n=0}^{n_{\max}} (U_7(E_{FBHD},n,\eta_g))^{\frac{1}{2}} \tag{11.19a}$$

The magneto EP in this case is given by

$$J_B = \frac{\alpha_0 e^2 g_v B k_B T}{2\pi^2\hbar^2} \sum_{n=0}^{n_{\max}} F_0(\eta_{5A}) \tag{11.19b}$$

where

$$\eta_{5A} = (k_B T)^{-1}[\mathrm{E}_{FHBD} - (\mathrm{E}_{L5} + W - h\upsilon)]$$

where E_{L5} is the Landau sub-band energy which can be obtained from (11.18a) by substituting $\mathrm{k}_z = 0$ and $E = E_{L5}$

(e) The model of Palik et al.

To the fourth order in effective mass theory and taking into account the interactions of the conduction, light hole, heavy-hole and split-off hole bands, the electron energy spectrum in III-V semiconductors in the presence of a quantizing magnetic field $\vec{B}$ can be written as

$$E = J_{31} + (n+\frac{1}{2})\hbar\omega_0 + \frac{\hbar^2 k_z^2}{2m_c} \pm \frac{1}{4}(\frac{m_c}{m_0})\hbar\omega_0 g_0^* \pm k_{30}\alpha(n+\frac{1}{2})(\hbar\omega_0)^2 \pm k_{31}\alpha\hbar\omega_0(\frac{\hbar^2 k_z^2}{2m_c}) + k_{32}\alpha\left[\hbar\omega_0\left(n+\frac{1}{2}\right) + \frac{\hbar^2 k_z^2}{2m_c}\right] \tag{11.20}$$

where

$$J_{31} = -\frac{1}{2}\alpha\hbar\omega_0\left[(1-y_{11})/(2+x_{11})^2\right]\cdot J_{32},$$

$$J_{32} = \left\{\left[\frac{1}{3}(1-x_{11})^2-(2+x_{11}^2)\right](2+x_{11})\cdot y_{11}+\frac{1}{2}(1-x_{11}^2)(1+x_{11})(1+y_{11})\right\},$$

$$g_0^* = 2\left\{1-\left[\frac{(1-x_{11})}{(2+x_{11})}\right]\left[\frac{(1-y_{11})}{y_{11}}\right]\right\},\ k_{30} = (1-y_{11})(1-x_{11})$$

$$\left\{\left[\left(2+\frac{3}{2}x_{11}+x_{11}^2\right)\cdot\frac{(1-y_{11})}{(2+x_{11})^2}\right]-\frac{2}{3}y_{11}\right\},$$

$$k_{31} = (1-y_{11})\left[\frac{(1-x_{11})}{(2+x_{11})}\right]\left\{\left[\left(2+\frac{3}{2}x_{11}+x_{11}^2\right)\cdot\frac{(1-y_{11})}{(2+x_{11})^2}\right]-\frac{2}{3}(1-x_{11})y_{11}\right\},$$

$$k_{32} = -\left[\left(1+\frac{1}{2}x_{11}^2\right)/\left(1+\frac{1}{2}x_{11}\right)\right](1-y_{11})^2, x_{11} = [1+(\frac{\Delta}{E_g})]^{-1} \text{ and } y_{11} = \frac{m_c}{m_0}$$

Under the condition of heavy doping, the (11.20) assumes the form

$$J_{34}k_z^4 + J_{35,\pm}(n)k_z^2 + J_{36,\pm}(n) - \gamma_3(\mathrm{E},\eta_g) = 0 \tag{11.21}$$

where

$$J_{34} = \alpha k_{32}(\hbar^2/2m_c)^2,\ J_{35,\pm}(n) = \left[\tfrac{\hbar^2}{2m_c} \pm \alpha k_{31}\hbar\omega_0\cdot\tfrac{\hbar^2}{2m_c} + \alpha k_{32}\hbar\omega_0\cdot\tfrac{\hbar^2}{2m_c}(n+\tfrac{1}{2})\right],$$

$$J_{36,\pm}(n) = \left[J_{31} \pm \frac{1}{4}(\frac{m_c}{m_0})\hbar\omega_0 g_0^* \pm k_{30}\alpha(\hbar\omega_0)^2(n+\frac{1}{2}) + k_{32}\alpha[(\hbar\omega_0)(n+\frac{1}{2})]^2\right]$$

The (11.21) can be written as

$$k_z^2 = A_{35,\pm}(E,n,\eta_g) \tag{11.22a}$$

where

$$A_{35,\pm}(E,n,\eta_g) = (2J_{34})^{-1}\left[-J_{35,\pm}(n)+\sqrt{(J_{35,\pm}(n))^2-4J_{34}[J_{36,\pm}(n)-\gamma_3(\mathrm{E},\eta_g)]}\right]$$

The EEM at the Fermi Level can be written from (11.22a) as

$$m_\pm^*(E_{FBHD},n,\eta_g) = \frac{\hbar^2}{2}A'_{35,\pm}(E_{FBHD},n,\eta_g) \tag{11.22b}$$

Thus, the EEM is a function of Fermi energy, Landau quantum number and the scattering potential.

The electron concentration is given by

$$n_0 = \frac{eBg_v}{2\pi^2 h} \sum_{n=0}^{n_{\max}} [Y_{34HD}(E_{FBHD}, n, \eta_g) + Z_{34HD}(E_{FBHD}, n, \eta_g)] \tag{11.23a}$$

where $Y_{34HD}(E_{FBHD}, n, \eta_g) = [\sqrt{A_{35HD,+}(E_{FBHD}, n, \eta_g)} + \sqrt{A_{35HD,-}(E_{FBHD}, n, \eta_g)}]$

and $Z_{34HD}(E_{FBHD}, n, \eta_g) = \sum_{\gamma=1}^{s_0} L_B(r)[Y_{34HD}(E_{FBHD}, n, \eta_g)]$

The magneto EP in this case is given by

$$J_B = \frac{\alpha_0 e^2 g_v B k_B T}{2\pi^2 \hbar^2} \sum_{n=0}^{n_{\max}} F_0(\eta_{6A}) \tag{11.23b}$$

where

$$\eta_{6A} = (k_B T)^{-1} [\mathrm{E}_{FHBD} - (\mathrm{E}_{L6} + W - h\upsilon)]$$

where E_{L6} is the Landau sub-band energy which can be obtained from (11.22a) by substituting $\mathrm{k}_z = 0$ and $E = E_{L6}$.

11.2.3 The EP from HD II-VI Semiconductors Under Magnetic Quantization

The magneto dispersion relation of the carriers in heavily doped II-VI semiconductors are given by

$$\gamma_3(E, \eta_g) = a_0' \frac{2eB}{\hbar}(n + \frac{1}{2}) + b_0' k_z^2 \pm \lambda_0' [\frac{2eB}{\hbar}(n + \frac{1}{2})]^{\frac{1}{2}} \tag{11.24}$$

The (11.24) can be written as

$$k_z^2 = U_{8\pm}(E, n, \eta_g) \tag{11.25a}$$

where $U_{8\pm}(E, n, \eta_g) = (\mathrm{b}_0')^{-1} [\gamma_3(E, \eta_g) - \frac{2eBa_0'}{\hbar}(n + \frac{1}{2}) \mp \bar{\lambda}_0 [\frac{2eB}{\hbar}(n + \frac{1}{2})]^{\frac{1}{2}}]$

The EEM at the Fermi Level can be written from (11.25a) as

$$m^*(E_{FBHD}, \eta_g) = \frac{\hbar^2}{2} U'_{8\pm}(E_{FBHD}, n, \eta_g) \tag{11.25b}$$

The electron concentration is given by

$$n_0 = \frac{eBg_v}{2\pi^2 h} \sum_{n=0}^{n_{max}} \left[Y_{35HD}(E_{FBHD}, n, \eta_g) + Z_{35HD}(E_{FBHD}, n, \eta_g) \right] \tag{11.26a}$$

where $Y_{35HD}(E_{FBHD}, n, \eta_g) = \left[\sqrt{\mathrm{U}_{8+}(E_{FBHD}, n, \eta_g)} + \sqrt{\mathrm{U}_{8-}(E_{FBHD}, n, \eta_g)} \right]$ and $Z_{35\,HD}(E_{FBHD}, n, \eta_g) = \sum_{r=1}^{s_0} L_B(r)[Y_{35\,HD}(E_{FBHD}, n, \eta_g)]$

The magneto EP in this case is given by

$$J_B = \frac{\alpha_0 e^2 g_v B k_B T}{2\pi^2 \hbar^2} \sum_{n=0}^{n_{max}} F_0(\eta_{7A}) \tag{11.26b}$$

where

$$\eta_{7A} = (k_B T)^{-1} [\mathrm{E}_{FHBD} - (\mathrm{E}_{L7} + W - h\upsilon)]$$

where E_{27} is the Landau sub-band energy which can be obtained from (11.25a) by substituting $k_z = 0$ and $E = E_{L7}$.

11.2.4 The EP from HD IV-VI Semiconductors Under Magnetic Quantization

The electron energy spectrum in IV-VI semiconductors are defined by the models of Cohen, Lax, Dimmock and Bangert and Kastner respectively. The Magneto EP in HD IV-VI semiconductors is discussed in accordance with the said model for the purpose of relative comparison.

(a) Cohen Model

In accordance with the Cohen model, the dispersion law of the carriers in IV-VI semiconductors is given by

$$E(1 + \alpha E) = \frac{p_x^2}{2m_1} + \frac{p_z^2}{2m_3} - \frac{\alpha E p_y^2}{2m'_2} + \frac{p_y^2(1 + \alpha E)}{2m_2} + \frac{\alpha p_y^4}{4m_2 m'_2} \tag{11.27}$$

where $p_1 \equiv \hbar k_i$, $i = x, y, z$, m_1, m_2, and m_3 are the effective carrier masses at the band-edge along x, y and z directions respectively and m_2' is the effective-mass tensor component at the top of the valence band (for electrons) or at the bottom of the conduction band (for holes).

The magneto electron energy spectrum in IV-VI semiconductors in the presence of quantizing magnetic field$B$ along z-direction can be written as

$$E(1+\alpha E) = \left(n+\frac{1}{2}\right)\hbar\omega(E) \pm \frac{1}{2}g^*\mu_0 B + \frac{3}{8}\alpha\left(n^2+n+\frac{1}{2}\right)\hbar^2\omega^2(E) + \frac{\hbar^2 k_z^2}{2m_3} \tag{11.28a}$$

where $\omega(E) \equiv \frac{|e|B}{\sqrt{m_1 m_2}}\left[1+\alpha E\left(1-\frac{m_2}{m_2'}\right)\right]^{1/2}$.

Therefore the magneto dispersion law in heavily doped IV-VI materials can be expressed as

$$\frac{\hbar^2 k_z^2}{2m_3} = U_{16,\pm}(E, n, \eta_g) \tag{11.28b}$$

where

$$\begin{aligned} U_{16,\pm}(E,n,\eta_g) = {} & [\gamma_2(E,\eta_g) - (n+\frac{1}{2})\frac{\hbar eB}{\sqrt{m_1 m_2}} \mp \frac{1}{2}g*\mu_0 B - \frac{3\alpha}{8}(n^2+n+\frac{1}{2})\left(\frac{\hbar eB}{\sqrt{m_1 m_2}}\right)^2 \\ & - \gamma_3(E,\eta_g)\left[\frac{\alpha}{2}(n+\frac{1}{2})\frac{\hbar eB}{\sqrt{m_1 m_2}}\left(1-\frac{m_2}{m_2'}\right) \right. \\ & \left. + \frac{3\alpha^2}{8}(n^2+n+\frac{1}{2})\left(\frac{\hbar eB}{\sqrt{m_1 m_2}}\right)^2\left(1-\frac{m_2}{m_2'}\right)\right]] \end{aligned}$$

The EEM at the Fermi Level can be written from (11.28b) as

$$m_\pm^*(E_{FBHD}, n, \eta_g) = m_3 U_{16,\pm}'(E_{FBHD}, n, \eta_g) \tag{11.28c}$$

Thus, the EEM is a function of Fermi energy, Landau quantum number and the scattering potential.

The carrier statistics under the condition of extreme degeneracy in this case can be expressed as

$$n_0 = \frac{g_v eB}{\pi^2\hbar^2}\sqrt{m_3}\sum_{n=0}^{n_{\max}}\left(U_{16,\pm}(\mathrm{E}_{FBHD}, n, \eta_g)\right)^{\frac{1}{2}} \tag{11.29a}$$

The magneto EP in this case is given by

$$J_B = \frac{\alpha_0 e^2 g_v B k_B T}{2\pi^2 \hbar^2} \sum_{n=0}^{n_{\max}} F_0(\eta_{8A}) \tag{11.29b}$$

where

$$\eta_{8A} = (k_B T)^{-1} [\mathrm{E}_{FHBD} - (\mathrm{E}_{L8} + W - h\upsilon)]$$

where E_{L8} is the Landau sub-band energy which can be obtained from (11.28b) by substituting $k_z = 0$ and $E = E_{L8}$

(b) Lax Model

In accordance with this model, the magneto dispersion relation assumes the form

$$E(1 + \alpha E) = (n + \frac{1}{2})\hbar\omega_{03}(E) + \frac{\hbar^2 k_z^2}{2m_3} \pm \frac{1}{2}\mu_0 g^* B \tag{11.30}$$

where $\omega_{03}(\mathrm{E}) = \frac{eB}{\sqrt{m_1 m_2}}$

The magneto dispersion relation in heavily doped IV-VI materials, can be written following (11.30) as

$$\gamma_2(E, \eta_g) = (n + \frac{1}{2})\hbar\omega_{03}(\mathrm{E}) + \frac{\hbar^2 k_z^2}{2m_3} \pm \frac{1}{2} g^* \mu_0 B \tag{11.31}$$

(11.31) can be written as

$$\frac{\hbar^2 k_z^2}{2m_3} = U_{17,\pm}(E, n, \eta_g) \tag{11.32a}$$

where

$$U_{17,\pm}(E, n, \eta_g) = \gamma_2(E, \eta_g) - (n + \frac{1}{2})\hbar\omega_{03}(\mathrm{E}) \pm \frac{1}{2} g^* \mu_0 B$$

The EEM at the Fermi Level can be written from (11.32a) as

$$m^*(E_{FBHD}, \eta_g) = m_3 U'_{17,\pm}(E_{FBHD}, n, \eta_g) \tag{11.32b}$$

The electron concentration under the condition of extreme degeneracy can be written as

$$n_0 = \frac{g_v eB}{\pi^2 \hbar^2} \sqrt{m_3} \sum_{n=0}^{n_{\max}} (U_{17,\pm}(\mathrm{E}_{FBHD}, n, \eta_g))^{\frac{1}{2}} \tag{11.33a}$$

The magneto EP in this case is given by

$$J_B = \frac{\alpha_0 e^2 g_v B k_B T}{2\pi^2 \hbar^2} \sum_{n=0}^{n_{\max}} F_0(\eta_{9A}) \tag{11.33b}$$

where

$$\eta_{9A} = (k_B T)^{-1}[\mathrm{E}_{FHBD} - (\mathrm{E}_{L9} + W - h\upsilon)]$$

where E_{L9} is the Landau sub-band energy which can be obtained from (11.32a) by substituting $k_z = 0$ and $E = E_{L9}$

(c) Dimmock Model

The dispersion relation under magnetic quantization in HD IV-VI semiconductors can be expressed in accordance with Dimmock model as

$$\begin{aligned} &\gamma_2(E,\eta_g) + \alpha\gamma_3(E,\eta_g)\frac{2eB}{\hbar}(n+\frac{1}{2})\frac{\hbar}{2}\left(\frac{1}{m_t^+} - \frac{1}{m_t^-}\right) + \alpha\gamma_3(E,\eta_g)x\frac{\hbar^2}{2}\left(\frac{1}{m_l^+} - \frac{1}{m_l^-}\right) \\ &= \frac{\hbar^2 k_s^2}{2m_t^*} + \frac{\hbar^2 k_z^2}{2m_l^*} + \frac{\hbar^2 k_s^2}{2m_t^-} + \frac{\hbar^2 k_z^2}{2m_l^-} + \alpha\left[\frac{\hbar^4 k_s^4}{4m_t^- m_t^+} + \frac{\hbar^4 k_s^2 k_z^2}{4m_l^- m_l^+} + \frac{\hbar^4 k_z^2 k_s^2}{4m_t^+ m_l^-} + \frac{\hbar^4 k_z^2}{4m_l^- m_l^+}\right] \\ &= \frac{2eB}{\hbar}(n+\frac{1}{2})\frac{\hbar^2}{2}\left(\frac{1}{m_t^*} - \frac{1}{m_t^-}\right) + x\frac{\hbar^2}{2}\left(\frac{1}{m_l^*} - \frac{1}{m_l^-}\right) \\ &\quad + \alpha\left[\frac{\hbar^4}{4m_t^- m_t^+}\left(\frac{2eB}{\hbar}(n+\frac{1}{2})\right)^2 + x\left[\frac{\hbar^4 eB}{2m_l^+ m_t^- \hbar} + \frac{\hbar^4 eB}{2m_t^+ m_l^- \hbar}\right](n+\frac{1}{2}) + \frac{\hbar^4}{4m_l^- m_l^+}x^2\right] \end{aligned} \tag{11.33c}$$

where $x = k_z^2$

Therefore the magneto dispersion relation in heavily doped IV-VI materials, whose unperturbed carriers obey the Dimmock Model can be expressed as

$$k_z^2 = U_{17}(E, n, \eta_g) \tag{11.34}$$

where

$$U_{17}(E,n,\eta_g) = [2p_9]^{-1}\left[-q_9(E,n,\eta_g) + \left[q_9^2(E,n,\eta_g) + 4p_9R_9(E,n,\eta_g)\right]^{\frac{1}{2}}\right] \tag{11.35}$$

$$p_9 = \frac{\alpha\hbar^4}{4m_l^- m_l^+}, q_9(\mathrm{E},n,\eta_g) = [\frac{\hbar^2}{2}(\frac{1}{m_l^*}+\frac{1}{m_l^-}) + \frac{\alpha\hbar^3 eB}{2}(n+\frac{1}{2})(\frac{1}{m_t^- m_l^+}+\frac{1}{m_l^- m_t^+}) - \alpha\gamma_3(\mathrm{E},\eta_g)(\frac{1}{m_l^+}+\frac{1}{m_l^-})$$

and

$$\begin{aligned} R_9(\mathrm{E},n,\eta_g) = [\gamma_2(\mathrm{E},\eta_g) + \alpha \mathrm{eB}\gamma_3(\mathrm{E},\eta_g)(n+\frac{1}{2})\hbar(\frac{1}{m_t^+}-\frac{1}{m_t^-}) \\ - \hbar \mathrm{eB}(n+\frac{1}{2})(\frac{1}{m_t^*}+\frac{1}{m_t^-}) - \frac{\alpha\hbar^2}{m_t^- m_t^+}[\mathrm{eB}(n+\frac{1}{2})]^2] \end{aligned} \tag{11.36a}$$

The EEM at the Fermi Level can be written from (11.34) as

$$m^*(E_{FBHD},n,\eta_g) = \frac{\hbar^2}{2}U_{17}'(E_{FBHD},n,\eta_g) \tag{11.36b}$$

Thus, the EEM is a function of Fermi energy, Landau quantum number and the scattering potential.

The electron concentration under the condition of extreme degeneracy can be written as

$$n_0 = \frac{g_v eB}{\pi^2\hbar}\sum_{n=0}^{n_{\max}}(U_{17}(\mathrm{E}_{FBHD},n,\eta_g))^{\frac{1}{2}} \tag{11.37a}$$

The magneto EP in this case is given by

$$J_B = \frac{\alpha_0 e^2 g_v B k_B T}{2\pi^2\hbar^2}\sum_{n=0}^{n_{\max}}F_0(\eta_{10A}) \tag{11.37b}$$

where

$$\eta_{10A} = (k_B T)^{-1}[\mathrm{E}_{FHBD} - (\mathrm{E}_{L10} + W - h\upsilon)]$$

where E_{L10} is the Landau sub-band energy which can be obtained from (11.34) by substituting $\mathrm{k}_z = 0$ and $E = E_{L10}$

(d) Model of Bangert and Kastner

In accordance with this model [73], the carrier energy spectrum in HD IV-VI semiconductors can be written following (3.68) as

$$\frac{k_s^2}{\rho_{11}^2(E,\eta_g)}+\frac{k_y^2}{\rho_{12}^2(E,\eta_g)}=1 \tag{11.38}$$

where

$$\rho_{11}(E,\eta_g)=\frac{1}{\sqrt{S_1(E,\eta_g)}},\ \rho_{12}(E,\eta_g)=\frac{1}{\sqrt{S_2(E,\eta_g)}},$$

$$\begin{aligned}S_1(E,\eta_g)=&\left[2\gamma_0(E,\eta_g)\right]^{-1}[\frac{(\bar{R})^2}{E_g}\{c_1(\alpha_1,E,E_g)-iD_1(\alpha_1,E,E_g)\}\\&+\frac{(\bar{S})^2}{\Delta_c'}\{c_2(\alpha_2,E,E_g)-iD_2(\alpha_2,E,E_g)\}\\&+\frac{(\bar{Q})}{\Delta_c''}\{c_3(\alpha_3,E,E_g)-iD_3(\alpha_3,E,E_g)\}]\end{aligned}$$

and

$$\begin{aligned}S_2(E,\eta_g)=&\left[2\gamma_0(E,\eta_g)\right]^{-1}[\frac{2(\bar{A})^2}{E_g}\{c_1(\alpha_1,E,\eta_g)-iD_1(\alpha_1,E,\eta_g)\}\\&+\frac{(\bar{S}+\bar{Q})^2}{\Delta_c''}\{c_3(\alpha_3,E,\eta_g)-iD_3(\alpha_3,E,\eta_g)\}]\end{aligned}$$

Since $S_1(E,\eta_g)$ and $S_2(E,\eta_g)$ are complex, the energy spectrum is also complex in the presence of Gaussian band tails.

Therefore the magneto dispersion law in the presence of a quantizing magnetic field B which makes an angle θ with k_z axis can be written as

$$k_z^2=U_{18}(E,n,\eta_g) \tag{11.39a}$$

where

$$\begin{aligned}U_{18}(E,n,\eta_g)=&\left[\rho_{11}^2(E,\eta_g)\sin^2\theta+\rho_{12}^2(E,\eta_g)\cos^2\theta\right]\\&-\left[\frac{2eB}{\hbar}(n+\frac{1}{2})\left[\left(\rho_{11}^2(E,\eta_g)\rho_{12}(E,\eta_g)\right)^{-1}\{\rho_{11}^2(E,\eta_g)\sin^2\theta\right.\right.\\&\left.\left.+\rho_{12}^2(E,\eta_g)\cos^2\theta\}^{\frac{3}{2}}\right]\right]\end{aligned}$$

The EEM at the Fermi Level can be written from (11.39a) as

$$m^*(E_{FBHD}, n, \eta_g) = \frac{\hbar^2}{2} \text{ Real part of } [U_{18}(E_{FBHD}, n, \eta_g)]' \tag{11.39b}$$

Thus, the EEM is a function of Fermi energy, Landau quantum number and the scattering potential and the orientation of the applied quantizing magnetic field.

The electron concentration under the condition of extreme degeneracy can be written as

$$n_0 = \frac{eg_v B}{\pi^2 \hbar} \text{ Real part of} \sum_{n=0}^{n_{\max}} (U_{18}(\mathrm{E}_{FBHD}, \mathrm{n}, \eta_g))^{1/2} \tag{11.40a}$$

The magneto EP in this case is given by

$$J_B = \frac{\alpha_0 e^2 g_v B k_B T}{2\pi^2 \hbar^2} \text{ Real part of} \sum_{n=0}^{n_{\max}} F_0(\eta_{11A}) \tag{11.40b}$$

where

$$\eta_{11A} = (k_B T)^{-1} [\mathrm{E}_{FHBD} - (\mathrm{E}_{L11} + W - h\upsilon)]$$

where E_{L11} is the complex Landau sub-band energy which can be obtained from (11.39a) by substituting $\mathrm{k}_z = 0$ and $E = E_{L11}$

(e) Model of Foley and Langenberg

The dispersion relation of the conduction electrons of IV-VI semiconductors in accordance with Foley et al. can be written as [74]

$$E + \frac{E_g}{2} = E_-(k) + [[E_+(k) + \frac{E_g}{2}]^2 + P_\perp^2 k_s^2 + P_\parallel^2 k_z^2]^{\frac{1}{2}} \tag{11.41}$$

where $E_+(k) = \frac{\hbar^2 k_s^2}{2m_\perp^+} + \frac{\hbar^2 k_z^2}{2m_\parallel^+}, E_-(k) = \frac{\hbar^2 k_s^2}{2m_\perp^-} + \frac{\hbar^2 k_z^2}{2m_\parallel^-}$ represents the contribution from the interaction of the conduction and the valence band edge states with the more distant bands and the free electrons term, $\frac{1}{m_\perp^\pm} = \frac{1}{2}\left[\frac{1}{m_{tc}} \pm \frac{1}{m_{tv}}\right], \frac{1}{m_\parallel^\pm} = \frac{1}{2}\left[\frac{1}{m_{lc}} \pm \frac{1}{m_{lv}}\right]$

Following the methods as given in Chap. 1, the dispersion relation in heavily doped IV-VI materials in the present case is given by

$$\begin{aligned} &\left[\left[\gamma_3(E, \eta_g) + \frac{E_{g_0}}{2}\right] - \left[\frac{\hbar^2 k_s^2}{2m_\perp^-} + \frac{\hbar^2 k_z^2}{2m_\parallel^-}\right]\right]^2 = \left[\frac{\hbar^2 k_s^2}{2m_\perp^+} + \frac{\hbar^2 k_z^2}{2m_\parallel^+}\right]^2 + \frac{E_{g_0}^2}{4} \\ &+ E_{g_0}\left[\frac{\hbar^2 k_s^2}{2m_\perp^+} + \frac{\hbar^2 k_z^2}{2m_\parallel^+}\right] + P_\parallel^2 k_z^2 + P_\perp^2 k_s^2 \end{aligned} \tag{11.42}$$

Therefore the magneto-dispersion relation in heavily doped IV-VI materials can be written as

$$\gamma_3^2(E,\eta_g)+\frac{E_g^2}{4}+E_g\gamma_3(E,\eta_g)+[\frac{\hbar eB}{m_\perp^-}(n+\frac{1}{2})+\frac{\hbar^2 x}{2m_\parallel^-}]^2-2[\gamma_3(E,\eta_g)+\frac{E_g}{2}][\frac{\hbar eB(n+\frac{1}{2})}{m_\perp^-}+\frac{\hbar^2 x}{2m_\parallel^-}]$$
$$=[\frac{\hbar eB(n+\frac{1}{2})}{m_\perp^+}+\frac{\hbar^2 x}{2m_\parallel^+}]^2+E_g[\frac{\hbar eB}{m_\perp^+}(n+\frac{1}{2})+\frac{\hbar^2 x}{2m_\parallel^+}]+P_\parallel^2 x+P_\perp^2\frac{2eB}{\hbar}(n+\frac{1}{2}) \quad (11.43)$$

where $k_z^2=x$.

Therefore the magneto dispersion relation in IV-VI heavily doped materials, where unperturbed carriers follow the model of Foley et al. can be expressed as

$$k_z^2=U_{19}(E,n,\eta_g) \quad (11.44a)$$

where

$$U_{19}(E,n,\eta_g)=[2p_{91}]^{-1}[-q_{91}(E,n,\eta_g)+\{q_{91}^2(E,n,\eta_g)+4p_{91}R_{91}(E,n,\eta_g)\}^{\frac{1}{2}}]$$

$$p_{91}=\frac{\hbar^4}{4}[\frac{1}{\left(m_\parallel^+\right)^2}-\frac{1}{\left(m_\parallel^-\right)^2}],\ q_{91}(E,n,\eta_g)$$
$$=[\frac{\hbar^3 eB}{m_\perp^- m_\parallel^+}(n+\frac{1}{2})+\mathrm{p}_\parallel^2+\frac{\hbar^3 E_g}{2m_\parallel^+}-\frac{\hbar^2 eB(n+\frac{1}{2})}{m_\perp^- m_\parallel^+}+\frac{\hbar^2}{m_\parallel^-}(\gamma_3(E,\eta_g)+\frac{E_g}{2})]$$

$$R_{91}(E,\eta_g,n)=[\gamma_3^2(E,\eta_g)+E_g\gamma_3(E,\eta_g)-\frac{2\hbar eB}{m_\perp^-}(\gamma_3(E,\eta_g)+\frac{E_g}{2})(n+\frac{1}{2})$$
$$-E_g\frac{\hbar eB}{m_\perp^+}(n+\frac{1}{2})-P_\perp^2\cdot\frac{2eB}{\hbar}(n+\frac{1}{2})]$$

The EEM at the Fermi Level can be written from (11.44a) as

$$m^*(E_{FBHD},n,\eta_g)=\frac{\hbar^2}{2}U_{19}'(E_{FBHD},n,\eta_g) \quad (11.44b)$$

Thus, as noted already in this case also the EEM is a function of Fermi energy, Landau quantum number and the scattering potential.

The electron concentration under the condition of extreme degeneracy can be written as

$$n_0=\frac{g_v eB}{\pi^2\hbar}\sum_{n=0}^{n_{\max}}(U_{19}(\mathrm{E}_{FBHD},n,\eta_g))^{\frac{1}{2}} \quad (11.45a)$$

The magneto EP in this case is given by

$$J_B = \frac{\alpha_0 e^2 g_v B k_B T}{2\pi^2 \hbar^2} \sum_{n=0}^{n_{\max}} F_0(\eta_{12A}) \tag{11.45b}$$

where

$$\eta_{12A} = (k_B T)^{-1} [\mathrm{E}_{FHBD} - (\mathrm{E}_{L12} + W - h\upsilon)]$$

where E_{L12} is the Landau sub-band energy which can be obtained from (11.44a) by substituting $k_z = 0$ and $E = E_{L12}$.

11.2.5 The EP from HD Stressed Kane Type Semiconductors Under Magnetic Quantization

The dispersion relation of the conduction electrons in heavily doped Kane type semiconductors can be written following (1.206) of Chap. 1 as

$$\frac{k_x^2}{a_{\|}^2(\mathrm{E}, \eta_g)} + \frac{k_y^2}{b_{\|}^2(\mathrm{E}, \eta_g)} + \frac{k_z^2}{c_{\|}^2(\mathrm{E}, \eta_g)} = 1$$

where

$$a_{\|}(\mathrm{E},\eta_g) = \frac{1}{\sqrt{P_{\|}(\mathrm{E}, \eta_g)}}, \ \mathrm{b}_{\|}(\mathrm{E}, \eta_g) = \frac{1}{\sqrt{Q_{\|}(\mathrm{E}, \eta_g)}} \text{ and } c_{\|}(\mathrm{E}, \eta_g) = \frac{1}{\sqrt{S_{\|}(\mathrm{E}, \eta_g)}} \tag{11.46}$$

The electron energy spectrum in heavily doped Kane type semiconductors in the presence of an arbitrarily oriented quantizing magnetic field B which makes an angle $\bar{\alpha}_1$, $\bar{\beta}_1$ and $\bar{\gamma}_1$ with k_x, k_y and k_z axes respectively, can be written as

$$(k_z')^2 = U_{41}(\mathrm{E}, n, \eta_g) \tag{11.47a}$$

where $U_{41}(\mathrm{E}, n, \eta_g) = I_2(\mathrm{E},\eta_g)[1 - I_3(\mathrm{E},n, \eta_g)]$

$$I_2(\mathrm{E}, \eta_g) = [[a_{11}(\mathrm{E}, \eta_g)]^2 \cos^2 \bar{\alpha}_1 + [b_{11}(\mathrm{E}, \eta_g)]^2 \cos^2 \bar{\beta}_1 + [c_{11}(\mathrm{E}, \eta_g)]^2 \cos^2 \bar{\gamma}_1]$$

and

$$I_3(\mathrm{E}, n, \eta_g) = \frac{2eB}{\hbar}(n+\frac{1}{2})[a_{11}(\mathrm{E}, \eta_g) b_{11}(\mathrm{E}, \eta_g)] c_{11}(\mathrm{E}, \eta_g)]^{-1}[I_2(\mathrm{E}, \eta_g)]^{1/2}$$

The EEM at the Fermi Level can be written from (11.47a) as

$$m^*(E_{FBHD}, n, \eta_g) = \frac{\hbar^2}{2} U'_{41}(E_{FBHD}, n, \eta_g) \tag{11.47b}$$

From (11.47b) we observe that the EEM is a function of Fermi energy, Landau quantum number, the scattering potential and the orientation of the applied quantizing magnetic field.

The electron concentration under the condition of extreme degeneracy can be written as

$$n_0 = \frac{g_v eB}{\pi^2 \hbar} \sum_{n=0}^{n_{\max}} (U_{41}(\mathrm{E}_{FBHD}, n, \eta_g))^{\frac{1}{2}} \tag{11.48a}$$

The magneto EP in this case is given by

$$J_B = \frac{\alpha_0 e^2 g_v B k_B T}{2\pi^2 \hbar^2} \sum_{n=0}^{n_{\max}} F_0(\eta_{13A}) \tag{11.48b}$$

where

$$\eta_{13A} = (k_B T)^{-1}[\mathrm{E}_{FHBD} - (\mathrm{E}_{L13} + W - h\upsilon)]$$

where E_{L13} is the Landau sub-band energy which can be obtained from (11.47a) by substituting $k_z = 0$ and $E = E_{L13}$

11.2.6 The EP from HD Tellurium Under Magnetic Quantization

The magneto dispersion relation of the conduction electrons in HD Te can be expressed following (1.231) of Chap. 1 as

$$k_z^2 = U_{42\pm}(E, n, \eta_g) \tag{11.49a}$$

where

$$U_{42,\pm}(\mathrm{E},n,\eta_g) = (2\Psi_1^2)^{-1}[\{2\gamma_3(\mathrm{E},\eta_g)\Psi_1 + \Psi_3^2 - 4\Psi_1\Psi_2\frac{eB}{\hbar}(n+\frac{1}{2})\} - \{\Psi_3^4 + 4\Psi_1\Psi_3^2\gamma_3(\mathrm{E},\eta_g) + \frac{8eB}{\hbar}(n+\frac{1}{2})\Psi_1^2\Psi_4^2 - \Psi_1\Psi_2\Psi_3^2)\}^{-1/2}]$$

The EEM at the Fermi Level can be written from (11.49a) as

$$m_{\pm}^*(E_{FBHD},n,\eta_g) = \frac{\hbar^2}{2}U'_{42\pm}(E_{FBHD},n,\eta_g) \tag{11.49b}$$

Thus from (11.49b) we note that the EEM is a function of three variables namely Fermi energy, Landau quantum number and the scattering potential.

The electron concentration under the condition of extreme degeneracy can be written as

$$n_0 = \frac{g_v eB}{2\pi^2\hbar}\sum_{n=0}^{n_{\max}}(U_{42,\pm}(\mathrm{E}_{FBHD},n,\eta_g))^{\frac{1}{2}} \tag{11.49c}$$

The magneto EP in this case is given by

$$J_B = \frac{\alpha_0 e^2 g_v B k_B T}{2\pi^2\hbar^2}\sum_{n=0}^{n_{\max}}F_0(\eta_{14A}) \tag{11.49d}$$

where

$$\eta_{14A} = (k_B T)^{-1}[\mathrm{E}_{FHBD} - (\mathrm{E}_{L14} + W - h\upsilon)]$$

where E_{L14} is the Landau sub-band energy which can be obtained from (11.49a) by substituting $k_z = 0$ and $E = E_{L14}$.

11.2.7 The EP from HD Gallium Phosphide Under Magnetic Quantization

The magneto dispersion relation in HD GaP can be written following (1.248) of Chap. 1 as

$$k_z^2 = U_{43}(\mathrm{E},n,\eta_g) \tag{11.50a}$$

where

$$U_{43}(E,\, n, \eta_g) = (2b^2)^{-1}[\{2\gamma_3(\mathrm{E},\eta_g)b + c - 2Db - 4ab\frac{eB}{\hbar}(n+\frac{1}{2})\} + \{[c^2 + 4bc\gamma_3(\mathrm{E},\eta_g) + 4D^2b^2 - 4cDb] - \frac{8eB}{\hbar}(n+\frac{1}{2})(2ab^2D + 4\gamma_3(\mathrm{E},\,\eta_g)b^2a + abc - 2b^2a\gamma_3(\mathrm{E},\,\eta_g) - b^2c)\}^{-1/2}]$$

The EEM at the Fermi Level can be expressed from (11.50a) as

$$m^*(E_{FBHD}, n, \eta_g) = \frac{\hbar^2}{2} U'_{43}(E_{FBHD}, n, \eta_g) \tag{11.50b}$$

Thus, from (11.50b) it appears that the EEM is the function of Fermi energy, Landau quantum number and the scattering potential.

The electron concentration under the condition of extreme degeneracy can be written as

$$n_0 = \frac{g_v eB}{\pi^2\hbar} \sum_{n=0}^{n_{\max}} (U_{43}(\mathrm{E}_{FBHD}, n, \eta_g))^{\frac{1}{2}} \tag{11.50c}$$

The magneto EP in this case is given by

$$J_B = \frac{\alpha_0 e^2 g_v B k_B T}{2\pi^2\hbar^2} \sum_{n=0}^{n_{\max}} F_0(\eta_{15A}) \tag{11.50d}$$

where

$$\eta_{15A} = (k_B T)^{-1}[\mathrm{E}_{FHBD} - (\mathrm{E}_{L15} + W - h\upsilon)]$$

where E_{L15} is the Landau sub-band energy which can be obtained from (11.50a) by substituting $k_z = 0$ and $E = E_{L15}$.

11.2.8 The EP from HD Platinum Antimonide Under Magnetic Quantization

The magneto dispersion relation in HD $PtSb_2$ can be written following (1.270) as

$$k_z^2 = U_{44}(\mathrm{E},\, n, \eta_g) \tag{11.51a}$$

where

$$U_{44}(\mathrm{E},n,n_g) = \frac{1}{2T_{41}}[T_{71}(\mathrm{E},n,n_g) + \sqrt{T_{71}^2(\mathrm{E},n,n_g) + 4T_{41}T_{71}(\mathrm{E},n,n_g)}],$$

$$T_{71}(\mathrm{E},n,n_g) = T_{51}(\mathrm{E},n_g) - T_{31}\frac{2eB}{\hbar}(n+\frac{1}{2}) \text{ and } T_{72}(\mathrm{E},n,n_g)$$

$$= [T_{61}(\mathrm{E},n_g) + \frac{2eB}{\hbar}(n+\frac{1}{2})T_{21}(\mathrm{E},n_g)]$$

The EEM at the Fermi Level can be written from (11.51a) as

$$m^*(E_{FBHD},n,\eta_g) = \frac{\hbar^2}{2}U'_{44}(E_{FBHD},n,\eta_g) \quad (11.51b)$$

Thus, from the above equation we infer that the EEM is a function of Landau quantum number, the Fermi energy and the scattering potential.

The electron concentration under the condition of extreme degeneracy can be written as

$$n_0 = \frac{g_veB}{\pi^2\hbar}\sum_{n=0}^{n_{\max}}(U_{44}(\mathrm{E}_{FBHD},n,\eta_g))^{\frac{1}{2}} \quad (11.52a)$$

The magneto EP in this case is given by

$$J_B = \frac{\alpha_0 e^2 g_v B k_B T}{2\pi^2\hbar^2}\sum_{n=0}^{n_{\max}}F_0(\eta_{16A}) \quad (11.52b)$$

where $\eta_{16A} = (k_BT)^{-1}[\mathrm{E}_{FHBD} - (\mathrm{E}_{L16} + W - h\upsilon)]$

where E_{L16} is the Landau sub-band energy which can be obtained from (11.51a) by substituting $k_z = 0$ and $E = E_{L16}$.

11.2.9 The EP from HD Bismuth Telluride Under Magnetic Quantization

The magneto dispersion relation in HD Bi_2Te_3 can be written following (1.285) as

$$k_x^2 = U_{45}(E,\eta_g,n) \quad (11.53a)$$

where

$$U_{45}(E,\eta_g,n) = \frac{\gamma_2(E,\eta_g) - (n+\frac{1}{2})\frac{e\hbar B}{M_{31}}}{\bar{\omega}_1} \text{ and } M_{31} = \frac{m_0}{(\bar{\alpha}_{22}\bar{\alpha}_{33} - \frac{(\bar{\alpha}_{23})^2}{4})^{\frac{1}{2}}}$$

The EEM at the Fermi Level can be written from (11.53a) as

$$m^*(E_{FBHD},\eta_g) = \frac{\hbar^2}{2} U'_{45}(E_{FBHD},n,\eta_g) \quad (11.53b)$$

The electron concentration under the condition of extreme degeneracy can be written as

$$n_0 = \frac{g_v eB}{\pi^2\hbar} \sum_{n=0}^{n_{\max}} (U_{44}(\mathrm{E}_{FBHD},n,\eta_g))^{\frac{1}{2}} \quad (11.54a)$$

The magneto EP in this case is given by

$$J_B = \frac{\alpha_0 e^2 g_v B k_B T}{2\pi^2\hbar^2} \sum_{n=0}^{n_{\max}} F_0(\eta_{17A}) \quad (11.54b)$$

where

$$\eta_{17A} = (k_B T)^{-1}[\mathrm{E}_{FHBD} - (\mathrm{E}_{L17} + W - h\upsilon)]$$

where E_{L17} is the Landau sub-band energy which can be obtained from (11.53a) by substituting $k_z = 0$ and $E = E_{L17}$.

11.2.10 The EP from HD Germanium Under Magnetic Quantization

(a) Model of Cardona et al.

The magneto dispersion relation in HD Ge can be written following (1.300) as

$$k_x^2 = U_{46}(E,\eta_g,n) \quad (11.55a)$$

where

$$U_{46}(\mathrm{E},n,\eta_g) = \frac{2m_{\parallel}^*}{\hbar^2}[\gamma_3(\mathrm{E},\eta_g) + \frac{E_{go}}{2} - [\frac{E_{go}^2}{4} + \frac{E_{go}\hbar^2}{m_{\perp}^*}\frac{2eB}{\hbar}(n+\frac{1}{2})]]^{1/2}$$

The EEM at the Fermi Level can be written from (11.55a) as

$$m^*(E_{FBHD}, n, \eta_g) = \frac{\hbar^2}{2} U'_{46}(E_{FBHD}, n, \eta_g) \tag{11.55b}$$

From (11.55b) it appears that the EEM is a function of Fermi energy and Landau quantum number due to band non-parabolicity.

The electron concentration under the condition of extreme degeneracy can be written as

$$n_0 = \frac{g_v eB}{\pi^2 \hbar} \sum_{n=0}^{n_{\max}} (U_{46}(\mathrm{E}_{FBHD}, n, \eta_g))^{\frac{1}{2}} \tag{11.56a}$$

The magneto EP in this case is given by

$$J_B = \frac{\alpha_0 e^2 g_v B k_B T}{2\pi^2 \hbar^2} \sum_{n=0}^{n_{\max}} F_0(\eta_{18A}) \tag{11.56b}$$

where $\eta_{18A} = (k_B T)^{-1}[\mathrm{E}_{FHBD} - (\mathrm{E}_{L18} + W - h\upsilon)]$where E_{L18} is the Landau sub-band energy which can be obtained from (11.55a) by substituting $k_z = 0$ and $E = E_{L18}$

(b) Model of Wang and Ressler

The magneto dispersion relation in HD Ge can be written following (1.321) as

$$k_2^2 = U_{47}(\mathrm{E}, n, \eta_g) \tag{11.57a}$$

where

$$U_{47}(\mathrm{E}, n, \eta_g) = \left(\frac{m_\parallel^*}{\hbar^2 \bar{\alpha}_6}\right)[1 - \bar{\alpha}_5(n + \frac{1}{2})\hbar\omega_\perp - \{\theta_7(n) - 4\bar{\alpha}_6\gamma_3(\mathrm{E}, \eta_g)\}^{1/2}],\ \omega_\perp = \frac{eB}{m_\perp^*} \text{ and}$$

$$\theta_7(n) = [1 + (\bar{\alpha}_5)^2\{(n + \frac{1}{2})\hbar\omega_\perp\}^2 - 2\bar{\alpha}_5(n + \frac{1}{2})\hbar\omega_\perp) + 4\bar{\alpha}_6(n + \frac{1}{2})\hbar\omega_\perp - 4\bar{\alpha}_6\bar{\alpha}_4 - 4\bar{\alpha}_6\gamma_3\{(n + \frac{1}{2})\hbar\omega_\perp\}^2]$$

The EEM at the Fermi Level can be written from (11.57a) as

$$m^*(E_{FBHD}, n, \eta_g) = \frac{\hbar^2}{2} U'_{47}(E_{FBHD}, n, \eta_g) \tag{11.57b}$$

From (11.57b) we note that the mass is a function of Fermi energy and quantum number due to band non-parabolicity.

The electron concentration under the condition of extreme degeneracy can be written as

$$n_0 = \frac{g_v eB}{\pi^2 \hbar} \sum_{n=0}^{n_{\max}} (U_{47}(\mathrm{E}_{FBHD}, n, \eta_g))^{\frac{1}{2}} \tag{11.58a}$$

The magneto EP in this case is given by

$$J_B = \frac{\alpha_0 e^2 g_v B k_B T}{2\pi^2 \hbar^2} \sum_{n=0}^{n_{\max}} F_0(\eta_{19A}) \tag{11.58b}$$

where

$$\eta_{19A} = (k_B T)^{-1} [\mathrm{E}_{FHBD} - (\mathrm{E}_{L19} + W - h\upsilon)]$$

where E_{L19} is the Landau sub-band energy which can be obtained from (11.57a) by substituting $k_z = 0$ and $E = E_{L19}$.

11.2.11 The EP from HD Gallium Antimonide Under Magnetic Quantization

The magneto dispersion relation in HD GaSb can be written following (1.338) as

$$k_z^2 = U_{48}(\mathrm{E}, n, \eta_g) \tag{11.59a}$$

where

$$U_{48}(\mathrm{E}, n, \eta_g) = [\frac{-2eB}{\hbar}(n + \frac{1}{2}) + (2\alpha_9^2)^{-1} [\{2\alpha_9 \gamma_3^2(\mathrm{E}, \eta_g) + \alpha_9 \bar{\mathrm{E}}'_{g0}) + \frac{\alpha_{10}(\bar{\mathrm{E}}'_{g0})^2}{4}\}$$
$$- \{\alpha_9^2 (\bar{\mathrm{E}}'_{g0})^2 + \frac{\alpha_{10}^2 (\bar{\mathrm{E}}'_{g0})^4}{16} + \alpha_9 \alpha_{10} \gamma_3(\mathrm{E}, \eta_g)(\bar{\mathrm{E}}'_{g0})^2 + \frac{\alpha_9 \alpha_{10} (\bar{\mathrm{E}}'_{g0})^3}{2}\}^{1/2}]],$$
$$\alpha_9 = \frac{\hbar}{2m_0} \text{and}\, \alpha_{10} = \frac{\hbar}{\bar{E}'_{g0}}(\frac{1}{m_c} - \frac{1}{m_0})$$

The EEM at the Fermi Level can be written from (11.59a) as

$$m^*(E_{FBHD}, \eta_g) = \frac{\hbar^2}{2} U'_{48}(E_{FBHD}, n, \eta_g) \quad (11.59b)$$

The electron concentration under extreme degeneracy can be written as

$$n_0 = \frac{g_v eB}{\pi^2 \hbar} \sum_{n=0}^{n_{max}} (U_{48}(\mathrm{E}_{FBHD}, n, \eta_g))^{\frac{1}{2}} \quad (11.60a)$$

The magneto EP in this case is given by

$$J_B = \frac{\alpha_0 e^2 g_v B k_B T}{2\pi^2 \hbar^2} \sum_{n=0}^{n_{max}} F_0(\eta_{20A}) \quad (11.60b)$$

where

$$\eta_{20A} = (k_B T)^{-1} [\mathrm{E}_{FHBD} - (\mathrm{E}_{L20} + W - h\upsilon)]$$

where E_{L20} is the Landau sub-band energy which can be obtained from (11.59a) by substituting $k_z = 0$ and $E = E_{L20}$.

11.2.12 The EP from HD II-V Semiconductors Under Magnetic Quantization

The dispersion relation of the holes are given by [75–77]

$$E = \theta_1 k_x^2 + \theta_2 k_y^2 + \theta_3 k_z^2 + \delta_4 k_x \mp [\{\theta_5 k_x^2 + \theta_6 k_y^2 + \theta_7 k_z^2 + \delta_5 k_x\}^2 + G_3^2 k_y^2 + \Delta_3^2]^{\frac{1}{2}} \pm \Delta_3 \quad (11.61)$$

where k_x, k_y and k_z are expressed in the units of 10^{10} m^{-1},

$$\theta_1 = \frac{1}{2}(a_1 + b_1),\ \theta_2 = \frac{1}{2}(a_2 + b_2),\ \theta_3 = \frac{1}{2}(a_3 + b_3),\ \delta_4 = \frac{1}{2}(A + B),$$
$$\theta_5 = \frac{1}{2}(a_1 - b_1),\ \theta_6 = \frac{1}{2}(a_2 - b_2),\ \theta_7 = \frac{1}{2}(a_3 - b_3),\ \delta_5 = \frac{1}{2}(A - B),$$
$$a_i (i = 1, 2, 3, 4), b_i, A, B, G_3 \textit{ and } \Delta_3 \textit{ are system constants}$$

The hole energy spectrum in HD II-V semiconductors can be expressed following the method of Chap. 1 as

$$\gamma_3(\mathrm{E},\eta_g) = \theta_1 k_x^2 + \theta_2 k_y^2 + \theta_3 k_z^2 + \delta_4 k_x \mp [\{\theta_5 k_x^2 + \theta_6 k_y^2 + \theta_7 k_z^2 + \theta_5 k_x\}^2 + G_3^2 k_y^2 + \Delta_3^2]^{\frac{1}{2}} \pm \Delta_3 \tag{11.62}$$

the magneto dispersion law in HD II-V semiconductors assumes the form

$$k_y^2 = U_{49,\pm}(\mathrm{E}, n, \eta_g) \tag{11.63a}$$

where

$$U_{49,\pm}(\mathrm{E}, n, \eta_g) = [I_{35}\gamma_3(\mathrm{E}, \eta_g) + I_{36,\pm}(n) \pm [\gamma_3^2(\mathrm{E}, \eta_g) + \gamma_3(\mathrm{E}, \eta_g) I_{38,\pm}(n) + I_{39,\pm}(n)]^{\frac{1}{2}},$$
$$I_{35} = \frac{\theta_2}{(\theta_2^2 - \theta_5^2)}, I_{36,\pm}(n) = \frac{I_{33,\pm}(n)}{2(\theta_2^2 - \theta_5^2)}, I_{38,\pm}(n) = (4\theta_5^2)^{-1}[4\theta_2 I_{33,\pm}(n) + 8\theta_2^2 I_{31,\pm}(n) - \theta_5^2 I_{31,\pm}(n)],$$
$$I_{39,\pm}(n) = (4\theta_5^2)^{-1}[I_{33,\pm}^2(n) + 4\theta_2^3 I_{34,\pm}(n) - 4\theta_5^3 I_{34,\pm}(n)], I_{33,\pm}(n) = [G_3^2 + 2\theta_5 I_{32}(n) - 2\theta_2 I_{31,\pm}(n)],$$
$$I_{34,\pm}(n) = [I_{32,\pm}^2(n) + \Delta_3^2 - I_{31,\pm}(n)], I_{31,\pm}(n) = [(n+\frac{1}{2})\hbar\omega_{31} - \frac{\delta_4^2}{4\theta_1} \pm \Delta_3], I_{32}(n) = [(n+\frac{1}{2})\hbar\omega_{32} - \frac{\delta_5^2}{4\theta_5}],$$
$$\omega_{31} = \frac{eB}{\sqrt{M_{31}M_{32}}}, \omega_{32} = \frac{eB}{\sqrt{M_{33}M_{34}}}, M_{31} = \frac{\hbar^2}{2\theta_1}, M_{32} = \frac{\hbar^2}{2\theta_3}, M_{33} = \frac{\hbar^2}{2\theta_5} \text{ and } M_{34} = \frac{\hbar^2}{2\theta_7}.$$

The EEM at the Fermi Level can be written from (11.63a) as

$$m_{\pm}^*(E_{FBHD}, n, \eta_g) = \frac{\hbar^2}{2} U'_{49,\pm}(E_{FBHD}, n, \eta_g) \tag{11.63b}$$

From (11.63b) we note that the EEM is a function of Fermi energy, Landau quantum number and the scattering potential.

The electron concentration under extreme degeneracy can be written as

$$n_0 = \frac{g_v eB}{2\pi^2\hbar} \sum_{n=0}^{n_{\max}} (U_{49,\pm}(\mathrm{E}_{FBHD}, n, \eta_g))^{\frac{1}{2}} \tag{11.64a}$$

The magneto EP in this case is given by

$$J_B = \frac{\alpha_0 e^2 g_v B k_B T}{2\pi^2\hbar^2} \sum_{n=0}^{n_{\max}} F_0(\eta_{21A}) \tag{11.64b}$$

where

$$\eta_{21A} = (k_B T)^{-1}[\mathrm{E}_{FHBD} - (\mathrm{E}_{L21} + W - h\upsilon)]$$

where E_{L21} is the Landau sub-band energy which can be obtained from (11.63a) by substituting $k_z = 0$ and $E = E_{L21}$.

11.2.13 The EP from HD Lead Germanium Telluride Under Magnetic Quantization

The dispersion relation of the carriers in n-type $Pb_{1-x}Ga_xTe$ with x = 0.01 can be written following Vassilev [78] as

$$\left[E - 0.606k_s^2 - 0.0722k_z^2\right]\left[E + \bar{E}_g + 0.411k_s^2 + 0.0377k_z^2\right] = 0.23k_s^2 + 0.02k_z^2 \pm \left[0.06\bar{E}_g + 0.061k_s^2 + 0.0066k_z^2\right]k_s \quad (11.65)$$

where $\bar{E}_g(= 0.21 \text{ eV})$ is the energy gap for the transition point, the zero of the energy E is at the edge of the conduction band of the Γ point of the Brillouin zone and is measured positively upwards, k_x, k_y and k_z are in the units of 10^9 m^{-1}.

The magneto dispersion law in HD $Pb_{1-x}Ge_xTe$ can be expressed following the methods as given in Chap. 1 as

$$\left[\frac{2\theta_0(E,\eta_g)}{1+Erf(E/\eta_g)}\right] + \gamma_3(E,\eta_g)\left[\overline{E_{g_o}} - 0.345x - 0.390\frac{eB}{\hbar}(n+\frac{1}{2})\right] = \frac{0.46eB}{\hbar}(n+\frac{1}{2}) + 0.02x \pm \left[0.06\overline{E_{g_o}} + 0.122\frac{eB}{\hbar}(n+\frac{1}{2}) + 0.0066x\right]\left(\frac{2eB}{\hbar}(n+\frac{1}{2})\right)^{\frac{1}{2}} + \left[\overline{E_{g_o}} + \frac{0.822eB}{\hbar}(n+\frac{1}{2}) + 0.377x\right]\left[\frac{1.212eB}{\hbar}(n+\frac{1}{2}) + 0.722x\right] \quad (11.66)$$

The (11.66) assumes the form

$$k_z^2 = U_{50,\mp}(E,n,\eta_g) \quad (11.67a)$$

where

$$U_{50,\mp}(E,n,\eta_g) = (2p_{10})^{-1}\left[q_{10}(E,n,\eta_g) - \left[q_{10}^2(E,n,\eta_g) + 4p_{10}R_{10,\mp}(E,n,\eta_g)\right]^{\frac{1}{2}}\right]$$

$$p_{10} = (0.377 \times 0.722), q_{10}(E,n,\eta_g) = [0.02 + 0.345\gamma_3(E,\eta_g) \pm 0.0066(\frac{2eB}{\hbar}(n+\frac{1}{2}))^{\frac{1}{2}} + 0.377 \times \frac{1.212eB}{\hbar}(n+\frac{1}{2}) + 0.722[\bar{E}_{go} + 0.822\frac{eB}{\hbar}(n+\frac{1}{2})]]$$

and

$$R_{10,\mp}(E,n,\eta_g) = [\frac{2\theta_0(E,\eta_g)}{1+Eof(E/\eta_g)} + \gamma_3(E,\eta_g)[\bar{E}_{go} - 0.390\frac{eB}{\hbar}(n+\frac{1}{2})] \mp (0.06\bar{E}_{go} + 0.122\frac{eB}{\hbar}(n+\frac{1}{2}))(\frac{2eB}{\hbar}(n+\frac{1}{2}))^{\frac{1}{2}} - (\bar{E}_{go} + 0.822\frac{eB}{\hbar}(n+\frac{1}{2}))\frac{1.212eB}{\hbar}(n+\frac{1}{2}) - \frac{0.46eB}{\hbar}(n+\frac{1}{2})].$$

The EEM at the Fermi Level can be written from (11.67a) as

$$m^*_{\mp}(E_{FBHD},\eta_g) = \frac{\hbar^2}{2} U'_{50,\mp}(E_{FBHD},n,\eta_g) \tag{11.67b}$$

Thus from (11.67b) we note that the EEM is a function of the Fermi energy, Landau quantum number and the scattering potential.

The electron concentration under extreme degeneracy can be written as

$$n_0 = \frac{g_v eB}{2\pi^2\hbar} \sum_{n=0}^{n_{\max}} (U_{50,\mp}(\mathrm{E}_{FBHD},n,\eta_g))^{\frac{1}{2}} \tag{11.68a}$$

The magneto EP in this case is given by

$$J_B = \frac{\alpha_0 e^2 g_v B k_B T}{2\pi^2\hbar^2} \sum_{n=0}^{n_{\max}} F_0(\eta_{22A}) \tag{11.68b}$$

where

$$\eta_{22A} = (k_B T)^{-1}[\mathrm{E}_{FHBD} - (\mathrm{E}_{L22} + W - h\upsilon)]$$

where E_{L22} is the Landau sub-band energy which can be obtained from (11.67a) by substituting $k_z = 0$ and $E = E_{L22}$.

11.3 Open Research Problems

(R.11.1) Investigate the EP in the presence of an arbitrarily oriented quantizing magnetic field for all the materials as given in problems in R.1.1 of Chap. 1 in the presence of the Gaussian type band tails.

(R.11.2) Investigate the EP in the presence of an arbitrarily oriented quantizing magnetic field in HD nonlinear optical semiconductors by including broadening and the electron spin. Study all the special cases for HD III-V, ternary and quaternary materials in this context.

(R.11.3) Investigate the EPs for HD IV-VI, II-VI and stressed Kane type compounds in the presence of an arbitrarily oriented quantizing magnetic field by including broadening and electron spin.

(R.11.4) Investigate the EP for all the materials as stated in R.1.1 of Chap. 1 in the presence of an arbitrarily oriented quantizing magnetic field by including broadening and electron spin under the condition of heavily doping.

(R.11.5) Investigate the EP in the presence of an arbitrarily oriented quantizing magnetic field and crossed electric fields in HD nonlinear optical semiconductors by including broadening and the electron spin. Study all the special cases for HD III-V, ternary and quaternary materials in this context.

(R.11.6) Investigate the EPs for HD IV-VI, II-VI and stressed Kane type compounds in the presence of an arbitrarily oriented quantizing magnetic field and crossed electric field by including broadening and electron spin.

(R.11.7) Investigate the EP for all the materials as stated in R.1.1 of Chap. 1 in the presence of an arbitrarily oriented quantizing magnetic field and crossed electric field by including broadening and electron spin under the condition of heavy doping.

(R.11.8) Investigate the 2D EP in QWs of HD nonlinear optical, III-V, II-VI, IV-VI and stressed Kane type semiconductors.

(R.11.9) Investigate the 2D EP for HD QWs of all the materials as considered in problems R.1.1.

(R.11.10) Investigate the 2D EP in the presence of an arbitrarily oriented non-quantizing magnetic field for the QWs of HD nonlinear optical semiconductors by including the electron spin. Study all the special cases for III-V, ternary and quaternary materials in this context.

(R.11.11) Investigate the EPs in QWs of HD IV-VI, II-VI and stressed Kane type compounds in the presence of an arbitrarily oriented non-quantizing magnetic field by including the electron spin.

(R.11.12) Investigate the 2D EP for HD QWs of all the materials as stated in R.1.1 of Chap. 1 in the presence of an arbitrarily oriented magnetic field by including electron spin and broadening.

(R.11.13) Investigate the EP for all the problems of R.1.1 under an additional arbitrarily oriented electric field in the presence of heavy doping.

(R.11.14) Investigate the EP for all the problems of R.1.1 under the arbitrarily oriented crossed electric and magnetic fields in the presence of heavy doping.

(R.11.15) Investigate the 2D EP for all the problems in R.1.1 the presence of finite potential well under the conditions of formation of band tails and applied external parallel magnetic field.

(R.11.16) Investigate the 2D EP for all the problems in R.1.1 the presence of parabolic potential well under the conditions of formation of band tails and applied external parallel magnetic field.

(R.11.17) Investigate the 2D EP for all the problems in R.1.1 the presence of circular potential well under the conditions of formation of band tails and applied external parallel magnetic field.

(R.11.18) Investigate the 2D EP for accumulation layers of HD nonlinear optical, III-V, IV-VI, II-VI and stressed Kane type semiconductors in the presence of an arbitrary electric quantization.

(R.11.19) Investigate the 2D EP in accumulation layers of all the materials as stated in R.1.1 of Chap. 1 under the condition of heavy doping and in the presence of electric quantization along arbitrary direction.

(R.11.20) Investigate the 2D EP in the presence of an arbitrarily oriented electric quantization for accumulation layers of HD nonlinear optical semiconductors. Study all the special cases for III-V, ternary and quaternary materials in this context.

(R.11.21) Investigate the 2D EPs in accumulation layers of HD IV-VI, II-VI and stressed Kane type compounds in the presence of an arbitrarily oriented electric quantization.

(R.11.22) Investigate the 2D EP in accumulation layers of all the materials as stated in R.1.1 of Chap. 1 in the presence of an arbitrarily oriented quantizing electric field under the conditions of formation of band tails and applied external parallel magnetic field.

(R.11.23) Investigate the 2D EP in the presence of an arbitrarily oriented magnetic field in accumulation layers of HD nonlinear optical semiconductors by including the electron spin. Study all the special cases for HD III-V, ternary and quaternary materials in this context.

(R.11.24) Investigate the 2D EPs in accumulation layers of HD IV-VI, II-VI and stressed Kane type compounds in the presence of an arbitrarily oriented non-quantizing magnetic field by including the electron spin.

(R.11.25) Investigate the 2D EP in accumulation layers of all the materials as stated in R.1.1 of Chap. 1 in the presence of an arbitrarily oriented non-quantizing magnetic field by including electron spin and heavy doping.

(R.11.26) Investigate the 2D EP in accumulation layers for all the problems from R.11.22 to R.11.26 in the presence of an additional arbitrarily oriented electric field.

(R.11.27) Investigate the 2D EP in accumulation layers for all the problems from R.11.22 to R.11.26 in the presence of arbitrarily oriented crossed electric and magnetic fields.

(R.11.28) Investigate the 2D EP in accumulation layers for all the problems from R.11.22 to R.11.26 in the presence of surface states.

(R.11.29) Investigate the 2D EP in accumulation layers for all the problems from R.11.22 to R.11.26 in the presence of hot electron effects.

(R.11.30) Investigate the 2D EP in accumulation layers for all the problems from R.11.22 to R.11.26 by including the occupancy of the electrons in various electric subbands.

(R.11.31) Investigate the 2D EP in Doping superlattices of HD nonlinear optical, III-V, II-VI, IV-VI and stressed Kane type materials.

(R.11.32) Investigate the 2D EP in Doping superlattices of all types of materials as discussed in problem R.1.1 as given in Chap. 1 under the conditions of formation of band tails and applied external parallel magnetic field.

(R.11.33) Investigate the 2D EP in the presence of an arbitrarily oriented non-quantizing magnetic field for Doping superlattices of HD nonlinear optical semiconductors by including the electron spin. Study all the special cases for HD III-V, ternary and quaternary materials in this context.

(R.11.34) Investigate the 2D EPs in Doping superlattices of HD IV-VI, II-VI and stressed Kane type compounds in the presence of an arbitrarily oriented non-quantizing magnetic field by including the electron spin.

(R.11.35) Investigate the 2D EP for Doping superlattices of all the materials as stated in R.1.1 of Chap. 1 in the presence of an arbitrarily oriented non-quantizing magnetic field by including electron spin under the conditions of formation of band tails and applied external parallel magnetic field.

(R.11.36) Investigate the 2D EP for all the problems from R.11.32 to R.11.35 in the presence of an additional arbitrarily oriented non-quantizing electric field.

(R.11.37) Investigate the 2D EP for all the problems from R.11.32 to R.11.35 in the presence of arbitrarily oriented crossed electric and magnetic fields.

(R.11.38) Investigate all the problems from R.11.1 to R.11.37, in the presence of arbitrarily oriented light waves and magnetic quantization.

(R.11.39) Investigate all the problems from R.11.1 up to R.11.37 in the presence of exponential, Kane, Halperin and Lax and Bonch-Bruevich band tails [79].

(R.11.40) Investigate all the problems of this chapter by removing all the mathematical approximations and establishing the uniqueness conditions in each case.

(R.11.41)
(a) Investigate the EP in all the bulk semiconductors as considered in this appendix in the presence of defects and magnetic quantization.
(b) Investigate the EP as defined in (R.11.2.1) in the presence of an arbitrarily oriented quantizing magnetic field including broadening and the electron spin (applicable under magnetic quantization) for all the bulk semiconductors whose unperturbed carrier energy spectra are defined in Chap. 1.

(R.11.42) (R.11.42) Investigate the EP as defined in (R.11.2.1) in the presence of quantizing magnetic field under an arbitrarily oriented (a) non-uniform electric field and (b) alternating electric field respectively for all the semiconductors whose unperturbed carrier energy spectra are defined in Chap. 1 by including spin and broadening respectively.

(R.11.43) Investigate the EP as defined in (R.11.2.1) under an arbitrarily oriented alternating quantizing magnetic field by including broadening and the electron spin for all the semiconductors whose unperturbed carrier energy spectra as defined in Chap. 1.

(R.11.44) Investigate the EP as defined in (R.11.2.1) under an arbitrarily oriented alternating quantizing magnetic field and crossed alternating electric field by including broadening and the electron spin for all the semiconductors whose unperturbed carrier energy spectra as defined in Chap. 1.

(R.11.45) Investigate the EP as defined in (R.11.2.1) under an arbitrarily oriented alternating quantizing magnetic field and crossed alternating non-uniform electric field by including broadening and the electron spin whose for all the semiconductors unperturbed carrier energy spectra as defined in Chap. 1.

(R.11.46) Investigate the EP as defined in (R.11.2.1) in the presence and absence of an arbitrarily oriented alternating quantizing magnetic field under exponential, Kane, Halperin, Lax and Bonch-Bruevich band tails [69] for all the semiconductors whose unperturbed carrier energy spectra as defined in Chap. 1 by including spin and broadening (applicable under magnetic quantization).

(R.11.47) Investigate the EP as defined in (R.11.2.1) in the presence of an arbitrarily oriented quantizing magnetic field for all the semiconductors as defined in (R.11.2.6) under an arbitrarily oriented (a) non-uniform electric field and (b) alternating electric field respectively whose unperturbed carrier energy spectra as defined in Chap. 1.

(R.11.48) Investigate the EP as defined in (R.11.2.1) under an arbitrarily oriented alternating quantizing magnetic field by including broadening and the electron spin for all semiconductors whose unperturbed carrier energy spectra as defined in Chap. 1.

(R.11.49) Investigate the EP as defined in (R.11.2.1) under an arbitrarily oriented alternating quantizing magnetic field and crossed alternating electric field by including broadening and the electron spin for all the semiconductors whose unperturbed carrier energy spectra as defined in Chap. 1.

(R.11.50) Investigate all the appropriate problems of this section under magnetic quantization after proper modifications introducing new theoretical formalisms for functional, negative refractive index, macro molecular, organic and magnetic materials.

(R.11.51) Investigate all the appropriate problems of this section for HD p-InSb, p-CuCl and stressed semiconductors under magnetic quantization having diamond structure valence bands whose dispersion relations of the carriers in bulk semiconductors are given by Cunningham [79], Yekimov et al. [80] and Roman [81] respectively.

(R.11.52) Investigate all the problems of this section by removing all the mathematical approximations and establishing the respective appropriate uniqueness conditions.

References

1. N. Miura, *Physics of Semiconductors in High Magnetic Fields*, Series on Semiconductor Science and Technology (Oxford University Press, USA, 2007)
2. K.H.J. Buschow, F.R. de Boer, *Physics of Magnetism and Magnetic Materials* (Springer, New York, 2003)
3. D. Sellmyer, R. Skomski (eds.), *Advanced Magnetic Nanostructures* (Springer, New York, 2005)
4. J.A.C. Bland, B. Heinrich (eds.), *Ultrathin Magnetic Structures III: Fundamentals of Nanomagnetism (Pt. 3)* (Springer, Germany, 2005)
5. B.K. Ridley, *Quantum Processes in Semiconductors*, 4th edn. (Oxford publications, Oxford, 1999)
6. J.H. Davies, *Physics of Low Dimensional Semiconductors* (Cambridge University Press, UK, 1998)
7. S. Blundell, *Magnetism in Condensed Matter*,Oxford Master Series in Condensed Matter Physics (Oxford University Press, USA, 2001)
8. C. Weisbuch, B. Vinter, *Quantum Semiconductor Structures: Fundamentals and Applications* (Academic Publishers, USA, 1991)
9. D. Ferry, *Semiconductor Transport* (CRC, USA, 2000)
10. M. Reed (ed.), *Semiconductors and Semimetals: Nanostructured Systems* (Academic Press, USA, 1992)
11. T. Dittrich, *Quantum Transport and Dissipation* (Wiley-VCH Verlag GmbH, Germany, 1998)
12. A.Y. Shik, *Quantum Wells: Physics and Electronics of Twodimensional Systems* (World Scientific, USA, 1997)
13. K.P. Ghatak, M. Mondal, Zietschrift fur Naturforschung A **41a**, 881 (1986)
14. K.P. Ghatak, M. Mondal, J. Appl. Phys. **62**, 922 (1987)
15. K.P. Ghatak, S.N. Biswas, Phys. Stat. Sol. (b) **140**, K107 (1987)
16. K.P. Ghatak, M. Mondal, J. Mag. Mag. Mat. **74**, 203 (1988)
17. K.P. Ghatak, M. Mondal, Phys. Stat. Sol. (b) **139**, 195 (1987)
18. K.P. Ghatak, M. Mondal, Phys. Stat. Sol. (b) **148**, 645 (1988)
19. K.P. Ghatak, B. Mitra, A. Ghoshal, Phys. Stat. Sol. (b) **154**, K121 (1989)
20. K.P. Ghatak, S.N. Biswas, J. Low Temp. Phys. **78**, 219 (1990)
21. K.P. Ghatak, M. Mondal, Phys. Stat. Sol. (b) **160**, 673 (1990)
22. K.P. Ghatak, B. Mitra, Phys. Lett. **A156**, 233 (1991)
23. K.P. Ghatak, A. Ghoshal, B. Mitra, Nouvo Cimento **D13D**, 867 (1991)
24. K.P. Ghatak, M. Mondal, Phys. Stat. Sol. (b) **148**, 645 (1989)
25. K.P. Ghatak, B. Mitra, Int. J. Elect. **70**, 345 (1991)
26. K.P. Ghatak, S.N. Biswas, J. Appl. Phys. **70**, 299 (1991)
27. K.P. Ghatak, A. Ghoshal, Phys. Stat. Sol. (b) **170**, K27 (1992)
28. K.P. Ghatak, Nouvo Cimento **D13D**, 1321 (1992)
29. K.P. Ghatak, B. Mitra, Int. J. Elect. **72**, 541 (1992)
30. K.P. Ghatak, S.N. Biswas, Nonlinear Opt. **4**, 347 (1993)
31. K.P. Ghatak, M. Mondal, Phys. Stat. Sol. (b) **175**, 113 (1993)
32. K.P. Ghatak, S.N. Biswas, Nonlinear Opt. **4**, 39 (1993)
33. K.P. Ghatak, B. Mitra, Nouvo Cimento **15D**, 97 (1993)
34. K.P. Ghatak, S.N. Biswas, Nanostruct. Mater. **2**, 91 (1993)
35. K.P. Ghatak, M. Mondal, Phys. Stat. Sol. (b) **185**, K5 (1994)
36. K.P. Ghatak, B. Goswami, M. Mitra, B. Nag, Nonlinear Opt. **16**, 9 (1996)
37. K.P. Ghatak, M. Mitra, B. Goswami, B. Nag, Nonlinear Opt. **16**, 167 (1996)
38. K.P. Ghatak, B. Nag, Nanostruct. Mater. **10**, 923 (1998)
39. D. Roy Choudhury, A.K. Choudhury, K.P. Ghatak, A.N. Chakravarti, Phys. Stat. Sol. (b) **98**, K141 (1980)
40. A.N. Chakravarti, K.P. Ghatak, A. Dhar, S. Ghosh, Phys. Stat. Sol. (b) **105**, K55 (1981)

41. A.N. Chakravarti, A.K. Choudhury, K.P. Ghatak, Phys. Stat. Sol. (a) **63**, K97 (1981)
42. A.N. Chakravarti, A.K. Choudhury, K.P. Ghatak, S. Ghosh, A. Dhar, Appl. Phys. **25**, 105 (1981)
43. A.N. Chakravarti, K.P. Ghatak, G.B. Rao, K.K. Ghosh, Phys. Stat. Sol. (b) **112**, 75 (1982)
44. A.N. Chakravarti, K.P. Ghatak, K.K. Ghosh, H.M. Mukherjee, Phys. Stat. Sol. (b) **116**, 17 (1983)
45. M. Mondal, K.P. Ghatak, Phys. Stat. Sol. (b) **133**, K143 (1984)
46. M. Mondal, K.P. Ghatak, Phys. Stat. Sol. (b) **126**, K47 (1984)
47. M. Mondal, K.P. Ghatak, Phys. Stat. Sol. (b) **126**, K41 (1984)
48. M. Mondal, K.P. Ghatak, Phys. Stat. Sol. (b) **129**, K745 (1985)
49. M. Mondal, K.P. Ghatak, Phys. Scr. **31**, 615 (1985)
50. M. Mondal, K.P. Ghatak, Phys. Stat. Sol. (b) **135**, 239 (1986)
51. M. Mondal, K.P. Ghatak, Phys. Stat. Sol. (b) **93**, 377 (1986)
52. M. Mondal, K.P. Ghatak, Phys. Stat. Sol. (b) **135**, K21 (1986)
53. M. Mondal, S. Bhattacharyya, K.P. Ghatak, Appl. Phys. **A42A**, 331 (1987)
54. S.N. Biswas, N. Chattopadhyay, K.P. Ghatak, Phys. Stat. Sol. (b) **141**, K47 (1987)
55. B. Mitra, K.P. Ghatak, Phys. Stat. Sol. (b) **149**, K117 (1988)
56. B. Mitra, A. Ghoshal, K.P. Ghatak, Phys. Stat. Sol. (b) **150**, K67 (1988)
57. M. Mondal, K.P. Ghatak, Phys. Stat. Sol. (b) **147**, K179 (1988)
58. M. Mondal, K.P. Ghatak, Phys. Stat. Sol. (b) **146**, K97 (1988)
59. B. Mitra, A. Ghoshal, K.P. Ghatak, Phys. Stat. Sol. (b) **153**, K209 (1989)
60. B. Mitra, K.P. Ghatak, Phys. Lett. **142A**, 401 (1989)
61. B. Mitra, A. Ghoshal, K.P. Ghatak, Phys. Stat. Sol. (b) **154**, K147 (1989)
62. B. Mitra, K.P. Ghatak, Sol. State Elect. **32**, 515 (1989)
63. B. Mitra, A. Ghoshal, K.P. Ghatak, Phys. Stat. Sol. (b) **155**, K23 (1989)
64. B. Mitra, K.P. Ghatak, Phys. Lett. **135A**, 397 (1989)
65. B. Mitra, K.P. Ghatak, Phys. Lett. **A146A**, 357 (1990)
66. B. Mitra, K.P. Ghatak, Phys. Stat. Sol. (b) **164**, K13 (1991)
67. S.N. Biswas, K.P. Ghatak, Int. J. Elect. **70**, 125 (1991)
68. P.R. Wallace, Phys. Stat. Sol. (b) **92**, 49 (1979)
69. B.R. Nag, *Electron Transport in Compound Semiconductors, Springer Series in Solid-State Sciences*, vol. 11 (Springer, Germany, 1980)
70. K.P. Ghatak, S. Bhattacharya, D. De, *Einstein Relation in Compound Semiconductors and Their Nanostructures*, vol. 116, Springer Series in Materials Science (Springer, Germany, 2009)
71. C.C. Wu, C.J. Lin, J. Low Temp. Phys. **57**, 469 (1984)
72. M.H. Chen, C.C. Wu, C.J. Lin, J. Low Temp. Phys. **55**, 127 (1984)
73. E. Bangert, P. Kastner, Phys. Stat. Sol (b) **61**, 503 (1974)
74. G.M.T. Foley, P.N Langenberg. Phys. Rev. B **15B**, 4850 (1977)
75. M. Singh, P.R. Wallace, S.D. Jog, E. Arushanov, J. Phys. Chem. Solids **45**, 409 (1984)
76. Y. Yamada, Phys. Soc. Japan **35**, 1600 (1973)
77. Y. Yamada, Phys. Soc. Japan **37**, 606 (1974)
78. L.A. Vassilev, Phys. Stat. Sol. (b) **121**, 203 (1984)
79. R.W. Cunningham, Phys. Rev. **167**, 761 (1968)
80. A.I. Yekimov, A.A. Onushchenko, A.G. Plyukhin, L. Al Efros, J. Expt. Theor. Phys. **88**, 1490 (1985)
81. B.J. Roman, A.W. Ewald, Phys. Rev. **B5**, 3914 (1972)

Chapter 12
Appendix B: The EP from Super-Lattices of HD Semiconductors Under Magnetic Quantization

12.1 Introduction

In this chapter, we shall study the EP under magnetic quantization in III-V, II-VI, IV-VI, HgTe/CdTe and strained layer, HDSLs with graded interfaces in Sects. 12.2.1–12.2.5 respectively. From Sects. 12.2.6 to 12.2.10, we shall investigate the same DMR in III-V, II-VI, IV-VI, HgTe/CdTe and strained layer, HD effective mass SLs. The last Sect. 12.2.3 contains open research problems.

12.2 Theoretical Background

12.2.1 The EP from HD III-V Superlattices with Graded Interfaces Under Magnetic Quantization

The magneto EP in this case is given by

$$J_B = \frac{\alpha_0 e^2 g_v B k_B T}{2\pi^2 \hbar^2} \text{ Real part of} \sum_{n=0}^{n_{\max}} F_0(\eta_{1SLHD}) \tag{12.1}$$

where

$$\eta_{1SLHD} = (k_B T)^{-1} [\mathrm{E}_{F1} - (\mathrm{E}_{SL1} + W - h\upsilon)]$$

where E_{SL1} is the complex Landau sub-band energy which can be obtained from (4.3) by substituting $k_z = 0$ and $E = E_{SL1}$ *is the Fermi energy in this case.*

It appears that the evaluation of EP in this case requires an expression of the electron concentration which is given by

K.P. Ghatak, *Einstein's Photoemission*, Springer Tracts in Modern Physics 262,
DOI 10.1007/978-3-319-11188-9_12

$$n_0 = \frac{g_v eB}{\pi^2 \hbar} \sum_{n=0}^{n_{\max}} [\phi_{1C}(E_{F1}, n) + \phi_{2C}(E_{F1}, n)] \tag{12.2}$$

where

$$\phi_{1C}(E_{F1}, n) = \left[\left(G_{8E_{F1},n} + \sqrt{G^2_{8E_{F1},n} - H_{8E_{F1},n}}\right)/2\right]^{1/2},$$

$$\phi_{2C}(E_{F1}, n) = \sum_{r=1}^{s} \theta_{2r,1}[\phi_{1C}(E_{F1}, n)],\ \theta_{2r,i} = 2(k_B T)^{2r}\left(1 - 2^{1-2r}\right)\zeta(2r)\frac{\partial^{2r}}{\partial E_{Fi}^{2r}}$$

and $i = 1, 2, 3, \ldots$

12.2.2 The EP from HD II-VI Superlattices with Graded Interfaces Under Magnetic Quantization

The magneto EP in this case is given by

$$J_B = \frac{\alpha_0 e^2 g_v B k_B T}{2\pi^2 \hbar^2} \text{Real part of} \sum_{n=0}^{n_{\max}} F_0(\eta_{2SLHD}) \tag{12.3}$$

where

$$\eta_{2SLHD} = (k_B T)^{-1}[E_{F2} - (E_{SL2} + W - h\nu)]$$

where E_{SL2} is the complex Landau sub-band energy which can be obtained from (4.10) by substituting $k_z = 0$ and $E = E_{SL2}$ *and* E_{F2} *is the Fermi energy in this case.*

The electron concentration in this case can be written as

$$n_0 = \frac{g_v eB}{\pi^2 \hbar} \sum_{n=0}^{n_{\max}} [\phi_{3C}(E_{F2}, n) + \phi_{4C}(E_{F2}, n)] \tag{12.4}$$

where

$$\phi_{3C}(E_{F2}, n) = \left[\left(G_{19E_{F2},n} + \sqrt{G^2_{19E_{F2},n} - H_{19E_{F2},n}}\right)/2\right]^{1/2}$$

and $\phi_{4C}(E_{F2}, n) = \sum_{r=1}^{s} \theta_{2r,2}[\phi_{3C}(E_{F2}, n)]$

12.2.3 The EP from HD IV-VI Superlattices with Graded Interfaces Under Magnetic Quantization

The magneto EP in this case is given by

$$J_B = \frac{\alpha_0 e^2 g_v B k_B T}{2\pi^2 \hbar^2} \sum_{n=0}^{n_{\max}} F_0(\eta_{3SLHD}) \tag{12.5}$$

where

$$\eta_{3SLHD} = (k_B T)^{-1} [E_{F3} - (E_{SL3} + W - h\upsilon)]$$

where E_{SL3} is the Landau sub-band energy which can be obtained from (4.16) by substituting $k_z = 0$ and $E = E_{SL3}$ *and* E_{F3} *is the Fermi energy in this case.*

The electron concentration is given by

$$n_0 = \frac{g_v e B}{\pi^2 \hbar} \sum_{n=0}^{n_{\max}} [\phi_{5C}(E_{F3}, n) + \phi_{6C}(E_{F3}, n)] \tag{12.6}$$

where

$$\phi_{6C}(E_{F3}, n) = \sum_{r=1}^{s} \theta_{2r,3} [\phi_{6C}(E_{F3}, n)]$$

12.2.4 The EP from HD HgTe/CdTe Superlattices with Graded Interfaces Under Magnetic Quantization

The magneto EP in this case is given by

$$J_B = \frac{\alpha_0 e^2 g_v B k_B T}{2\pi^2 \hbar^2} \text{Real part of} \sum_{n=0}^{n_{\max}} F_0(\eta_{4SLHD}) \tag{12.7}$$

where

$$\eta_{4SLHD} = (k_B T)^{-1} [E_{F4} - (E_{SL4} + W - h\upsilon)]$$

where E_{SL4} is the complex Landau sub-band energy which can be obtained from (4.21) by substituting $k_z = 0$ and $E = E_{SL4}$ *and* E_{F4} *is the Fermi energy in this case.*

The electron concentration is given by

$$n_0 = \frac{g_v eB}{\pi^2 \hbar} \sum_{n=0}^{n_{\max}} [\phi_{7C}(E_{F4}, n) + \phi_{8C}(E_{F4}, n)] \tag{12.8}$$

where

$$\phi_{7C}(E_{F4}, n) = \left[\left(G_{192E_{F4},n} + \sqrt{G^2_{192E_{F4},n} - H_{192E_{F4},n}}\right) \Big/ 2\right]^{1/2}$$

$$\text{and } \phi_{8C}(E_{F4}, n) = \sum_{r=1}^{s} \theta_{2r,4}[\phi_{7C}(E_{F4}, n)]$$

12.2.5 The EP from HD Stained Layer Superlattices with Graded Interfaces Under Magnetic Quantization

The magneto EP in this case is given by

$$J_B = \frac{\alpha_0 e^2 g_v B k_B T}{2\pi^2 \hbar^2} \sum_{n=0}^{n_{\max}} F_0(\eta_{6SLHD}) \tag{12.9}$$

where

$$\eta_{6SLHD} = (k_B T)^{-1}[E_{F6} - (E_{SL6} + W - h\upsilon)]$$

where E_{SL6} is the Landau sub-band energy which can be obtained from (4.26) by substituting $k_z = 0$ and $E = E_{SL6}$ *and* EF_6 *is the Fermi energy in this case.*

The electron concentration is given by

$$n_0 = \frac{g_v eB}{\pi^2 \hbar} \sum_{n=0}^{n_{\max}} [\phi_{9C}(E_{F6}, n) + \phi_{10C}(E_{F6}, n)] \tag{12.10}$$

where

$$\phi_{8C}(E_{F6}, n) = \sum_{r=1}^{s} \theta_{2r,6}[\phi_{7C}(E_{F6}, n)]$$

12.2.6 The EP from HD III-V Effective Mass Superlattices Under Magnetic Quantization

The magneto EP in this case is given by

$$J_B = \frac{\alpha_0 e^2 g_v B k_B T}{2\pi^2 \hbar^2} \text{Real part of} \sum_{n=0}^{n_{\max}} F_0(\eta_{7SLHD}) \tag{12.11}$$

where

$$\eta_{7SLHD} = (k_B T)^{-1}[\mathrm{E}_{F7} - (\mathrm{E}_{SL7} + W - h\upsilon)]$$

where E_{SL7} is the complex Landau sub-band energy which can be obtained from (4.31) by substituting $k_z = 0$ and $E = E_{SL7}$ *and* E_{F7} *is the Fermi energy in this case.*

The electron concentration is given by

$$n_0 = \frac{g_v eB}{\pi^2 \hbar} \sum_{n=0}^{n_{\max}} [\phi_{11}(E_{F7}, n) + \phi_{12}(E_{F7}, n)] \tag{12.12}$$

where

$$\phi_{11}(E_{F7}, n) = \left[\left(\delta_{7E_{F7},n} + \sqrt{\delta^2_{7E_{F7},n} - \delta_{8E_{F7},n}}\right)\Big/2\right]^{1/2}$$

$$\text{and } \phi_{12}(E_{F7}, n) = \sum_{r=1}^{s} \theta_{2r,7}[\phi_{11}(E_{F7}, n)].$$

12.2.7 The EP from HD II-VI Effective Mass Superlattices Under Magnetic Quantization

The magneto EP in this case is given by

$$J_B = \frac{\alpha_0 e^2 g_v B k_B T}{2\pi^2 \hbar^2} \text{Real part of} \sum_{n=0}^{n_{\max}} F_0(\eta_{8SLHD}) \tag{12.13}$$

where

$$\eta_{8SLHD} = (k_B T)^{-1}[E_{F8} - (E_{SL8} + W - h\upsilon)]$$

where E_{SL8} is the complex Landau sub-band energy which can be obtained from (4.35) by substituting $k_z = 0$ and $E = E_{SL8}$ *and* E_{F8} *is the Fermi energy in this case.*

The electron concentration is given by

$$n_0 = \frac{g_v eB}{\pi^2 \hbar} \sum_{n=0}^{n_{\max}} [\phi_{13}(E_{F8}, n) + \phi_{14}(E_{F8}, n)] \tag{12.14}$$

where

$$\phi_{13}(E_{F8}, n) = \left[\left(\Delta_{13E_{F8},n} + \sqrt{\Delta^2_{13E_{F8},n} - \Delta_{14E_{F8},n}} \right) \Big/ 2 \right]^{1/2},$$

$$\text{and } \phi_{14}(E_{F8}, n) = \sum_{r=1}^{s} \theta_{2r,7}[\phi_{13}(E_{F8}, n)]$$

12.2.8 The EP from HD IV-VI Effective Mass Superlattices Under Magnetic Quantization

The magneto EP in this case is given by

$$J_B = \frac{\alpha_0 e^2 g_v B k_B T}{2\pi^2 \hbar^2} \sum_{n=0}^{n_{\max}} F_0(\eta_{9SLHD}) \tag{12.15}$$

where

$$\eta_{9SLHD} = (k_B T)^{-1} [\mathrm{E}_{F9} - (\mathrm{E}_{SL9} + W - h\upsilon)]$$

where E_{SL9} is the Landau sub-band energy which can be obtained from (4.39) by substituting $k_z = 0$ and $E = E_{SL9}$ *and* E_{F9} *is the Fermi energy in this case.*

The electron concentration is given by

$$n_0 = \frac{g_v eB}{\pi^2 \hbar} \sum_{n=0}^{n_{\max}} [\phi_{15}(E_{F9}, n) + \phi_{16}(E_{F9}, n)] \tag{12.16}$$

where

$$\phi_{16}(E_{F9}, n) = \sum_{r=1}^{s} \theta_{2r,6}[\phi_{15}(E_{F9}, n)]$$

12.2.9 The EP from HD HgTe/CdTe Effective Mass Superlattices Under Magnetic Quantization

The magneto EP in this case is given by

$$J_B = \frac{\alpha_0 e^2 g_v B k_B T}{2\pi^2 \hbar^2} \text{Real part of} \sum_{n=0}^{n_{\max}} F_0(\eta_{10SLHD}) \tag{12.17}$$

where

$$\eta_{10SLHD} = (k_B T)^{-1} [E_{F10} - (E_{SL10} + W - h\upsilon)]$$

where E_{SL10} is the is the complex Landau sub-band energy which can be obtained from (4.43) by substituting $k_z = 0$ and $E = E_{SL10}$ *and* E_{F10} *is the Fermi energy in this case.*

The electron concentration is given by

$$n_0 = \frac{g_v eB}{\pi^2 \hbar} \sum_{n=0}^{n_{\max}} [\phi_{17}(E_{F10}, n) + \phi_{18}(E_{F10}, n)] \tag{12.18}$$

where

$$\phi_{17}(E_{F10}, n) = \left[\left(\Delta_{13HE_{F10},n} + \sqrt{\Delta^2_{13HE_{F10},n} - \Delta_{14HE_{F10},n}}\right) \Big/ 2\right]^{1/2}$$

and $\phi_{18}(E_{F10}, n) = \sum_{r=1}^{s} \theta_{2r,7}[\phi_{17}(E_{F10}, n)]$.

12.2.10 The EP from HD Stained Layer Effective Mass Superlattices Under Magnetic Quantization

The magneto EP in this case is given by

$$J_B = \frac{\alpha_0 e^2 g_v B k_B T}{2\pi^2 \hbar^2} \sum_{n=0}^{n_{\max}} F_0(\eta_{11SLHD}) \tag{12.19}$$

where

$$\eta_{11SLHD} = (k_B T)^{-1} [\mathrm{E}_{F11} - (\mathrm{E}_{SL11} + W - h\upsilon)]$$

where E_{SL11} is the Landau sub-band energy which can be obtained from (4.48) by substituting $k_z = 0$ and $E = E_{SL11}$ *and* E_{F11} *is the Fermi energy in this case.*

The electron concentration is given by

$$n_0 = \frac{g_v eB}{\pi^2 \hbar} \sum_{n=0}^{n_{\max}} [\phi_{19}(E_{F11}, n) + \phi_{20}(E_{F11}, n)] \tag{12.20}$$

where

$$\phi_{20}(E_{F11}, n) = \sum_{r=1}^{s} \theta_{2r,6}[\phi_{19}(E_{F11}, n)].$$

12.3 Open Research Problem

(R.12.1) Investigate the EP for all types of HD super-lattices under alternating magnetic field and alternating non uniform electric field applied simultaneously in arbitrary directions.

Chapter 13
Appendix C: The EP from HDS and Their Nano-structures Under Cross-Fields Configuration

13.1 Introduction

The influence of crossed electric and quantizing magnetic fields on the transport properties of semiconductors having various band structures are relatively less investigated as compared with the corresponding magnetic quantization, although, the cross-fields are fundamental with respect to the addition of new physics and the related experimental findings. In 1966, Zawadzki and Lax [1] formulated the electron dispersion law for III-V semiconductors in accordance with the two band model of Kane under cross fields configuration which generates the interest to study this particular topic of semiconductor science in general [2–38].

In Sect. 13.2.1 of theoretical background, the EP in HD nonlinear optical materials in the presence of crossed electric and quantizing magnetic fields has been investigated by formulating the electron dispersion relation. The Sect. 13.2.2 reflects the study of the EP in HD III-V, ternary and quaternary compounds as a special case of Sect. 13.2.1. The Sect. 13.2.3 contains the study of the EP for the HD II-VI semiconductors in the present case. In Sect. 13.2.4, the EP under cross field configuration in HD IV-VI semiconductors has been investigated in accordance with the models of the Cohen, the Lax nonparabolic ellipsoidal and the parabolic ellipsoidal respectively. In the Sect. 13.2.5, the EP for the HD stressed Kane type semiconductors has been investigated. The Sect. 13.3 contain three open research problems.

K.P. Ghatak, *Einstein's Photoemission*, Springer Tracts in Modern Physics 262,
DOI 10.1007/978-3-319-11188-9_13

13.2 Theoretical Background

13.2.1 The EP from HD Nonlinear Optical Semiconductors Under Cross-Fields Configuration

The (1.2) of Chap. 1 can be expressed as

$$T_{22}(\mathrm{E},\eta_g)=\frac{p_s^2}{2m_\perp^*}+\frac{p_z^2}{2M_{||}}T_{22}(\mathrm{E},\eta_g)[T_{21}(\mathrm{E},\eta_g)]^{-1} \tag{13.1}$$

where, $p_s=\hbar k_s$ and $p_z=\hbar k_z$

We know that from electromagnetic theory that,

$$\vec{B}=\nabla\times\vec{A} \tag{13.2}$$

where, $\vec{A}$ is the vector potential. In the presence of quantizing magnetic field B along z direction, the (13.2) assumes the form

$$0\hat{i}+0\hat{j}+B\hat{k}=\begin{vmatrix}\hat{i} & \hat{j} & \hat{k}\\ \frac{\partial}{\partial x} & \frac{\partial}{\partial y} & \frac{\partial}{\partial z}\\ A_x & A_y & A_z\end{vmatrix} \tag{13.3}$$

where $\hat{i}$, $\hat{j}$ and $\hat{k}$ are orthogonal triads. Thus, we can write

$$\begin{aligned}\frac{\partial A_z}{\partial y}-\frac{\partial A_y}{\partial z}&=0\\ \frac{\partial A_x}{\partial z}-\frac{\partial A_z}{\partial x}&=0\\ \frac{\partial A_y}{\partial x}-\frac{\partial A_x}{\partial y}&=B\end{aligned} \tag{13.4}$$

This particular set of equations is being satisfied for $A_x=0$, $A_y=Bx$ and $A_z=0$.

Therefore in the presence of the electric field E_o along x axis and the quantizing magnetic field B along z axis for the present case following (13.1) one can approximately write,

$$T_{22}(\mathrm{E},\eta_g)+|e|E_o\hat{x}\rho(E,\eta_g)=\frac{\hat{p}_x^2}{2m_\perp^*}+\frac{(\hat{p}_y-|e|B\hat{x})^2}{2m_\perp^*}+\frac{\hat{p}_z^2}{2a(E,\eta_g)} \tag{13.5}$$

where

$$\rho(E) \equiv \frac{\partial}{\partial E}[\mathrm{T}_{22}(E,\eta_g)] \text{ and } a(E,\eta_g) \equiv m_{||}^*[\mathrm{T}_{22}(E,\eta_g)]^{-1}[\mathrm{T}_{21}(E,\eta_g)]$$

Let us define the operator $\hat{\theta}$ as

$$\hat{\theta} = -\hat{p}_y + |e|B\hat{x} - \frac{m_\perp^* E_o \rho(E,\eta_g)}{B} \tag{13.6}$$

Eliminating the operator $\hat{x}$, between (13.5) and (13.6) the dispersion relation of the conduction electron in tetragonal semiconductors in the presence of cross fields configuration is given by

$$\mathrm{T}_{22}(E,\eta_g) = \left[\left((n+\frac{1}{2})\hbar\omega_{01}\right) + \left(\frac{[\hbar k_z(E)]^2}{2a(E,\eta_g)}\right) - \left(\frac{E_o \hbar k_y \rho(E,\eta_g)}{B}\right) - \left(\frac{M_\perp \rho^2(E,\eta_g)E_o^2}{2B^2}\right)\right] \tag{13.7}$$

where,

$$\omega_{01} = \frac{|e|B}{m_\perp^*} \tag{13.8a}$$

The EEMs along Z and Y directions can, respectively be expressed from (13.7) as

$$\begin{aligned} m_z^*(\bar{E}_{FBHD},\eta_g,n,E_0) = {} & \text{Real part of } [a'(\bar{E}_{FBHD},\eta_g)[T_{22}(\bar{E}_{FBHD},\eta_g) - (n+\frac{1}{2})\hbar\omega_{01} \\ & + \frac{M_\perp \rho^2(\bar{E}_{FBHD},\eta_g)E_0^2}{2B^2}] + a(\bar{E}_{FBHD},\eta_g)[T_{22}'(\bar{E}_{FBHD},\eta_g) \\ & + \frac{M_\perp \rho(\bar{E}_{FBHD},\eta_g)\rho'(\bar{E}_{FBHD},\eta_g)E_0^2}{B^2}]] \end{aligned} \tag{13.8b}$$

and

$$\begin{aligned} m_y^*(\bar{E}_{FBHD},\eta_g,n,E_0) = {} & (\frac{B}{E_0})^2 \text{Real part of } [\rho(\bar{E}_{FBHD},\eta_g)]^{-3}[T_{22}(\bar{E}_{FBHD},\eta_g) \\ & - (n+\frac{1}{2})\hbar\omega_{01} + \frac{M_\perp \rho^2(\bar{E}_{FBHD},\eta_g)E_0^2}{2B^2}][[\rho(\bar{E}_{FBHD},\eta_g) \\ & [T_{22}'(\bar{E}_{FBHD},\eta_g) + \frac{M_\perp \rho(\bar{E}_{FBHD},\eta_g)\rho'(\bar{E}_{FBHD},\eta_g)E_0^2}{B^2}] \\ & - [T_{22}(\bar{E}_{FBHD},\eta_g) - (n+\frac{1}{2})\hbar\omega_{01} \\ & + \frac{M_\perp \rho^2(\bar{E}_{FBHD},\eta_g)E_0^2}{2B^2}]\rho'(\bar{E}_{FBHD},\eta_g)]] \end{aligned} \tag{13.8c}$$

where $\bar{E}_{FBHD}$ is the Fermi energy in the presence of cross-fields configuration and heavy doping as measured from the edge of the conduction band in the vertically upward direction in the absence of any quantization.

When $E_0 \to 0$, $m_z^*(\bar{E}_{FBHD}, \eta_g, n, E_0) \to \infty$, which is a physically justified result. The dependence of the EEM along y direction on the Fermi energy, electric field, magnetic field and the magnetic quantum number is an intrinsic property of cross fields together with the fact in the present case of heavy doping, the EEM exists in the band gap. Another characteristic feature of cross field is that various transport coefficients will be sampled dimension dependent. These conclusions are valid for even isotropic parabolic energy bands and cross fields introduce the index dependent anisotropy in the effective mass.

The formulation of EP requires the expression of the electron concentration which can, in general, be written excluding the electron spin as

$$n_o = \frac{-g_v}{L_x \pi^2} \sum_{n=0}^{n_{\max}} \int_{\bar{E}_0}^{\infty} I(E, \eta_g) \frac{\partial f_o}{\partial E} dE \tag{13.9}$$

where L_x is the sample length along x direction, $\bar{E}_0$ is determined by the equation

$$I(\bar{E}_0, \eta_g) = 0$$

where

$$I(E, \eta_g) = \int_{x_l(E,\eta_g)}^{x_h(E,\eta_g)} k_z(E) dk_y \tag{13.10}$$

in which, $x_l(E, \eta_g) \equiv \frac{-E_0 M_\perp \rho(E, \eta_g)}{\hbar B}$ and $x_h(E, \eta_g) \equiv \frac{|e| B L_x}{\hbar} + x_l(E, \eta_g)$

Thus we get

$$\begin{aligned} I(E, \eta_g) &= \frac{2}{3}\left[\frac{B\sqrt{2a(E, \eta_g)}}{\hbar^2 E_0 \rho(E, \eta_g)}\right]\left[\left[T_{22}(E, \eta_g) - \left(n + \frac{1}{2}\right)\frac{h|e|B}{m_\perp^*} + |e| E_0 L_x \rho(E, \eta_g) - \frac{m_\perp^* E_0^2 [\rho(E, \eta_g)]^2}{2B^2}\right]^{\frac{3}{2}}\right. \\ &\left. - \left[T_{22}(E, \eta_g) - \left(n + \frac{1}{2}\right)\frac{h|e|B}{m_\perp^*} - \frac{m_\perp^* E_0^2 [\rho(E, \eta_g)]^2}{2B^2}\right]^{\frac{3}{2}}\right] \end{aligned} \tag{13.11}$$

Therefore the electron concentration is given by

$$n_0 = \left(\frac{2g_v B\sqrt{2}}{3L_x\pi^2\hbar^2 E_o}\right) \text{Real part of} \sum_{n=0}^{n_{\max}} \left[T_{41HD}\left(n, \bar{E}_{FBHD}, \eta_g\right) + T_{42HD}\left(n, \bar{E}_{FBHD}, \eta_g\right)\right] \tag{13.12}$$

where

$$T_{41}\left(n, \bar{E}_{FBHD}, \eta_g\right) \equiv \frac{\sqrt{a\left(\bar{E}_{FBHD}, \eta_g\right)}}{\rho\left(\bar{E}_{FBHD}, \eta_g\right)}\left[\left[T_{22}\left(\bar{E}_{FBHD}, \eta_g\right) - \left(n+\frac{1}{2}\right)\frac{\hbar|e|B}{M_\perp} + |e|E_0 L_x \rho\left(\bar{E}_{FBHD}, \eta_g\right) - \frac{m_\perp^* E_0^2\left[\rho\left(\bar{E}_{FBHD}, \eta_g\right)\right]^2}{2B^2}\right]^{3/2}\right.$$
$$\left. - \left[T_{22}\left(\bar{E}_{FBHD}, \eta_g\right) - \left(n+\frac{1}{2}\right)\frac{\hbar|e|B}{m_\perp^*} - \frac{m_\perp^* E_0^2\left[\rho\left(\bar{E}_{FBHD}, \eta_g\right)\right]^2}{2B^2}\right]^{3/2}\right]$$

where $\overline{E}_{FBHD}$ is the Fermi energy in this case and

$$T_{42HD}\left(n, \bar{E}_{FBHD}, \eta_g\right) \equiv \sum_{r=1}^{s}\left[L(r) T_{41HD}\left(n, \bar{E}_{FBHD}, \eta_g\right)\right]$$

The EP in DMR this case can be written as

$$J(E_0, B) = \left[\frac{-\alpha_0 e g_v}{4L_x\pi^2\hbar}\text{Real part of}\sum_{n=0}^{n_{\max}} \int_{e''+W-h\upsilon}^{\infty} I(E, \eta_g)\frac{\partial E}{\partial k_z}\frac{\partial f(E)}{\partial E} dE\right] \tag{13.13}$$

where e'' is the Landau sub-band energy under cross-fields configuration.

Thus using (13.12), (13.13) and the allied definitions we can study the EP in this case.

13.2.2 *The EP from HD Kane Type III-V Semiconductors Under Cross-Fields Configuration*

(a) Under the conditions $\delta = 0$, $\Delta_{\|} = \Delta_{\perp} = \Delta$ and $m_{\|}^* = m_{\perp}^* = m_c$, (13.7) assumes the form

$$T_{33}\left(E, \eta_g\right) = \left(n+\frac{1}{2}\right)\hbar\omega_0 + \frac{\left[\hbar k_z(E)\right]^2}{2m_c}$$
$$- \frac{E_0}{B}\hbar k_y\left\{T_{33}\left(E, \eta_g\right)\right\}' - \frac{m_c E_0^2\left[\left\{T_{33}\left(E, \eta_g\right)\right\}'\right]^2}{2B^2} \tag{13.14a}$$

where $T_{33}(E, \eta_g) = T_{31}(E, \eta_g) + iT_{32}(E, \eta_g)$

The use of (13.14a) leads to the expressions of the EEM's along z and y directions as

$$m_z^*(\bar{E}_{FBHD},\eta_g,n,E_0) = m_c \text{ Real part of } [\{T_{33}(\bar{E}_{FBHD},\eta_g)\}'' + \frac{m_c E_0^2\{T_{33}(\bar{E}_{FBHD},\eta_g)\}'\{T_{33}(\bar{E}_{FBHD},\eta_g)\}''}{B^2}] \quad (13.14b)$$

$$m_y^*(\bar{E}_{FBHD},\eta_g,n,E_0) = (\frac{B}{E_0})^2 \text{ Real part of } [\{T_{33}(\bar{E}_{FBHD},\eta_g)\}']^{-1}[T_{33}(\bar{E}_{FBHD},\eta_g) - (n+\frac{1}{2})\hbar\omega_0 + \frac{m_c E_0^2[\{T_{33}(\bar{E}_{FBHD},\eta_g)\}']^2}{2B^2}][\frac{-\{T_{33}(\bar{E}_{FBHD},\eta_g)\}''}{[\{T_{33}(\bar{E}_{FBHD},\eta_g)\}']^2}[T_{33}(\bar{E}_{FBHD},\eta_g) - (n+\frac{1}{2})\hbar\omega_0 + \frac{m_c E_0^2[\{T_{33}(\bar{E}_{FBHD},\eta_g)\}']^2}{2B^2}] + 1 + \frac{m_c E_0^2\{T_{33}(\bar{E}_{FBHD},\eta_g)\}''}{B^2}] \quad (13.14c)$$

The Landau energy $(\bar{E}_{n_1})$ can be written as

$$T_{33}(\bar{E}_{n_1},\eta_g) = (n+\frac{1}{2})\hbar\omega_0 - \frac{m_c E_0^2[\{T_{33}(\bar{E}_{n_1},\eta_g)\}']^2}{2B^2} \quad (13.15)$$

The electron concentration in this case assumes the form

$$n_0 = \frac{2g_v B\sqrt{2m_c}}{3L_x\pi^2\hbar^2 E_0} \text{ Real part of } \sum_{n=0}^{n_{\max}} [T_{43HD}(n,\bar{E}_{FB},\eta_g) + T_{44HD}(n,\bar{E}_{FB},\eta_g)] \quad (13.16)$$

where,

$$T_{43HD}(n,\bar{E}_{FBHD},\eta_g) \equiv \Big[[T_{33}(\bar{E}_{FB},\eta_g) - (n+\frac{1}{2})\hbar\omega_0 - \frac{m_c E_0^2}{2B^2}[\{T_{33}(\bar{E}_{FB},\eta_g)\}']^2 + |e|E_0L_x[\{T_{33}(\bar{E}_{FB},\eta_g)\}']]^{3/2} - [T_{33}(\bar{E}_{FB},\eta_g) - (n+\frac{1}{2})\hbar\omega_0 - \frac{m_c E_0^2}{2B^2}[\{T_{33}(\bar{E}_{FB},\eta_g)\}']^2]^{3/2}\Big]\frac{1}{[\{T_{33}(\bar{E}_{FB},\eta_g)\}']}$$ and

$$T_{44HD}(n,\bar{E}_{FBHD},\eta_g) \equiv \sum_{r=1}^{s}[L(r)T_{43HD}(n,\bar{E}_{FBHD},\eta_g)].$$

Thus using (13.13), (13.16) and the allied definitions we can study the EP in this case.

(b) Under the condition $\Delta \gg E_g$, (13.14a–c) assumes the form

$$\gamma_2(E,\eta_g) = \left(n+\frac{1}{2}\right)\hbar\omega_0 - \frac{E_0}{B}\hbar k_y \gamma_2'(E,\eta_g) - \frac{m_c E_0^2}{2B^2}\left(\gamma_2'(E,\eta_g)\right)^2 + \frac{[\hbar k_z(E)]^2}{2m_c} \quad (13.17a)$$

The use of (13.17a) leads to the expressions of the EEM's along z and y directions as

$$m_z^*(\bar{E}_{FBHD},\eta_g,n,E_0) = m_c[\{\gamma_2(\bar{E}_{FBHD},\eta_g)\}'' + \frac{m_c E_0^2\{\gamma_2(\bar{E}_{FBHD},\eta_g)\}'\{\gamma_2(\bar{E}_{FBHD},\eta_g)\}''}{B^2}] \quad (13.17b)$$

$$\begin{aligned} m_y^*(\bar{E}_{FBHD},\eta_g,n,E_0) = {} & (\frac{B}{E_0})^2 \frac{1}{[\{\gamma_2(\bar{E}_{FBHD},\eta_g)\}']}[\gamma_2(\bar{E}_{FBHD},\eta_g) - (n+\frac{1}{2})\hbar\omega_0 \\ & + \frac{m_c E_0^2[\{\gamma_2(\bar{E}_{FBHD},\eta_g)\}']^2}{2B^2}][\frac{-\{\gamma_2(\bar{E}_{FBHD},\eta_g)\}''}{[\{\gamma_2(\bar{E}_{FBHD},\eta_g)\}']^2}[\gamma_2(\bar{E}_{FBHD},\eta_g) \\ & - (n+\frac{1}{2})\hbar\omega_0 + \frac{m_c E_0^2[\{\gamma_2(\bar{E}_{FBHD},\eta_g)\}']^2}{2B^2}] + 1 + \frac{m_c E_0^2\{\gamma_2(\bar{E}_{FBHD},\eta_g)\}''}{B^2}] \end{aligned} \quad (13.17c)$$

The Landau energy $(\bar{E}_{n_2})$ can be written as

$$\gamma_2(\bar{E}_{n_2},\eta_g) = \left(n+\frac{1}{2}\right)\hbar\omega_0 - \frac{m_c E_0^2}{2B^2}\left(\gamma_2'(\bar{E}_{n_2},\eta_g)\right)^2 \quad (13.18)$$

The expressions for n_0 in this case assume the forms

$$n_0 = \frac{2g_v B\sqrt{2m_c}}{3L_x\pi^2\hbar^2 E_0}\sum_{n=0}^{n_{\max}}\left[T_{47HD}\left(n,\bar{E}_{FBHD},\eta_g\right) + T_{48HD}\left(n,\bar{E}_{FBHD},\eta_g\right)\right] \quad (13.19)$$

where

$$\begin{aligned} T_{47HD}(n,\bar{E}_{FBHD},\eta_g) \equiv {} & \Big[[\gamma_2(\bar{E}_{FBHD},\eta_g) - (n+\frac{1}{2})\hbar\omega_0 + |e|E_0 L_x(\gamma_2'(\bar{E}_{FBHD},\eta_g)) - \frac{m_c E_0^2}{2B^2}(\gamma_2'(\bar{E}_{FBHD},\eta_g))^2]^{3/2} \\ & - [(\gamma_2(\bar{E}_{FBHD},\eta_g)) - (n+\frac{1}{2})\hbar\omega_0 - \frac{m_c E_0^2}{2B^2}(\gamma_2'(\bar{E}_{FBHD},\eta_g))^2]^{3/2}\Big][\gamma_2'(\bar{E}_{FBHD},\eta_g)]^{-1} \end{aligned}$$

and

$$T_{48HD}\left(n, \bar{E}_{FBHD}, \eta_g\right) \equiv \sum_{r=0}^{s} L(r)\left[T_{47HD}\left(n, \bar{E}_{FBHD}, \eta_g\right)\right].$$

Thus using (13.13), (13.19) and the allied definitions we can study the EP in this case.

(c) For $\alpha \to 0$ and we can write,

$$\gamma_3(E, \eta_g) = \left(n + \frac{1}{2}\right)\hbar\omega_0 - \frac{E_0}{B}\hbar k_y \gamma_3'(E, \eta_g) - \frac{m_c E_0^2}{2B^2}\left(\gamma_3'(E, \eta_g)\right)^2 + \frac{[\hbar k_z(E)]^2}{2m_c} \tag{13.20a}$$

The use of (13.20a) leads to the expressions of the EEM's along z and y directions as

$$m_z^*(\bar{E}_{FBHD}, \eta_g, n, E_0) = m_c\left[\{\gamma_3(\bar{E}_{FBHD}, \eta_g)\}'' + \frac{m_c E_0^2\{\gamma_3(\bar{E}_{FBHD}, \eta_g)\}'\{\gamma_3(\bar{E}_{FBHD}, \eta_g)\}''}{B^2}\right] \tag{13.20b}$$

$$\begin{aligned} m_y^*(\bar{E}_{FBHD}, \eta_g, n, E_0) = &\left(\frac{B}{E_0}\right)^2 \frac{1}{\left[\{\gamma_3(\bar{E}_{FBHD}, \eta_g)\}'\right]}[\gamma_3(\bar{E}_{FBHD}, \eta_g) - \left(n + \frac{1}{2}\right)\hbar\omega_0 \\ &+ \frac{m_c E_0^2\left[\{\gamma_3(\bar{E}_{FBHD}, \eta_g)\}'\right]^2}{2B^2}][\frac{-\{\gamma_3(\bar{E}_{FBHD}, \eta_g)\}''}{[\{\gamma_3(\bar{E}_{FBHD}, \eta_g)\}']^2}[\gamma_3(\bar{E}_{FBHD}, \eta_g) \\ &- (n + \frac{1}{2})\hbar\omega_0 + \frac{m_c E_0^2[\{\gamma_3(\bar{E}_{FBHD}, \eta_g)\}']^2}{2B^2}] + 1 + \frac{m_c E_0^2\{\gamma_3(\bar{E}_{FBHD}, \eta_g)\}''}{B^2}] \end{aligned} \tag{13.20c}$$

The Landau energy $(\bar{E}_{n_3})$ can be written as

$$\gamma_3(\bar{E}_{n_3}, \eta_g) = \left(n + \frac{1}{2}\right)\hbar\omega_0 - \frac{m_c E_0^2}{2B^2}\left(\gamma_3'(\bar{E}_{n_3}, \eta_g)\right)^2 \tag{13.21}$$

The expressions for n_0 in this case assume the forms

$$n_0 = \frac{2g_v B\sqrt{2m_c}}{3L_x \pi^2 \hbar^2 E_0}\sum_{n=0}^{n_{max}}\left[T_{49HD}\left(n, \bar{E}_{FBHD}, \eta_g\right) + T_{50HD}\left(n, \bar{E}_{FBHD}, \eta_g\right)\right] \tag{13.22}$$

where

$$T_{49HD}(n, \bar{E}_{FBHD}, \eta_g) \equiv \left[\left[\gamma_3(\bar{E}_{FBHD}, \eta_g) - \left(n+\frac{1}{2}\right)\hbar\omega_0 + |e|E_0 L_x(\gamma_3'(\bar{E}_{FBHD}, \eta_g)) - \frac{m_c E_0^2}{2B^2}(\gamma_3'(\bar{E}_{FBHD}, \eta_g))^2\right]^{3/2} - \left[(\gamma_3(\bar{E}_{FBHD}, \eta_g)) - \left(n+\frac{1}{2}\right)\hbar\omega_0 - \frac{m_c E_0^2}{2B^2}(\gamma_3'(\bar{E}_{FBHD}, \eta_g))^2\right]^{3/2}\right][\gamma_3'(\bar{E}_{FBHD}, \eta_g)]^{-1}$$

and

$$T_{50HD}(n, \bar{E}_{FBHD}, \eta_g) \equiv \sum_{r=0}^{s} L(r)[T_{49HD}(n, \bar{E}_{FBHD}, \eta_g)].$$

Thus using (13.13), (13.22) and the allied definitions we can study the EP in this case.

13.2.3 The EP from HD II-VI Semiconductors Under Cross-Fields Configuration

The electron energy spectrum in HD II-VI semiconductors in the presence ofelectric field E_0 along x direction and quantizing magnetic field B along z direction can approximately be written as

$$\gamma_3(E, \eta_g) = \beta_1(n, E_0) - \frac{E_0}{B}\hbar k_y \gamma_3'(E, \eta_g) - \frac{m_\parallel^* E_0^2}{2B^2}(\gamma_3'(E, \eta_g))^2 + \frac{[\hbar k_z(E)]^2}{2m_\parallel^*} \quad (13.23a)$$

where

$$\beta_1(n, E_0) \equiv \left[\left(n+\frac{1}{2}\right)\hbar\omega_{02} - \left(\frac{E_0^2 m_\perp^*}{2B^2}\right) + D\left\{\left(n+\frac{1}{2}\right)\hbar\omega_{02} + \left(\frac{E_0^2 m_\perp^*}{2B^2}\right)\right\}^{1/2}\right], \quad \omega_{02} \equiv \frac{|e|B}{m_\perp^*}.$$

and

$$D \equiv \pm\frac{\bar{\lambda}_0\sqrt{2m_\perp^*}}{\hbar}$$

The use of (13.23a) leads to the expressions of the EEM's along z and y directions as

$$m_z^*(\bar{E}_{FBHD},\eta_g,n,E_0) = m_\parallel^*[\{\gamma_3(\bar{E}_{FBHD},\eta_g)\}'' + \frac{m_\parallel^* E_0^2\{\gamma_3(\bar{E}_{FBHD},\eta_g)\}'\{\gamma_3(\bar{E}_{FBHD},\eta_g)\}''}{B^2}] \tag{13.23b}$$

$$m_y^*(\bar{E}_{FBHD},\eta_g,n,E_0) = (\frac{B}{E_0})^2 \frac{1}{[\{\gamma_3(\bar{E}_{FBHD},\eta_g)\}']}[\gamma_3(\bar{E}_{FBHD},\eta_g) - \beta_1(n,E_0) + \frac{m_\parallel^* E_0^2[\{\gamma_3(\bar{E}_{FBHD},\eta_g)\}']^2}{2B^2}][\frac{-\{\gamma_3(\bar{E}_{FBHD},\eta_g)\}''}{[\{\gamma_3(\bar{E}_{FBHD},\eta_g)\}']^2}[\gamma_3(\bar{E}_{FBHD},\eta_g) - \beta_1(n,E_0) + \frac{m_\parallel^* E_0^2[\{\gamma_3(\bar{E}_{FBHD},\eta_g)\}']^2}{2B^2}] + 1 + \frac{m_\parallel^* E_0^2\{\gamma_3(\bar{E}_{FBHD},\eta_g)\}''}{B^2}] \tag{13.23c}$$

The Landau energy $(\bar{E}_{n_4})$ can be written as

$$\gamma_3(\bar{E}_{n_4},\eta_g) = \beta_1(n,E_0) - \frac{m_\parallel^* E_0^2}{2B^2}\left(\gamma_3'(\bar{E}_{n_4},\eta_g)\right)^2 \tag{13.24}$$

The expression for n_0 in this case assumes the form

$$n_0 = \frac{2g_v B\sqrt{2m_\parallel^*}}{3L_x\pi^2\hbar^2 E_0}\sum_{n=0}^{n_{\max}}\left[T_{53HD}(n,\bar{E}_{FBHD},\eta_g) + T_{54HD}(n,\bar{E}_{FBHD},\eta_g)\right] \tag{13.25}$$

where

$$T_{53HD}(n,\bar{E}_{FBHD},\eta_g) \equiv \left[\left[\gamma_3(\bar{E}_{FBHD},\eta_g) - \beta_1(n,E_0) + |e|E_0L_x(\gamma_3'(\bar{E}_{FBHD},\eta_g)) - \frac{m_\parallel^* E_0^2}{2B^2}\left(\gamma_3'(\bar{E}_{FBHD},\eta_g)\right)^2\right]^{3/2} - \left[(\gamma_3(\bar{E}_{FBHD},\eta_g)) - \beta_1(n,E_0) - \frac{m_\parallel^* E_0^2}{2B^2}\left(\gamma_3'(\bar{E}_{FBHD},\eta_g)\right)^2\right]^{3/2}\right]\left[\gamma_3'(\bar{E}_{FBHD},\eta_g)\right]^{-1}$$

and

$$T_{54HD}(n,\bar{E}_{FBHD},\eta_g) \equiv \sum_{r=0}^{s} L(r)\left[T_{53HD}(n,\bar{E}_{FBHD},\eta_g)\right].$$

Thus using (13.13), (13.25) and the allied definitions we can study the EP in this case.

13.2.4 The EP from HD IV-VI Semiconductors Under Cross-Fields Configuration

The (3.68) can be written as

$$\frac{p_s^2}{2M_1^*(E,\eta_g)}+\frac{p_z^2}{2M_3^*(E,\eta_g)}=g(E,\eta_g) \tag{13.26a}$$

where

$$M_1^*(E,\eta_g)=[\frac{(\overline{R})^2}{E_g}\{c_1(\alpha_1,E,E_g)-iD_1(\alpha_1,E,E_g)\}$$
$$+\frac{(\overline{S})^2}{\Delta_c'}\{c_2(\alpha_2,E,E_g)-iD_2(\alpha_2,E,E_g)\}$$
$$+\frac{(\overline{Q})^2}{\Delta_c''}\{c_3(\alpha_3,E,E_g)-iD_3(\alpha_3,E,E_g)\}]^{-1}$$
$$M_3^*(E,\eta_g)=[\frac{2(\overline{A})^2}{E_g}\{c_1(\alpha_1,E,E_g)-iD_1(\alpha_1,E,E_g)\}$$
$$+\frac{(\overline{S}+\overline{Q})^2}{\Delta_c''}\{c_3(\alpha_3,E,E_g)-iD_3(\alpha_3,E,E_g)\}]^{-1}$$

and

$$g^*(\mathrm{E},\eta_g)=2\hbar^2\gamma_0(\mathrm{E},\eta_g)$$

In the presence of quantizing magnetic field B along z direction and the electric field along x-axis, from above equation one obtains

$$\frac{\hat{p}_x^2}{2M_1^*(E,\eta_g)}+\frac{(\hat{p}_y-|e|B\hat{x})^2}{2M_1^*(E,\eta_g)}+\frac{\hat{p}_z^2}{2M_3^*(E,\eta_g)}=g^*(E,\eta_g)+|e|E_0\hat{x}\rho_1^*(E,\eta_g) \tag{13.26b}$$

where $\rho_1^*(E,\eta_g)=\frac{\partial}{\partial E}[g^*(E,\eta_g)]$

Let us define the operator $\hat{\theta}$ as

$$\hat{\theta}=-\hat{p}_y+|e|B\hat{x}-\frac{\rho_1^*(E,\eta_g)E_0\left[M_1^*(E,\eta_g)\right]}{B} \tag{13.27}$$

Eliminating $\hat{x}$, between the above two equations, the dispersion relation of the conduction electrons in HD stressed Kane type semiconductors in the presence of cross fields configuration can be expressed as

$$g^*(E,\eta_g) = (n+\frac{1}{2})\hbar\overline{\omega_{i1}}(E,\eta_g) + \frac{\hbar^2 k_z^2}{2M_3^*(E,\eta_g)} - \frac{E_0}{B}\rho_1^*(E,\eta_g)\hbar k_y - \frac{E_0^2}{2B^2}[\rho_1^*(E,\eta_g)]^2 M_1^*(E,\eta_g) \quad (13.28a)$$

where $\overline{\omega_{i1}}(E,\eta_g) = eB[M_1^*(E,\eta_g)]^{-1}$

The use of (13.28a) leads to the expressions of the EEM's along z and y directions as

$$m_z^*(\bar{E}_{FBHD},\eta_g,n,E_0) = \text{Real part of } [[M_3^*(\bar{E}_{FBHD},\eta_g)]'[g^*(\bar{E}_{FBHD},\eta_g) - (n+\frac{1}{2})\hbar\overline{\omega_{i1}}(\bar{E}_{FBHD},\eta_g) + \frac{E_0^2}{2B^2}[\rho_1^*(\bar{E}_{FBHD},\eta_g)]^2 M_1^*(\bar{E}_{FBHD},\eta_g)] + [M_3^*(\bar{E}_{FBHD},\eta_g)][[g^*(\bar{E}_{FBHD},\eta_g)]' - (n+\frac{1}{2})\hbar[\overline{\omega_{i1}}(\bar{E}_{FBHD},\eta_g)]' + \frac{E_0^2}{2B^2}[[\rho_1^*(\bar{E}_{FBHD},\eta_g)]^2[M_1^*(\bar{E}_{FBHD},\eta_g)]'] + 2[M_3^*(\bar{E}_{FBHD},\eta_g)][\rho_1^*(\bar{E}_{FBHD},\eta_g)][\rho_1^*(\bar{E}_{FBHD},\eta_g)]']] \quad (13.28b)$$

and

$$m_y^*(\bar{E}_{FBHD},\eta_g,n,E_0) = (B/E_0)^2 \text{ Real part of } [\rho_1^*(\bar{E}_{FBHD},\eta_g)]^{-3}[g^*(\bar{E}_{FBHD},\eta_g) - (n+\frac{1}{2})\hbar\overline{\omega_{i1}}(\bar{E}_{FBHD},\eta_g) + \frac{E_0^2}{2B^2}[\rho_1^*(\bar{E}_{FBHD},\eta_g)]^2 M_1^*(\bar{E}_{FBHD},\eta_g)][[\rho_1^*(\bar{E}_{FBHD},\eta_g)][[g^*(\bar{E}_{FBHD},\eta_g)]' - (n+\frac{1}{2})\hbar[\overline{\omega_{i1}}(\bar{E}_{FBHD},\eta_g)]' + \frac{E_0^2}{2B^2}[[\rho_1^*(\bar{E}_{FBHD},\eta_g)]^2[M_1^*(\bar{E}_{FBHD},\eta_g)]'] - [\rho_1^*(\bar{E}_{FBHD},\eta_g)]'[g^*(\bar{E}_{FBHD},\eta_g) - (n+\frac{1}{2})\hbar\overline{\omega_{i1}}(\bar{E}_{FBHD},\eta_g) + \frac{E_0^2}{2B^2}[\rho_1^*(\bar{E}_{FBHD},\eta_g)]^2 M_1^*(\bar{E}_{FBHD},\eta_g)]] \quad (13.28c)$$

The Landau level energy (E_{n9}) in this case can be expressed through the equation

$$g^*(E_{n_9},\eta_g) = (n+\frac{1}{2})\hbar\overline{\omega_{i1}}(E_{n_9},\eta_g) - \frac{E_0^2}{2B^2}[\rho_1^*(E_{n9},\eta_g)]^2 M_1^*(E_{n9},\eta_g) \quad (13.28d)$$

The electron concentration can be written as

$$n_0 = \frac{2B}{3L_x\pi^2\hbar^2E_0}\text{Real part of}\sum_{n=0}^{n_{\max}}[T_{4131HD}(n,\overline{E}_{FBHD},\eta_g)+T_{4141HD}(n,\overline{E}_{FBHD},\eta_g)] \tag{13.28e}$$

where

$$\begin{aligned}
T_{4131HD}(n,\overline{E}_{FBHD},\eta_g) &= [\frac{\sqrt{2M_3^*(\overline{E}_{FBHD},\eta_g)}}{\rho_1^*(\overline{E}_{FBHD},\eta_g)}][[T_{51}(n,\overline{E}_{FBHD},\eta_g)\\
&\quad+\frac{E_0}{B}\rho_1^*(\overline{E}_{FBHD},\eta_g)\hbar x_{hHD1}(\overline{E}_{FBHD},\eta_g)\rho_1^*(\overline{E}_{FBHD},\eta_g)]^{\frac{3}{2}}\\
&\quad-[T_{51}(n,\overline{E}_{FBHD},\eta_g)\\
&\quad+\frac{E_0}{B}\rho_1^*(\overline{E}_{FBHD},\eta_g)\hbar x_{lHD1}(\overline{E}_{FBHD},\eta_g)\rho_1^*(\overline{E}_{FBHD},\eta_g)]^{\frac{3}{2}}],\\
T_{51}(n,\overline{E}_{FBHD},\eta_g) &= [g^*(\overline{E}_{FBHD},\eta_g)-(n+\frac{1}{2})\hbar\overline{\omega_{i1}}(\overline{E}_{FBHD},\eta_g)\\
&\quad+\frac{M_1^*(\overline{E}_{FBHD},\eta_g)E_0^2}{2B^2}[\rho_1^*(\overline{E}_{FBHD},\eta_g)]^2]\\
x_{lHD1}(\overline{E}_{FBHD},\eta_g) &= \frac{-M_1^*(\overline{E}_{FBHD},\eta_g)E_0[\rho_1^*(\overline{E}_{FBHD},\eta_g)]}{B},\\
x_{hHD1}(\overline{E}_{FBHD},\eta_g) &= \frac{|e|BL_x}{\hbar}+x_{lHD1}(\overline{E}_{FBHD},\eta_g)
\end{aligned}$$

and $T_{4141HD}(n,\bar{E}_{FBHD},\eta_g)\equiv\sum_{r=1}^{s}L(r)T_{4131HD}(n,\bar{E}_{FBHD},\eta_g)$

Thus using (13.13), (13.28c) and the allied definitions we can study the EP in this case.

13.2.5 The EP from HD Stressed Semiconductors Under Cross-Fields Configuration

The use of (2.48) can be written as

$$\frac{p_x^2}{2m_1^*(E,\eta_g)}+\frac{p_y^2}{2m_2^*(E,\eta_g)}+\frac{p_z^2}{2m_3^*(E,\eta_g)}=G^*(E,\eta_g) \tag{13.28f}$$

where

$$m_1^*(E,\eta_g) = [2\hbar^2[\gamma_0(E,\eta_g) - I(1)T_{17}]]^{-1},$$
$$T_{17} \equiv \left[E_g - C_1\varepsilon - (\bar{a}_0 + C_1)\varepsilon + \frac{3}{2}\bar{b}_0\varepsilon_{xx} - \frac{\bar{b}_0}{2}\varepsilon + \left(\sqrt{3}/2\right)\varepsilon_{xy}\bar{d}_0\right],$$
$$m_2^*(E,\eta_g) = [2\hbar^2[\gamma_0(E,\eta_g) - I(1)T_{27}]]^{-1},$$
$$T_{27} \equiv \left[E_g - C_1\varepsilon - (\bar{a}_0 + C_1)\varepsilon + \frac{3}{2}\bar{b}_0\varepsilon_{xx} - \frac{\bar{b}_0}{2}\varepsilon - \left(\sqrt{3}/2\right)\varepsilon_{xy}\bar{d}_0\right]$$
$$m_3^*(E,\eta_g) = [2\hbar^2[\gamma_0(E,\eta_g) - I(1)T_{37}]]^{-1},$$
$$T_{37} \equiv \left[E_g - C_1\varepsilon - (\bar{a}_0 + C_1)\varepsilon + \frac{3}{2}\bar{b}_0\varepsilon_{zz} - \frac{\bar{b}_0}{2}\varepsilon\right]$$

and the other symbols are written in (1.196a).

In the presence of quantizing magneticfield B along z direction and the electric field along x-axis, from (13.28d) one obtains

$$\frac{\hat{p}_x^2}{2m_1^*(E,\eta_g)} + \frac{(\hat{p}_y - |e|B\hat{x})^2}{2m_2^*(E,\eta_g)} + \frac{\hat{p}_z^2}{2m_3^*(E,\eta_g)} = G^*(E,\eta_g) + |e|E_0\hat{x}\left[\frac{m_1^*(E,\eta_g)}{m_2^*(E,\eta_g)}\right]^{\frac{1}{2}}\rho^*(E,\eta_g) \tag{13.29}$$

where $\rho^*(E,\eta_g) = \frac{\partial}{\partial E}[G^*(E,\eta_g)]$

Let us define the operator $\hat{\theta}$ as

$$\hat{\theta} = -\hat{p}_y + |e|B\hat{x} - \frac{\rho^*(E,\eta_g)E_0\left[m_1^*(E,\eta_g)m_2^*(E,\eta_g)\right]^{\frac{1}{2}}}{B} \tag{13.30}$$

Eliminating $\hat{x}$, between the above two equations, the dispersion relation of the conduction electrons in HD stressed Kane type semiconductors in the presence of cross fields configuration can be expressed as

$$G^*(E,\eta_g) = (n+\frac{1}{2})\hbar\overline{\omega_i}(E,\eta_g) + \frac{\hbar^2k_z^2}{2m_3^*(E,\eta_g)} - \frac{E_0}{B}\rho^*(E,\eta_g)[\frac{m_1^*(E,\eta_g)}{m_2^*(E,\eta_g)}]^{\frac{1}{2}}\hbar k_y - \frac{E_0^2}{2B^2}[\rho^*(E,\eta_g)]^2 m_1^*(E,\eta_g) \tag{13.31a}$$

where

$$\overline{\omega_i}(E,\eta_g) = eB[m_1^*(E,\eta_g)m_2^*(E,\eta_g)]^{-\frac{1}{2}}$$

The use of (13.31a) leads to the expressions of the EEM's along z and y directions as

$$\begin{aligned} m_z^*(\bar{E}_{FBHD},\eta_g,n,E_0) = {} & [[m_3^*(\bar{E}_{FBHD},\eta_g)]'[G^*(\bar{E}_{FBHD},\eta_g) \\ & - (n+\frac{1}{2})\hbar\overline{\omega_i}(\bar{E}_{FBHD},\eta_g) + \frac{E_0^2}{2B^2}[\rho^*(\bar{E}_{FBHD},\eta_g)]^2 m_1^*(\bar{E}_{FBHD},\eta_g)] \\ & + [m_3^*(\bar{E}_{FBHD},\eta_g)][[G^*(\bar{E}_{FBHD},\eta_g)]' - (n+\frac{1}{2})\hbar\overline{[\omega_i}(\bar{E}_{FBHD},\eta_g)]' \\ & + \frac{E_0^2}{2B^2}[2[\rho^*(\bar{E}_{FBHD},\eta_g)][\rho^*(\bar{E}_{FBHD},\eta_g)]'[m_1^*(\bar{E}_{FBHD},\eta_g)] \\ & + [m_1^*(\bar{E}_{FBHD},\eta_g)]'[\rho^*(\bar{E}_{FBHD},\eta_g)]^2]]] \end{aligned} \tag{13.31b}$$

$$\begin{aligned} m_y^*(\bar{E}_{FBHD},\eta_g,n,E_0) = {} & (B/E_0)^2\,[m_4^*(\bar{E}_{FBHD},\eta_g)]^{-3}[G^*(\bar{E}_{FBHD},\eta_g) \\ & - (n+\frac{1}{2})\hbar\overline{\omega_i}(\bar{E}_{FBHD},\eta_g) \\ & + \frac{E_0^2}{2B^2}[\rho^*(\bar{E}_{FBHD},\eta_g)]^2 m_1^*(\bar{E}_{FBHD},\eta_g)][[m_4^*(\bar{E}_{FBHD},\eta_g)][[G^*(\bar{E}_{FBHD},\eta_g)]' \\ & - (n+\frac{1}{2})\hbar[\overline{\omega_i}(\bar{E}_{FBHD},\eta_g)]' + \frac{E_0^2}{2B^2}[[\rho^*(\bar{E}_{FBHD},\eta_g)]^2[m_1^*(\bar{E}_{FBHD},\eta_g)]'] \\ & - [m_4^*(\bar{E}_{FBHD},\eta_g)]'[G^*(\bar{E}_{FBHD},\eta_g) - (n+\frac{1}{2})\hbar\overline{\omega_i}(\bar{E}_{FBHD},\eta_g) \\ & + \frac{E_0^2}{2B^2}[\rho^*(\bar{E}_{FBHD},\eta_g)]^2 m_1^*(\bar{E}_{FBHD},\eta_g)]] \end{aligned} \tag{13.31c}$$

where

$$m_4^*(\bar{E}_{FBHD},\eta_g) = [[\rho^*(\bar{E}_{FBHD},\eta_g)][\frac{m_1^*(\bar{E}_{FBHD},\eta_g)}{m_2^*(\bar{E}_{FBHD},\eta_g)}]^{\frac{1}{2}}]$$

The Landau level energy(E_{n_8}) in this case can be expressed through the equation

$$G^*(E_{n_8},\eta_g) = (n+\frac{1}{2})\hbar\overline{\omega_i}(E_{n_8},\eta_g) - \frac{E_0^2}{2B^2}[\rho^*(E_{n_8},\eta_g)]^2 m_1^*(E_{n_8},\eta_g) \tag{13.31d}$$

The electron concentration can be written as

$$n_0 = \frac{2B}{3L_x\pi^2\hbar^2 E_0}\sum_{n=0}^{n_{\max}}[T_{413HD}(n,\overline{E}_{FBHD},\eta_g) + T_{414HD}(n,\overline{E}_{FBHD},\eta_g)] \tag{13.31e}$$

where

$$
\begin{aligned}
T_{413HD}(n, \overline{E}_{FBHD}, \eta_g) = {} & [\frac{\sqrt{2m_3^*(\overline{E}_{FBHD}, \eta_g)}}{\rho^*(\overline{E}_{FBHD}, \eta_g)}][[T_5(n, \overline{E}_{FBHD}, \eta_g) \\
& + \frac{E_0}{B}\rho^*(\overline{E}_{FBHD}, \eta_g)\hbar x_{hHD}(\overline{E}_{FBHD}, \eta_g)\rho^*(\overline{E}_{FBHD}, \eta_g)]^{\frac{3}{2}} \\
& - [T_5(n, \overline{E}_{FBHD}, \eta_g) + \frac{E_0}{B}\rho^*(\overline{E}_{FBHD}, \eta_g)\hbar x_{lHD}(\overline{E}_{FBHD}, \eta_g)\rho^*(\overline{E}_{FBHD}, \eta_g)]^{\frac{3}{2}}], \\
T_5(n, \overline{E}_{FBHD}, \eta_g) = {} & [G^*(\overline{E}_{FBHD}, \eta_g) - (n + \frac{1}{2})\hbar\overline{\omega_l}(\overline{E}_{FBHD}, \eta_g) \\
& + \frac{m_1^*(\overline{E}_{FBHD}, \eta_g)E_0^2}{2B^2}[\rho^*(\overline{E}_{FBHD}, \eta_g)]^2] \\
x_{lHD}(\overline{E}_{FBHD}, \eta_g) = {} & \frac{-m_1^*(\overline{E}_{FBHD}, \eta_g)E_0[\rho^*(\overline{E}_{FBHD}, \eta_g)]}{B}, \\
x_{hHD}(\overline{E}_{FBHD}, \eta_g) = {} & \frac{|e|BL_x}{\hbar} + x_{lHD}(\overline{E}_{FBHD}, \eta_g)
\end{aligned}
$$

and

$$
T_{414HD}(n, \bar{E}_{FBHD}, \eta_g) \equiv \sum_{r=1}^{s} L(r)T_{413HD}(n, \bar{E}_{FBHD}, \eta_g)
$$

Thus using (13.13), (13.31e) and the allied definitions we can study the EP in this case.

13.3 Open Research Problems

R.13.1 Investigate the EP in the presence of an arbitrarily oriented quantizing magnetic and crossed electric fields in HD tetragonal semiconductors by including broadening and the electron spin. Study all the special cases for HD III-V, ternary and quaternary materials in this context.

R.13.2 Investigate the EPs for all models of HD IV-VI, II-VI and stressed Kane type compounds in the presence of an arbitrarily oriented quantizing magnetic and crossed electric fields by including broadening and electron spin.

R.13.3 Investigate the EP for all the ultrathin of HD materials whose bulk dispersion relations are stated in R.1.1 of Chap. 1 in the presence of an arbitrarily oriented quantizing magnetic and crossed electric fields by including broadening and electron spin.

References

1. W. Zawadzki, B. Lax, Phys. Rev. Lett. **16**, 1001 (1966)
2. M.J. Harrison, Phys. Rev. A **29**, 2272 (1984)
3. J. Zak, W. Zawadzki, Phys. Rev. **145**, 536 (1966)
4. W. Zawadzki, Q.H. Vrehen, B. Lax, Phys. Rev. **148**, 849 (1966)
5. Q.H. Vrehen, W. Zawadzdi, M. Reine, Phys. Rev. **158**, 702 (1967)
6. M.H. Weiler, W. Zawadzki, B. Lax, Phys. Rev. **163**, 733 (1967)
7. W. Zawadzki, J. Kowalski, Phys. Rev. Lett. **27**, 1713 (1971)
8. C. Chu, M.S. Chu, T. Ohkawa, Phys. Rev. Lett. **41**, 633 (1978)
9. P. Hu, C.S. Ting, Phys. Rev. B **36**, 9671 (1987)
10. E.I. Butikov, A.S. Kondratev, A.E. Kuchma, Sov. Phys. Solid State **13**, 2594 (1972)
11. K.P. Ghatak, J.P. Banerjee, B. Goswami, B. Nag, Nonlin. Opt. Quantum Opt. **16**, 241 (1996)
12. M. Mondal, K.P. Ghatak, Phys. Status Solidi (b) **133**, K67 (1986)
13. M. Mondal, N. Chattopadhyay, K.P. Ghatak, J. Low Temp. Phys. **66**, 131 (1987)
14. K.P. Ghatak, M. Mondal, Zeitschrift fur Physik B **69**, 471 (1988)
15. M. Mondal, K.P. Ghatak, Phys. Lett. A **131A**, 529 (1988)
16. M. Mondal, K.P. Ghatak, Phys. Stat. Sol. (b) Germany **147**, K179 (1988)
17. B. Mitra, K.P. Ghatak, Phys. Lett. **137A**, 413 (1989)
18. B. Mitra, A. Ghoshal, K.P. Ghatak, Phys. Stat. Sol. (b) **154**, K147 (1989)
19. B. Mitra, K.P. Ghatak, Phys. Stat. Sol. (b) **164**, K13 (1991)
20. K.P. Ghatak, B. Mitra, Int. J. Electron. **70**, 345 (1991)
21. S.M. Adhikari, D. De, J.K. Baruah, S. Chowdhury, K.P. Ghatak, Adv. Sci. Focus **1**, 57 (2013)
22. S. Pahari, S. Bhattacharya, D. De, S.M. Adhikari, A. Niyogi, A. Dey, N. Paitya, S.C. Saha, K. P. Ghatak, P.K. Bose, Phys. B **405**, 4064 (2010)
23. S. Bhattaacharya, S. Choudhary, S. Ghoshal, S.K. Bishwas, D. De, K.P. Ghatak, J. Comp. Theo. Nanosci. **3**, 423 (2006)
24. M. Mondal, K.P. Ghatak, Annalen der Physik **46**, 502 (1989)
25. K.P. Ghatak, B. De MRS Proceedings **242**, 377 (1992)
26. K.P. Ghatak, B. Mitra, Int. J. Theor. Phys. **70**, 343 (1991)
27. B. Mitra, K.P. Ghatak, Phys. Lett. A **141**, 81 (1989)
28. B. Mitra, K.P. Ghatak, Phys. Lett. A **137**, 413 (1989)
29. M. Mondal, N. Chattopadhyay, K.P. Ghatak, J. Low Temp. Phys. **73**, 321 (1988)
30. M. Mondal, K.P. Ghatak, Phys. Lett. A **131**, 529 (1988)
31. K.P. Ghatak, N. Chattopadhyay, S.N. Biswas, International society for optics and photonics, In: Proceedings of the society of photo-optical instrumentation engineers, USA, 203 1987
32. K.P. Ghatak, M. Mitra, B. Goswami, B. Nag, Nonlin. Opt. **16**, 167 (1996)
33. K.P. Ghatak, D.K. Basu, B. Nag, J. Phys. Chem. Sol. **58**, 133 (1997)
34. S. Biswas, N. Chattopadhyay, K.P. Ghatak, International society for optics and photonics, In: Proceedings of the society of photo-optical instrumentation engineers, USA, 836, 175 1987
35. K.P. Ghatak, M. Mondal, S. Bhattacharyya, SPIE **1284**, 113 (1990)
36. K.P. Ghatak, SPIE, Photonic Mater. Opt. Bistability **1280**, 53 (1990)
37. K.P. Ghatak, S.N. Biswas, SPIE, Growth Charact. Mater. Infrared Detectors Nonlin. Opt. Switches **1484**, 149 (1991)
38. K.P. Ghatak, SPIE, Fiber Opt. Laser Sens. IX **1584**, 435 (1992)

Chapter 14
Appendix D: The EP from HD III-V, Ternary and Quaternary Semiconductors Under Strong Electric Field

14.1 Introduction

In the investigation of transport properties of nano-devices under electric field, we assumed that the electron energy spectrum becomes an invariant quantity, which is not true especially in the presence of strong electric field. In nano-devices the in-built electric field is so large that the electron dispersion relation changes fundamentally and in this chapter we shall investigate the influence of intense electric field on the EP under various physical conditions in III-V, ternary and quaternary materials. In Sect. 14.2.1, of theoretical background Sect. 14.2, we shall study the EP under strong electric field in HD said semiconductors. The Sect. 14.2.2, explores the EP in the presence of quantizing magnetic field under strong electric field in HD said materials. In Sect. 14.2.3, we study the EP in QWs of HD III-V, ternary and quaternary materials under strong electric field. In Sect. 14.2.4, the EP has been investigated in NWs of HD III-V, ternary and quaternary materials. In Sect. 14.2.5, the EP has been investigated in QBs of HD III-V, ternary and quaternary materials In Sect. 14.2.6, we study the magneto EP in QWs of HD III-V, ternary and quaternary materials under strong electric field. In Sect. 14.2.7, the magneto EP in effective mass superlattices of HD said materials under strong electric field has been investigated. The last Sect. 14.3, contains 43 open research problems.

14.2 Theoretical Background

14.2.1 The EP Under Strong Electric Field in HD III-V, Ternary and Quaternary Materials

In the presence of strong electric field F_s along x direction, the electron energy spectrum in Kane type III-V semiconductors whose unperturbed conduction electrons obey the three band models of Kane can be expressed following [1] as

K.P. Ghatak, *Einstein's Photoemission*, Springer Tracts in Modern Physics 262,
DOI 10.1007/978-3-319-11188-9_14

$$\frac{\hbar^2 k^2}{2m_c} = [e_1 E^4 + e_2 E^3 + e_3 E^2 + e_4 E + e_5 - \frac{e_6}{1 + CE} + e_7 (1 + CE)^{-2}] \quad (14.1)$$

where

$$e_1 = Q_f \omega_1, \quad Q_f = \frac{m_c}{m_r} E_g^{-4} [5 e_f E_g^{-2} - 6G_f + 7 h_f E_g^{-4}], \quad m_r (= (\frac{1}{m_c} + \frac{1}{m_v})^{-1})$$

is the reduced mass, m_v is the effective heavy hole mass at the edge of the valance band $e_f = A_f P_f$,

$$A_f = [F + E_g (E_g - \delta')]^2 m_c (6 m_\gamma^2 (\delta')^4)^{-1},$$

$F = eF_s$, F_s is the electric field along x direction.

$$\delta' = \frac{\Delta E_g^2}{\chi}, \chi = 6E_g^2 + 9\Delta E_g + 4\Delta^2, \frac{1}{m_r} = (\frac{1}{m_c} + \frac{1}{m_v}), G_f = e_f (4\delta' + C_f),$$

$$C_f = (2E_g Q^2 + PQ(E_g - E_g') - 2p^2 E_g), P = \frac{r_0^2}{2} (\frac{E_g - \delta'}{E_g + \delta'}),$$

$$r_0 = [\frac{6}{\chi} (E_g + \Delta)(E_g + \frac{2}{3}\Delta)]^{\frac{1}{2}}$$

$$Q = \frac{t^2}{2}, t = [\frac{6}{\chi} (E_g + \frac{2}{3}\Delta)]^{\frac{1}{2}}, h_f = (4\delta' e_f C_f)(B_f)^{-1}, B_f = (P + Q)^2$$

$$P_f = E_g^{-3} (e_f E_g^{-2} - G_f + h_f E_g^{-4}), \omega_1 = a_1^2, a_1 = \frac{ab}{c},$$

$$a = \frac{1}{E_g}, b = \frac{1}{E_g + \Delta}, c = (E_g + \frac{2}{3}\Delta)^{-1},$$

$$e_2 = Q_f \omega_2, \omega_2 = 2a_1 b_1, b_1 = (c)^{-2} (ac + bc - ab),$$

$$e_3 = (1 - P_f) a_1 + Q_f \omega_3, \omega_3 = (b_1^2 + 2a_1 c_1),$$

$$c_1 = [\frac{1}{c} (1 - \frac{a}{c})(1 - \frac{b}{c})],$$

$$e_4 = [(1 - P_f) b_1 + Q_f \omega_4], \omega_6 = \frac{2c_1 b_1}{c} (1 - \frac{cc_1}{b_1}), \omega_4 = 2b_1 c_1,$$

$$e_5 = [(1 - P_f) c_1 + Q_f \omega_5],$$

$$\omega_5 = (c_1^2 - 2c_1 b_1), e_7 = Q_f \omega_7, \omega_7 = c_1^2, e_6 = [(1 - P_f) c_1 - Q_f \omega_6],$$

Using (14.1) and (1.4) we get

$$\begin{aligned}\frac{\hbar^2k^2}{2m_c}\int_{-\infty}^{E} F(v)dv =& e_1\int_{-\infty}^{E}(E-v)^4F(v)dv + e_2\int_{-\infty}^{E}(E-v)^3F(v)dv\\ &+ e_3\int_{-\infty}^{E}(E-v)^2F(v)dv + e_4\int_{-\infty}^{E}(E-v)F(v)dv\\ &+ e_5\int_{-\infty}^{E}F(v)dv - e_6\int_{-\infty}^{E}\frac{F(v)dv}{1+c(E-v)}\\ &+ e_7\int_{-\infty}^{E}F(v)dv[1+c(E-v)]^2\end{aligned} \tag{14.2}$$

Let us put

$$\begin{aligned}I(11) =& \int_{-\infty}^{E}(E-v)^4F(v)dv\\ =& E^4\int_{-\infty}^{E}F(v)dv + \int_{-\infty}^{E}v^4F(v)dv + 6E^2\int_{-\infty}^{E}v^2F(v)dv\\ &- 2E^3\int_{-\infty}^{E}vF(v)dv - 4E\int_{-\infty}^{E}v^3F(v)dv\end{aligned} \tag{14.3}$$

Now

$$\begin{aligned}\int_{-\infty}^{E} v^4(F(v))dv =& [\frac{3\eta_g^4}{8\pi}[1+Erf(\frac{E}{\eta_g})] - \frac{\eta_g^5}{4E\pi}\exp(\frac{-E^2}{\eta_g^2})]\\ =& \frac{3\eta_g^4}{8\pi}[1+Erf(\frac{E}{\eta_g}) - \frac{2\eta_g}{3E}\exp(\frac{-E^2}{\eta_g^2})] = \delta_0'(E,\eta_g)\end{aligned} \tag{14.4}$$

$$\int_{-\infty}^{E} v^2F(v)dv = \frac{\eta_g^2}{4}[1+Erf(\frac{E}{\eta_g})] \tag{14.5}$$

$$\int_{-\infty}^{E} v(F(v))dv = \frac{-\eta_g}{2\sqrt{\pi}}\exp(\frac{-E^2}{\eta_g^2}) \tag{14.6}$$

$$\int_{-\infty}^{E} v^3F(v)dv = \frac{-\eta_g^3}{2\sqrt{\pi}}\exp(\frac{-E^2}{\eta_g^2})[1+\frac{E^2}{\eta_g^2}] \tag{14.7}$$

Thus

$$\begin{aligned} I(11) &= \frac{E^4}{2}[1+Erf(\frac{E}{\eta_g})] + \frac{3\eta_g^4}{8\pi}[1+Erf(\frac{E}{\eta_g}) - \frac{2\eta_g}{3E}\exp(\frac{-E^2}{\eta_g^2})] \\ &+\frac{3}{2}(E\eta_g)^2[1+Erf(\frac{E}{\eta_g})] + \frac{E^3\eta_g}{\sqrt{\pi}}\exp(\frac{-E^2}{\eta_g^2}) \\ &+\frac{2E\eta_g^3}{\sqrt{\pi}}\exp(\frac{-E^2}{\eta_g^2})[1+\frac{E^2}{\eta_g^2}] \\ &= \varphi_0(E,\eta_g) \end{aligned} \tag{14.8}$$

In Chap. 1 we have proved that

$$I(\alpha, E, \eta_g) = \int_{-\infty}^{E} \frac{F(v)dv}{1+e(E-v)} = \frac{2}{c\eta_g\sqrt{\pi}}e^{-u^2}\left[\sum_{p=1}^{\infty}\frac{e^{\frac{-p^2}{4}}}{p}\sinh(pu)\right] - i\frac{\sqrt{\pi}}{c\eta_g}e^{-u^2} \tag{14.9}$$

where $u = \frac{1+cE}{c\eta_g}$.

The theorem of differentiation under the sign of integration tells us

$$\begin{aligned} \frac{\partial}{\partial x}\int_{A(x)}^{B(x)} F(x,y)dy = &\int_{A(x)}^{B(x)} \frac{\partial}{\partial x}[F(x,y)]dy + F(x,B(x)\frac{\partial B(x)}{\partial x}) \\ &- F(x,A(x)\frac{\partial A(x)}{\partial x}) \end{aligned} \tag{14.10}$$

where the notations have their usual meaning and the integrals are convergent.

Using (14.9) and (14.10) and differentiating (14.9) with respect to E we get

$$
\begin{aligned}
&-c\int_{-\infty}^{E}\frac{F(v)dv}{[1+c(E-v)]^2}+\frac{1}{\sqrt{\pi\eta_g^2}}\exp(\frac{-E^2}{\eta_g^2})\\
&\quad=\frac{2}{c\eta_g\sqrt{\pi}}e^{-u^2}2u\frac{1}{\eta_g}\left[\sum_{p=1}^{\infty}\frac{e^{\frac{-p^2}{4}}}{p}\sinh(pu)\right]\\
&\quad+\frac{2}{c\eta_g\sqrt{\pi}}e^{-u^2}\frac{1}{\eta_g}\sum_{p=1}^{\infty}e^{\frac{-p^2}{4}}\cosh(pu)-i\frac{\sqrt{\pi}}{c\eta_g}e^{-u^2}2u\frac{1}{\eta_g}
\end{aligned}
\tag{14.11}
$$

$$
\therefore\int_{-\infty}^{E}\frac{F(v)dv}{[1+c(E-v)]^2}=c_3(E,\eta_g,c)+iD_3(E,\eta_g,c)
$$

where

$$
\begin{aligned}
c_3(E,\eta_g,c)&=[\frac{1}{\pi c\eta_g}\exp(\frac{-E^2}{\eta_g^2})-\frac{4ue^{-u^2}}{c^2\eta_g^2\sqrt{\pi}}[\sum_{p=1}^{\infty}\exp(\frac{-p^2}{4})p^{-1}\sinh(pu)]\\
&\quad-\frac{2}{c^2\eta_g^2\sqrt{\pi}}e^{-u^2}\sum_{p=1}^{\infty}e^{\frac{-p^2}{4}}\cosh(pu)]\\
D_3(E,\eta_g,c)&=\frac{2u}{c^2\eta_g^2}\exp(-u^2),\quad u=\frac{1+cE}{c\eta_g}
\end{aligned}
$$

Again (14.9) can be written as

$$
\int_{-\infty}^{E}\frac{F(v)dv}{1+c(E-v)}=c_1(c,E,\eta_g)-ic_2(c,E,\eta_g)
\tag{14.12}
$$

where $c_1(c,E,\eta_g)=\frac{2}{c\eta_g\sqrt{\pi}}e^{-u^2}\left[\sum_{p=1}^{\infty}\frac{e^{\frac{-p^2}{4}}}{p}\sinh(pu)\right]$ and $c_2(c,E,\eta_g)=\frac{\sqrt{\pi}}{c\eta_g}e^{-u^2}$.

We know that

$$
\begin{aligned}
\int_{-\infty}^{E}(E-v)^3F(v)dv&=[\frac{E}{2}[1+Erf(\frac{E}{\eta_g})][E^2+\frac{3}{2}\eta_g^2]\\
&\quad+\frac{\eta_g}{2\sqrt{\pi}}e^{-\frac{E^2}{\eta_g^2}}(4E^2+\eta_g^2)]\\
&=\varphi_1(E,\eta_g)
\end{aligned}
\tag{14.13}
$$

Therefore the dispersion relation in HD III-IV semiconductors whose unperturbed conduction electrons obey the three band models of Kane in the presence of an electric field along x axis can be expressed as

$$\frac{\hbar^2 k^2}{2m_c} = J_4(E, c, \eta_g) \quad \text{where } J_4(E, c, \eta_g) = J_1(E, c, \eta_g) + iJ_2(E, c, \eta_g), \tag{14.14}$$

$$\begin{aligned} J_1(E, c, \eta_g) = {} & 2[1 + Erf(\frac{E}{\eta_g})]^{-1}[e_1\varphi_0(E, \eta_g) + e_2\varphi_1(E, \eta_g) \\ & + e_3\theta_0(E, \eta_g) + e_4\gamma_0(E, \eta_g) \\ & + e_{g_{\frac{1}{2}}}[1 + Erf(\frac{E}{\eta_g})] - e_6 c_1(E, c, \eta_g) + e_7 c_3(E, c, \eta_g)] \end{aligned}$$

and $J_2(E, c, \eta_g) = 2[1 + Erf(\frac{E}{\eta_g})]^{-1}[e_6 c_2(E, c, \eta_g) + e_7 D_3(E, c, \eta_g)]$.

For two band model of Kane, the dispersion relation in the presence of electric field F_s along x direction is given by

$$\frac{\hbar^2 k^2}{2m_c} = P_{11f}E(1 + \alpha E) - Q_{11f} \tag{14.15}$$

where $P_{11f} = [1 + (Q_{11f})(\frac{5m_c}{m_r E_g})], Q_{11f} = \frac{(\hbar F)^2}{12 m_r E_g^2}$.

Therefore under the condition of heavy doping (14.15) assumes the form

$$\frac{\hbar^2 k^2}{2m_c} = J_5(E, \eta_g) \tag{14.16a}$$

where $J_5(E, \eta_g) = P_{11f}\gamma_2(E, \eta_g) - Q_{11f}$.

Thus (14.14) and (14.16a) are key equations for investigating the electronic properties in III-V Kane type heavily doped semiconductors in the presence of a strong electric field.

The EEM in III-V Kane type HD semiconductors in the presence of a strong electric field whose energy band structures in the absence of any perturbation are defined by three and two band models of kane can be written from (14.14) and (14.16a) as

$$m^*(E_{FE}, F) = m_c \,\text{Real part of } [J_4(E_{FE}, c, \eta_g)]' \tag{14.16b}$$

and

$$m^*(E_{FE}, F) = m_c[J_5(E_{FE}, \eta_g)]' \tag{14.16c}$$

where E_{FE} is the Fermi energy in the present case.

Thus following (14.14) electron concentration is given by

$$n_0 = \frac{g_v}{3\pi^2}(\frac{2m_c}{\hbar^2})^{\frac{3}{2}} \text{ Real part of } [J_6(E_{FE}, c, \eta_g) + J_7(E_{FE}, c, \eta_g)] \tag{14.17}$$

where $J_6(E_{FE}, c, \eta_g) = [J_4(E_{FE}, c, \eta_g)]^{\frac{3}{2}}$ and $J_7(E_{FE}, c, \eta_g) = \sum_{r=1}^{s} L(r)J_6(E_{FE}, c, \eta_g)$

For the dispersion relation (14.16a) the corresponding electron concentration can be written as

$$n_0 = \frac{g_v}{3\pi^2}(\frac{2m_c}{\hbar^2})^{\frac{3}{2}}[J_8(E_{FE}, \eta_g) + J_9(E_{FE}, \eta_g)] \tag{14.19a}$$

where $J_8(E_{FE}, \eta_g) = [J_5(E_{FE}, \eta_g)]^{\frac{3}{2}}$ and $J_9(E_{FE}, \eta_g) = \sum_{r=1}^{s} L(r)J_8(E_{FE}, \eta_g)$.

The velocity along z direction and the density of states function in this case for HD optoelectronic Kane type materials under intense electric field whose conduction electrons in the absence of perturbation obey the three band model of Kane can respectively be written as

$$v_z(\mathrm{E}_1') = \sqrt{\frac{2}{m_c}\frac{[J_4(\mathrm{E}_{1E_0}', c, \eta_g)]^{1/2}}{J_4'(\mathrm{E}_{1E_0}', c, \eta_g)}} \tag{14.19b}$$

$$N(\mathrm{E}_1') = 4\pi g_v(\frac{2m_c}{h^2})^{3/2}\sqrt{J_4(\mathrm{E}_{1E_0}', c, \eta_g)}[J_4'(\mathrm{E}_{1E_0}', c, \eta_g)] \tag{14.19c}$$

where $E_{1E_0}' = E - E_{01HDE_0}, E_{01HDE_0} = \Omega_1 + W - hv$, Ω_1 is the root of the equation

$$J_4(\Omega_1, c, \eta_g) = 0 \tag{14.19d}$$

The EP in this case is given by

$$J_{LHD} = \frac{4\pi\alpha_0 e m_c g_v}{h^3} \text{Real part of} \int_{E_{01HDE_0}}^{\alpha} J_4(\mathrm{E}_{1E_0}', c, \eta_g)f(E)d\mathrm{E}_{1E_0}' \tag{14.19e}$$

Similarly the EP for perturbed two band model of Kane and that of parabolic energy bands can respectively be expressed as

$$J_{LHD} = \frac{4\pi\alpha_0 e m_c g_v}{h^3} \int\limits_{E_{02HDE_0}}^{\alpha} J_5(E'_{2E_0}, c, \eta_g) f(E) d\,E'_{2E_0} \tag{14.19f}$$

where $E'_{2E_0} = E - E_{02HDE_0}, E_{02HDE_0} = \Omega_2 + W - h\nu$, Ω_2 is the root of the equation

$$J_5(\Omega_2, c, \eta_g) = 0. \tag{14.19g}$$

14.2.2 The EP from the Presence of Quantizing Magnetic Field Under Strong Electric Field in HD III-V, Ternary and Quaternary Materials

The electron energy spectrum under magnetic quantization can be written as

$$\frac{\hbar^2 k_z^2}{2m_c} + (n + \frac{1}{2})\hbar\omega_0 = J_4(E, c, \eta_g) \tag{14.20}$$

$$\frac{\hbar^2 k_z^2}{2m_c} + (n + \frac{1}{2})\hbar\omega_0 = J_5(E, \eta_g) \tag{14.21a}$$

The EEM in this case can be written using (14.20) and (14.21a) as

$$m^*(E_{FEB}, F) = m_c \text{ Real part of } [J_4(E_{FEB}, c, \eta_g)]' \tag{14.21b}$$

and

$$m^*(E_{FEB}, F) = m_c [J_5(E_{FEB}, \eta_g)]' \tag{14.21c}$$

where E_{FEB} is the Fermi energy in the present case.

The electron concentration for the dispersion relation (14.20) is given by

$$n_0 = \frac{eBg_v\sqrt{2m_c}}{\pi^2\hbar^2} \text{ Real part of } \sum_{n=0}^{n_{\max}} [J_{10}(E_{FEB}, c, \eta_g, n) + J_{11}(E_{FEB}, c, \eta_g, n)] \tag{14.22}$$

where $J_{10}(E_{FEB}, c, \eta_g, n) = \sqrt{J_4(E_{FEB}, c, \eta_g, n) - (n + \frac{1}{2})\hbar\omega_0}$ and $J_{11}(E_{FEB}, c, \eta_g, n) = \sum_{r=1}^{s} L(r) J_{10}(E_{FEB}, c, \eta_g, n)$.

The magneto EP in this case is given by

$$J_{HBE_0} = \frac{\alpha_0 e^2 B g_v k_B T}{2\pi^2\hbar^2} \text{ Real part of } \sum_{n=0}^{n_{\max}} F_0(\eta_{DB4,1}) \tag{14.23}$$

where $\eta_{DB4,1} = [E_{FEB} - (L_{DB4,1} + W - h\upsilon)][k_B T]^{-1}$ and $\mathrm{L}_{DB4,1}$ is the Landau subband energy which can be obtained from (14.20) by substituting $k_z = 0$ and $E = \mathrm{L}_{DB4,1}$.

The electron concentration for the dispersion relation (14.2.1) is given by

$$n_0 = \frac{eBg_v\sqrt{2m_c}}{\pi^2\hbar^2}\sum_{n=0}^{n_{\max}}[J_{12}(E_{FEB},\eta_g,n) + J_{13}(E_{FEB},\eta_g,n)] \tag{14.24a}$$

where $J_{12}(E_{FEB},\eta_g,n) = \sqrt{J_5(E_{FEB},\eta_g,n) - (n+\frac{1}{2})\hbar\omega_0}$ and $J_{13}(E_{FEB},\eta_g,n) = \sum_{r=1}^{s} L(r)J_{12}(E_{FEB},\eta_g,n)$.

The magneto EP in this case is given by

$$J_{HBE_0} = \frac{\alpha_0 e^2 B g_v k_B T}{2\pi^2\hbar^2}\sum_{n=0}^{n_{\max}} F_0(\eta_{DB4,2}) \tag{14.24b}$$

where $\eta_{DB4,2} = [E_{FEB} - (\mathrm{L}_{DB4,2} + \mathrm{W} - h\upsilon)][\mathrm{k}_B\mathrm{T}]^{-1}$ and $\mathrm{L}_{DB4,2}$ is the Landau subband energy which can be obtained from (14.21a) by substituting $k_z = 0$ and $E = \mathrm{L}_{DB4,2}$.

14.2.3 The EP from QWs of HD III-V, Ternary and Quaternary Materials Under Strong Electric Field

For QWs the 2D dispersion laws for (14.14) and (14.16a) assume the forms

$$\frac{\hbar^2 k_s^2}{2m_c} + \frac{\hbar^2}{2m_c}\left(\frac{n_z\pi}{d_z}\right)^2 = J_4(E,c,\eta_g) \tag{14.25}$$

and

$$\frac{\hbar^2 k_s^2}{2m_c} + \frac{\hbar^2}{2m_c}\left(\frac{n_z\pi}{d_z}\right)^2 = J_5(E,\eta_g) \tag{14.26a}$$

The EEM in this case can be written using (14.25) and (14.26a) as

$$m^*(E_{FESQ}, F) = m_c \text{ Real part of } [J_4(E_{FESQ}, c, \eta_g)]' \tag{14.26b}$$

and

$$m^*(E_{FESQ}, F) = m_c[J_5(E_{FESQ}, \eta_g)]' \tag{14.26c}$$

where E_{FESQ} is the Fermi energy in this case.

The 2D electron concentration for (14.25) can be written as

$$n_0 = \frac{m_c g_v}{\pi \hbar^2} \text{Real part of} \sum_{n_z=1}^{n_{zmax}} [J_{17}(E_{FESQ}, c, \eta_g, n_z) + J_{18}(E_{FESQ}, c, \eta_g, n_z)] \tag{14.27}$$

where $J_{17}(E_{FESQ}, c, \eta_g, n_z) = [J_4(E_{FESQ}, c, \eta_g) - \frac{\hbar^2}{2m_c}(\frac{n_z \pi}{d_z})^2]$ and $J_{18}(E_{FESQ}, c, \eta_g, n_z) = \sum_{r=1}^{s} L(r)[J_{17}(E_{FESQ}, c, \eta_g, n_z)]$.

The EP in this case is given by

$$J_{2DE_0} = [\frac{\alpha_0 e m_c g_v}{2\pi d_z \hbar^2}(\frac{m_c}{2})^{1/2}] \text{ Real part of} \sum_{n_{zmin}}^{n_{zmax}} [\frac{\sqrt{J_4(E_{nzE_0,1}, c, \eta_g)}}{J_4'(E_{nzE_0,1}, c, \eta_g)}][J_{17}(E_{FESQ}, c, \eta_g, n_z) + J_{18}(E_{FESQ}, c, \eta_g, n_z)] \tag{14.28}$$

where $E_{nzE_0,1}$ is the sub-band energy in this case and is obtained by substituting $k_s = 0$ and $E = E_{nzE_0,1}$ in (14.25).

The 2D electron concentration for (14.26a) can be written as

$$n_0 = \frac{m_c g_v}{\pi \hbar^2} \sum_{n_z=1}^{n_{zmax}} [J_{19}(E_{FESQ}, c, \eta_g, n_z) + J_{20}(E_{FESQ}, c, \eta_g, n_z)] \tag{14.29a}$$

where $J_{19}(E_{FESQ}, c, \eta_g, n_z) = [J_5(E_{FESQ}, c, \eta_g) - \frac{\hbar^2}{2m_c}(\frac{n_z \pi}{d_z})^2]$ and $J_{20}(E_{FESQ}, c, \eta_g, n_z) = \sum_{r=1}^{s} L(r)[J_{19}(E_{FESQ}, c, \eta_g, n_z)]$.

The EP in this case is given by

$$J_{2DE_0} = [\frac{\alpha_0 e m_c g_v}{2\pi d_z \hbar^2}(\frac{m_c}{2})^{1/2}] \sum_{n_{zmin}}^{n_{zmax}} [\frac{\sqrt{J_5(E_{nzE_0,2}, c, \eta_g)}}{J_5'(E_{nzE_0,2}, c, \eta_g)}][J_{19}(E_{FESQ}, c, \eta_g, n_z) + J_{20}(E_{FESQ}, c, \eta_g, n_z)] \tag{14.29b}$$

where $E_{nzE_0,2}$ is the sub-band energy in this case and is obtained by substituting $k_s = 0$ and $E = E_{nzE_0,2}$ in (14.26a).

14.2.4 The EP from NWs of HD III-V, Ternary and Quaternary Materials Under Strong Electric Field

The dispersion relations of the 1D electrons in NWs of HD optoelectronic materials in the presence of light waves can be expressed from (14.14) and (14.16a) as

$$k_y^2 = \frac{2m_c J_4(E, c, \eta_g)}{\hbar^2} - \left(\frac{\pi n_{z71}}{d_z}\right)^2 - \left(\frac{\pi n_{x71}}{d_x}\right)^2 \tag{14.29c}$$

$$k_y^2 = \frac{2m_c J_5(E, c, \eta_g)}{\hbar^2} - \left(\frac{\pi n_{z72}}{d_z}\right)^2 - \left(\frac{\pi n_{x72}}{d_x}\right)^2 \tag{14.30}$$

where n_{x7J} $(J = 1, 2, 3)$ is the size quantum number.

The electron concentration per unit length are respectively given by

$$n_{1DL} = \frac{2g_v\sqrt{2m_c}}{\pi\hbar} \text{ Real part of } \sum_{n_{x71}=1}^{n_{x71\max}} \sum_{n_{z71}=1}^{n_{z71\max}} \left[\phi_{77,1}(E_{F1DLE_0}, n_{x71}, n_{z71}) + \phi_{78,1}(E_{F1DLE_0}, n_{x71}, n_{z71})\right] \tag{14.29e}$$

$$n_{1DL} = \frac{2g_v\sqrt{2m_c}}{\pi\hbar} \sum_{n_{x72}=1}^{n_{x72\max}} \sum_{n_{z72}=1}^{n_{z72\max}} \left[\phi_{79,2}(E_{F1DLE_0}, n_{x72}, n_{z72}) + \phi_{80,2}(E_{F1DLE_0}, n_{x72}, n_{z72})\right] \tag{14.31}$$

where E_{F1DLE_0} is the Fermi energy in NWs in the present case.

$$\phi_{77,1}(E_{F1DLE_0}, n_{x71}, n_{z71}) = \left[J_4(E_{F1DLE_0}, c, \eta_g) - G_{71}(n_{x71}, n_{z71})\right]^{\frac{1}{2}},$$

$$G_{7i}(n_{x7i}, n_{z7i}) = \frac{\hbar^2}{2m_c}\left[\left(\frac{\pi n_{x7i}}{d_x}\right)^2 + \left(\frac{\pi n_{z7i}}{d_z}\right)^2\right],$$

$$\phi_{78,1}(E_{F1DLE_0}, n_{x71}, n_{z71}) = \sum_{r=1}^{s_0} L(r)\left[\phi_{77,1}(E_{F1DLE_0}, n_{x71}, n_{z71})\right],$$

$$\phi_{79,2}(E_{F1DLE_0}, n_{x72}, n_{z72}) = \left[J_5(E_{F1DLE_0}, c, \eta_g) - G_{72}(n_{x72}, n_{z72})\right]^{\frac{1}{2}} \text{ and}$$

$$\phi_{80,2}(E_{F1DLE_0}, n_{x72}, n_{z72}) = \sum_{r=1}^{s} L(r)\left[\phi_{79,2}(E_{F1DLE_0}, n_{x72}, n_{z72})\right].$$

The generalized expression of photo current in this case is given by

$$I_L = \frac{\alpha_0 e g_v k_B T}{\pi \hbar} \sum_{n_{x7i}=1}^{n_{x7i\max}} \sum_{n_{z7i}=1}^{n_{z7i\max}} F_0\left(\eta_{7iE_0}\right) \tag{14.32}$$

where, $\eta_{7iE_0} = \frac{E_{F1DLE_0} - \left(E'_{7iE_0} + W - h\upsilon\right)}{k_B T}$ and E'_{7iE_0} are the sub-band energies in this case and are defined through the following equations

$$\left.\begin{aligned} J_4\left(E'_{71E_0}, c, \eta_g\right) &= G_{71}\left(n_{x71}, n_{z71}\right) \\ J_4\left(E'_{72E_0}, c, \eta_g\right) &= G_{72}\left(n_{x72}, n_{z72}\right) \end{aligned}\right\} \tag{14.33}$$

Real part of the (7.22) should be used for computing the EP from NWs of HD optoelectronic materials under intense electric field whose unperturbed energy band structures are defined by the three-band model of Kane.

14.2.5 The EP from QBs of HD III-V, Ternary and Quaternary Materials Under Strong Electric Field

The dispersion relations of the electrons in QBs of HD optoelectronic materials in the presence of intense electric field can respectively be expressed from (14.14) and (14.16a) as

$$\frac{2m_c J_4\left(E_{Q1E_0}, c, \eta_g\right)}{\hbar^2} = H_{71}\left(n_{x71}, n_{y71}, n_{z71}\right) \tag{14.34}$$

$$\frac{2m_c J_5\left(E_{Q2E_0}, c, \eta_g\right)}{\hbar^2} = H_{72}\left(n_{x72}, n_{y72}, n_{z72}\right) \tag{14.35}$$

where E_{QiE_0} is the totally quantized energy and $H_{7i}\left(n_{x7i}, n_{y7i}, n_{z7i}\right) = \left(\frac{\pi n_{x7i}}{d_x}\right)^2 + \left(\frac{\pi n_{y7i}}{d_y}\right)^2 + \left(\frac{\pi n_{z7i}}{d_z}\right)^2$.

The electron concentration can, in general, be written as

$$n_{ODL} = \left(\frac{2g_v}{d_x d_y d_z}\right) \sum_{n_{x7i}=1}^{n_{x7i\max}} \sum_{n_{y7i}=1}^{n_{y7i\max}} \sum_{n_{z7i}=1}^{n_{z7i\max}} F_{-1}\left(\eta_{7i0DE_0}\right) \tag{14.36}$$

where $\eta_{7i0DE_0} = \frac{E_{F0DLE_0} - E_{QiE_0}}{k_B T}$ and E_{F0DLE_0} is the Fermi energy in QBs of HD optoelectronic materials in the presence of intense electric field as measured from the edge of the conduction band in the vertically upward direction in the absence of any quantization.

Real part of the (7.26) should be used for computing the carrier density from QBs of HD optoelectronic materials whose unperturbed energy band structures are defined by the three-band model of Kane.

The photo-emitted current densities in this case are given by the following equations

$$J_{0DL} = \frac{(\alpha_0 e g_v)}{d_x d_y d_z}\left(\frac{m_c}{2}\right)^{-1/2} \text{Real part of} \sum_{n_{x71}=1}^{n_{x71\max}} \sum_{n_{y71}=1}^{n_{y71\max}} \sum_{n_{z71\min}}^{n_{z71\max}} \left[\frac{\sqrt{J_4\left(E_{n_{z71}}, c, \eta_g\right)}}{J_4'\left(E_{n_{z71}}, c, \eta_g\right)}\right] F_{-1}\left(\eta_{710DE_0}\right) \tag{14.37}$$

$$J_{0DL} = \frac{(\alpha_0 e g_v)}{d_x d_y d_z}\left(\frac{m_c}{2}\right)^{-1/2} \sum_{n_{x72}=1}^{n_{x72\max}} \sum_{n_{y72}=1}^{n_{y72\max}} \sum_{n_{z72\min}}^{n_{z72\max}} \left[\frac{\sqrt{J_5\left(E_{n_{z72}}, c, \eta_g\right)}}{J_5'\left(E_{n_{z72}}, c, \eta_g\right)}\right] F_{-1}\left(\eta_{720DE_0}\right). \tag{14.38}$$

14.2.6 The Magneto EP from QWs of HD III-V, Ternary and Quaternary Materials Under Strong Electric Field

Under magnetic quantization, (14.14) and (14.16a) assume the forms

$$(\mathrm{n} + \frac{1}{2})\hbar\omega_c + \frac{\hbar^2}{2m_c}(\frac{n_z\pi}{d_z})^2 = J_4(\overline{E}_{E_0 1}, c, \eta_g) \tag{14.39a}$$

and

$$(\mathrm{n} + \frac{1}{2})\hbar\omega_c + \frac{\hbar^2}{2m_c}(\frac{n_z\pi}{d_z})^2 = J_5(\overline{E}_{E_0 2}, c, \eta_g) \tag{14.39b}$$

where $\overline{E}_{E_0 1}$ and $\overline{E}_{E_0 2}$ are totally quantized energy in the respective cases.

The energy along z direction for both the cases can be expressed as

$$\frac{\hbar^2}{2m_c}(\frac{n_z\pi}{d_z})^2 = J_4(\overline{E}_{n_z E_0 1}, c, \eta_g) \tag{14.39c}$$

and

$$\frac{\hbar^2}{2m_c}\left(\frac{n_z\pi}{d_z}\right)^2 = J_5(\overline{E}_{n_zE_0 2}, c, \eta_g) \tag{14.39d}$$

where $\overline{E}_{n_zE_0 1}$ and $\overline{E}_{n_zE_0 2}$ are the energies along z direction for (14.39a) and (14.39b) respectively.

The electron concentration for (14.39a) is given by

$$n_0 = \frac{g_v eB}{\pi\hbar} \text{ Real part of } \sum_{n=0}^{n_{\max}} \sum_{n_z=1}^{n_{z\max}} F_{-1}(\eta_{EOD,1}) \tag{14.39e}$$

where $\eta_{EOD,1} = [k_B T]^{-1}[E_{FTD} - \overline{E}_{E_0 1}]$, E_{FTD} is the Fermi energy in this case.

The EP in this case can be expressed as

$$J_{TDE_0} = \frac{\alpha_0 e^2 B g_v}{h d_z} \frac{\sqrt{2}}{\sqrt{m_c}} \sum_{n_{z\min}}^{n_{z\max}} \sum_{n=0}^{n_{\max}} \frac{\sqrt{J_4(E_{nzE_0 1}, c, \eta_g)}}{J_4'(E_{nzE_0 1}, c, \eta_g)} F_{-1}(\eta_{EOD1}) \tag{14.39f}$$

The electron concentration for (14.39b) is given by

$$n_0 = \frac{g_v eB}{\pi\hbar} \sum_{n=0}^{n_{\max}} \sum_{n_z=1}^{n_{z\max}} F_{-1}(\eta_{EOD,2}) \tag{14.39g}$$

where $\eta_{EOD,2} = [k_B T]^{-1}[E_{FTD} - \overline{E}_{E_0 2}]$.

The EP in this case can be expressed as

$$J_{TDE_0} = \frac{\alpha_0 e^2 B g_v}{h d_z} \frac{\sqrt{2}}{\sqrt{m_c}} \sum_{n_{z\min}}^{n_{z\max}} \sum_{n=0}^{n_{\max}} \frac{\sqrt{J_5(E_{nzE_0 2}, c, \eta_g)}}{J_5'(E_{nzE_0 2}, c, \eta_g)} F_{-1}(\eta_{EOD2}). \tag{14.39h}$$

14.2.7 The Magneto EP from Effective Mass Superlattices of HD III-V, Ternary and Quaternary Materials Under Strong Electric Field

The electron dispersion law in III-V effective mass super lattices can be written as

$$k_x^2 = \left[\frac{1}{L_0^2}[\cos^{-1}\{f_{HD}(E, c, \eta_g, k_y, k_z)\}]^2 - k_\perp^2\right] \tag{14.40}$$

where

$$f_{HD}(E, c, \eta_g, k_y, k_z) = [[a_{1HD} cos[a_0 C_{1HD}(E, \eta_{g1}, c_1, k_\perp) + b_0 D_{1HD}(E, \eta_{g2}, c_2, k_\perp)]]$$
$$- [a_{2HD} \cos[a_0 C_{1HD}(E, \eta_{g1}, c_1, k_\perp) - b_0 D_{1HD}(E, \eta_{g2}, c_2, k_\perp)]]],$$

$$a_{1HD} = [\sqrt{\frac{m_{c2} J_4'(0, c_2, \eta_{g2})}{m_{c1} J_4'(0, c_1, \eta_{g1})}} + 1]^2 \cdot [4\sqrt{\frac{m_{c2} J_4'(0, c_2, \eta_{g2})}{m_{c1} J_4'(0, c_1, \eta_{g1})}}]^{-1},$$

$$a_{2HD} = [\sqrt{\frac{m_{c2} J_4'(0, c_2, \eta_{g2})}{m_{c1} J_4'(0, c_1, \eta_{g1})}} - 1]^2 \cdot [4\sqrt{\frac{m_{c2} J_4'(0, c_2, \eta_{g2})}{m_{c1} J_4'(0, c_1, \eta_{g1})}}]^{-1},$$

$$C_{1HD}(E, \eta_{g1}, c_1, k_\perp) = [\frac{2m_{c1}}{\hbar^2} J_4(E, \eta_{g1}, c_1) - k_\perp^2]^{1/2} \text{ and}$$

$$D_{1HD}(E, \eta_{g1}, c_1, k_\perp) = [\frac{2m_{c2}}{\hbar^2} J_4(E, \eta_{g2}, c_2) - k_\perp^2]^{1/2}.$$

In the presence of a quantizing magnetic field B along k_x direction, the magneto electron energy spectrum can be written as

$$k_x^2 = \omega_{HD}(E, \eta_g, c, n) \tag{14.41a}$$

where $\omega_{HD}(E, \eta_g, c, n) = [\frac{1}{L_0^2}[cos^{-1}\{f_{HD}(E, c, \eta_g, n)\}]^2 - \frac{2eB}{\hbar}(n + \frac{1}{2})]$

$$f_{HD}(E, c, \eta_g, n) = [[a_{1HD} cos[a_0 C_{1HD}(E, \eta_{g1}, c_1, n) + b_0 D_{1HD}(E, \eta_{g2}, c_2, n)]]$$
$$- [a_{2HD} cos[a_0 C_{1HD}(E, \eta_{g1}, c_1, n) - b_0 D_{1HD}(E, \eta_{g2}, c_2, n)]]],$$

$$C_{1HD}(E, \eta_{g1}, c_1, n) = [\frac{2m_{c1}}{\hbar^2} J_4(E, \eta_{g1}, c_1) - \frac{2eB}{\hbar}(n + \frac{1}{2})]^{1/2}$$

and $D_{1HD}(E, \eta_{g1}, c_1, n) = [\frac{2m_{c2}}{\hbar^2} J_4(E, \eta_{g2}, c_2) - \frac{2eB}{\hbar}(n + \frac{1}{2})]^{1/2}$.

The EEM in this case can be written from (14.41a) as

$$m^*(E_{SL}, \eta_g, c, n) = \frac{\hbar^2}{2} \text{Real part of } [\omega_{HD}(E_{SL}, \eta_g, c, n)]' \tag{14.41b}$$

where E_{SL} is the Fermi energy in this case.

The electron concentration is given by

$$n_0 = \frac{g_v eB}{\pi^2 \hbar} \text{Real part of } \sum_{n=0}^{n_{\max}} [J_{40}(E_{SL}, \eta_g, n) + J_{41}(E_{SL}, \eta_g, n)] \tag{14.42}$$

where $J_{40}(E_{SL}, \eta_g, n) = [\omega_{HD}(E_{SL}, \eta_g, c, n)]^{\frac{1}{2}}$ and $J_{41}(E_{SL}, \eta_g, n) = \sum_{r=1}^{s} L(r)$ $[J_{40}(E_{SL}, \eta_g, n)]$.

The EP in this case is given by

$$J_{ESE_0} = \frac{\alpha_0 e^2 B g_v k_B T}{2\pi^2 \hbar^2} \sum_{n=0}^{n_{\max}} F_0(\eta_{SLL1}) \tag{14.43}$$

where $\eta_{SLL1} = (k_B T)^{-1}[\mathrm{E}_{SL} - (\omega_{SL1} + W - hv)]$ and ω_{SL1} is the sub-band energy in this case.

The electron dispersion law in III-V effective mass super lattices whose constituent materials obey (14.16a) can be expressed as

$$k_x^2 = [\frac{1}{L_0^2}[\cos^{-1}\{f_{HD5}(E, \eta_g, k_y, k_z)\}]^2 - k_\perp^2] \tag{14.44}$$

where

$$\begin{aligned} f_{HD5}(E, \eta_g, k_y, k_z) &= [[a_{1HD5}\cos[a_0 C_{1HD5}(E, \eta_{g1}, k_\perp) + b_0 D_{1HD5}(E, \eta_{g2}, k_\perp)]] \\ &\quad - [a_{2HD5}\cos[a_0 C_{1HD5}(E, \eta_{g1}, k_\perp) - b_0 D_{1HD5}(E, \eta_{g2}, k_\perp)]]], \\ a_{1HD5} &= [\sqrt{\frac{m_{c2}J_5'(0, \eta_{g2})}{m_{c1}J_5'(0, \eta_{g1})}} + 1]^2 \cdot [4\sqrt{\frac{m_{c2}J_5'(0, \eta_{g2})}{m_{c1}J_5'(0, \eta_{g1})}}]^{-1}, \\ a_{2HD5} &= [\sqrt{\frac{m_{c2}J_5'(0, \eta_{g2})}{m_{c1}J_5'(0, \eta_{g1})}} - 1]^2 . [4\sqrt{\frac{m_{c2}J_5'(0, \eta_{g2})}{m_{c1}J_5'(0, \eta_{g1})}}]^{-1}, \\ C_{1HD5}(E, \eta_{g1}, k_\perp) &= [\frac{2m_{c1}}{\hbar^2} J_5(E, \eta_{g1}) - k_\perp^2] \text{ and} \\ D_{1HD5}(E, \eta_{g1}, k_\perp) &= [\frac{2m_{c2}}{\hbar^2} J_5(E, \eta_{g2},) - k_\perp^2] \end{aligned}$$

In the presence of a quantizing magnetic field B along k_x direction, the magneto-electron energy spectrum can be written as

$$k_x^2 = \omega_{HD5}(E, \eta_g, n) \tag{14.45a}$$

where

$$\begin{aligned} \omega_{HD5}(E, \eta_g, n) &= [\frac{1}{L_0^2}[cos^{-1}\{f_{HD5}(E, \eta_g, n)\}]^2 - \frac{2eB}{\hbar}(n + \frac{1}{2})], \\ f_{HD5}(E, \eta_g, n) &= [[a_{1HD5}cos[a_0 C_{1HD5}(E, \eta_{g1}, n) + b_0 D_{1HD5}(E, \eta_{g2}, n)]] \\ &\quad - [a_{2HD5}cos[a_0 C_{1HD5}(E, \eta_{g1}, n) - b_0 D_{1HD5}(E, \eta_{g2}, n)]]], \\ C_{1HD5}(E, \eta_{g1}, n) &= [\frac{2m_{c1}}{\hbar^2} J_5(E, \eta_{g1}) - \frac{2eB}{\hbar}(n + \frac{1}{2})] \end{aligned}$$

$$\text{and } D_{1HD5}(E, \eta_{g1}, n) = [\frac{2m_{c2}}{\hbar^2} J_5(E, \eta_{g2}) - \frac{2eB}{\hbar}(n + \frac{1}{2})]$$

The EEM in this case can be written from (14.45a) as

$$m^*(E_{SL},\eta_g,n) = \frac{\hbar^2}{2}[\omega_{HD5}(E_{SL},\eta_g,n)]' \quad (14.45b)$$

The electron concentration is given by

$$n_0 = \frac{g_v eB}{\pi^2\hbar}\sum_{n=0}^{n_{\max}}[J_{50}(E_{SL},\eta_g,n) + J_{51}(E_{SL},\eta_g,n)] \quad (14.46)$$

where $J_{50}(E_{SL},\eta_g,n) = [\omega_{HD5}(E_{SL},\eta_g,n)]^{\frac{1}{2}}$ and $J_{51}(E_{SL},\eta_g,n) = \sum_{r=1}^{s} L(r)$ $[J_{50}(E_{SL},\eta_g,n)]$

The EP in this case is given by

$$J_{ESE_0} = \frac{\alpha_0 e^2 B g_v k_B T}{2\pi^2\hbar^2}\sum_{n=0}^{n_{\max}} F_0(\eta_{SLL2}) \quad (14.47)$$

where $\eta_{SLL2} = (k_B T)^{-1}[\mathrm{E}_{SL} - (\omega_{SL2} + W - h\nu)]$ and ω_{SL2} is the sub-band energy in this case.

14.3 Open Research Problems

(R.14.1) Investigate the EP for the HD bulk materials whose respective dispersion relations of the carriers in the absence of any field is given in Chap. 1 in the presence of intense electric field which change the original band structure and consider its effect in the subsequent study in each case.

(R.14.2) Investigate the EP as defined in (R.14.1) in the presence of an arbitrarily oriented non-uniform light waves for all the HD materials as considered R.14.1.

(R.14.3) Investigate the EP as defined in (R.14.1) in the presence of an arbitrarily oriented non-quantizing alternating non-uniform electric field for all the cases of R.14.1.

(R.14.4) Investigate the EP as defined in (R.14.1) for all the HD materials in the presence of arbitrarily oriented non-quantizing non-uniform electric field for all the appropriate cases.

(R.14.5) Investigate the EP as defined in (R.14.1) for all the HD materials in the presence of arbitrarily oriented non-quantizing alternating electric field for all the appropriate cases of problem R.14.4.

(R.14.6) Investigate the EP as defined in (R.14.1) for the negative refractive index, organic, magnetic and other advanced optical HD materials in the presence of arbitrarily oriented electric field.

(R.14.7) Investigate the EP as defined in (R.14.1) in the presence of alternating non-quantizing electric field for all the problems of R.14.6.

(R.14.8) Investigate the EP as defined in (R.14.1) for all the multiple quantum confined HD materials whose unperturbed carrier energy spectra are defined in R.14.1 in the presence of arbitrary oriented quantizing magnetic field by including the effects of spin and broadening respectively.

(R.14.9) Investigate the EP as defined in (R.14.1) in the presence of an additional arbitrarily oriented alternating quantizing magnetic field respectively for all the problems of R.14.8.

(R.14.10) Investigate the EP as defined in (R.14.1) in the presence of arbitrarily oriented alternating quantizing magnetic field and arbitrary oriented non-quantizing non-uniform electric field respectively for all the problems of R.14.8.

(R.14.11) Investigate the EP as defined in (R.14.1) in the presence of arbitrary oriented alternating non-uniform quantizing magnetic field and additional arbitrary oriented non-quantizing alternating electric field respectively for all the problems of R.14.1.

(R.14.12) Investigate the EP in the presence of arbitrary oriented and crossed quantizing magnetic and electric fields respectively for all the problems of R.14.8.

(R.14.13) Investigate the EP for all the appropriate HD low-dimensional systems of this chapter in the presence of finite potential wells.

(R.14.14) Investigate the EP for all the appropriate HD low-dimensional systems of this chapter in the presence of parabolic potential wells.

(R.14.15) Investigate the EP for all the appropriate HD systems of this chapter forming quantum rings.

(R.14.16) Investigate the EP for all the above appropriate problems in the presence of elliptical Hill and quantum square rings respectively.

(R.14.17) Investigate the EP for multiple carbon nano-tubes.

(R.14.18) Investigate the EP for multiple carbon nano-tubes in the presence of non-quantizing non-uniform alternating light waves.

(R.14.19) Investigate the EP for multiple carbon nano-tubes in the presence of non-quantizing non-uniform alternating magnetic field.

(R.14.20) Investigate the EP for multiple carbon nano-tubes in the presence of crossed electric and quantizing magnetic fields.

(R.14.21) Investigate the EP for all types of HD semiconductor nano-tubes for all the HD materials whose unperturbed carrier dispersion laws are defined in Chap. 1.

(R.14.22) Investigate the EP for HD semiconductor nano-tubes in the presence of non-quantizing alternating light waves for all the materials whose unperturbed carrier dispersion laws is defined in Chap. 1.

(R.14.23) Investigate the EP for HD semiconductor nano-tubes in the presence of non-quantizing alternating magnetic field for all the materials whose unperturbed carrier dispersion laws are defined in Chap. 1.

(R.14.24) Investigate the EP for HD semiconductor nano-tubes in the presence of non-uniform light waves for all the materials whose unperturbed carrier dispersion laws are defined in Chap. 1.

(R.14.25) Investigate the EP for HD semiconductor nano-tubes in the presence of alternating quantizing magnetic fields for all the materials whose unperturbed carrier dispersion laws are defined in Chap. 1.

(R.14.26) Investigate the EP for HD semiconductor nano-tubes in the presence of crossed electric and quantizing magnetic fields for all the materials whose unperturbed carrier dispersion laws are defined in Chap. 1.

(R.14.27) Investigate the EP for all the appropriate nipi structures of the HD materials whose unperturbed carrier energy spectra are defined in Chap. 1.

(R.14.28) Investigate the EP for all the appropriate nipi structures of the HD materials whose unperturbed carrier energy spectra are defined in Chap. 1, in the presence of an arbitrarily oriented non-quantizing non-uniform additional electric field.

(R.14.29) Investigate the EP for all the appropriate nipi structures of the HD materials whose unperturbed carrier energy spectra are defined in Chap. 1 in the presence of non-quantizing alternating additional magnetic field.

(R.14.30) Investigate the EP for all the appropriate nipi structures of the HD materials whose unperturbed carrier energy spectra are defined in Chap. 1 in the presence of quantizing alternating additional magnetic field.

(R.14.31) Investigate the EP for all the appropriate nipi structures of the HD materials whose unperturbed carrier energy spectra are defined in Chap. 1 in the presence of crossed electric and quantizing magnetic fields.

(R.14.32) Investigate the EP for HD nipi structures for all the appropriate cases of all the above problems.

(R.14.33) Investigate the EP for the appropriate accumulation layers of all the materials whose unperturbed carrier energy spectra are defined in Chap. 1 in the presence of crossed electric and quantizing magnetic fields by considering electron spin and broadening of Landau levels.

(R.14.34) Investigate the EP for quantum confined HD III-V, II-VI, IV-VI, HgTe/CdTe effective mass super-lattices together with short period, strained layer, random, Fibonacci, poly-type and saw-tooth super-lattices.

(R.14.35) Investigate the EP in the presence of quantizing magnetic field respectively for all the cases of R.14.34.

(R.14.36) Investigate the EP in the presence of non-quantizing non-uniform additional electric field respectively for all the cases of R.14.34.

(R.14.37) Investigate the EP in the presence of non-quantizing alternating electric field respectively for all the cases of R.14.34.

(R.14.38) Investigate the multiphoton EP in the presence of crossed electric and quantizing magnetic fields respectively for all the cases of R.14.34.

(R.14.39) Investigate the EP as defined in (R.14.1) for HD quantum confined super-lattices for all the problems of R.14.35.

(R.14.40) Investigate the EP as defined in (R.14.1) in the presence of quantizing non-uniform magnetic field respectively for all the cases of R.14.34.

(R.14.41) Investigate the EP as defined in (R.14.1) in the presence of crossed electric and quantizing magnetic fields respectively for all the cases of R.14.34.

(R.14.42) Investigate the EP as defined in (R.14.1) for all the systems in the presence of alternating strain.

(R.14.43) Investigate all the problems of this chapter by removing all the mathematical approximations and establishing the respective appropriate uniqueness conditions.

Reference

1. K.P. Ghatak, S. Bhattacharya, *Debye Screening Length; Effects of Nanostructured Materials*, vol. 255. Springer Tracts in Modern Physics (Springer, Berlin, 2014)

Materials Index

K.P. Ghatak, *Einstein's Photoemission*, Springer Tracts in Modern Physics 262,
DOI 10.1007/978-3-319-11188-9

Subject Index

K.P. Ghatak, *Einstein's Photoemission*, Springer Tracts in Modern Physics 262,
DOI 10.1007/978-3-319-11188-9

F

G

MIX
Papier aus verantwortungsvollen Quellen
Paper from responsible sources
FSC® C105338
FSC
www.fsc.org

If you have any concerns about our products,
you can contact us on
ProductSafety@springernature.com

In case Publisher is established outside the EU,
the EU authorized representative is:
Springer Nature Customer Service Center GmbH
Europaplatz 3, 69115 Heidelberg, Germany

Printed by Libri Plureos GmbH
in Hamburg, Germany